Sustainable Built Environment Conference 2019 Wales

Policy to Practice

IOP Conference Series: Earth and Environmental Science
Volume 329

Cardiff, United Kingdom
24 - 25 September 2019

ISBN: 978-1-5108-9665-9
ISSN: 1755-1307

Printed from e-media with permission by:

Curran Associates, Inc.
57 Morehouse Lane
Red Hook, NY 12571

Some format issues inherent in the e-media version may also appear in this print version.

Institute of Physics
Dirac House, Temple Back
Bristol BS1 6BE UK

Phone: 44 1 17 929 7481
Fax: 44 1 17 920 0979

techtracking@iop.org

Additional copies of this publication are available from:

Curran Associates, Inc.
57 Morehouse Lane
Red Hook, NY 12571 USA
Phone: 845-758-0400
Fax: 845-758-2633
Email: curran@proceedings.com
Web: www.proceedings.com

TABLE OF CONTENTS

SUSTAINABLE NEIGHBOURHOODS

ENGAGEMENT AND BEHAVIOUR

BUILDING ENERGY PERFORMANCE

BUILDING THERMAL PERFORMANCE

RESIDENTIAL RETROFIT

CLIMATE AND BUILDINGS

URBAN HEAT & GREEN ARCHITECTURE

INNOVATION

POLICY TO PRACTICE (2)

Foreword

On 24th and 25th September 2019 Cardiff University and Cynnal Cymru hosted the Sustainable Built Environment Conference: SBE19Wales Policy to Practice, at the Principality Stadium, Cardiff.

The aim of SBE19Wales Policy to Practice was to bring together climate and energy researchers, built environment professionals, policy experts and government officials to focus on the transition of energy and low carbon policies into practice. The conference topics allowed delegates to increase their knowledge and skills related to climate change policies and actions, building design and retrofit issues and the performance of buildings and materials. The papers have been grouped according to themes.

The International Sustainable Built Environment Conference Series began in 2000 and operates on a three year cycle starting with a year of planning, followed by a year of national or regional conferences, culminating in a single global event in year three. The series places a core emphasis on peer-reviewed papers, presentations of regional policy papers, displays of pre-evaluated projects and a small number of high-quality commercial exhibits. These events also provide fast-track inclusion of the best technical papers from national/regional conferences into the global event, as well as registration policies designed to encourage the attendance of students and delegates from developing countries. In all of these conferences, one or more local research-based organisations take on the financial and organizational responsibility while the international series owners, CIB, iiSBE, FIDIC and UN Environment provide organisational advice and use their networks to ensure a large multi-disciplinary audience for the events.

The Organisers would also like to thank the Scientific Committee for their time and support:

Professor Vincent Buhagiar, University of Malta
Professor Joe Clarke, University of Strathclyde
Professor Malcolm Cook, Loughborough University
Dr Sarah Darby, Environmental Change Institute, University of Oxford
Dr Hu Du, Welsh School of Architecture, Cardiff University
Dr Heba Elsharkawy, University of East London
Dr Tina Fawcett, Environmental Change Institute, University of Oxford
Simon Foxell, The Architects Practice
Professor Rajat Gupta, Oxford Brookes University
Dr Shan Shan Hou, Welsh School of Architecture, Cardiff University
Dr Kathryn Janda, UCL Energy Institute
Professor Phillip Jones, Welsh School of Architecture, Cardiff University
Ingrid Kaltenegger, Joanneum Research, Austria
Professor Ian Knight, Welsh School of Architecture, Cardiff University
Werner Lang, Technical University of Munich
Dr Simon Lannon, Welsh School of Architecture, Cardiff University
Nils Larsson, iiSBE
Gudrun Lettmayer, Joanneum Research, Austria
Professor Conghong Liu, Tianjin University
Professor Yanfeng Liu, Xian University of Architecture and Technology
Richard Lorch, Buildings & Cities
Dr Alice Moncaster, Open University
Professor Tadj Oreszczyn, UCL Energy Institute
Dr Joanne Patterson, Welsh School of Architecture, Cardiff University
Assistant Professor Sofie Pelsmakers, Aarhus School of Architecture (Denmark) and Tampere University Technology (Finland)
Professor Deo Prasad, UNSW, Sydney

IOP Conf. Series: Earth and Environmental Science **329** (2019) 011002 doi:10.1088/1755-1315/329/1/011002

Peer review statement

All papers published in this volume of *IOP Conference Series: Earth and Environmental Science* have been peer reviewed through processes administered by the proceedings Editors. Reviews were conducted by expert referees to the professional and scientific standards expected of a proceedings journal published by IOP Publishing.

Understanding the decarbonisation of housing: Wales as a case study

E Green[1,2], S Lannon[1], J Patterson[1], F Variale[1] and H Iorwerth[1]

[1] Welsh School of Architecture, Cardiff University
[2] Corresponding Author: GreenE11@cardiff.ac.uk

Abstract. International targets requiring UK nations to reduce carbon emissions by at least 80% of 1990 levels by 2050 were made obsolete in 2019. The Climate Change Act now commits UK government to reducing greenhouse gas emissions by at least 100% of 1990 levels by 2050. This includes more stringent emissions targets for the devolved administrations. Welsh Government is responsible for the delivery of decarbonisation in Wales. The Welsh housing stock, among the oldest and least efficient in Europe, produces 21% of national carbon emissions. The CCC report recommends that Wales target no less than a 95% reduction in carbon emissions by 2050. This research explores how policy could deliver this scale of emissions reduction, by predicting the impact of established and emerging 'best practice' retrofit solutions on the existing housing stock. Fourteen recurring dwelling 'types' are identified within the Welsh housing stock (based on physical metrics, condition and tenure) using large data sets. Appropriate retrofit 'narratives' are established and then simulated for each of the fourteen dwelling types. Their impact is measured in terms of capital cost, primary energy use, fuel bills, and carbon emissions. By exploring the impact of a range of approaches to retrofit, and by considering different scenarios for clean energy supply, this research has informed the ongoing development of a route map for delivering decarbonisation targets in Wales.

1. Background

Longstanding international targets required UK nations to reduce greenhouse gas emissions by at least 80% of 1990 levels by 2050. Welsh Government committed to these targets via the Environment (Wales) Act 2016. However, following widespread acknowledgement of a climate emergency in 2019, and subsequent publication of a report by the Committee for Climate Change [1], UK government is now committed to reducing greenhouse gas emissions by at least 100% of 1990 levels (net zero) by 2050, significantly increasing the scale of the challenge.

The CCC report recommends that Wales deliver a 95% reduction in greenhouse gases by 2050. The Welsh housing stock produces 21% of national carbon emissions (compared with 29% of UK carbon emissions, and reflecting the relatively high level of emissions from industry in Wales). Key factors affecting the challenge of decarbonising the Welsh housing stock include:

- Cleaner electricity generation is a clear achievement of the last decade. However, progress in this sector masks a failure to decarbonise other sectors, including housing. The UK is not on course to meet the legally binding fourth and fifth UK carbon budgets [2], and will not do so unless risks to the delivery of existing policies are reduced significantly and new policies go beyond achievements to date.
- The composition of the Welsh housing stock has been stable for the last two decades, with a significant increase only in the private rented sector. Low rates of replacement and an

underperforming housebuilding sector mean that more than 90% of existing homes are predicted to remain in use by 2050.

- Welsh housing is among the oldest and least efficient in Europe. One third of homes were built before 1919. Just 6% were built in the last 30 years (as of 2016), increasing energy demand for heating and reducing comfort. Despite energy efficiency initiatives, almost a quarter of households currently experience fuel poverty.
- Welsh housing is varied in terms of type, age, physical form and construction. Dwellings have been modified over time to create a diverse stock of varying quality and condition. There is no single 'solution' for such a varied housing stock. However, a study of the potential to retrofit the more common dwelling 'archetypes' reveals appropriate pathways for decarbonisation.

2. Method

The research underpinning this paper was commissioned by Welsh Government in two stages: Stage 1 produced the 'what works' scoping review (see Section 3). This review of academic and grey literature identified forty domestic retrofit case studies where there was clear evidence of energy and emission reductions. For discussion purposes, retrofit actions were classified according to key themes: Thinking strategically; Building fabric; Renewables; Services; Financial; Supply chain; People. An assessment was made of each retrofit action in terms of confidence and timescale. This enabled a further classification of retrofit actions into 'what works', 'needs exploring' and 'big challenge' categories.

Stage 2 began with a statistical review of existing data (including the VOA data set, the EPC dataset 2016 [3] and the Welsh House Condition Survey 2018 [4]) to establish dwelling archetypes that collectively represent the Welsh housing stock. Modelling the housing stock using archetypes is a well-established method [5], and is particularly useful when modelling retrofit options [6]. Fourteen dominant dwelling types were identified and modelled in SAP [7] (see Section 4: Modelling the Welsh housing stock). These models were used to simulate the Welsh housing stock in 1998, 2018 and 2050. To simulate the condition of the stock in 2050, the Stage 1 review was used to establish four distinct retrofit 'narratives' for each dwelling type. The four narratives simulated were 1) retrofit limited by historic building fabric, 2) retrofit to a standard commensurate with current UK Building Regulations, 3) retrofit to a standard commensurate with SAP90, and 4) retrofit to SAP90 in an off-grid situation. To assess the significance of decarbonisation of energy supply via the national grid, three energy supply 'scenarios' were also considered: 40% clean energy supply, 60% clean energy supply and 80% clean energy supply.

The resulting research (see Section 5: findings) provides a collection of retrofit simulations which broadly reflect the Welsh housing stock. Each dwelling model describes the impact of the four retrofit strategies in terms of capital cost, energy use, fuel bills and carbon emissions. Each model also considers the three distinct energy supply scenarios. These models have been used to draw conclusions about the potential to decarbonise the Welsh housing stock through retrofit. The conclusions have been used to shape recommendations regarding the ongoing development by Welsh Government of a route-map for decarbonisation. Key recommendations are outlined and explained in Section 6.

3. 'What works': the Stage 1 review

The key objective of the Stage 1 scoping review was to begin understanding possible pathways towards decarbonisation of the existing Welsh housing stock by establishing 'what works': a database of retrofit actions with an empirical basis for assessing their effectiveness in terms of capital costs and impact on energy efficiency and carbon emissions, alongside an understanding of the challenges likely to reduce or resist their effective implementation. Key components of the review include:

- A review of Welsh policy and housing statistics, to develop understanding of context.
- A database of 40 'best practice' retrofit case studies with empirical evidence.
- A review of relevant literature from academic, governmental, industry and advisory sources
- A questionnaire survey, directed towards a wide range of stakeholders.
- The assembled literature was scrutinised to ensure relevance (to Wales) and replicability.
- Additional focussed searches were carried out when a gap was found in the literature.

40 case studies were selected based on their relevance to the decarbonisation of Welsh housing, and on the quality of underpinning empirical evidence, with the aim of finding at least one case study to represent each known retrofit action. Many of the case studies are located in Wales because this increases their relevance. Most case studies are located in the UK, with only a small number overseas. A number of the case studies were associated with retrofit programmes or funding streams, notably the Retrofit for the Future and WEFO funding programmes. Each case study also explains the key challenges that were encountered.

To develop appropriate strategies for decarbonisation over the next thirty years, pathways will need to combine tried and tested retrofit actions ('what works') with actions that are not currently fully understood. Literature was therefore reviewed from a wide range of sources, with the intention of capturing more speculative 'blue sky' thinking. 46 published pieces of literature were reviewed, including academic papers, industry standards and best practice advice, stakeholder reports and guidance for policy-makers.

Analysis of case studies and literature identified recurring retrofit actions, and discussed the underpinning understanding and applicability of each, in terms of 'timescale' (short term, medium term, long term) and 'confidence' (anecdotal, documented and understood). Combined within a simple matrix, the retrofit actions were grouped according to 'what works', 'needs exploring' and 'big challenges':

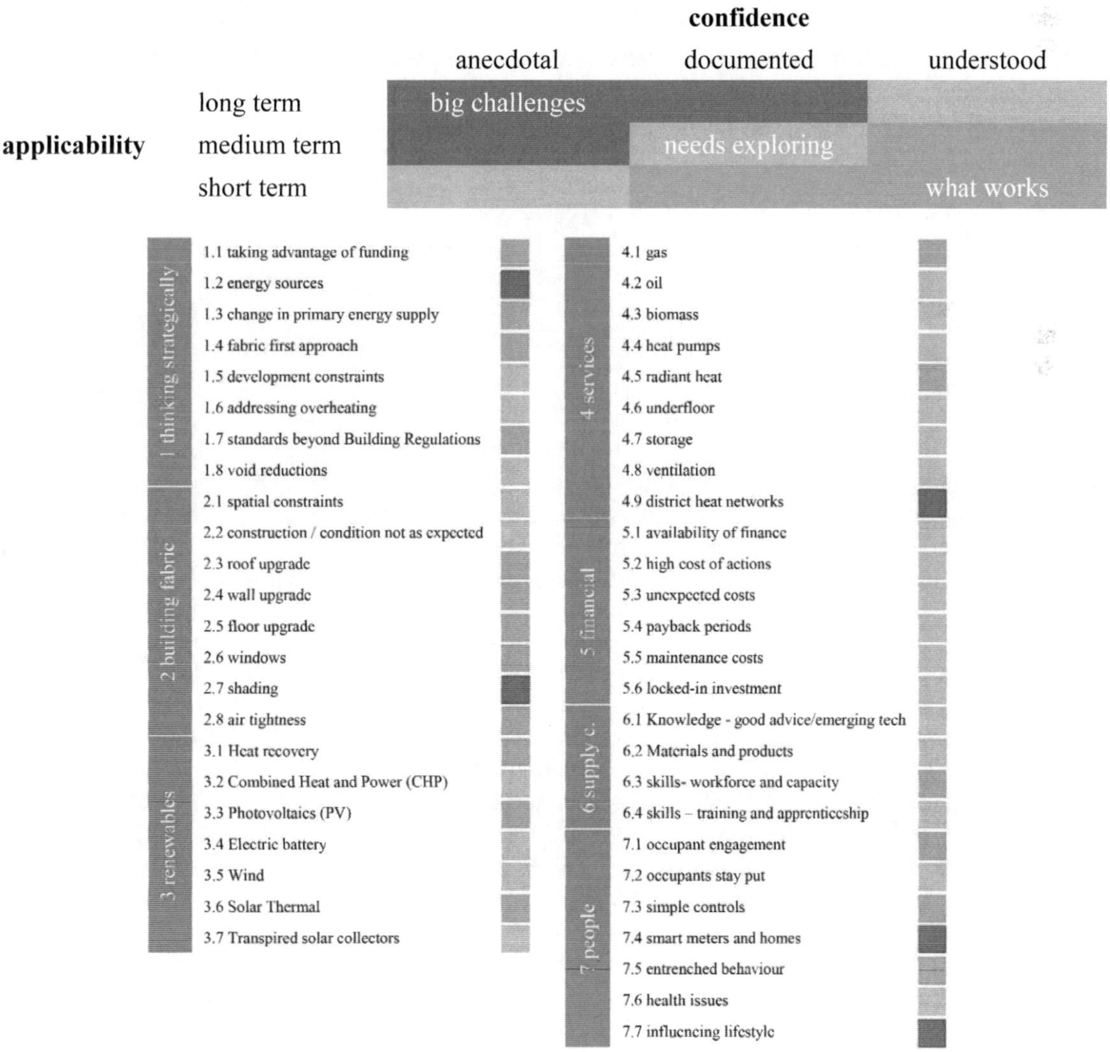

Figure 1. Grouping common retrofit actions into a 'what works' classification

While the Stage 1 scoping review [8] cannot be considered to be comprehensive, it captures the breadth of current, relevant practice and exploration in the professional and academic domains. However, for this resource to remain useful, it must be updated regularly with emerging best practice.

4. Modelling the Welsh housing stock

The primary aim of the Stage 2 research was to understand how the nature of the Welsh housing stock informs the development of a strategy for reducing carbon emissions from the housing sector, and establishes reasonable limits for decarbonisation, by connecting the lessons learnt in the Stage 1 review to the specific context of the Welsh housing stock.

To do so, it was necessary to develop a taxonomy of commonly occurring dwelling archetypes that collectively represent the makeup and condition of the existing Welsh housing stock. SAP models of these archetypes were then used to simulate the decarbonisation of the housing stock using understood approaches to retrofit. Implications for energy efficiency, fuel costs and affordable warmth were considered to be an important part of the modelling process.

Two data sets were used to establish the dwelling archetypes. A large, low level detail dataset from the Valuation Office Agency (VOA) was combined with a smaller, high detail dataset from the Welsh Housing Condition Survey (WHCS), related to circa 2,500 dwellings within the Welsh housing stock (collected 2016-2018), to provide a breakdown by dwelling type and age, as outlined below:

Table 1. Fourteen archetypes collectively represent 84% of the Welsh housing stock.

	End terrace	Mid terrace	Semi-det.	Detached	Flat	Total
Pre 1919	type 1, 3%	type 2, 9%	type 3, 4%	type 4, 7%		23%
1919- 1944			type 5, 5%			5%
1945- 1964			type 6, 10%			10%
1965 - 1990	type 7, 4%	type 8, 6%	type 9, 10%	type 10, 9%	type 11, 4%	33%
Post 1990			type 12, 5%	type 13, 7%	type 14, 1%	13%
Total	7%	15%	34%	23%	5%	84%

The physical characteristics associated with each archetype are considered to be constant, and enabled digital models to be developed for each dwelling archetype. The modelling techniques used in this report are based on the Government's Standard Assessment Procedure for Energy Rating of Dwellings (SAP). The outcomes of SAP are based on the energy required for space heating, water heating, lighting and contributions from renewable technologies. This energy requirement is then used to predict the energy costs and associated Carbon emissions. (A spreadsheet version of the SAP 2012 9.92 worksheet has been developed by the WSA and tested against approved software Stroma FSAP 2012. The spreadsheet version allows the modelling of different fuel costs and emissions rates, which is critical to understand the different outcomes for 1990 and the three future grid scenarios.)

The fourteen dwelling models were then used to evaluate the impact of future retrofit on carbon emissions by comparing the performance of each dwelling as it was in 1990, as it currently performs, and as it might perform in 2050. Further information was needed to simulate the condition of each archetype at these three points in time. This data came from three sources:

- Historic data for the 1990 'baseline' condition came from the United Kingdom housing energy fact file [10]
- Data for the current condition (2016) came from the EPC data for Wales (2016) [3], which represents circa 60% of the total Welsh housing stock (collected over a 9 year period).
- To develop a model for 2050, four retrofit narratives were developed. Retrofit actions were selected based on lessons learnt from the Stage 1 review, developing a specification for each narrative. Each narrative was designed to represent a viable approach for part, or all, of the Welsh housing stock.

Table 2. Four retrofit narratives developed from the Stage 1 review.

Good practice	Actions are driven by best value – in terms of affordability, cost effectiveness, and availability of skills and resources in the current marketplace. Meets current Building regulations, indicative SAP rating of 87.6.
Best practice	Assumes an aspirational client or owner occupier, likely to be more concerned with long term quality than short term cost. Environmental impact is a priority. Exceeds current Building regulations, indicative SAP rating of 89.4.
Heritage	Actions are constrained, e.g. as a result of listed building status or location within a conservation area. Impact on dwelling fabric is assumed to be challenging. Indicative SAP rating of 71.
Rural	Location is assumed to dictate off grid energy solutions. The focus is on energy conservation and use of locally viable renewables. Indicative SAP rating 89.8.

The retrofit of each of the fourteen dwelling types was explored in terms of capital costs, and its impact on energy demand, fuel costs, and carbon emissions. Overheating was also investigated. The capital cost data associated with each retrofit action were provided by cost consultant Lee Wakemans. Predicted costs are based on current cost data and recent, relevant case studies. Some costs are assumed to vary with dwelling size, and others are assumed to be static.

By combining the results arising from each individual dwelling type in their representative proportions, it was possible to illustrate the degree to which potential modifications to the Welsh housing stock could achieve carbon emission reductions, along with associated impacts on energy use and capital cost. The results of this modelling enabled a balanced discussion of approaches to decarbonisation that considers carbon emissions alongside capital costs and impact on fuel bills for occupants.

5. Findings

This paper explores retrofit actions that directly affect the housing stock. However, the impact of key actions external to the housing sector – particularly decarbonisation of energy supply – should not be under-emphasised. To date, emissions from the housing sector are estimated to have reduced by more than 40% from baseline 1990 levels. Three quarters of this improvement comes from cleaner primary energy supply (from changes to the mains gas and mains electricity grid) rather than changes to the housing stock itself.

The degree to which energy supply continues to decarbonise will significantly affect decarbonisation of the housing stock, and could influence the selection and effectiveness of dwelling-based retrofit actions. For this reason, it was deemed necessary to consider decarbonisation of energy supply within this piece of work. Three distinct energy supply scenarios were allowed to influence the models, to explore the impact of potential future changes to energy supply on decarbonisation of the existing housing stock, as follows:

Scenario 1 – minor future improvements to the national grid (40% clean energy supply)
Scenario 2 – significant future improvements to the national grid (60% clean energy supply)
Scenario 3 – transformational change to the national grid (80% clean energy supply).

The resulting discussion established the significance of each of the three energy supply scenarios in informing the development of a route map towards decarbonisation, in the context of current uncertainty around future changes to clean energy supply. Results are observed at the scale of individual dwelling archetypes (subsection 5.1), and in terms of the Welsh housing stock as a whole (subsection 5.2):

5.1 Findings: the dwelling archetypes

The Stage 1 scoping review suggests that there is considerable scope to develop appropriate retrofit strategies for the Welsh housing stock by employing actions that are well understood, and skills and products that are widely available.

Retrofit actions affecting dwelling fabric are the most thoroughly understood. The effectiveness of renewables and systems-based actions is less well understood, being more influenced by innovation and emerging technologies. The influence of people is the least understood aspect of retrofit, and introduces the greatest level of uncertainty around effectiveness, making future work around lifestyle choices and behaviour change particularly important.

The physical size and shape of a dwelling are not necessarily factors that change the approach taken to retrofit, apart from purpose built flats which are prone to overheating (see below). However, the physical size and shape of a dwelling do have considerable impact on capital cost and energy costs.

The selection of 'appropriate' retrofit actions is more likely to be informed by the current condition and character of the dwelling and its surrounding neighbourhood, by prior retrofit actions, and to a certain extent by the personal choice of the occupant or owner.

The predicted reduction in carbon emissions resulting from the retrofit of each dwelling archetype is highly dependent on the decarbonisation of energy supply, as described in table 3, below [8].

Table 3. Reduction in carbon emissions achieved by retrofit of the dwelling archetypes, for four retrofit strategies and three energy supply scenarios

	Carbon reduction by energy supply scenario:		
Retrofit strategy:	1. 40% clean energy	2. 60% clean energy	3. 80% clean energy
Heritage	58-66%	78-83%	100%+
Good practice	64-76%	81-87%	100%+
Best practice	83-89%	92-95%	100%+
Rural	86-96%	93-98%	100%+

After retrofit, 86% of the housing stock are predicted to experience an acceptable risk of overheating (a slight to medium risk, during peak summer months only). 7% of dwellings (mid terraced houses built 1965-1990) are likely to experience an elevated risk of overheating, which would need to be addressed (e.g. through improved ventilation or cooling). 7% of homes (flats built post 1965) would experience a serious increase in overheating (severe risk throughout peak summer months). This would need to be dealt with through adoption of an appropriate retrofit strategy, such as electric heating / cooling via a heat pump or air conditioning package.

Baseline capital costs are predictable for the four retrofit narratives, as described by the ranges below. Low costs are consistently for older, smaller mid terraced properties and high costs are for older, larger detached dwellings, as follows:

Good practice narrative	£17k to £32k
Best practice narrative	£33.5k to £63.3k
Heritage narrative	£10.8k to £25.5k
Rural narrative	£39.4k to £66.8k

The detailed specification of individual retrofit actions can have a considerable further impact on capital cost – particularly decisions around the use of materials or products that are ethically sourced, environmentally sustainable or have related health benefits.

For all narratives other than 'rural', *annual energy costs are predicted to rise considerably when strategies focussing on cleaner energy supply and retrofit of heating systems are employed.* (The 'rural' narrative predicts high existing energy costs by assuming heating is via Economy 7.) The increase in energy costs for the 'best practice' narrative are as follows:

Scenario 1 (which assumes no increase in fuel tariff) average increase 47% (26% to 59%)

Scenario 2 (which assumes a 50% increase in fuel tariff) average increase 120% (89 to 138%)

It is therefore reasonable to suppose that strategies focusing on cleaner energy supply and retrofit of heating systems will impact negatively on fuel bills for occupants. In contrast, when a holistic retrofit strategy is implemented, the percentage *reduction* in energy costs compared to current energy costs are:

Good practice narrative	average of 33% (14 to 49%)
Best practice narrative	average of 29% (20 to 42%)

Heritage narrative	average of 71% (58 to 78%)
Rural narrative	average of 11% (1 to 28%)

In the likely event that the costs associated with cleaner energy supply are passed on to households, there could be considerable increases in energy costs for residents, and therefore in fuel poverty. A low level of dwelling fabric retrofit will leave occupants highly vulnerable to future changes in fuel tariffs.

5.2 Findings: the Welsh housing stock as a whole

To model the housing stock as a whole in 2050, some assumptions were made. It was assumed that 15% of homes must be retrofitted to the 'heritage' standard because their retrofit is constrained by character (typically these would be listed buildings or located in a conservation area). It was assumed that 10% of homes must be retrofitted to the 'rural' narrative because they are situated in off-grid locations. It was assumed that the social housing and private rented sectors (30% of all housing) will be retrofitted to 'best practice' because housing standards provide a mechanism for enforcing holistic retrofit. It was also assumed that a further 15% of the stock, deemed to be in fuel poverty, will be retrofitted to 'best practice', to mitigate against the real risk that decarbonisation increases fuel poverty.

Initially, no assumptions were made regarding the remaining 30% of the housing stock (owner occupied homes, which do not currently have a mechanism for enforcing holistic retrofit). However, the research clearly indicated that unless clean energy supply goes considerably beyond 60%, carbon emissions can only be reduced beyond 90% if all homes are improved to the same standard:

Table 4. Assumptions made regarding retrofit strategy adopted by housing sector

Housing sector	% of stock	Narrative	SAP equivalent
Older protected homes	15%	Heritage	71
Off-grid homes	10%	Rural	90
Social housing + PRS	30%	Best practice	90
Fuel poor households	15%	Best practice	90
Remainder (owner occupied)	30%	Best practice	90

The results of modelling the housing stock as a whole in this way are as follows:

- Under scenario 1 (40% clean energy supply) it is not currently tenable for the existing housing stock to achieve >90% reduction in carbon emissions using established retrofit practice.
- Under scenario 2 (60% clean energy supply) >90% reduction in carbon emissions is only achieved when the entire existing housing stock (excepting 15% assumed to be older protected homes) is retrofitted to SAP90 or higher, with a transition to electricity as a source of heat (i.e. best practice or rural narratives).
- Under Scenario 3 (80% clean energy supply) target levels of decarbonisation are achieved through improvements in energy supply and services alone. However, fabric-based retrofit remains an important part of any decarbonisation strategy. It is critical that fabric-based retrofit is employed to reduce demand in primary energy use (as heating systems, hot water supply and transportation all transition to electricity as a primary energy source) and to minimise increases in fuel bills for occupants, and a consequent increase in fuel poverty.

6. Conclusions: key recommendations drawn from this research

Cleaner primary energy supply from the national grid is an essential part of any decarbonisation route map, and cannot be ignored. UK Government must be lobbied to ensure that energy supplied by the national grid exceeds 60% clean energy by 2050. Smaller scale generation of clean energy at a local or regional level must also be encouraged.

Cleaner primary energy supply is likely to increase fuel tariffs. Action must be taken to reduce household energy use and monitor future trends in fuel poverty, to protect vulnerable households and ensure that a consequence of cleaner energy supply is not an acute increase in fuel poverty.

Dwelling retrofit strategies that upgrade heating systems must also uplift dwelling fabric to an acceptable standard, to diminish increases in energy costs and fuel poverty.

There should be no distinction between standards for retrofit and newbuild, which are confusing and divisive. Similarly, there should be no distinction between standards based on tenure or housing type.

Some housing has retrofit constraints, notably pre-1919 housing with an established character or identity that would be diminished by extensive retrofit of the physical fabric. However the justification for such 'acceptable fails' must be carefully defined, so as not to jeopardise decarbonisation targets.

All other housing must be retrofitted beyond SAP90, which equates to current Building Regulations. The most straightforward way to achieve this currently is to insist all housing targets an EPC A rating.

Phased targets for retrofit of the entire housing stock (in line with carbon budgets) may appear to offer stepping stones for decarbonisation. However, retrofit of most fabric-based actions will be more cost effective if only undertaken once. Other options, such as boiler replacement, are likely to be undertaken more than once before 2050. There is a risk that phased targets will diminish, or obviate, the benefit of work carried out in early phases. In any case, actions that avoid lock-in should be utilised, and economies of scale should be explored.

Prescriptive retrofit approaches can limit opportunities for sensitive retrofit and reduce effectiveness.

To meet stringent decarbonisation targets, retrofit actions must overcome the performance gap (i.e. the results should be measured as delivered, not as predicted). A skilled, trained supply chain is needed to ensure that retrofit is appropriately conceived and properly implemented.

Retrofit standards are easier to enforce for social housing and the PRS sectors. Work must be undertaken exploring how to initiate this level of retrofit in the owner occupied sector. Point of sale represents an opportunity to drive new performance standards.

The impact of people on retrofit remains the least understood part of the process, and the one most likely to impact (positively or negatively) on the effectiveness of any retrofit strategies. Evidence indicates that occupants must be included in the retrofit process, and understand the retrofit actions being undertaken, if the anticipated carbon savings are to be realised.

A flexible approach that pushes all housing to achieve appropriate standards by 2050 is the only way to anticipate achieving >90% reductions in carbon emissions under assumed energy supply scenarios.

References

[1] Committee for Climate Change (2019), Net Zero – The UK's contribution to stopping global warming

[2] BEIS (2017), Updated energy and emissions projections

[3] MHCLG (2016), Energy Performance of Buildings Data England and Wales EPC Data

[4] Welsh Government, Welsh House Condition Survey 2018

[5] Jones, P. J., Patterson, J. L. and Lannon, S. C. 2007. *Modelling the built environment at an urban scale - Energy and health impacts in relation to housing*. Landscape and Urban Planning 83(1), pp. 39-49.

[6] Swan, L.G., Ugursal, V.I. 2009. Modeling of end-use energy consumption in the residential sector: A review of modeling techniques, Renewable and Sustainable Energy Reviews, 13(8): 1819-1835.

[7] BRE, 1998. The Government's Standard Assessment Procedure for Energy Rating of Dwellings. Building Research Establishment for DETR, Garston.

[8] Green, E. Lannon, S. Patterson, J. and Variale, F. (2018). *Homes of Today for Tomorrow STAGE 1: Decarbonising Welsh Housing between 2020 and 2050 - scoping review*. Cardiff University

[9] Green, E. Lannon, S. Patterson, J. and Iorwerth, H. (2019). *Homes of Today for Tomorrow STAGE 2: Exploring the potential of the Welsh housing stock to meet 2050 decarbonisation targets.* Cardiff University

[10] DECC (2013), United Kingdom housing energy fact file

Considering embodied energy and carbon in heritage buildings – a review

F Wise*, A Moncaster, D Jones and E Dewberry

Department of Engineering and Innovation, The Open University, Walton Hall, Milton Keynes, MK7 6AA, UK

*Freya.wise@open.ac.uk

Abstract. Approximately 20% of UK buildings can be defined as 'heritage buildings', offering unique values that should be preserved. They tend to use more energy than newer buildings, creating a strong case for energy retrofits to reduce energy use, greenhouse gas emissions, and improve thermal comfort. However, few studies of heritage retrofits examine embodied impacts, which are the energy and carbon impacts required to manufacture, transport and construct materials and components. This study considers the whole life (embodied plus operational) impacts of retrofitting heritage buildings, through a systematic literature review and thematic analysis. It concludes that; both embodied and operational impacts should be considered in retrofitting projects, retrofitting is better than demolish and rebuild in lifecycle terms, there is a lack of policy mandating for the measurement of lifecycle impacts and low impact retrofitting can be better for conserving heritage values and reducing embodied carbon.

1. Introduction

Heritage buildings have important cultural roles and help to shape the character of our rural and urban areas [1]. Heritage buildings are variously and broadly defined, and they can have a range of historical, aesthetic and communal values [2, 3]. Estimates suggest that approximately 20% of UK buildings could be considered to have heritage value and many of these have varying levels of protection/exemptions in planning law [4].

Reducing energy use and greenhouse gas emissions (referred to herein as carbon) is vital for mitigating climate change. The building sector is responsible for approximately 36% of global energy consumption and 39% of carbon emissions each year [5]. In the UK the building sector is responsible for around 30% of annual carbon emissions, with similar figures for other European countries [6]. There is growing evidence that heritage buildings can have better energy performance than standard models predict and there are strong arguments for their preservation [7]. It is also recognised however that most of these buildings should still be retrofitted to reduce their energy and carbon, while taking their heritage values into account [8, 9].

Currently energy retrofits of heritage buildings, and indeed buildings more generally, often only consider operational energy and carbon savings [10], that is, the energy/carbon spent during the building's use. At present the embodied energy and carbon impacts of retrofit, from the manufacture, transport and construction of retrofit components and their maintenance and replacement throughout a building's lifetime, are rarely examined [10]. This is part of a lifecycle assessment (LCA) approach that considers both embodied and operational energy/carbon to determine lifecycle impacts [11] The potential value of taking a lifecycle perspective on heritage building retrofits has occasionally been identified but not explored in detail [12]. The use of LCA for buildings is becoming more mainstream but is still not commonplace and the need to take this approach is currently not recognised in national UK policy, or used to any great extent in practice [13].

This paper explores the literature on embodied energy/carbon in heritage buildings and addresses the following question: What academic research has been carried out into the embodied carbon and energy of heritage buildings to date, and what are the main conclusions and gaps? This study will inform a forthcoming research project that will include a consideration of any research gaps identified.

2. Methodology

This paper is based on a systematic literature review of embodied energy/carbon in heritage buildings followed by an identification of themes that emerge from the literature. This review offers a thorough and structured search of the literature with clear inclusion criteria [14]. A thematic analysis is then used to identify key areas in the literature that are pertinent to the research question; this is deemed an effective tool for synthesising papers with a broad diversity of methods [15]. In addition, the system boundaries and scope of the papers that undertake LCA is identified. This study will follow a similar method to that of Pomponi and Moncaster [15], who systematically reviewed the literature on embodied carbon in the built environment; this study however will focus on heritage buildings.

2.1. The literature review

The review searched: title, abstract and keywords, in Scopus, Science Direct, Web of Science and Google Scholar, as these represent four of the largest scholarly databases covering this topic. Because LCA in buildings is still a developing field the search was limited to the last decade and included only peer reviewed journal articles in English. The terms in Figure 1 were searched as indicated below, in individual combinations for thoroughness.

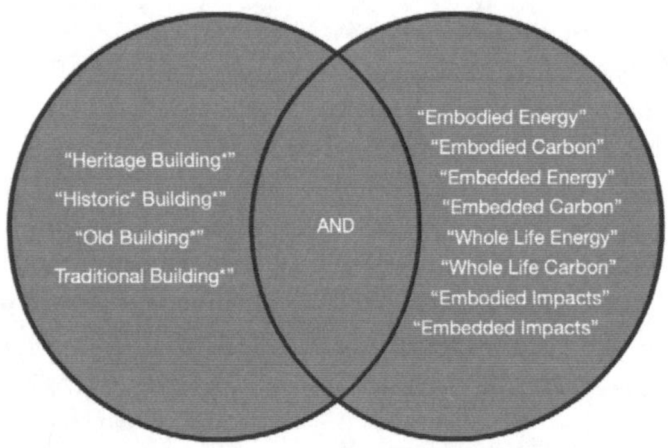

Figure 1. Venn diagram of Search Strings.

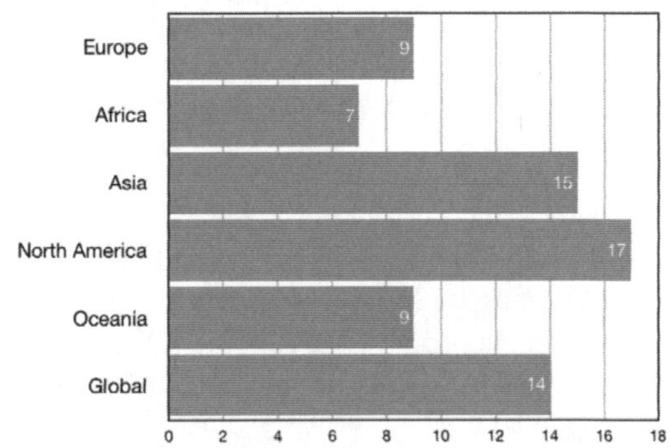

Figure 2. Research distribution by continent.

An initial 948 publications were returned by the search and their titles and abstracts were reviewed to identify meaningful use of the search terms. If there was any uncertainty about a paper's relevance it was retained at this stage. After this process and the removal of duplicates 141 papers remained, which were then read in full and further irrelevant papers were removed, leaving 87 papers relevant for inclusion. It is worth noting that not all of the papers included in this review deal specifically with heritage buildings. 34 of the 87 studies are LCA studies, of which 38% examine heritage buildings while the others assess particular traditional materials or older buildings more generally. These, non-heritage building, LCA studies have been included in the review because they may still have relevant information, transferrable to a heritage setting.

2.2. Thematic Analysis

Summaries of the papers were then created, grouped into broad themes and then refined through several iterations consistent with the principles of thematic analysis as described in [16]. Six key themes were identified through this process. In addition, a quantitative breakdown of the studies by country and publication date was undertaken. The system boundaries, which define the scope of the study, and data sources of the 34 LCA studies were identified, as were the methodologies of the non LCA studies.

2.3. Limitations

Potentially relevant publications from conferences, in books or the technical literature were excluded from this study. Initial examination suggested that conference papers were similar in scope and content to the journal articles reviewed. The effectiveness of the review process is also determined by the search terms meaning that some relevant papers may have been missed.

3. Results and Discussion

3.1. Breakdown of papers reviewed

Figures 2 and 3 show research by continent and country. Research is represented from all continents showing broad global interest in this topic, although almost 40% of papers explore European contexts.

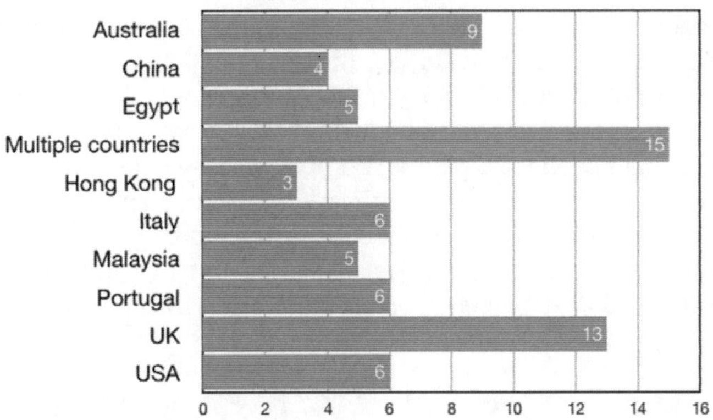

Figure 3. Top ten countries for research.

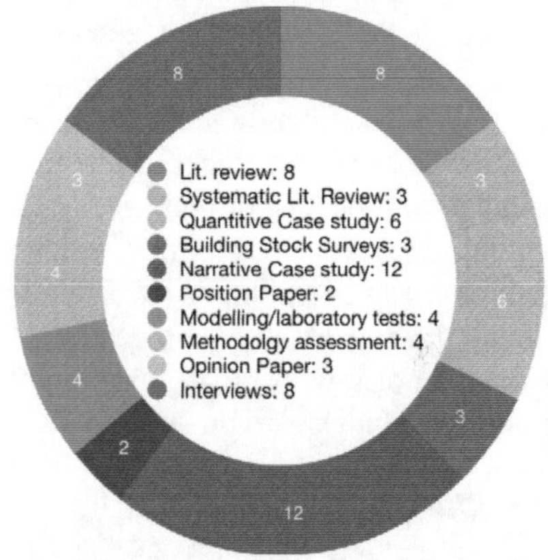

Figure 4. Methodologies of non LCA studies.

The methodologies of the 53 non LCA studies are shown in Figure 4. 34% of papers use narrative or quantitative case studies, making this the most common method, followed by narrative and systematic literature reviews at 20%. This appears consistent with other research identifying the prevalence of the case study method for heritage building research [12].

The publication trends of the papers studied appear to show an overall increase during the period reviewed, Figure 5. There is also an apparent step change in 2012 which, given the preponderance of European studies, could be linked to the publication of the European TC350 LCA standards in 2011-2012 as identified in [17]. This trend follows a general increase in heritage buildings research, evident in recent years. The diversity of journals in which these studies were published is broad, totalling 55 different publications. This indicates the lack of a dedicated journal for heritage buildings and embodied carbon, with relevant articles spread across different fields and disciplines.

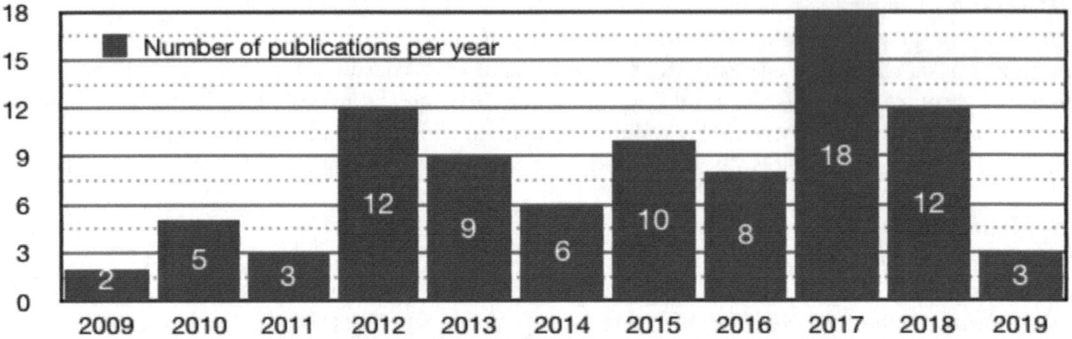

Figure 5. Publications per year during review period.

3.2. System Boundaries of LCA Studies

The system boundaries, data sources and scope of the 34 LCA studies were identified and are shown in Table 1 below. Many of these studies reference the ISO 1440 LCA standards and, in line with this, 88% had clear system boundaries [18,19], only 4 studies (12%) were deemed too unclear to categorise [20-23]. Figure 6 shows the different lifecycle stages. System boundary variation can mean the results of different studies are not directly comparable and as can be seen in Table 1 a range of lifecycle stages are included by different studies. 23% use cradle to grave (A1-C4), 23% cradle to site (A1-A4) and 12% cradle to gate (A1-A3). Cradle to grave is considered to provide the most comprehensive picture of lifecycle impacts, while studies with narrower boundaries are likely to underestimate the full impacts [11]. Studies used a number of different databases for materials' data, with several studies augmenting this with local data to provide a more accurate and contextual picture. For example, by considering the carbon intensity of a specific country's fuel mix [24] or the use of primary data where possible as in [25]. In other studies, however database information has been transposed to different national contexts, such as the use of UK average transport data in a Malaysian project [26], and this may reduce the accuracy of results. A broad variation of methods, system boundaries and data sources are therefore used in the LCA studies and this may reduce the accuracy and comparability of results.

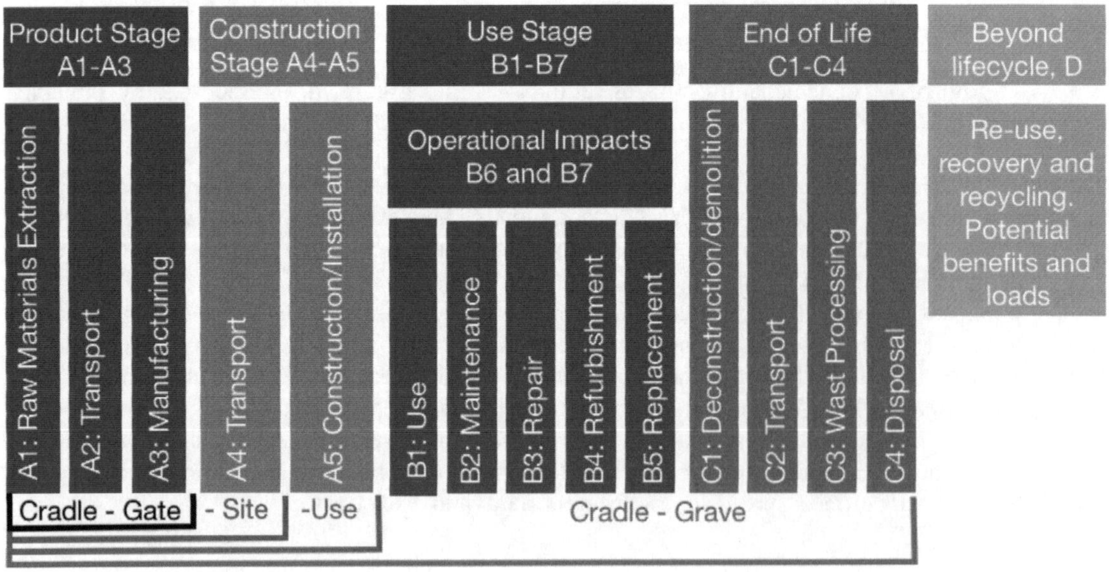

Figure 6: Lifecycle stages relating to buildings adapted from EN 15978 [27].

Table 1: System boundaries and scope of LCA studies.

Author	Lifecycle stages	LCA Database	Building type and scope. ie material, retrofit packages, repair options, etc.*
Almeida et al	Unclear	Multiple	**EBs**: 3 retrofit packages
Asdrubali et al	A1-B5	Eco-Invent	**EB**: 4 retrofit packages
Balleri and Marini	A1-A3	Bath	**EB**: Retrofit package + or - seismic strengthening
Bin Marsono and Balasbaneh	A1-A5	Unclear	**RB**: 7 wall material combinations
Brandão et al	A1-A3	Bath	**M**: Partition wall for HB
Burg and Fuglseth	A1-C4	Eco-Invent	**HB**: Retrofit package, demolish and replace
Chiang et al	A1-A4	Augmented database	**EB**: 32 internal finish options
D'Alessandro et al	A1-C4	Augmented database	**M**: straw bale building
Ding	A1-A4	Augmented database	**HBs**: Main construction materials
Ferreira et al	A1-A5	GaBi	**HB**: Retrofit package or demolish and replace
Franzoni et al	A1-A3	Eco-Invent	**HB**: 52 surface cleaning options
Gong et al	A1-C4	Unclear	**M**: 3 building frame structures
Hu	A1-C4	Athena	**HB**: complete building
Ip and Miller	A1-A5	Eco-Invent	**M**: Hemp line wall system
Iyer-Raniga and Wong	A1-C4	Eco-Invent	**HBs**: range of retrofit options
Kayan	A1-A4	Bath	**HBs**: 3 paint repair options
Kayan et al 2017 (a)	A1-A4	Bath	**HB**: 4 stone repair options
Kayan et al 2018, 2017, 2016, 2011	A1-A4	Bath	No specific building
Khorassani et al	A1-C4	Eco-Invent	**HB**: 1 retrofit package
Kyriakidis et al	A1-A3	Bath	**M**: 3 wall types, 1 traditional
Loussos et al	Unclear	NIBE	**EB**: 3 facade retrofits
Marique and Rossi	A, B6 -C4	Custom tool.	**HB**: Complete building
Mastali et al	A1-A3	Eco-Invent	**M**: Mortar
Pineda et al	A1-A5	Eco-Invent	**HB**: 6 grouts for stonework repair
Pomponi and D'Amico	A1-D	Multiple	**RB**: Timber double skinned facade
Pulselli et al	Unclear	Unclear	**RB**: complete buildings
Rodrigues and Freire	A1-B5	Unclear	**HB**: Insulation variants
Sharif and Hammad	Unclear	Multiple	**EB**: Retrofit package
Shen et al	A1-C4	Multiple	**HBs**: 4 whole buildings
Weiler et al	A1-C4	Eco-Invent	**EB**: 2 retrofit packages, demolish and replace
Zhang and Wang	A1-C4	Unclear	**NB**: complete building

*__HB/s__: Heritage building/s, **EB/s:** Existing Building/s, **NB**: New Building, **RB**: Reference Building, **M**: Materials

3.3. Thematic Analysis

3.3.1 In lifecycle terms, preserving and retrofitting heritage buildings is better than their demolition and replacement with new energy efficient buildings (35 papers). Avoiding demolition waste and the embodied carbon investment of new build is frequently cited as a major benefit of preservation in heritage building adaptive re-use literature [28-31]. This is often stated with only limited detail but seems to be a widely accepted view [32-35]. A number of LCA studies, however, actually assess re-use and retrofit versus demolition and rebuild. A Belgian heritage building study concluded that a retrofit package was 57% better in lifecycle carbon terms than demolish and rebuild [36]. [37] identified that a high level retrofit to a building constructed in 1975 was 15% better than a new construction built to 2016 German standards and [38] identified that re-use could be 4%-46% better in a Portuguese context. [38] identified 13% savings for their actual case study because of extensive structural support requirements due to seismic activity. 2 studies in this review [39, 40] suggested that the embodied carbon already extant in heritage buildings should be included in calculations as wastage if they are demolished, but most studies consider this a sunk cost and only include the embodied carbon required for demolition and new construction. An additional benefit of preservation identified in [41,42] is the temporal element to emissions. Carbon emissions must be reduced now to avoid climate change tipping points and uncontrollable feedback loops. Demolish and rebuild requires a large investment of embodied carbon in the present for future operational savings, which can have a highly negative impact, especially as energy mixes are expected to decarbonise in the future.

3.3.2 Traditional materials can perform better in lifecycle terms than highly manufactured modern solutions but research on more materials in diverse contexts is needed (25 papers). Studies in China and Malaysia assessed the lifecycle carbon of building frame materials, concluding that wood is significantly better than concrete and brick and slightly better than steel [43-44]. [19] examined hemp-lime walls in a UK context, concluding that these walls can have negative embodied carbon due to the potential for carbon sequestration. [33] also showed the carbon sequestration potential of hemp fibres in mortars with recycled aggregates but found that a maximum of 6% fibres was possible for structural applications. Other studies developed novel products; a low embodied carbon, wood and cork partition wall for Portuguese heritage buildings [45], or a double-skinned timber façade for a UK building [18]. Several of these LCA studies only measure initial embodied carbon and do not consider operational carbon or material lifespans [44-47]. As with studies that only look at operational carbon this is an issue because a full lifecycle picture is not developed, making assessments of benefits problematic. Considering the lifespan of different retrofit solutions is also important, to aid comparison of options. Other studies, [48- 51], identify the carbon benefits of traditional materials qualitatively, often in contrast to materials such as concrete, but do not provide any quantitative evidence for these benefits.

3.3.3 The embodied carbon of maintenance and repair should be included in the LCA of heritage buildings (12 papers). It is worth noting that 5 of these papers, by Kayan et al, are markedly similar in content, examining 4 options for stone repair in heritage buildings [26,52,53]. The authors' model identifies the importance of considering the durability and recurrent embodied carbon of different repair options as well as their initial embodied carbon and, in this context, identifies stone replacement as better than plastic repair or repointing. The importance of embodied carbon and the benefits of durability are noted by the same author in [54] when examining roof paint maintenance on Malaysian heritage buildings. A study from Hong Kong [24] explored the embodied carbon, financial cost and labour intensity of different internal finishes. A study of different grouting mortars in a Spanish heritage building [55] identified that hydrated lime grout is better than cement in embodied carbon and structural terms in a repair context. Finally, an LCA on 52 different surface cleaning products for use in Italian heritage buildings identified the high carbon impact of the cotton wool used in many cleaning processes [56]. There is a lack of studies on the maintenance of wooden components and elements such as windows which conservation organisations identify as having high heritage value [57]. This highlights that the durability and maintenance of retrofit options should be considered.

3.3.4 Lack of policy requirements to measure embodied carbon in heritage buildings (10 papers). This is seen as a key challenge to encouraging the calculation of embodied carbon. Two literature reviews identify the lack of, and need for, legislation on embodied energy in heritage buildings [58] and in buildings more generally [59]. [58] also specifically identifies the importance of comparing the lifecycle carbon of different retrofit options. Interviews with Australian building developers identify little incentive for them to invest time and effort in calculating embodied carbon and the majority say they would not undertake this unless it was mandatory [60]. A study from Hong Kong also identified a lack of embodied impact consideration in practice and suggested that governments mandate for the routine availability of better-quality lifecycle data [61]. [62] cites a lack of embodied carbon policies in Australia, Spain and Europe. This lack of policy is particularly important for heritage buildings, where retrofits which affect heritage values must make lifecycle, not just operational, carbon savings.

3.3.5 Lack of tools and certification schemes that facilitate embodied carbon measurement in heritage buildings (12 papers). This is identified in numerous countries, particularly because key schemes such as LEED and BREEAM, which encourage the use of low carbon materials, are generally deemed poorly suited to a heritage building context [63,64]. Meanwhile the Portuguese ECS scheme and other similar decision support tools, often only consider operational carbon [65-68,60]. This is also identified in an Australian context [64,68] where the mandatory Green Star certification system was shown to not fully consider embodied carbon and to be poorly suited to heritage buildings [68]. Two studies [12,41] identify that some tools are being developed, such as a Swedish project covering lifecycle assessment and heritage values impacts and the EU's EFFESUS project. EFFESUS, completed in 2016, created a decision support tool assessing the lifecycle carbon and heritage impact of retrofits; however, the tool is not yet available as the business model is still in development.

3.3.6 Including embodied carbon in a lifecycle perspective significantly effects the most appropriate retrofit options and may be complementary to sensitive retrofit of heritage buildings (13 papers). A key finding is that smaller retrofit interventions are often better in lifecycle terms than more intrusive interventions [57]. An Australian study of different retrofits for eight heritage buildings identified that double glazing with ultra violet film actually increased the lifecycle energy by an average of 2% [62]. In contrast secondary glazing had average savings of 2%, while the addition of thermal curtains to the original windows led to average savings of 3% with a much lower impact on heritage values and reduced financial costs. This links to findings reported in [57], that curtains, shutters and secondary glazing had better lifecycle carbon than most double-glazing options in Scottish heritage buildings. An Italian study of retrofit packages to an 80-year old school identified that the cost optimal package, with smaller interventions, was significantly better in lifecycle carbon terms than retrofitting to government standards or as a nearly zero energy building [69]. [47] showed the low embodied carbon and significant thermal benefits of adding thermal lime plaster to a traditional adobe wall in southern Europe. [41] identified the benefits of smaller interventions and also noted that environmentally conscious user behaviour can significant affect operational carbon in heritage buildings, with almost 50% less energy used in the case study than predicted by modelling. The effect of user behaviour was also considered in insulation retrofits to a Portuguese heritage building where different temperature set points were modelled [70]. A final study, a heritage complex in Spain, considered the project's social and economic impacts in addition to the lifecycle carbon, identifying these as broadly positive [71].

4. Conclusions

This study has identified a developing global interest in this topic, particularly in Europe, and shown that relevant articles are spread across a broad range of journals from different disciplines.

In response to the research question 'What academic research has been carried out into the embodied carbon and energy of heritage buildings to date and what are the main conclusions and gaps?' There are few LCA studies dealing with heritage buildings specifically. There is a broad diversity in system boundaries for the LCA studies and this, and the out of context use of LCA databases, reduces accuracy and makes comparisons challenging. Preserving heritage buildings has been shown by several authors to have lifecycle carbon benefits over demolition and rebuild, with temporal aspects of carbon emissions providing additional support for preservation. Some evidence was found for the carbon benefits of traditional materials and these would also help to preserve heritage values. The importance of including recurrent carbon from repair and maintenance activities was identified, as was the need to consider retrofit lifespans. The literature showed a global lack of policy mandating for the calculation of embodied carbon and identified that the majority of practitioners are unlikely to undertake these measurements without regulation or incentives. A lack of support tools and certification schemes for embodied impacts in heritage buildings was also seen, although there appears to be some development in this area. Finally, the importance of considering the embodied and operational carbon of heritage building retrofits in a lifecycle perspective was identified and it was seen that this approach might be complementary to heritage sensitive retrofitting by providing support for lower impact interventions.

The literature therefore strongly suggests that embodied carbon affects the best retrofit choices for heritage buildings and should be considered in these projects. Embodied carbon influences whether to retrofit or replace and which materials and retrofit options are best in lifecycle terms. More evidence is needed to help build the case to policy makers that the calculation of lifecycle carbon should be included in regulation. Further research on the lifecycle performance of different retrofit options and traditional materials is required to reduce carbon from heritage buildings and help mitigate climate change.

References

[1] Deliotte R E 2017 *Heritage Works: A toolkit of best practice in heritage regeneration* (Historic England)
[2] Mazzarella L 2015 *Energy Build.* **95** 23–31
[3] CADW 2011 *Criteria for assessing historic buildings for the purposes of grant assistance* (Welsh Government)
[4] Pickles D and McCaig I 2017 *Energy Efficiency & Historic Buildings: Application of Part L of the Building Regulations to Historic & Traditional* (Historic England)
[5] Global Alliance for Buildings and Construction 2017 *Global Status Report 2017* (UN Environment)
[6] UK Green Building Council and Aldersgate Group 2017 *Energy efficiency in the UK's buildings: key priorities for the new government* (London)
[7] Pracchi V 2014 *Hist. Environ. Policy Pract.* **5** 210–25
[8] Cabeza L F, de Gracia A and Pisello A L 2018 *Energy Build.* **177** 96–111
[9] Ascione F, Bianco N, De Masi R F, de'Rossi F and Vanoli G P 2015 *Energy Build.* **95** 172–83
[10] May N and Griffiths N 2015 *Planning Responsible Retrofit* (Sustainable Traditional Buildings Alliance)
[11] Birgisdottir H, Moncaster A, Houlihan Wiberg A, Chae C, Yokoyama K, Balouktsi M, Seo S, Oka T, Lützkendorf T and Malmqvist T 2017 *Energy Build.* **154** 72–80
[12] Webb A L 2017 *Renew. Sustain. Energy Rev.* **77** 748–59
[13] UK Green building Council 2015 *Tackling Embodied Carbon in Buildings* (London: UK Green Building Council)
[14] Sawyer R K 2017 *Educ. Res. Rev.* 15
[15] Pomponi F and Moncaster A 2016 *J. Environ. Manage.* **181** 687–700
[16] Braun V and Clarke V 2006 *Qual. Res. Psychol.* **3** 77–101
[17] Moncaster A M and Symons K E 2013 *Energy Build.* **66** 514–23
[18] Pomponi F and D'Amico B 2017 *Build. Environ.* **124** 42–56
[19] Ip K and Miller A 2012 *Resour. Conserv. Recycl.* **69** 1–9
[20] Almeida M, Ferreira M and Barbosa R 2018 *Buildings* **8** 103
[21] Sharif S A and Hammad A 2019 *J. Build. Eng.* **21** 429–45
[22] Loussos P, Konstantinou T, Van den Dobbelsteen A and Bokel R 2015 *Buildings* **5** 622–49
[23] Pulselli R M, Niccolucci V and Marchettini N 2009 *Int. J. Des. Nat. Ecodynamics* **4** 11–5
[24] Chiang Y H, Li J, Zhou L, Wong F K W and Lam P T I 2015 J. *Clean. Prod.* **109** 326–35
[25] D'Alessandro F, Bianchi F, Baldinelli G, Rotili A and Schiavoni S 2017 *J. Build. Eng.* **11** 56–68
[26] Kayan B A, Zaid N S M and Mahmud N S 2017 *J. Des. Built Environ.* 15
[27] European Committee for Standardization, 2012b. *EN 15978: Sustainability of construction*

works - Assessment of environmental performance of buildings - Calculation method
[28] Allen A D 2012 *Transylv. Rev. Adm. Sci.* **8** 11–8
[29] Thornton B J 2011 *J. Green Build.* **6** 1–7
[30] Matthews J, Dyson K and Love P E D 2016 *Built Environ. Proj. Asset Manag.* **6** 44–57
[31] Khalil A M R, Hammouda N Y and El-Deeb K F 2018 *Heritage* **1** 57–87
[32] Throsby D 2016 *City Cult. Soc.* **7** 81–6
[33] Ragheb G, Ragheb A A and Ragheb R A 2017 *Int. J. Curr. Eng. Technol.* **7** 8
[34] Firzan M, Keumala N, Zawawi R and Ahmad A G 2015 *Built. Environ.* **12** 13
[35] Craun Z 2012 *J. Int. Aff. N. Y.* **65** 157–XII
[36] Marique A-F and Rossi B 2018 *J. Environ. Manage.* **224** 396–405
[37] Weiler V, Harter H and Eicker U 2017 *Energy Build.* **134** 319–28
[38] Ferreira J, Duarte Pinheiro M and de Brito J 2015 *J. Build. Eng.* **3** 114–26
[39] Akande O K, Odeleye D and Coday A 2014 *Int. J. Sustain. Dev. Plan.* **9** 237–50
[40] Merlino K R 2014 *Public Hist.* **36** 70–85
[41] Berg F and Fuglseth M 2018 *J. Archit. Conserv.* **24** 152–67
[42] Zhang X and Wang F 2017 *Energy Build.* **155** 404–13
[43] Gong X, Nie Z, Wang Z, Cui S, Gao F and Zuo T 2012 *J. Ind. Ecol.* **16** 576–87
[44] Bin Marsono A K and Balasbaneh A T 2015 *Constr. Build. Mater.* **85** 100–8
[45] Brandão F J da S, Paio A, Sousa J P and Rato V 2016 *Gest. Tecnol. Proj* **11** 9–24
[46] Mastali M, Abdollahnejad Z and Pacheco-Torgal F 2018 *Constr. Build. Mater.* **160** 48–56
[47] Kyriakidis A, Michael A, Illampas R, Charmpis D C and Ioannou I 2018 *Energy* **161** 1016–27
[48] Quagliarini E, Stazi A, Pasqualini E and Fratalocchi E 2010 *Sustainability* **2** 3291–308
[49] Kolokotsa D, Maravelaki-Kalaitzaki P, Papantoniou S, Vangeloglou E, Saliari M, Karlessi T and Santamouris M 2012 *Sol. Energy* **86** 1648–59
[50] Ede A N, Olofinnade O M, Edwin E-A and Bamigboye G 2017 *Int. J. Appl. Eng. Res.* **12** 7873–83
[51] Keable R 2010 *Carbon Clim. Law Rev.* **2010** 357–63
[52] Kayan B A, Halim, I A and Mahmud N S 2017 *Chem. Eng. Trans.* 337–342
[53] Kayan B A, Banfill P F G and Forster A M 2016 *Smart Sustain. Built Environ.* **5** 143–64
[54] Kayan B A 2017 *Int. J. Build. Pathol. Adapt.* **35** 63–89
[55] Pineda P, García-Martínez A and Castizo-Morales D 2017 *Constr. Build. Mater.* **138** 528–47
[56] Franzoni E, Volpi L, Bonoli A, Spinelli R and Gabrielli R 2018 *Energy Build.* **165** 92–105
[57] Curtis R 2010 *J. Archit. Conserv.* **16** 7–27
[58] Lidelöw S, Örn T, Luciani A and Rizzo A 2019 *Sustain. Cities Soc.* **45** 231–42
[59] Zeng R and Chini A 2017 *Energy Build.* **155** 172–84
[60] Wilkinson S J and Remoy H 2017 *Sustain. Build.* **2** 1–10
[61] Langston C, Chan E H W and Yung E H K 2018 *Sustainability* **10** 3229
[62] Iyer-Raniga U and Wong J P C 2012 *Build. Environ.* **47** 138–49
[63] Loli A and Bertolin C 2018 *Buildings* **8** 22
[64] Balderstone S 2012 *Hist. Environ.* **24** 6
[65] Nunes P, Lerer M and Graça G C da 2013 *Energy Eng.* **110** 7–34
[66] Ferreira J, Duarte Pinheiro M and de Brito J 2013 *Constr. Build. Mater.* **49** 425–47
[67] Baker H, Moncaster A and Al Tabbaa A 2017 *Proc. Inst. Civ. Eng. - Forensic Eng.* **170** 144–56
[68] Conejos S, Langston C, Chan E H W and Chew M Y L 2016 *Build. Res. Inf.* **44** 507–19
[69] Asdrubali F, Ballarini I, Corrado V, Evangelisti L, Grazieschi G and Guattari C 2019 *Build. Environ.* **147** 461–72
[70] Rodrigues C and Freire F 2017 J. *Clean. Prod.* **157** 94–105
[71] Khorassani S M, Ferrari A M, Pini M, Blundo D S, Muiña F E G and García J F 2018 *Int. J.* Life Cycle Assess. 1–22

Innovations for a carbon free Dutch housing stock in 2050

Henk Visscher

Architecture and the Built Environment, Delft University of Technology, Julianalaan 134, 2628 BL Delft. Email: h.j.visscher@tudelft.nl

Abstract. In 2017 the new Dutch coalition government agreed on high ambitions for the reduction of CO2 emissions to meet the Paris agreement and policies to achieve this were detailed in a Climate Agreement between the government and many stakeholder organisation in June 2019. In 2030 CO2 emissions have to be reduced to 49% compared to 1990. In 2050 we should have a CO2 emission free and energy neutral built environment. The existing housing stock plays a major role in the realisation of these goals. The Dutch housing stock comprises of 7.5 million dwellings. The majority of which has a rather bad thermal energy condition and are mainly heated by natural gas. All these houses have to be renovated to a nearly zero carbon performance. By 2021 yearly more than 50,000 new homes per year will have to be delivered on nZEB level without a natural gas connection and at least 50,000 existing dwellings will have to be made gas-free per year. These are steps towards 200,000 zero carbon renovations per year, a pace that is needed to make the entire stock carbon free in the 30 years up to 2050. To support these goals a large innovation programme is currently being developed. This paper presents the required innovations. Research on the impact of the policies and measures applied in the recent years point to the challenges that have to be overcome. A particular challenge is to realize the intended performance and really save energy and reduce CO2 emissions on a large scale.

Keywords: Nearly zero energy housing renovation; CO2 free built environment, Guaranteeing high performance houses

1. Introduction

In 2017 the new Dutch government set some severe goals to meet with the Paris agreement of limiting the global heating to a maximum of 2 degrees Celsius. To operationalise the goals the government installed some 'round tables' with many stakeholder groups to negotiate a new national Climate Agreement. In June 2019 the Agreement was a fact and it contains many detailed goals, policies and regulations to achieve the goals [1].

In the 1960's natural gas was discovered in the northern province Groningen and was heavily exploited. Ever since then most Dutch houses were heated by this gas. From 2010 however, earthquakes started to happen in Groningen and geological investigations proved a direct causal relation to the extraction of gas form the earth. Since then the earthquakes appeared more and more often and became heavier, causing cracks in the houses and creating unsafe situations. The pressure on the government to reduce the extraction of the gas, increased. The production is already largely reduced. Finally, recently the government decided to completely stop the gas extraction in 2030. The gas from Groningen has another caloric composition than e.g. Norwegian or Russian gas and the existing heaters in Dutch homes cannot use the other gas. So, besides the climate argument, also the simple fact that Dutch houses have to find another energy source is a very strong argument for the energy transition in the Dutch housing stock. The currently available solutions are either connecting the houses to a district heating network, with heat from e.g. a power plant, or heat from geothermal source, find alternative gas solutions (bio gas, hydrogen), or go to all electrical solutions with a heat pump. At the moment such a solution is only sensible if the house is very well insulated.

According to the Climate Agreement CO2 emissions have to be reduced to 49% in 2030 compared to 1990. In 2050 we should have a CO2 emission free and energy neutral built environment. The existing housing stock plays a major role in the realisation of these goals. The Dutch housing stock comprises of 7.5 million dwellings. The majority of which has a rather bad thermal energy condition and are mainly heated by natural gas. All houses have to be renovated to a zero-carbon performance. By 2021 yearly more than 50,000 new homes per year will have to be delivered nZEB without natural gas connection and heating and at least 50,000 existing dwellings will have to be made gas-free per year. These are steps towards 200,000 zero carbon renovations per year, a pace that is needed to make the entire stock carbon free in the 30 years up to 2050.

2. Climate agreement

The goals have been worked out into detailed proposals for policies and measures laid down in the Climate Agreement [1]. This is the result of months of negotiations by the so called 'sector round tables' with representatives of all relevant stakeholder groups like home owners, construction companies, building service companies, other relevant industries, energy companies, the financial sector and many more. In order to achieve the target of 3.4 Mton CO2 reduction in the built environment in 2030, approximately 1.5 million existing homes have to be made sustainable and the CO2 emissions in existing utility buildings must be reduced by 1 Mton extra in 2030.

In order to make the challenge manageable, the sector table proposes a phased and programmatic approach, on the one hand working on an energetic start and on the other hand on the conditions for innovations and later up-scaling and rollout. It is important that offering parties quickly learn, through, among other things, the so called 'Starter motor' projects (projects of housing associations with large subsidies to make the investments budget neutral), how to efficiently (more) make large numbers of homes more sustainable. The lessons from this first larger bundled demand offers opportunities for technical and organizational innovations. This allows providers to develop a more efficient and cheaper offer. Clients gain experience with tending larger numbers of renovations. These are necessary conditions for actual up-scaling. There will have to be broad experience with what is cost-effective and practically feasible in different situations, before tackling scaling up and being able to roll out on a large scale. That also gives time to work out the conditions for later scaling up and deployment carefully.
For homes, the approach is a combination of seduction and direction through the 'Neighbourhood-oriented approach'. Building owners can also be tempted at an individual level to become more sustainable. This approach succeeds if sustainability can be recouped through the (decreased) energy bill. In order to finance these investments through energy saving and lower energy costs and make them affordable, much innovation and cost reduction is still required. That is why will be started with experiments and with an innovation program to systematically learn and experiment, so that after some years a cost-effective up-scaling and roll out can take place. The conditions for up-scaling, innovation, greater efficiency and cost reduction will be realized by, making funding available and making proposals to take away the remaining unprofitable summit of building-related measures, infrastructure and sustainable sources through pricing and subsidizing. This involves a slider in the energy tax, which stimulates investments in sustainability and attractive financing methods, which means that investments can also be paid. Further incentives and action perspectives are offered by making known through standardization which (final) situation buildings have to be brought in order to be able to be heated without natural gas in the future.

To summarize: A broad spectrum of policies, measures and financial support will be applied to boost the renovations of homes. In the first years his won't be cost effective, but because of innovations and up-scaling the cost should be reduced to a cost-effective level. By then regulations and financial arrangements will more or less force all home owners to take part in the renovation programme.

IOP Conf. Series: Earth and Environmental Science **329** (2019) 012003 doi:10.1088/1755-1315/329/1/012003

This paper will focus on new and improved regulations for the energy transition in houses. Research on the impact of the policies and measures applied in the recent years point to the challenges that have to be overcome. A particular challenge is to realize the intended performance and really save energy and reduce CO2 emissions on a large scale.

3. Goals and policy measures

According to the European Energy Performance of Buildings Directive (EPBD) all new buildings will have to meet an Energy performance standard. By 2021 the standard will be on the level of nearly zero energy (nZEB – nearly Zero Energy Building). Currently member states still have their own calculation tool and performance level for the interpretation of 'nearly zero'. In the Netherlands this will be: total energy demand 25 kWh/m2.yr; primary fossil energy use 25 kWh/m2.yr and share renewable energy 50%. For existing buildings an Energy Performance Certificate is required at the moment of selling or renting out. The EPC, or energy label, is a presentation of the theoretical energy performance of the building. Only for large renovations the standard for newly built buildings is required. For existing buildings there is currently no tool to force improvement.

The new Climate Agreement [1] proposes the development of a new Standard to better facilitate and to force the renovation to fossil free houses. The ideas for this new standard are proposed as follows: It is possible to determine 'sensible' sustainability renovation option for dominant or characteristic types of housing in the Netherlands, based on cost benefits and desired reduction of heat demand. A so-called 'regret-free' refurbishment with which the building owner is assured that several times within the technical lifespan a radical adaptation to the same building parts is not necessary, in anticipation of the alternative to natural gas that is chosen in the neighbourhood-oriented approach. This standard is formulated at the level of the entire house (net heat demand in kWh / m2 / year, following the NTA8800 determination method). By determining the standard, the sensitivity of the standard for the (later) definitive choice of the heat alternative is explored. Not everyone will be able to renovate the whole house. For renovations, where only one or a few building parts are being tackled (such as roof, façade, floor), target values are given for insulation (in Rc, or U values) and required ventilation. The standard for the entire house is leading, the target values for building parts contribute to this. The standard and the target values will be set no later than 1 July 2019. The standard can be considered as one of the inputs for the guideline and therefore for choosing the most suitable heat source for a district. The 'regret-free' standard is a means to achieve the intended purpose of a low-carbon built environment. The intended standard for the existing building is not mandatory for owner-occupiers yet, but gives interpretation about the desired energy performance of existing homes in advance of the neighbourhood-oriented approach. In 2025, the standard will be evaluated on the basis of criteria to be determined, in conjunction with other instruments and the neighbourhood-oriented approach. After that, the standard can be tightened up if necessary, better supported or it can get a more binding character.

Unlike owner-occupiers, tenants do not have the freedom to decide for themselves how the houses will be adapted to meet the minimum requirements imposed on the dwelling from the alternative heat source. In order to give landlords an action perspective and to protect tenants against high energy costs, the standard in 2050 is therefore mandatory for homes intended for letting. The obligation for landlords does not mean that they disproportionately share in collective costs (volatility risk). Landlords are responsible for adapting a home, so that it meets the standard at the time that the housing is connected to the new infrastructure via the neighbourhood-oriented approach. Tenants will lend their cooperation to the necessary adjustments to prevent them or subsequent tenants from being left out in the cold or receiving very high energy costs. In order to optimally inform homeowners of all options to renovate their homes, validated information on sustainability measures and the accompanying indicative energy savings and investment profiles will be provided on a website as of 1 January 2020. This is linked to financing and subsidy opportunities.

4. Research findings: impact of policies in recent years

While developing new policy tools for the up-scaling or renovations and energy saving measures we should take in to account some lessons we learned from the applied regulations and measures in the recent years.

One of the most important policy tools required by the EPBD in the European Member States is the issuing of Energy Performance Certificates (or EPCs). These EPCs give an indication of the required energy to provide a certain average temperature in the building and depend on physical characteristics of the building. The certificate has no mandatory implications in the sense that owners could be forced to improve their buildings to certain levels. Nonetheless it is a crucial instrument for benchmarking and formulating policy goals. Building owners in all EU Member States have to obtain an EPC for a building at the moment it is sold or rented out. This is not yet current practice everywhere, mostly due to lack of enforcement. This especially applies to the private housing stock.

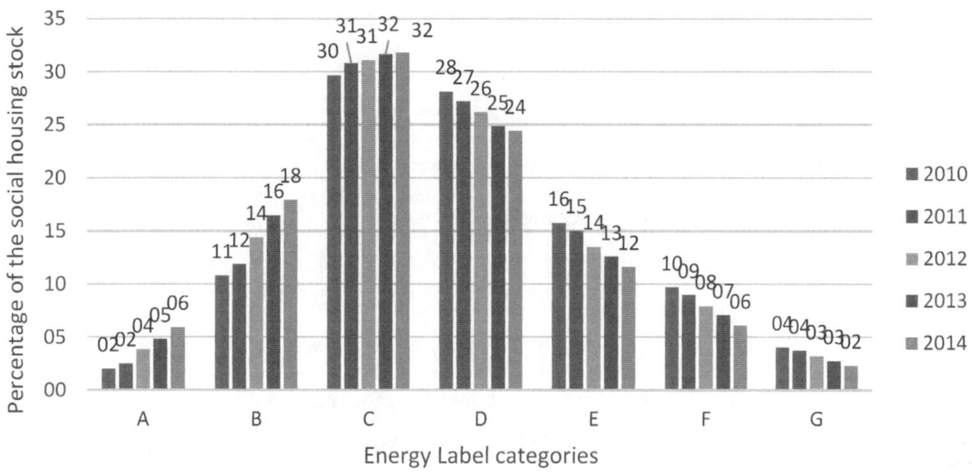

Figure 1: Distribution of the energy labels of the non-profit rented housing sector in SHAERE database [2]

The social housing sector in the Netherlands agreed with the government and the national tenants' union to a covenant about energy renovation goals. This was initiated in 2008 and updated in 2012. Most important goal is to initiate renovations that lead to an average energy label B in the beginning of 2021 for the whole sector, which comprises 2.3 million dwellings (35% of the total housing stock in the Netherlands). All social housing is already labelled for many years. Since 2010 nearly all dwellings have a fairly reliable label. Every year AEDES, the national umbrella organisation of the housing associations, collects the energy label data of all housing associations. Every year more associates really do so. In 2015 the data of more than 60 % of the whole social housing stock was collected in this way. This database is called SHAERE. Research with the SHAERE data base enables showing the progress in renovation. Figure 1 demonstrates the label steps over the years 2010 to 2014. It can be noted that most of the renovations lead to small improvements. If the current figures are extrapolated to 2021, we can see that the goals of an average label B will not be reached (Figure 2). The label indexes relate the calculation of the Energy Index, which is for label B 1,25. In 2015, the calculation method for the EI changed with the purpose to simplify it. Due to the changes the labels form before and after the change are difficult to compare. Therefor also the continuation of the improvement line can't be drawn and it is also hard to conclude about whether or not the targets of the covenant will be met in 2021. On the other hand, the progress between 2010 and 2015 shows clearly that the formulated goal won't be reached. At this moment the actual goals for the built environment are reconsidered. Also, the housing association are debating about the goals for 2050 and the scenarios for reaching these.

IOP Conf. Series: Earth and Environmental Science **329** (2019) 012003 doi:10.1088/1755-1315/329/1/012003

Stimulated by government policies in 2013 an agreement was made by a group of 27 housing associations and buildings and the Dutch government to work on a programme for 111,000 houses to Net Zero Energy (E=0) levels. This was called Energiesprong (Energy leap) [3]. E=0 means, annually a house does not consume more energy for heating, hot water, lights and appliances than it produces. The concept is based on refurbishments financed by the energy cost savings; a refurbishment executed within 10 days and a 40-year energy performance warranty from the builder. The approach is further based on organizing a massive demand for a Net Zero Energy (E=0) refurbishments, making financers and governments tune their financing offerings and regulations towards this product and challenging the construction sector to start an ambitious innovation process. Since the take of of the programme several projects have started and some houses are yet already taken into use and even some monitoring results are yet available [4]. Recently Energiesprong is working also in the UK and France to get an independent market development team set up to drive a similar approach as in the Netherlands.

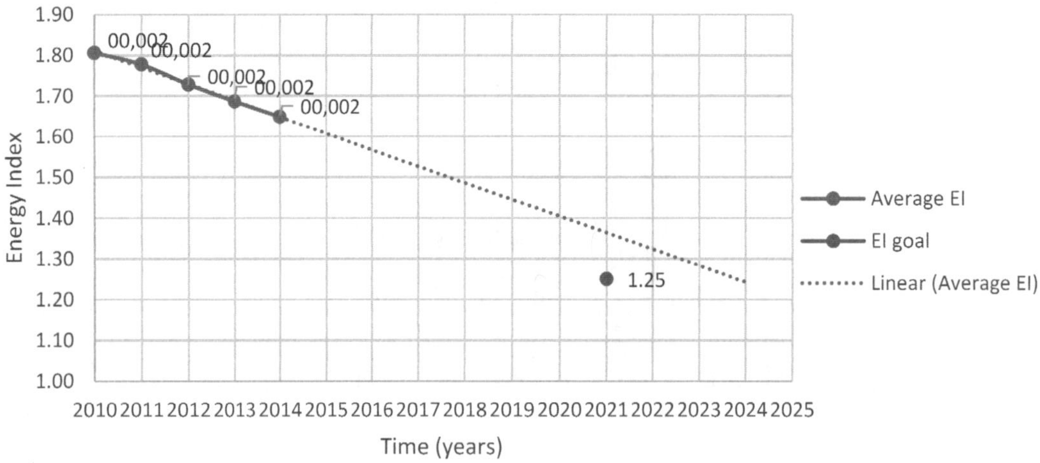

Figure 2: Development of the EI in the Dutch non-profit housing sector since 2010. [5]

The costs for Net Zero Energy refurbishment are substantially more than the net present value of today's energy bill. Therefore, an intensive innovation process in the construction sector is needed focused on reduction of costs and process time and energy performance guarantees. This will require moving from doing one-off projects towards developing mass-produced refurbishment packages with a performance standard. And then the idea is that these packages will be sold in shops, which are cosy and have helpful staff. To catalyse that innovation process, a large demand volume this new product is needed.

The financing of the refurbishment is done by giving the tenants an "energy plan" including a guarantee for a hot house (22o), warm water (certain amount of shower time per day) and an electricity bundle for electric appliances for which they together pay a fixed monthly fee. If they exceed that agreed amount of energy performance, they pay the additional electricity consumption to the utility. The bundle fee is paid to the housing association (HLM). This pays of the investment. In the private sector it works similarly, with an increase in mortgage instalments to finance the investment covered by the reduction in energy costs. In the Dutch case, the cost of living stays therefore the same for tenants or private homeowners. Investing in factories are required to pre-fabricate these refurbishment solutions with a view to scale. It helped the government to change regulation (i.e. ability for associations to collect the energy plan money) and it made the financier revalue the properties that would be refurbished in order to free extra room to borrow money. The first 200 prototypes and test houses have been built. By the end of 2016, there should have been 10000 Net Zero Energy refurbishments completed. Following that deal, Energiesprong is now working to bring these refurbishment packages to the private homeowner sector.

A common question is whether this only works for certain houses that all look the same. In fact, it is mass customization. Houses are all 3D scanned and refurbishment packages are individually produced because even a set of row houses are not all exactly the same, even if it looks like they are. The focus until now has been on the most prevalent typologies of houses in the Dutch market of which 2.3 million houses in the Netherlands exist, but flexibility will increase over time with the number of possible solutions. The first designs will establish whole new supply chains based on prefabricated industrially production methods providing new platforms for further, interactive innovation process (very much like the car industry).

The programme is running now for a few years and we can see that the planned number of projects and dwellings are not yet renovated and / or in development. The investment costs are still too high: about €70.000. It will only become a feasible business case if the cost could be reduced to about €40.000. The housing associations that invested in the start-up of the programme with the expectation that the cost would drop if the numbers would increase. Currently the progress seems to slow down.

The actual domestic energy use is besides the physical characteristics of a dwelling, largely influenced by the use and behaviour of the tenants. Some preliminary figures demonstrate the difficulty in 'forcing' reduced energy use by improvements of dwellings. The dwellings with the worst EPC (G) in practise use far less energy as expected, while the most advanced dwellings (A) use much more. This is probably due to a combination of the rebound effect and an increase in comfort level of the dwellings [6,7,8] and underperformance of the buildings and installations. The large difference between theory and practice is called the performance gap and is recognised in more and more international studies. In a research project by [6,7] the actual energy consumption was compared with the theoretical use according to the EPC's (Figure 3).

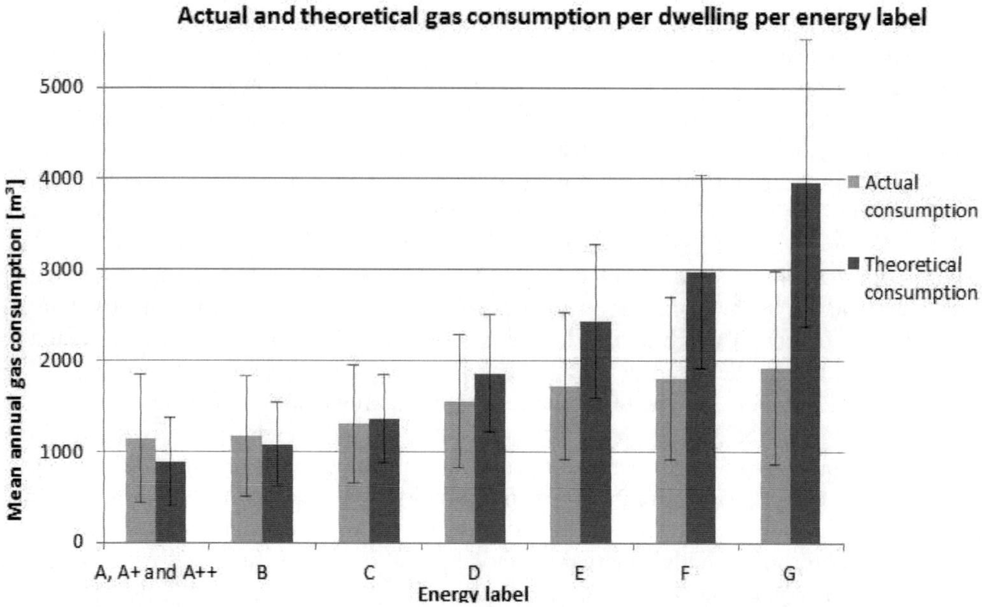

Figure 3 Actual and theoretical gas consumption in Dutch dwellings - per m2 dwelling area. [6]

This research was based on the Dutch energy labels issued in 2010 - a total of over 340,000 cases with 43 variables (regarding building location and technical characteristics, the properties of the label itself etc.). This data set was derived from the publicly available database of the EPCs. This data was, on the basis of the addresses of the households, linked to actual energy use data. The energy data was provided by the CBS (Statistics Netherlands), which collected this data from the energy companies. The combined data file was then cleaned by deleting incomplete or obvious incorrect EPCs. This resulted in

193,856 usable cases. This still large sample proved to be representative for all housing types and energy label classes.

To understand how the energy label relates to the discrepancies, the gas and electricity consumption in various label categories were examined and analysed. The actual and theoretical gas use per dwelling was compared and then analysed per m2 of dwelling (Figure 5). Little difference exists between the actual and theoretical energy use calculated per dwellings and per m2 , except the difference in actual gas use between label A and label B. At the level of individual dwellings, the actual consumption was identical, but at the level of m2 the dwellings in category A use less gas than dwellings in category B. This may relate directly to the fact that dwellings in label category A were found to be considerably larger than all other dwellings. From these figures it is clear that although better labels lead to higher actual gas consumption, there is a clear difference between the mean theoretical and mean actual gas consumption for each label. For the most energy-efficient categories (A, A+ and A++) and for category B, Figure 5 shows that the theoretical calculation underestimated the actual annual gas consumption. This is in contrast to the rest of the categories for which the theoretical calculation largely overestimated the actual annual gas consumption. This research indicates that the energy label has some predictive power for the actual gas consumption. However, according to the labels, dwellings in a better label category should use on average significantly less gas than dwellings with poorer labels, which is not the case [9, 10, 11]. This finding also has a major impact on the way the improvement policies for the energy use in the housing stock are formulated [12] and what can be expected from the renovations.

5. Discussion

Without any doubt there is a necessity to drastically reduce the use of fossil-fuel energy sources by reducing the demand for energy and switching from fossil to renewable sources. Buildings account for 40% of Europe's energy consumption and three-quarters of the floor area of the building stock is residential. The targets are clear and the technical solutions are available. Good insulation and product innovations can reduce the energy demand for heating and cooling for a large part. The remaining energy demand can be delivered by renewables like sunlight and heat, district heating, heat pumps, etc. The remaining electricity demand for appliances can in the first place be reduced by further product innovation and then be provided by photovoltaic panels. There are no reasons not to apply these solutions in new buildings at a large scale on the short term. Evaluations of the current practice show however that there is a lot to be gained here. To improve this situation, it has to be assured that constructions and installations are installed properly and in such way that they are not vulnerable for unpredictable or misuse by the occupants. This will set demands on both the construction industry as on the control and enforcement process (and the parties responsible). The improvement of the existing building stock is a huge challenge. The potential energy savings are large, but the barriers to overcome are also high. As stated before, almost three quarters of the future housing stock (2050) has already been built. Studies show however that it is hard to increase the rate and depth of energy renovations of the existing stock. Actual energy (and financial) savings in renovated dwellings stay behind expectations because of rebound effects. There are important barriers. Many owners believe that the benefits of the measures do not outweigh the costs. Besides that, the cost of improving the energy performance of a dwelling does not (proportionally) increase the value of the dwelling. The new Dutch Climate Agreement should deal with these challenges.

6. Conclusion

The Paris agreement in combination with the fact that the availability of own natural gas in the Netherlands will stop in 2030 urges for a policy and innovation plans with very high ambitions. The government will apply a large arrange of policies, measures and financial support. The standards to calculate the energy demand of buildings will be renewed to better facilitate a step wise, no-regret improvement to gas free solutions. Technical and process innovations and up-scaling should make renovations cheaper. In the next years many new approaches have to be developed and implemented. By all of this we should keep monitoring the real impact of the policies and regulations and also measure the real reduction of CO2 emissions. Previous research has shown the failure of calculation tools and renovation agreements. Furthermore, the plans and renovation standards might have to be adapted to external developments like price developments and developments of the availability of renewable energy and the possibility of storage of renewable energy.

7. References

[1] Rijksoverheid, 2019, Klimaatakkoord, https://www.klimaatakkoord.nl/
[2] Filippidou, F, Nieboer, N, Visscher, H, 2017, Are we moving fast enough? The energy renovation rate of the Dutch non-profit housing stock using the national energy labelling database, Energy Policy 109, 488-498.
[3] Rovers, R., (2014), New energy retrofit concept: 'renovation trains' for mass housing, Building Research and Information, 42 (6) 757-767.
[4] Energiesprong, 2016, Eerste ervaringen met prestatiegarantie voor nul op de meter woningen (First experiences with performance guarantees for zero on the meter dwellings). http://energielinq.nl/resources/monitoringprestatiegarantie
[5] Filippidou, F., Nieboer, N. & Visscher, H. 2019, Effectiveness of energy renovations: a reassessment based on actual consumption savings, Energy Efficiency. 12, 1, p. 19-35.
[6] Majcen, D., Itard, I. and Visscher, H.J. 2013a, Theoretical vs. actual energy consumption of labelled dwellings in the Netherlands: Discrepancies and policy implications, Energy Policy 54, 125 – 136.
[7] Majcen, D., Itard, L., Visscher, H.J. 2013b, Actual and theoretical gas consumption in Dutch dwellings: What causes the differences? Energy Policy, 61, 460-471.
[8] Majcen, D., Itard, L., Visscher, H.J. 2015, Statistical model of the heating prediction gap in Dutch dwellings: Relative importance of building, household and behavioural characteristics, Energy and Buildings, 105, 43-59.
[9] Brom, P van den, Meijer, A. and Visscher, H., 2018, Performance gaps in energy consumption: household groups and building characteristics, Building Research and Information, 46-1, 54-70
[10] Brom, P van den., Meijer, A. and Visscher, H. (2019), Actual energy saving effects of thermal renovations in dwellings—longitudinal data analysis including building and occupant characteristics, Energy and Buildings. 182, p. 251-263
[11] Brom, P. van den, Hansen, A. R., Gram-Hanssen, K., Meijer, A. & Visscher, H. (2019), Variances in residential heating consumption – Importance of building characteristics and occupants analysed by movers and stayers, Applied Energy. 250, p. 713-728
[12] Visscher H., Meijer F., Majcen D., Itard L., 2016, Improved governance for energy efficiency in housing, Building Research & Information, 44:5-6, 552-561, DOI: 10.1080/09613218.2016.1180808

What buildings policy might look like if we took climate change seriously

Tina Fawcett, Marina Topouzi
Environmental Change Institute, University of Oxford, South Parks Road, Oxford, OX1 3QY, UK
Corresponding Author Email: tina.fawcett@eci.ox.ac.uk

Abstract. In 2018, the Intergovernmental Panel on Climate Change released its report on the impacts of global warming of 1.5°C. It called for "rapid, far-reaching and unprecedented changes in all aspects of society" to reduce the risks of increasing climate change. Energy use in buildings is one of the key contributors to greenhouse gas emissions in the UK and EU. Detailed policies to support high standards in new build and acceleration of ambitious retrofit are a necessary part of any serious response to climate change.

This paper begins with a description of key UK policies on energy use in the buildings, and future UK government plans as expressed in the Clean Growth Strategy. It analyses the direction of travel, and looks at the rate of change compared with what will be required to rapidly reduce carbon emissions from the sector. Current buildings policies are insufficient to meet the challenge of climate change. Both positive and negative examples of changes in policy are presented.

A new conceptualization of the contribution of energy demand to delivering net zero goals, developed by CREDS, is introduced. In addition, key characteristics of the buildings sector and its energy use are put forward. Bringing together the role of energy demand, these characteristics of buildings, and the broader literature on buildings and energy policy, we have set out an exploratory set of guidelines for developing policy to reduce carbon emissions from buildings further, faster and more flexibly. These are classified into the guiding principle, approaches to policy creation, expanding the boundaries of policies and focus on quality. These should form the basis for further discussion.

If policy were designed as if we took climate change seriously, according to the guidelines developed, what would this mean for buildings? Fundamentally it would require a momentous change in the culture of construction, so that what is valued moves beyond design, aesthetics and functionality, to include low carbon and highly efficient performance. New targets and approaches to delivering them are required – and this paper contributes to thinking about both.

Key words: Climate change; Buildings; Policy

1. Introduction

In 2018, the Intergovernmental Panel on Climate Change released its report on the impacts of global warming of 1.5°C [1]. It called for "rapid, far-reaching and unprecedented changes in all aspects of society" to reduce the risks of increasing climate change. While there are many governmental statements in support of taking action on climate change, the evidence is that change is not happening at anything like the rate required. In November, the United Nations Environment Programme released its 'Emissions

IOP Conf. Series: Earth and Environmental Science **329** (2019) 012004 doi:10.1088/1755-1315/329/1/012004

Gap Report 2018' [2], which concluded that nations need to 'triple efforts to reach the 2C target' (which is less demanding than the 1.5°C target in the IPCC report). The UK is not currently on track to meet intermediate targets leading up to its 2050 target for 80% GHG emissions reductions [3].

In recent months, three civil society movements demanding substantial action in response to the threat of climate change have emerged. The most high-profile is the global 'school strike for climate' movement, inspired by Greta Thunberg who began her weekly strike outside the Swedish parliament in autumn 2018. On 15th February 2019, British school pupils and their supporters joined this growing world-wide movement, with the first UK-wide climate strike. In other European countries, school strikes are a weekly event, attracting thousands, or even tens of thousands. Secondly, the international Extinction Rebellion movement has taken direct action to highlight climate change and species and habitat loss, and calls for radical economic, social and political change. Thirdly, local councils are being urged to declare a 'climate emergency' and as of 27 March 2019, councils representing more than 36 million citizens in Australia, America, Canada, the UK and Switzerland have done so. This includes councils covering 20 million UK citizens [4]. These movements are all demanding immediate, significant and sustained reductions in greenhouse gas emissions.

In 2010, the world's buildings accounted for 32% of global final energy use and 19% of energy-related greenhouse gas (GHG) emissions. Under business-as-usual projections, use of energy in buildings globally could double or even triple by 2050 [5]. Detailed policies to support high standards in new build and acceleration of ambitious retrofit are a necessary part of any serious response to climate change.

The paper is structured as follows. First, we discuss what we mean by taking climate change seriously. This is followed by an overview of UK buildings policy. A new approach to thinking about energy demand is introduced. Principles for guiding development are developed and we illustrate what these could mean at the individual building level. The paper closes with discussion and conclusions.

2. Taking climate change seriously

What do we mean by taking climate change seriously? In a nutshell, it means setting (and achieving) carbon reduction goals for the UK that are consistent with both global justice and limiting temperature rise to 2C or less. To know what this would mean for the buildings sector, this goal then needs to be shared across sectors, acknowledging the constraints on other sectors. Clearly, there are a lot of judgements to be made – each of which could be questioned and debated. The aim of this paper is not to suggest a set of numbers and dates, but to raise key issues and to propose a set of guidelines for developing future policy.

First, a brief review of current targets. The UK Climate Change Act (2008) establishes a long-term target of reducing greenhouse gas (GHG) emissions to 80% below 1990 levels by 2050. To meet this target, via the Committee on Climate Change, the government has established five-yearly carbon budgets which restrict the amount of greenhouse gases UK can legally emit in each five-year period. In October 2017, the government published its Clean Growth Strategy [6]. This set its approach to meeting the 4th (2023–2027) and 5th (2028–2032) carbon budgets and to ensuring that the UK maximizes the economic and environmental benefits of green technology. The IPCC 1.5C report was published after the Clean Growth Strategy, and the government has asked the Climate Change Committee to provide advice on revised long-term targets for greenhouse gas emissions and the UK's transition to a net zero-carbon economy. While the UK is expected to meet targets up to 2023, as mentioned previously, it is not currently on track to meet later carbon budgets.

In May 2019, the Committee on Climate Change recommended that the UK adopt a target of 'net zero' greenhouse gas emissions by 2050. The following month the government announced that the net zero target would be adopted into law. This is a major step forward for the UK and increases the level of ambition considerably. Ambitious targets have also been adopted by elsewhere. Within the EU, for example, France aims to have net zero emissions by 2050, and Germany aims to reduce 2050 emissions by 95% compared with 1990 levels. However, these targets do not necessarily end the discussion. For the UK, the Centre for Alternative Technology has a 'Zero Carbon Britain' programme suggest net zero greenhouse gas emissions can be achieved for the by 2030. Extinction Rebellion are calling for net zero emissions to be achieve by 2025. So while the UK's net zero target by 2050 currently looks ambitious in relation to past policy, the increasingly worrying scientific news about the impacts of climate change, and growing public appetite for greater change, makes it like to be superseded in future.

In thinking about future targets, a number of issues would need to be considered, at various levels. At the highest level, to set the UK's national target, global justice is important. Within the national envelope, the targets for different sectors need consideration. For buildings, residential and non-residential buildings face different challenges, as do new and existing buildings. Should more devolved targets be set in terms of carbon or energy or both? Should targets apply to buildings, individuals, organisations or all of these? There is evidence that there is still plenty of scope for reducing emissions considerably using conventional energy efficiency measures [7], but we will need to go beyond these. In essence, taking climate change seriously means achieving close to zero net carbon emissions well before 2050 in the buildings sector, with more ambitious interim targets than are currently in place.

3. Current policy on buildings

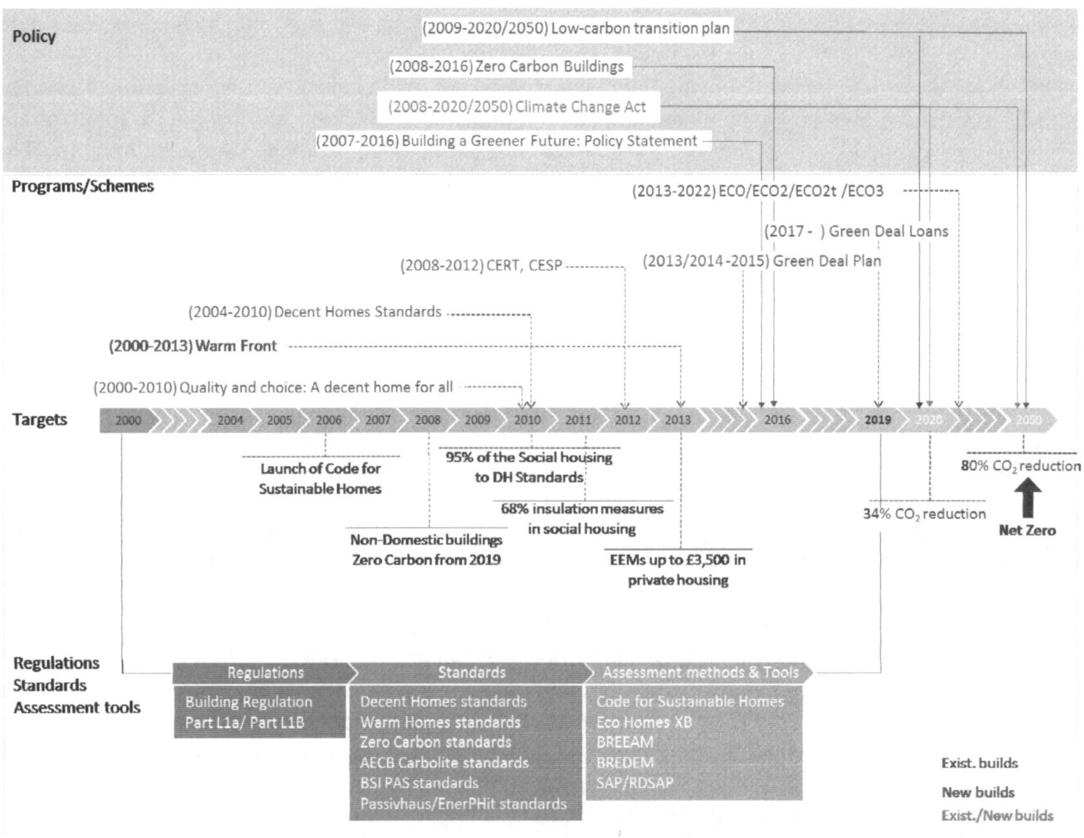

Figure 1: Overview of UK energy policy in buildings, 2000 onwards.

Policy on buildings and energy in the UK responds both to EU requirements and to domestic priorities. In this summary we have focused on the policies which are UK level only (so not describing the different approaches in Scotland and Wales, where devolved powers permit), and have not name-checked the relevant EU directives. Figure 1 sets out key policies from 2000, individual programmes and schemes, targets (up to 2050), regulations, standards and assessment tools. Policies and measures have been separated into those for new buildings, existing buildings and for all buildings. There have been various policy interventions across the period – some more successful than others, a myriad of standards, and the underpinning of Building Regulations, which most powerfully affect new buildings. Several important policies have been removed (e.g. Warm Front, Code for Sustainable Homes, Green Deal Plan) or redesigned to deliver lower energy and carbon savings (e.g. ECO). UK buildings policy at present does not have the policy mix necessary to meet its top level goals [3].

Here we describe briefly how lack of continuity of policy in the new build residential sector has adversely affected carbon emissions, development and delivery of innovative, low-carbon market solutions, and decision-making by local authorities on standards. The Code for Sustainable Homes, introduced in 2006 was to have been the framework leading to a low or zero carbon new building stock, with 2016 as the date by which high standards - 'Code 6'- were required. However, the housing standards review in 2015 resulted in scrappage of the zero carbon buildings policy and the mandatory 2016 target. This decision has been widely criticized. At the same time as this national standard was removed, the situation was made more complex for local authorities, in terms of whether their planning powers permitted them to require higher standards for developments. Simultaneously, the Building Act 1984 was changed to allow new optional standards to be introduced and required; while an amendment to the Planning and Energy Act 2008 constrained the ability of local authorities to go beyond the Building Regulations energy efficiency standards for new homes. This ambiguity has led to some authorities requiring higher energy efficiency standards (where particular regulatory tests are met), but others have understood the situation to mean that they cannot go beyond current basic energy efficiency requirements as outlined in Part L1A of the Building Regulations, thus holding back long-term zero carbon goals.

A more positive example of change relates to professional standards, training and quality. In 2016, the Each Home Counts review outlined 27 ground-breaking recommendations to improve the government's and consumers' confidence in domestic retrofit [8]. The quality mark for residential retrofit suggested will be delivered via a government-endorsed quality scheme supported by an industry Code of Conduct, a Consumer Charter and a framework of technical standards for retrofit (PAS2035 – to be launched in 2019). This scheme will set the requirements for assessing a building's existing condition, identifying and evaluating the options of improvement and the risk involved in the design and specification of the energy efficiency measures, and set the monitoring and evaluation requirements and the skills of the actors involved in all stages of a retrofit process.

4. A new approach to energy demand policy – the big picture

The Centre for Research into Energy Demand Solutions (CREDS) is developing thinking on the place of energy demand in the UK energy transition (to net zero emissions). Demand for energy, in buildings and other sectors, is determined by the energy-using activities undertaken and the efficiency with which these occur. However, in the context of the transition, reducing demand for energy is no longer the only issue. There are at least two other demand-side issues which need to be addressed – demand flexibility and decarbonisation of energy sources used at the point of demand [9]. These multiple aims for demand change in the energy transition – efficiency, reduction, flexibility and a switch to sustainable fuels – cannot effectively be analysed separately. A sustainable, affordable and secure energy system will requirs all of these elements in combination. Figure 2 sets out a simple representation of how CREDS sees them contributing to energy system transformation.

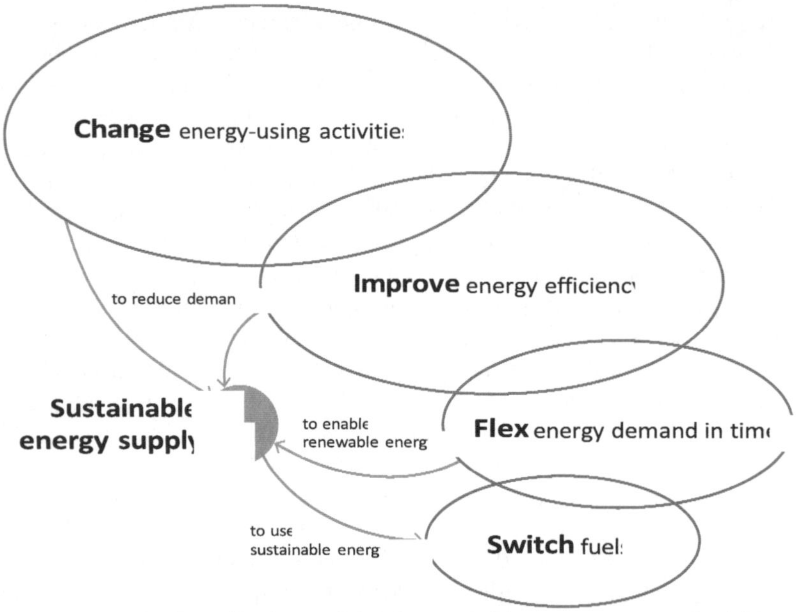

Figure 2: Contributions of the demand side to energy sustainability (source [9]).

The diagram shows what is necessary on the demand side, and the links between that and a sustainable energy supply, to give a (simple) whole systems overview. Changing energy-using activities and improving energy efficiency are critical to ensuring demand is low enough to be met by sustainable fuels. In addition, it is important that demand is flexible in time (the 'flex' bubble) ensuring that the sustainable energy supply can meet peak as a well as average demand, and at lower cost. Most buildings-related policy has focused on increasing energy efficiency as a means of reducing demand, with much less focus on changing energy-using activities. Delivering changes in energy use could range from overall approaches such as passive design in retrofit (e.g. cooling and heating), to detailed requirements like the provision of outdoor clothes drying facilities in new residential developments. While this diagram is not presented as hierarchical, as all steps are necessary, the 'bubble' sizes indicate something about the relative importance of different elements of demand side action.

5. Developing principles to guide buildings and energy policy

When thinking about how to create policy for the buildings sector, it is important to take into account the nature of buildings and the energy services they provide. There are two critical characteristics of buildings in this context:

- Buildings are unique and change/ age differently (even factory-manufactured buildings);
- Energy use from buildings is a result of interactions between the buildings, the external environment, their energy-using equipment and controls, and their users, to achieve or maintain their comfort preferences, safety and security.

These characteristics matter for policy because they tell us important things about what technologies, control systems and organisation- and people-focused interventions are likely to be most effective. The first characteristic suggests that 'one size does not fit all' – certainly not for more ambitious, deep renovation strategies. So, for example, while the idea of off-site manufacturing of insulation panels may be helpful in some situations, it cannot be a universal answer. The second characteristic reflects evidence that energy used in very similar homes and offices can vary considerably, with evidence that this can be due to a myriad of small differences in hours of occupation, habits and technologies [10]. This characteristic also points perhaps to the limitations of building energy modelling (for some purposes), and the distinction between performance of buildings according to standard assumptions, and their real-life performance.

Bringing together the CREDS approach to energy demand change, these characteristics of buildings, and the broader literature on buildings and energy policy, we have set out an exploratory set of guidelines for developing policy to reduce carbon emissions from buildings further, faster and more flexibly. These are classified into the guiding principle, approaches to policy creation, expanding the boundaries of policies and focus on quality. The guidelines chosen are very briefly justified.

Table 1: Exploratory set of guidelines for developing building and energy policy.

Theme	Key guidelines
Guiding principle	**Contribute to the UK's net-zero goals** taking account of all aspects of energy demand change (Figure 2)
Approaches to policy creation	**Use a policy mix:** A complex set of problems requires a complex response. Policy is needed at different scales, focused on different actors, technologies, decisions etc. [11]. It also needs to go beyond energy efficiency.
	Allow flexibility - this is critical given the variability of buildings. We need flexible/adaptable approaches for some aspects of policy, without compromising overall carbon reduction goals. For example, heritage buildings might not face strict low carbon standards, or deep residential renovation could happen over extended time periods.
	Recognise multiple benefits: Energy saving or shifting energy use in time (flexibility) has multiple benefits – environmental, economic and social. These must be recognised within policy making, at all scales from the single building, to city to national, and multiple actors engaged around their own values.
	Embed policy experimentation and learning: (includes support for new technologies) Learning by doing will be essential, as will being allowed to tell 'learning stories' [12]. Establish feedback loops between different scales and levels as a reflexive responsive mechanism to respond to implementation failures avoiding/reducing the design-performance gap.
Expand the boundaries of policy	**Recognise policy connections.** To make effective policy we need to go beyond the boundaries of current policy – to include renewables, flexibility / demand response, planning, heat networks etc. For example, there needs to be legal reform of governance for buildings in multiple ownership, if they are to have energy retrofits [13].
	Adapt to climate change: Buildings need to adapt to existing and expected climate change – particularly higher summer temperatures, more heat waves and extreme weather events. These are likely to affect the health of building occupants and increase the deterioration of buildings.
	Include embodied energy: As the whole economy must decarbonise, the embodied energy in buildings materials, services and equipment (for new build and retrofit) will become more significant.
Focus on quality	**Improve skills & training.** Without a skilled construction sector, with sustainable values, change cannot be delivered. A new standard under development, PAS 2035, should help upskilling by changing the culture of responsibility and trust between multiple actors, For example it embeds the notion that a project doesn't end at the handover stage, but ensures quality of performance continues throughout its lifespan.
	Focus on evaluation & enforcement This is required to close the design-performance gap, to ensure renovation work is done to specification and to improve the policy process. We need feedback loops that reach from middle actors to top level actors (from the designers, builders, suppliers etc to the policy makers).

6. What would buildings look like if we took climate change seriously?

If we could design policy as if we took climate change seriously, according to the guidelines above, what would this mean for buildings? Thinking particularly about new buildings, the culture of construction needs a momentous change in which the prominence of a building will not be limited primarily to the design (aesthetics) and functionality but will put low carbon and highly efficient performance at the top of the hierarchy. This requires a whole cultural transformation, change of the values and principles towards demand reduction not only of the building designers, and other professions in the construction sector and supply chain market, but also of the building users. Also a move beyond linear processes in which the finish line is at the delivery or completion of a building, installation of technology or a system.

It would mean, at single building level, improving a building's energy performance to net zero carbon emissions by including:

- high performance building envelope;
- energy efficient services in heating, cooling/ ventilation and lighting;
- control systems and onsite generation flexible enough to respond to user preferences/lifestyle and changing conditions;
- passive building design or solutions (that minimise or eliminate the need for mechanical heating, cooling and ventilation) as single or complex of buildings;
- flexible design and adaptive solutions to technical, social and climate changes;
- passive/active interaction of the users with the building system (user behaviour and awareness of energy use practices);
- repair and maintenance services that follow a building's ageing and life-span changes, allowing for short term/long term interventions.

7. Discussion

In this paper we've started the discussion from setting the broader picture of the role of demand in the energy transition, followed by a set of guidelines that can drive building energy policy design in order to deliver the low carbon transition and zero or low carbon emissions buildings that have the characteristics discussed above.

The challenge of responding to climate change cannot be over-stated for the buildings sector or for society as a whole. While there have been impressive improvements in buildings energy efficiency and a switch towards renewable sources of energy in recent years, there is a huge distance to go and little time in which to do it. The aim of this paper has been to try and link this high-level concern with the more detailed world of energy use in buildings and policy design. The table of guidelines which has been developed is very much open for discussion. It has attempted to compress important insights from across the literature into a rather simple format. Clearly this risks overlooking important linkages between topics, and all sorts of nuances. Hopefully the advantage is that by taking an overview approach, we remember the big picture, and avoid focusing all our attention on policy details, important as these are. The next step is to discuss the contents of this table with colleagues (including at the conference) and discover whether it is thought to be a useful approach, and a basis for building policy in response to climate change.

8. Conclusions

The unprecedented challenge of decarbonising energy means that, while we can and should learn from past UK, EU and international policy experience, we are likely to need new approaches to the design, types and mixes of policy, institutions and delivery mechanisms. We will need to re-think governance and expand the ambition and reach of policy. The energy transition will require changes in technologies, practices and choices for every household and business, many of which we do not currently know how to organise technically, cost-effectively or in a socially acceptable way.

This paper has set out key elements of current policy, and shown how these are inadequate to meet the challenges we currently face. It has used the recently developed CREDS energy demand conceptualisation, to re-think how policy is formed, and specifically how it should be formed for

IOP Conf. Series: Earth and Environmental Science **329** (2019) 012004 doi:10.1088/1755-1315/329/1/012004

buildings taking into account key characteristics of the buildings sector and its energy use. This has led to development of an exploratory set of guidelines for creating effective buildings policy. These should form the basis for further discussion.

References

[1] IPCC, *Global Warming of 1.5 °C*. 2018, Intergovernmental Panel On Climate Change: Available: http://www.ipcc.ch/report/sr15/ [Accessed November 2018].

[2] UNEP, *Emissions gap report 2018*. 2018, United Nations Environment Programme: http://www.unenvironment.org/emissionsgap [Accessed December 2018].

[3] CCC, 2018. *Reducing UK emissions - 2018 Progress report to parliament.* The Committee on Climate Change, London.

[4] Climate Emergency Declaration, 2019. Climate emergency declarations in 423 councils cover 36 million citizens. Available: https://climateemergencydeclaration.org/climate-emergency-declarations-cover-15-million-citizens/ [Accessed April 2019]

[5] Chalmers, P., 2014. Climate Change: Implications for Buildings. Key Findings from the Intergovernmental Panel on Climate Change Fifth Assessment Report, in *Climate: Everyone's Business*. ECF, BPIE, GBPN, CJBS and CISL.

[6] BEIS, *The clean growth strategy: Leading the way to a low carbon future*. 2017, Department for Business, Energy and Industrial Strategy: Available: https://www.gov.uk/government/publications/clean-growth-strategy [Accessed Oct 2017].

[7] Rosenow, J., et al., *The remaining potential for energy savings in UK households.* Energy Policy, 2018. 121: p. 542-552.

[8] Bonfield, P., 2016. *Each Home Counts: An Independent Review of Consumer Advice, Protection, Standards and Enforcement for Energy Efficiency and Renewable Energy*. Department for Business, Energy and Industrial Strategy and the Department for Communities and Local Government: www.gov.uk/government/publications/bonfield-review-terms-of-reference.

[9] Eyre, N. and Killip, G. (Eds.), 2019. *Shifting the focus: energy demand in a net-zero carbon UK.* Centre for Research into Energy Demand Solutions. www.creds.ac.uk.

[10] Topouzi, M., 2015. *Occupants' interaction with low-carbon retrofitted homes and its impact on energy use.* Unpublished DPhil, Environmental Change Institute, University of Oxford, Oxford.

[11] Rosenow, J., Fawcett, T., Eyre, N., Oikonomou, V., 2016. *Energy efficiency and the policy mix.* Building Research & Information 44, 562-574.

[12] Janda, K.B. and M. Topouzi, 2015. *Telling tales: using stories to remake energy policy.* Building Research & Information,. 43(4) 516-533.

[13] Bright, S. and Weatherall, D. 2017. *Framing and mapping the governance barriers to energy upgrades in flats.* Journal of Environmental Law 29(2) 203-229

Acknowledgments

This work was funded by UK Research and Innovation, grant agreement number EP/R035288/1 (Centre for Research into Energy Demand Solutions).

Energy-related policy in the housing sector of Bulgaria: in search for a meeting point of social and technical dimensions

E Dimitrova[1]*, M Petrova-Tasheva[2], I Mutafchiiska[2], A Burov[2]

[1] ECHOES Project, University of Architecture, Civil Engineering and Geodesy, 1, Hristo Smirnenski Boulevard, 1046 Sofia, Bulagria
[2] Urban Planning Department, Faculty of Architecture, University of Architecture, Civil Engineering and Geodesy, 1, Hristo Smirnenski Boulevard, 1046 Sofia, Bulagria
*Corresponding author: eldim2002@gmail.com

Abstract. Understanding the way in which a policy framework put into action supports or hampers stakeholders' energy-related action in the field of housing renovation could provide meaningful insight on needed respect for stakeholders' needs, expectations, motivations and capacity for action. The paper presents a study undertaken on the case of Bulgaria as part of ongoing ECHOES research project (EU-funded under the Horizon 2020 Program). It discusses the complex challenges faced in linking policy and technical to social and cultural considerations in renovating the existing residential stock in Bulgaria while meeting energy-efficiency requirements with an explicit focus on predominant apartment buildings in multiple ownership. The refurbishment process requires, on one hand, considerable investment, which is in most cases hardly accessible for the homeowners; on the other hand, collective decision-making with numerous stakeholders involved, often with diverse social and economic status. The presented two national case studies illustrate two alternative approaches applied – one searching to benefit from the national energy market peculiarities, and the other taking the opportunities provided by European and national grant schemes introduced in the country. Based on contacts and semi-structured interviews with various actors involved in the two initiatives, the dynamics of the situation throughout the projects' life cycle sand and in the longer term and the changing personal estimations of costs and benefits are analysed. Conclusions are drawn about the importance of continuity in implemented energy-related policies. The authors claim that specific paths to EU energy policy goals under varying economic, technical and sociocultural context should be well continually informed by comprehensive interdisciplinary research.

1. Introduction

Energy transition has been acknowledged as a key policy area at the EU level for more than three decades [1]. It has been also among the major research challenges addressed by the EC framework research programmes [2]. Research has repeatedly addressed technical as well as social dimensions of the process; yet understanding and influencing stakeholders' energy-related behaviour still seems a challenge that needs a deeper study of interacting technical, social and cultural factors. ECHOES research project, funded under Horizon 2020 programme, [3] provided a chance for studying stakeholders' needs, motivations and capacity for action under the multidimensional context in eight participating countries. Undertaken within the common methodological framework of the project, the research presented hereafter addresses the peculiarities of the process in Bulgaria and discusses the complex challenges faced in linking economic and technical to social and cultural considerations during the implementation of the national policy on the renovation of multi-family residential buildings with an explicit focus on energy efficiency (EE). The renovation process required, on one hand, considerable investment, which was not in most cases accessible for the homeowners, and, on the other hand, collective decision-making of numerous stakeholders, often with diverse social and economic status.

The developed two case studies illustrate different approaches chosen and applied by two homeowners' collectives when addressing energy issues in a multifamily building in multiple ownership – one searching to benefit from the national energy market development, and the other taking the opportunities provided by the introduced EU-funded and national grant schemes. Based on literature review, personal contacts and semi-structured interviews with actors involved in the two initiatives, the study traces the practical steps undertaken by institutions, homeowners' collectives and individuals, their outcomes and follow ups, and the actors' estimations of costs and benefits throughout the process. It aims at identifying general and peculiar enabling and hindering factors to the ongoing process in the country but also at drawing the attention of policy makers on the importance of contextually sensitive research in formulating and implementing effective policies for energy transformation.

2. Research questions and methodology

Residential buildings are nowadays responsible for about 25% of the overall energy consumption in Europe [4]. Their energy efficiency is, therefore, of key importance to the process of energy transition aimed at. European policy documents acknowledge that enhancing EE goes beyond the implementation of technical measures [5]. At a city and regional scale, large-scale housing retrofit programmes have the potential to significantly reduce energy demand and CO2 emissions, whilst at the same time, having other positive impacts such as affordable warmth, improving health and quality of life; therefore, a more systematic and holistic approach integrating technological and environmental issues with social, economic and political ones has been recommended to 'low carbon' research [6]. The interpretation of energy behaviours from the point of view of social practice theory has outlined that the behavioural patterns of a group of people sharing common culture or lifestyle can only be properly understood and changed, if also the material and meaning component of this behaviour is understood [7]. Research in the field has outlined that enhanced inhabitants' participation, decreasing power imbalances and awareness rising on individual and group level have been also identified as important motivating factors in implementing energy efficiency measures in a residential building [8, 9]. The barriers rooted in the socio-cultural context in multifamily residential buildings are often related to the urgency of action required and the limited owners' capacity to invest and to expected inconveniences caused by the refurbishment [10]; the choice of technical options is often limited by the length of the administrative approval procedures or raises conflicts between various stakeholders [11,12]. The processes of mutual learning of the actors involved in a housing renovation initiative is outlined as an important peculiarity of a dynamic sociotechnical system [13]. All these aspects are, however, largely uncovered yet by systematic research under the Bulgarian context despite the acknowledged urgency of undertaking EE measures in the housing sector [14] and the practical efforts undertaken since mid-1990s [15].

The research questions addressed by the current study concern the meeting points of three types of factors in the housing renovation process: policy and management (PM), spatial and technical (ST), and social and cultural (SC) ones, with an explicit focus on their meeting points under the Bulgarian context. The case study-based approach followed ECHOES general research methodology; it was expected to provide a deeper understanding of the sociocultural context of the implemented innovations and the enabling/hindering factors rooted in actors' values and attitudes, their perceptions of benefits and uncertainties, while responding to technical challenges and active renovation rules (figure 1).

Three groups of stakeholders were identified in the two case studies, and their representatives contacted: homeowners, public authorities and energy exerts/ service providers. Information on the process in case study 1 (CS1) was retrieved through personal contacts with a member of a specialized NGO providing methodological support to EE initiatives, and with an expert from the Energy Efficiency Agency in Sofia. Five semi-structured interviews were conducted with individuals involved in the two consecutive projects shaping case study 2 (CS2) - one with an ex deputy mayor of the administrative district concerned (R1), one with an energy agency expert (R2) and three with flat owners in two different buildings renovated through the project (R3, R4 and R5).

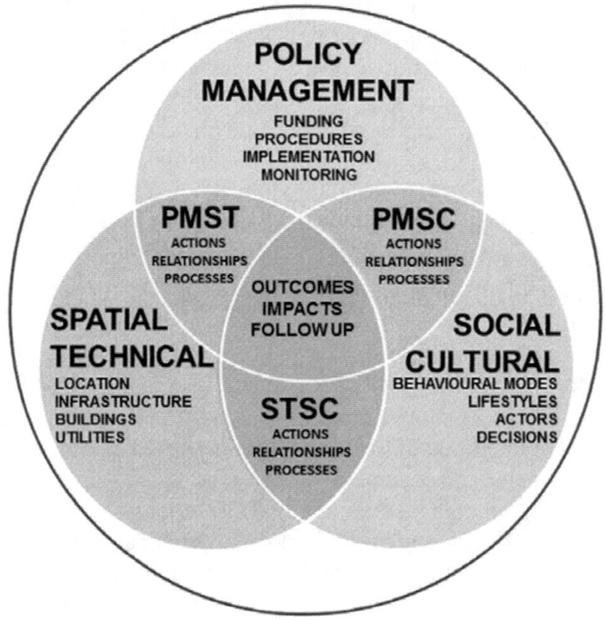

Figure 1. Research concept and focus [source: authors]

3. Case studies description

The two case studies follow the same structure, outlining as follows: location and technical parameters of the buildings; technological challenges and innovations considered; type of funding chosen, actors involved, undertaken action and interactions among the respondents, their personal estimations of outcomes and follow-up - achievements and failures, success factors and bottlenecks, lessons learned.

3.1. Case study 1 (CS1): Homeowners' energy cooperative

Location and technical parameters of the building and the installation: A 28-kW photovoltaic system for electricity generation with an estimated output of 35.00 MWh/year, was installed in 2013 on the roof of a recently constructed 15-storey multifamily building with two entrances and a total of 120 apartments, located in a dynamically developing neighbourhood of Sofia.

Technological challenges and innovations. The electricity to be produced by the solar system installed on the roof of the building was to be sold to the energy supplying company and expected to bring profit to the homeowners as the market price of green energy was considerably higher than the subsidized energy provided by the company. The revenue was to be distributed among the owners and used for covering the maintenance costs for the common parts of the building.

Funding scheme. The construction of the building had been jointly funded by the owners through a multi-step collection of needed resources. After the construction was over, part of the money remained unspent and the owners had to decide on how to utilize it. They voted for installing a photovoltaic system for electricity generation on the roof of their building.

Actors involved. The investment idea came from one of the owners with expertise in the renewable energy field, who was member of the Bulgarian Solar Association. The Head and the Managing Board members of the Homeowners' Association (HOA) had the major contribution in collecting all the requested 120 homeowners' signatures under the formal agreement. They initially acted as a volunteering team and were later authorized by the HOA to officially take the responsibilities.

Actions and procedures. The final decision on the investment in the photovoltaic system came after a long and heated debate with numerous clarifications and proofs provided on possible benefits and risks. The information campaign was followed by lengthy negotiations; in line with the legal requirements all the owners had to sign the formal agreement; finding the missing owners (some of them being abroad) was a next challenge. The owner with the relevant technical competence in the energy field took the responsibility for defining the installation technical parameters, obtaining the permission for installing it and preparing the documents for connecting it to the national electricity grid.

Relations between the actors. The considerable number of owners with various socio-cultural and economic profiles was a first difficulty in initiating the process. Yet, the major bottleneck proved to be signing the contract with the private company providing electricity to Sofia. The process of linking the building to the energy supply system was considerably delayed in the last stage of negotiations. The delay coincided with a change in the national regulatory framework on the conditions for generating energy from renewable sources and, thus, with the reduction of the purchase price per MWh. According to the signed contract, the final price of the sold energy was over 40% lower than the initially estimated one. Thus, the investment payback period was extended from 5 to 10 years, which was highly disappointing for the owners and had a negative effect on their relations with HOA Managing Board.

Outcomes and follow-up. After the photovoltaic installation was set into operation, the project was broadly promoted in media and received considerable publicity. There was considerable interest in realizing such an urban project and many inquiries about practical implementation requirements. The need for HOA's own funding was, however, considered a serious barrier by most of the interested homeowners. After the photovoltaic modules started bringing revenue to the owners' association from the electricity produced by the solar system and sold to the energy supply company, a second collective decision was taken, and the money was invested in energy saving windowpanes for the staircases of the two entrances. The owners realized that the benefits of a collective investment bringing guaranteed revenue in future would be far higher than the ones of distributing small amounts of already devaluated money among numerous individual stakeholders.

3.2. Case study 2 (CS2): Grant-based complex renovation of residential buildings

Location and technical parameters of the building. Three residential buildings in a traditional quarter at a closer proximity to the city centre of Sofia were successfully renovated through two consecutive projects providing 100% grants in support of EE measures implementation. The in-depth study was focused on a 5-storey building with 10 apartments, built in 1930s under less restrictive regulations on energy efficiency than the current ones.

Technological challenges and innovations. The renovation responded to the estimated EE deficiencies in the building as a result of undertaken energy audit; the full refurbishment of the building included upgrading of the building installations, the underground parts and the roof structure; a solar system for water heating was installed on the roof and the building was certified as one of energy class A+. Analyses on the real-life functioning were then carried out in order to identify needed adjustments and improve the operational efficiency of the building. Trainings were organized, and guidance was provided on ways for reducing energy and resource consumption in everyday households' activities.

Funding scheme. Funding was provided through two grant-based projects: *Demonstration Project for the Renovation of Multifamily Residential Buildings* (DPRMRB) [16], the grant of which covered 100% of the construction work expenses for the renovation of the common parts of the buildings and 80% for each dwelling (20% of the expenses were funded by the owners and the municipal authorities funded the upgrading of the adjacent area); and *Staccato* project (an international project funded under the 6th EU Framework Program) [17] for installing solar systems for hot water in residential buildings.

Actors involved. The person elected by the homeowners to head the association, was the real driving factor in applying for the DPRMRB grant - he enabled the collective decision for joining the project and took the responsibility for communicating with the local and central administration responsible for the project. He also played a leading role during the construction process by providing investor's control

IOP Conf. Series: Earth and Environmental Science **329** (2019) 012005 doi:10.1088/1755-1315/329/1/012005

and acting as a mediator between the construction company and the homeowners. Other owners with specific expertise also contributed in different phases the project.

Actions and procedures. In 2010 The Ministry of Regional Development and Public Works (MRDPW) signed a Memorandum with *Staccato* project consortium on reconciling the activities of the DPRMRB and *Staccato* Project. As a result, the homeowners' associations of three residential buildings, renovated under the Demonstration Project, could be involved in the *Staccato* project and receive solar hot water installations on the building roofs for free. In Phase 1 of the project the homeowners made a collective decision for joining the project; in Phase 2 the solar system was installed, and the building was certified as energy class A+. In Phase 3 the real-life operation of all the installations was analysed and needed precise adjustments to improve its operational efficiency were identified; work continued with the HOAs on capacity-building for reducing everyday energy and resource consumption; two elderly homeowners - engineers with expertise in the energy field, took the leadership in this phase.

Relations between the actors. The small number of owners who had known each other for many years was a key enabling factor in the collective decision-making on joining the project. Due to the persistent efforts of the HOA head, all the signatures for the contract and the financial contribution requested were collected in quite a brief time span; the preparation of the application was based on solidarity with the people lacking the financial resources needed for signing the contract.

Outcomes and follow-up. Owners' satisfaction with the renovation results in CS2 continued until the end of the first heating period when the unwillingness of the local monopolist company, responsible for the district heating in the city of Sofia, to change the calculation methodology resulted in a prolonged debate on the method of calculating the energy prices for A+ energy class buildings. Looking for a solution, the owners introduced smart technology for distance reporting of the actual energy consumption. Regretfully, that did not solve the problem. Additional spending caused by a failure of the water heating system at the end of the first operating year brought next disappointment among the homeowners. The poor maintenance and the resulting improper operation of the installation raised a next challenge to the inhabitants: either to acquire additional knowledge and skills themselves or hire a specialized maintenance company. The first option was expected to place additional burden on some of the owners; the second questioned the economic feasibility of the installation. The repair of a circulation pump was organised with the help of the specialised NGO involved in the initiative, yet the search for acceptable long-term solution was still going on by the time of the interviews.

4. Discussion

The enabling and hindering factors identified in the two case studies were related to the meeting points linking the three aspects of the housing renovation process through ten topics, as follows: (a) Policy & Management with Spatial & Technical (PMST) aspects - availability of funding; relevance of the legal and regulative framework; grant applications and institutional channels; communication and dissemination of technical competence; (b) Policy & Management with Social and Cultural (PMSC) aspects - the role of local administration; homeowners' proactive role in implementing EE measures; HOAs empowerment; and (c) Spatial and Technical with Social and Cultural (STSP) aspects - homeowners' diversity of relations; HOA's decision-making: awareness, technical expertise and risk estimation; HOAs' communications with service providers (see Fig. 1).

4.1. Policy and Management to Spatial and Technical (PMST) aspects

Availability of funding for energy saving measures. Funding programmes proved to be crucial factors for the implementation of EE policy in the Bulgarian housing sector with more than 92% private ownership and minor financial capacity of the homeowners for investing in energy efficiency. In CS1 the availability of 'own' funding by the homeowners was a key enabling factor as it eliminated loan-related risks for people, who were capable and willing to take the initiative with all the risks and benefits provided by the just emerging energy market in the country. In CS2, the proactive attitude of the homeowners to implementing energy efficiency measures helped them in consecutively capturing different opportunities provided by the variety of grant funding schemes in the country.

Relevance of active legislation and regulations. CS1 proved the importance of good knowledge about the legislation and regulation changes under the dynamically changing policy context in Bulgaria. Relying on that, the homeowners were able to sign the contract for selling the electricity produced at a favourable price. In parallel, the currently active market-based financial mechanisms of implementing EE measures, although limited in scope, encountered serious implementation difficulties, e.g. concerning the contracts with the energy supply companies. CS2 outlined the need for simplifying the administrative procedures when installing the solar collectors for water heating. Both case studies provided insights that synchronizing all the rules and regulations on the implementation of EE measures but also the long-term management of the buildings and installations afterwards would result in a more positive attitude and greater confidence among the owners implementing their own collective decisions.

Grant applications and institutional channels. The availability of an energy expert among the homeowners in CS1 was estimated as a key success factor for the initiation of the project as information and knowledge on project opportunities, application procedures and profitability of measures in the energy field is currently neither widespread nor easily accessible and understandable.

Communication and dissemination of technical competences. Attracting expert non-profit organizations working in the field of energy efficiency proved to be a strong supportive factor in CS1.The renovation process went smoothly in CS2, due to the long-term personal commitment and responsibilities voluntarily taken by homeowners with technical and other relevant expertise.

4.2. Policy and Management to Social and Cultural (PMSC) aspects

The role of local administration. The support of local administration was of core importance in the preparatory stage of the renovation process in CS2 – it comprised dissemination of information, awareness raising and training. The visibility and accessibility of public experts and employees was commented as very important for building and maintaining trust throughout all the stages of the process. The further empowerment of local authorities and the real governance decentralization could be expected to provide better conditions for collective EE-related action in residential buildings.

Homeowners' proactive role in implementing EE measures. In both case studies well-educated active members of the owners' collectives put a lot of voluntary effort and time in the process. In CS1 a team of initiators also took various roles in implementing the collective decisions; actors with access to needed legal information on required procedures joined in promoting the initiative among the homeowners and organizing the communication during the installation of the solar system. Homeowners with specific expertise actively joined in the preparation and submission of the application documents in CS2. The experience there proved that communicating possible benefits and building a clear homeowners' vision on desired outcomes provides for continuity in implementing a complex model for overall energy renewal of the residential buildings by consecutively using different funding schemes.

Empowerment of homeowners' associations. The establishment of homeowners' associations followed the legal requirements introduced in the country; in both case studies HOAs played crucial role in initiating the renovation and in legally communicating with all the institutions on behalf of the owners. In CS2 the HOA registration provided for formalizing the existing long-term informal relationships. A further empowerment of homeowners' collectives throughout the whole lifecycle of an energy-related project could be achieved by clearly defining their rights and obligations in selecting the contractors for the construction works and the building materials, and in controlling the interventions.

4.3. Spatial and Technical to Social and Cultural (STSC) aspects

Homeowners' diversity and rules of interaction. An identified major challenge to the EE-related housing renovation in the country was embedded in the housing typology - multifamily high-rise buildings with numerous privately-owned dwellings and shared ownership on the common parts of the buildings - roofs, entrances, staircases and basements. The collective decision-making on undertaking a renovation initiative required that homeowners with a variety of social and economic profiles could come to a shared vision on future action in rather short period of time. The estimated difficulties for coming to a decision in CS1 stemmed from the large number of owners but also from the lack of community spirit

IOP Conf. Series: Earth and Environmental Science **329** (2019) 012005 doi:10.1088/1755-1315/329/1/012005

caused by the frequent change of homeowners and the feeling of the uninhabited of anonymous neighbours next door. *"Apartments are continuously sold and rented in these blocks we do not even know half of the people there". (R1).* The smaller number of inhabitants in CS2 and the established long-term relations contributed for their sense of belonging to a small community: *"Some people could not provide the money, and the neighbours helped through the fund accumulated among themselves"* (R3).

HOAs' decision-making. Information, technical expertise and adequate risk estimation proved to be important enabling factors in both case studies. The uncertainty about future energy-related conditions negatively influenced the homeowners' investment intentions. All the reforms in the energy field should be therefore introduced in a coherent and transparent manner and clearly communicated in advance; a comprehensive preliminary study of their potential short, medium, and long-term effects on all stakeholders should be requested.

Communication between HOAs and service providers. The HOA's initial trust to the specialised service providers in CS2 due to the good quality of the accomplished construction works was expected to guarantee the continuity of their positive attitude to EE measures. Yet, the owners then accumulated negative experience with the companies providing maintenance services for the common parts of the building and were often reluctant to using their services. Through the improper exploitation resulting in generated additional expenses, the initial effort for achieving higher energy efficiency often turned to be useless as either the saved money was used to cover repair expenses. Thus, although the renewal required complex technical measures and the meeting of high building standards, signing a maintenance contract with requested too high payment was difficult to accept for the association (R2).

4.4. Relevance and effectiveness of the research methodology applied

Relevance to the research objectives. The two case studies illustrate two different proactive approaches in addressing energy efficiency in the multifamily residential buildings in Bulgaria. Proactive voluntary action and collective decision-making contributed in both cases to building urban social capital. Despite currently being of rather limited applicability in the country, CS1 approach provided an insight that activating private investment resources and expertise on energy efficiency in the housing sector could provide for motivating well-educated communities of relatively high-income individuals to take collective decisions on increasing the energy efficiency of their homes. The process in CS2 addressed the current complex social and technological challenges in the country. Due to the applied grant scheme it contributed to improving the living conditions of a broad target group, while preventing social exclusion. Regretfully, the experienced difficulties resulted in the next national programme on energy efficiency of multifamily residential buildings [18] only in the introduction of 100% grants for the dwellings refurbishment at the expense of reduced interventions covering minimum EE measures.

The research approach of the current study could be considered an important step in identifying the mutual influences among numerous factors in the EE field and in enhancing the national policy sensitivity to the social and cultural dimensions of the process. It should be further combined with expert qualitative and quantitative estimations on the overall mid- and long-term effects of the implemented EE measures on energy savings and GHG emissions at the urban and national level.

5. Conclusion

Functioning residential buildings are complex socio-technical systems. Throughout their lifecycle they undergo numerous changes depending on multiple factors. The technical and technological aspects of the EE-aimed interventions there need to be related to the variety of inhabitants' needs and behaviour, changing ownership, economic capacity and perceptions of comfort and discomfort. The combination of private and collective ownership in multifamily buildings in Bulgaria raises the challenges of varying economic capacity of different households and the resulting different estimations of costs and benefits by the owners. Relevant technological expertise is required throughout the life of the building in support of an informed decision-making process at the level of the owners' collectives; it is not always available or practically accessible. Providing such an expertise needs an explicit well-focused effort. Activating the available expert knowledge and skills among the owners themselves could be the way to a continuous learning process but also to more sustainable local communities. The coordination and continuity of efforts for the implementation of effective energy-related housing policies need to be also sensitive to the peculiarities of an evolving economic, technical and sociocultural local and regional context.

Acknowledgment
The study was undertaken in 2018 within ECHOES project (contract 727470, H2020-EU.3.3.6. Robust decision making and public engagement, Topic LCE-31-2016-2017).

References
[1] Langsdorf S 2011 EU Energy Policy: from the ECSC to the Energy Roadmap 2050. Green European Foundation; http://archive.gef.eu/uploads/media/History_of_EU_energy_policy.pdf. Accessed: 28/05/2019.
[2] EC, COM (2006) 847 final. Towards a European strategic energy technology plan. Brussels, 10.1.2007.
[3] ECHOES project, Horizon 2020 programme 2016-19), www. https://echoes-project.eu/.
[4] Eurostat. https://epthinktank.eu/2016/07/08/energy-efficiency-in-buildings/.
[5] Friege, J. & Chappin, E. (2014). Modelling decisions on energy-efficient renovations: A review. Renewable and Sustainable Energy Reviews, 39, 196-208.
[6] COST Action TU1104 – Smart Energy Regions 2014 ISBN 978-1-899895-14-4.
[7] Hargreaves T 2011 Practicing behavioural change: applying social practice theory to pro-environmental behavioural change, *J Consumer Culture*, **11**,1
[8] Šahović N and da Silva P P 2016 Community renewable energy - research perspectives. 1st Energy Economics Iberian Conference, EEIC|CIEE, Lisbon; *Energy Procedia* **106**, pp 46-58.
[9] Heuts E and Versele A 2016. RenoseeC: Renovating with a social, ecological and economic benefit through a collective approach. SBE16 Tallinn and Helsinki Conference; Build Green and Renovate Deep, 5-7 October, Tallinn and Helsinki. In *Energy Procedia*, **96**, pp 540-550.
[10] van Doren D, Driessen PPJ, Runhaar Hens and Giezen M 2016 Scaling-up low-carbon urban initiatives: Towards a better understanding. *Urban Studies*. **55** 1, pp 175-194
[11] Cabinet Office Behavioral Insights Team 2011 Behaviour change and energy use, Ref:06537/0711; Available at: www.nationalarchives.gov.uk/doc/open-government-licence/, Accessed: 01042019
[12] World Business Council for Sustainable Development (WBCSD), 2007; Energy Efficiency in Buildings (EEB) project, Summary Report
[13] Andersen HT, Dimitrova, E and Schmeidler K 2013 Urban knowledge and large housing estates in Europe. eds HT Andersen and R Atkinson *Production and use of urban knowledge. European experiences* chapter 6 (Springer) pp 103-132.
[14] The World Bank 2018 Bulgaria: National Program for Energy Efficiency in Residential Buildings. Program Design Report for the Second Phase.
[15] Nakova K 2007 Energy Efficiency Networks in Eastern Europe and Capacity Building for Urban Sustainability. The experience of two municipal networks. *Indoor and Built Environment*, **16**, 3 pp 248-254
[16] Ministry of regional development and public works (MRRB) 2007 Demonstration project for renovation of multi-family residential buildings (DPRMFRB) (2007-2011). Available at: https://www.mrrb.bg/en/energy-efficiency/proon/demonstration-project-for-renovation-of-multi-family-buildings-2007-2011/. Accessed: 20/05/2019
[17] Refurbishment of multifamily buildings, *Staccato* project. Available at: https://www.concertoplus.eu/reducing-co2/sofia/ Accessed: 20/05/2019.
[18] Energy Efficiency of Multi-Family Residential Buildings National Programme. Available at: https://www.mrrb.bg/en/energy-efficiency/energy-efficiency-of-multi-family-residential-buildings-national-programme/. Accessed: 05/07/2019

IOP Conf. Series: Earth and Environmental Science **329** (2019) 012006 doi:10.1088/1755-1315/329/1/012006

Prediction, comparison and analysis of comfort level and energy consumption of a passively remodeled dwelling based on BP neural network computation

Yiming Du[1]; Qiong Huang[2,*], Anxiao Zhang[3]; Qi Zhang[4]; Hao Jing[5]

[1]Tianjin University, Tianjin, China, yiming.du@tju.edu.cn
[2]Tianjin University, Tianjin, China, qhuang@tju.edu.cn
[3]Tianjin University, Tianjin, China, zhanganxiao9012@gmail.com
[4]Tianjin University, Tianjin, China, zhangqi_arch@vip.163.com
[5]Tianjin Zhaojiang Co. Ltd, Tianjin, China, 334796305@qq.com
[*]Corresponding Author

Abstract. Many ordinary dwellings are not well built in the vast rural areas of China, with poor comfort level and high energy consumption. The China Resources Charity Foundation funded the renovation of more than 200 farmhouses. Compared with the BP neural network model established with the measured data, the BP neural network model established with the simulated data can more quickly estimate the energy saved by the renovation measures of rural dwellings and the improved indoor environment, which is very necessary for the government and enterprises to make decisions to help. In this paper, one of the renovated houses (No. 285) was selected for tracking measurement before and after renovation to obtain real indoor environment and energy consumption data. Simulation software (DesignBuilder) was used to simulate the other 36 renovated houses in order to train the BP neural network model. The energy consumption of No. 285 was calculated by using the trained BP neural model, and the calculated results were compared and analyzed with the simulated data and measured data. Results show that the BP neural network model after training is better than the simple energy consumption simulation software in energy consumption prediction, but there are still errors from the actual measurement values. Regardless of software predictions or actual measurements, the results show that the renovation measures have indeed exerted a very positive impact. As to the overall trend, the software predictions are the same as of the actual measurements, and the variance of day and night temperature variance and the annual extreme values are greatly reduced.

1. Introduction

Energy consumption has become an important social issue in China. In recent years, a large number of energy-saving building practices have been carried out in China, including various types of public buildings, such as gymnasiums, libraries and office buildings. However, the prediction and research of residential building energy consumption have more potential for energy saving, especially in China, a country with a large population. In China, the main forms of residential buildings are divided into two types: one is dwelling house in the city, and the other is single house in the countryside. At present, the research on residential building energy consumption mainly focuses on urban dwelling house, and some achievements have been made. Due to poor construction conditions and imperfect public facilities, single houses distributed in rural areas have a serious waste of energy, which should be paid more attention to.

The prediction and research of building energy consumption are helpful to reduce building energy consumption. However, the energy consumption of buildings is a complex issue. Different climatic environments, building types, construction methods, materials used to maintain structures, orientation of buildings, cooling requirements and other factors have a significant impact on the energy consumption of buildings.

At present, the research methods of building energy consumption can be roughly divided into two kinds, physical modeling method and data prediction method [1]. Physical modeling methods rely on thermodynamic rules for detailed modeling and analysis. Common software includes Designbuilder, EnergyPlus, eQuest, and Ecotect. The software calculates building energy consumption based on detailed building and environmental parameters, such as construction details, operating time, heating, ventilation and air conditioning (HVAC) design information and climate information [2]. However, some of the detailed data may not be available to users, which can lead to a decrease in prediction accuracy. Instead, the data algorithm method is based on a large number of operations of existing data and uses the reasoning of mathematical model to predict future energy consumption. In contrast, it is more efficient and operable. In recent years, with the constant modification of software parameters and the continuous evolution of algorithm methods, the two energy consumption calculation methods have different advantages in different buildings.

In China, residential buildings can be roughly divided into urban aggregative housing and rural detached housing. Compared to urban housing, rural housing usually needs to be self-sufficient in heating, power supply and water supply. However, the countryside in China is usually backward. The ordinary rural houses are badly built and the thermal environment is often neglected, with no any insulation for walls. Backward architectural cognition and unskilled construction skills make these houses with poor comfort level but high energy consumption.

Unfortunately, these backward villages are often located in the most scenic areas, and inefficient energy use can wreak havoc on the beautiful landscapes. Therefore, the energy-saving reconstruction of these houses can not only improve the living comfort of farmers, but also protect the beautiful environment.

To achieve fast and reliable building energy conservation prediction, we must rely on the computational method, namely the meta-model method. Meta-model, or agent model, refers to the agent model calculated based on the computer model, which simulates more complex design variables with a large amount of known building data. Compared with prediction models, the prediction calculation of meta-models is faster. Common meta-models include Adaptive Regression Splines (MARS), Kriging (KG), radial basis function (RBF), artificial neural network (ANNs) and support vector Regression (SVR) [7].

There is no optimal meta-model technology, but the accuracy of prediction results depends on the most suitable technology selection and technical input, and ANNs model is still the most commonly used building energy consumption prediction technology. In many studies, ANNs model has shown excellent performance, providing a regression coefficient (R) higher than 0.9 in most cases, which is more accurate than many physical, statistical and regression methods [15]. It is worth noting that ANNs model is the most commonly used alternative model for building performance simulation (BPS), because it has stable and satisfactory performance in solving large problems [7]. Neural networks are especially suitable for studying problems with the influence of large variables, such as this study.

An ANN (see Ref. [16] for details) is a processing data system that learns the relationship between inputs and outputs by studying recorded data, obtained from the original model. It is a network of computation units, called neurons, which are connected by weighted links (called synaptic connections or synapses), over which information is transmitted and manipulated. Each neuron receives input data from the previous ones by means of synapses, handles these data and combines them through a transfer function so as to generate output data that are sent to the following neurons. The network learns from provided inputs and outputs (coming from the original model) by means of training. In particular, this is an iterative procedure that aims to properly set the weights of the synaptic connections by minimizing a certain parameter, such as the sum of squared errors (SSE) or the root mean squared error (RMSE) [12] [13].

The performance of an ANN significantly depends on input and output data, as well as its architecture and parameters, which must be set carefully [17]. The ANN model used in this study is a feed-forward MLP, composed of three layers and thus with only one hidden layer. The number of hidden neurons is detected by means of 'trial-and-error'. This parameter highly affects the ANN performance: when it increases, the training data set error decreases at the cost of compromising the model generalization ability. The network is trained with Levenberg-Marquardt back-propagation algorithm coupled with Bayesian regularization. A sigmoidal function for the hidden layer and a linear function for the output layer are used as transfer functions. The training is stopped when the RMSE stabilizes or the maximum number of epochs, set equal to 1000 as in Ref. [15], is reached. Then, the network is tested on a second sample of input and output data by considering the coefficient of regression (R) and the distribution of the relative error between the ANN outputs and the original model targets as performance indicators.

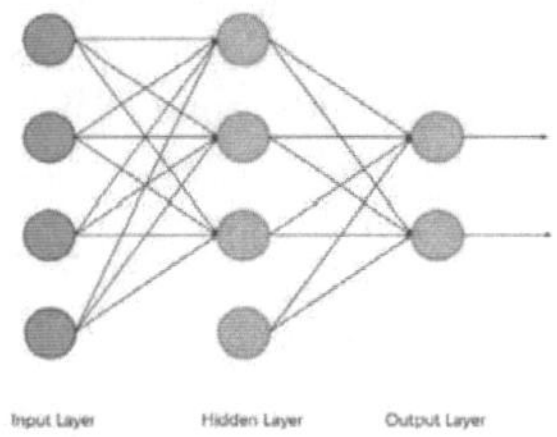

Figure 1. Architecture of a feed-forward multi-layer perceptron (MLP) ANN.

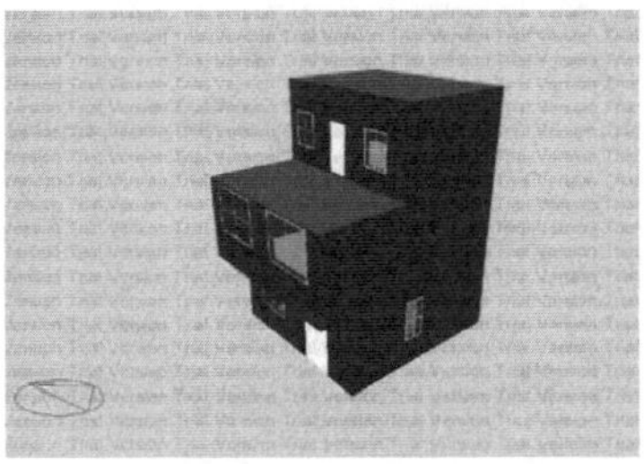

Figure 2. Model of No. 285.

Figure 3. Image of No. 285

China Resources Charitable Foundation is a charitable organization of China Resources Group. Every year, companies and employees of China Resources Group make donations to the foundation, and the annual fund of about RMB 100 million Yuan is used to improve the living environment of poor rural residents. In 2013, China Resources Charitable Foundation funded the renovation of rural houses in three natural villages of WuDian Town, JinZhai District, AnHui Province to improve the living environment of farmers. In consideration of durability and economy, the renovation mainly adopts the method of passive building energy saving instead of the addition of active equipment.

This paper selected one house (No. 285) of the modified houses as the study case. The indoor environment parameters and energy consumption in a year were recorded through the field measurement, the measured data were compared with the annual simulation results obtained by the energy consumption simulation software and the data calculation result using BP network, and then the result was used to improve the energy saving prediction, so as to predict the energy saving potential of the local rural house more accurately.

Simulation data which input BP neural network are used to quickly know the value of renovation measures to improve the indoor environmental quality of rural houses and reduce energy consumption. This rapid estimation helps local governments formulate renovation measures. In addition, most rural houses are built by villagers themselves, which has a lot of irrationality and randomness, and both construction and use exist at the same time. The houses are in the process of constant change for a long time, and it is necessary to keep the basic construction situation of the houses stable for a long time according to field measurement. The measured data for a long time cannot accurately reflect the actual situation of rural buildings.

2. Methodology

No. 285 house is located in Gutang Village, Wudian Town, Jinzhai County, Anhui Province, which is 175 kilometers away from the provincial capital Hefei. Jinzhai's climate is subtropical monsoon climate characterized by hot and rainy summers and gentle and less rainy winters. The Gutang Village is located in the depths of the Dabie Mountain, effected by the mountain microclimate, and there is large temperature difference in the morning and evening with high humidity: the average temperature is 33 °C in the summer, and the average temperature is - 1 °C in winter.

No. 285 is a three-layer house with flat roof, which is one of many detached houses along the street of Gutang Village. The left and right walls are connected with other houses. No. 285 covers an area of 140.89 m^2 in total, with the use space mainly on Layers 1 and 2, including a living room, two bedrooms, one kitchen and one dining room. No. 285 has a poor construction quality. The wall is a single-layer red brick wall with a thickness of 240 mm, the windows are made of aluminum alloy single-layer glass, the roof deck is 150-mm concrete single-layer floor, and all enclosures have simple structures and maintenance functions and do not have any insulation and energy saving measures. The interior decoration is simple with only painting of the wall surface, and there is no insulation measures such as suspended ceiling.

No. 285 is mainly used by a couple around 50 years old. The couple make a living by farming and do some small handicrafts. They are basically at home except in the busy farming seasons, so the housing usage rate is very high. The couple has a son who works outside all year round with his wife. They come back to live during the Spring Festival every year, which is almost negligible. The elderly couple cook three times a day. Most of the energy used for cooking is on fire, with a small amount of electricity and about 3.2 tons of coal each year. There are no cooling ways in the summer, and in the winter they rely on electric blankets (filaments heated by resistance wires) for heating. The couple said they did not feel hot in summer, but felt colder in winter. They have to heat their rooms with charcoal brazier in a month or so in winter.

According to the construction situation of No. 285 house and its use characteristics, the energy-saving transformation was carried out to reduce its energy consumption and improve the indoor living environment. The renovation methods include: 1. Wall renovation: add 100-mm thick polystyrene board insulation on the basis of the original wall, and externally brush the thermal mortar; 2. Window renovation: replace single-layer glass with double-layer hollow glass, and replace aluminum alloy window frame with plastic steel window frame with better tightness; 3. Kitchen renovation: use electrical appliances instead of open fire.

To accurately record the indoor environment of No. 285 before and after the renovation, the air temperature and humidity recorder was installed in the main space of the farm house.

The air temperature and humidity recorder is a HOBO U10-003 type temperature and humidity data collector produced by Onset. The recorder can be used to monitor and record the temperature and relative humidity in the indoor environment, and can be widely used for monitoring people's indoor environmental comfort.

Finally, the digital model of No. 285 house was built using DesignBuilder software to simulate the energy consumption and indoor environment before and after the renovation. The envelope setting parameters before renovation are shown below.

Table 1. Opaque exterior parameters of envelope summary before renovation.

	Construction	Reflectance	U-Factor with Film (W/m2-K)	U-Factor without Film (W/m2-K)	Gross Area(m2)	Net Area(m2)
Block1	150mm brick	0.30	2.297	3.500	105.70	93.17
Block2	150mm brick	0.30	2.297	3.500	128.43	115.09
Block3	150mm brick	0.30	2.297	3.500	127.01	120.3

Table 2. Exterior fenestration parameters of envelope summary before renovation

	Glass Construction	Glass Area(m2)	Frame Area(m2)	Divider Area (m2)	Area of One Opening (m2)
Block1	single-layer	4.86	0.65	0.15	5.66
Block2	single-layer	7.59	0.77	0.19	8.55
Block3	single-layer	4.09	0.59	0.14	4.82

After renovation, the U-Factor with Film was set as 0.337 W/m2-K, and the U-factor without Film was set as 0.355 W/m2-K in opaque exterior. In exterior fenestration, the glass U-factor was set as 3.835, the glass solar heat gain coefficient (SHGC) was set as 0.768, and the glass visible transmittance was set as 0.744. In HVAC system, the thermostat setpoint temperature at peak load is 24 ℃ in the sensible cooling and 22 ℃ in sensible heating. The calculated design load is 10116 kWh in the sensible cooling and 9241 kWh in sensible heating of total three zones.

In Ref. [18], the implementation of simulation-based large-scale uncertainty/sensitivity analysis of building energy performance (SLABE) has allowed to qualify the existing building stock, which constitutes the examined category, by defining 46 characteristic parameters related to thermal energy demand for space conditioning and thermal comfort. These parameters are related to geometry, thermal envelope, building operation and type of in-room terminals for space conditioning. Based on the actual situation of rural houses in Jinzhai, we have sorted out 34 parameters, as shown in Table 3.

Table 3. Characteristic parameters of the existing building stock taken from Ref. [16].

Geometry	p1	Orientation (angle between North axis and building North)
	p2	Area of each floor (m2)
	p3	Form ratio
	p4	Floor height (m)
	p5	Window to wall ratio: South
	p6	Window to wall ratio: East
	p7	Window to wall ratio: North
	p8	Window to wall ratio: West
	p9	Number of floors
Envelope	p10	Type of window glasses
	p11	Type of window frames
	p12	Clay (floor) thickness (m)
	p13	Clay (floor) thermal conductivity (W/m K)
	p14	Clay (floor) specific heat (J/kg K)
	p15	Expanded clay (roof) thickness (m)
	p16	Expanded clay (roof) thermal conductivity (W/m K)
	p17	Expanded clay (roof) specific heat (J/kg K)
	p18	External bricks' (walls) thickness (m)
	p19	External bricks' (walls) thermal conductivity (W/m K)
	p20	External bricks' (walls) specific heat (J/kg K)
	p21	Floor block thickness (m)
	p22	Floor block specific heat (J/kg K)
	p23	Internal bricks' (walls) thickness (m)
	p24	Internal bricks' (walls) thermal conductivity (W/m K)
	p25	Internal bricks' (walls) specific heat (J/kg K)
	p26	Roof block thickness [m]
	p27	Roof block thermal conductivity (W/m K)
	p28	Roof block specific heat (J/kg K)
Operation	p29	People density (people/m^2)
	p30	Artificial light load (W/m^2)
	p31	Infiltration rate (h^{-1})
HVAC	p32	Heating set point temperature (° C)
	p33	Cooling set point temperature (° C)
Energy	p34	total energy

To train the BP neural network model more accurately, 69 other houses were measured in the village, No. 213-284. These houses vary widely in construction, from original rammed earth structure to multistory brick structure built in recent years. For training a model more similar to No. 285, we selected the above 69 houses surveyed and selected 36 houses of 39 households for detailed mapping, including the size, construction and use of the buildings. According to the mapping results, we use DesignBuilder simulation software to build the digital models. Using software simulation results and mapping statistical results, the information summary of the 36 houses were completed.

Rural housing has a small scale, simple and crude construction, simple use condition, and no too many parameters for input, so the calculation time of BP neural network is very short.

3. Results

According to the information in figure 2, the BP neural network model was built. According to BP neural network theory and operation steps, the prediction of building energy consumption is realized by programming in MATLAB.

- Data import and normalization

  Firstly, the data T was imported with the output of numeric matrix type. Input data and output data were determined, 26 groups (70%) of the 36 groups of data were randomly selected as training data and the rest as test data, and then the normalization processing to the data was carried out.

- Set up BP neural network
  BP neural network was constructed and parameter configuration was set. The number of iterations (epoch=1000). Learning rate lr = 0.1; the target goal = 0.00004;
- Test the training model
  Test the model with the remaining 30% data, and output the graph of network results.

The BP neural network was ended by the 8th generation, and the gradient was 0.00461. The best performance occurred in the 6th generation, with a performance value of 0.0658. The calculation results are shown in figure 4: it can be seen from the figure that the fitting degree of BP model for training data is very good, reaching 0.99777, and it is generally considered that R < 0.05 is a small error. However, the fitting degree of test data is poor, only 0.55947, with a comprehensive performance of 0.81519, which cannot provide very accurate prediction.

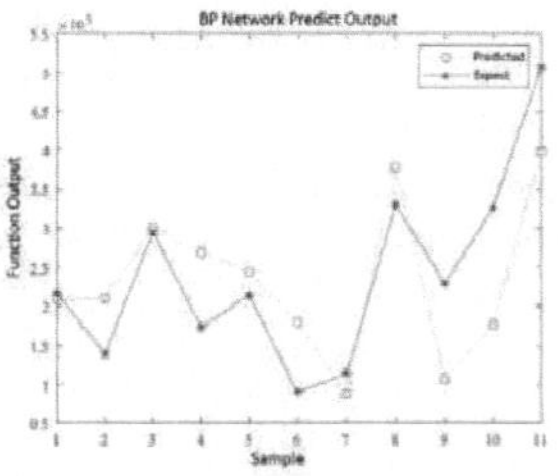

Figure 4. BP Network Output.

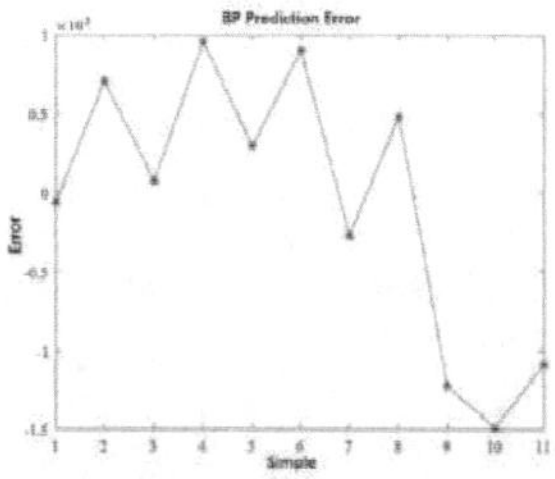

Figure5. BP Network Error.

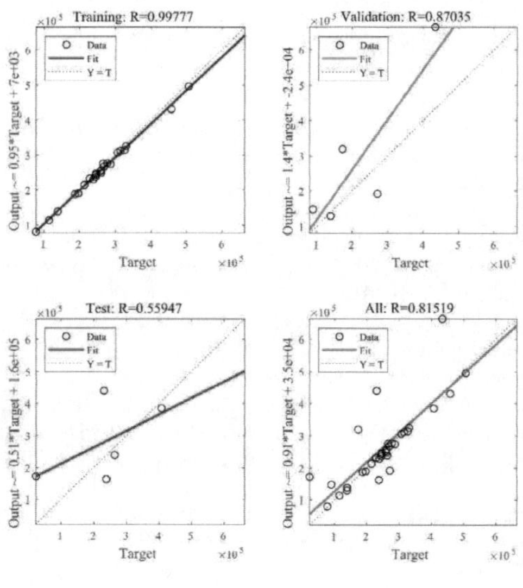

Figure 6. Results of the BP Network Calculation.

DesignBuilder software simulation results show that before modification, 8791 kWh of electricity is used in summer and the total energy is 13171 kWh.

Using the BP neural network model and the information of mapping, the energy consumption of No. 285 was calculated. The calculation procedure is the same as that in the test of BP neural network model. Firstly, input data were loaded and normalized, then the BP neural network model was used for calculation, and finally the results were anti-normalized. The final calculation result of No. 285 is 13911 (kWh).

Results show that before the modification, 3220 kWh of electricity and 3.2 tons of coal are used in summer, equal to 13300 energy, but after the modification, 4203 kWh of electricity and 2.4 tons of coal are used in winter, equal to 11763 energy.

Table 4. Energy consumption results

	Simulation	BP neural net work	Measurement
No. 285 Total energy	13864	13911	14122

4. Conclusions

The BP neural network model after training is better than the simple energy consumption simulation software in energy consumption prediction. This means that the BP neural network model based on simulated data is suitable for rapid estimation of energy saving in rural houses, which is helpful for the government to formulate specific policies to help. Compared with the simulation of rural housing by single energy consumption software, the calculation of energy consumption by adding BP neural network training model is more accurate. There are still errors from the actual measurement values, mainly because the quality of the houses actually built is not perfect, especially the rural houses, and there will be some differences between the actual usage and the software simulation. In addition, the data show that the fitting degree of the trained BP neural network model is not very ideal, which may be related to the insufficient amount of data. Some houses that are quite different from other houses contribute points farther from the fitting curve.

Regardless of software predictions or actual measurements, the results show that the renovation measures have indeed exerted a very positive impact. As to the overall trend, the software predictions are the same as the actual measurements, and the variance of day and night temperature and the annual extreme values are greatly reduced.

Acknowledgements
This research was funded by the National Key Research and Development Program of China (Grant No. 2016YFC0700201) and the National Science Foundation of China (Grant No. 51338006).

References

[1] Ascione, N. Bianco, C. De Stasio, G. M. Mauro, and G. P. Vanoli 2017 Artificial neural networks to predict energy performance and retrofit scenarios for any member of a building category: A novel approach, *Energy*, vol. **118**, pp. 999–1017
[2] S. Ramakrishnan, X. Wang, J. Sanjayan, and J. Wilson 2017 Thermal performance of buildings integrated with phase change materials to reduce heat stress risks during extreme heatwave events, *Appl. Energy*, vol. **194**, pp. 410–421
[3] Lee SH, Hong T, Piette MA and Taylor-Lange SC 2015 Energy retrofit analysis toolkits for commercial buildings: a review. *Energy*, vol. **89**, pp. 1087–1100
[4] Poel B, Cruchten GV, Balaras CA 2007 Energy performance assessment of existing dwellings. *Energy Build*, vol.**39**, pp. 393–403
[5] US Department of Energy. Energy Efficiency and Renewable Energy Office, Building Technology Program (2013), EnergyPlus 8.0.0. Available at: http://apps1.eere.energy.gov/buildings/energyplus/
[6] TRNSYS. Transient system simulation program. University of Wisconsin; 2000.
[7] Li YF, Ng SH, Xie M and Goh TN 2007 A systematic comparison of metamodeling techniques for simulation optimization in decision support systems. *Appl Soft Comput*, vol.**4**, pp. 1257–1273, Oct.
[8] Hopfe CJ, Emmerich MT, Marijt R and Hensen JLM 2012 Robust multi-criteria design optimisation in building design. In: Proceedings of the building simulation and optimization conference, Loughborough (UK)
[9] Eguía P, Granada E, Alonso JM, Arce E and Saavedra A 2016 Weather datasets generated using kriging techniques to calibrate building thermal simulations with TRNSYS. J *Build Eng* vol.**7**, pp. 78–91, 2016
[10] Dong B, Cao C and Lee S 2005 Applying Support Vector Machines to predict building energy consumption in tropical region. *Energy Build*, vol.**37**, pp. 545–553
[11] Eisenhower B, O'Neill Z, Narayanan S, Fonoberov VA and Mezi c I 2012 A methodology for meta-model based optimization in building energy models. *Energy Build*, vol.**47**, pp. 292–301
[12] Agnier L, Haghighat and F. Multiobjective 2010 optimization of building design using TRNSYS simulations, genetic algorithm, and artificial neural network. *Build Environ*, vol.**45**, pp. 739–746
[13] Asadi E, da Silva MG, Antunes CH, Dias L and Glicksman L 2014 Multi-objective optimization for building retrofit: a model using genetic algorithm and artificial neural network and an application. *Energy Build*, vol.**81**, pp. 444–456
[14] Kalogirou SA 2000 Applications of artificial neural-networks for energy systems. *Appl Energy* vol.**67**, pp. 17–35
[15] Paudel S, Elmtiri M, Kling WL, Le Corre O and Lacarriere B 2014 Pseudo dynamic transitional modeling of building heating energy demand using artificial neural network. *Energy Build*vol.**70**, pp. 81–93

[16] Haykin SS. Neural networks and learning machines (third ed.). Upper Saddle River: Pearson Education.

[17] Kalogirou SA and Bojic M 2000 Artificial Neural Networks for the prediction of the energy consumption of a passive solar building. *Energy* vol.**25**, pp. 479–491

[18] Mauro GM, Hamdy M, Vanoli GP and Bianco N, Hensen JLM 2015 A new methodology for investigating the cost-optimality of energy retrofitting a building category. *Energy Build,* vol.**107**, pp. 456–478

IOP Conf. Series: Earth and Environmental Science **329** (2019) 012007 doi:10.1088/1755-1315/329/1/012007

Adapting Traditional Passive Strategies within Contemporary House to Decrease High energy consumption Impact in Nejd Region, Saudi Arabia

Ali Abdulmohsen Aldersoni, David Hou Chi Chow
School of Architecture, University of Liverpool. Abercromby Square, Liverpool L69 7ZN, UK

Abstract. Saudi Arabia has been identified by the Intergovernmental Panel on Climate Change (IPCC) as one of the strongest warming projected regions which indicates a serious future impact if no radical change is made in energy consumption [1]. The Saudi Electricity Company indicated that electricity consumption for 2017 is 204,597 GW [2]. Residential buildings have used 55% of the annual energy consumed in Saudi Arabia [3]. Therefore, this research aims to investigate the efforts that can be made to reduce this extreme consumption of energy by using simple traditional design techniques that were used in traditional buildings in the Nejd region, Saudi Arabia. The Nejd region was chosen because it represents the majority of Saudi's topography and weather, and is recognized as the largest region in the country with the biggest population size [4]. Two houses were monitored during the summer of 2018. One represents a traditional old house while the other represents a contemporary house built by the Ministry of Housing (M.of.H) in Riyadh City. The two case studies provided valuable data for further analysis as the architectural design for the houses were analysed to distinguish the architectural features and weaknesses. Two models for the two cases were created using DesignBuilder and validated against measured data. The computational models were then used for dynamic thermal simulation, and used to estimate the energy consumption for the selected cases. The integrated passive strategies in traditional house were analysed, and compared it with a contemporary house. The result shows a 52.3% differences in annual energy consumption between the traditional house with contemporary house. In addition, the simulation result illustrates a 24.3% reduction in the annual energy consumption for the contemporary house is achieved by applying the passive strategies to the house design.

1. Introduction

Saudi Arabia recognises the issues surrounding energy sustainability caused by the drastic rise in energy demand over the last ten years. There has been an increasing in electricity demand in Saudi Arabia leading to a consumption of approximately 9.2 kilowatt hours per year per capita in 2015, while it was approximately 7.7 kilowatt hours per year in 2010. Currently, residential buildings consume 55% of the total national energy consumption [3]. The residential building sector's energy consumption will continue to increase, and statistics imply that 70% of current construction projects being carried out currently are linked to housing [6]. Saudi Arabia's residential buildings use an excessive amount of electrical energy, even more so throughout summer when approximately 60% of energy consumed is due to air conditioning [7]. Adding to the seriousness of this problem the Ministry of Housing in KSA is working to establish 781,636 residential units which will be completed in 2020 [8]. To fulfil increasing demands, the Saudi government is planning to expand the capacity of electric generation from 55GW to 120GW by 2020. This will be the Middle East's greatest increase of electricity generation [9]. Responding to the high energy consumption crisis causing the building sector's substantial growth, another initiative was introduced in 2012, the Royal Decree no.6927 and it introduced a minimal standard for thermal insulation in every building sector, to enhance the efficiency of energy in buildings. In 2017, the building code in Saudi was designed to be more efficient[10].

The rapid increase in usage of heating and cooling systems in buildings had numerous disadvantages as well as higher consumption of energy, including a growth in peak electricity demand, indoor air quality problems and more significantly, environmental problems like ozone depletion and global warming [7]. Consequently, alternative solutions were presented to eliminate or lower usage of systems for conventional mechanical heating and cooling. For the study presented in this paper, five traditional passive strategies were investigated to assess their effectiveness in reducing the energy consumption in a traditional house, and evaluated their negative and positive impact on the energy consumption for contemporary houses in Saudi Arabia. The five strategies are: window to wall ratio, exterior wall thermal mass, shading device, ceiling height, and the inclusion of greenery spaces.

2. Literature Review

Many researchers have shown that implementing an energy code and standard can perform a significant role in improving the energy efficiency in buildings. The Saudi government recent initiative has been the basis of an obligatory building code for new build residential buildings. It was established to minimise energy consumption as of January 2017. According to the SEEC (Saudi Energy Efficiency Centre), a reduction of 30-40% of energy has been expected to be accomplished in the residential sector [12].

Table *1* explains the building code requirements for the thermal building envelope. The building code's effect on energy consumption was tested by Mohammed Abuhussain on a typical house in the city of Jeddah [10]. The study concluded that by adhering to the building code, the existing villa accomplished a significant calculated reduction of 40% from the annual total of cooling demand. Abuhussain advised that further passive strategies to be included in the energy code for residential buildings in varying climate zones of Saudi so that the effects of potential climate change can be neutralised.

Table 1: Building code requirement for the thermal transmittance (U-Values) for low-rise / residential buildings

Opaque Elements	Climate Zone 1 (Riyadh, Jeddah, Dammam cities)	Climate Zone 2 (Ha'il, Tabuk cities)	Climate Zone 3 (Asir City)
Roof	0.31	0.37	0.42
Exterior Walls	0.53	0.61	0.7
Opaque Doors –All Assemblies	2.84	2,84	2,84
Vertical Glazing , 25% of wall			
All Assemblies	2.67 SHGC - 0.25	2.67 SHGC - 0.25	2.67 SHGC - 0.25
0% -3% All Types	4.26 SHGC- 0.35	4.26 SHGC- 0.35	4.26 SHGC- 0.35

Other studies concentrated on the theoretical aspect of traditional building characteristics. For example, Farzaneh Soflaei studied architectural features in the Iranian traditional central courtyard [14]. The study illustrated the analysis of the environmental physics for central courtyards which cover four important aspects. The dimension and orientation of the courtyard perform a significant role for maximising radiation during the cold seasons towards passive heating as well as daylighting, including maximum amiable airflow for natural ventilation and passive cooling for indoor areas were possible throughout the warmer seasons. On the other hand, Albert Malama studied the thermal performance of traditional and contemporary housing in the cool season of Zambia. The statistics of the research on the traditional house show they are more uncomfortable in the cooler seasons as compared to the contemporary house.

Traditional buildings with passive strategies in Saudi Arabia were not involved in saving energy consumption strategies, and there is no study that researches the difference between the energy consumption in traditional and contemporary building. Therefore, this paper aims to bridge this gap and take advantage of traditional passive strategies in new contemporary houses

3. Methodology

In order to distinguish the traditional passive strategies and draw comparisons between traditional and contemporary houses, two typical traditional and contemporary houses located in Riyadh city were selected for this study. Riyadh city is the capital of Saudi Arabia, and it has the highest population in the country, which increases the demand on the residential facilities [4]. The first case study is on a traditional house located in the Aldhoo neighbourhood, one of the most traditional areas in the centre of Riyadh city which received special attention from the Riyadh Development Authority. The second case is a contemporary house that was built by M.of.H in Al-Uyaynah, Riyadh city, and it was completed in April 2018. On-site measurements for both cases were collected including exterior and interior temperature, exterior and interior humidity, wind speed and direction, and solar radiation. It is important to mention that data collection was based on climate analysis that was conducted in August, which is a favourable period as the hottest month in 2018. This data was validated by using a simulation software program, DesignBuilder. 3D models were built based on architectural and constructional drawings for both houses. The drawings and data were obtained from three sources: (i) review of building drawing filed with projects' consulters, (ii) site visits to buildings under construction, (iii) informal interviews with contractors. The energy consumption for cooling and heating systems was calculated using the temperature data collected. In addition, standard Saudi family occupancy was taken into consideration for energy analysis[16].

Traditional passive strategies have been investigated in order to see the negative and positive impact on energy consumption in the traditional and M.of.H houses. This study applies to the contemporary house features and compares them to the traditional building to investigate the size of the negative impact. Five passive strategies were investigated in the two types of houses. The passive strategies include window to wall ratio, walls and window shading device, height ceiling, thermal mass, and outside green area.

3.1. Case Studies

3.1.1. Contemporary House (M.of.H House)

In 2018, 12161 houses were built by M.of.H in Saudi Arabia which represent the typical contemporary house for a single Saudi family. According to Opoku and Abdul-Muhmin, 40% of the residential building stock is classified as single family house [17]. As a result, enhancing the energy performance in this form of residential buildings can positively reduce the overall energy consumption in Saudi Arabia's building sector.

The total land area for this house is 500 m^2, and it comprises of two storeys with 272 m^2 total building area. This house was designed to inhabit 6 people. M.of.H projects were built by using a concrete construction system with adherence to the minimum standard amount of thermal insulation. Ground floor, intermediate floor, roof, interior walls, and exterior walls construction and thermal insulation details were illustrated in Table 2. However, the architectural drawing clarifies that environmental effects have not been taken in to consideration when buildings are designed.

3.1.2. Traditional House

The traditional house with one central courtyard is the most popular traditional house type in the Middle East. The courtyard is the main element that all spaces connect with directly. Therefore, most of the house opens (windows and doors) are directed to the courtyard which contains 15.1% of the total land area. The total land area is 155 m2, and the total building area is 270 m^2.

The Riyadh Development Authority and the General Authority for Tourism and National Heritage (SAUDI) are very careful about the architectural sensitivity of traditional houses. Therefore,

the house has been renovated with the same original natural materials that were used in the original building with minor additions of artificial material used to protect and preserve the natural materials and the building's construction (Table 2). Clay block, mud plaster, and Athl joist (local tree trunk) are the main construction element that distinguished the traditional house in Saudi Arabia.
Interestingly, the building property of traditional house is compatible with the new Saudi Building Code requirements. The building envelope insulation fulfils the new Saudi building code requirements which has a significant impact on the current and future climate [10].

Table 2: Specification of M.of.H and traditional Buildings' Materials and Their Thermal Properties.

	M.of.H House (Contemporary House)		Traditional House	
Categories	Thickness (m)	U-value (w/m2-k)	Thickness (**M**)	U-value (w/m2-k)
Ground floor	0.29	2.40	0.30	1.080
Exterior walls	0.25	0.82	0.57	0.391
interior walls	0.25	1.40	0.42	0.513
Intermediate floor	0.36	2.10	0.50	0.234
Roof	0.48	0.43	0.67	0.199

3.2. Modeling validation

3.2.1. On site measurement & Simulation tools

In order to collect weather data for the selected case, iButtons data loggers (DS1921H-F5 Thermochron) were used in 5-6 of the most important spaces such as bedrooms, living room, and kitchen for collecting hourly interior temperature data. The loggers were installed in constant positions that were distant from any heat source. Davis weather station (Pro2) was used to collect hourly exterior weather data. Weather stations rose 2m from the rooftop with a custom stainless steel stand to prevent any outside climate obstacles (Figure 1). It's important to mention that all data loggers were tested in closed room in the basement of the architecture school building at university of Liverpool to calibrate the accuracy of the loggers' reading. The result illustrated a good agreement between the data loggers' records.

3.2.2. Validation

To examine the different options for energy efficiency, baseline energy for the traditional house and the M.of.H house were developed, to identify the passive strategies which could be implemented on the residential building in Saudi Arabia. Using the architectural drawings with DesignBuilder software (version 5.0.1) a 3D model for a standard villa was developed as depicted in Figure 2. A validated computer model has been used to principally assess the energy consumption for different types of residential building in Riyadh.

Figure 1: Equipment used for on-site measurements

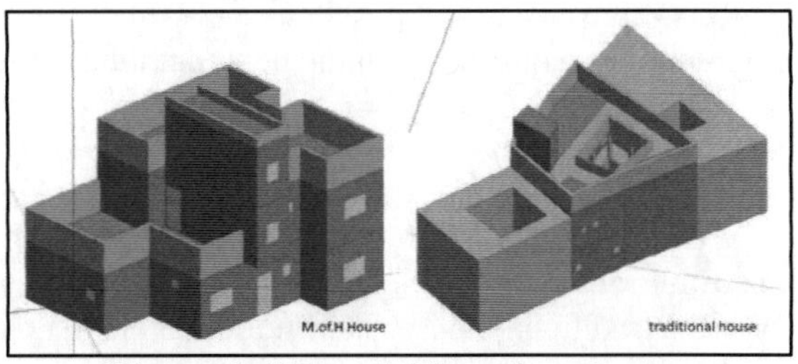

Figure 2: 3D model for selected cases

Figure 3 depicts the result of hourly temperature calibrations between the actual interior temperature affected by architectural characteristics for measured rooms. The software temperature calibration gave near identical result with a 1.02 % average difference for five spaces during a one-month period.

According to Abuhussain the researcher can consider the model to be valid as long as the difference between simulated and measured results is less than 5% [10].

Figure 4 illustrated the hourly temperature calibration between the on-site measured and DesignBuilder simulation results for room 2 in the traditional house. The validation result gave a 3.16% average difference for six spaces during a one-month period. The actual reading for interior temperature was affected by airflow inside the room because the windows and doors were open. The airflow is fluctuated from room to other, and this was caused by opening size and direction. That means the interior temperature was affected by the exterior temperature when the natural wind passes through the building's openings which increases the amount of airflow's velocity.

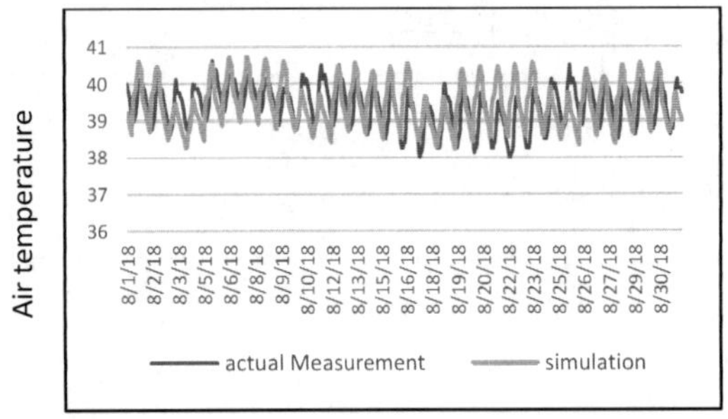

Figure 3: Hourly calibration between on-site measured data and simulation for dining room in the M.of.H house.

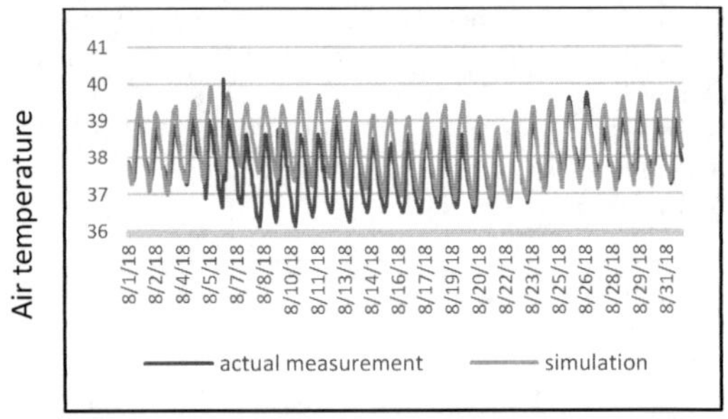

Figure 4: Hourly calibration between on-site measured data and simulation for room 2 in the traditional house

3.3. Energy consumption evaluation

In order to calculate the cooling and heating annual energy consumption, EnergyPlus has been chosen to demonstrate the analysis for the whole-building energy simulation. EnergyPlus is relatively straightforward simulation program which does not require extensive time to preform simulations. It has a user-friendly interface. For this study, four aspects were considered to assume a close result of actual energy consumption which are: building occupancy, energy use, HVAC system, and AC set point. The first aspect is the building's occupancy information which describes Saudi's standard number of family members and behaviour. Jubran Alshahrani found that 45% of Saudi families have up to 5-6 members [16]. In addition, Alshahrani described Saudi family behaviour and energy use which are important aspects of energy consumption. Six different zones are examined to study the energy used and people's habits of consumption in the sleeping zone, kitchen zone, dining zone, living zone, family zone, and guest zone. The third aspect is the HVAC system, which is the window AC, and is requested by the Ministry of Housing for such houses in the specification sheet for M.of.H projects. Significantly, the AC has a cooling set point of 25.5C and a heating set point of 20C base on the minimum requirement in the new Saudi building code [18].

3.4. Assessment process for traditional passive strategies

3.4.1. Assessment of passive strategies in contemporary house

The traditional construction and design techniques applied to the contemporary house were inspired form the traditional house. The additional design characteristics are described in the following:

- Window to wall ratio is an important variable affecting energy performance in residential buildings in a hot, dry climate. The ratio was decreased from 6.5% to 2.5%, and the window size became 0.5m X 1m.
- The overhang shading was added to shade the windows and solid walls. The vertical shading extends 1.6m overhang which is 50% of the ceiling height. This percentage was determined from the traditional cloister extension into the courtyard. In the traditional house, the cloister overhangs 1.85m, which is 50% of ceiling height.

- Contemporary materials were used on the basis of an Al-Sanea study which advises using melded polystyrene and heavyweight concrete blocks [19], the exterior walls are to be of thermal mass, and the wall thickness should be increased from 0.25 m to 0.4 m. This suggestion gave the contemporary house similar thermal mass properties to the traditional house.
- The green area covered all of the threshold between the edges of the house and the fence. This addition was inspired by the main function of courtyard, which is expected to have a direct impact into the ground temperature. A recent study conducted by Limor Shashua demonstrated that landscape decreased 4°C of the ground temperature [20].

4. Result and Discussion

4.1. Energy consumption

4.1.1. Contemporary House (M.of.H House)

The study results indicate that the average annual energy consumption of the typical M.of.H House is approximately 31800 kWh. It is axiomatic that cooling accounts for 93.8% of total energy consumption especially in a hot dry city like the Alyuaynah region. An extremely hot summer requires 5655 kWh in August alone when the outside temperature reaches 43.7 C°. On the other hand, the heating system needs 1960 kWh a year which is 6.2% of the of annual total energy consumption.

4.1.2. Traditional House

The study results indicate that the average annual energy consumption of the typical traditional house is proximately 15155 KWh, despite the temperatures in the Riyadh region reaching 48.7C°, which is 5C more than the Alyuaynah region. The cooling system requires about 13215 Kwh, which is 87.2% of the total energy consumption, and heating takes up 12.8% of total annual energy consumption which is around 1940 KWh.

4.2. traditional passive Strategies

As can be seen in Figure 10, during hot summer months, the traditional house requires less cooling energy than the contemporary house. It is apparent that passive cooling strategies reduced the cooling loads for the traditional building in comparison to the modern building, which has been built without consideration for passive techniques. By referring to the simulation results, the traditional passive cooling strategies can provide significant reduction to cooling energy consumption for M.of.H housing in Riyadh. The M.of.H Houses consume approximately 52.3% more energy than the traditional house, although the traditional house is located in a relatively hotter climate (Figure 5). Furthermore, the research established the impact of passive strategies on the traditional house design by including characteristics of the contemporary house in the design. A rise of 85.7% total cooling energy use and total annual cooling compared to the base case was shown in the results (Figure 6).

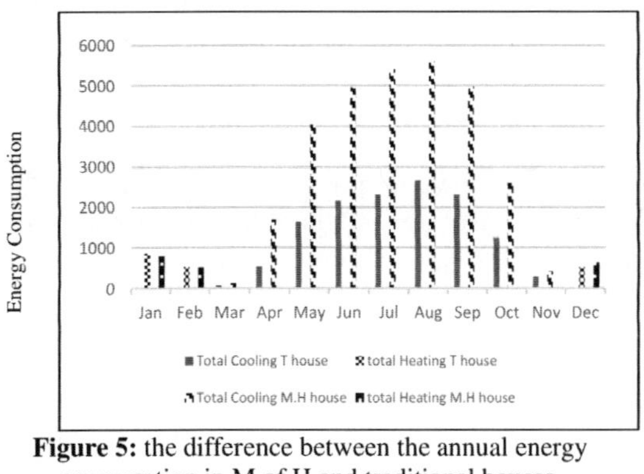

Figure 5: the difference between the annual energy consumption in M.of.H and traditional houses.

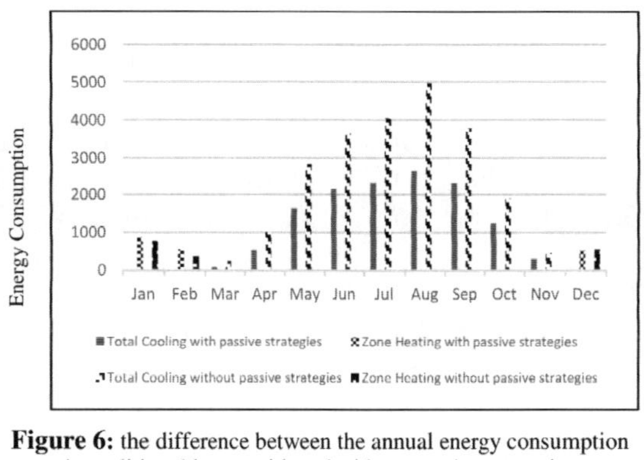

Figure 6: the difference between the annual energy consumption in traditional house with and without passive strategies

Since the construction and design concepts of traditional and modern houses are completely different, the energy consumption in these houses are also different. The objective of this research is to demonstrate the significant impact of traditional passive cooling strategies into M.of.H contemporary houses. The energy saving that were achieved by additional traditional design technique, is explained in the following:

- The simulation result shows roughly 5.4% saving of the total annual cooling energy consumption by decreasing 4% of base case's window to wall ratio.
- The overhang shading for windows and solid walls, achieved 4.8% reduction in cooling annual loads.
- The thermal mass technique saved approximately 7.4% of the annual cooling and heating energy consumption
- The exploitation of the open green areas around the house has a significant impact on annual energy demand with a reduction of 12.4% of total annual cooling energy demand.

Traditional passive strategies improved the energy efficiency for the contemporary house by reducing the total annual cooling energy demand by approximately 29%. Figure 7 shows that 24.3% reduction in the annual cooling and heating energy consumption for the contemporary house is achieved by adding the traditional passive techniques to the house design.

The effectiveness of applying passive cooling techniques was negative on the performance of the improved case during the winter season, where the total annual heating energy consumption was increased by 58%. This is linked to two main reasons including minimizing direct solar radiation, and decreasing the ground temperature. However, this only represents 5.7% of total annual energy consumption for improved case.

High ceiling technique was neglected of traditional passive techniques included in the analysis of this study. It was found that raising the height of the ceiling will increase cooling loads by roughly 23% due to the increase of the space volume. However, this strategy could provide further energy reduction during natural ventilation modes.

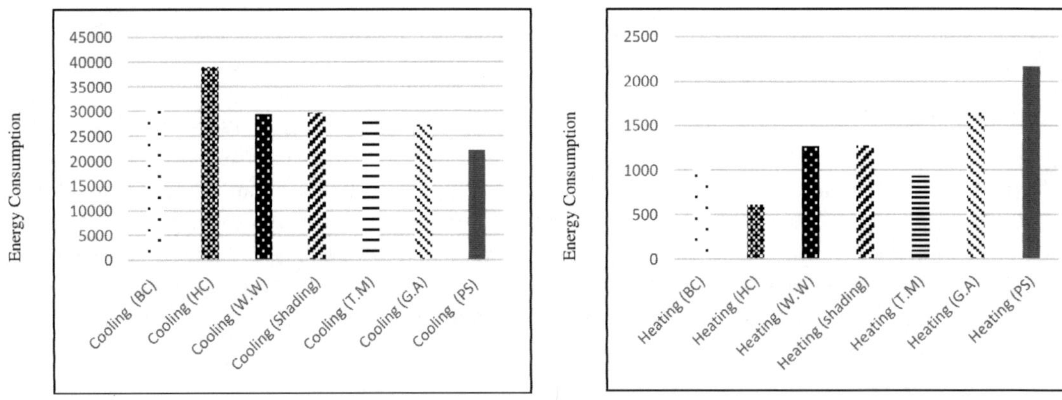

Figure 7: the different between the annual cooling and heating energy consumption for M.of.H house with five passive strategies.

Key: BC (Base case), HC(Ceiling Height), WW (Window to wall ratio), shading (shading device), T.M (thermal mass), G.A (Green area), PS (Passive strategies case)

5. Conclusion and recommendation

In Saudi Arabia, contemporary residential buildings significantly contribute to high demands of electrical energy. Specifically, energy consumption is caused by space cooling, with summer time being a major cause of peak demand and therefore high levels of electricity consumption. In this paper, five traditional passive strategies are investigated to assess their effectiveness in reducing the energy consumption in contemporary Saudi houses. The simulation results show that traditional houses without passive strategies request double energy consumption. In addition, the effectiveness of these strategies has been investigated and analysed to reduce the cooling and heating energy demand of M.of.H residential buildings. The five evaluated traditional passive techniques include window to wall ratio, walls and window shading device, height ceiling, thermal mass, and outside green area. These five techniques are attempted in traditional and contemporary houses. It was found that traditional passive techniques can achieve a significant reduction in cooling and heating energy use for a typical residential building located in Riyadh with a rate of 28.7 % for the annual cooling demand and with 24.3 % from the total annual energy consumption. The results of this study fulfil the aim of the new 2030 vision of Saudi Arabia by reducing the total energy demand for the residential sector in the country.

One of the limitations of this study is the parametric analysis for traditional passive strategies to distinguish the optimum design solution that is compatible with a contemporary house in Saudi Arabia. In addition, other design principles in a traditional house such as central courtyard should be studied to investigate their ability to implement these techniques in a future version of contemporary houses in Saudi Arabia.

6. References

[1] IPCC, "Climate Change 2014: Synthesis Report. Contribution of Working Groups I, II and III to the Fifth Assessment Report of the Intergovernmental Panel on Climate Change," IPCC,Geneva, Switzerland, 151 pp, 2014.

[2] SEC, "Annual Report 2017," Saudi Arabia, 2017.

[3] I. Mezghani and H. Ben Haddad, "Energy consumption and economic growth: An empirical study of the electricity consumption in Saudi Arabia," *Renew. Sustain. Energy Rev.*, vol. 75, no. November 2016, pp. 145–156, 2017.

[4] G. A. for S. KSA, "Population and Housing Census General Organization for Statistics," *STATS*, 2017. [Online]. Available: https://www.stats.gov.sa/ar/13. [Accessed: 20-Nov-2018].

[5] A. M. Al-ibrahim and A. Varnham, "A review of inlet air-cooling technologies for enhancing the performance of combustion turbines in Saudi Arabia," *Appl. Therm. Eng.*, vol. 30, no. 14–15, pp. 1879–1888, 2020.

[6] F. Alrashed and M. Asif, "Prospects of Renewable Energy to Promote Zero-Energy Residential Buildings in the KSA," *Energy Procedia*, vol. 18, pp. 1096–1105, 2012.

[7] A. Alaidroos and M. Krarti, "Evaluation of Passive Cooling Systems for Residential Buildings in the Kingdom of Saudi Arabia," vol. 138, no. June 2016, pp. 1–11, 2019.

[8] Saudi Ministry of Housing, "Housing projects record," *Saudi Ministry of Housing*, 2018. [Online]. Available: https://sakani.housing.sa/all-projects-map. [Accessed: 10-Apr-2018].

[9] F. Alrashed and M. Asif, "Analysis of critical climate related factors for the application of zero-energy homes in Saudi Arabia," *Renew. Sustain. Energy Rev.*, vol. 41, pp. 1395–1403, 2015.

[10] M. A. Abuhussain, D. HC, and S. Sharples, "Assessing the adaptability of the Saudi residential building ` s energy code for future climate change scenarios," *PLEA*, 2018.

[11] M. Santamouris, K. Pavlou, A. Synnefa, K. Niachou, and D. Kolokotsa, "Recent progress on passive cooling techniques Advanced technological developments to improve survivability levels in low-income households," vol. 39, pp. 859–866, 2007.

[12] SASO, "Thermal Transmittance Values for Residential Building," Saudi Arabia, 2014.

[13] M. M. Bajwa, "Effectiveness of Nocturnal ventialative passive cooling strategy in the Maritime desert climate of the Arabian Gulf region," *Renew. Energy*, vol. 2, no. 3, pp. 237–254, 1992.

[14] F. Soflaei, M. Shokouhian, S. Majid, and M. Shemirani, "Investigation of Iranian traditional courtyard as passive cooling strategy (a field study on BS climate)," *Int. J. Sustain. Built Environ.*, vol. 5, no. 1, pp. 99–113, 2016.

[15] A. Malama and S. Sharples, "Thermal Performance of Traditional and Contemporary Housing in the Cool Season of Zambia," vol. 1323, no. 96, pp. 69–78, 1997.

[16] J. Alshahrani, "Reducing High Energy Demand Associated with Air-Conditioning Needs in Saudi Arabia," vol. 12, no. 1, pp. 1–57, 2018.

[17] R. A. Opoku and A. G. Abdul-muhmin, "Housing preferences and attribute importance among low-income consumers in Saudi Arabia," *Habitat Int.*, vol. 34, no. 2, pp. 219–227, 2010.

[18] S. B. code national Committee, "Saudi Code for Energy Conservation (SBC 601-602)," *SbC*, 2018. [Online]. Available: https://www.sbc.gov.sa/Ar/Feedback/Pages/SBC_601.aspx. [Accessed: 17-Mar-2019].

[19] S. A. Al-sanea, M. F. Zedan, and S. N. Al-hussain, "Effect of thermal mass on performance of insulated building walls and the concept of energy savings potential," vol. 89, pp. 430–442, 2012.

[20] L. Shashua-bar, D. Pearlmutter, and E. Erell, "Landscape and Urban Planning The cooling efficiency of urban landscape strategies in a hot dry climate," vol. 92, pp. 179–186, 2009.

IOP Conf. Series: Earth and Environmental Science **329** (2019) 012008 doi:10.1088/1755-1315/329/1/012008

Analysing the benefits and challenges of retrofitting rural dwellings in Hunan, China to the Passivhaus EnerPHit standard

C Liu, H Mohammadpourkarbasi and S Sharples
School of Architecture, University of Liverpool, Liverpool L69 7ZN, UK

Abstract. Chinese rural housing largely consists of uninsulated reinforced-concrete apartment blocks with poor energy performance. These dwellings are structurally sound, costly to demolish and challenging to recycle. Retrofit is, therefore, potentially more worthwhile than new build. Currently, there is no Chinese standard for retrofitting dwellings. This research examined the viability of applying the German EnerPHit retrofit standard to Chinese rural dwellings in Hunan, southern China (hot summers and cold winters). Dwellings were evaluated in terms of building structure, materials and systems, and a common type of apartment was selected. The real-world thermal performance of the dwelling was monitored and then the dwelling was modelled using the dynamic thermal simulation software DesignBuilder and Passivhaus Planning Package (PHPP). Monitoring data were used to validate the software's predicted values. Next, energy-efficient EnerPHit retrofitting measures were incrementally applied to the dwelling model. Simulation results indicated that it was possible for the apartment to meet the EnerPHit standard if an optimised combination of thermal measures were applied. Good ventilation heat recovery was essential for winter comfort and minimum energy consumption; in summer, using adjustable shading and a high efficiency humidity recovery ventilation system was important. Appropriate natural ventilation schedules contributed in lowering cooling energy demand.

1. Introduction

The German Passivhaus standard, with its emphasis on super insulated and super airtight building envelopes, is a relatively new concept for China. The Hamburg House at the 2010 World Expo in Shanghai was the first certificated Passivhaus in China [1]. Currently, there are 21 certified Passivhaus buildings in China [2], with most of them located in the cold climate of northern China. However, the hot summer/cold winter climate zone covers 14 provinces in China, has one-third of the country's population and is the most economically and culturally developed region. The development of passive buildings in this region could make a great contribution to reducing China's building energy consumption. Furthermore, this region is not included in the centralized heating area plan and most buildings use high power electric heaters or split air conditioners for winter heating, resulting in high carbon emissions and energy costs and a difficulty in achieving comfort [3]. As the most densely populated area in China, many existing residential buildings have been built in recent decades, even in rural areas. Those buildings are largely uninsulated reinforced-concrete flats with no insulation applied. They are structurally sound, costly to demolish and not easy to recycle. Consequently, retrofit is a more reasonable solution than new builds.

This paper investigates the modelled energy performance of a rural dwelling in the hot summer/cold winter area of Hunan, China that was retrofitted to the Passivhaus EnerPHit standard. The real weather condition and indoor thermal comfort of this area were analysed by measured data, and the possible retrofit strategies were tested using the dynamic simulation software DesignBuilder and Passivhaus Planning Package (PHPP). The results suggest that the main challenges for rural Hunan dwellings to meet the EnerPHit standard are related to summer cooling demand rather than winter heating.

2. Literature review

Passivhaus buildings have been built in cold European climate since the 1990s and have been successful in significantly reducing heating energy demand whilst maintaining high levels of comfort. Recent Passivhaus research has been examining cooling energy savings in new and retrofit buildings for warm and hot climates [4][5]. A study of a Greek house, retrofitted to the Passivhaus EnerPHit standard, where the main challenge was to achieve summer comfort and low cooling demand, suggested that a 2kW mini split unit for space cooling, combined with shading and night ventilation cooling, could achieve indoor temperatures in summer that did not exceed 26°C [6]. In a Bulgarian study, a building used a 3 kWh air-to-air heat pump to supply active cooling and lower cooling demand. The careful design of shading for south windows and shutters and exterior blinds helped limit solar gain [7]. In China, the Passive House Database suggests that there are currently no retrofitted buildings that meet the EnerPHit standard and only five Passivhaus certificated new builds in the hot summer/cold winter region [2]. One of these certificated buildings, Lychee Garden, is a detached single-family house with an air tightness of 0.57 h^{-1} with good thermal protection and careful thermal bridge design. It uses an air-to-air heat pump for cooling and the monitored cooling load is $11W/m^2$. However, its architect acknowledged that the airtightness made the house very sensitive to occupancy activities like cooking, which could create overheating problems [8]. In conclusion, Passivhaus has been shown to provide comfortable indoor conditions at an extremely low heating and cooling load in different climates [9], while flexible design and solutions according to the exact cases and realities is the key to achieving passive standards

3. Methodology

The property investigated in this study was a semi-detached 4-storey building situated in Hunan, China. The building has multiple mixed usages, like most residential buildings in Chinese towns. The ground floor is for commercial use and the top storeys are individual flats; thus, only the flats were considered for EnerPHit retrofitting. Hunan has a hot summer/cold winter climate. The summer is hot and humid, with peak outside temperatures frequently reaching 38°C and 1175 cooling degree days against a baseline of 18°C [10]. On the other hand, the winter minimum outdoor temperature is around 0°C with 1621 heating degree days. Residents will commonly use individual radiators to heat a small area of a room while the remaining space remains cold. Figure 1 shows the floor plan and dimensions (in mm) of the flat used in the study, while Table 1 summarizes the properties.

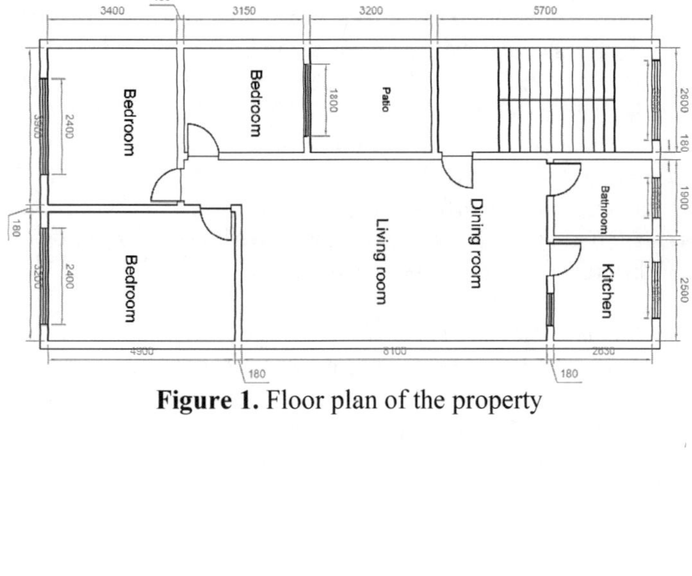

Figure 1. Floor plan of the property

Table 1. General information about the property

Location	Hunan, south China
Number of flat	3
Total treat floor area	252m^2
Floor to floor height	2.8m (1st and 2nd floors) 3.4m (3rd floor)
Total glazing area	51.8 m^2
Net volume	772.8 m^3

The property has a reinforced-concrete structure with no insulation. The building's major axis faces east-west, with the main east façade facing the main road. The three flats of this property are designed in an exact same layout, as shown in Figure 1. Each of them has a treated floor area (TFA) of 84 m^2, which does not include the patio and staircase area. Constructional and material information about the building is given in Table 2

Table 2. Properties of the building envelope

Outside wall	5mm putty paint + 10mm cement mortar + 180mm clay brick + 10mm cement mortar + 10mm outside porcelain tiles
Floors	10mm porcelain tiles + 10mm cement+ 50mm cement mortar + 100m reinforced concrete raft + 3mm putty paint
Roof	50mm cement + 100m reinforced concrete raft + 400mmair gap + 10mm wood board + 3mm putty paint
Windows	Aluminium window frame + 4mm single glass

3.1 Monitoring of the property
Two periods of indoor environmental condition measurements were made in different rooms. The indoor temperature and relative humidity in the living room and second bedroom were recorded by Rotronic TL-1D devices; a Rotronic CL11 was used to record the CO_2 levels, temperature and relative humidity in the master bedroom; finally, the outside temperature and relative humidity adjacent to the property were measured by an EasyLog EL-GFX-2. All loggers were logged at 15-minute intervals, not exposed to direct sunlight and kept away from heat sources. The two recording periods were 23rd January to 21st February 2018 (one month) and 1st July to 31st December 2018 (six months).

3.2 Software simulation
Building energy simulation of retrofitting the case study property to EnerPHit standard was done using the dynamic simulation software DesignBuilder (DB) and Passive House Planning Package (PHPP) software. The weather files used for simulation were generated by the climate database software Meteonorm. The logged data were used to calibrate the DesignBuilder base model for the different seasonal climates. Then, a series of energy efficient measures were applied on the calibrated base model to meet the standard. PHPP is an Excel based software carefully developed by the Passivhaus Institute specifically for calculating and certifying the new buildings and refurbished buildings to achieve the passive standards. It has been proven that the PHPP is an appropriate design tool in all Chinese climates [11]. Version 9 was used in this study. PHPP's interface, designPH, was used to build the model of the actual property. Then, the same energy efficient retrofit measures which were applied in DesignBuilder were simulated in PHPP, though there are some parameter settings which are different between the two software because their different calculation methods.

3.3. Proposed strategies to meet the PH standard
To retrofit the case study property to the EnerPHit standard in the model, it was important to follow the 'fabric first' approach, which prioritises heat retention and air tightness, followed by efficient heating and ventilation systems. Insulation was placed on the interior of the bricks, and this eliminated thermal bridges. Rockwool was chosen for the whole envelope because it is a common insulation material in China. Window type 'Passive 130' from Hebei Orient Sundar Co. was adopted as it is a certified passive

component produced in China. Airtightness was assumed to be 0.6 ach and a mechanical ventilation with heat recovery (MVHR) system, with a heat recovery efficiency of 85% and humidity recovery efficiency of 77%, was modelled Controlled night-time natural cooling in summer was used to avoid overheating. Finally, extensive shading to exclude solar gain was employed.

4. Results and Discussion

4.1. Temperature and relative humidity data recorded in the building

Table 3 presents an overview of the monthly mean average temperatures and relative humidities for both outdoors and indoors for the two data logging periods. The indoor temperature and humidity shown are the average value of living-room and bedroom.

Table 3. Monthly temperature T (°C) and relative humidity RH (%) values

	Outside				Inside			
Month	Ave T	Max T	Min T	Ave RH	Ave T	Max T	Min T	Ave RH
23Jan-21Feb	6.8	10.9	4	70%	8.4	9.3	7.7	71%
Jul	31.0	38.1	26.5	69%	30.7	31.6	29.6	62%
Aug	28.8	34.6	25.5	76%	30.5	31.4	30.4	66%
Sep	25.8	30.8	22.4	75%	27.4	28.3	26.6	66%
Oct	18.5	23.6	15.1	75%	20.3	21.0	19.7	62%
Nov	13.8	18.4	10.8	80%	15.3	16.3	15.0	73%
Dec	7.4	9.6	5.6	87%	9.2	10.0	8.9	78%

The measured outside values show the seasonal climatic variations between winter and summer months. The measured inside values were used to assess thermal comfort in the pre-retrofit property, and Figure 2 shows the results of the analysis based on the Passivhaus standard of comfort being in the range 20°C to 25°C. February, December, July and August had no temperature between 20°C to 25°C, and September, October and November had 14.3%, 43.2% and 5.6% of the time in the Passivhaus standard comfort temperature range.

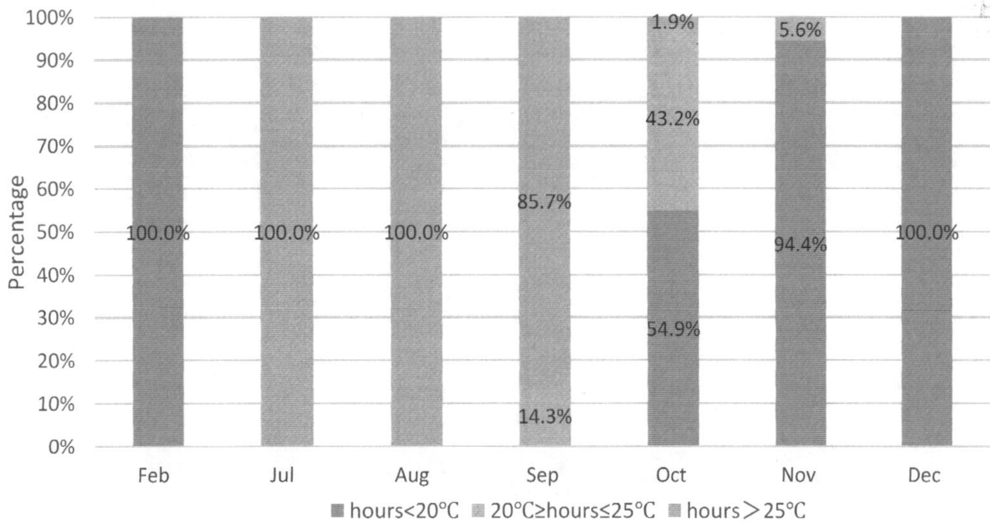

Figure 2. Percentage of comfort hours: existing property during both occupied and unoccupied hours

4.2. DesignBuilder base model calibration

The DesignBuilder base dwelling modelled the characteristics of the property, including construction materials, lights, equipment, occupancy and activity schedules. Recorded data from the two measurement periods were used to validate the DesignBuilder base model. Figure 3 displays average temperature comparisons of each hour from the recorded temperature and DesignBuilder simulated temperature in the living room. The blue dashed line shows the measured temperature data, whilst the three solid lines represent the DesignBuilder simulated results with three different envelope airtightness level: 5ach (purple); 3ach (green) and 1ach (red) respectively. Analysis of the differences between the measured and modelled data indicated a variation no greater than ±1.5°C, suggesting that the DesignBuilder model was credible.

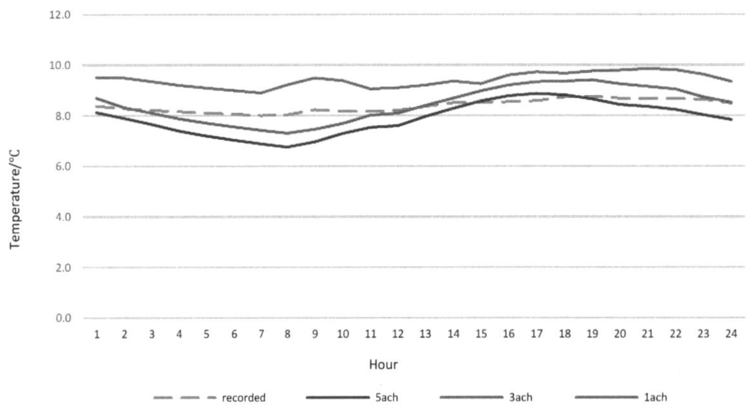

Figure 3. Hourly calibration between measured temp and DB simulated average temp in living room, based on 1, 3 and 5 ach, 23rd Jan and 21st Feb 2018

Figure 4 draws on a 6-month period (1st July – 31 Dec 2018) of measured and modelled values. The results show that a maximum temperature gap of 3°C degrees in July, which may be because the measured outside temperature was about 3°C degree higher than the DB weather file imported from Meteonorm. In the winter months, the two outside temperatures were closer, and the two inside temperature values converge. The two calibration processes were employed principally to validate the reliability of the DB base model and give confidence in the simulation of Passivhaus retrofitting.

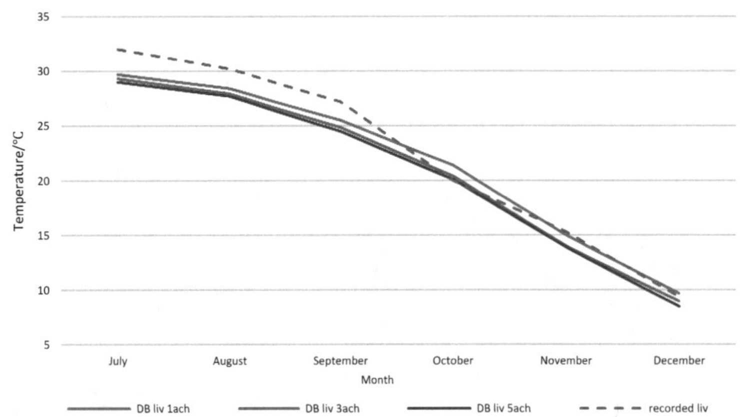

Figure 4. Monthly calibration between measured temp and DB simulated temp in living room, based on 1, 3 and 5 ach, 1st July to 31st Dec 2018

4.3. Simulation results

Table 4 shows the comparison of U-values of the dwelling's components before and after applying the Passivhaus EnerPHit standard. DesignBuilder and PHPP showed slightly different results of U-values with the exact same materials because of their different calculate methods. But both of their calculated U-values after applying 250mm rock wool were much lower than the EnerPHit standard. For the window unit, PHPP could directly use the certified 'Passive 130' window (U-value of 0.8 W/m^{2}K). A similar U-value (0.78 W/m^{2}K) was found in the DesignBuilder menu by selecting an argon-filled triple low emissivity (LoE) glazing.

The mechanical system settings of DesignBuilder and PHPP simulation are shown in Table 5. For the mechanical ventilation with heat recovery (MVHR) system, the heat recovery efficiency was set at 85% in both DesignBuilder and PHPP simulation, while the humidity recovery efficiency was 85% in DesignBuilder and 77% in PHPP. For the cooling system, the DB simulation used its default cooler with a coefficient of 2.5 controlled by room activity schedule. PHPP cooling combines recirculation cooling with additional dehumidification as the summer humidity in this area is very high, and so dehumidification is important. Both DB and PHPP have summer night-time ventilation cooling but controlled by different methods. DB controls by night-time cooling schedule and PHPP controls by temperature difference and window opening gap, so the total night ventilation value is controlled around 1.77ach. Similarly, the shading method is different between the two programmes as well - DB uses a 2 metre overhang for summer simulation and removes the overhang for winter simulation. PHPP can add a shading reduction factor only for summertime, a 10% factor is applied which could prevent 90% of solar heat from passing through the glazing.

Table 4. Comparison of pre-retrofit and retrofitted envelope U-values and the EnerPHit standard

	Existing U-value W/m^{2}K			Retrofitted U-value W/m^{2}K		EnerPHit standard
	DB	PHPP	Insulation material	DB	PHPP	W/m^{2}K
Outside walls	2.30	2.54	250mm Rockwool	0.125	0.125	0.3
Floors	2.85	3.69	250mm Rockwool	0.126	0.127	0.5
Roof	1.76	1.88	250mm Rockwool	0.123	0.125	0.3
Glazing	5.85	5.80	DB: triple glazing LoE PHPP: Orient-Passive 130	0.78	0.80	1.05

Table 5. Mechanical system inputs for DesignBuilder and PHPP

	DesignBuilder	PHPP
MVHR	Sensible heat recovery efficiency: 85% Latent heat recovery efficiency: 85% Outside air definition by zone; Outside air 1.25ac/h; Heat by MVHC	Heat recovery efficiency 85% Humidity recovery efficiency 77% Supply air: 30 m^3/ h Supply by zone; Heat by MVHC
Airtightness	0.6ach	0.6ach
Cooling	Default DB cooler with system seasonal CoP 2.5. Schedule set by room activity	Recirculation cooling with max 2.0kW cooling capacity and seasonal energy efficiency ratio 2.0. Additional dehumidification with seasonal energy efficiency ratio 3.2
Natural ventilation	Vent by zone with outside air speed 5ac/h. Night-time cooling schedule	Additional night ventilation for cooling controlled by ΔT difference. Total night ventilation value 1.77ach
Shading	Outside windows with 2m overhang in summer months; No overhang in winter months	Additional reduction factor summer shading: 10% for all outside windows

Running the PHPP software after retrofitting the Hunan dwelling to the specification shown in Tables 4 and 5 confirmed that the EnerPHit standard had been met. The EnerPHit standard criteria for Hunan's climate require a yearly energy demand for space heating of less than 20 kWh/m^2; for space cooling less than 27 kWh/m^2 and a total primary energy demand (heating, cooling, hot water and electrical appliances) of less than 120 kWh/m^2. Figure 5 shows the heating and cooling demand of the dwelling before and after the retrofit as calculated by DB and PHPP. The pre-retrofit dwelling consumed 273 kWh/m^2 of energy for heating, which is 18.9 to 21 times greater than the DB and PHPP simulated amount of the retrofitted building. Cooling energy demand pre-retrofit was 198 kWh/m^2 compared to 18.1 kWh/m^2

and 25 kWh/m^2 for the DB and PHPP retrofitted simulations respectively. Figure 6 compares the demands and compares them to the EnerPHit criteria. For the heating demand, the DB value (14.4 kWh/m^2) and PHPP (13.0 kWh/m^2) are 28% and 35% lower respectively than the criteria of 20 kWh/m^2. For the cooling demand, the DB and PHPP cooling demands are 32% and 7% lower than the criterion. For the primary energy demand, DB and PHPP both used a factor of 2.6 for electricity and factor of 1.1 for gas. The DB model just meets the primary energy criterions, whilst PHPP shows a lower demand of 108 kWh/m^2. This difference in the two calculation methods for primary energy consumption is interesting given that the same source factors were used in both models.

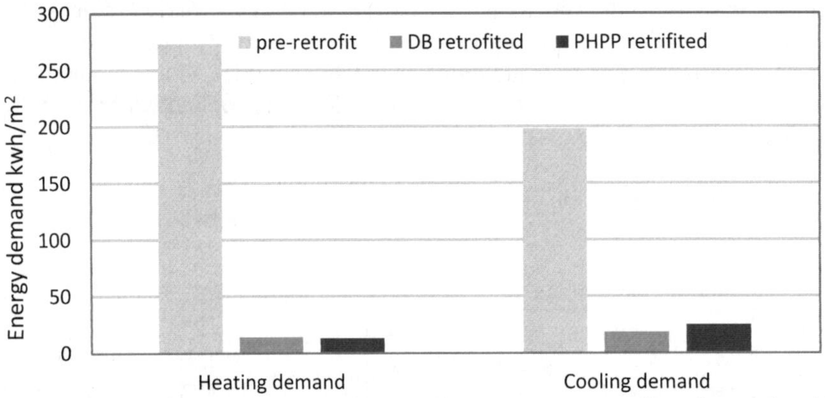

Figure 5. Comparison of modelled energy demand between pre-retrofit and retrofitted building

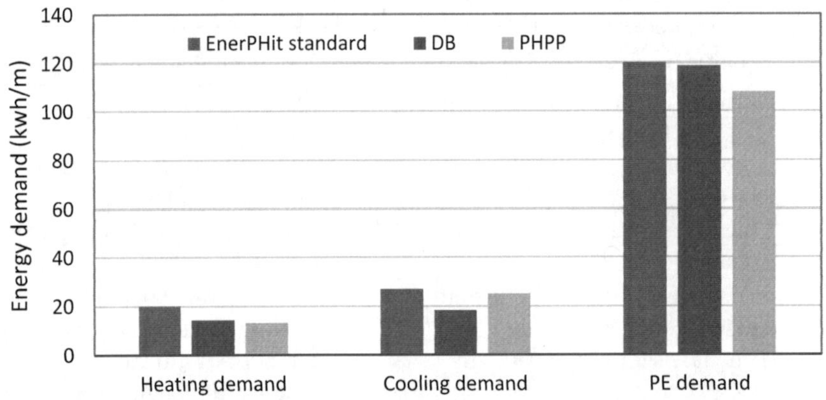

Figure 6. Comparison of predicted energy performance of the retrofitted building simulated by DB and PHPP with the EnerPHit criteria.

4.4 Main options to reduce energy demand in PHPP simulation

PHPP was used to examine the impact on energy demand of the various retrofit measures once the insulation was in place. Figure 7 shows how each measure performed. Firstly, the MVHR system, with no dehumidification, shading and summer night-time ventilation applied in the simulation, was considered. Next, summer night-time ventilation was added to the MVHR system. This produced a small drop in cooling and primary demand, but the heating demand was unchanged. Following this, dehumidification was added, which produced a considerable drop in cooling demand and primary demand. Finally, the option of shading was applied, which brought a large decrease in cooling and primary demand. From these results it appears that insulation combined with a high efficiency MVHR system are the core retrofit strategies for this property to meet the EnerPHit heating demand criterion. For the cooling demand, shading is the most efficient strategy, while dehumidification and summer night-time ventilation strategies are also important.

5. Conclusion

The results presented in this paper suggest that retrofitting an existing rural low-rise flat block in Hunan's hot summer/cold winter climate, under the constraint of the EnerPHit standard, is an achievable task. However, a series of strategies must be followed as it is necessary to deal with not only the cold winter but also the more challenging summer weather. In fact, in Hunan's climate, it is relatively easy to achieve the heating EnerPHit criterion by applying sufficient insulation materials, good windows and a high efficiency MVHS system. The simulation results from DB and PHPP both show that the heating demand is lower than the criterion after applying those measures. For summer, insulation helps to lower the cooling demand. The results show that the shading is the most efficient strategy to decrease the cooling demand. The high moisture recovery efficiency of the ventilation system is an option which will not lower the energy demand very much but will significantly increase the comfort level, as the indoor humidity is quite high. In addition, the role of summer night-time ventilation also needs attention, as it helps to reduce the cooling demand by 33% based on DB simulation.

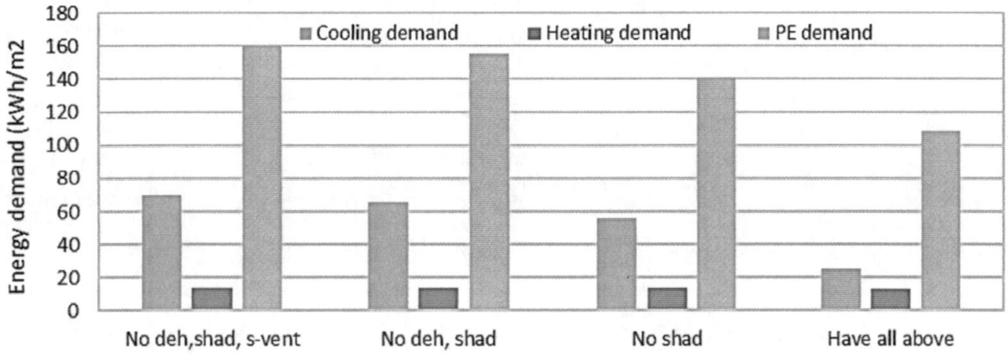

Figure 7. PHPP testing of retrofit parameters on cooling, heating and primary energy demand.

DB and PHPP simulated results for the retrofitted building for heating demand were very close, (14.4 kWh/m^2 and 13 kWh/m^2 respectively). The DB simulation showed a lower cooling demand (18.1 kWh/m^2) than PHPP (25 kWh/m^2), which may be because in the PHPP calculations there was additional dehumidification equipment. Although the retrofitted building has been investigated by two robust and extensively validated tools, DesignBuilder and PHPP, and simulations have reached the EnerPHit standard, the assessment about comfort indoor temperature and overheating risk after retrofit is inadequate. Moreover, this paper only focused on one dwelling, so the results are not sufficient to represent the feasibility of retrofitting rural housing in Hunan to EnerPHit standard. Further research, such as investigating different dwellings and post-retrofit operational performance in real-life conditions, are required to fully investigate and understand the effects of retrofitting rural housing to the EnerPHit standard on energy consumption and thermal comfort for Hunan and other hot summer/cold winter climate areas of China. Life-cycle assessment, capital expenditure and operational performance of each energy efficient retrofit measure will also be studied in the future.

6. References

[1] Tao Y 2015 Practice of demonstration and construction of passive houses in Shandong Province; Building Technology
[2] Passive House Database; https://passivehouse-database.org/
[3] Chen D 2014 Practice and thinking on the construction of "passive house" in hot summer and cold winter area; Building Technology
[4] James M and James B 2016 *Passive house in different climates: the path to net zero* (New York: Routledge)
[5] Schnieders J 2009 *Passive houses in south west Europe: a quantitative investigation* (Darmstadt: Passivhaus Institut)

[6] Pallantzas S 2018 Can a single mini-split unit heat and cool a 100m^2 residence in Athens, Greece? *22nd International Passive House Conference* Munich
[7] Dobrevski S 2018 How to get cost-effective summer comfort *22nd International Passive House Conference* Munich
[8] Miles P 2016 Divine insulation: Passivhaus architecture https://www.ft.com/content/85449ed0-f729-11e5-96db-fc683b5e52db
[9] Schnieders J, Feist W and Rongenc L 2015 Passive Houses for different climate zones *Energ Buildings* **105** 71-87
[10] Klimaat Consulting 2018 http://www.klimaat.ca/
[11] Schnieders J, Schulz T, Feist W, Sichen B K, Jiang S H, Winkel S, Buteikyte E and Sifferlen C 2016 Passive house in Chinese climates (Darmstadt: Passivhaus Institut)

IOP Conf. Series: Earth and Environmental Science **329** (2019) 012009 doi:10.1088/1755-1315/329/1/012009

Passive Strategies for Buildings in Hot and Dry Climates: Optimisation of Informal Apartment Blocks in Cairo

Gionatan Vignola[1,3], Ryan Kiracofe[2] and Udo Dietrich[1]

[1] Faculty of Resource Efficiency in Architecture and Planning, HafenCity University Hamburg, Überseeallee 16, 20457 Hamburg, Germany

[2] Faculty of School for Engineering of Matter, Transport & Energy, Arizona State University, 501 E Tyler Mall, Tempe, AZ 85287 U.S.A.

[3] Author to whom any correspondence should be addressed (gionatan.vignola@hcu-hamburg.de)

Abstract. Cairo's dry and hot climate leads to challenges related to health, thermal comfort and energy consumption that its urban dwellers encounter daily. The informal apartment blocks in which most of the Cairene households live are generally poorly insulated against external heat gains, hardly shaded and poorly ventilated. Furthermore, especially during summer, about one-fifth of households experience constant power interruptions. This research shows to what extent it is possible, by using passive strategies, to optimise the architecture of informal apartment blocks, to meet the end user's thermal comfort expectations while being energy and environmentally sound. By using Primero, an EnergyPlus based software, existing informal and traditional Cairenes buildings have been modelled, and thermal comfort and energy performance simulated and compared. Performance of informal buildings has been optimised and improved by using passive strategies. This has been possible by changing the structure of the building (adding an overhang), and by experimenting with the use of different materials for external wall constructions (all of which available in Egypt).

1. Introduction

Cairo, with its 18 million inhabitants, is one of the cities of the African continent in which the processes of urban transformation are more visible. With a population growth of about 2% per year – meaning that each year a city as big as Cardiff is adding up with its 360.000 inhabitants to Cairo - the city struggles to supply its inhabitants with adequate housing, urban infrastructure and energy without interruptions [1][2][3]. The city tries since years to distress the city centre and redirect its population towards new satellite cities (e.g. New Cairo), as well as to find solutions to deal with informality, the latter being still a significant characteristic in the pedigree of the Egyptian Capital, but still, there is much uncertainty about how effective and to what extent the situation is going to change for better.

About 65 per cent of Cairo´s urban dwellers live in informal settlements. The informal apartment blocks in which most of the Cairenes households live are generally poorly insulated against external heat gains, hardly shaded and poorly ventilated [2]. Thinking about Cairo´s climate, being hot and dry (BWh by Köppen-Geiger climate classification), it is deducible that living in reinforced-concrete skeleton buildings with external walls filled with one layer of 12 centimetres fired bricks affects the thermal comfort of its inhabitants negatively.

Most of the people that are not affected negatively by this – a tiny minority – are the ones that can afford to buy and run mechanical ventilation. As reported by the Ministry of Electricity and

Renewable Energy in 2015, the residential sector in Egypt is responsible for about 44 per cent of the share of energy consumption (64.546 GWh). This number, which is also consistently growing, is due both to the "expansion of residential compounds [...] and the widespread use of domestic appliances, especially the air conditioners in the summer" [4][5].

This paper aims to give an insight into the research done at the REAP faculty of the HafenCity University Hamburg in which contemporary and traditional (and vernacular) buildings in different climates are studied and analysed, and thanks to energy performance simulation, they are optimised to increase end-user thermal comfort while decreasing energy consumption. This is the third paper that summarises results obtained with buildings in Cairo. The first paper focuses on the indicator for passive optimisation as well as a method for optimising the performance of a building (*Passive adaptive strategies and indicators for the optimisation of comfort and energy demand in buildings in hot climates,* [6]). The second one focuses on materials and thermal mass that are adequate for Cairo´s climate (*Optimised external and internal constructions in buildings in hot and dry climates to support thermal comfort without air conditioning,* [7]).

This paper has the goal to show the thermal performance of buildings built in Cairo and the potential for optimising them by using passive strategies. After having explained in detail the methodology used, in the chapter *Simulation and Performance* 18 rooms found in five informal buildings and one traditional building that was built by Hassan Fathy following traditional strategies have been analysed, and results on thermal performance are given. In the chapter *Optimisation of Informal Buildings and Results*, four rooms have been optimised to see to what extent it is possible to increase thermal comfort while reducing energy consumption only by using passive strategies. In the last chapter *Discussion and Conclusion,* the outcomes of this research are discussed as well as the further research that should follow up to have a holistic view of the matter.

2. Methodology and Assumptions

2.1. Simulation Process and Assumptions

The purpose of this report is to examine the thermal performance of 18 rooms in buildings found in Cairo, Egypt, modelled in Primero, an EnergyPlus based software that, based on room characteristics and usage, delivers hourly information about indoor temperature, comfort and energy [8]. There are two distinct categories of rooms analysed: informal and traditional ones. The five informal buildings were taken from a series of case studies researched by the Eidgenössische Technische Hochschule Zürich [9], and the traditional one is a building designed and built by Hassan Fathy. Rooms selected for analysis focused on Cairo, Egypt; thus, the geographic and weather file selected was for the same region. From here, the floor plans were measured using computer software, and the measurements of the room and components were put into Primero so that the room could be analysed. Several assumptions were necessary to develop a working model. The followings are the constant assumptions for all the buildings:

- Rooms modelled with the usage of 'residential (bedroom) profile
- Usage is 6935 hours per year (19 hours/day)
- One or two external sun-exposed walls and no external shading from surrounding buildings
- Cross-Ventilation possible at up to 3 air changes per hour
- Single glazing window with internal roller blinds
- Shading control on blinds-closed when sun shining on façade
- Low equipment and lighting loads
- Internal constructions and obstructions modelled adiabatically
- Artificial light (on/off) according to daylight access

IOP Conf. Series: Earth and Environmental Science **329** (2019) 012009 doi:10.1088/1755-1315/329/1/012009

For all internal walls and barriers, adiabatic conditions were assumed. The short-coming in this assumption is that it expects all adjoining rooms (especially those without exterior exposure) to have the same temperature. In reality, there is a temperature difference between rooms, which could increase or decrease the overall comfort in the desired room. The process for Cross-Ventilation at various temperatures is a necessary assumption, as well as an understanding of the limitations on how Primero calculates temperature. This is done through a 1D-steady state flow, where air changes (rather than volumetric pressures) are simulated based on a typical user behaviour; below 10° C windows are closed, between 10 and 20° C windows are more and more opened, 20 till 39° C windows are fully open (maximum air change), above 39° C windows closed to avoid that external heat comes in. There were some specific assumptions necessary to model the informal rooms. Based on preliminary research, each of the informal rooms had the same building envelope composition: structural steel-reinforced concrete with fired brick infill. Since research indicates that most builds are five to eight floors, with the ground floor primarily used for commercial purposes, the rooms modelled were modelled above ground height. Additionally, the rooms modelled from buildings by Hassan Fathy – that follow traditional principles - have specific assumptions that should be outlined clearly (footnote 4). Based on literary information, it was assumed that all the constructions would have a uniform envelope composition, with adobe as a basis. The only difference from model to model was the thickness of the walls, which was derived from the architectural drawings. Although many of his buildings feature no glazing, or more traditional wooden structures, to get comparable results, the standard glazing assumption was used as the parameter in each of the constructions. When north was not provided on the drawings, it was assumed based on building orientation.

2.2. Indicator and Optimisation Process

The following sections summarise the methodology for "*Indicator Topt*" and optimisation strategy, which was previously published [6][7]. Passive buildings are optimised to reach thermal comfort using only passive systems. In a climate like Cairo´s (Hot and Dry) it is a challenge to design such buildings in which the operative indoor temperature (*Top*) is always below outdoor temperature as well as also below the comfort temperature (*Tcomf*) – in our case set at 28° C. As shown in figure 1 the indicator *Topt* counts exceeding hours in which *Top>Tcomf* only in case of insufficient passive optimisation, therefore, only if the indoor temperature (*Top*) exceeds outdoor temperature (*Te*). *Topt* is determined at all hours of use and summed up to a yearly value in Kelvin hours (Kh). By dividing *Topt* by the number of hours the room is used (in our case 6935) a yearly average in Kelvin is generated (*Toptw*). To keep comfort – as a rule of thumb – the difference between *Top* and *Te* should not exceed 1 K. The results exceeding this threshold, are considered an opportunity for optimisation.

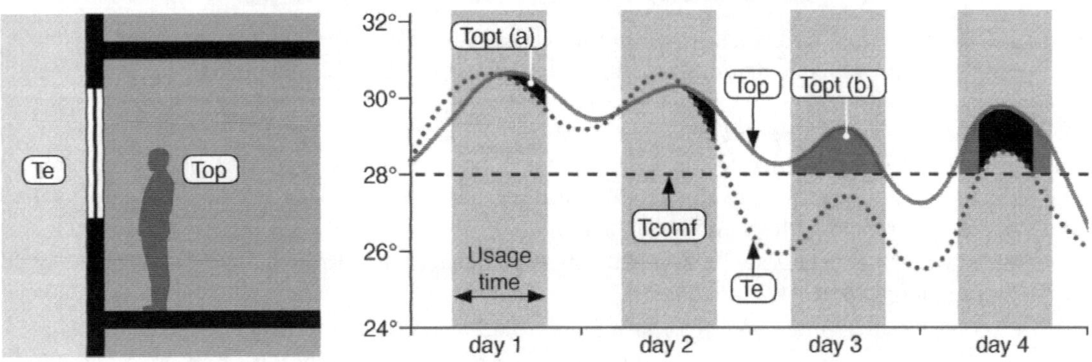

Figure 1. Visualisation of *Topt* (potential of passive optimisation). (a) indicates the situation in which *Top* and *Te* are above *Tcomf*, and *Top* is above *Te* (*Top>Tcomf*; *Te>Tcomf*; *Top>= Te*). (b) indicates when *Top* is above *Tcomf* and *Te* is below *Tcomf* (*Top>Tcomf*; *Te<Tcomf*). [6]

The optimisation strategy used follows the basic rules that are required when building in this kind of climate. First, priority is given to the reduction of solar heat gains (by shading windows and facade, and adding a window screening with the optimal grade of transmissivity), then priority is given to heat removal (different ventilation strategies as well as by optimising the building envelope). In this paper, the phases of optimisation are visualised to simplify the readability as well as to understand which phases are the most important for thermal performance.

3. Simulation and Performance

3.1. Characteristics of Buildings

Especially thinking about the buildings built from the seventies on - thanks to the Egyptian economic boom – there is a typology that covers almost 90% of all dwellings in Cairo: the multi-story apartment block. These are reinforced-concrete skeleton buildings with walls filled with one layer of 12 centimetres fired bricks (figure 2). On average, they are five to eight floors high - but can reach exceptionally more than 15 [1][2] While all floors above the ground are residential, typically on the ground floor commercial activities take place. Due to the high built density, most of the buildings have only one exposed façade. Ventilation, as well as light, for rooms situated in the centre of the building is possible through shafts – a typical element that can be found in Old Cairo or or 19th-century buildings [9]. Traditional buildings on the other way, are more difficult to be found in Cairo, and depending on the age they have, they can be traced back to Pharaonic, Islamic, Ottoman, as well as Colonial periods. Hamed Said building, built in two phases between 1942-1945 by architect Hassan Fathy, is one of the buildings that tries to show the legacy to traditional Islamic architecture, be it by choice of materials, or the design itself. The floorplan evolves around a central courtyard in which high trees find their space. Construction is massive with thick walls in adobe and small windows. Most of the ceilings are domes, and apart from corridors and a gallery that are to find around the courtyard, all rooms have at least one exposed wall [10].

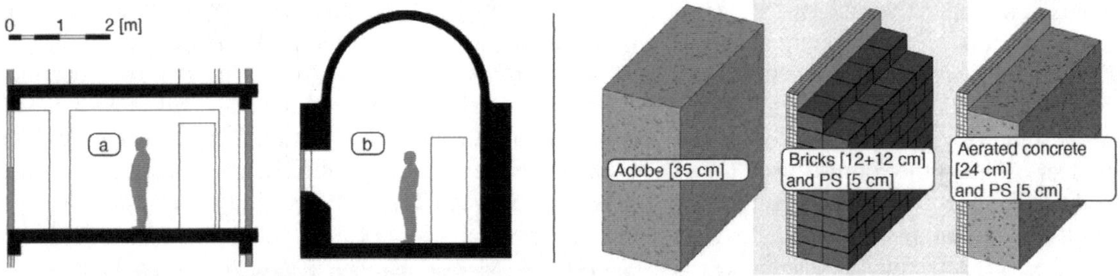

Figure 2. Left: simplified floor sections of informal (a) and traditional building (b) based on [9] and [10]. Right: external wall constructions used in the optimisation (Phase 3, Chapter 4).

3.2. Performance of Existing Buildings

Table 1 summarises the results of the simulations on the 18 simulated rooms. Across the rooms modelled, there is a wide range of values for comfort metrics. The immediate cause for the differences is not evident. Nevertheless, this is what can be observed:

- Most of the rooms have a *Toptw* below the threshold of 1 Kelvin. Only two rooms exceed the threshold of 1 K.
- Five rooms in three buildings have a *Toptw* below 0.16 Kelvin (Rooms 1, 2, 3, 4, 5).
- The best buildings are comfortable >75% and the worst buildings are comfortable < 50%.
- The only room in a traditional building has a *Toptw* below 0.5.
- When grouped by [K] 0.10 bins, the height of the room has a direct relationship with the comfort metrics. As of height increases, comfort increases.
- The larger the surface area of the floor plan, the higher the overall performance of the room.

The primary outcome of this part of the analysis shows that some of the most apparent correlations are the ones between *Toptw* and either the height or the surface area of the floor plan. We have been looking at correlations with the orientation of the room, as well as the importance of having a window on the façade or a shaft, and we realised that more research should be done on that matter – at least with the set of buildings and rooms we have at disposal. When looking at differences in the performance between a traditional and an informal design, we can observe that the room number 8 performs within the best 50% of the analysed rooms.

Table 1. Rooms Characteristics and Simulation Results (as Built).

		Characteristics					Results	
Room Nr.	**Building Name**	**Area [m²]**	**Height [m]**	**Window Area [m²] / Sun Exposed [Y/N]**	**Sun Exposed Window Area / Area of Usage [ratio]**	**Sun Exposed Walls (Nr. / Orientation)**	**Topt [Kh]**	**Toptw [K]**
1	Mahmoud's	29,0	3,3	1,25 (Y)	0,04	1 (S)	559	0,08
2	Hanaa's	28,1	3,0	1,5 (Y)	0,05	1 (S)	697	0,1
3	Mahmoud's	48,6	3,3	1,25 (Y)	0,03	1 (N)	699	0,1
4	Kadry's	16,0	2,9	1(N)	0,06	2 (E & S)	1007	0,15
5	Hanaa's	16,1	3,0	1,5 (Y)	0,09	1 (S)	1066	0,15
6	Hanaa's	20,2	3,0	1,5 (N)	0,07	1 (W)	2482	0,36
7	Hanaa's	19,4	3,0	0,0	0,0	1 (S)	2599	0,37
8	Hamed Said	11,3	3,2	1,45 (Y)	0,13	1 (S)	3188	0,46
9	Hanaa's	15,8	3,0	0,0	0,0	1 (N)	3690	0,53
10	Mahmoud's	18,9	3,3	1,25 (N)	0,07	1 (N)	3980	0,57
11	Hanaa's	12,3	3,0	0,0	0,0	1 (E)	5315	0,77
12	Abdullah's	14,4	2,8	0,0	0,0	2 (N & E)	5584	0,81
13	Hamdy's	27,3	3,0	0,0	0,0	2 (N & E)	5922	0,85
14	Abdullah's	14,4	2,8	1,33 (Y)	0,09	2 (N & W)	6603	0,95
15	Hamdy's	34,3	3,0	3,0 (Y)	0,09	2 (E & S)	6621	0,95
16	Kadry's	16,0	2,9	1,4 (N)	0,09	2 (N & E)	6897	0,99
17	Hanaa's	19,3	3,0	0,0	0,0	2 (N & E)	6953	1
18	Kadry's	11,8	2,9	1,41 (N)	0,12	1 (W)	9384	1,35

Informal Building / External walls in fired bricks (12 cm) [no background colour in table]
Traditional Building (Grey) / External walls in adobe (60 cm)
Informal Building used for Optimisation in Chapter 4 (Dark Grey) / External walls in fired bricks (12 cm)

To generalise conclusions, more buildings should be included in the set [4]. With an increased number of modelled rooms, it may become clearer what can be sources of such results. While there are definitive parameters, such as building envelope composition and external shading, that improve the room profile, it is not currently clear yet – out of simulations - if there is a noticeable difference in comfort between the informal buildings and those designed by Fathy.

4. Optimisation of Informal Buildings and Results

In this section, the four models of rooms belonging to the two buildings that consequently performed worst – all rooms are in the *Toptw* range of between 0,85 and 0,95 Kelvin – have been simulated and optimised. Starting with the status quo (Phase 01), in Phase 02 an overhang has been designed and dimensioned to keep heat gains away from façade and windows. In Phase 03, simulations were carried out by experimenting with different external wall constructions.

[4] Added to the selected traditional building, two other buildings designed by Hassan Fathy– and that follow traditional passive strategies - have been simulated, namely the Shri Zaher Building and Carr Building (two rooms each). Including Hamed Said Building, all traditional rooms performed in the top half of the overall table. They were not shown as part of the results because they are not in Egypt.

In Phase 01 the same assumptions are made as in *Chapter 2.1*. Both buildings share similar characteristics: each building has a room without window (the one oriented towards North-West), has one room with external walls facing North and East, and has one room with external walls facing North and West (with window). The main differences can be seen in the size of the rooms and windows. Hamdy's building rooms have about the double of the room surface as Abdullah´s, and the window has most of the double of the surface than Abdullah´s.

As can be seen in figure 3, for each room, three extracts of results the different phases of optimisation are shown. While in the first two phases the three results are identical (Phase 01 is the status quo with the standard construction, in Phase 02 the reduction of solar gains is obtained by

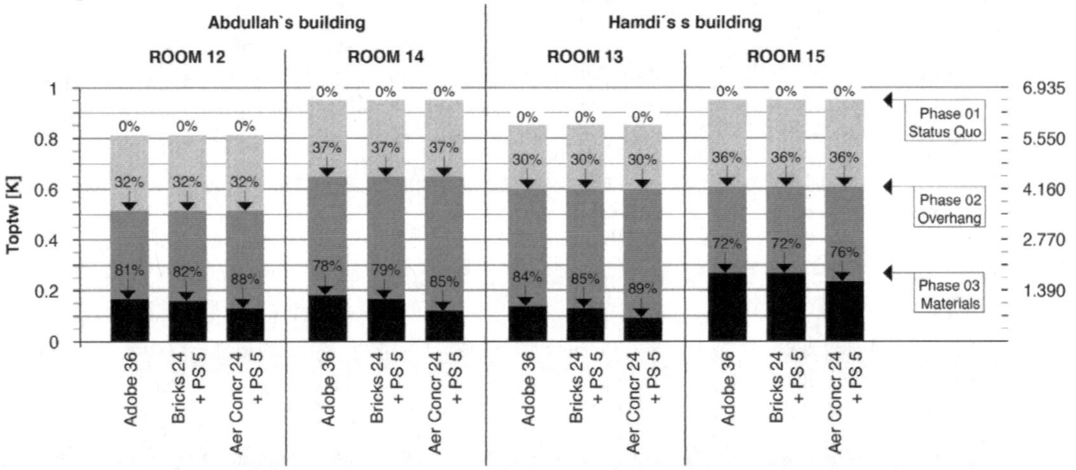

Figure 3. Simulation and visualisation of results (*Topt*) of four rooms at different stages of optimisation. The numbers in the coloured bars show the percentage of optimisation related to the status quo (Phase 01).

adding an overhang) the difference between the three results are due to the last phase of optimisation (external wall construction). Here, even though we have simulated 12 different external wall constructions, we have decided to show results of materials that are either traditional (adobe), or that can be found in Cairo (bricks, aerated concrete, polystyrene). As we have explained in Chapter 2.2, there are still two optimisation processes that could take place: in Phase 02 the optimisation of grade of transmissivity of the window, and between Phase 2 and 3 the optimisation of the ventilation strategy.

Regarding the grade of transmissivity of the window, we realised that by adding a screen in front of a window, the performance is not increasing – at least with the two analysed rooms with a window. Previous simulations show that with the room having another orientation (window towards South façade) a window screen can increase the performance ([6]). Concerning the optimisation of the ventilation strategy, since simulations done for status quo implied already cross ventilation possible whenever needed – in other words, best case possible –, no results showing other ventilation strategies are shown.

The first thing we can easily observe by looking at the status quo is that the rooms that receive fewer heat gains during the day (room 12 and 13) are the ones that in Phase 01 perform better than the other two. This is because first, they have no windows, and second, they face East – meaning that they get solar heat gains only during the morning.

For Phase 02, different overhang depths were simulated for each room. The best result in *Toptw*, have been reached by dimensioning the overhang as it follows: Room 12 with 1,75 meters, Room 14 with 2 meters, Room 13 with 1,45 meters, Room 15 with 2 meters. To note: both rooms facing West, measure 2 meters. By looking at results after having dimensioned the external shading, we see that by keeping away direct sun radiations from the façades, it is possible to increase the room performance in a range between the 30 and 37 per cent. Moreover, it is precisely by looking at the rooms that are more exposed to sun radiations (room 14 and 15) that it is observable the highest improvement with results of around 37 per cent.

After Phase 03, by changing the external wall constructions, the highest percentage of performance improvement is to be found in rooms 12 and 13 – with a range of between 51-59 per cent of improvement from the previous phase, while the rooms 14 and 15 gain between 36 - 49 per cent. Most probably, the situation is due to the windows as well as the window to wall ratio. Room 15 - with a bigger window - is the one that, even though is performing well, has got the lower increase in optimisation.

Looking at external wall construction of the four rooms a clear trend is noticeable: while adobe (36 cm depth) is consistently performing well – as well as fired bricks with polystyrene (24 + 5 cm) - a construction wall of 24 centimetres aerated concrete with the addition of 5 centimetres' polystyrene is the one that of the three performs best. This is due to the highest U-value as well as the thermal mass of the construction [7].

5. Discussion and Conclusion

Cairo's dry and hot climate leads to challenges related to health, thermal comfort and energy consumption that its urban dwellers encounter daily. This is also due to the apartment blocks in which most of the Cairenes live; they are generally hardly shaded, poorly insulated against external heat gains and poorly ventilated.

In chapter *3.2 Performance of Existing Buildings* this has been partially confirmed by the simulations that were done on 18 buildings. Two buildings have a *Toptw* exceeding 1 K (meaning that on average the operative temperature is 1 K higher than comfort temperature or outdoor temperature). Eight buildings are in the range of between 0,5 – 1 K, and the best-performing ones find themselves in a range between 0,08 and 0,5 K. Part of the analysis indicates that there are correlations between *Toptw* and either surface area of the floor plan or room height. Added to that, potentially (footnote 4), traditional buildings find themselves within the best performing one. There are definitive parameters, such as building envelope composition and external shading, that improve the room profile. Nevertheless, due to the number of other room's characteristics that play a role in thermal comfort (e.g. orientation, windows, window to wall ratio, connection to a shaft) more research and a broader dataset are needed to make a general conclusion.

In chapter *4. Optimisation of Informal Buildings and Results*, an optimisation strategy that focuses first to reduce heat gains and then to remove the heat has been used. By following this strategy, it has been possible to increase thermal comfort by an average of 80%. In the best case by reducing *Toptw* by 0,83 K (Room 14). In the worst case (Room 15 – facing West, with an exposed window area of three meters) by reducing *Toptw* by 0,73 K. This has been possible especially by changing the structure of the building – adding an overhang, and by experimenting with the use of different materials for external wall constructions (adobe, bricks, aerated concrete, polystyrene - all of which available in Egypt).

With this research, it was possible to demonstrate with digital model simulations to what extent – and in which range - it would be possible to optimise the thermal performance of informal buildings by using passive strategies only. The methods used for doing this has proven again to be effective, and, when thinking about the time needed to get results, it has also proven to be efficient [6] [7].

Nevertheless, questions regarding correlations and comparison of different rooms remain partially unanswered. We have already some hints about windows size, orientation, size of the room and windows to wall ratios. However, probably, a bigger and diverse dataset together with simulations done on a building scale (building simulation), could help to find out how the geometry of rooms and buildings can be improved, as well as to find out what are the "no go" in this climatic condition. Such work could also help to understand differences between traditional and contemporary buildings, and it would be helpful especially to learn how the traditional passive strategies could be applied in contemporary and informal architecture while meeting the needs of end-users that generate this kind of architecture.

While thinking further about the improvements in the performance of buildings, this research also raises new questions that need an answer. *How could be such improvements integrated into the Cairenes building culture?* would be an important one. However, inevitably, before answering questions that look at the implementation of strategies only, questions that look at a sustainable future should be addressed. Assuming that all means have been found to improve the performance of buildings, what would be the consequences of such interventions when we think about social acceptance and economic power of informal settlers, for example? We know how important are social structures, as well as that most of the people living in informality are also weaker in terms of economic power. So, how could we integrate this new knowledge to be accepted and implemented with the economic means at disposal without becoming a burden? Also, what would be the consequences by improving performance in buildings related to the ecological dimension? If we are going, for example, to use materials such as polystyrene in informal buildings – for the one hand side we would increase the thermal performance of buildings, but on the other side, we would generate a problem related to disposal and recycling that right now is not there.

We remain sure, that the most innovative solutions that can be found in the field of building performance and building efficiency, if thought and implemented using sustainable principles, can lead to an improvement to health conditions, as well as responsible use of resources and energy.

Acknowledgements: The research was partially made possible thanks to funding through the DAAD RISE Undergraduate Fellowship, which provided student resources towards the preliminary findings presented in the paper.

References

[1] Horwood C 2011 *Cairo: A City in Transition. Cities & Citizens Series: Study 2* (Nairobi Kenya: United Nations Human Settlement Programme)

[2] Kraas F, Leggewie C, Lemke P, Matthies E, Messner D, Nakicenovic, N, ... Schneidewind, U 2016 *Humanity on the Move: Unlocking the Transformative Power of Cities: Flagship Report.* (Berlin: Wissenschaftlicher Beirat d. Bundesregierung Globale Umweltveränderungen)

[3] CLA - Cardiff Local Authority 2018 *2017 Mid Year Estimate* retrieved from https://www.cardiff.gov.uk (01.03.2019)

[4] ARE - Arab Republic of Egypt, Ministry of Electricity and Energy 2011 *Annual Report 2010 / 2011*

[5] ARE - Arab Republic of Egypt, Ministry of Electricity and Renewable Energy 2015 *Annual Report 2014 / 2015*

[6] Dietrich U and Vignola G 2019 Passive adaptive strategies and indicators for the optimisation of comfort and energy demand in buildings in hot climates. In *The Cities and Climate Change Complex. Proc. of Cities and Climate Conference 2017* Potsdam Germany (Heidelberg: Springer, to be published in 2019)

[7] Dietrich U and Vignola G 2018 Optimised external and internal constructions in buildings in hot and dry climates to support thermal comfort without air conditioning. *Conf. Proc. of the International Conference for Sustainable Design of the Built Environment 2018 London UK*

[8] Primero Comfort 2018 *What is primero* retrieved from http://www.primerosoftware.de/what-primero-english/ (01.03.2019)

[9] Angélil M and Malterre-Barthes C 2016 *Housing Cairo - the [informal] response* (Berlin: Ruby Press)

[10] Steele J 1989 *The Hassan Fathy Collection. A Catalogue of Visual Documents at the Aga Khan Award for Architecture* (Bern, Switzerland: The Aga Khan Trust for Culture)

IOP Conf. Series: Earth and Environmental Science **329** (2019) 012010 doi:10.1088/1755-1315/329/1/012010

Modelling the impact of ground temperature and ground insulation on cooling energy use in a tropical house constructed to the Passivhaus Standard

R C Sigalingging[1], D Chow[1] and S Sharples[1]

[1]University of Liverpool, Liverpool L69 3BX, United Kingdom

Corresponding Author, R.C.Sigalingging@liverpool.ac.uk

Abstract. In tropical climates ventilation introduces high levels of moisture into dwellings, leading to raised values of indoor relative humidity and air temperatures, creating occupant thermal discomfort. The Passivhaus standard, by creating a super insulated and air-tight envelope, reduces heat losses and provides low energy comfort. This approach in tropical housing might be effective but could potentially increase mechanical cooling demand. This research investigated the effect of ground temperatures on thermal comfort and energy-saving in a tropical dwelling. Terraced houses in Jakarta, Indonesia were chosen as this a popular house type. IES VE software was used to study building energy use, and the Passivhaus software PHPP examined the application of Passivhaus criteria. Field measurements of air temperature and relative humidity were used to validate the software model. Analysis revealed that the building's predicted air temperatures were affected by the assumed ground temperature models in IES. Good agreement between the measured and modelled values was achieved only when a particular ground model was chosen. Having validated the IES model, dwelling insulation and air-tightness levels were incrementally changed until they met the Passivhaus standard. Finally, the feasibility of meeting the Passivhaus energy standard for cooling in the modified tropical house was tested.

1. Introduction

Low-density housing demand in Indonesia has contributed significantly to uncontrolled urban sprawl around the edges of Indonesia's major cities [1]. This unplanned development is characteristic of contemporary Asian Urbanization [2]. Many of these dwellings are badly constructed, being poorly insulated and not well-sealed. The air conditioning (AC) that is employed to achieve thermal comfort for residents is used inefficiently and expensively. Indonesia's hot and humid climate, coupled with rapid urbanisation and economic growth, means the demand for cooling energy in residential buildings will increase sharply in the coming decades [3]. In 2017, the three largest energy user sectors in Indonesia were transport (47%), industry (30%) and housing (16%) [4]. It is forecast that, by 2020, building energy consumption in South East Asia countries will exceed that of developed countries, with 56% of total building energy demand being used for air-conditioning [5]. Therefore, it is desirable to investigate low energy cooling systems that could help reduce the need for electric air conditioning.

Radiant cooling is one such strategy to cool buildings. Concrete Core Temperature Control (CCTC), for example, pumps cool water around pipes cast inside the floor [6]. In tropical climates, ground soil temperatures range from around 15°C to 25°C, and so can be much cooler than the ambient air temperature [7]. Floor slabs in contact with the ground can, therefore, be effective in cooling a building.

The use of ventilation to try and cool buildings can transfer high amounts of moisture in to rooms, resulting in poor internal thermal and health conditions in which the optimum humidity levels of 40% to 60% [5] are exceeded. It is, therefore, necessary to create design approaches to building design that can keep indoor air humidity low while still reducing cooling energy consumption. One radical alternative solution is to apply the German Passivhaus standard [8], with its very high levels of insulation and air tightness, to Indonesian dwellings.

This paper presents the validation and testing of a house model developed using the commercial dynamic thermal simulation software IES. Monitored air temperature and relative humidity data from a real dwelling were used to validate the IES software model before Passivhaus concepts were applied to the house model. This paper describes the development of a Passivhaus standard for housing built in Indonesia that works for both thermal comfort and energy efficiency. The energy advantages from cool ground temperatures and no floor insulation were observed in the development of the house model.

1.1. The Passivhaus standard

The requirements for the Passivhaus standard, which originated in Germany, are shown in Table 1 [8]. These requirements are principally performance-based, using passive measures, minimal thermal bridging and a whole house mechanical ventilation system with highly efficient heat recovery [9]. Even though the initial developments were made within the relatively mild climate of Central and Northern Europe, studies have suggested that Passivhaus could be a feasible option in other climate types [10]. The Passive-On study forecast a number of issues related to Passivhaus criteria for warmer climate, including the introduction of a limit for summer cooling energy demand, a higher infiltration rate and an indoor comfort temperature that coincided with adaptive thermal comfort standards [11]. The application of the Passivhaus standard must also properly consider moisture balances and latent loads for buildings in a hot and humid climate.

Table 1. Passive House Criteria.

			Criteria[1]			Alternative Criteria[2]
Heating						
Heating demand	[kWh/(m²a)]	≤	15			-
Heating load[3]	[W/m²]	≤	-			10
Cooling						
Cooling + dehumidification demand	[kWh/(m²a)]	≤	15 + dehumidification contribution[4]			variable limit value[5]
Cooling load[6]	[W/m²]	≤	-			10
Airtightness						
Pressurization test result n_{50}	[1/h]	≤	0.6			
Renewable Primary Energy (PER)[7]			Classic	Plus	Premium	
PER demand[8]	[kWh/(m²a)]	≤	60	45	30	±15 kWh/(m²a) deviation from criteria...
Renewable energy generation[9] (with reference to projected building footprint)	[kWh/(m²a)]	≥	-	60	120	...with compensation of the above deviation by different amount of generation

1.2. Thermal Comfort

Thermal comfort assessment is an important tool to assess energy usage and measurement thermal satisfaction achievement. The term 'thermal envelope' used here refers to the shell of the dwelling as a barrier to unwanted heat or mass transfer between the interior of the building and the outside conditions [12]. In hot climates, the effect of air movement and humidity are particularly important because heat lost by evaporation dominates [13]. The impact of the external temperatures on perceived comfort has to some extent also been incorporated in to design standards such as ASHRAE Standard 55. The finding on the ASHRAE adaptive comfort standard study was the difficulty of creating generalization for areas that have mean outdoor temperatures above 23°C. The research indicated that buildings in this zone were unable to maintain thermal comfort for many hours of the day [14]. This is consistent with other research that found some climatic differences between cities in the lowland and highland in Indonesia, which could lead difference in people's comfort temperature due to physical adaptation [15]. However, the Indonesia National Standardization Agency (SNI) still issues standards whereby the comfort temperature is set at 25°C ± 1 °C and comfort relative humidity at 60% ± 10% [16]. The result from a study of comfort temperatures for people in the Depok area (a Jakarta satellite town) indicated that the comfort temperature was higher than this national standard, at 27.6°C [17].

2. Research methodology

This paper investigated the impact of ground temperature model on a simulated Passivhaus building. A row (terraced) house of the type that is commonly found in the Jakarta Metropolitan Region was chosen for the study. IES VE 2018 software was used to predict the dwelling's indoor temperatures, relative humidity and energy use. The building model in IES VE was built by inserting building information such as building materials, cooling systems, lights and appliances, and a presupposed occupancy schedule. The selected monitored house had a floor area between 50m^2 to 69m^2 floor area in which form most of the existing urban housing stock for housing in the Jakarta Metropolitan Region [18]. Empirical validation compared the computer simulation results with monitored data from the house. Analysis was done by investigating the typical housing characteristic, building performance and energy consumption.

3. Case study monitoring

The dwelling was in the major housing development area of Bogor, Jakarta Metropolitan Region. The dwelling measured 6m x 10m in plan, with a total floor area of 55m^2, and a floor-to-ceiling height of 2.85m. The building's main axis orientation was northerly; it was not insulated, being constructed from a single layer of brick, and with single glazing windows (Figure 1 and Figure 2).

Figure 1. Exterior of the row house.

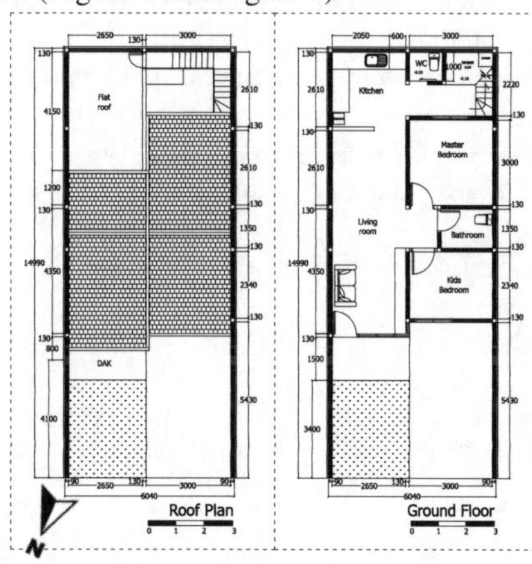

Figure 2. Floor plans of the row house.

The monitoring of the case study building was undertaken during a period of consistent occupation over two selected periods, one in January – February for the rainy season, and the other in September – November for the hot season. Air temperature and relative humidity values were collected at 30 minute intervals using Tinytag and Rotronic data loggers placed at a height of 1.5m in the living room, master bedroom and a sheltered outdoor area. Because of space limitations, only the living room results are presented in this paper.

4. Validation of the IES VE 2018 software

A three-dimensional dwelling model in IES VE software was built using the data obtained from field measurements. The IES VE simulation result was compared with measured period data, in this paper in mid-September, which has comparatively high weekly air temperature and a larger gap of temperature differences between maximum and minimum temperature compared to all the measured data. The

IOP Conf. Series: Earth and Environmental Science **329** (2019) 012010 doi:10.1088/1755-1315/329/1/012010

dwelling model in this software was using plan provided by the home owner as a guidance, building materials information was based on the contractor specification, and the occupant activity schedule was gained from field observations. The building elements used to build the base model in IES VE 2018 can be seen in Table 2.

Table 2. Building elements.

Building Element	Constructional layers
External and internal walls	25 mm thick cement plaster + 100 mm thick clay brick + 25 mm thick cement plaster
Party wall	25 mm thick cement plaster + 100 mm thick clay brick + 25 mm thick cement plaster
Floor	8 mm thick ceramic tile + 22 mm thick cement screed + 100 mm thick concrete slab + soil layer
Window	6 mm thick single layer glass
Ceiling	6 mm thick gypsum board
Pitched roof	20 mm thick roof tile + 25 mm thick timber batten
Ceiling slab	22 mm thick cement screed + 100 mm thick concrete slab + 20 mm thick cement plaster

4.1. The ground temperature model used in the simulation

Figure 3 shows the detailed analysis of the air temperature and relative humidity measured data compared with simulation results for the living room. A large divergence between measured and modelled values was evident for daytime results, with simulation data showing internal air temperatures similar to the external temperatures. Undertaking a series of parametric studies of the IES model indicated that the assumed ground temperatures were creating the divergence. The default setting in IES VE sets the ground temperature equal to the external air temperature, but this may not be appropriate for tropical weather and soil conditions. To find out the setting for the ground temperature, the research continued by looking at ground temperatures in tropical countries, particularly Jakarta, Indonesia.

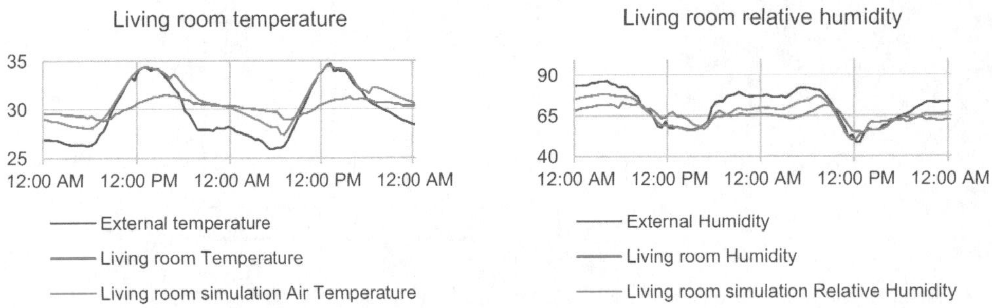

Figure 3. Comparison of measured and modelled air temperatures and relative humidities.

Soil properties are difficult to generalize because of local heterogeneity and the lack of broad based data; also, soil condition is not a standard variable collected at most weather stations [20]. Available literature indicates that underground soil temperatures can vary from 8°C to 27°C in some parts of the world, especially in cold-dominated areas (North America), and between 15°C and 25°C in tropical climates [7]. The average monthly ground temperature in Jakarta, based on the weather analysis software Climate Consultant, is 26°C [20]. Because soil temperature responds to the net effect of the daily surface energy balance, it can be estimated by computing a running average of air temperature, with progressively longer integration times as soil depth increases [21].

By using the literature information about ground temperature above, the setting on the IES VE simulation was set using outside air with an offset temperature profile. The analysis method for ground construction setting in IES VE software is based on EN-ISO 13370 [22]. The software analysis indicated that an offset temperature profile set at - 5°C from the external temperature was appropriate for this simulation. The graph in Figure 4 is ground temperature that was estimated by offsetting measured external temperature by -5°C following analysis above. The graph indicated that there was more than 75% of the time when the ground temperatures were below the adaptive comfort level.

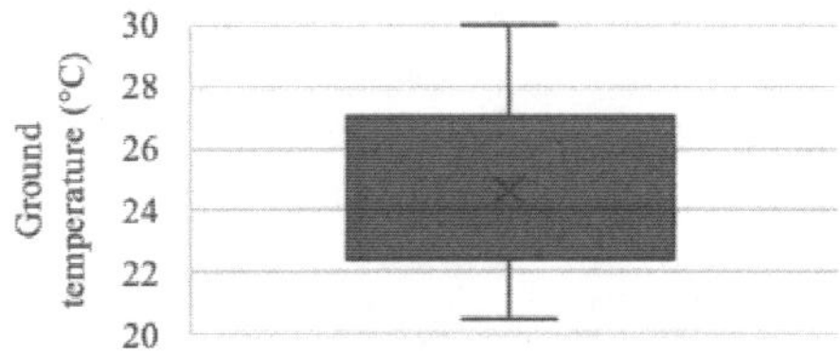

Figure 4. Box plot of ground temperatures during a one-week period.

4.2. The validation results

Figure 5 shows the air temperature and relative humidity for the living room after applying the adjusted ground temperature model. The results indicate that the simulations were satisfactory in displaying the same trends as the measured data. Living room results in Figure 5 suggest that the differences between the measured and modelled air temperature values were around 1 to 2°C, and the differences for relative humidity were around 10%.

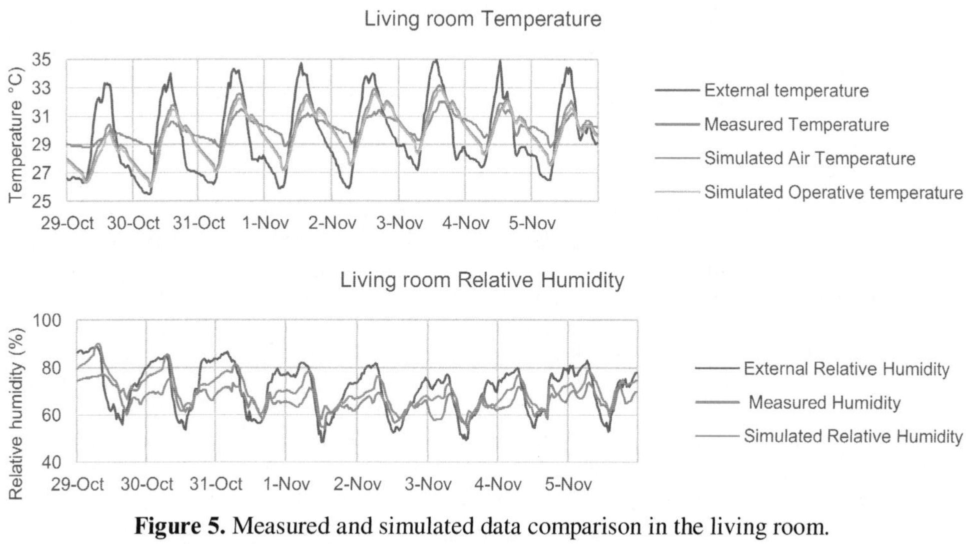

Figure 5. Measured and simulated data comparison in the living room.

To ensure that the IES dwelling model was statistically valid, an analysis was conducted in accordance with ASHRAE Standard 14-2002 [23]. As an indication of the mean ratio of relative error between two values (measured and modelled), the mean bias error MBE and Coefficient of Variation of the Root Mean Square Error CVRMSE were calculated. ASHRAE Standard 14 defines the acceptable limits for calibration to hourly data as within ±10% MBE (hourly) and ≤30% CVRMSE (hourly) measured at a utilities level [23]. The evaluation results for this study showed relative humidity and temperature MBE values for the living room of -5.9% and +0.3% respectively and CVRMSE values of +7.5% and +3.1% respectively. Therefore, it can be concluded that the IES simulation can model satisfactorily the studied house in the tropical climate and produce relatively similar results compared with the field measurement data.

5. The impact of ground temperature impact on a Passivhaus building in a tropical climate

To study the effect of ground temperature on a tropical Passivhaus building, the validated IES building model had Passivhaus elements applied to it. Table 3 indicates the materials that were used in the building model to follow the Passivhaus concept. This performance of the Passivhaus model will be compared with a Passivhaus model that had the floor insulation removed to let the ground floor act as a thermal sink and potentially provide radiant cooling. The house with its original building elements was also included in the comparison to look at the performance of the building with the Passivhaus standard to study the energy saving to reach thermal comfort. All three scenarios used the same HVAC system (AC and dehumidifier) that was located in the living room, master bedroom, and children's bedroom. For the simulation, the AC temperature set point in IES was 26°C, and relative humidity controller was set at 60%.

Table 3. Passivhaus building elements.

Building Element	Constructional layers
External and internal walls	25 mm thick cement plaster + 100 mm thick clay brick + 100 mm XPS Extruded Polystyrene + 25 mm thick cement plaster
Party wall	25 mm thick cement plaster + 100 mm thick clay brick + 100 mm XPS Extruded Polystyrene + 25 mm thick cement plaster
Floor	8 mm thick ceramic tile + 22 mm thick cement screed + 100 mm thick concrete slab + Urea Formaldehyde Foam + soil layer
Window	6 mm thick double layer glass
Ceiling	6 mm thick gypsum board
Pitched roof	20 mm thick roof tile + 25 mm thick timber batten + 100 mm MW Glass Wool (rolls)
Ceiling slab	22 mm thick cement screed + 100 mm thick concrete slab + 20 mm thick cement plaster

The loggers used in this study recorded air temperature. Operative temperature is often considered a better indicator of thermal comfort as it combines both air and mean radiant temperatures in a space but is not easy to measure in real building situations. The IES software was run to test for any significant differences between predicted air and operative temperatures in the house. However, as shown in Figure 5, there were only minor difference between simulated air temperatures with simulated operative temperature and so it was feasible to use air temperature for the validation and the comfort analysis.

6. Result and discussion

With the application of the Passivhaus standard to the IES dwelling model, simulation results for the living room showed that the Passivhaus building model provided stable air temperatures (Figure 6), although the air temperatures were mostly above 27.6°C (the adaptive comfort level). For the original building elements scenario, the air temperature was still fluctuating and following the external air temperature, with a few hours being above 27.6°C. Meanwhile, the Passivhaus application without floor insulation always maintained room air temperatures below 27.6°C. Relative humidity for the original house scenario tracked outside relative humidity and, for most of the time, was above 60%. On the other hand, the Passivhaus model and Passivhaus without floor insulation indicated low relative humidities, with relative humidity staying below 60% by using the AC + Dehumidifier system. With the same building HVAC system and occupancy schedule, the IES simulation predicted an annual cooling energy use of 11.41 MWh for the original layout, 10.89 MWh for the house with Passivhaus application, and 8.61 MWh for the Passivhaus without floor insulation (Figure 7).

From this study, the application of Passivhaus into a tropical dwelling provided stable interior air temperatures and relative humidities. The building airtightness reduced cooling energy slightly, and by removing floor insulation significant energy savings were made. The floor can be seen to act as a radiant cooling source for the dwelling built in tropical climate since the ground temperature that below room adaptive comfort level.

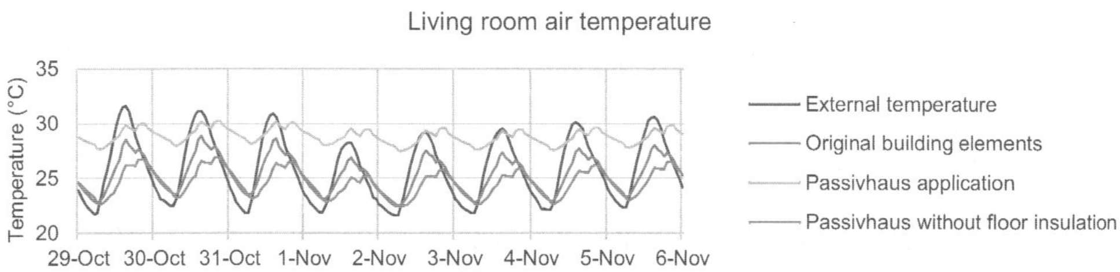

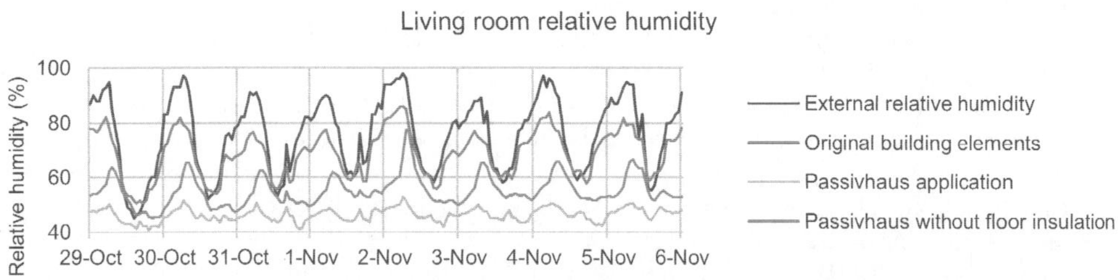

Figure 6. Hourly air temperatures comparison in living room for Passivhaus approaches and original building elements.

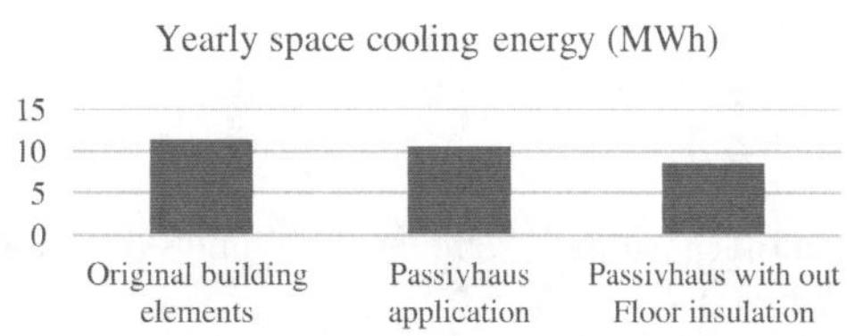

Figure 7. Yearly space cooling energy use for the original building elements, the Passivhaus and the Passivhaus without floor insulation.

7. Conclusions and discussion

This paper explored the effect of ground temperature on the performance of a Passivhaus-enhanced dwelling in a tropical climate. The original dwelling was monitored to gain air temperature, relative humidity and activity schedule data that were used to create a validated IES building model. The validation process identified ground temperature as an important variable in the model. The impact of coupling the building to the ground without using insulation was also examined and found to reduce cooling energy demand whilst maintaining good comfort levels. The application of the Passivhaus concept in tropical climates still needs further analysis. Besides the additional construction cost for the building, the skill of construction workers needs to be improved, especially residential construction workers, because the use of insulation, double glazing, and constructing air-tight building is not common practice in Indonesia.

References

[1] Rahadi RA, Wiryono SK, Koesrindartoto DP, et al. Factors influencing the price of housing in Indonesia. *Int J Hous Mark Anal* 2015; 8: 169–188.

[2] Mookherjee D, Hoerauf E. Is It Sprawling Yet? A Density-Based Exploration of Sprawl in the Urban Agglomeration Region Around the Mega City of Delhi. Springer, Tokyo, pp. 49–60.

[3] Levine M, Ürge-Vorsatz D, Blok K, et al. Residential and commercial buildings. *Clim Chang 2007 Mitig* 2007; 387–446.
[4] Adi AC, Ajiwihanto NE, Lasnawatin F, et al. *Handbook of Energy and Economic Statistics of Indonesia*. Jakarta: Ministry of Energy and Mineral Resources, https://www.esdm.go.id/assets/media/content/content-handbook-of-energy-and-economic-statistics-of-indonesia.pdf (2018, accessed 9 March 2019).
[5] Katili AR, Boukhanouf R, Wilson R. Space Cooling in Buildings in Hot and Humid Climates—A Review of the Effect of Humidity on the Applicability of Existing Cooling Techniques. *Proccedings 14th Int Conf Sustain Energy Technol* 2015; 25–27.
[6] Oettl F. *Austrian Embassy Jakarta*, http://www.ecreee.org/event/regional-workshop-and-training-energy-efficiency-buildings-praia-cabo-verde (2014, accessed 23 March 2019).
[7] Alam MR, Zain MFM, Kaish ABMA, et al. Underground soil and thermal conductivity materials based heat reduction for energy-efficient building in tropical environment. *Indoor Built Environ* 2015; 24: 185–200.
[8] Passive House Institute. *Criteria for the Passive House, EnerPHit and PHI Low Energy Building Standard*. Darmstadt, Germany, www.passivehouse.com (2016, accessed 11 June 2019).
[9] Passivhaus Trust. Passivhaus : Basic Principles, http://www.passivhaus.org.uk/page.jsp?id=17 (2017, accessed 6 December 2018).
[10] Schnieders J, Hermelink A. CEPHEUS results: measurements and occupants' satisfaction provide evidence for Passive Houses being an option for sustainable building. *Energy Policy* 2006; 34: 151–171.
[11] eERG. The Passive-On Project. *Passive-On*, http://www.eerg.it/passive-on.org/en/index.php (2018, accessed 14 November 2018).
[12] Nicol JF, Humphreys MA. Adaptive thermal comfort and sustainable thermal standards for buildings. *Energy Build* 2002; 34: 563–572.
[13] Nicol F. Adaptive thermal comfort standards in the hot–humid tropics. *Energy Build* 2004; 36: 628–637.
[14] de Dear RJ, Brager GS. Thermal comfort in naturally ventilated buildings: revisions to ASHRAE Standard 55. *Energy Build* 2002; 34: 549–561.
[15] Karyono TH. Thermal Comfort in Indonesia. In: *Sustainable Houses and Living in the Hot-Humid Climates of Asia*. Singapore: Springer Singapore, pp. 115–121.
[16] SNI. *Konservasi energi sistem tata udara pada bangunan gedung*, https://anzdoc.com/konservasi-energi-sistem-tata-udara-pada-bangunan-gedung.html (2000).
[17] Santy, Matsumoto H, Susanti L. Development of a Passive House Standard for Tropical Climates (Indonesia) - The Initial Stage. *CLIMA 2016 - Proc 12th REHVA World Congr*, http://vbn.aau.dk/files/233778874/paper_753.pdf (2016).
[18] Badan Pusat Statistic. 2010 Population Census - Household by Region and Floor Area of Dwelling Unit - Indonesia, http://sp2010.bps.go.id/index.php/site/tabel?tid=333&wid=0 (2018, accessed 1 October 2018).
[19] Waring RH, Running SW. Spatial Scaling Methods for Landscape and Regional Ecosystem Analysis. *For Ecosyst* 2008; 225–V.
[20] Milne M. Climate Consultant. *University of California*, http://www.energy-design-tools.aud.ucla.edu/climate-consultant/ (2019, accessed 17 March 2019).
[21] Hu QI, Feng S. A Daily Soil Temperature Dataset and Soil Temperature Climatology of the Contiguous United States. *J Appl Meteorol* 2003; 42: 1139.
[22] IES-VE. Ground Contact Constructions, https://www.iesve.com/support/faq/pdf/groundcontactconstructions_faq307.pdf (accessed 1 February 2019).
[23] ASHRAE. Ashrae Guide and Data Book(Fundamentals). *Ashrae*.

Problems and barriers affecting sustainable commuting: Case study of people's daily commute to Kasetsart University, Bangkok, Thailand

P Wethyavivorn and N Sukwattanakorn
Department of Civil Engineering, Kasetsart University, 50 Ngamwongwan Road, Bangkok 10900

Abstract. Bangkok is one of the cities with the worst road traffic congestion in the world. The city needs to shift toward a more sustainable commuting mode as soon as practicable. This study aims to explore modes of transportation that students and staff of Kasetsart University, Bangkok, use to commute to the university. Data was collected from 315 students and staff through online questionnaires. Interviews with selected respondents were also conducted to gather more insights relating to the perceived barriers to sustainable commuting. Observation of surrounding infrastructure was also conducted. The daily commuting were divided into 4 categories: pure active commuting including walking and biking (19.05%), pure active commuting and public transportation (18.41%), combination modes including private motorized vehicles (25.40%), and the least sustainable mode, pure private motorized vehicles (37.14%). The study further investigated the perceived barriers to 5 modes of sustainable transportation available to most respondents, namely, walking, biking, city bus, public van, and rapid transit. The most significant barriers to pure active commuting was the hot weather and the lack of walking and biking infrastructure. The major barrier to using city buses was the waiting time, together with the longer travel time especially during rush hours. For public vans, the barrier was found to be the safety issue due to careless drivers as well as the van condition. Most of the respondents would like to use the rapid transit, however, they thought the cost was a little high for daily commute.

1. Introduction

It is widely recognized that sustainable transportation plays important role in reducing energy consumption and greenhouse gas emission, especially in big cities [1, 2]. Sustainable transportation can be achieved through good active commuting infrastructure together with attractive public transportation system. The role of active commuting in increasing physical activity of both children and adult have been emphasized in many countries since it lowers the risk of chronic diseases [3]. The role of public transport is extended beyond reducing energy to creating economic equity and establishing good public value [4].

Bangkok is the capital city of Thailand, with population of 9.4 million according to the last census held in 2015 by the National Statistical Office of Thailand. In 2018, the number of registered vehicles in Bangkok was 9,778,661, which was one fourth of total vehicles registered in Thailand. Around 37% of the vehicles in Bangkok are motorbikes. Approximately 65 % of the city's residents use private vehicles to get to work and return home, while only 35 % use rapid transit, city bus, taxis and other modes of public transport [5]. The city has been consistently ranked as one with the worst traffic in the world for many years. Moreover, Thailand has the second highest road traffic fatality rate in the world with an annual estimate of 66 deaths every day and more than two third of those resulting from motorbikes [6].

IOP Conf. Series: Earth and Environmental Science **329** (2019) 012011 doi:10.1088/1755-1315/329/1/012011

University students have been reported in many studies as the group that use the most sustainable commuting [7, 8, 9]. The university has an important role in promoting, reinforcing, as well as reshaping sustainable development behaviour and awareness, especially in sustainable transportation. The university education helps craft the right attitude and behaviour necessary to develop positive changes in the society as they grow [7, 10]. It is evident that Bangkok has a problem in developing sustainable transportation. It is necessary to understand how people commute daily and what the barriers to sustainable commuting are. This study aims to investigate the issues by using Kasetsart University as a case study. The university is situated in the northern part of Bangkok where access to public transport is between less than 12 minutes to half an hour. The area has been classed as high density urban area as shown in Figure 1.

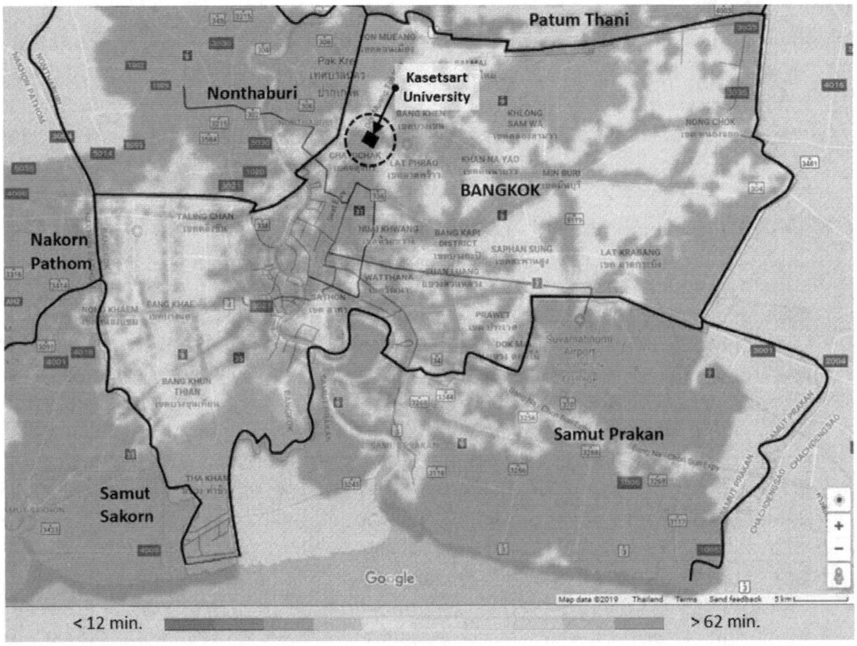

Figure 1. Access to road public transportation of Bangkok and vicinity area adapted from Guntamuanglee thesis at Chulalongkorn University [11].

2. Factors affecting mode of transportation

Factors influencing travel behaviour are divided into 4 categories: physical environment, social & cultural environment, intra-individual factors, and policy and regulatory environment [12, 13]. The physical environment includes factors such as safety, accessibility, neighbourhood environment, weather, and distance. In the study conducted in Glasgow, the distance to workplace, bad weather, and danger from motor traffic have been found to be the barriers to walking and cycling to workplace [14]. Lack of adequate cycling infrastructure which was associated with safety was the major barrier found in Brazil [15]. Bus passengers in Edinburgh reported feeling unsafe due to drunk people and the behaviour of other passengers [16]. Study of public transport in Portugal identified discomfort in public bus due to hot weather, crowding and smell. Light rail in Portugal was perceived as a more reliable and comfortable mode than bus [17]. Santos et.al. also found that numbers of days of rain affected the use of public transportation [9].

Social & cultural environment includes factors related to social norms. In countries with high economic inequality both crime and traffic safety are higher than in more equal societies. Therefore, parents in high economic inequality societies often restricted their children's independent mobility [18]. Stradling et.al. also indicated that travelling by bus did not create right impression to others [16].

Intra-individual factors include factors such as health condition, personal attitude, economic status, and lifestyle. The study of students and staff of University of Western Australia found that the

IOP Conf. Series: Earth and Environmental Science **329** (2019) 012011 doi:10.1088/1755-1315/329/1/012011

time it took to travel, the forward planning required and also the need to pick up and drop off others on the way prohibit them from travelling more sustainably [7].

Policy and regulatory environment refers to government interventions to support sustainable commuting such as investment in high quality infrastructure and public transportation, tax incentives, development of smart cards or applications to facilitate sustainable commuting.

3. Transportation mode in Bangkok

There are several modes of transportation available in Bangkok and its vicinities as shown in Figure 2. For this study, these modes were divided into 4 categories ranging from the most sustainable to the least sustainable.

Figure 2 Modes of transportation in Bangkok.

3.1 Active commuting: the most sustainable way to commute. This consisted of 2 modes, namely walking and biking.

3.2 Active commuting with public transportation: this was also considered sustainable commuting since it utilizes mass transportation services provided in the neighbourhood. Mass transportation available in Bangkok and its vicinities included city bus, passenger train, rapid transit, public van, public boat, and Songthaew, a modified pickup truck used to transport people in a neighbourhood.

3.3 Sustainable commuting in combination with private motorized vehicles was considered partially sustainable. The private motorized vehicles included 4 modes: private car, private motorbike, car taxi, and motorbike taxi, very popular in Bangkok because of the heavy congestion.

3.4 Pure private motorized vehicles: this was classed in not sustainable category. Single person transportation with private motorized vehicles was considered the most inefficient mode to travel in the city.

4. Research Approach

The design of the study was a cross-sectional descriptive survey using a self-reported online questionnaire. The questionnaire was developed based on the study by The University of Western Australia [7] due to its availability and in great details with the aim of gathering comprehensive information regarding daily travel patterns and barriers to sustainable commuting. Barriers to sustainable commuting were obtained from previous literatures and discussions with commuters. The questionnaire went through several revisions due to complexity of travel patterns in Bangkok before adoption into the actual survey. The population for the study consisted of students and staff who routinely commute to Kasetsart University. A total of 400 questionnaires were distributed and 315 questionnaires were returned.

The questionnaire consisted of three main sections: section 1 elicited personal demographic information (11 questions); section 2, existing travel patterns (18 questions); section 3 was concerned with perceived barriers toward 5 modes of sustainable transportation, namely, walking (31 factors), biking (27 factors), public bus (29 factors), public van (27 factors), and rapid transit (29 factors). The respondents had to rate these factors on the Five-point Likert scales. These 27 – 31 factors for each mode were divided into 4 subgroups: physical environment, social & cultural environment, intra-individual factors, and policy and regulatory environment.

The qualitative study was conducted by observation of the neighbourhood within 1 kilometre around the campus. Face-to-face interview of staff and students was also conducted with those who indicated willingness in the online questionnaire to gain further insight on the barriers to sustainable commuting.

5. Case study

Kasetsart University is a public research university with the area of 335 acres as shown in Figure 3. There are approximately 35,000 students and 7,500 employees. The university is flanked by three major roads on the east, west, and south sides. There are city bus stops around the university along 5 main gates, which are also used as public van stops. Passenger train station is on the west side across the road. Rapid transit line and stations are under construction on the east and west sides of the university. Monorail system on the south side is expected in the near future. The nearest rapid transit station is 3.6 kilometres from the university. Motorbike taxis are widely available in the neighbourhood, in which most of the students reside.

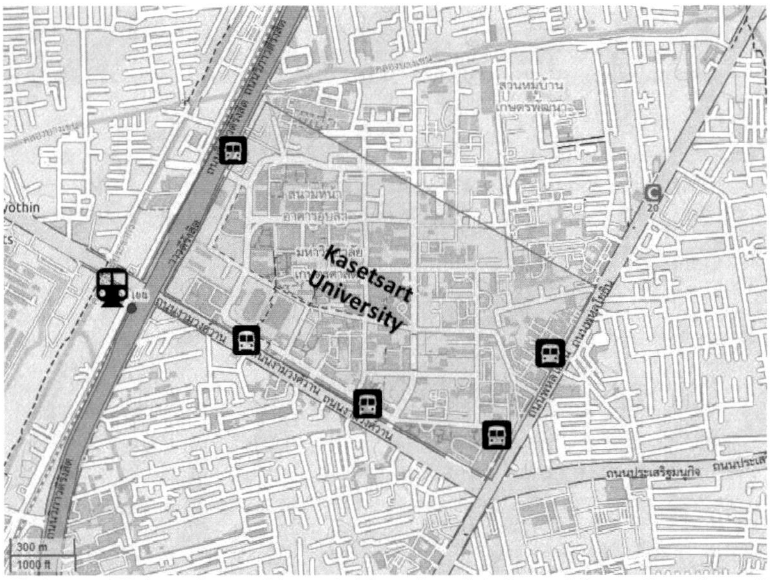

Figure 3. Kasetsart University and its neighbourhood (adapted from OpenStreetMap.com)

6. Kasetsart University current mode of transport

Most of the respondents are students between 21 – 25 years old living within 2 kilometres from the campus. Almost half of the respondents use walking as part of their travelling. One fourth of the respondents reported that they used motorbike taxi, city bus, and private car as part of their daily commute. Only 15% used private motorbike. Biking was used by less than 10% of the respondents.

From the four categories, the least sustainable mode, pure private motorized vehicles, was used by most, at 37.14%. Followed by the group that used combination mode with private motorized vehicles at 25.04%. The most sustainable mode, walk and bike, was used by 19.05% and the public transportation in combination with walk and bike was used by 18.41% of the respondents. Table 1 shows the adoption percentage of each mode of transportation by each class of respondents. It is evident from the table that the higher the monthly salary was the less they actively commute, and the more they used private motorized vehicles. Also, the greater the distance of residence from the university, and also of the residence to the main road the less they use active commuting.

Table 1. Modes of transportation divided by different groups of respondents.

	Persons	Walk & Bike	Public with Walk & Bike	Combi with Private Motorized	Private Motorized
		60 (19.05%)	58 (18.41%)	80 (25.40%)	117 (37.14%)
Sex					
Male	197	19.80%	19.80%	24.37%	36.04%
Female	118	17.80%	16.10%	27.12%	38.98%
Age					
Less than 20 years old	44	36.36%	15.91%	25.00%	22.73%
21 – 25 years old	212	17.92%	18.87%	27.83%	35.38%
26 – 30 years old	24	16.67%	20.83%	25.00%	37.50%
More than 30 years old	35	5.71%	17.14%	11.43%	65.71%
Student or Staff					
Student	234	23.08%	17.95%	27.78%	31.20%
Staff	81	7.41%	19.75%	18.52%	54.32%
Monthly Salary					
Less than 5,000 Baht (120 GBP)	35	34.29%	20.00%	20.00%	25.71%
5,000 – 10,000 Baht	149	22.15%	21.48%	26.17%	30.20%
10,000 - 20,000 Baht	85	12.94%	16.47%	30.59%	40.00%
More than 20,000 Baht	46	8.70%	10.87%	17.39%	63.04%
Distance from the Campus					
Less than 1 kilometer	84	48.81%	8.33%	23.81%	19.05%
1 – 2 kilometers	58	29.31%	12.07%	29.31%	29.31%
2 – 10 kilometers	90	2.22%	27.78%	20.00%	50.00%
More than 10 kilometers	83	0.00%	22.89%	30.12%	46.99%
Distance from Residence to the Main Road					
Less than 0.5 kilometer	136	33.82%	17.65%	20.59%	27.94%
0.5 – 1.0 kilometer	92	8.70%	19.57%	35.87%	35.87%
More than 1 kilometer	87	6.90%	18.39%	21.84%	52.87%

The percentage of active mode found in this study was comparable to the University of Western Australia's students [19] and District of Columbia working people [20]. However, it was significantly lower than 55% found in McMaster University's students [21], 47% found in Netherland adolescents [22], 35% found in University of Albama's students [23], 31% found in Hochimin adolescents [24], 28% found in England [25], and 25% found in University of California Los Angeles's students [8]. It was also found that if the people living within 1 kilometre around Kasetsart University is excluded from study, the active mode drop significantly down to only 8% as shown in Figure 4.

Moreover, within the walking range of 1 kilometre from the university, the results indicated that only half of this group actively commuted to the Kasetsart University compared to 90% of University of Western Australia's students, 83% of people in England [25], and 58% of staff in University of Western Australia [19].

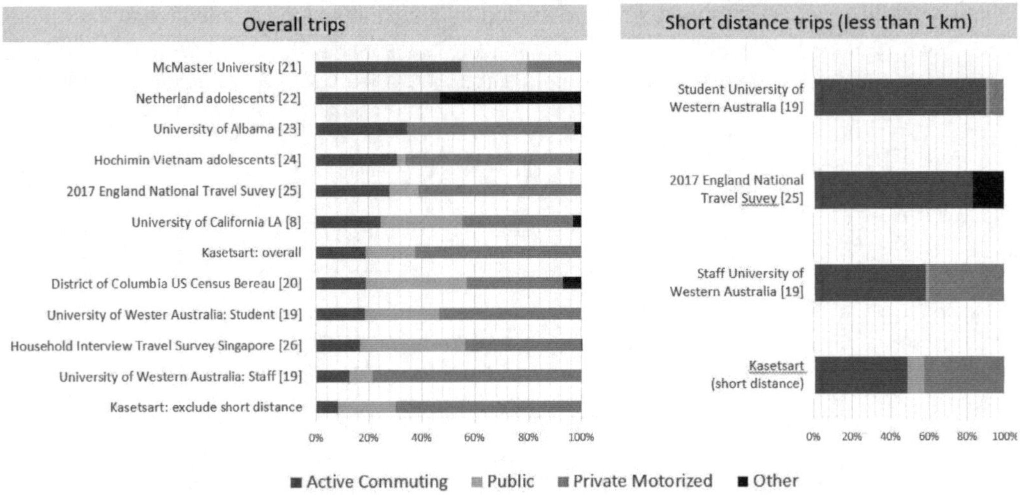

Figure 4. Compare Kasetsart University and other studies

7. Barriers to sustainable commuting

The questionnaire also asked the respondents who used private motorized vehicles whether they ever consider changing to the more sustainable mode of transportation, only one fourth of this showed their willingness to change. The barriers to five mode of sustainable transportation available to most of the respondents: walking, biking, city bus, public van, and rapid transit, were explored.

Even though walking is conceivably accessible to all able-bodied students and was used by half of the respondents, the other half did not use walking as part of their daily commute. The perceived barriers were the distance, the inconvenience including lack of time. Hot and rainy weather also discouraged them from walking. The respondents also showed concerns regarding safety issues including both safety from crime as well as traffic safety. The respondents indicated that the walk path was not well-connected and not separated from motor vehicles. From observation, there were often no footpath around residences. Walkways to the main road were often shared with cars, motorbikes, bikes, parked vehicles and street food stalls. Street lights are often dim, affording little comfort to passers-by. Where there were footpaths, they were often intermittently obstructed by power line poles, trees, street food stalls and motorbike taxis stations. The respondents also mentioned that there was no policy to promote or regulate the foot path in Bangkok. Moreover, the existing law and regulations related to the sidewalks had not been enforced effectively. The motorbikes were seen running on the sidewalks everywhere in Bangkok especially during rush hours due to traffic congestion.

Most of the respondents reported that they could ride a bike but only one fourth of them used it daily. The barriers to biking in the study were found similar to walking. The interview with respondents found that they were afraid of traffic accidents since most of the time they had to share the road with cars and motorbikes. They also mentioned that there was no dedicated bike crossing to the university, making reaching the university a dangerous undertaking. These issues were common in large cities in Southeast Asia such as Hochimin, Vietnam [24] and Kota Bharu, Malaysia [26] where there were no active commuting infrastructure and well enforced regulations.

A barrier to the use of city bus was the waiting time, since there was no bus schedule. When one arrives, it was always too crowded, especially during rush hour. Since the bus has to stay on the bus lane and stop at bus stops, it was perceived as the slowest way to commute during rush hours. However, it was perceived as a more convenient mode compared to walking and biking due to the hot and rainy weather in Bangkok. The respondents also felt that it was safer than walking and biking to the university.

Public van is like a minibus, taking people from the Bangkok's vicinities into the city. Public van is faster than the bus since it does not stop as frequently as the city bus and it can take toll road and fast lane depending on the lines. However, due to the driver and van conditions, public van often gets into accidents resulting in passenger casualties. Female respondents also mentioned their fear of criminal act from the driver and fellow passengers.

There was no major barriers to the use of rapid transit, except for the high fare, ranging from 20 to 100 baht, depending on the length of the travel. The respondents felt the rapid transit was the only public transport that offered convenience, comfort and safety.

8. Conclusions

Most people commuted to Kasetsart University via non-sustainable mode including private motorized vehicles along or in combination with other public transports. The most private motorized vehicles used were motorbike taxis, especially when the residence was far from the main road, or when it was too hot or when safety was an issue. It was found that as much as 39% of the alleys in Bangkok were dead end, compared to 19% in London, around 7% in Singapore and Tokyo, and only 3.6% in New York [28]. Instead of walking from the residences situated in these dead end alleys as observed around university, they were willing to pay the minimum charge of 20 baht per trip for the motorbike taxi twice a day. Although walking distance of up to 800 meters has been acceptable in many countries, the lack of appropriate walking infrastructure made this distance untenable in Bangkok.

Most transportation infrastructure developed in Bangkok during the last few decades seems dedicated to cars only. Many roads, tollways, bridges, and even tunnels have been developed and expanded continuously with the hope of alleviating the traffic congestion in the city. Instead, the development has worsened the traffic situation in Bangkok. The development which concentrates solely on cars has forced people living in the city to get a car of their own prior to buying a house so they can commute safely. The infrastructure has shaped the attitude and behaviour of the people in a way that has not been realized before.

Commuting in as big a city as Bangkok consumes substantial energy and emits large amount of carbon dioxide. If the policy is not directed toward more sustainable commuting, the energy consumed and the resulting pollution will harm the population in this city. Priority has to be shifted toward non-motorized mode of transportation while restricting the number of private motorized vehicles. Rapid transit is currently developed to cover most part of Bangkok. Government subsidies to lower the rapid transit fare should be considered. Good walking and biking infrastructure must be designed and developed to ensure the connectivity and the safety of people from their residence to the public transportation. City bus should be improved in terms of number of buses as well as application to ascertain the availability as well as the bus location. Moreover, car ownership costs should be increased including excise tax, vehicle registration tax, parking costs, etc. to discourage people from buying motorized vehicles.

Measures need to be implemented to incentivize people to commute more sustainably. Investment in walking and biking infrastructure is less capital intensive than those directed at building roadways, but can yield greater benefits when environmental cost is factored in. Sustainable commuting is no longer a choice for Bangkok, it is rather a necessity if Bangkok is to remain a liveable capital city of Thailand.

9. References

[1] Åkerman J and Höjer M 2006 How much transport can the climate stand?—Sweden on a sustainable path in 2050 *Energy Policy* **34**(14), 1944-57

[2] Kennedy C, Steinberger J, Gasson B, Hansen Y, Hillman T, Havránek M, Pataki D, Phdungsilp A, Ramaswami A and Mendez GV 2009 Greenhouse Gas Emissions from Global Cities *Environmental Science & Technology* **43**(19), 7297-302

[3] Andersen LB 2017 Active commuting is beneficial for health *BMJ* **357** j1740

[4] Stjernborg V and Mattisson O 2016 The Role of Public Transport in Society—A Case Study of General Policy Documents in Sweden *Sustainability* **8**(11), 1120

[5] Breuer H 2017 Urban Mobility: Urban Mobility: A Master Plan to Counteract non-stop Traffic Congestion Siemens Company stories Available at: https://new.siemens.com/global/en/

IOP Conf. Series: Earth and Environmental Science **329** (2019) 012011 doi:10.1088/1755-1315/329/1/012011

[6] World Health Organization 2019 Strengthening Road Safety in Thailand. Available online at: http://www.searo.who.int/thailand/areas/roadsafety/en/
[7] Smith T, Giles-Corti B, Pikora T, Bulsara M, Bull F, Stark K and Shilton T 2004 Commuting habits and potential for change. Physical Activity Research Group, School of Population Health and School of Human Movement and Exercise Science, The University of Western Australia, Perth, Australia
[8] Zhou J 2012 Sustainable commute in a car-dominant city: Factors affecting alternative mode choices among university students *Transportation Research Part A: Policy and Practice* **46**(7), 1013-29
[9] Santos G, Maoh H, Potoglou D and Brunn T 2013 Factors influencing modal split of commuting journeys in medium-size European cities *Journal of Transport Geography* **30**, 127-37
[10] Carlos JLB 2003 Sustainable transportation planning on college campuses *Transport Policy* **10** (1), 35–49
[11] Guntamuanglee A 2013 The level of accessibility to public transportation systems in the Bangkok Metropolitan region, Master Thesis, Chulalongkorn University [In Thai]
[12] Garrard J 2009 Active transport: Children and young people, Melbourne, VicHealth
[13] Gebel K, King L, Bauman A, Vita P, Gill T, Rigby A and Capon A 2005 Creating healthy environments: A review of links between the physical environment, physical activity and obesity. Sydney, NSW Health Department and NSW Centre for Overweight and Obesity
[14] Crawford F, Mutrie N and Hanlon P 2001 Employee attitudes towards active commuting *International Journal of Health Promotion & Education* **39**(1), 14-20
[15] Souza AA, Sanches SP and Ferreira, MAG 2014 Influence of Attitudes with Respect to Cycling on the Perception of Existing Barriers for Using this Mode of Transport for Commuting *Procedia - Social and Behavioral Sciences* **162,** 111-20
[16] Stradling S, Carreno M, Rye T and Noble A 2007 Passenger perceptions and the ideal urban bus journey experience *Transport Policy* **14,** 283-92
[17] Beirão G and Cabral SJA 2007 Understanding attitudes towards public transport and private car: A qualitative study *Transport Policy* **14**(6), 478-89.
[18] Davison KK, Werder JL and Lawson CT 2008 Children's active commuting to school: current knowledge and future directions *Preventing chronic disease* **5**(3), A100
[19] Shannon T, Giles-Corti B, Pikora T, Bulsara M, Shilton T, Bull F 2006 Active commuting in a university setting: Assessing commuting habits and potential for modal change Transport Policy, 13(3), 240-53. doi:10.1016/j.tranpol.2005.11.002
[20] US Census Bureau 2013 Community Survey Available at: https://www.census.gov/
[21] Whalen KE, Páez A, Carrasco JA 2013 Mode choice of university students commuting to school and the role of active travel *Journal of Transport Geography* **31,** 132-42
[22] Bere E, Horst K, Oenema A, Prins R, Brug J 2008 Socio-demographic factors as correlates of active commuting to school in Rotterdam, the Netherlands Preventive Medicine, 47(4), 412-6
[23] Lundberg B and Weber J 2014 Non-motorized transport and university populations: an analysis of connectivity and network perceptions *Journal of Transport Geography* **39,** 165-78
[24] Trang N, Hong TK, Dibley MJ 2012 Active Commuting to School Among Adolescents in Ho Chi Minh City, Vietnam *American Journal of Preventive Medicine*, **42**(2), 120 – 8
[25] National Travel Survey: England 2017 Available at: https://assets.publishing.service.gov.uk/
[26] Household Interview Travel Survey (HITS) 2016 Available at: https://www.lta.gov.sg/
[27] Mat Yazid MR, Ismail R, Atiq R 2011 The Use of Non-Motorized For Sustainable Transportation in Malaysia, Procedia Engineering, **20**(2011), 125-34
[28] Yupho, S 2018 Dead End Analysis Available at: https://www.facebook.com/innithi/posts/10156246058728369

IOP Conf. Series: Earth and Environmental Science **329** (2019) 012012 doi:10.1088/1755-1315/329/1/012012

3D built-environment attributes and household road traffic noise exposure in Hong Kong

Mengdi Guo[1,3], Jianxiang Huang[1], Michael Yuxuan Ni[2]

[1] Department of Urban Planning and Design, 8/F Knowles Building, the University of Hong Kong, Pokfulam Road, Hong Kong
[2] School of Public Health, G/F, Patrick Manson Building (North Wing), 7 Sassoon Road, Pokfulam, Hong Kong
[3] Corresponding Author Email: mdguo@hku.hk

Abstract. Road traffic noise is an environmental health hazard in Hong Kong and other high-density cities. The dense built environment modifies noise propagation, reflects, absorbs or diffracts sound depending on building morphology, road configuration, and open space layout, etc. Mitigation of urban traffic noise is of growing concerns to planning and design practitioners. Existing assessment methods are limited in reliably accounting for localized variations in noise exposure associated with a high-density city. The aim of this research is to 1) develop a 3D database of road traffic noise exposure for a large number of households in a high-density city; 2) explore the linkages between built environment attributes and road traffic noise exposure. A 3D built environment database was constructed for Hong Kong using building geometries, topography and urban traffic noise data. Window coordinates for each household were extracted using address and building floor plans. Computer simulation was conducted to determine traffic noise exposure at window locations using CadnaA for a random sample of 8,158 households across the city. Results revealed that 76.3% of the households are exposed to excessive road traffic noise by WHO standards. Household traffic noise exposure are significantly associated with proximity to secondary road, story-level of the flat, and other urban form attributes. The 3D database is of value for public health research in relation to noise and urban noise mitigation measures. The next step is to develop an efficient and reliable simulation tool to support planning and design decisions in traffic noise mitigation.

1. Introduction

Noise is an environmental hazard associated with health issues such as cardiovascular diseases, diabetes, sleep disturbances, etc. [1], [2]. Residents in high-density cities are especially vulnerable to noise due to the agglomeration of noise sources, built-up structures, and urban residents in a limited area. Road traffic is a ubiquitous source of environmental noise that is widely spread across the city. Around one million residents in Hong Kong were subjected to excessive road traffic noise according to the Environmental Protection Department (EPD) [3], [4].

Population noise exposure in urban area was found correlated with a variety of built environment attributes, including yet not limited to traffic conditions, density of buildings and roads, building morphology, material characters, microclimate, provision of open space, etc. [5]–[7]. Wang and Kang found significant impacts of morphological attributes on noise exposure with a study of Greater Manchester and Wuhan [5]. The impact of individual attribute varies in the two cities. For instance, building coverage and road traffic noise exposure are positively correlated in Greater Manchester, yet negatively correlated in Wuhan. A study in Hong Kong found dwelling scale and peripheral street canyon index were linked to road traffic noise exposure in large street block groups [6]. Ryu et al. found ground space index, floor space index, traffic volume, traffic speed and road coverage significantly correlated with noise exposure of gridded urban area in Cheongju, Korea [7]. The stuides of Silva, Oliveira and Silva and Guedes, Bertoli and Zannin stated that environmental noise was associated with construction density, provision of open space, and spatial configuration in Braga, Portugal [8], [9].

In existing studies, understandings of the relationship between population noise exposure and built environment attributes were mixed. Due to data limitation, not many studies have investigated the noise impacts of built environment attributes; even less effort was put into high-density cities. Existing studies either limit in resolution of data provision or representativeness of selected cases, which may lead to the loss of statistical power or misclassification of household or individual noise exposure [5], [6], [9], [10]. In regional scale researches, individual noise exposure is usually calculated using averaged noise levels of the most exposed façade [11], [12]. A less accurate approach is to calculate noise exposure at building level, using noise mapping in combination with interpolation methods [13], [14]. Precise simulation of individual or household noise exposure is usually conducted in neighborhood or smaller scale studies [8], [9]. It can hardly be applied to a large area due to limits in data provision and high assessment cost.

This study aims to advance the spatial resolution of household traffic noise exposure simulation in dense urban area of Hong Kong, in order to enhance the understanding of traffic noise exposure in high-density cities and inform noise mitigation measures in urban planning practices. In detail, we will 1) develop a database of annual averaged road traffic noise exposure of households from public health cohorts in Hong Kong, 2) and evaluate the linkages between built environment attributes and household road traffic noise exposure. This study is supported by a comprehensive built environment database of Hong Kong for precise identification of built environment attributes for each household address. We use the state-of-the-art noise simulation tools to calculate household road traffic noise exposure with high spatial resolution and reliable accuracy. Section two introduces the methods and data used in this study; section three presents result and discussion; the study is concluded in section four.

2. Methods and Data

For participants of public health cohorts in Hong Kong, we identified the geo-coordinates of their household window locations in 3D urban space, calculated the annual averaged road traffic noise exposure for selected households, linked the noise exposure with built environment attributes, and analyzed the relationship between traffic noise exposure and built environment characters.

Data input of this study includes: 1) iB1000 digital topographic map that covers a majority of Hong Kong urban area for the year 2015, obtained from Lands Department [15], 2) household addresses of FAMILY Cohort [16] and Birth Cohort [17] participants, 3) flat-unit information fetched from CentaData [18], 4) built environment information obtained from OpenStreetMap, an open map data platform [19], 5) territory-wide traffic noise database of road network in Hong Kong for the year 2015, calculated with road traffic database of Hong Kong [20], average traffic speed obtained from Google Map, and topography data [15], using the CRTN algorithm [21] as endorsed by the Hong Kong Environmental Protection Department (EPD) [22], and 6) data obtained from other open data platform such as Google and Housing Authority. The cleaning, analyses and linkages of the above databases were executed in the ArcGIS 10.4, using programming languages Python 2.7 and R 3.3.1. The spatial reference was "Hong Kong 1980 Grid System" [23].

2.1. *Identifying major window locations of households in 3D urban space*

A semi-manual method is used to identify 3D geo-coordinates of the centre point of the living or bed room window for apartments of FAMILY Cohort and Birth Cohort participants. A sum of 8,158 window coordinates was identified from the household addresses of both cohorts: 6,358 from FAMILY Cohort and 1,800 from Birth Cohort.

Firstly, address data of the cohort participants was paired with flat-unit and floor plan information and cleaned in the meanwhile. Flat-unit information of households living in private housing estates was automatedly fetched from CentaData (Centadata Company Limited, no date). Most flat-unit information of public housing estates was obtained by field study, except for the few with available data in open data platform. A pilot field visit to 48 buildings of 6 public housing estates in Shek Kip Mei, Hong Kong was conducted in Jan 2017 (Figure 1 left & middle), in order to obtain flat-unit information. Field investigation of a large number of public housing estates were carried out in early June of 2017. The pairing process added flat-unit attributes such as floor plan, apartment information, etc. to the address database and excluded the samples without valid flat-unit information.

Next, 3D coordinates were added to the address database using floor plan and flat-unit information obtained in the pairing process. A floor plan was matched with the building outline in ArcGIS space (Figure 1 right); the window location was pinpointed on the building outline, at the middle of the largest window of living or bed room; X-Y coordinates of the point was then calculated in ArcGIS space. Height attribute (Z coordinate) of each window point was calculated using the base and roof level of the building, total storeys of the building and the actual storey of household.

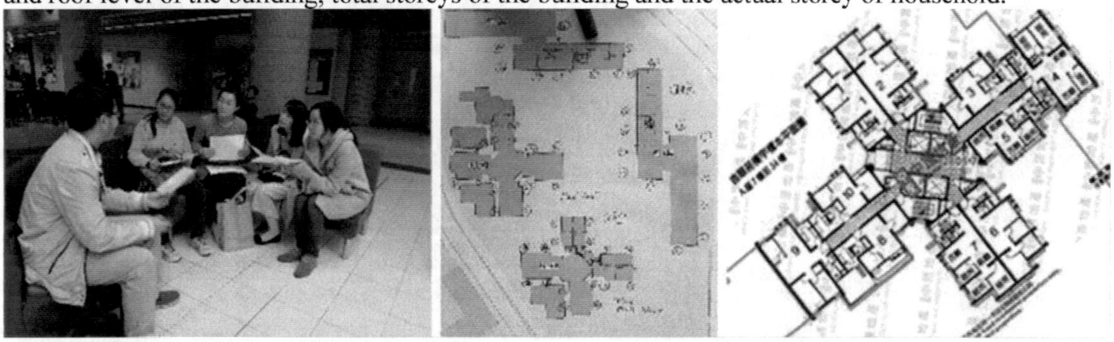

Figure 1. left, briefing of pilot field visit to public housing estates in Shek Kip Mei in January 2017; middle, flat unit information obtained in the visit; right, match floorplan with building outlines in GIS database and pinpoint window location in ArcGIS platform.

2.2. *Simulation of household road traffic noise exposure*

We use annual-averaged road traffic noise exposure (L_{den}), a metric recommended by WHO and mandated by European Noise Directive [24], [25] for health studies, to evaluate long-term noise exposure for the 8,158 households identified in 2.1. We use CadnaA, an extensively validated software that has been widely applied to simulate noise exposure in high-density cities, such as Wuhan, Tianjin, and Singapore, to simulate the traffic noise exposure at the identified household window locations [5], [26], [27]. The input data consists of a territory-wide database of buildings, topography, road centrelines with annual averaged road traffic noise levels (L_{Aden}), and 3D coordinates of receiver points located one metre away from the identified window location in perpendicular direction to the building façade [28]. Urban area of Hong Kong was divided into 237 sub-areas (blue frames in **Figure 2** left), in order to fit in the calculation capacity of CadnaA. The 8,158 identified household addresses were located in 211 sub-areas. For each sub-area, built environment and traffic data for simulation was assembled within a 300-metre outer buffer (red frames in **Figure 2** left) [22].

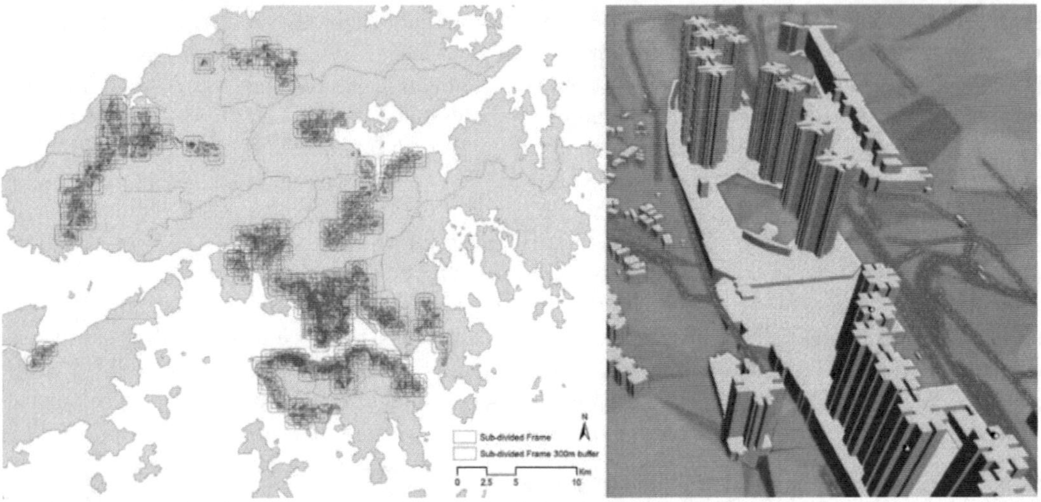

Figure 2. left, Hong Kong urban area was divided into 237 sub-areas (blue frames) in ArcGIS platform for acoustic simulation in CadnaA space, the built environment and traffic data for calculation was collected within a 300-metre outer buffer (red frames) of the sub-area; right, 3D view of noise simulation in CadnaA.

2.3. Statistical analysis

We examined the linkages between household road traffic noise exposure and built environment attributes using multi-variant regression analysis. Variables of household road traffic noise exposure, flat-unit information and built environment attributes were assembled in previous sections and linked with cohort address data. A total of 6,363 households were selected with variables summarized in **Table 1**.

Table 1. Summary of household road traffic noise exposure, flat-unit information, and built environment attributes linked with 6,363 households in the FAMILY Cohort and the Birth Cohort.

Variables	Unit	Mean	Std. Dev.	Min	Max
Household exposure to road traffic noise	dB(A)	61.18	9.57	27.3	86
Household proximity to the nearest highway or major road	100 Metre	1.80	1.60	0.06	11.39
Household proximity to the nearest secondary road	100 Metre	0.32	0.23	0	2.00
Traffic noise level of the nearest road	dB(A)	68.1	6.16	0	84.68
Density of all roads in a 500m buffer	1/km	3.96	6.47	0	31.06
Building coverage in a 500m buffer	%	28.85	12.3	3.23	94.49
Average building height in a 500m buffer	Metre	28.86	12.33	6.59	73.83
Number of flat units per floor of the building	count	9.62	5.25	1	60
Actual storey of the flat	count	16.2	10.54	1	58
Total storeys of the building	count	31.82	9.64	1	65

3. Results and Discussion

Territory-wide distribution of road traffic noise exposure (L_{Aden}) at window location of the 8,158 households was shown in **Figure 3**, where each column represented one household. Traffic noise level of road centre lines was presented in coloured lines. Average household road traffic noise exposure was 62.91 dB(A) in Hong Kong Island, 63.87 dB(A) in Kowloon, and 59.77 dB(A) in New Territories. A high variation in traffic noise exposure was observed between households living close by. Around 76.3% of the studied households have a road traffic noise exposure higher than 55 dB(A), the WHO day time noise standard [29]; while around 36.7% exceeded the Hong Kong's local standard [30]. The database improves the spatial resolution of household road traffic noise exposure and is of value to advance researches on health impacts of road traffic noise in high-density cities like Hong Kong.

Figure 3. Territory-wide distribution of road traffic noise exposure simulated at the window locations of the 8,158 households selected from the FAMILY Cohort and the Birth Cohort, and traffic noise levels of road centrelines.

The result of regressing household road traffic noise exposure on built environment attributes was presented in Table 2, where the dependent variable was household road traffic noise exposure; independent variables were flat-unit and built-environment attributes stated in Table 1. Results showed correlations between household road traffic noise exposure and road traffic attributes, built environment density, and flat locations in 3D urban space.

- Road-traffic variables, such as proximity of highways or secondary roads, were associated with household road traffic noise exposure ($p < 0.001$). Noise impacts of secondary roads on households are higher than those of the highways or major roads: regression results showed that being 100m closer to the nearest highway or major road was linked with an increase of 1.55 dB in household road traffic noise exposure; the increment was 7.78 dB for a nearest secondary road, although major roads have higher traffic noise levels. A possible reason is that households in Hong Kong are most likely to live close to a secondary road in a dense neighborhood; while a highway, although generating higher levels of traffic noise in itself, is often located further away and equipped with sound barriers.
- Correlations were found between household traffic noise exposure and urban form. Household noise exposure increases 0.1dB if the building coverage increases one percent; it decreases 0.17dB if average building height of the area increases one meter (both $p<0.001$). Peak traffic noise exposure appears at the height of 20 stories (**Figure 4**): for households below 20 story, living one story higher is linked to higher road traffic noise exposure ($p<0.001$). A higher residential building is negatively associated with household noise exposure ($p<0.01$).
- Results suggest a need for 3D noise simulation at household level, since household noise exposure depend significantly on the orientation of the apartment, location of windows, presence of nearby buildings, and proximity of road traffic, etc. Much of the variability will be lost if a 2D simulation is used as it is currently employed by the Hong Kong EPD [3].

Table 2. Regressing household road traffic noise exposure on flat-unit and built environment attributes with a subset of 6,363 households living in private housing estates.

Dependent Variable: Household Road Traffic Noise Exposure (dB(A))		
Number of obs.	6,363	
F (9, 6353)	254.66	
Prob > F	0	
R-squared	0.2651	
Adjusted R-squared	0.2641	
Root MSE	8.2011	
Household exposure to road traffic noise (dB(A))	***β*** (95% CI)	p-value
Independent Variables: Road-Traffic Variables		
Household proximity to nearest highway or major road (100m)	-1.547 (-1.678, -1.416)	(0.000***)
Household proximity to the nearest secondary road (100m)	-7.779 (-8.697, -6.861)	(0.000***)
Traffic noise level of the nearest road (dB(A))	0.282 (0.249, 0.316)	(0.000***)
Road density within 500m radius (1/km)	0.055 (0.023, 0.086)	(0.001***)
Control Variables		
Building coverage within 500m radius (%)	0.103 (0.085, 0.122)	(0.000***)
Average building height within 500m radius(m)	-0.167 (-0.186, -0.147)	(0.000***)
Number of flat units per floor of the building (count)	-0.084 (-0.124, -0.043)	(0.000***)
Actual storey of the flat (count)	0.116 (0.095, 0.137)	(0.000***)
Total storeys of the building (count)	-0.037 (-0.064, -0.009)	(0.01**)
Intercept	48.894 (46.304, 51.484)	(0.000***)

* 90% significance level, ** 95% significance level, *** 99% significance level

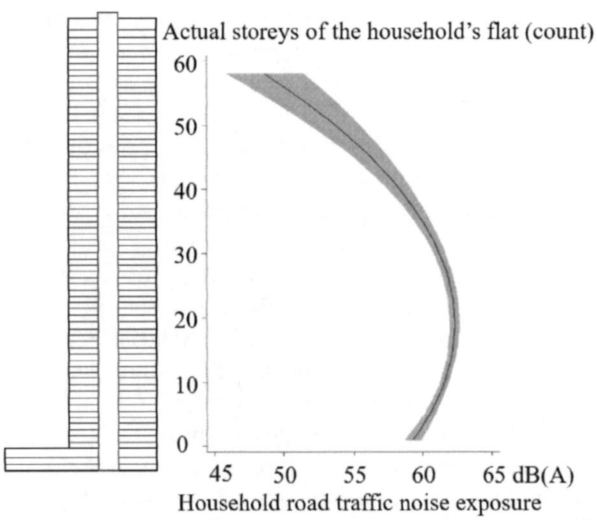

Figure 4. The vertical dimension of household road traffic noise exposure, n = 6,363. Average peak noise exposure appears at the height of 20 storey.

Results have practical implications for urban planning and design practices. Prevailing mitigation measures either work on road network or building design. Road network measures such as low noise road surfaces (LNRS) is mainly applied to major roads and highway. Yet findings of this study suggest that more attention shall be paid to the mitigation of traffic noise on secondary roads. The mitigation effect of building scale measures, including fins, acoustic balcony & window, building setback, etc., may not work as expected considering the highly complex noise propagation in high-density built environment.

The study is delimitated by road traffic noise, while other noise sources are not included such as construction, domestic activities, or aviation. The next step is to evaluate simulation data with those obtained from field measurement and to study the health impact of road traffic noise exposure. A long-term goal is to develop a convenient and reliable software tool for planning and design practitioners. Such tool can provide timely and reliable feedback to decision makers in order to improve urban acoustic environment. Designers shall be able to optimize building forms and layouts, organize floor plans, design façades, etc. based on the noise exposure of buildings in 3D space.

4. Conclusion

A 3D database consisted of household road traffic noise exposure and built environment attributes is developed for a large number of households in Hong Kong; 8,158 households are linked to two public health cohorts (6,358 households from the FAMILY Cohort and 1,800 households from the Birth Cohort) with comprehensive health records. Significant correlations are found between household road traffic noise exposure and urban attributes, i.e. proximity to highways, secondary roads, and floor level of apartments, etc. Results also suggest the necessity to have a 3D noise simulation tool in order to support assessing acoustic impacts of urban planning and design projects.

Acknowledgements

This paper is supported by Health and Medical Research Fund, Food and Health Bureau (HMRF Grant #13143421), HKUrbanLab Seed Funding of the Faculty of Architecture and the University of Hong Kong, and URC Small Equipment Grant 2016/2017 of the University of Hong Kong.

References

[1] Basner M *et al.* 2014 Auditory and non-auditory effects of noise on health *Lancet* **383** 9925 pp 1325-32
[2] Babisch W and Kim R 2011 Environmental noise and cardiovascular disease *Burden of Disease from Environmental Noise* (Denmark: WHO Regional Office for Europe) pp 15-42.
[3] EPD 2016 Noise [Online] Available: https://www.gov.hk/en/residents/environment/noise/noisepollutioninhk.htm. [Accessed: 09-Mar-2018].
[4] GovHK 2019 Traffic noise [Online] Available: https://www.gov.hk/en/residents/environment/noise/trafficnoise.htm. [Accessed: 09-Mar-2018].
[5] Wang B and Kang J 2011 Effects of urban morphology on the traffic noise distribution through noise mapping: A comparative study between UK and China *Appl. Acoust.* **72** 8 pp 556-568
[6] Lam KC *et al. 2013* Relationship between road traffic noisescape and urban form in Hong Kong *Environ. Monit. Assess.* **185** 12 pp 9683-95
[7] Ryu H, Park IK, Chun BS, and Chang SI 2017 Spatial statistical analysis of the effects of urban form indicators on road-traffic noise exposure of a city in South Korea *Appl. Acoust.* **115** pp 93-100
[8] Silva LT, Oliveira M and Silva JF 2014 Urban form indicators as proxy on the noise exposure of buildings *Appl. Acoust.* **76** pp 366-376
[9] Guedes ICM, Bertoli SR and Zannin PHT 2011 Influence of urban shapes on environmental noise: A case study in Aracaju - Brazil *Sci. Total Environ.* **412** pp 66-76
[10] de Souza LCL and Giunta MB 2011 Urban indices as environmental noise indicators *Comput. Environ. Urban Syst.* **35** 5 pp 421-430
[11] Sørensen M *et al.* 2013 Long-term exposure to road traffic noise and incident diabetes: a cohort study *Environ Health Perspec* **121** 2 pp 217-222

[12] Roswall N *et al.* 2015 Residential exposure to traffic noise and health-related quality of life: A population-based study *PLoS One* **10** 3 pp 1-13
[13] Dratva J *et al.* 2012 Transportation noise and blood pressure in a population-based sample of adults *Environ. Health Perspect.* **120** 1 pp 50-55
[14] Selander J *et al.* 2009 Long-term exposure to road traffic noise and myocardial infarction *Epidemiology* **20** 2 pp 272-279
[15] Survey and Mapping Office Lands Department 2015 Digital Map Products [Online]. Available: www.landsd.gov.hk/mapping/en/digital_map/mapprod.htm. [Accessed: 07-Mar-2018].
[16] Leung GM *et al.* 2017 Cohort Profile: FAMILY Cohort *Int. J. Epidemiol.* **46** 2 pp 1-13
[17] Schooling CM, Hui LL, Ho LM, Lam TH and Leung GM 2012 Cohort Profile: 'Children of 1997': a Hong Kong Chinese birth cohort *Int. J. Epidemiol.* **41** 3 pp 611-620
[18] Centadata Company Limited 2019 CentaData [Online]. Available: hk.centadata.com. [Accessed: 09-Mar-2018].
[19] OSMF 2019 OpenStreetMap [Online]. Available: https://openstreetmap.org/. [Accessed: 09-Mar-2018].
[20] Transport Department 2016 *The Annual Traffic Census 2015* pp 1-204
[21] Department of Transport Welsh Office 1988 *Calculation of road traffic noise* (London: H.M.S.O.) pp 1-95
[22] EPD 2011 *Road traffic noise impact assessment under the environmental impact assessment ordinance* (Hong Kong) pp 1-20
[23] Survey and Mapping Office Lands Department 1995 *Explanatory Notes on Geodetic Datums in Hong Kong* pp 1-18
[24] European Commission 2002 Directive 2002/49/EC of The European Parliament and of the Council of 25 June 2002 relating to the assessment and management of environmental noise *Off. J. Eur. Communities* **189** pp12-26
[25] WHO 2011 Burden of disease from environmental noise *World Health* pp 1-105
[26] Zhao WJ *et al.* 2017 3D traffic noise mapping using unstructured surface mesh representation of buildings and roads *Appl. Acoust.* **127** pp 297-304
[27] Zhou Z, Kang J, Zou Z and Wang H 2017 Analysis of traffic noise distribution and influence factors in Chinese urban residential blocks *Environ. Plan. B Urban Anal. City Sci.* **44** 3 pp 570-587
[28] ISO 2007 *ISO 1996-2 Acoustics-Description, measurement and assessment of environmental noise - Part2: Determination of sound pressure levels* pp 1-40
[29] Berglund B, Lindvall T and Schwela D 2000 New WHO guidelines for community noise *Noise Vib. Worldw.* **31** 4 pp 24–29
[30] EPD 1997 *Technical memorandum on environmental impact assessment process (environmental impact assessment ordinance, Cap.499, S.16)* pp 1-83

IOP Conf. Series: Earth and Environmental Science **329** (2019) 012013 doi:10.1088/1755-1315/329/1/012013

Preliminary Analysis on the Status Quo and Problems of Land Use of Educational Space in Primary and Secondary Schools in the Downtown Area, Tianjin- Results of data analysis based on information entropy and equilibrium

Shouren Zhang[1,2], Qing Ye[1], Yawei Du[1],Kun Song[1]

[1]School of Architecture, Tianjin University, China
[2]Corresponding Author: 179499690@qq.com

Abstract. In this article, in order to have a better understanding of the current situation and supply-demand challenges with regard to the land use of educational space in primary and secondary schools in the downtown area, Tianjin and address the related problems, the comparative analysis is conducted on the land-use layout of 30 ordinary primary and secondary schools. In addition, for these primary and secondary schools, based on the information entropy, balance and dominance of land-use structure, and in conjunction with the actual conditions, spatial location and time of construction, efforts are made to analyze the diversity and balance of educational space. Generally speaking, the land-use layout of the surveyed schools falls into three types: ring-shaped with a center, linear, and group-based. Besides, thus, the main factors which exert certain influence on the pattern of spatial layout and the land-use structure of educational space in primary and secondary schools in the downtown area, Tianjin are identified by data analysis. It provides reference for the intensive use of educational space in primary and secondary schools, and lays a solid foundation for formulating a series of land-use planning and management policies targeted at primary and secondary schools.

1. Introduction

In recent years, many cities suffer the ever-growing shortage of both educational resources of primary and secondary schools and educational land, which has become a bottleneck posing threat to the coordinated development of primary and secondary schools. In this article, 30 primary and secondary schools in the downtown area of Tianjin are selected to conduct a comparison of their spatial layout and land-use structure, in order to get a detailed picture of the overall educational land of these schools, the supply-demand contradictions, and those pressing problems to be addressed.

According to the statistics of the National Education Development Statistics Bulletin issued by the Ministry of Education, the statistics of the number of primary schools, junior high schools and high schools in Tianjin for the past five years, the number of students in school, the building area of school buildings and the floor area per student are calculated as follows:

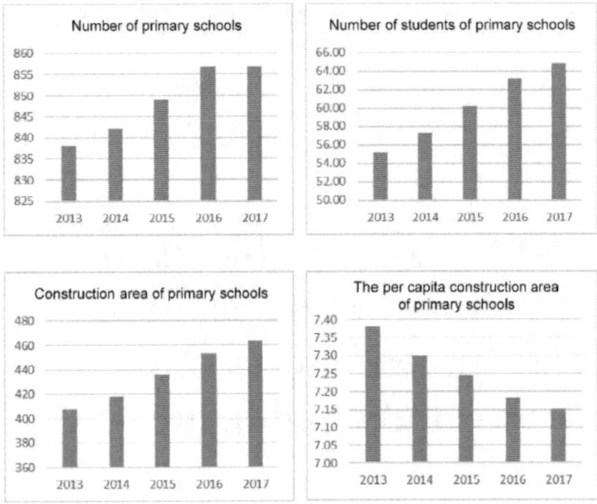

Figure 1. The Elementary School Campus

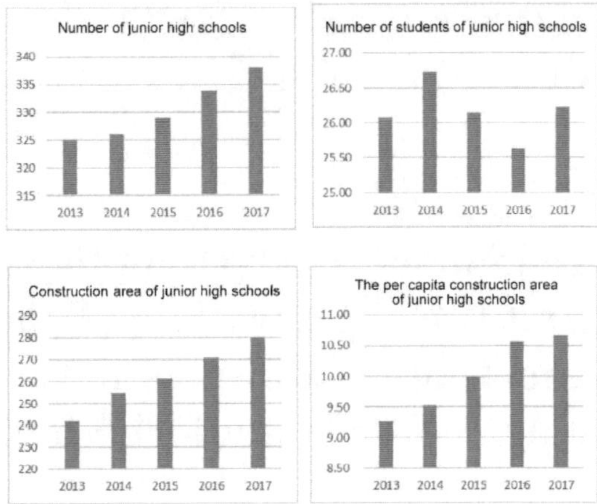

Figure 2. The Junior High School Campus

In 2013-2015, the number of primary schools in Tianjin increased rapidly while in 2016-2017 it has not changed. As of 2017, the number of primary schools was 857. During this period, the number of students and the floor space of school buildings showed an upward trend. In 2017, the total number of students in school reached 648,800, and the construction area of school buildings reached 463 hectares. The final calculated primary school students' per capita construction area showed a downward trend. In 2017, it was 7.15 m2/person.

In 2013-2015, the number of junior high schools in Tianjin increased slowly, and the growth in 2015-2017 accelerated. As of 2017, the number of junior high schools was 338. During this period, the number of students in junior high schools did not change significantly, and the overall number remained at 25-Between 270,000 people, the building area of school buildings has been increasing year by year. The final calculated average construction area of junior high school students has shown an increasing trend, reaching 10.67 m2/person in 2017.

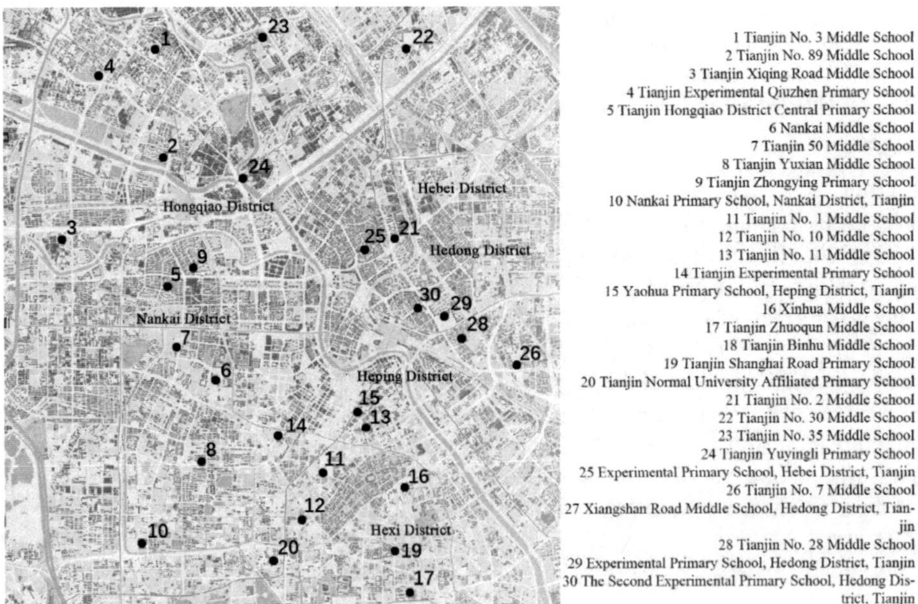

Figure 3. The spatial distribution of 30 primary and secondary schools in the downtown area of Tianjin

This paper takes the six districts in Tianjin as an example. According to the list of schools provided by the Tianjin Education Commission website, 5 districts are selected in each district of Hebei, Heping, Hexi, Nankai, Hongqiao and Hedong districts. The information collection of the primary and secondary schools with different school establishment time, school scale and campus environment are shown in Figure 4.

2. The Analysis on the Spatial Layout of Educational Land of Primary and Secondary Schools in the Downtown Area of Tianjin

The spatial layout, mainly focusing on the geographical location, relationship of combination and distribution characteristics of each functional space, is one of the key topics of human geography and urban planning. This study straightens out the spatial layout of 30 primary and secondary schools in Tianjin, and on the basis of this, the analysis is carried out on the overall combination relationship of primary and secondary schools. With the aid of the analysis on the plane composition of functional space, the combination of land forms, and so on, such insights are gained: the distribution characteristics and the combination law of functional elements in the geographical space. The model of spatial layout falls into three categories: center-ring type, line-type, and group-type, which are shown in Table 2.

Table 1. Schematic Map of Typical Land Use Layout Patterns for Primary and Secondary Schools in Tianjin

Pattern	Schematic illustrations of typical land use layout patterns
line-type	
center-ring type	
group-type	

	construction land		sports land
	green land		other land

3. The Analysis on the Land-use Structure of Educational Land of Primary and Secondary Schools in the Downtown Area of Tianjin

3.1. The quantization method of land-use structure - information entropy

It is supposed that the total area of the region with a fixed administrative boundary is A. The land type here falls into n categories. A (xi) represents the area of each type, and the probability of the occurrence of a certain type is expressed as：

$$p(x_i) = \frac{A(x_i)}{A}, i = 1, 2, \cdots, n$$

The normalization conditions are met. Based on the principles of information theory, for this region, the information entropy of the land-use structure can be defined as:

$$H(x) = -\Sigma p(x_i) \ln p(x_i)$$

In order to make the information-entropy values of the land-use structure more comparable, such concepts as the equilibrium and dominance of the land-use structure are introduced. The equilibrium E and the dominance D are respectively expressed as

$$E = \frac{H}{H_{max}} = -\sum_{i=1}^{N} p(x_i) \ln p(x_i) / \ln(n)$$

$$D = 1 - E$$

The equilibrium E of the land-use structure is the ratio of the actual entropy value to the maximum entropy value; H is the information entropy of the land-use structure; Hmax is the information-entropy maximum; the dominance D is the land-use concentration, showing that the region is dominated by one advantageous land type or several advantageous land types.

According to the principles of information theory, if the region is home to different land types, the area of each type is much closer, with the larger information-entropy value. In a region where its potential is untapped, the information-entropy value is 0, that is to say, Hmin=0; conversely, when the land types in the region tend to be proportionate and stable as well as the conditions of maximizing the information entropy are met, namely, A（x1）=A（x2）=...=A（xn）=A/n, the information-entropy value reaches the maximum, that is to say, Hmax=ln(n). Since H≤Hmax, the change interval of E is [0,1]. When E=1, the region enjoys the optimal land-use equilibrium; on the contrary, when E=0, the region where the land is not developed is in the most unbalanced state.

3.2. *The analysis on the land-use structure*

According to the above formula, for the 30 universities and colleges in Tianjin, the information entropy and the equilibrium of the land-use structure are calculated. The results are shown in the table.

Table 2. Data Measurement of Primary and Secondary Schools in Tianjin

District	Number	Information entropy	Average	Variance	Equilibrium
	1	1.18			0.86
	2	1.20			0.87
Hongqiao District	3	1.26	1.21	0.0939	0.91
	4	1.33			0.97
	5	1.05			0.76
	6	1.16			0.84
	7	1.21			0.88
Nankai District	8	1.07	1.20	0.8612	0.78
	9	1.24			0.90
	10	1.33			0.96
	11	1.33			0.96
	12	1.25			0.90
Heping District	13	1.18	1.22	0.0664	0.86
	14	1.17			0.85
	15	1.15			0.83

	16	1.06			0.77
	17	1.21			0.88
Hexi District	18	1.31	1.21	0.0875	0.95
	19	1.28			0.93
	20	1.20			0.87
	21	1.26			0.92
	22	1.24			0.90
Hebei District	23	1.38	1.23	0.0981	1.00
	24	1.07			0.78
	25	1.22			0.88
	26	1.26			0.91
	27	1.35			0.98
Hedong District	28	1.01	1.18	0.1195	0.73
	29	1.19			0.86
	30	1.10			0.80

It can be seen from the table that for the 30 primary and secondary schools in Tianjin, the average information entropy and the average equilibrium of the land-use structure are 1.21 and 0.88 respectively. The maximum is 1.38 and 1.00 respectively - Tianjin No. 35 Middle School; the minimum is 1.01 and 0.73 respectively - Tianjin No. 28 Middle School. Based on the mean, extreme and distribution of the information-entropy values of the selected primary and secondary schools, we can see that universities and colleges, with different geographical locations and at different stage of development have different information entropy of the land-use structure, and accordingly, the land-use diversity and equilibrium are also different.

According to the status of various functional space in primary and secondary schools (Table 1), the land-use structure is statistically measured. The results are shown in Figure 2.

Based on the results of Figure 2, we can see that the land-use structure of the selected 30 universities and colleges in Tianjin has the following characteristics:

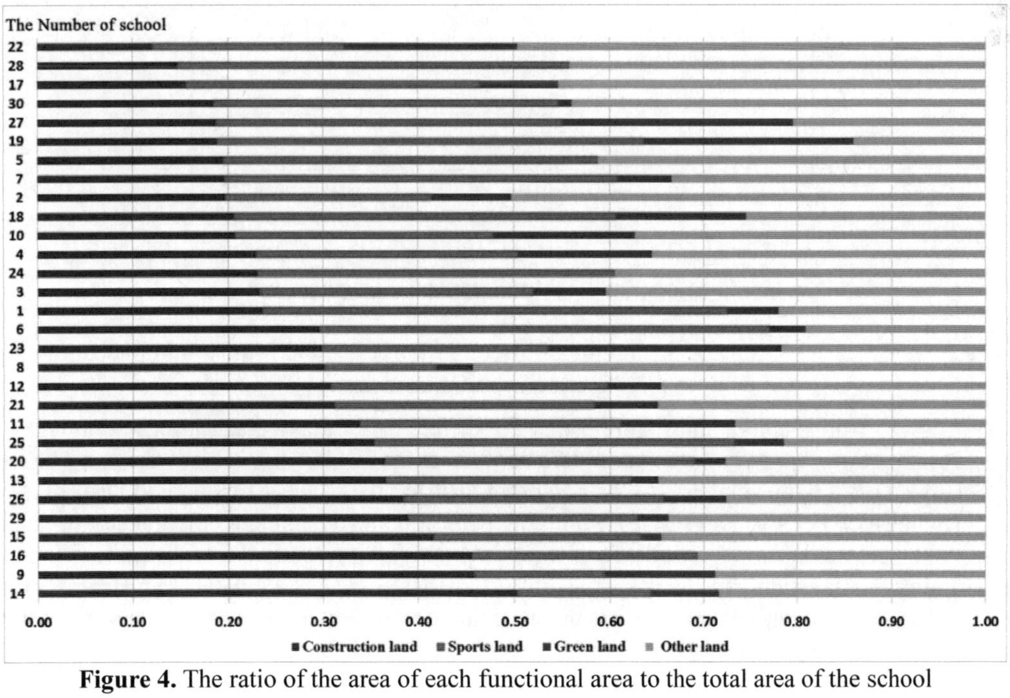

Figure 4. The ratio of the area of each functional area to the total area of the school

(1) Construction land

Construction land, as the core type of the functional space of primary and secondary schools makes up a proportion of 28% on average in the land-use structure of the 30 primary and secondary schools. However, there are certain gaps in this respect. For instance, these key schools such as Tianjin Experimental Primary School and Tianjin Zhongying Middle School and Xinhua Middle School, affected by such factors as geographical location, land price, school fund and basic conditions, have a fairly high proportion of construction land in the land-use structure, which accounts for 50%, 46% and 46% respectively. Tianjin No. 30 Middle School has a relatively large area of land, where some space is idle. Thus, the construction land accounts for a fairly low proportion of merely 12%, while the other roads and squares occupy a large area, about 50%.

(2) Sports land

Sports land is the second largest functional land in primary and secondary schools in Tianjin. Different schools, due to such factors as the time of establishment as well as the land area and scale, have remarkable difference in the proportion of sports land in the land-use structure. The sports land accounts for 49% in Tianjin No. 3 Middle School, which is the highest; the lowest is 12% Tianjin Yuxian Middle School. There is a negative correlation between the proportion of construction land and sports land. These key schools such as Tianjin No. 3 Middle School and Nankai Middle School with a time-honored history, sufficient sources of fund, and wide social attention have been updated in order to meet the requirements of modern education, with the sports land accounting for 49% and 48% respectively. By contrast, the sports land in such schools as Tianjin Yuxian Middle School, Tianjin Zhongying Middle School and Tianjin Experimental Primary School accounts for merely 12%, 14% and 14% respectively, due to their unreasonable campus planning and spatial layout.

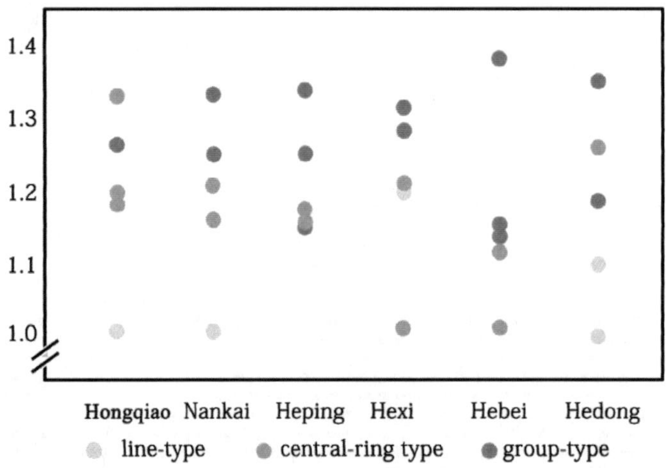

Figure 5. Information entropy of the land-use structure of primary and middle schools based on the differences in spatial location

It can be seen from Table 2 and Figure 3 that the difference in the average information entropy of the land-use structure of the primary and middle schools in the six districts is not obvious. The largest is 1.23 - Hebei District, the smallest is 1.18 - Hedong District; the values of Hongqiao District, Nankai District, Heping District and Hexi District are between 1.20 and 1.22. Thus, the gap in the land-use structure of the selected schools is not large, which is mainly attributed to the undiversified functional space and simple allocation of school land. No obvious data deviates from the standard. Thereby, a preliminary conclusion can be reached that the planning of primary and secondary schools in the six districts is under way in an orderly manner.

As is shown in the table of the land-use structure, the information-entropy variance, and the information-entropy distribution of the selected schools in the six districts of Tianjin, the minimum information-entropy variance is found in Heping District - 0.0664, and the values are more proportionate, one following another. Since in Heping District the education projects have a solid ground and the district is home to quality students, teachers and schools, the campus there is more sophisticated. In Hedong District, the maximum information-entropy variance is 0.11946, and the values are far from proportionate. Thus, the campus planning there is still in its infancy, and a set of campus system which is well-defined and reasonable has not been formed.

It can be seen from the distribution of the information-entropy values that the line-type spatial layout has a small information-entropy value, and value of the group type is fairly large. These values are attributed to the following facts: for one thing, since the line-type spatial layout is limited by the land conditions, it is not easy to integrate the functional space; for another, the corners and transferring space are used in an improper manner. On the contrary, the group-type space, more flexible as it is, has its moderate area and reasonable planning, and as a result, the information-entropy value is larger and the land-use structure is more balanced.

4. The main problems to be addressed in terms of the education land of primary and secondary schools in Tianjin

(1) There is an insufficient source of land, especially the outdoor space. Much effort should be made for outdoor landscape.

Firstly, the land in primary and secondary schools is generally tense. In the campus, the teaching and administrative buildings play a dominant role, which complicates the problem of insufficient outdoor space. Besides, in the outdoor area, some space must be reserved for the playground, and as a result, the land for landscape construction and extracurricular activities for students is limited. Secondly, the campus suffers its undiversified landscape design. In many schools, landscape design is merely defined as a sign of boundary or activity space, without any creative and commemorative landscape elements. In addition, some landscape design is manifested by merely several seats or flower beds, in the absence of any landscape axis, which fails to bring the initiative of students into full play.

(2) There is no specific design for on-campus functional space, and the land-use structure needs to be optimized.

The campuses in different regions suffer their unbalanced planning and development, and there is no guiding strategy. The campuses of multi-campus schools are simplistic in function, without any connection with each other; most of single-campus schools have their relatively compact land-use structure, which compresses the area of other functional space. For instance, the entrance space is small, thus causing inconvenience in the personnel coordination when the traffic flow is high; due to the lack of outdoor space, the time for extracurricular activities of students is cut; the lack of courtyard space poses threat to the sustainable development of "green campus". Besides, some campus-renovation plans cannot adapt to the requirements of modern education, and a series of campus-construction projects are launched in a passive manner, thus resulting in such problems as the disorder of both functional space and activity space for teachers and students.

(3) The campus is homogenized, and the on-campus spatial layout suffers the lack of personality.

Compared with universities and colleges featuring the diversified spatial function, primary and secondary schools have their simple and pure land-use structure. Therefore, the spatial layout and the land-use patterns are more or less similar. In terms of the on-campus interior space, some schools are merely home to necessary teaching and administrative buildings, without such public buildings as libraries and school-history museums; in most schools, stadiums, swimming pools, and so on cannot be found. With regard to the on-campus outdoor space, most schools are home to such space as the entrance square, the playground and the fragmented negative space surrounding the buildings, and there is no comfortable outdoor space for daily activities, which provides convenience and enjoyable experience for students. In some middle schools, the on-campus outdoor landscape lacks the people-oriented design, in order to meet the needs of students.

5. Conclusions

Based on the comparison and analysis of the spatial layout and land-use structure of 30 primary and middle schools in the downtown area of Tianjin, this article, proceeding from the four land-use functions of construction land, green land, sports land and other land for roads and squares concludes that the spatial layout of primary and secondary schools in Tianjin falls into three types: line-type, central-ring type, and group-type. On the basis of measuring the information entropy, equilibrium and dominance of the educational land in these schools, the comparative analysis is conducted on the diversity and equilibrium of the spatial layout and the land-use structure, in conjunction with the historical development and location characteristics of the six districts in Tianjin. The results show that universities and colleges, with different geographical locations and at different stage of development, have different information entropy values of the land-use structure, as well as the land-use diversity and equilibrium. (1) The gap in the land-use structure of primary and secondary schools in the six districts of Tianjin is not obvious, and the campus planning is under way in an orderly manner. (2) Since in Heping District the education projects have a solid ground and the district is home to quality students, teachers and schools, the campus there is more sophisticated, and the information-entropy values are more proportionate, one following another. In Hedong District, with these values far from proportionate, the campus planning is still in its infancy, and a set of well-defined and reasonable campus system has not been formed. (3) The line-type spatial layout is limited by the land conditions, and therefore the information-entropy value of the land use is small; on the contrary, the group-type space, more flexible as it is, has its moderate area and reasonable planning, and as a result, the information-entropy value is larger and the land-use structure is more balanced. Therefore, such factors as the type, construction period and geographical locations exert certain influence upon the spatial layout and the land-use structure of the educational land of primary and secondary schools.

This study, identifying the problems in the spatial layout and the land-use structure of primary and secondary schools in Tianjin, aims to provide reference for campus planning and to render decision-making basis for the intensive land utilization of these schools.

References

[1] Coulter J E, Shi L and Jenkins S. 2007 Environment as the Stage for Economic Actors. Chinese Journal of Population, Resources and Environment. p3

[2] Ma Zheng. and Wang Yongwang. 2010 Appreciating the Development Strategy of Hongqiao District based on the Comparison of Economic Indicators in the Six Districts of Tianjin. Tianjin Economy p 5

[3] Li Pengfei. 2013 Optimization of the Land-use Layout in Primary and Secondary Schools in the Downtown Area of Zhengzhou

[4] Shao Yan. 2012 Research on the Balanced Development of Basic Education in Tianjin Based on GIS

IOP Conf. Series: Earth and Environmental Science **329** (2019) 012014 doi:10.1088/1755-1315/329/1/012014

Contribution of Space Factors to Decisions on Comfort of Healthy Building Design

Qi Zhen[1], Qiong Huang[1] and Qi Zhang[1]

[1] Department of Architecture, Tianjin University, 92 Weijin Street, Tianjin, China

zhenqi@tju.edu.cn; qhuang@tju.edu.cn; zhangqi_arch@vip.163.com

Abstract. Healthy buildings are closely related to human health and comfort. Based on the literature survey, this paper explores the definition of comfort in healthy buildings. Through literature analysis and expert interviews, 16 building space factors affecting building indoor environment were identified. To introduce the building space factors into the research on the comfort evaluation of healthy buildings, 152 questionnaires were sent out and the relationship between these factors and building indoor environment comfort evaluation and their rankings were discussed. Data analysis includes the reliability analysis, normal distribution test, Pearson correlation coefficient and MANOVA. The results have shown that the perceived comfort is strongly influenced by the building space factors, and the relationship is complicated. The rankings from professionals and non-professionals were different. Several factors have been identified as key influencing factors. Besides, the perceived comfort is also related to other factors such as age, gender, etc. Such discovery primarily provides an ordering of building space design factors, which will contribute to the buildings' health, comfort and sustainable development through architectural design from an architect's perspective.

1. Introduction

The building design aims to provide a healthier and more comfortable environment for people. People stay indoors for more than 90% of the time [1], so most countries believe that the indoor environment has a major impact on human health and well-being [2][3]. Studies have shown that healthier buildings can reduce occupants' stress [4], increase productivity, and reduce health care costs by 1-5% [5]. Therefore, to determine the relationship between building environment and comfort is conducive to the design and operation of healthy buildings.

In the indoor environment, the literature review examining the comfort of building occupants mainly focuses on the impact of physical environmental conditions on overall IEQ satisfaction, such as noise, light, and temperature [6][7][8]. The influence of architectural design on the indoor environment is mainly reflected in the space characteristics, and the building space factors can interact with occupants [9]. Studies have shown that the architectural decoration and light color have little effect on thermal comfort [10][11], but the building space factors exert a significant influence on overall IEQ satisfaction [12]. Calculating the impact of spatial factors on human comfort and their importance index can help architects design healthier buildings.

1.1. Connotation and elements of healthy buildings

In 2000, the healthy building was explained at the Healthy Building Conference in Espoo, Finland. In addition to temperature, humidity, ventilation and light, the healthy building is also seen as a way to build an indoor environment, including layout, color and personal psychological needs [13]. In

October 2014, the International Well Building Institute (IWBI) in the United States released the world's first healthy building assessment standard, *The Well Building Standard V1 (WELL)*. The *WELL* is divided into seven health concepts, i.e. air, water, nutrition, light, fitness, comfort and spirit [14]. In March 2015, the standard was officially introduced into China by the Green Business Certification (GBCI) and IWBI. The Architectural Society of China released the *Healthy Building Evaluating Standard T/ASC02-2016 (HBES)* on January 6, 2017. The evaluation index system includes six categories: water, air, comfort, fitness, humanities and services [15]. The *ASHB* first defines a healthy building as a building that provides a healthier environment, facilities and services for building users, promotes physical and mental health, and improves health performance, based on the realization of building functions.

The concept of a healthy building is believed to be a 'built environment that encourages positive well-being of human beings' [16]. For decades, researchers have conducted studies to investigate the relationship between human health and the built environment. Different factors in a building affect its safety and hygiene. For example, factors such as lighting, quality of air, thermal comfort, aural comfort, colors, and textures are known to have a positive relationship with a healthy built environment [17]. Apart from these physical dimensions, some immeasurable aspects, such as aesthetics, job satisfaction, and social relationships, play important roles in the state of general well-being [18].

1.2. Comfort of healthy buildings

The comfort specified in the *WELL* focus on significantly reducing the most common causes of physiological disturbances, distraction and irritation, and largely enhancing acoustics, ergonomics, smell and thermal comfort to avoid stress and injury and improve individual comfort, work efficiency and health level. *HBES* incorporates natural light and lighting into comfort evaluation. As shown in Table 1, the concept of comfort in the standard includes three elements of indoor environmental quality (sound, heat, vision) and ergonomics. However, even if the requirements of existing standards are met, not all occupants are satisfied with the indoor environment [19].

A healthy building should provide a comfortable indoor environment for occupants, which makes them healthy, both physically and psychologically [20].The importance of indoor environment has reached a consensus among many researchers. Comfort perceived by human is disrupted by many (individual, social, and architectural) factors. Jaakola believes that different determinants simultaneously affect both human health and comfort, and sometimes the effects may be synergistic [21]. They combine to form an overall comfort experience, which is the concept of comfort that is discussed in this paper, that is, the occupants' overall satisfaction with the indoor environment.

Table 1. Comfort characteristics of WELL and HBES.

Standard	Comfort characteristics	Comfort characteristics	Standard
WELL	Accessibility design standard in *ADA*	Indoor artificial cold/heat source	HBES
	Ergonomics: vision and physiology	Ergonomics	
	Outdoor noise intrusion	Acoustic environment	
	Indoor noise	Thermal and humid environment	
	Reverberation time	Light environment	
	Sound masking		
	Silencing surface		
	Sound barrier		
	Thermal comfort		
	Independent thermal control		
	Radiant thermal comfort		
	Olfactory comfort		

1.3. Hypotheses

The purpose of this paper is to confirm the impact of building space factors on human comfort, to enrich the terms of comfort characteristics in standard of healthy buildings. Obviously, it is important for architects to make decisions to optimize building design and thus to improve space comfort. The lack of space quality standards has led architects to rely entirely on personal experience in optimizing space design, and unfortunately their preferences are not the same as those of occupants. After interviews and questionnaires, this paper discussed whether all spatial factors contribute to comfort and counted the rankings of space factors that occupants consider important. This paper also attempted to identify several factors that unrelated to space but affect the comfort experience. These factors include occupants' age, gender, etc.

We attempted to make three hypotheses:

Hypothesis 1: There are several building space factors that are important for people to evaluate whether a room is comfortable or not.

Hypothesis 2: When physical environmental conditions meet the standards, assessing whether the indoor environment is comfortable mostly depends on the spatial conditions that are of high importance to people's comfort.

Hypothesis 3: Factors unrelated to space such as age, gender, etc., also have an impact on people's overall comfort evaluation.

2. Determination of building space factors

The internal space of the building is separated from the natural space for some purpose by people with the use of a certain material and technical means. It has the closest relationship with people and has the greatest impact on people. It should have a beautiful form under the premise of meeting functional requirements, and thus meet people's mental and aesthetic needs.

Table 2. Building space factors related to comfort.

	Code	Item	Grade 1	Grade 7
Dimension	A1	Floor dimension [b]	Too small	Too large
	A2	Height of ceilings [b]	Too short	Too high
	A3	Shape of rooms [a]	Unsatisfactory	Satisfactory
Enclosure	B1	Openness of spaces [b]	Too closed	Too open
	B2	Size of windows [b]	Too small	Too large
	B3	Orientation of windows [a]	Unsatisfactory	Satisfactory
	B4	Presence of natural images [a]	Unsatisfactory	Satisfactory
	B5	Indoor/outdoor sunshades [a]	Unsatisfactory	Satisfactory
	B6	Entrance/exit location [a]	Unsatisfactory	Satisfactory
Layout and separation	C1	Separation in the room (column, mezzanine, etc.) [a]	Unsatisfactory	Satisfactory
	C2	Layout [a]	Unsatisfactory	Satisfactory
	C3	Function districts [a]	Unsatisfactory	Satisfactory
Color and texture	D1	Color of ground/wall/ceiling [a]	Unsatisfactory	Satisfactory
	D2	Material of ground/wall/ceiling [a]	Unsatisfactory	Satisfactory
	D3	Texture of ground/wall/ceiling [a]	Unsatisfactory	Satisfactory
	D4	Other decorations [a]	Unsatisfactory	Satisfactory

[a] These items have a 7-point scale: 1-being not satisfactory, 7-being satisfactory.

[b] These items are two-sided, the best score is 4, both grade 1 and 7 being most negative. For analysis, we re-encode them into a negative-positive 4-point scale (1 and 7: mark 1; 2 and 6: mark 2; 3 and 5: mark 3; 4: mark 4)

In the 21st century, study about spatial attribute perception started [22]. Most studies examined single space attribute, such as size, window-wall ratio, and color. The perception content of these space attributes includes space forms, such as spaciousness, closure, complexity, and organization, and general preferences, such as like and satisfaction [23]. Other studies have examined the relationship between certain human emotions (like pleasure, stress and anxiety) or their work efficiency and architectural design features [24]. It is believed that architectural design factors are important for human experience [25]. We summarized the building space factors discussed in the literature and interviewed 50 professionals. According to their comprehensive results, we classified the building space factors, which are divided into four categories (dimension, enclosure, layout and separation, color and texture) and a total of 16 factors, as shown in Table 2. For the convenience of further study, these impact factors were coded.

3. Methodology

3.1. Questionnaire development

The purpose of the questionnaire survey is to investigate different views on the importance of the above spatial factors, and to determine people's satisfaction with different building space factors and overall environmental satisfaction. Besides, there are people's social contact, performance and emotion, and this paper also verified their relevance. Questionnaires with a 7-point semantic differential scale were used to further analyze these identified influencing factors.

There were four main parts in the questionnaire. Section 1 was the background information of the respondents. Five questions (Table 3) were designed to gather basic information about respondents (gender, age, occupation, daily working hours and work experience) which was believed to affect the overall comfort evaluation and help ensure the authenticity and validity of the data collected at the same time. In Section 2, 16 questions corresponding to 16 spatial factors (Table 2) were designed to determine the importance of each factor affecting healthy buildings and their rankings from its response. Section 3 was about other factors, including related indoor environmental quality (IEQ) factors and personal control degree, and the respondents can add other important factors that have not been mentioned before (Table 3). Section 4 was about the space atmosphere, performance and emotion of respondents (Table 3).

Table 3. Other aspects.

Basic information	Personal control	space atmosphere	Performance	Emotions
Gender	Flexible selection of stations [c]	Privacy [d]	Work enthusiasm [d]	Fatigue [e]
Age	Sunshade [c]	Crowding [e]	Work efficiency [d]	Anxiety [e]
Occupation	Window opening [c]			Depression [e]
Working hours	Temperature [c]			Laziness [e]
Work experience	Light [c]			
	Noise [c]			
	Air quality [c]			

[c] These items have a 7-point scale: 1-being no control at all, 7-full control.
[d] These items have a 7-point scale: 1-very low, 7-very high.
[e] These items have a 7-point scale: 1-being agree, 7-being disagree.

3.2. Data collection

The questionnaires were distributed to respondents by email or QR code. According to the purpose of the questionnaire survey, the target samples from several cities in different regions such as Tianjin, Beijing, etc. were selected, including experienced professionals such as architects and non-professionals. Therefore, to determine the key factors that affect the comfort of the building space environment from their response are more scientific and reasonable.

A total of 152 questionnaires were sent out from March to April 2019, and 151 valid responses were collected with an effective response rate of 99%. The background information about the respondents is shown in Table 4.

Table 4. Distribution of the returns from different groups.

Male/Female ratio	Age		Daily working hours		Work experience (year)		Professional / non-professional ratio
1:1.07	under 18	1.99%	3~5	3.31%	Less than 1	22.52%	1:2.02
	18~25	23.18%	5~8	27.81%	1~3	33.77%	
	26~30	41.72%	8~11	58.94%	3~5	14.57%	
	31~40	14.57%	Over 11	9.93%	Over 5	29.14%	
	41~50	11.26%					
	51~60	7.28%					

3.3. analysis

Data analysis was divided into three steps. First, reliability analysis and normal distribution test were carried out. Then, Pearson correlation coefficient was used to analyze the relationship between building space factors and the overall comfort of indoor environment. Finally, the effects of other factors on indoor environmental comfort were analyzed. SPSS 22.0 was used to analyze the data and research the principal components of impact factors influencing the comfort of healthy building.

4. Results and discussion

4.1. Reliability analysis and normal distribution test

Table 5. Reliability Statistics of the factors.

Cronbach's Alpha	Cronbach's Alpha Based on Standardized Items	N of Items
0.898	0.896	16

Table 6. Normal distribution test of the factors.

		A1	A2	A3	B1	B2	B3	B4	B5
N	Valid	151	151	151	151	151	151	151	151
	Missing	0	0	0	0	0	0	0	0
Mean		3.72	3.98	4.49	3.87	3.62	4.50	4.03	4.21
Std. Deviation		1.167	1.023	1.356	1.287	1.258	1.708	1.645	1.490
Skewness		-0.154	0.002	-0.101	0.162	-0.556	-0.181	-0.006	-0.150
Std. Error of Skewness		0.197	0.197	0.197	0.197	0.197	0.197	0.197	0.197
Kurtosis		1.636	2.572	0.197	0.852	0.054	-0.676	-0.653	-0.226
Std. Error of Kurtosis		0.392	0.392	0.392	0.392	0.392	0.392	0.392	0.392
		B6	C1	C2	C3	D1	D2	D3	D4
N	Valid	151	151	151	151	151	151	151	151
	Missing	0	0	0	0	0	0	0	0
Mean		4.32	4.05	4.12	3.88	4.17	4.20	4.11	3.99
Std. Deviation		1.378	1.496	1.469	1.553	1.375	1.361	1.329	1.369
Skewness		-0.017	0.054	-0.017	0.169	0.215	0.003	0.084	-0.051
Std. Error of Skewness		0.197	0.197	0.197	0.197	0.197	0.197	0.197	0.197
Kurtosis		0.039	-0.049	0.050	-0.421	0.040	0.310	0.258	0.056
Std. Error of Kurtosis		0.392	0.392	0.392	0.392	0.392	0.392	0.392	0.392

Cronbach's alpha is often used to test the internal consistency of collected data. The reliability is acceptable if the Cronbach's alpha value is more than 0.7 [26]. Cronbach's alpha was used in this study to test the internal consistency among the impact factors. The Cronbach's coefficient was more than 0.7, as shown in Table 5.

The normality test of space comfort evaluation was conducted by using the kurtosis and skewness coefficients. As shown in Table 6, all the skewness and kurtosis coefficients except A1 and A2 were less than 1, so it could be considered that the results are approximately normally distributed.

4.2. Rankings of factors and correlation analysis

With collected data, the correlation rank of factors was calculated. As shown in the Table 7, rankings from professionals and that from non-professionals showed inconsistent performance. Layout and separation, color and texture were regarded as important to the indoor environment comfort by all the respondents. However, compared with non-professionals, professionals think that texture had little impact on comfort. Professionals believed that all three sub-factors of dimension were very important, while non-professionals believed floor dimension and height of ceilings were not so important. As far as Enclosure was concerned, the size of the window was considered important by professionals, followed by the openness of the space and interior and exterior shading measures, while non-professionals considered the openness of the space to be more important, followed by shading and window size. The reason for the difference in ranking was believed to be that professionals were more sensitive to factors such as space scale and window-wall ratio because they were engaged in architectural design. Of course, it was also influenced by personal preference. Based on the overall ranking, some factors (floor dimension / orientation of windows, etc.) were considered to have a weak relationship with the overall comfort of the room.

Pearson correlation coefficient was used to measure the relationship between the satisfaction degree of 16 factors and the overall satisfaction degree of overall comfort. As shown in the Table 8, there was a significant correlation between the evaluation of 16 factors and the overall comfort evaluation. In general, factors with obvious correlation were also considered important except four (shape of rooms, openness of spaces, size of windows and texture of ground/wall/ceiling). Dimension (A1-A3) was related to crowding degree of the room, work enthusiasm and work efficiency. Enclosure (B4-B6), layout and separation (C1-C3) and color and texture (D1-D4) were related to privacy and crowding of the room and work enthusiasm, while only B4 was related to work efficiency. Building space factors had little influence on the evaluation results of emotion.

Table 7. Correlation index and ranking of the factors.

Groups	Code	Item	Correlation index and ranking						
			Professional		non-professional		All		
Dimension	A1	Floor dimension	0.88	7	0.71	16	0.77	15	
	A2	Height of ceilings	0.92	3	0.77	13	0.82	10	*
	A3	Shape of rooms	0.86	9	0.84	4	0.85	3	*
Enclosure	B1	Openness of spaces	0.84	10	0.85	3	0.85	3	*
	B2	Size of windows	0.90	5	0.80	11	0.83	9	*
	B3	Orientation of windows	0.80	15	0.78	12	0.79	12	
	B4	Presence of natural images	0.82	13	0.77	13	0.79	12	
	B5	Indoor/outdoor sunshades	0.84	10	0.81	10	0.82	10	*
	B6	Entrance/exit location	0.82	13	0.72	15	0.75	16	
Layout and separation	C1	Separation in the room	0.96	1	0.87	1	0.90	1	*
	C2	Layout	0.92	3	0.82	9	0.85	3	*
	C3	Function districts [a]	0.90	5	0.83	7	0.85	3	*
Color and texture	D1	Color of ground/wall/ceiling	0.96	1	0.87	1	0.90	1	*
	D2	Material of ground/wall/ceiling	0.84	10	0.84	4	0.84	8	*
	D3	Texture of ground/wall/ceiling	0.72	16	0.83	7	0.79	12	
	D4	Other decorations	0.88	7	0.84	4	0.85	3	*

* Factors with high correlation.

Table 8. Pearson correlation coefficient of the factors with the overall comfort.

		A1	A2	A3	B1	B2	B3	B4	B5
Overall comfort	Pearson Correlation	.338**	.231**	.391**	.180*	.191*	.380**	.503**	.451**
	Sig. (2-tailed)	.000	.004	.000	.027	.019	.000	.000	.000
	N	151	151	151	151	151	151	151	151
		B6	C1	C2	C3	D1	D2	D3	D4
Overall comfort	Pearson Correlation	.506**	.614**	.642**	.600**	.564**	.536**	.550**	.518**
	Sig. (2-tailed)	.000	.000	.000	.000	.000	.000	.000	.000
	N	151	151	151	151	151	151	151	151

** Correlation is significant at the 0.01 level (2-tailed).
* Correlation is significant at the 0.05 level (2-tailed).

4.3. Evaluation of factors unrelated to space

Multivariate Analysis of Variance was used to analyze the effect of factors unrelated to space (age, gender, etc.) on the overall comfort evaluation. As shown in Table 9, all the eleven factors had little influence on the evaluation of overall comfort in the statistical sense. However, after statistical analysis, compared with women, men had a higher tolerance for the environment. With the increase of age, daily working hours and work experience, the evaluation of environmental comfort decreased.

Table 9. MANOVA of the factors that affect the degree of satisfaction with the overall comfort.

Source	Type III Sum of Squares	Mean Square	F	Sig.	Partial Eta Squared
Gender	1.538	1.538	1.118	.292	.008
Age	13.511	2.702	1.965	.088	.067
Daily working hours	9.662	3.221	2.343	.076	.049
Work experience	4.711	1.570	1.142	.334	.024
Flexible selection of stations	9.579	1.368	1.317	.250	.083
Control of sunshade	6.370	1.062	1.021	.416	.057
Control of window opening	2.577	.368	.354	.926	.024
Control of temperature	7.744	1.106	1.064	.392	.068
Control of light	5.682	.812	.781	.605	.051
Control of noise	1.925	.275	.265	.966	.018
Control of air quality	9.074	1.296	1.247	.284	.079

5. Conclusion

In this study, questionnaire survey was used to check whether the building space factors had influence on comfort evaluation, thus contributing to the comfort decision-making in healthy building design. The results proved the three previous hypotheses. First, 11 of the 16 spatial factors selected had significant influence on the overall comfort evaluation. Secondly, in general, when the physical environmental conditions met the standards, the factors considered to be more important had a higher correlation with overall comfort. Some of the anomalies were considered to come from three reasons: professional experience, sample size, and personal preference. Finally, there were some factors (personal characteristics, personal control) that interfere with comfort assessment, although not obviously.

In this study, only questionnaires were used for investigation and analysis, the indoor environment quality of the buildings was not specifically measured, which was more convenient in the laboratory. Therefore, it is suggested that future research can quantify the impact of building space factors on comfort evaluation based on laboratory studies.

Acknowledgments
This research was financially supported by the National Key Research and Development Program of China (Grant No. 2016YFC0700200).

References

[1] Hidayetoglu M L, Yildirim K and Cagatay K 2010 The effects of training and spatial experience on the perception of the interior of buildings with a high level of complexity *Sci. Res. Essays* 5 (5) pp 428–39
[2] CEC 2004 The European environment and health action plan 2004-2010 COM (2004) 416 final vol I SEC (2004)729
[3] WHO 2004 Regional office for Europe Declaration EU/04/5046267/6
[4] Salleh M R 2008 Life event, stress, and illness *Malays. J. Med. Sci.* **15** (4) pp 9–18
[5] Morton B and Ramos J 2014 The drive toward healthier buildings 2014: the market drivers and impact of building design and construction on occupant health, well-being and productivity *Smart Mark. Rep.* Dodge Data Anal
[6] Choi J H and Aziz A 2009 Loftness V. Decision support for improving occupant environmental satisfaction in office buildings: The relationship between subset of IEQ satisfaction and overall environmental satisfaction Proceedings of the 9th International Conference Healthy Buildings *Syracuse NY USA paper* p 747
[7] Lai A C K, Mui K W, Wong L T and Law L Y 2009 An evaluation model for indoor environmental quality (IEQ) acceptance in residential buildings. *Energy Build* 41(9) pp 930-6
[8] Astolfi A and Pellerey F 2008 Subjective and objective assessment of acoustical and overall environmental quality in secondary school classrooms. *J Acoust Soc Am* 123(1) pp 163-73
[9] Van Leeuwen J P and Wagter H 1997 *Architectural design-by-features* in: R. Junge (Ed.) CAAD Futures Springer Dordrecht
[10] Fanger P O, Breum NO and Jerking E 1997 Can colour and noise influence man's thermal comfort? *Ergonomics* **20**(1) pp 11-8
[11] Rohles F H and Wells W V 1977 The role of environmental antecedents on subsequent thermal comfort *ASHRAE Trans* **83**(2) pp 21-9
[12] Xu H, Huang Q and Zhang Q 2018 A study and application of the degree of satisfaction with indoor environmental quality involving a building space factor *Building and Environment* 143 pp 227–39
[13] Loftness V, Hakkinen B, Adan O and Nevalainen A 2007 Elements that contribute to healthy building design *Environmental Health Perspectives* 115(6) pp 965–70
[14] The WELL building standard. v1 with Q2 2017. [EB /OL]
[15] Wang Q Q, Meng C and Li G Z 2017 Introduction to the preparation of T/ASC 02—2016 Assessment Standard for Healthy Building *Building Science* **33**(02) pp 163-6
[16] Ho D C W, et al. 2004 Assessing the health and hygiene performance of apartment buildings *Facilities* **22** (3/4) pp 58–69
[17] Rousseau D and Wasley J 1997 *Healthy by design, Building and Remodeling Solutions for Creating Healthy Homes* Hartley & Marks Publishers Inc. WA
[18] Samuelsson L 2000 Quality assurance of the indoor environment in schools *Offices and dwellings*
[19] Frontczak M and Wargocki P 2011 Literature survey on how different factors influence human comfort in indoor environments *Building and Environment* 46(4) pp 922–37
[20] Grawitch M J and Ballard D W 2016 *The Psychologically Healthy Workplace: Building a Win-win Environment for Organizations and Employees* American Psycho- logical Association
[21] Jaakola J J K 1998 The office environment model: a conceptual analysis of the sick building syndrome *Indoor Air Journal (suppl. 4)* pp 7-16
[22] Stamps A E III 2010 Effects of permeability on perceived enclosure and spaciousness *Environ. Behav.* 42 (6) pp 864–86
[23] Stamps A E III 2011 Effects of boundary height and horizontal size within boundary on

perceived enclosure *Percept. Mot. Skills* 113 (3) pp 995–8

[24] Kwallek N, Lewis C M and Robbins A S 1988 Effects of office interior color on workers' mood and productivity *Percept. Mot. Skills* 66 pp 123–8

[25] Ergan S, Shi Z and Yu X 2018 Towards quantifying human experience in the built environment: A crowdsourcing based experiment to identify influential architectural design features *Journal of Building Engineering* 20(May) pp 51–9

[26] Aigbavboa C and Thwala W 2013 Confirmatory Factor Analysis of Neighbourhood Features Amongst South African Low-income Housing Occupants *American Society of Civil Engineers* pp 1107-22

IOP Conf. Series: Earth and Environmental Science **329** (2019) 012015 doi:10.1088/1755-1315/329/1/012015

Study concerning the conception and construction of Long-Lasting Flexible Pavements

M Condurat[1] and C Ursanu[2]

[1] Regional Directorate of Roads and Bridges Iasi (D.R.D.P. Iasi), Romanian National Company of Road Infrastructure Management (C.N.A.I.R. S.A Romania)
[2] Astaldi Spa Italy, Romania Branch, Bucharest Office

Abstract. The paper presents the current state of Long Lasting Flexible road pavement design, which is characterized by permanent development, both in terms of the calculation scheme that can reproduce as accurately as possible the actual operating behaviour and also in terms of more efficient materials usage. The research has shown that these types of flexible, well-designed and executed road pavements could have a longer lifetime than classical ones. In the same context, the energy performances and the Global Warming Potential, related to the construction and maintenance of Long-Lasting Flexible Pavements, calculated using a Life Cycle Assessment approach, are given, in order to substantiate their ecological advantages compared with a classical road pavement. This paper emphasizes the necessity to adopt appropriate constructions alternatives for road pavements, also within the initial construction phase, as well as within their maintenance and intervention stages, in order to have a sustainable build environment. In conclusion, roadwork and recent research have shown that Sustainable road pavements, known also as Long Lasting Pavements could be designed and built, resulting in significant material savings and, therefore, low maintenance costs with high performance and significantly longer service life compared to classical road pavements.

1. Introduction

The transportation system represents a significant key component of the current development process. Besides the positive effects correlated with the continuous development of transport system a wide diversity of, mostly negative impacts are also produced. Taking into account the actual high levels of pollutions worldwide, the current goal is to achieve a sustainable development for the transport systems. Given the accelerated increase of vehicle fleet and the extensive development of the road network, the paper concentrates on the construction of road transport infrastructure. The development processes within this area implies the productions of a wide range of externalities. These refer to ecological impact, lithographic materials depletion, energy and fuel consumption or ecosystem equilibrium disruption, social and economic impact [1]. In order to tackle these externalities, sustainable and eco-friendly transportation policies and practices have to be promoted worldwide. Recent research has shown a direct link between transport development and climate change, high levels of pollution and depletion of lithographic resources. In this context, it is essential to make changes, take appropriate measures and promote low carbon mobility, given that these are fundamental features for a sustainable future and competitive urban areas. Taking into account the actual high levels of pollution and the fact that, based on the studies performed by the European Commission, transport activities have been proved to be responsible for almost 32% of Europe's energy consumption and 28% of total CO_2 emissions [2], the development of sustainable road construction technologies and processes has become more and more important.

In addition to the initial construction of the road pavement, the maintenance and rehabilitation phases, are major factors in the increase of greenhouse effect. The aim of this paper is to study the approaches, which could be taken for the transition to low carbon mobility, in regards with the new technologies, which support this transition, namely the construction of Long-Lasting Flexible Pavement. The results obtained are envisaged to be assimilated and implemented in the actual policies and strategies adopted for the North East Region of Romania.

Low carbon mobility is characterized through specific actions correlated to the transportation system, which results in significantly reduced consumption of non-renewable materials and smaller quantities of pollutant emissions released in atmosphere and also, to an increased service life. These actions or measures refer to the new technologies and processes encouraging this transition, specific policies developed for their implementation, as well as the human behaviour and daily personal choices [3].

Thus, the paper presents the new developments correlated to the transport infrastructure towards low carbon mobility, which consist in designing and construction of Long-Lasting Pavements, road pavements characterized by an initial service life of forty or more years. These road pavements are a safer, an efficient and more environmentally friendly road structures alternatives and are capable to withstand unusual actions without suffering major faults and without losing functionality [3]. Furthermore, they are characterized by low maintenance costs with high performance and significantly longer service life compared to classical road pavements. These new concepts and construction technologies, developed recently, in the frame of Technical University Gheorghe Asachi of Iasi, such as Long Lasting Flexible Pavements (LLFP) [4], [5] involving new concepts of road pavements, will extend the life of asphalt roads from twenty-five to fifty years, thus contributing to significantly reduced costs and emissions, during the construction and maintenance processes. In Romania, the design and the construction of durable pavements, started with the direct involvement the Strategic Highway Research Program [6] designed to evaluate the performance of Romanian pavements and to enhance its pavement performance prediction models, being over 30 LTPP test section in the Romanian program, including asphalt and concrete pavements situated in specific rural and urban locations. One of such RO-LTPP sector, considered as a typical Romanian pavement with extended life, its envisaged design life being of forty to fifty years, is localized on the national road DN 12A – Miercurea Ciuc – Onesti, km. 30+000 – 30+150 and it has been built twenty-seven years ago [7]. Another example of a Long-Lasting Pavement constructed in the North East Region of Romania is represented by the sector build on the national road DN 17 Suceava – Vatra Dornei using a sustainable rigid pavement reinforced with steel fibers, recovered from post-consumed tires, as a part of the EU collaborative research project EcoLanes [3].

As a response to the actual ecologic and economic problems arisen from the use of fossil fuel and lithographic materials, the Long-Lasting Flexible Pavements contributes to the following:

- An efficient and effective transport network to make the North East Region of Romania internationally competitive;
- Urban congestion solutions;
- Increased service life and low costs for maintenance in addition to materials saving and overall a smaller carbon footprint.

The implementation of Long-Lasting Flexible Pavement could be achieved progressively starting with research and experimental road pavement construction (end user) and moving forward to the regional level. This could be performed based on appropriate management decisions through close collaboration between research institutes, technology progress, decision-makers and people, the development of norms and regulations, and training courses regarding the design, construction and maintenance of these types of road pavements. At regional level, the low carbon mobility can be implemented through research programmes, standards and accreditation for carbon reduction products and services regulations, experimental road pavement construction, planning guidance and incentives.

IOP Conf. Series: Earth and Environmental Science **329** (2019) 012015 doi:10.1088/1755-1315/329/1/012015

2. The design of Long-Lasting Flexible Pavements

The current state of flexible road pavement design is characterized by permanent development, both in terms of calculation scheme that can reproduce as accurately as possible the actual operating behaviour and in terms of more efficient usage of materials.

The main role of road pavements is to support the traffic loads and to transmit them to the foundation soil without suffering irreversible deformations during their exploitation period, followed by distresses that affect their functionality. Consequently, the road pavement design requires to establish the stresses and strains of the actions to which the structure is subjected in exploitation in order to assess the structural response and to ensure high reliability. Compared to other types of construction, road pavements have the particularity of being designed to fail after a certain period of operation of the road. Nowadays, due to the possibilities offered by automated methods of road pavement design, a number of simplifications can be used. The calculation objectives are to determine the strains and displacements in a spatial computing model, taking into account both the non-linear behaviour of the materials determined by experimental tests and the degree of cooperation/adhesion between the component layers of the road pavement. It also takes into account the dynamic loads, the optimization of the road pavement as a whole and its component elements [8].

Sustainable road pavement ("Long Lasting" or "Perpetual Pavements") have been designed since the 1960s. Research has shown that these types of flexible, well-designed and executed road pavements, can have a longer lifespan than conventional ones [9]. Many of these road pavements are the product of "full-depth" design of asphalt pavement that are built directly on foundation or "deep-strength" soil that are placed on a layer of natural aggregates and it is very important to know that they carry a very high traffic [10]. One of the main advantages of this road pavement is that the whole section of the structure is smaller than those commonly used, hence, the material consumption is significantly reduced. Recent advances in material selection, mix design, performance testing, and road pavement design have enabled the development of an appropriate methodology for sustainable road pavements with an increased technical performance and a lifetime of about 50 years [11], [12], [13].

3. Composition of flexible road structures

Flexible road pavements are road structures in which there are no layers containing hydraulic or puzzolanic binders, and the road wearing course is bituminous. The composition of the road pavement will be determined taking into account the following factors:

- the minimum thicknesses of the different layers of the road pavement;
- the maximum thicknesses of the different road layers, taking into consideration certain constraints specific to the execution technologies;
- reducing the number of layers and interfaces, in order to minimize the risk of adhesion layers failures;
- establishing the shape of the layer so that its thickness can be taken into account in the road pavement design regarding the action at the freeze-thaw phenomenon.

4. Case study I. Sustainable road pavement design. Intermediate solutions made from asphalt macadam, respectively, asphalt mix AB2

4.1. Scheme of the research program

The following types of road pavements have been studied, as follows:

A) The Alternative 1 takes into consideration the design of a sustainable road pavement having as an intermediate layer, the asphalt macadam, which has the asphalt courses as shown in figure 1.

According Alternative 1, the Sustainable road pavement is composed of a stone matrix asphalt with the maximum size of aggregates of 16 mm (SMA16) and an asphalt macadam course laid on another stone matrix asphalt layer, supported by a ballast foundation course, which is designed on an improved subgrade course, as shown in figure 1.

B) The Alternative 2 analyses a sustainable road pavement with an intermediate layer of asphalt mix AB2, as shown in figure 1, presented below.

According Alternative 2, in this case, the Sustainable road pavement is composed of a stone matrix asphalt and an asphalt mix AB2 course laid on another stone matrix asphalt layer, supported by a ballast foundation course, which is designed on an improved subgrade course, as shown in figure 1.

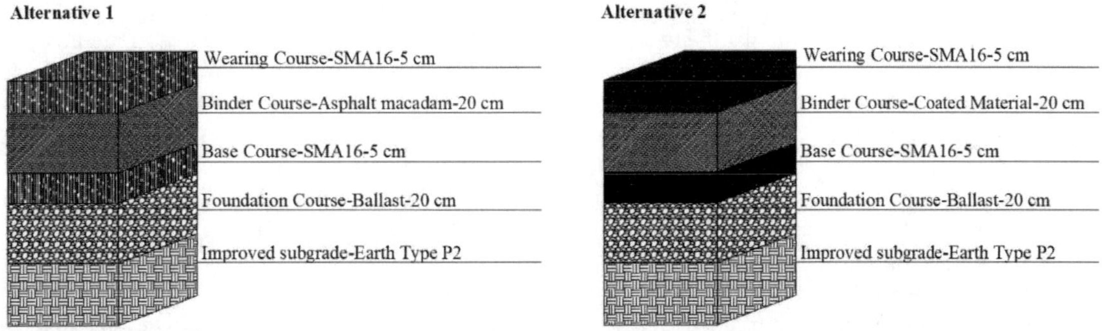

Figure 1. Sustainable road pavements with: A) asphalt macadam as an intermediate layer and B) asphalt mix AB2 as an intermediate layer

4.2. Assessment of the ecological impact associated with a Long-Lasting Flexible Pavement

In order to assess the ecological performances of Long-Lasting Flexible Pavement, a Life Cycle Assessment Analysis, incorporated in the asPECT software, developed by TRL Laboratories U.K., has been performed on the Alternatives previously presented.

asPECT software, Version 3.1, developed by Transport Research Laboratory - TRL UK, gives a methodology for computing greenhouse gas emissions on the pavement life cycle correlated with using asphalt materials on roads construction. The asPECT software enables the assessment of pollutant emissions based on data collected concerning materials, transport and mixture plant characteristics [14]. The software database contains the necessary formulas and emission factors for calculating CO_2e emissions associated with the production, laying and maintenance of bituminous layers by taking into consideration all the stages of materials and energy production and all the processes from raw material extraction, production, transport and use phase of the asphaltic mixture to the end of their life, shown in figure 2.

The process for calculating the carbon footprint associated with flexible road pavement consists in three main phases [3], namely:

- The introduction of raw materials used in the asphalt mixture (total annual energy consumption for the acquisition, broken down by type of fuel and operation);
- Data introduction regarding the asphalt mixes plant characteristics (plant type, annual production, energy consumption and asphalt mix composition);
- Data introduction regarding installation of bituminous mix and visualization of the results.

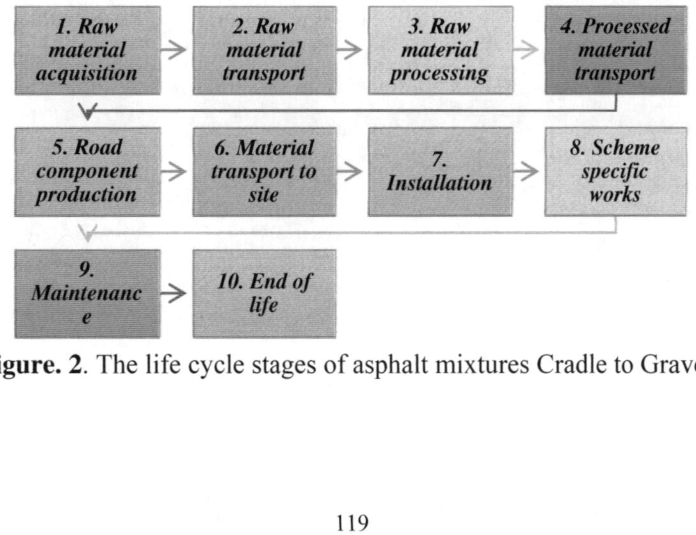

Figure. 2. The life cycle stages of asphalt mixtures Cradle to Grave [14]

The Life Cycle Assessment Analysis has been conducted on a built section of 1000 ml long and 7.0 m wide based a CRADLE TO GRAVE perspective, presented above, in figure 2.

4.3. Results obtained

A) Alternative 1. Sustainable road pavement with intermediate layer - Asphalt Macadam

For this assumption, it is found that the two verification criteria, namely the admissible horizontal tensile stress at the bottom of the bituminous layers and the vertical strain at the level of subgrade are met for medium and heavy traffic categories. For these criteria to be fulfilled for the other types of computational traffic categories, such as very heavy and exceptional traffic, it would be necessary to modify the proposed sustainable road structure by increasing the total thickness of the asphalt layers from 30 cm to higher values, namely 34 cm, either retaining the initial thicknesses for the SMA layers and increasing the thickness of the intermediate layer to 24 cm, either by retaining the thickness of the intermediate layer and correspondingly increasing the thickness of the SMA layers. Another change leading to the verification of the two criteria for the very heavy and exceptional categories of traffic could be the use of materials characterized by a high elasticity (SMA asphalt mixes for the upper and lower layers of the durable road pavement with bitumen modified with polymers).

In terms of ecological impact associated with the construction of a Long-Lasting Flexible Pavement, using Alternative 1, the results of the quantitative assessment of greenhouse gas emissions, are presented in Table 1.

Table 1. CO_2e emissions associated with Alternative 1: Sustainable road pavements with asphalt macadam as an intermediate layer

No.	Life Cycle Stage	kg CO_2e/t	Total kg CO_2e
1-3	Material extraction and processing	14.91	75152.93
4	Transport to plant	28.08	141515.24
5	Asphalt production	42.28	213081.12
6	Transport to site	40.23	202748.00
7	Laying and compacting	4.70	23688.00
8	Project works	106.46	15969.64
9	Maintenance	34.00	5100.00
10	Reconstruction	14.80	74607.00
Project results summary (5340.00 t material)		**141.79**	**751861.93**

B) Alternative 2 - Sustainable road pavement with the intermediate layer of an asphalt mix AB2

For this alternative, given their modulus of deformation significantly greater than the asphalt macadam, the study has been conducted in a similar manner. For the subgrade type P2, as expected, given the higher elasticity modulus of the asphalt mix AB2 it has been found that the two verification criteria are met.

From an ecological perspective, the results of the quantitative assessment of greenhouse gas emissions associated with Alternative 2, are presented in Table 2.

IOP Conf. Series: Earth and Environmental Science **329** (2019) 012015 doi:10.1088/1755-1315/329/1/012015

Table 2. CO_2e emissions associated with Alternative 2 - Sustainable road pavements with asphalt mix AB2 as an intermediate layer

No.	Life Cycle Stage	kg CO_2e/t	Total kg CO_2e
1-3	Material extraction and processing	49.50	232178.28
4	Transport to plant	36.49	171120.49
5	Asphalt production	42.28	198283.82
6	Transport to site	30.17	141501.21
7	Laying and compacting	4.70	22043.00
8	Project works	106.46	15969.64
9	Maintenance	34.00	5100.00
10	Reconstruction	16.76	78623.00
Project results summary (4990.00 t material)		**175.38**	**864819.44**

In order to justify the ecological benefits associated with the construction of Long-Lasting Flexible Pavements, a classical road pavement has been proper analysed, taking into consideration the same assumptions defined for the two alternatives studied above. The difference between the classical road pavement and the sustainable versions consist in the fact that the classical ones are usually designed over an initial service life of 15 years. In this regard, a flexible road pavement consisting in the following layers, has been analysed:

- ✓ Wearing course (asphalt concrete with the maximum size of the aggregate of 16 mm, 4 cm);
- ✓ Binder course (asphalt concrete with the maximum size of the aggregate of 25 mm, 6 cm);
- ✓ Base course (asphalt base, 15 cm);
- ✓ Foundation layer (ballast, 20 cm);
- ✓ Subgrade (20 cm).

The results of the quantitative assessment of greenhouse gas emissions associated with road construction using a classical alternative are given in Table 3.

Table 3. CO_2e emissions associated with a classical road pavement over its initial service life

No.	Life Cycle Stage	kg CO_2e/t	Total kg CO_2e
1-3	Material extraction and processing	52.88	203434.62
4	Transport to plant	36.52	140500.34
5	Asphalt production	48.28	162643.47
6	Transport to site	6.48	24946.30
7	Laying and compacting	4.70	18080.90
8	Project works	241.23	36184.10
9	Maintenance	34.27	5140.00
10	Reconstruction	15.89	61120.78
Project results summary (4147.00 t mixture)		**154.08**	**652050.51**

In order to be properly compared these case studies, both from an ecological and technical perspectives, a structural overlay has been taken into account, which is envisaged to be applied at the end of initial service life of the classical road pavement, which will extend the service life of the road pavement and increase its mechanical resistance. Structural overlays represent a current road repair work which is meant to increase the pavement load bearing capacity. The bituminous road pavement is composed from two asphalt layers with a minimum thickness of 8 cm. This reinforcement strategy can be performed using two alternatives, namely: two-layer bituminous road pavement with a thickness of less than 13 cm or a road pavement with two-layer bituminous wearing course and an asphalt base course with a thickness greater than 13 cm, but not exceeding 18 cm. For the assessment of the ecological impact associated with the structural overlays, the reinforcement layers have been designed using the first assumptions (asphalt wearing course of 4 cm and a binder course of 5 cm) and the associated quantities of CO_2e are presented in Table 4. Also, the final results, calculated over the same service life as a Long Fasting Flexible Pavement, are given in Table 5.

Table 4. CO_2e emissions associated with structural overlays

No.	Life Cycle Stage	kg CO2e/t	Total kg CO2e
1-3	Material extraction and processing	44.76	64070.37
4	Transport to plant	35.41	50687.96
5	Asphalt production	42.28	60520.96
6	Transport to site	30.17	43189.55
7	Laying and compacting	4.70	6728.05
8	Project works	0.00	0.00
9	Maintenance	34.27	5140.00
10	Reconstruction	16.91	24206.05
Project results summary (1581.50 t mixture)		**160.95**	**254542.93**

Table 5. Final results of CO_2e emissions associated with a classical road pavement over its total service life

No.	Life Cycle Stage	kg CO_2e/t	Total kg CO_2e
1-3	Material extraction and processing	97.64	267504.99
4	Transport to plant	71.93	191188.30
5	Asphalt production	90.11	144518.35
6	Transport to site	36.65	68135.85
7	Laying and compacting	9.40	24808.95
8	Project works	241.23	36184.10
9	Maintenance	68.54	10280.00
10	Reconstruction	32.80	85326.83
Project results summary (5718.50 t mixture)		**315.03**	**906593.44**

5. Conclusions

The most challenging environmental issue facing humankind nowadays is represented by the acceleration of global warming and climate change associated with this phenomenon as a direct result of an increase quantities of pollutants emanated in the atmosphere. In this context, the paper presents the results of recent research undertaken for the assessment of the environmental indicators, associated with the construction and maintaining of an asphalt road pavement, based on a Cradle to Grave perspective, which takes into account all the stages of materials and energy production and also all the processes from lithographic material extraction, production, transportation and use phase of the products to their end of life. The Life Cycle Assessment Analysis has been conducted on a built road section of 1000 ml long and 7.0 m wide using a Cradle to Grave approach, incorporated in the asPECT software.

Within this paper two specific alternatives for the construction of a Long Lasting Flexible Pavement using different materials, have been analysed in an environmental perspective as well as a technical approach. These type as well-designed structures are characterized by an initial service life of forty or more years. The Long Lasting Flexible Pavements are safer, more efficient and environmentally friendly road pavements alternatives and are capable to withstand unusual actions without suffering major faults and without losing functionality [3].

The first alternative deals with the construction of a durable pavement, composed by a stone matrix asphalt wearing course and an asphalt macadam layer laid on another stone matrix asphalt layer, supported by a ballast foundation course, which is designed on an improved subgrade course. The alternative 2 analyses a sustainable road pavement with an intermediate layer of asphalt mix AB2. From a technical point of view both alternatives met the verification criteria used in the design phase. In terms of environmental perspective, the first alternative represents a more sustainable choice for the construction of the road pavement, with a reduction of the total quantities of CO_2e emissions of 15.02%, given the fact that the asphalt macadam requires smaller quantities of bitumen in the mix, while providing an increased bearing capacity.

When comparing the most ecological alternative of Long Lasting Flexible Pavement, namely Alternative 1, with a classical road pavement, one may notice that, in this case, Alternative 1 has an increased value of pollutant emission, with 15.31% higher than the classical version of road pavement. The difference is explained through the fact that Long Lasting Flexible Pavements are designed with an initial service life of forty or more years, compared with the classical one, where its service life is 15 years. In order to proper compare these two technical solutions, specific strategies for regular maintenance and structural reinforcement of the classical road flexible pavement have to be performed. In this context, a structural overlay, meant to extend the road service life, has been taken into account within the study. When the total service life is considered, one may notice that a Long Lasting Flexible Pavements, besides low maintenance costs and works, represents a more sustainable alternative for the construction, rehabilitation and modernisation of road transport infrastructure, with a total reduction of CO_2e emissions of 20.58%.

Therefore, looking at the results under an engineering and ecological consideration, the optimal construction strategy to be applied, consists in a sustainable road pavement with intermediate layer - asphalt macadam. The worst-case scenario studied in the frame of this research is represented by the construction of a classical pavement, where the specific intervention strategies ought to be applied when the pavement condition of the road is poor and the user's safety is endangered. In this case the only viable solution is reinforcing the pavement, the ecological impact being significantly higher.

In the context of improving the regions sustainably, the Long-Lasting Flexible Pavement represents a step forward achieving low carbon mobility based on their initial service life of forty or more years, low maintenance costs and reduced energy and materials consumptions. These road pavements represent a safer, an efficient and a more environmentally friendly alternatives of road structures and are capable to withstand unusual actions without suffering major faults and without losing their functionality. Moreover, it is imperative for the decision-making factors to have a decisive role in implementing these technologies in practice for fulfilling the sustainability criteria.

References

[1] Condurat, M., Nicuţă, A.M., Andrei, R. 2016. *Environmental Impact of Road Transport Traffic. A Case Study for County of Iaşi Road Network.* Proceedings of the 10th International Conference on Interdisciplinarity in Engineering, INTER-ENG 2016, Tg. Mureş, Volume 181, 2017, Pages 123-130, ISSN 1877-7058, Elsevier Ltd

[2] Eurostat. Statistics explained Consumption of energy http://ec.europa.eu/eurostat/statisticsexplained/index.php/Consumption_of_energy

[3] Condurat M and Patterson J 2016 *Welsh and Romanian policies for transition towards low carbon mobility* (Cardiff, UK, February 2016 Proceedings of the International Conference "Smart Energy Regions') pp. 59-70, ISBN: 978-1-899895-23-6

[4] Tanasele I 2012 *Conception and Design of Long Lasting Flexible Road Pavements* (PhD Thesis. Iasi: unpublished Technical University Gh. Asachi of Iasi, Faculty of Civil Engineering and Building Services) (in Romanian)

[5] Ursanu C 2014 *Studiul privind dimensionarea structurala a unor imrbacaminti rutiere flexibile durabile* (Disertation. Iasi: unpublished Technical University Gh. Asachi of Iasi, Faculty of Civil Engineering and Building Services). (in Romanian)

[6] SHRP 1994 *The Superpave Mix Design Manual for new Constructions and Overlays SHRP Report A-410.* [Online] Available at: http://onlinepubs.trb.org/onlinepubs/shrp/shrp-a-410.pdf

[7] Boboc V et al 2002 *Realizarea investigatiilor RO-LTTP de teren periodice in raza DRDP Brasov, A44/T4 Research Report* (Iasi, Technical University "Gh. Asachi"). (in Romanian).

[8] Garber N J Hoel L A *Traffic And Highway Engineering, Chapter 20 Design of Flexible Pavements* PWS Publishing, ISBN 0-534-36537-X, 1999, Page 925-989
[9] Newcomb D E and K R Hansen 2006 *Mix Type Selection for Perpetual Pavements. Proceedings. International Conference on Perpetual Pavements* (Ohio University, Columbus)
[10] Newcomb D E, M Buncher and I J Huddleston 2000 *Concepts of Perpetual Pavements. Transportation Research Circular No. 503. Perpetual Bituminous Pavements. Transportation Research Board* (Washington, DC)
[11] Timm D H, D E Newcomb and I Selvaraj. 2006 *A Practical Guide to Low-Volume Road Perpetual Pavement Design. Proceedings. Intl. Conf. on Perpetual Pavements* (Ohio University, Columbus)
[12] Thompson M R and S H Carpenter 2004 *Design Principles for Long Lasting HMA Pavements. Proceedings. Intl. Symp. on Design and Construction of Long Lasting Asphalt Pavements* (National Center for Asphalt Technology. Auburn University, Alabama) pp. 365-384
[13] Tarefoler, Rafiqul and Damien Bateman 2009 De*termining the Optional Perpetual Pavement Structure. Proceedings. Intl. Conf. on Perpetual Pavements* (Ohio University, Columbus) CD-ROM.
[14] TRL. *Asphalt Pavement Embodied Carbon Tool.* Transport Research Laboratory. online at http://www.sustainabilityofhighways.org.uk, 2014.

IOP Conf. Series: Earth and Environmental Science **329** (2019) 012016 doi:10.1088/1755-1315/329/1/012016

Urban building energy modelling and urban design for sustainable neighbourhood development-A China perspective

Feng Yang and Zhidian Jiang

College of Architecture and Urban Planning, Tongji University, 1239, Siping Road, Shanghai, 200092, China

yangfeng@tongji.edu.cn

Abstract. Urban design at the neighbourhood scale has profound effects on urban microclimate, and thus on urban buildings energy consumption (in particular heating, cooling and lighting), extent of applicability of passive heating/cooling strategies, indoor/outdoor thermal comfort and street vehicle emissions dispersal (air quality). Previous studies, especially recent explorations in the emerging field of urban building energy modelling (UBEM), suggest clear neighbourhood-scale energy impact from building type, density, layout/orientation and facade/fenestration treatment. The purpose of research is to inform design, but how the research findings in urban building energy inform urban design practice is not straight-forward, in that real-world urban development involves all stakeholders and needs to take environmental, social and economical factors into account, in addition to energy and urban climate. If to support more healthy, comfortable and energy-efficient urban district/ neighbourhood is one goal of green neighbourhood development (ND) rating tools, we should investigate possibilities to further and more efficiently integrate findings of UBEM into guidelines and credits system that inform climate-responsive urban design. This paper, based on a review on recent advancement in UBEM, investigates a more comprehensive and systematic prescriptive approach in ND to optimizing urban building energy performance by moderating urban form, fabric and land cover. Although UBEM as a scientific tool is diagnostic in evaluating urban design scenarios on a case basis, prescriptive design guidelines based on good science are suitable and useful for architects in integrating urban building energy issues in practice.

1. Introduction

Energy consumption in the building sector of China currently accounts for ~25% of the overall national energy consumption (not included is the energy used for construction material, which is included in the industrial sector). However, due to continuous urbanization and improvement on living standards, this component is increasing rapidly, closing the gap between China and developed countries.

While energy conservation for individual buildings has been (relatively) well studied and codes and standards for low-energy design has been issued and implemented in China, urban building energy conservation has been receiving increasing research interests. Because it has been increasingly aware that urban form of contemporary large cities, through affecting sun and wind, has clear effects on urban building operational energy usage. Although the effect on a single building may be small, the aggregation of effects at district/urban scale can be significant. For low-energy urban design and urban energy management, it is important to develop useful tools for modelling urban building energy.

IOP Conf. Series: Earth and Environmental Science **329** (2019) 012016 doi:10.1088/1755-1315/329/1/012016

Aimed at urban building energy efficiency and passive and climate-responsive urban design, this paper tries to examine the technical advancement in urban building energy modeling (UBEM) and the potentials it has in supporting more effective application of sustainable neighborhood development (ND) rating tools.

2. Urban building energy modelling

Urban design works on tens to thousands buildings. Accordingly, the energy modelling on the scale, i.e., urban district, neighborhood down to street block level, is considered urban building energy modelling (UBEM) [1].

Generally, UBEM can be divided into physics-based modeling and data-driven modeling (black or grey box) modeling. Below is a brief review on some of the published tools developed either specifically or with a clear and strong focus on UBEM.

2.1. A brief review of UBEM tools

2.1.1. Light and thermal (LT) method [2]: Developed by Nick Baker et al. Cambridge, LT method is innovative in association of building form/fabric (passive zone, orientation, w/w/ratio, obstruction angle etc.) with heating/cooling/lighting energy demand firstly through a set of correlations (visualized as a set of energy curves) and was later embedded in DEM-based image processing extension of GIS platform to estimate urban energy demand [3]. LT uses stationary modelling based on a set of assumptions. It is not a transient thermos-dynamic model, and may not work very well for warm/hot climates under which cooling is important building energy demand.

2.1.2. CitySIM [4]: developed by Robinson from his previous integrated urban sustainability early-stage assessment model SUNtool [5], and focuses on urban building energy.

2.1.3. Urban modelling interface (UMI)[6]: developed by Reinhart et al., at MIT, UMI is a Rhino plug-in to connect the CAD software with EnergyPlus simulation engine. It models operational energy, transportation (walkability/bikability) and daylight availability at the neighborhood scale. Essentially a multi-zone dynamic model, it quantifies solar/daylight effect of urban building adjacency based on a series of technical innovations such as a “shoe boxer” algorithm[7] an “AutoZoner” algorithm [8], to extrapolate for the energy use intensity (EUI) of all buildings in a district/neighbourhood.

2.1.4. Building block energy estimation (BBEE) [9]: developed by Yanda et al. from Tsinghua University, BBEE is inspired by the idea of Shoeboxer, and has two components-a BBEE algorithm and an energy database. Suitable for heating-dominated climate, the application would have considerable error when modelling under warm/hot climate, e.g., 27% biases can occur for Guangzhou.

2.1.5. EEP model [10]: The EEP and the followed two tools are all based on GIS platform and aimed as city-scale energy auditing tools. EEP aims for the whole city and uses GIS for compiling and organizing surveyed urban data and estimating building stock energy usage by stationary model SAP. GIS platform greatly facilitate the workflows of urban and building attributes information needed to carry out segmentation and characterization (archetyping) [11].

2.1.6. City energy analyst (CEA) [12], built as an extension of a GIS platform. It allows analysing the energy, carbon and financial benefits of urban design scenarios.

2.1.7. City building energy saver (CityBES) [13] is also GIS-based, and aims to support urban energy management.

2.1.8. Urban energy index for buildings (UEIB) [14]: The UEIB uses a unique form-based methodology, to abstract urban form while retaining its essential characteristics of energy demand.

According to the author it provide an innovative approach in addition to the extrapolation of thermal-physics modelling and statistical inference. It reduces actual urban districts into a representative notional grid by key morphological parameters e.g. building density, compactness index and floor area ratio, and calculates energy use intensity of the grid model.

There are many other UBEM tools not listed above, that adopt similar analytical structure and have been validated by various approaches. In addition, the urban effect indirectly associated with form factor but more with general urbanization such as UHI effect has counter-impact on heating and cooling energy. UWG is a set of algorithms intended to estimate UHI [15], it uses the EPW format weather file as input and output. It has been used by other research on context-dependency of urban buildings for instance the UHI and mutual-shading effects on high-rise commercial building in compact urban context in hot-humid climate of Brazil [16].

3. UBEM tools for low energy neighbourhood design

3.1. UBEM to inform low energy urban design?

Urban management on energy and carbon emission reduction seems to be one of the major field of application for UBEM tools. This is understandable for developed countries/regions where urbanization progression has mainly on renovation and redevelopment. But in developing countries e.g., China, new urban districts or even an entire new city are being constructed to accommodate the increasing urbanized population. Towards the common goal for emission reduction towards the 2030 mission, can UEBM tools support low carbon urban design for new build? Architects in large cities of China e.g. Shanghai always work overload at fast pace, for them an added step of (iterative) urban energy simulation to feedback design improvement seems rather difficult to accept. So is it possible to develop for them the prescriptive provisions in the form of diagrammatic design guide, regarding urban building density, layout, orientation, spacing and aspect ratio, window/wall ratio, etc.?

3.2. Towards mixed-use walkable living circles

Mixed-use neighborhood as an urban functional unit is considered a sustainable and pedestrian-friendly typology, and the mixed functionality and flexible block layout require deeper understanding on function/form and energy demand.

Urban housing developments in Chinese cities used to be barrack-like apartment slabs with main façade orientated towards South, with building spacing regulated by winter irradiation requirement. It is a common belief particularly among Southern Chinese citizens that this is the best layout for indoor/outdoor comfort and building energy saving. This overly-simplified layout pattern, despite its merits in utilizing passive solar and cross-ventilation, tends to cause many problems such as dull and faceless street, gated communities, and highly homogeneous urban space landscape.

This belief is in fact not well-grounded in a holistic perspective of sustainable neighborhood development, and our design research indicates that a more thoughtful urban design may improve the urban landscape, street walkability, indoor/outdoor microclimate as well as building energy demand. In the QianJiang NewTown (Hangzhou, ~160km away from Shanghai) planning scheme, an urban structure with smaller street block and denser street grids is developed (Fig.1. Research design team: Sun, Xu, Yang, et al.). Numerical simulation on the environmental performances of three urban form indicates that the new urban grid 1 and 2, which are smaller and denser while more diverse in orientation/height variation, show improved spatial microclimate and thermal comfort: percentages of strong and very strong thermal stress area (quantified by the universal thermal climate index UTCI) in the gird 1 and 2 are lower than that of grid 3, i.e. traditional urban form. The building energy demand simulation indicates a descending order of EUI (kwh/m^2.yr) of grid 3 (89), grid 1 (87) and grid 2(83). The improvement may be due to the mutual shading effect by adjacent buildings in the (semi-) enclosed layout. Note that the advantages in energy and comfort of grid 2 and gird 1 over grid 3 are marginal. Nevertheless, the results support that the enclosed and semi-enclosed neighborhood patterns are no inferior to the linear layout with respect to building energy demand and outdoor thermal comfort.

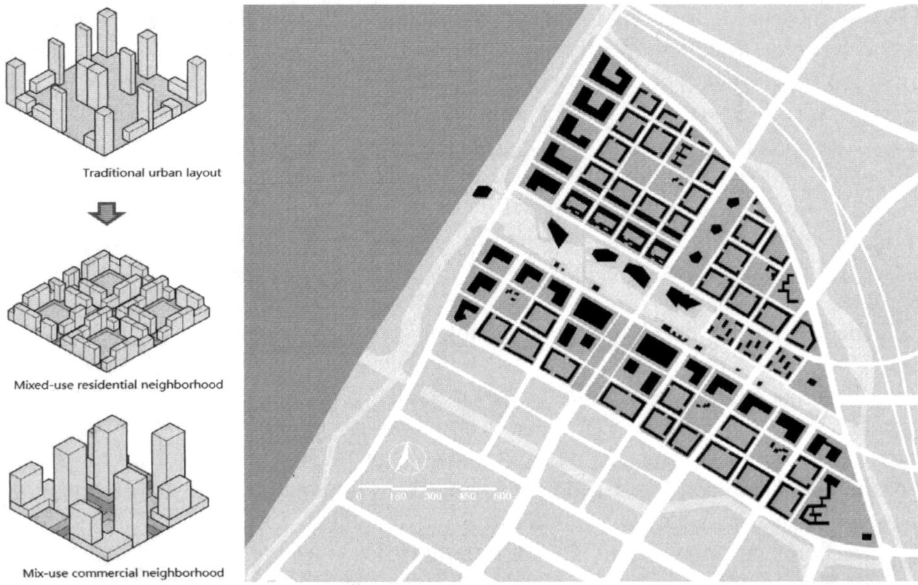

(a)

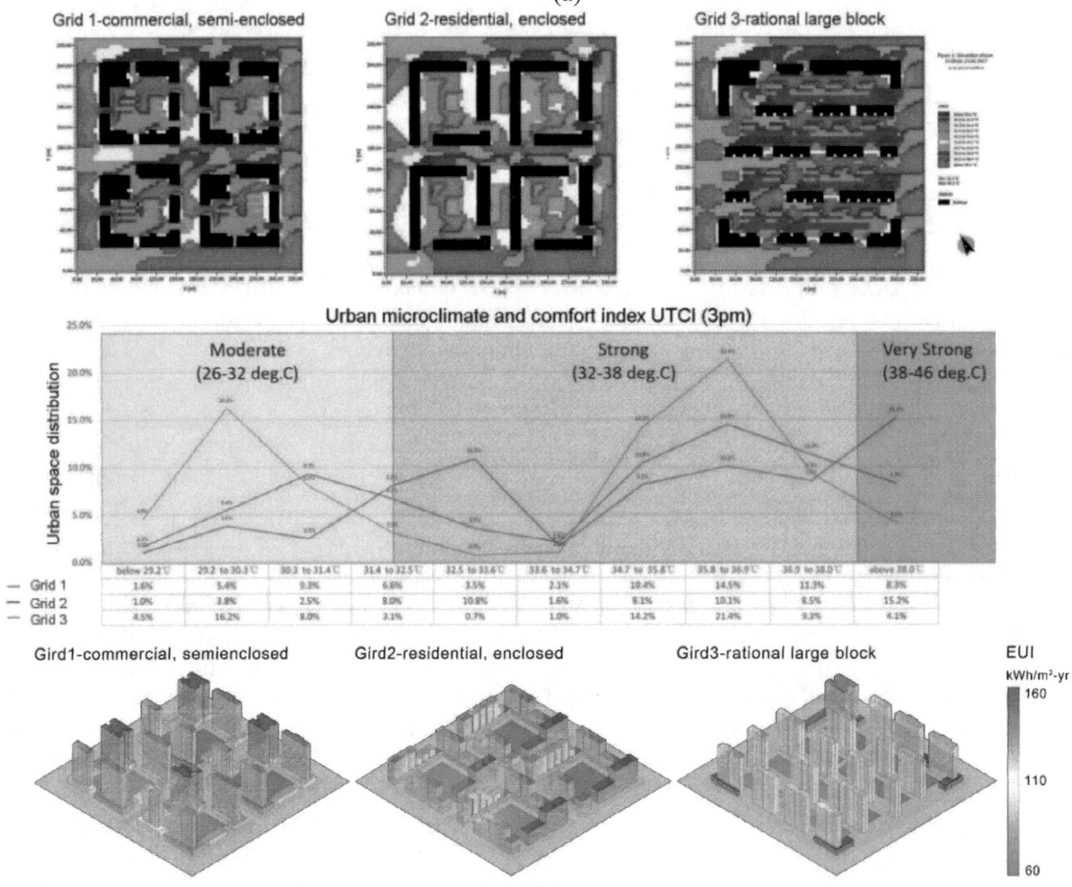

(b)

Figure 1: The Qianjiang NT urban design (a) and comparative analysis on microclimate and EUI (b)

Today new policy and design code in China are being issued to promote development based on unit of neighborhood (the so-called "5-10-15 min walk living circle" is in fact residential neighborhoods at different scales). Neighborhood embraces street life, walkable/bikable streets with commercial/retail street shops, and flexible building layout creates more diverse microclimate. The new Planning standard for urban residential area [17] allows higher density (footprint), lower green space whereas imposes stricter building height limit. The impact of urban form on UBE can be higher, and UBEM can be more meaningful and important in future low-carbon urban design.

4. UBEM and its integration with ND rating tools

The neighborhood development (ND) rating tools publicized in a number of countries/regions invariably include microclimate& energy-related credits, these include BREEAM Communities, LEED-ND; Bioclimatic Design Section in HKBEAM; CASBEE-UD and CASBEE-HI, etc. Some provisions are performance-targeted (i.e., what results to be achieved, 1°C or 2°C?), some are prescriptive (i.e., what design should do to achieve that 2°C reduction?). As for China, the Green Eco-District Evaluation Standard (GB/T51255-2017) includes credits regarding district-scale energy efficiency, such as solar orientation, wind corridor, greenery ratio and quality, heat island reduction, renewable energy and certified green buildings.

The two neighborhood development (ND) rating tools from the US and China, i.e., the LEED ND (V4, 2018) [18] and China Evaluation Standard of Green and Eco Urban District (GB/T-51255-2017) [19], or China green building label system, are compared as below regarding the provisions of code compliances that are associated with urban building energy:

4.1. Building orientation (for passive and active solar):

- LEED US: Two options: Opt 1-block orientation; Opt 2-building orientation. 75% of building main façade towards within plus/minus15° off due south, awarded 1 pt (Green infrastructure and building GIB, LEED-ND V4, 2018).
- GBL China: Building orientation that benefits UBE saving: 80%, 85%, 90% of all the ND residential buildings orient the main façade towards the direction that is deemed benefit building thermal energy saving under the local climate. (*The beneficial direction is unclear, in terms of passive/active solar, or natural ventilation? Probably South?*), awarded 6,8,10 pts. (4.2.8, GB 51255-2017).

4.2. Site ventilation:

- LEED US: No provisions for site ventilation is identified.
- GBL China: Provision of wind corridor wider than 50m in accordance with prevailing (*which season?*) wind direction, awarded 10 pts (4.2.9, GB 51255-2017).

4.3. UHI mitigation:

- LEED US: Three options. Opt 1 - nonroof, 50% shaded or high solar reflectance (SRI, three-year aged value at least 0.28) or open grid pavement (at least 50% unbound); Opt 2 – Roof, 75% of high SRI roof material, or vegetated (green) roof, or combination; Opt 3 – Mixed roof and nonroof). GIB credit, 1 pt (LEED-ND V4).
- GBL China: UHI intensity lower than 3°C, 2.5°C, awarded 3 or 5 pts. (*Detailed specification is not provided, e.g., summer or winter, day or night, instrumented experiment or numerical simulation?)* (5.2.10, GB 51255-2017).

4.4. Percentages of certified green buildings:

- LEED US: Percentage of certified GB areas meets 10%-20%, 20%-30%, 30%-40%, 40%-50%, >50%, awarded 1-5 pts (LEED-ND V4).

- GBL China: percentage of GB certified New builds with 2-start Label or higher, 35%-40%, >40%, awarded 10, 15 pts. Percentage of certified GB renovation areas 10-20%, >20%, awarded 5, 10 pts (6.6.1, 6.2.2. GB 51255-2017)

4.5. Renewable energy:

- LEED US: Percentages of RE-5%, 12.5%, 20%, awarded 1, 2, 3 pts (LEED-ND V4).
- GBL China: Percentages of RE-2.5%, 5%, 7.5%, awarded 5, 8, 10 pts (7.2.2, GB 51255-2017).

4.6. Building energy saving:

- LEED US: 90% building floor area meets Opt1- Energy performance simulation. Percentage improvement new builds of 12% (1 pt) or 20% (2 pts) over ANSI/ASHRAE/IESNA Standard 90.1–2010; Opt2-prespriptive compliance ASHRAE 50% Advanced Energy Design Guide (2 pts) (LEED-ND V4).
- GBL China: Percentage of new builds with design energy use intensity (EUI) 10% or lower than national standard (GB/T 51161) - 25%, 50%, 75%, awarded 5,7,10 pts (*A detailed simulation setting-up is lacking as compared to LEED*) (7.2.4, GB 51255-2017).

5. Challenges and suggestions for code compliance with some ND rating tools

5.1. Prescriptive vs. performative

In terms of time-saving and design support, prescriptive provisions seems better than performance indices provisions. Also, the performance indices compliance without detailed regulation on simulation setting-up (e.g., software and model chosen, boundary and initial conditions, workflow and protocol, etc.), for instance, the UHI mitigation provision in GB51255-2017, would allow large room of manipulation on simulation results.

5.2. Need for benchmarking

For case-based performance assessment, there is lack of a benchmarking procedure for ND rating tools, i.e. a proper base-case with which a proposed urban neighborhood scenario (new design or renovation) under energy modeling can be compared with. The urban base-case model is analogous to the base-case model for energy efficiency compliance simulation of individual building rating, e.g., Energy credits in LEED for BD+C as per required by Appendix G, ASHRAE 90.1.

6. Potentials of UBEM tools in support ND rating

6.1. Support prescriptive code-compliance provisions

As discussed before, prescriptive urban design guides would better suit the needs and workflow of architects. At its present state, some prescriptive provisions of ND tools are qualitative instead of quantitative, e.g., the air ventilation assessment (AVA) in Chapter 11 of the Hong Kong Planning and Design Guidelines. Some are quantitative, but how the threshold values are determined is unclear: is it scientifically tested, or just based on limited expert judgement? For that, well-developed and validated UBEM tools can be applied in a systemic way in supporting better-grounded prescriptive provisions.

6.2. Support base-case urban building energy model

While urban physicists and building scientists focus on technical improvement of their UBEM tools, architects and planners are (or should be) interested in possible implication of these tools in their practice. One such possibility would be to help build the base-case urban building energy models, depending on geography, climate, and typology (e.g., commercial, residential, or mixed). In new town building, key urban parameters are predetermined before urban design such as volumetric density FAR, footprint density BCR, green space ratio, land use and building functions, etc. A base-case urban model will comply with all the parameter setting, probably in an idealized grid setting to minimize manipulation on space and form (which is what architects are for). Design scenarios can be modelled by UBEM and the energy performance compared with the base-case.

A base-case urban model may help benchmarking the energy assessment workflow in ND rating and clearing out ambiguities and uncertainties. One interesting case is the UEIB that is introduced previously, which manages to "average" complex urban morphology into standardized "grid" with predefined key morphological parameters. The model, when used in assessing real urban areas, may risk averaging out the problematic spots where bad urban form/fabric adversely affect nearby building energy demand. UEIB however demonstrates the association of real-world urban areas, especially regarding the mean energy use intensity (EUI) and carbon emission rate, with a carefully-defined ideal urban grid model by a given set of planning indices. For the model itself, further validation using authoritative, thermo-dynamic energy modelling tools may be needed, as the UEIB results were compared with simulation results by Climatelite[14], a tool developed by BRE in 2013 and seems not accessible for the time being.

Acknowledgments

Authors acknowledge financial support from the National Natural Science Foundation of China (NSFC) Project (No.: 51678413), the Fundamental Research Funds for the Central Universities and the National Key R&D Program of China (No.: 2016YFC0700200).

References

[1] Reinhart CF, Cerezo Davila C. Urban building energy modeling – A review of a nascent field. *Building and Environment* 2016; **97**: 196-202.

[2] Baker N, Guedes MC, Shaikh N, Calixto L, Aguiar R. The LT-Portugal software: A design tool for architects. *Renewable Energy* 2013; **49**: 156-160.

[3] Ratti C, Baker N, Steemers K. Energy consumption and urban texture. *Energy and Buildings* 2005; **37**: 762-776.

[4] Robinson D. *Computer Modelling for Sustainable Urban Design: Physical Principles, Methods and Applications*. Earthscan, 2012.

[5] Robinson D, N. Campbell, W. Gaiser, K. Kabel, A. Le-Mouel, N. Morel, et al. SUNtool – A new modelling paradigm for simulating and optimising urban sustainability. *Solar Energy* 2007; **81** (9): 1196-1211.

[6] Reinhart CF, Timur Dogan, J Alstan Jakubiec, Rakha T, Sang A. Umi-an urban simulation environment for building energy use, daylighting and walkability. *13th Conference of International Building Performance Simulation Association*. Chambéry, France, 2013.

[7] Dogan T, Reinhart C. Shoeboxer: An algorithm for abstracted rapid multi-zone urban building energy model generation and simulation. *Energy and Buildings* 2017; **140**: 140-153.

[8] Dogan T, Reinhart C, Michalatos P. Autozoner: an algorithm for automatic thermal zoning of buildings with unknown interior space definitions. *Journal of Building Performance Simulation* 2016; **9** (2): 176-189.

[9] Zhu P, Yan D, Sun H, An J, Huang Y. Building Blocks Energy Estimation (BBEE): A method for building energy estimation on district level. *Energy and Buildings* 2019; **185**: 137-147.

[10] Jones P, Patterson J, Lannon S. Modelling the built environment at an urban scale—Energy and health impacts in relation to housing. *Landscape and Urban Planning* 2007; **83** (1): 39-49.

[11] Caputo P, Costa G, Ferrari S. A supporting method for defining energy strategies in the building sector at urban scale. *Energy Policy* 2013; **55**: 261-270.

[12] Fonseca JA, Nguyen T-A, Schlueter A, Marechal F. City Energy Analyst (CEA): Integrated framework for analysis and optimization of building energy systems in neighborhoods and city districts. *Energy and Buildings* 2016; **112**: 234-253.

[13] Chen Y, Hong T. Impacts of building geometry modeling methods on the simulation results of urban building energy models. *Applied Energy* 2018; **215**: 717-735.
[14] Rodríguez-Álvarez J. Urban Energy Index for Buildings (UEIB) : A new method to evaluate the effect of urban form on buildings' energy demand. *Landscape and Urban Planning* 2016; **148**: 170-187.
[15] Bueno B, Roth M, Norford L, Li R. Computationally efficient prediction of canopy level urban air temperature at the neighbourhood scale. *Urban Climate* 2014; **9:** 35-53.
[16] Lima I, Scalco V, Lamberts R. Estimating the impact of urban densification on high-rise office building cooling loads in a hot and humid climate. *Energy and Buildings* 2019; **182**: 30-44.
[17] MOHURD (Ministry of Housing and Urban-Rural Development). *GB50180-2018-Standard for urban residential area planning and design GB50180-2018*. 2018.
[18] USGBC. *LEED v4 for Neighborhood development*. U.S. Green Building Council, 2018.
[19] MOHURD (Ministry of Housing and Urban-Rural Development). *GB/T-51255-2017 Evaluation Standard of Green and Eco Urban District*. 2017.

IOP Conf. Series: Earth and Environmental Science **329** (2019) 012017 doi:10.1088/1755-1315/329/1/012017

Urban Policy as a Main Key towards Sustainable Built Environment of Popular Neighbourhood: Case Study of "*El Cabanyal*" Valencia Spain

Randa A. Mahmoud[1], Ahmed Wael[2]

[1]Ain Shams University, Department of Urban Planning and Design Cairo, Egypt
UFE Université Française d'Egypte, Département d'Architecture, Cairo, Egypt
INSA de Strasbourg, Département d'Architecture, Strasbourg, France. Email: randaabdelaziz@yahoo.fr, randa.mahmoud@insa-strasbourg.fr
[2] UFE Université Française d'Egypte, Département d'Architecture, Cairo, Egypt

Abstract. *El Cabanyal*, a popular neighbourhood and old fishermen's village, in the *Valencian* Community, eastern Spain, is a very explicit example on how popular neighbourhoods can be drastically transformed in line with urban policies' evolution. This neighbourhood -that has a strong cultural identity expressed by its vernacular architecture, local art and social life- kept resisting gentrification during the last 30 years. As a result, this neighbourhood has shown a great deal of deterioration as a price that had to pay in order to keep its own identity within the Mediterranean city of Valencia.

This paper demonstrates the evolution of local and popular neighbourhoods according to adopted urban policies by taking the neighbourhood of *El Cabanyal* as a case study. Therefore, it will firstly scrutinize the proposed urban policies/strategies by the regional government in order to develop this neighbourhood. Secondly, it will shed light on the participatory design solutions based on the local community involvement, to develop the area without destruction and getting use of its potentials. Finally, it will propose a diagnostic tool and ecological urban solutions that can be elaborated in order to tackle such complex problems.

We believe that the change of urban policies can give different spectrums to sustainable development strategies by addressing social problems and heritage conservation questions. The delicacy of the issue is a typical 21st Century problem encountering different contexts. It can be an obstacle for sustainable urban/social development if not addressed correctly. This is why it must be integrated in the strategies proposed by urban policies' decision makers.

1. Introduction

Urban integration of popular neighbourhoods in the modern world is a complex process, especially when sustainable built environment is taken into consideration. This process takes other composite dimensions for neighbourhoods with strong identities that related to patrimonial and social practices. In many cases, urban policies are the main factor determining whether these neighbourhoods are considered as potential keys or threats to sustainable development in the modern world. *El Cabanyal*, a popular neighbourhood and old fishermen's village in the *Valencian* region, eastern Spain is a very explicit example of this complex process.

1.1. El Cabanyal As A Fishermen Village

'*El Cabanyal'* is a historical district of the city of Valencia. Its origins, as an old fishermen's village, go back to the 15th century. The profession and the lifestyle of its inhabitants has shaped the urban tissue into distinctive parallel coastal grid. *El Cabanyal* is distinct with a specific vernacular architecture and a strong cultural identity that made of the neighbourhood a patrimonial historical pearl, declared in 1993 as an Asset of Cultural Interest (BIC)[1].

1.2. El Cabanyal in a Contemporary Context

Despite its historical importance, *El Cabanyal* has been suffering from a state of degradation due to political pressure, put by the previous municipal government in order to execute an operation of regenerative urban agenda, known as PEPRI[1], without taking neither the historical background nor the cultural identity into consideration. Consequently, this neighbourhood kept resisting gentrification during the last 30 years. As a result, *El Cabanyal* has shown a great deal of deterioration as a price that had to be paid in order to keep its own identity within the Mediterranean city of Valencia.

Nevertheless, the neighbourhood is relatively saved thanks to opposite urban movements and associations that represent the local community while taking part in the decision making process and strongly influencing the urban policy making.

The paper is based on a thesis of a graduation project, by the co-author that proposes somehow a method to repairing the damage done in the maritime part of the *Valencian* capital. It aims at providing a global solution of rehabilitation and restoration of the neighbourhood while focusing on the cultural heritage as a key to develop sustainably the built environment of the popular district without putting its authentic identity at risk. The goal is to frame a method to follow to help popular neighbourhoods becoming sustainable, especially those suffering from disparities compared to the urban fabric and socio-cultural structures of their neighbouring zones.

2. Paper's Aim and Interests

The research's main aim is to develop a diagnostic tool that can be used in the case of popular neighbourhoods at risk of losing their identities due to political or local policies. These zones suffer from disparities compared to their environing urban fabric, hence an imbalance of their socio-cultural structure and a threat to their neighbouring areas. The research's ultimate goal is to draw both viable analytical and intervention stratagems that will help these particular popular neighbourhoods evolving in order to become sustainably built environments coherent with the surrounding community.

3. Research Approach & Framework

This paper is based on a research work done by the co-author for his graduation project under the supervision of the author at the School of Architecture at the French University of Egypt and the INSA de Strasbourg for the year 2017/2018. The main objective of the project was to develop an urban stratum of *El Cabanyal* as a tool towards rehabilitation and restoration of the damaged maritime part of Valencia, while focusing on its cultural heritage as a potent factor of its development and the conservation of its popular identity. Consequently, the *Valencian* social and cultural aspects were studied for over a year in order to understand the local values and traditions.

3.1. Field Survey & Site Visits

The surveying procedure aimed at observing the city of Valencia in general and *El Cabanyal* in particular to apprehend a global understanding of the context. It considered the two possible points of view. The first, adapted by the past regional government and the PP party, is based on the need of radical interventions plus a "once for all" integration of the entire neighbourhood within the city. This integration will act on both physical and moral levels via changing the city's actual structure and is completely controlled by the local authorities. The second point of view -backed up by *El Cabanyal*

[1] *Plan Especial Protección y Reforma Interior* approved in 2001 by the Regional Government.

citizens, *Valencian* nationalists and the current regional government- adopts a softer approach of intervention and promotes for a bottom-up dialogue with the local inhabitants, yet without a real agreement on the action method. Despite their differences, both point of view agreed on the existence of an urban problem that needs an intervention strategy to end the deterioration affecting the zone.

Both point of views were deeply scrutinized through three site visits during the spring and summer of 2017and 2018. During these visits, the co-author focused on the observation of the neighbourhood's condition while summarizing and taking notes on the socio-cultural context and backgrounds of the inhabitants. The visits also coincided with special cultural manifestations of the area such as "*La semana santa marinera*" (Easter celebration) and the local market held every Thursday. Furthermore, field survey included the observation of some virtual platforms such as the website-forum *Salvem El Cabanyal* and the Association of neighbours of *El Cabanyal* (*Associació de veins i veines del Cabanyal*). The "*Salvem*" forum was and still is the main defender of the neighbourhood's identity against the destructive urban policy. These forums offers an extensive documentation and justified explanations in English, Spanish and *Valencian*, three languages mastered by the authors.

3.2. Finding a Diagnostic Tool

The research had to take the conclusions of the proposed projects and the survey outcomes into an evaluative phase. These proposed projects represent an example of the impact of urban policies on actual grounds and illustrate how the change of these policies can have a drastic influence on the urbanization process. Our aim was mainly to find a diagnostic method that leads eventually to a key intervention for sustainable built environment in popular neighbourhoods with strong identity and public opinion. This goal is to be reached through urban associations representing the local community while taking part in the decision making process and the urban policy application. This method comes in harmony with "UN Habitat III" principles of planned urbanization according to Dr Joan Clos's three pillars: Rules and Regulations, Urban Design, and a Financial Plan [2]. Together they form a three legged approach towards sustainable urbanization.

3.3. Application on El Cabanyal as Research Object

An adapted diagnostic approach was needed to attain viable solutions and action plans for the core problem. This adapted approach will also ensure a sustainable development for *El Cabanyal* within the same urban agenda of the city of Valencia. Hence, our diagnostic tool will rely on three main pillars:

Understanding *El Cabanyal*'s Identity via literature review and filed work. The field work is done physically by the co-author, also involved in the online platforms of urban movements such as "*Salvem El Cabanyal*" (Save *El Cabanyal*). The latter emerged to defend *El Cabanyal* against destruction proposed by the local authorities;

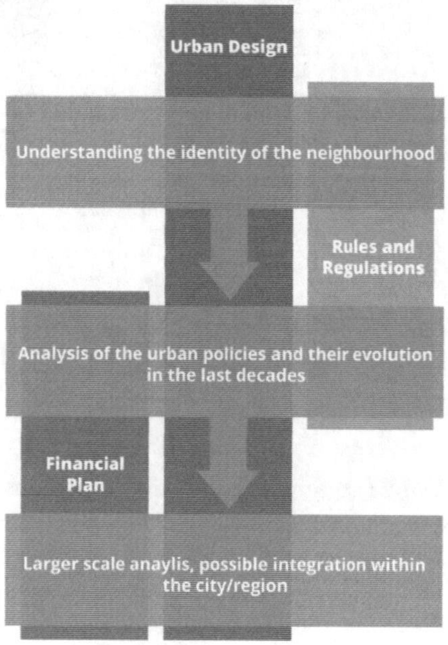

Figure 1: The diagnostic plan and the three pillars of urban organization. Source: Authors

2- Analysis of *Valencian* urban policies and their evolution in the last three decades. The main key to understand and contextualize popular actions -that fought against *El Cabanyal*'s destruction- is to trace urban policies of the Spanish authorities since the annexation of the fishermen village to the city in the late 19th century. These policies have been influenced by political opinion of the governing conservative party before the rise of nationalist *Valencian* party in 2015. Urban policies, that hooked up on commodification of the built environment worldwide, geared the city of Valencia towards mega projects integrating the sea as a main factor while neglecting the identity of *El Cabanyal*;

3- Possible integration of *El Cabanyal* within the City whilst conserving its identity. It is very important to shift our perspective to see from inside the *Valencian* community. This paradigm shift, also more compatible with current political policies, will cater for the determination of the most suitable zones for intervention on the urban design level. The three pillars of the diagnostic method are equally important and in the same order (Figure 1). The first step aims to completely apprehend the studied area via a variety of field work and literature review and analysis of the nowadays status. The second pillar links the current status to its origins and questions the urban policy in action. Finally, the third led to studying the area as a part of larger urban mechanism. This is to ensure a more holistic approach and to find practical interventions adapted to the well-studied context.

4. *El Cabanyal*: A Neighbourhood Facing A Brutal Gentrification Process

4.1. Valencian Urban Politics: An Ambition to Link the City to the Mediterranean Sea

Figure 2: Analysis of the current state of *El Cabanyal*, with the destructed buildings shown in black, buildings owned by the city council in white. Source: Bordas+Peiro Architects (2011) for an AJAC prize winning urban project.

El Cabanyal community in Valencia is considered recently the most struggling civil society in Spain against its municipal government [3]. Effectively, the political will behind *El Cabanyal* has always been consistent in a destructive approach as the popular neighbourhood represented an urban obstacle for the city council [1]. The latter was aiming at extending the road network through the neighbourhood to the

Mediterranean Sea. The future project would result in a complete metamorphosis of the neighbourhood's urban fabric and a loss of its identity.

Despite the non-execution of the municipal project, *El Cabanyal* has already started to suffer from an accelerated gentrification process, especially after the destruction of many typical local houses (Figure 2). The gentrification process manifested itself under various forms: the most remarkable one started at the late 19th century by the progressive imitation of the old city centre's social practices and way of life in *El Cabanyal* community in opposition to its urban fabric socio-cultural structure. Consequently, the *Valencian* lifestyle invaded literally *El Cabanyal* social and cultural identities. This process was accelerated in the last decades, by the recent actions of the local *Valencian* authorities, and reflected particularly on the evolution of architectural typology that went from the old historic *barraca* model to small three-four story buildings and lately to high rise buildings, especially in the zones closer to the city. This evolution was also accentuated by the demolition of more than 400 buildings, containing more than 1600 apartments/houses, after transferring their property to the city council, sometimes by forced agreements. More explicitly, buildings that survived are those whose locations weren't obstacles in the expansion of the city roads to the sea. This action is actually an evidence that the radical political plan wasn't aiming at linking the two parts of the city. On the contrary, it was about melting the neighbourhood within the city of Valencia to make use of the seaside as an attractive touristic spot that was an obstacle for investors (backed up by liberal-conservative local authorities).

The PEPRI project reflected the vision of the neo-liberal political urban policy practiced by the governing party at the time. Physical actions in that project would result in more radical gentrification, not only by changing the local architecture, urban fabric or public space structure, but also by replacing the local community with another one socially and culturally different.

4.2. The Part of El Cabanyal in the Strategic Plan

El Cabanyal is a living example of what a local resistance can be. This action started when some groups of local neighbours structured an antagonistic power against the municipal urban policy. The opposition actors managed to raise legitimate and confronting visions concerning the area's future that resulted in the stoppage of the PEPRI project in December 2009. This urban opponent power has actually shown an alternative creative way on how to re-imagine a city while integrating neighbourhoods with authentic identities. The local power succeeded in re-politicizing the urban agenda as a collective decision against the arrangements of the neo-liberal governance in a profound de-politicization of urban policies [1]. The case of *El Cabanyal* reveals how the collective power can be translated into political urban strategies. Nevertheless, and despite the growing power of the opposite action, the urban void silenced the area and made it less populated and less socially interactive. As a result, many local residents left the neighbourhood for the uncertainty of its future. The area was labelled "*Zona cero*" or (Zero Zone) for its un-safety. This sense of insecurity, especially at night, is also interpreted by drug dealing that got installed in the neighbourhood. The newly elected government made attempts to encourage people returning to the neighbourhood. Consequently, as visited in 2018, the local market is full of people, Thursday market takes place and is sufficiently populated. Students from near-by university campuses recently opted for the neighbourhood as a possible place to live in. However, many people still remain sceptical about moving back. We believe that there is a need for stronger and more organized interventions of contemporary adaptation.

5. The Rescue Of An Urban Identity Via The Reproduction Of Sustainable Spatial Solutions

5.1. The Urban Identity Conservation: The Urban Knowledge

El Cabanyal inhabitants' opposition is completely legitimate, as they were initially fighting for their right to the city. The need for re-appropriating private and public spaces that used to belong to them is obvious. This spatial necessity linked different social categories of the *El Cabanyal* neighbourhoods under the same cause. We believe that the key solution for this urban conflict between them and municipalities of Valencia inheres in the sustainable development of the built environment of *El*

Cabanyal. This can be achieved via producing alternative urban futures other than those previously proposed by the authorities. The identity of the neighbourhood, its cultural image and its evolution must be integrated as potential keys in the future urban plan. Moreover, the fusion of the local community and their strong perseverance are also assets to be considered. The identity conservation of *El Cabanyal* refers particularly to a collective living memory of the population, to different ways of life that took place in the heart of the *Valencian* Capital. It marks a popular resistance against a negative gentrification process and destruction intents. *El Cabanyal* identity expresses the *Valencian* uniqueness conserved along centuries and distinct compared to the historic city centre (influenced by the central Spanish urban model and mechanism). This is why, it has become an icon and a model to follow, especially in local community organisation, social cohesion, the sense of responsibility and strong will to protect popular values. This is why a successful urban intervention can be a proof of effectiveness of the diagnostic method and an exemplary tool, to reach sustainable built environment, adapted to a popular neighbourhood that can be mimicked and adapted to other similar cases.

5.2. Urban Spatial re-Production Policy: The Urban Project

The action plan for urban intervention is about limited physical interference in a chosen urban stratum of the parallel grid of *El Cabanyal*. The chosen part is actually the mostly affected by demolitions according to the "PEPRI". The intervention will take place in the heart of the area and everyday-life space that is connected to the local church, the municipal market and the *Plaça de Flor* square. This place is easily accessible from the *Cabanyal* Train Station and *Blasco Ibañez* Avenue.

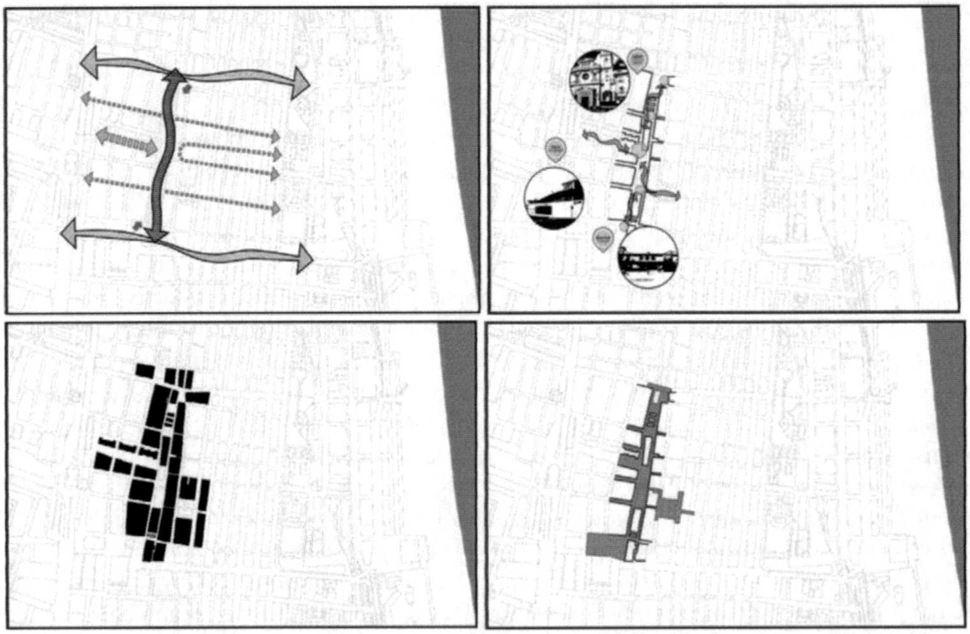

Figure 3: (a) Analysis of the circulation around the stratum, (b) Analysis of the nodes and important elements around the stratum, (c) Analysis of the current built environment (solids) around the stratum & (d) Analysis of the current public space considered in the project (voids) around the stratum. Source: Co-Author, 2018.

The project consists of transforming this stratum into a cultural hub for the *Valencian* population. This will be materialized through the management of the public space and the pedestrianization of two streets. Consequently, the proposed project will be making use of *El Cabanyal* lifestyle according to which public space is considered as semi-private where neighbours meet and do most of their activities. Moreover, the proposal includes the installation culture-related buildings that provides activities linked to the popular heritage. These culturally-based buildings will be in contact with proposed intermediate public spaces to host all kinds of cultural manifestations or events. The potential of the chosen area resides in its strong association to the neighbourhood, the easy accessibility from the rest of the city as it can offer an example for a sensible urban intervention. The latter will be done by redesigning new urban spaces based on the diagnostic tool (Figure 3). It will help generating an attractive zone enhanced by the exiting social interrelations and the urban dynamics of this Mediterranean context. The dynamics created by this intervention will act as igniters for an urban action or an example to follow that can reach the rest of the neighbourhood. Accordingly, on a larger scale, the rehabilitation of such an area will have also other consequences concerning the urban mechanism governing the relation between *El Cabanyal* and the municipalities since the annexation of the old fisherman village to the city The proposed project shows how an urban policy can save a popular neighbourhood by using a correct diagnostic tool. It illustrates the articulation between two opposed urban models. The first doesn't take the area's cultural power into consideration while the second adds more values to the heritages in the area. The project seeks developing social quality/cohesion in both outdoor and indoor spaces, as the conceptual space and the material space will be fused in the everyday "lived space"[4]. Therefore, the rehabilitation alone of old houses will not be enough as the addition of some buildings integrated with *El Cabanyal* identity will complete the action to add value to public spaces (Figures 4). We believe that the proposed project is capable of adding values to the current neighbourhood. The newly managed public spaces will act as a magnet and gain more the attention of visitors and foreigners who will rediscover the area through an itinerary of well-placed cultural nodes connecting pedestrian pathways through *El Cabanyal* ant all the way to the sea.

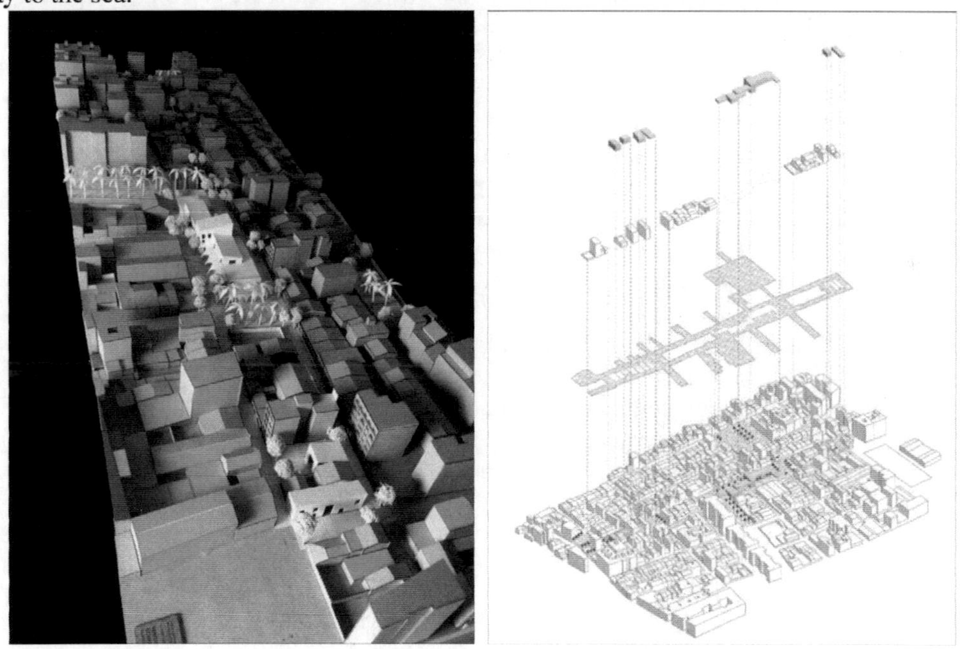

Figure 4 : (a) Physical model of the proposal in the scale of 1/200 (Left) showing the urban stratum and the added buildings in the voids (buildings in white are the proposed addition), (b) Exploded axonometric view of the proposal within the City (Right), showing the added volumes (in Cyan), the existing voids and the public space in question. Source: Authors

6. Conclusion

We live in a world where the "lived space" became politicized [5]; a symbol of domination where the user has no say on how his liveable space or his own spatial practices should be or will be. Imagination and creativity are killed and buried in the planning of the modern city where spaces become more generic and the notion of "city life" is rather commercialized; "lived space" is no longer existent as there is no room and no interest for it anymore.

Urban policies became an evident mechanism to control the social space/practices, to determine social exclusion or to limit social inclusion. These same urban policies could work, however, on restoring identities or creating new social platforms for expressing creativity. This continuous tensions gave birth to subversive socio-spatial movements that link various social categories who got united around the same cause which is: gaining back their everyday lived space. Social re-appropriation of the lived space by locals is like the light at the end of the tunnel that will guide them through to rescue their identity and to create their own space [6].

El Cabanyal neighbourhood is a part of city among others that struggled against being metamorphosed by neo-liberal forces. Urban social movements, struggles, frictions and resistances by community associations are the soldiers behind this resistance. In fact the famous neighbourhood showed the real potential of *dessensus* -as a source of creativity and original actions- against a hegemonic urban policy that neglects its cultural identity [7]. *El Cabnyal*'s "lived space" has finally witnessed a sacred platform from which phoenix has finally rose to contest against spatial injustice. This is how a new urban agenda is imposed by the people and for the people, as diverse social actors and citizens cooperatives worked together to represent an urban alternative that allows *El Cabnyal*'s inhabitants to live without being soulless.

Our research proved through filed study together with literature review how strongly the local identity is present in the everyday-life space, in the narrow streets of the neighbourhood of *El Cabanyal*. This is why, the proposed intervention was concentrating on how physically we can repair the damages while being very sensible to the social and cultural aspects when dealing with this popular neighbourhood. Moreover, it was very important to be well aware of the different factors that were behind the change of the urban policy of intervention in *El Cabanyal*:

1. The most important factor is the strong popular resistance that reflects a high sense of citizens' responsibility to defend their own identity and the desire to keep it intact;
2. The rule of laws and legal protection over patrimonial heritage and culture, materialized particularly in the sentence of the Spanish courts ordering the stoppage of the project;
3. The political shift, with the rise of *Valencian* Nationalists (*Compromís*) who replaced the Spanish conservatives (PP) which lead to the change of the municipality's plan *vis-à-vis* popular neighbourhoods.

In this particular context since 2015, with the politically guided shift in urban policy, *El Cabanyal* is no longer seen as an obstacle to economically efficient projects and the *Valencian* speculative maritime façade plan. It is rather seen as a local community of iconic importance with potentials of an interesting cultural heritage and geographical distinctiveness that should be protected. In addition, the growing attention towards achieving sustainable built environment made it clear that the urban tissue should be tackled differently. This is why our diagnostic tool, with its three pillars, was of a great benefits to be able to propose a project that adds values to the neighbourhood without oppressing its identity. This is diagnostic tool that worked beneficially with *El Cabanyal* can also be applied to other popular neighbourhoods with strong identities at risk of brutal gentrification.

References

[1] Eslava, N., L., (2016), Urban Regeneration And The Transformative Potential of Dissensus The Case Of "El Cabanyal" In Valencia, in: *International Conference from Contested Cities to Global Urban Justice*, Stream 1, Article nº 1-01.

[2] Clos, J., (2015), Guidelines On Urban and Territorial Planning, in: UN Habitat, 40 P.

[3] Díaz Orueta, F., 2006. Spain: Local Democracy and Citizen Participation. *Space and Polity*, 10(3), pp.263–278. Available at: http://www.tandfonline.com/doi/abs/10.1080/13562570601110666

[4] Purcell, M., (2001), Neighborhood Activism among Homeowners as a Politics of Space. in: *The Professional Geographer*, Issu:2- Volume 53, pp.178–194. Available at: http://www.tandfonline.com/doi/abs/10.1111/0033-0124.00278.

[5] Dikeç, M., (1987), Space, politics, and the political. Environment and Planning, *in: Society and*

IOP Conf. Series: Earth and Environmental Science **329** (2019) 012017 doi:10.1088/1755-1315/329/1/012017

Space, SAGE Publications.

[6] Stavrides, S., (2009), The December 2008 youth uprising in Athens: Glimpses of a possible city of thresholds, in : *Annual Meeting of International Sociological Association – Research Committee on Cities and Regions (RC21)*.
Available at: http://www.centrodametropole.org.br/ISA2009-BR/assets/papers/05-A-3.pdf.

[7] Agamben, G., (2014), The Power of Thought. in: *Critical Inquiry*, Volume 40-N°2, pp.480–491.

IOP Conf. Series: Earth and Environmental Science **329** (2019) 012018 doi:10.1088/1755-1315/329/1/012018

Strategies for Urban Residual Areas as a Planning Tool: Towards Achieving Sustainable Built Environment by Connecting Neighbourhoods

Alexander Kader[1] and Randa A Mahmoud[2,3,4]

[1] German University of Technology, Department of Urban Planning and Architectural Design, Muscat, Oman. e-mail: alexkader@gmx.de, alexander.kader@gutech.edu.om
[2]Ain Shams University, Department of Urban Planning and Design, Cairo, Egypt
[3]UFE Université Française d'Egypte, Département d'Architecture, Cairo, Egypt
[4]INSA de Strasbourg, Département d'Architecture, Strasbourg, France. e-mail: randaabdelaziz@yahoo.fr, randa.mahmoud@insa-strasbourg.fr

Abstract. In large parts of the Gulf countries, neighbourhoods' mono-functional structures with their low density and their lack of social functions, lively public spaces, public transportation and climate responsive design features have catered for an environmentally unfriendly development pattern. Borders between neighbourhoods are mostly socially repulsive as they usually host large residual and underutilized spaces such as oversized main roads and other infrastructure elements. Consequently, these residual areas constitute large breaks in the neighbourhoods' building pattern and physical barriers for the inhabitants. Since we see a great potential in these residual areas, we have elaborated a planning tool which is generally applicable for sustainable development of transitional spaces between neighbourhoods. This study demonstrates different strategies on how these urban thresholds between neighbourhoods can be transformed into central active mixed-use public spaces that will be reachable for all residents by walking. Accordingly, they can promote for social interaction between different neighbourhoods and thus social sustainability. Within this paper, at first, the thresholds between different residential neighbourhoods' structures are examined to better understand their qualities and deficits. Secondly, various sustainable development options for the threshold areas are demonstrated. An exemplary site in the Capital Area of Muscat, Oman serves as a case study. The third part consists of an evaluation of the elaborated proposals according to their ecological, social and economical efficiency in order to create adequate implementation strategies. Finally, we propose that the ecological solutions for residual areas as developed in this paper for the GCC region can be expanded to other hot-climate countries as a planning tool to achieve sustainable built environments by connecting different neighbourhoods.

1. Introduction

Ecological footprints and sustainability with all its aspects are preoccupying urban planners and designers nowadays. However, residential neighbourhoods in a large part of the Golf countries have shown the other way round. In most cases, neighbourhoods are developing as separated entities with mono-functional structures which usually are of low density, lack social functions, lively public spaces and public transportation. Moreover, only few showed climate responsive design features.

IOP Conf. Series: Earth and Environmental Science **329** (2019) 012018 doi:10.1088/1755-1315/329/1/012018

1.1. Thresholds between neighbourhoods as a real problem that needs to be resolved

The separation between neighbourhoods creates residual areas or thresholds that are mostly socially repulsive as they don't provide neither economic nor social activities. They usually host infrastructure elements, oversized main roads and underutilized spaces that could host potential economic/social poles. These areas are usually of low density and free of public transportation. This is why there is a high ratio of car dependency. Furthermore, the lack of climate adaptation and public utilities plus the mono-functionality in these thresholds have established a sense of social repulsiveness. Consequently, these residual spaces constitute large breaks in both the urban fabric and the social structures, as well as physical barriers for the inhabitants.

1.2. Thresholds as urban tool connecting neighbourhoods and achieving social sustainability/resilience

The social part to ultimately achieve sustainability was mostly neglected in most of urban planning trends, especially in the middle east. However, this approach which received very little attention is a main key to achieve sustainability in general [1]. Social sustainability is about resilience and stability of different social groups, which form one population, and the way they practice their social lives in the urban space. The more these groups are blended and tied together through social practices, the more the community will be resilient. This is why decisions and urban policies which consist of producing connected/loose urban fabrics contribute largely in creating or not strong/weak social bonds in the community. In order to achieve sustainability of the built environment, the key is to aim for social cohesion as a common goal in the new growing communities [2]. On the one hand the thresholds, between neighbourhoods in the Capital Area of Muscat, can be of a great potential and the residual areas can be used as a planning tool to connect neighbourhoods instead of separating them. Thresholds can be the bonding nodes in the urban fabric. They may attract different population to come together and to exchange. These nodes can be considered to some extent as a reinforcement tool for the urban fabric and the social structure of the community, hence they will promote social resilience and cohesion in these neighbourhoods. On the other hand, these nodes can be planned in a way to offer new propositions for sustainability via using ecological building materials and renewable energy sources suitable for the environment. We believe that finally thresholds can be a key to minimize exclusion and to achieve social diversity and self sufficient resources.

2. Research Background and Challenges

Urban development of neighbourhoods in the Gulf area is currently faraway from ecological notions. Each neighbourhood represents an enclave by itself where inhabitants live and practice their urban space separately. The lack of social interaction between neighbourhoods is a threat to social/urban sustainability of the whole city. Consequently, the threshold areas between these enclaves, if left without a clear planning strategy, will be a threat to these neighbourhoods' sustainability in general. We believe that there is a great potential in these unplanned areas as they could be transformed from barriers to adhesive zones and they could explore new economic opportunities. These thresholds could become the key to social/urban sustainability in the neighbourhoods, as every neighbourhood will be a part of a bigger tissue which integrates all its cells together in a balanced way. This proposition will reinforce the social cohesion of the population as well as the community resilience.

It will be as if we imagine that each neighbourhood represents a cell from a bigger tissue and each cell functions on its own vs. cells that function in harmony all together in one unifying organism.

Social sustainability is a must in order to manage urban and economic sustainability. The re-management of thresholds – which are existing between the neighbourhoods in Muscat, Oman – could contribute to achieve this social sustainability.

Figure 1. The challenges suburban residential neighbourhoods in the GCC countries are facing today.

3. Paper's Aim and Interests

The current mode of urbanization in the Gulf Countries, doesn't really take into consideration their ecological footprints. The problem is deeply aggravated by the existence of threshold spaces between different neighbourhoods. This study aims at highlighting planning strategies which could be used as a tool when it comes to the problem of residual spaces resulting from different neighbourhoods' implantation. These strategies take into consideration the existing urban fabric as an asset. They can be igniters to make the highest profit from the existing urban tissue via reducing the urban threshold inconveniences and maximizing the spatial liberty and their interactions. We believe that these strategies could be of a help to the gulf cities in shifting towards more climate resilient awareness and sustainable built environment development [3].

4. Neighbourhoods Today and Their Threshold Spaces

Urban policy in Oman promotes for the spread of detached mono-functional neighbourhoods. In fact every Omani individual is assigned at the age of 25 a piece of land on which she or he is supposed to build their own houses. The problem is not all the Omanis are ready to have their houses built at this age, which led to urban vacancies through neighbourhoods[4]. The fact that each individual is assigned a piece of land while he is not ready for the construction, plus the Gulf culture of having stand alone house even on small piece of land for intimacy reasons, cater for the creation of vacant areas between neighbourhoods. The microclimate of these residual spaces is especially harsh not only due to the high temperatures, but also because of unprotected sun and wind exposure. Furthermore, the hot climate makes it even harder to develop or to plant the resulted residual areas. Consequently, the neighbourhoods are disconnected from each other. All these factors are part of the reason for the current unsustainable development. This phenomenon was also nourished due to the socio-economic rapid growth of the gulf countries in general. Accordingly, the common residential building type is mostly the wide-spread single family house. The Capital area of Muscat plus the development of its suburbs, stretched for more than 100 km along the coastal plateau of Al Batina, is a living example of what a loose urban fabric looks like [5]. The loose urban fabric in Muscat, especially in the residual areas between neighbourhoods, has manifested the following symptoms within various levels:

4.1. Functional aspects

Most areas have mono-functional structures. The density of buildings and population is low. Buildings and public street spaces are completely exposed to the sun. Buildings, streets and the large amount of sealed surfaces contribute to urban overheating which causes the urban heat island effect. Buildings are arranged in an unsustainable way, many plots remain inbuilt. Climate responsive urban and architectural design strategies are missing.

4.2. Social aspects

The walkability within the neighbourhoods is very limited. Social functions are lacking. Lively public spaces are not existing. There is no public transport. Due to the lack of mobility, several population groups are socially excluded (e.g. elderly people, children, handicapped, low income expats). Social interaction is absent. Social repulsiveness has been established.

4.3. Threshold spaces
The borders between neighbourhoods consist of large residual and underutilized spaces, often as oversized main roads and other infrastructure. They constitute physical and mental barriers. Due to their large size they can be seen as breaks in the building pattern.

Figure 2 Al Khoud, Capital Area of Muscat / Oman – typical example of a residential neighbourhood with large areas of unused street space (threshold and residual space indicated with red texture), GPS 23.604316, 58.164498 (Source: Map data © 2018 Google, Digital Globe).

5. Case Study Area in the Capital Area of Muscat Oman
The area of Al Mabeila as indicated in figure 4 served as an exemplary site for the case study analysis.

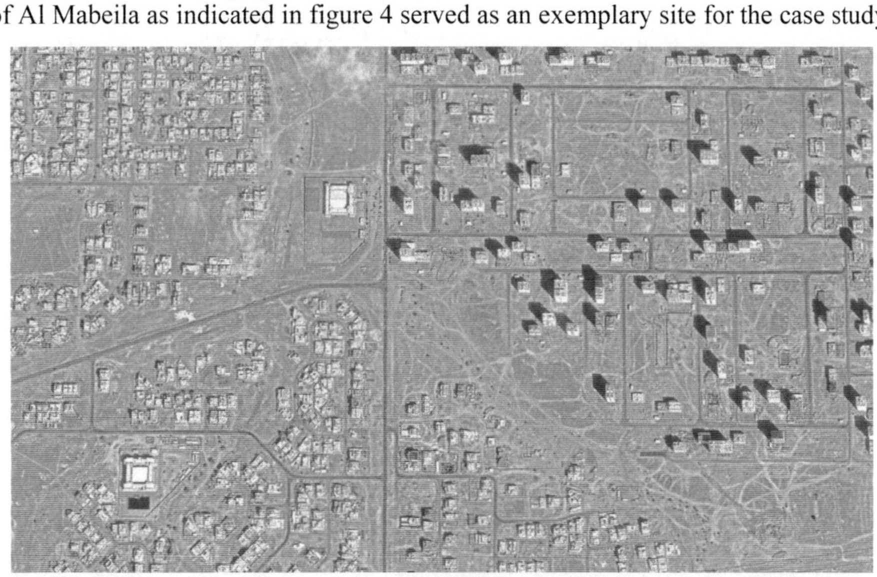

Figure 3. Satellite image of the case study area with diverse neighbourhoods and large unused spaces between them in Al Mabeila, Capital Area of Muscat, GPS 23.380301, 58.074790 (Source: Map data © 2018 Google, Digital Globe).

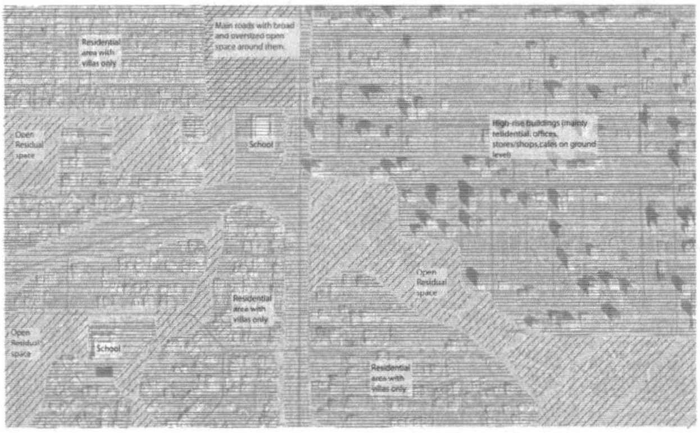

Figure 4. Satellite image of the case study area with indication of threshold and residual spaces.

6. Methods of Integrative Urban Development of Threshold Areas

We have developed a strategy of how the areas can be transformed into an integrative urban development with central active mixed-use built environments with lively public spaces which will be reachable for the residents of the adjacent neighbourhoods by walking. Thus the former threshold areas can promote for social interaction between different neighbourhoods and create social sustainability. Among others, key indicators for the success of the new urban areas are a re-densification of the neighbourhoods with a minimum of ca. 10,000 inhabitants/km^2 and public functions on the ground level of mixed-use buildings. Also the public exterior spaces are to be re-developed. The proposed urban green is to be irrigated with recycled wastewater from the households. Figures 5-8 are demonstrating this in regard to the individual interventions.

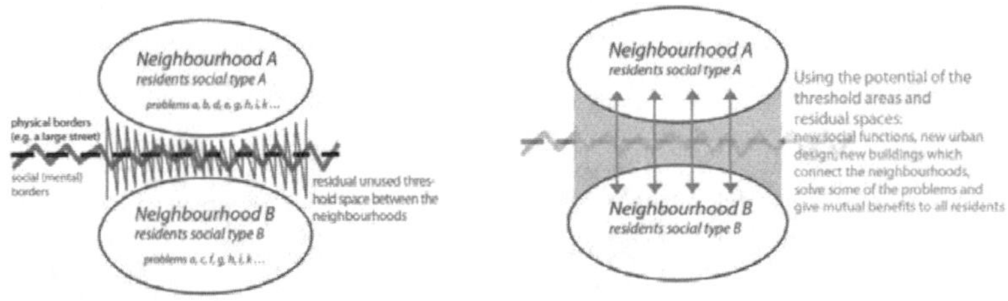

Current Situation Improvement Proposal

Figure 5. Diagrams of current situation (left) with separated neighbourhoods and unused / wasted residual threshold spaces and proposed transformation of the threshold areas into connecting areas.

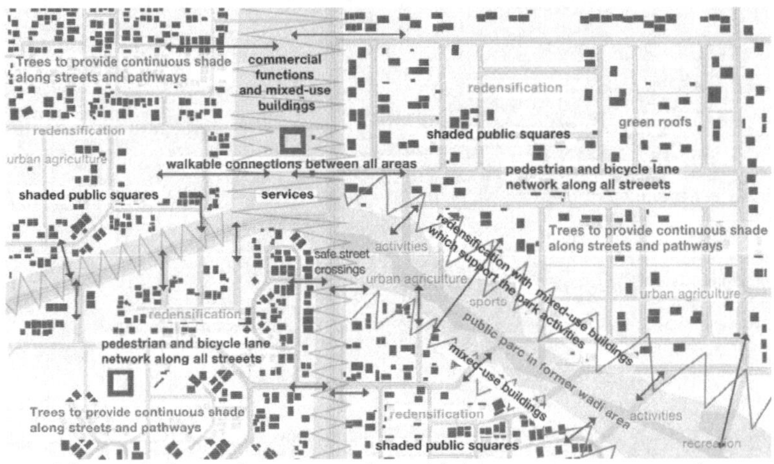

Figure 6. Strategic concept plan with improvement proposals.

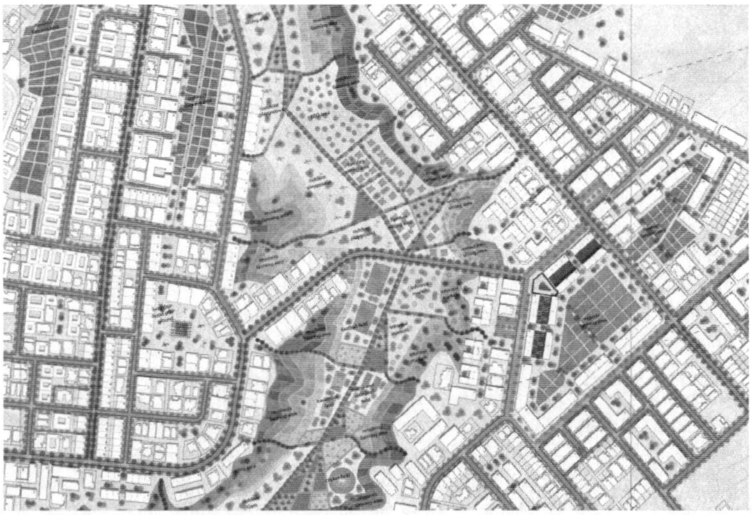

Figure 7. Design example from a student project supervised by one of the authors (Bachelor Thesis 2018, student: Hajir Al Alawi)

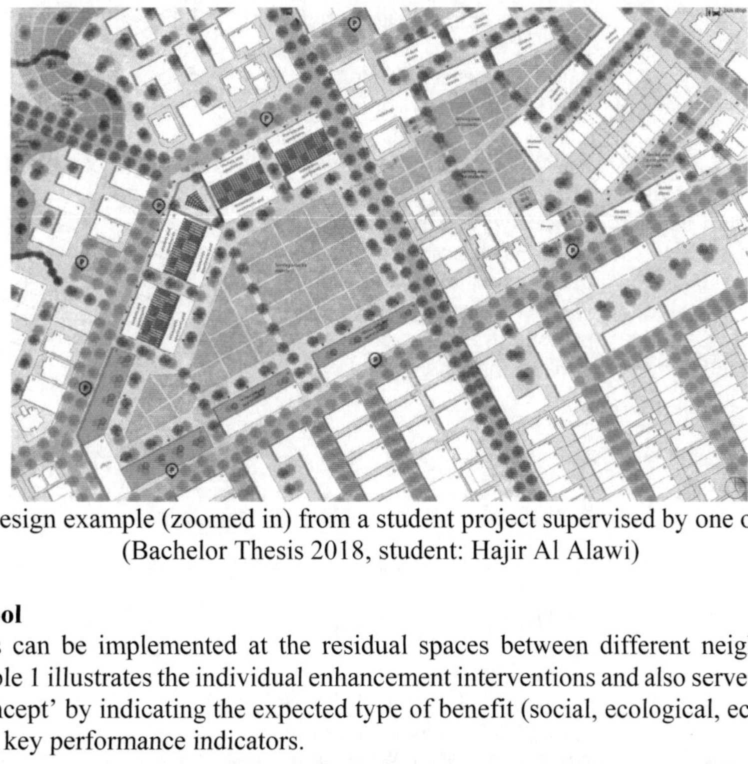

Figure 8. Design example (zoomed in) from a student project supervised by one of the authors (Bachelor Thesis 2018, student: Hajir Al Alawi)

7. Planning Tool

Urban nodes can be implemented at the residual spaces between different neighbourhoods. The following table 1 illustrates the individual enhancement interventions and also serves as a preliminary 'proof of concept' by indicating the expected type of benefit (social, ecological, economical) which also serve as key performance indicators.

Table 1. Toolbox of enhancement interventions for a more sustainable and neighbourhood connecting design

	Goal	Intervention	Type of Benefit Social	Ecological	Economical
1	Breaking physical barriers (e.g. roads), thus increasing the area's walkability	Creating safe pedestrian crossings	√	√	

2	Create enhanced liveliness and social interaction within a neighbourhood and between different neighbourhoods, increased number of homes, more services	Increasing population and building density by adding new mixed-use buildings in a compact arrangement	√	√	√
3		Creating urban nodes with services and public squares / plazas, all daily needs will thus be placed in walkable distance for all residents	√	√	√
4	Better adaptation to climate, reduction of urban overheating (urban heat island), improved air quality, increased walkability	Implementation of local plants, creation of shaded pathways with trees (irrigated with recycled household water)	√	√	
5		Green roofs, green facades with the use of local plants (irrigated with recycled household water)	√	√	√
6	Reduction of urban heat island by improving the solar reflectance of surfaces	For all sun exposed surfaces: use of materials with bright colours and smooth surfaces which have a high solar reflectance index (SRI). Use of pavement only when absolutely necessary, unsealed areas could be covered with green (irrigated with recycled household water)	√	√	
7	Better interconnections between the neighbourhoods for pedestrians and cyclists, mobility improvements, reduction of car use	Implementation of net of pedestrian and bicycle paths which connect the neighbourhoods and newly developed former residual areas	√	√	√
8		New network of pedestrian and bicycle paths to connect with existing network of residential streets. Existing connections to be extended for pedestrians and bicyclists	√	√	
9	Sustainable and healthy food production, reduced dependency on food import, improved micro-climate	Urban agriculture	√	√	√
10	Increase liveliness of neighbourhoods	Playground areas and outdoor activities	√		
11	Schools and kindergartens accessible by walking	New schools and kindergartens to be placed on former residual areas	√	√	√
12	Less emissions, less dependency on fossil fuels	Renewable energy generation (e.g. with building integrated photovoltaic, shading devices with integrated photovoltaic, small wind turbines)		√	√
13	Enhanced outdoor comfort, better microclimate	Wind corridors enable natural ventilation of the neighbourhoods	√	√	
14		Recreational green public spaces	√	√	
15	Mobility improvements	Public Transport Connections	√	√	
16		Providing public E-Bikes / E-Scooters	√	√	
17	Strengthened local economy, offering sustainable products	Local markets	√	√	√

18	Participatory development approach	Community-based participation in the design and management process (e.g. place making)	√	
19	Energy efficient building techniques	All new buildings are to be built in an energy efficient way		√
20	Increasing the population density	Use of adequate building typologies (e.g. apartment blocks, townhouses, courtyard houses; no villa nor twin-villa typology)	√	√

8. Conclusion

Thresholds spaces between neighbourhoods aggravate the urban evolution and diminish the eco-friendly development of residential areas in the Gulf Countries in general. Muscat in Oman is a living example for this pattern. This current mode of urbanization increases ecological footprints and plays against the sustainability of the built environment with all its aspects. The paper suggest the introduction of a planning tool which will connect separated neighbourhoods while targeting a wide spectrum of the population. This will offer the inhabitants a safe, secured climatically adapted mixed-use public urban space which strengthens exchanges and interrelated social practices. New knitted social relation between different neighbourhoods will promote especially social sustainability. The latter will be a key to achieve sustainability on a larger level. More explicitly, on urban level, the suggested urban nodes can be inserted to connect and enhance the urban qualities of the loose urban fabric in the void areas between neighbourhoods. They will introduce new attractive social activities which were lacking before. Consequently, they will also present new economic potentials and opportunities, as each neighbourhood will be more anchored in the urban tissue while having a new specific sociological and economic role, the social cohesion and community resilience will be reinforced. On architecture level, new urban nodes will be an opportunity to architectural development with local eco-friendly construction materials in a contemporary trend in order to support the built environment sustainability [6]. Finally we believe that the analyses, research and application on Muscat as a case study, proved that connecting neighbourhoods – via implementing urban nodes in the existing thresholds between residential areas – can be used as a planning tool to lead suburban residential districts in the Gulf Countries towards sustainable built environment. Furthermore, many aspects of the tool are also valid for other hot-climate countries.

References

[1] Hopwood, B., Mellor, M., O'Brien, G., (2005). *Sustainable Development: mapping different approaches*, in: Sustainable Development, 13: 38–52, 2005.

[2] Rashed, R., Mahmoud, R., (2018), Urban Social Sustainability in the Era of Digital Technology: the Case of NEOM, the World's Future Global Hub across Three Countries in: *the International conference of Sustainable Design of the Built Environment SDBE18,* London UK.

[3] Calderon, F., Stern, N. et al., (2014) *Better Growth, Better Climate – The New Climate Economy Report 2014*, The Global Commission on the Economy and Climate, World Resources Institute, Washington D.C.

[4] Kader, A. (2015). Design Proposals for a More Sustainable Urban Development of Residential Quarters in Oman. *Trialog A Journal for Planning and Building in the Global Context, 114*, pp 21-26

[5] Kader, A. (2018). Upgrading the Energy Performance of Existing Middle-Class Suburban Residential Buildings in the Gulf Region, *International Conference for Sustainable Design of the Built Environment SDBE 2018, Conference Proceedings*, pp 171-180

[6] Mahmoud, R., Rashed, R., (2019), Gentrification of Vernacular Materials as a Key Towards Sustainable Built Environment, in: *The International Conference "ArchCairo8, Building the Future Now"*, Cairo / Egypt, 2019

Guidelines for designing a sustainable neighbourhood based on TND theory in Egypt.

A Bondok[1], R Nasreldin[2] and A Shalaby[3]

[1] Researcher in urban design, Department of Architectural Engineering, Faculty of Engineering, Cairo University Giza, Egypt. eng.ahmedbondok2@gmail.com.
[2] Assistant professor, Department of Architectural Engineering, Faculty of Engineering, Cairo University Giza, Egypt. raniaibrahim@eng1.cu.edu.eg, raniaibrahim222@gmail.com.
[3] Professor of Architecture and Urban Development, Department of Architectural Engineering, Faculty of Engineering, Cairo University Giza, Egypt. And Visiting Assistant Professor Department of City and Regional Planning, Cornell University ,USA, amshalaby@gmail.com

Abstract. Neighbourhood is a particular geographic area that encloses a social network in which people interact socially. Therefore, the most appealing aspect of sustainable urbanism is certain to be the sustainable neighbourhood. This paper focusing on the urban development which contributes to the sustainable development of neighbourhood. Moreover, the problem of housing in Egypt is not a quantitative problem but lack of achieving the elements of sustainability and the new neighbourhoods are designed with the same basis as the Egyptian cities and thus will show the same results as in the existing cities. Consequently, there is absence of modern principles, lack of awareness for the importance of using new sustainable urban planning strategies in Egypt. This paper proposes a set of guidelines based on TND principles for the implementation of sustainable neighbourhood criteria of development. The methodology of this paper was literature review to analyse previous researches of sustainable neighbourhood development based on TND theory as well as, comparative analysis between national and international case studies. Furthermore, comparison between European and Egyptian case studies which used for presenting and analysing sustainable development criteria that used in real neighbourhood contexts. The outcome of this paper expected to influence decision-making in the area of sustainable urban development. Besides, helping to build a new and sustainable relationship between urban dwellers and urban space, and to increase the value of urban land and make our communities more sustainable.

Keywords: Sustainable Development, Traditional neighbourhood, Modern principles

1. Introduction

In the last few years, the neighbourhood's shape has extremely changed because of the quick urban growth. Urban sprawl is considered as an essential feature of fast growing neighbourhood, it is generally go along with various problems containing Inadequate land use, Low density rate, and car dependency as well as, Isolation usage. The future neighbourhoods should construct a diverse type of urban space, in which neighbourhood life flourishes and the most common issues of modern urbanization are treated. This paper suggest a set of guidelines for the application of sustainable neighbourhood development in order to assist in the link between sustainable urban areas and residents of neighbourhood toward the increase of land value

Research Problem: there is a great tendency by the Egyptian governorates to build new urban communities, but unfortunately these cities are being built in the same old urban design standards. Therefore, these new cities and communities have not proved to fulfil their designated goals without concerning the usage of these resources in the future. It is known that the problem of housing in Egypt is not a quantitative problem but, the problem of achieving the elements of sustainability in residential neighbourhoods.

Research Objective: the objective of this paper is to extract criteria of designing a sustainable neighbourhood development in urban aspect sector through proposing a set of guidelines based on TND principles for the implementation of these criteria of development.

2. Neighbourhood and Sustainable Urbanism.

Neighbourhoods are extremely important in the planning field, and neighbourhood-based planning grows constantly. Neighbourhood planning has more than one purpose, (Wendelyn, 1995) selecting and defining target neighbourhoods is a highly political and negotiable process, Defining neighbourhood for programmatic ends in any given case is always challenging. The title "neighbourhood" is a term given for the study purposes described, to form the relationship of a family life in a community. Investigations have shown that residential communities, when they meet the universal needs of family life, have similar parts performing similar functions. Those parts are used together in the neighbourhood unit system as a whole [1].

2.1 The Traditional Neighbourhood Definition

The basic increment of town planning is the traditional neighbourhood. The neighbourhood idea still strong however the size increases to city scale. Actual traditional neighbourhoods fulfil all requirements such as the housing, offices places, shopping spaces, city functions, and more facilities in sustainable and satisfying way by making neighbourhood more complete, compact, and connected. Even though the differ of numbers, five essential points of design are agreed to provide a common thread connecting great neighbourhoods [1].

Table 1. Neighbourhood attributes which agreed through the Congress charter of New Urbanism [2].

Neighbourhood Attributes	
Identifiable centre and Edge	The centre of the neighbourhood should be able to inform when any one has reached neighbourhood and its centre.
Walkable Size	The size of the neighbourhood should be from 40-200 acres and also should be suitable for walking.
Mix of Land Uses	The larger neighbourhoods have an accurate mixture of land uses and kinds of housing. Any mixture of uses significantly decreases the number of outer car trips wanted by dwellers, At least three dwelling types are necessary to create architectural diversity.
Integrated and Walkable Streets	A street network allows cyclists, pedestrians and drivers to move safely and restfully in a neighbourhood.
Special Sites Are Reserved for Civic Purposes	Sites are prepared significant by the geometry of the town design. Prominent locations, such as a terminated view seen down street or at a hill top, should be kept for landmark constructions. These sites are with intent chosen for building sites that will seal the extended view over a street or for anchoring neighbourhood square or a prominent street corner.

2.2 Traditional Neighbourhood Development (TND)

A human scale, walkable community with average to extreme residential densities and a mixed use centre is a Traditional Neighbourhood Development (TND). TND has a wide range of possibilities to improve modal split in order to accommodate other means of transport. It is essential in TND to design a connected network of narrow alleys with less curb radios. Traffic will be reduced by this network and new facilities for pedestrians will be created. Those guidelines are presented by NCDOT as the general design shows that non-vehicular trips are going to be available which in the same time wouldn't influence the safety considerations [3]. **The goal** of TND is to push residence to walk or bike more, decrease the vehicle external trips and the travelled distance as well, by 15% or more from a commercial insight, into a walkable community, Heavy traffic should be less [3].

2.2.1 The principals of Traditional Neighbourhood Development (TND)

- The density of the neighbourhoods increased
- There is a noticeable core in the neighborhood. This is a green or a square on a regular basis and a busy or memorable road corner sometimes. There would be a transit stop in this centre.
- Most of the dwellings are within 5 minutes walking distance from the center, an average of about 2,000 feet.
- There are a range of kinds of dwellings generally houses, rowhouses and flats so that younger and older people, individuals, and families, the poor and the rich can find places to live.
- Lot sizes and kinds of house differ in the same block.
- A tiny ancillary building is allowed inside each house's backyard. It can be used as a rental unit or workplace.
- A primary school is near enough to allow most kids to walk from home.
- There are small playgrounds that are not more than a tenth of a mile away from every dwelling.
- Neighbourhood streets form a linked network that disperses traffic by offering a variety of a multitude of pedestrian and vehicle paths.
- The roads are shaded by rows of trees and. This slows down traffic and creates a appropriate atmosphere for pedestrians and bicycles
- Parking is relegated to the back of houses, which are generally accessed through alleys.
- The neighbourhood should to be self-governing is structured. A formal association is debating and deciding on maintenance, safety and physical change issues.

2.3 LEED For Neighbourhood Developments (LEED-ND)

Green neighbourhoods are a fundamental part of the clarification to the environmental encounters dealing with the planet. Speedy urbanization requires community planning techniques which might be perfect for green intervention and transformation, as an opportunity for sustainable solutions including a variety of a small green enterprise beginning to a massive-scale of environmental services. LEED is a certificate program and internationally acknowledged standard for construction and design procedures for powerful green buildings. Produced by the USGBC in 2000 via a consensus based mostly process, It is a market-driven tool as well as, a third-party established standard of sustainability and process through the whole lifecycle of the building [4]. The degree of neighbourhood, those "drawing forums" can catalyse LEED for "Neighbourhood Development" (ND) policies, inclusive of low-priced housing, climate safety, and stepped forward public fitness. The country wide affiliation of Realtors lately located that -thirds of households might prefer a smaller domestic within taking walks distance of eating places, stores, and schools over a huge-lot belongings farther away with equipment like LEED ND, neighbourhoods can come to be prototypes of revolutionary management in achieving sustainability objectives [5]. Growing evidence points to a robust public choice for sustainable built environments, an inclusive of LEED ND's essential factors of traditional neighbourhood design as well as, multimodal tour.

3. Method

The research topic adopted on a scientific methodology, as shown in figure 1, The current study relies on a wide theoretical background and literature review based on an Inductive approach through collection, arrangement and classification of information about neighbourhoods and sustainable urbanism, TND theory and mention to LEED-ND . Comparative analysis of National and International Case In order to select the best strategy of sustainable neighbourhood development. In which successful narratives in redeveloping neighbourhoods are presented. The BO01 in Malmö, Sweden and Kronsberg in Hannover, Germany were chosen as international case studies due to their success in sustainable neighbourhood development. To get recommended strategies of the aforementioned urban projects will be highlighted; thus, establishing their contribution towards successful sustainable neighbourhood to suggest a set of guidelines can be applied on planning other new cities in Egypt.

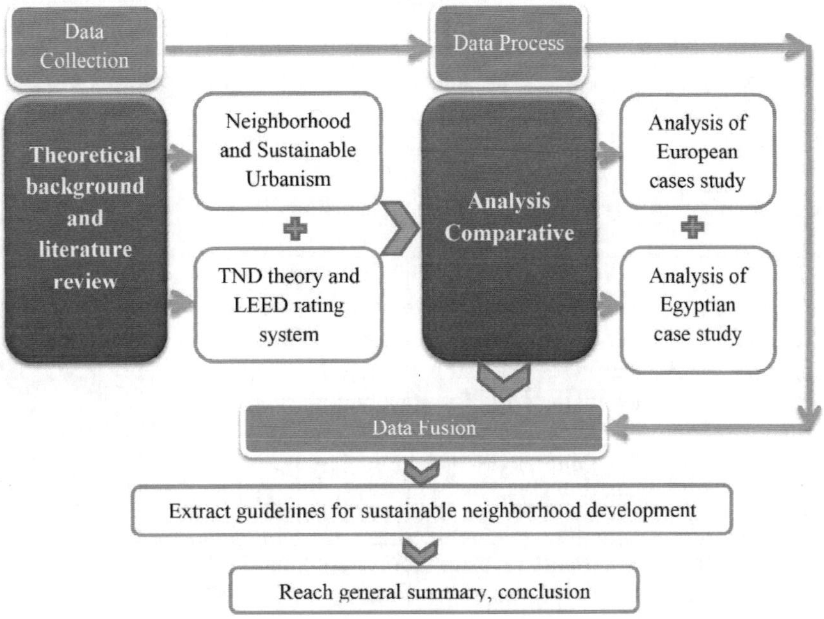

Figure 1. The figure illustrates the methodology of Research- Developed by Authors

The first step is to define the location and the administrative borders of the study area which is the criteria for analysis represented below in Figure 2 for any neighbourhood. However, this paper focuses on the urban aspects, through analysing the context based specifically on the following urban aspect criteria as shown in figure 2. In order to extract guidelines for the neighbourhood development from the perception of urban aspects.

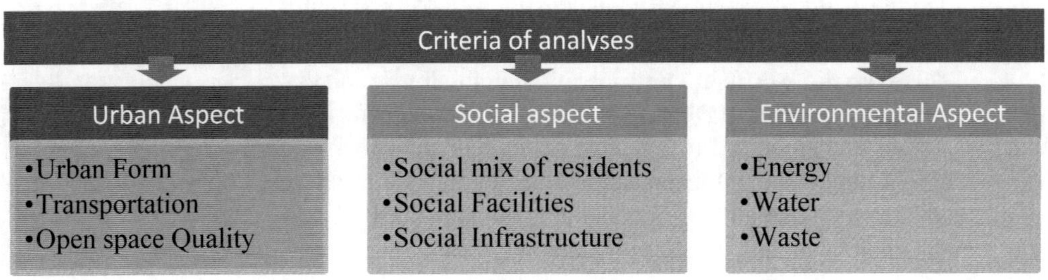

Figure 2 . The figure shows the criteria of analyse and the selected criteria of this research Developed by Authors.

3.1. Six October city -National Case study.

The Six October City was chosen as a case study for many reasons: It has an area of about 98842.2 acres. Has a potential that make it one of the most important new cities in Egypt. Moreover, it is an industrial, educational, commercial and residential city as shown in figure 4. Through these potentials, if the city is dealt with using the modern planning theories such as the theory of TND, it will produce sustainable neighbourhoods and accordingly sustainable cities [6].

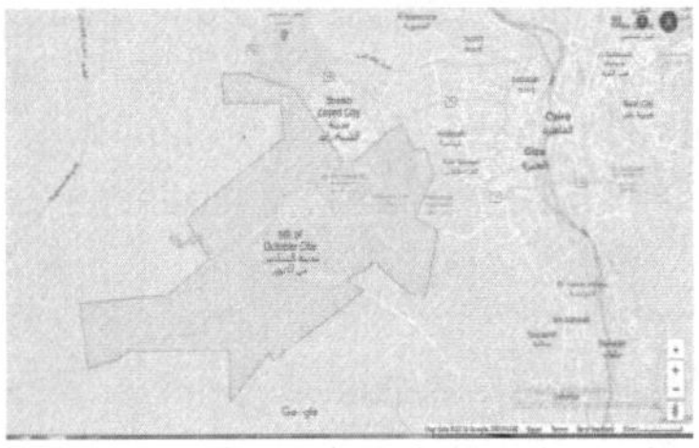

Figure 3. The 6th Of October city, location and its administrative boundaries in Google map.

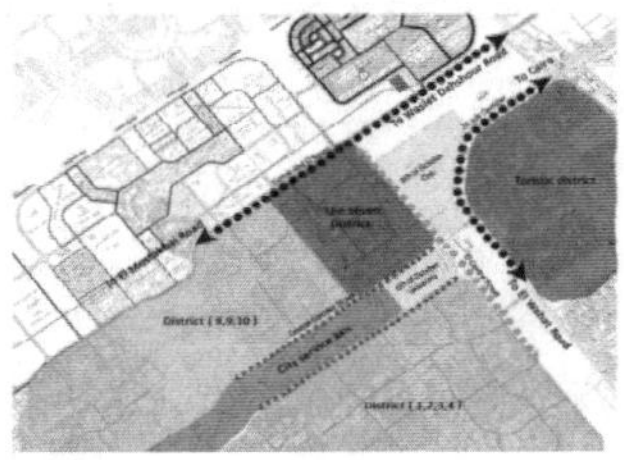

Figure 4. The Seventh District location in city

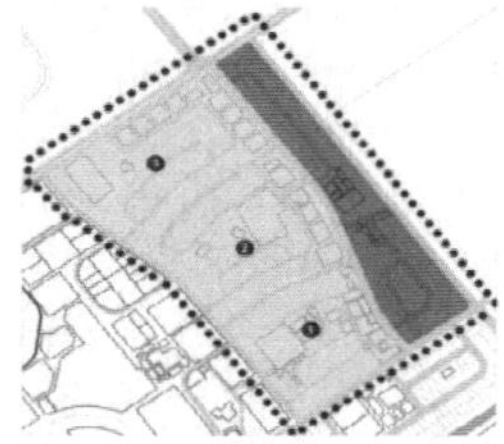

Figure 5. The Selected neighbourhood.

Table 2. Collective analysis for the National case study in urban aspect sector [6]

Urban aspect		
Urban Form		
Location	The area of study is located on the north of the districts area of the 6th of October and its boundaries are as follows: The North boundary: the future extension of the 6th of October city and the north extension area. The East boundary: the 6th of October Club. The West boundary: common service area for the Seventh and Eighth Districts followed by the Eighth District. The South boundary: The 6th of October University.	
Master Plan	The master plan consists of four neighbourhoods with total area 121.8 acres, has two types of housing (apartments and privet homes) and has a mosque and one school for each neighbourhood**.**	
Density	This case has area 121.8 acres and 402 number of dwellings so the density is 3.3% as there is still a shortage in the exact definition of true neighbourhood density. Referring to figure	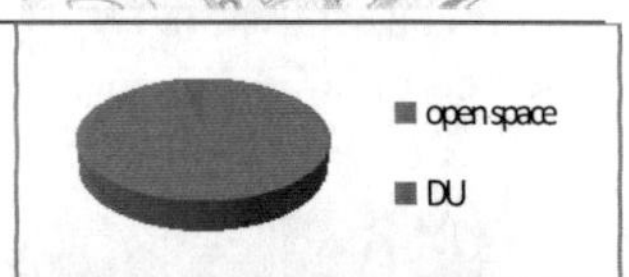
Scale	There are three different and specific heights for the blocks in the area of study: Blocks consisting of a basement floor and ground floor and two typical floors, Blocks consisting of a ground floor and three typical floors, Blocks consisting of a ground floor and four typical.	

Land use	Mixed and different types of land use.as follows: Residential 87.4%, Residential-commercial 6.5%, Residential-Administrative 2.7%, Commercial 0.5%, Educational 0.75%, Religious 0,75%, Public Service 1.25%.	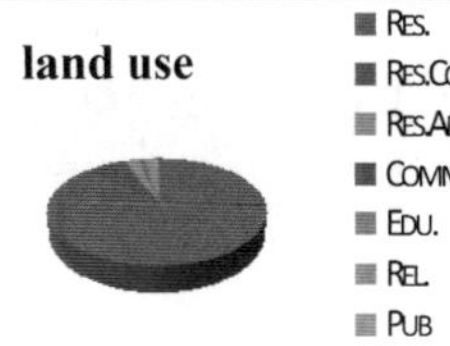
	Transportation	
Streets	There are four types of streets in the study area. The first one is the main axis to the city with width 30 meters. The second type of streets that boundary the area and its width is about 20 meters. The third type is the main streets within the district and its width is about 16 meters. The last type is the inner streets between the residential blocks and the width is about 10 meters.	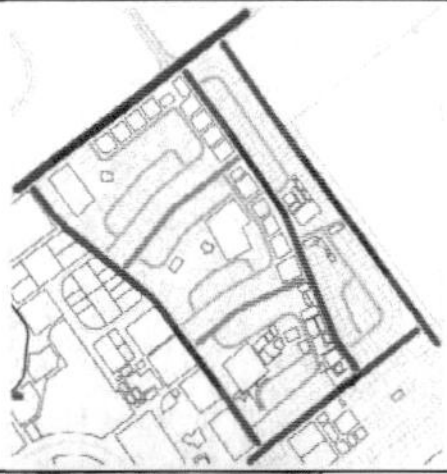
Parking space	There are no specific parking spaces on the District and depended on the parking on streets beside the buildings.	
	Open Space Quality	
Court yard	The most important potential of the case study is the courtyard there are many of them exist between buildings but without design or function is clear	
Avenue	There is no street that can be called this name in the study area but there are streets with the same standards to become avenue.	
Park	There are green spaces neglected within the district and not exploited and there is no use or classification.	
Green Corridor	There are corridors between the buildings but not designed and there is no specific network that can be used to connect the neighbours to each other	

3.2. *International Case studies.*

In this part we will analyse two successful case studies of sustainable neighbourhood development. The case is Bo01, Malmö, Sweden. Known as the "City of tomorrow". The second case is Kronsberg, Hannover, Germany. It is a complete case of visionary urban arranging and development with area 172 acres. The aim of analysis is to compare the similar problems between the national and international case studies to extract criteria that used in successful cases.

Table 3. Similar problems of Sixth of October, Sweden and Hannover [7].

Common issues	The Seventh District in Sixth of October	Bo01 in Malmö	Kronsberg in Hannover
Urban Design	-There is no development plan for the district. -Low density. -Lack of mixed use.	Shows that the formal characteristics of the planning system only provide limited insights into how spatial planning is conducted,	Lack of joint regional planning and development incentives,
Transportation	-There are no pedestrians pathes. -Lack of public transit. -Priority for cars	-The proportion of car journeys is high. - Traffic jam.	Lack of public transport and regional roads.
Open spaces	There are unused open spaces.	Lack of green outdoor environment.	Needs to recreation areas and zoological garden.

IOP Conf. Series: Earth and Environmental Science **329** (2019) 012019 doi:10.1088/1755-1315/329/1/012019

In each of two neighbourhoods, the agency planned, and it's played a serious role through the political leaders incorporating the process. The cities created special development committees, usually with their planning agencies, outsource consultants, citizen's groups as well as the authority services.

Table 4. Similar strategies of two international case studies [7].

Similar strategies	Bo01, Malmö, Sweden.	Kronsberg, Hannover, Germany.
Urban Form	Area: 54 acres, Density: 29 dwelling units/acre, parking ratio: 0.5%, -Has given priority to pedestrian and bike circulation, with frequent pedestrian shortcuts through blocks. -Has a mixed uses including shops, schools, services with 275% jobs per housing ratio. Has used a larger block type with a strong contrast between inside and outside, intimate spaces with expansive water views. -Has diversity in block types, architecture ideas and developed by a different architect developer team.	-Area: 172 acres, Density: 18 dwelling units/acre, parking ratio: 0.8%. -Also has particular significance to pedestrian as well as bike circulation, with frequent pedestrian shortcuts through blocks. Also has a mixed uses including shops, schools, services with 67% jobs per housing ratio. -Has developed a progression of block types from closed to open, from a one sided commercial strip to a more open landscape edge. Also has diversity in block types, developed by a different architect developer team.
Transportation	Each of two neighbourhoods transit system require a public (bus in addition to tram) as a crucial factor of the development, with the subsequently main criteria's: • 300-400 meters has a convenient location of transit stop for dwellings. • 6-8 minutes of intervals Headways. • Schedule Knowledge as well as Climate shelter was offered in Transit stops. • Connections are provided to desired destinations. • Each has street designs through traffic comforting measure in addition to street networks which gives importance to pedestrian then bikes.	
Green Space	Each neighbourhood authorize how the urban landscape can entirely alter the sensory knowledge of the city, consuming a favourable result not only energy in addition to, carbon dioxide emissions, besides environment however, it's further significant, on health and well-being. For the two neighbourhoods it has been a straight contact sharing of an extended urban landscape while, separating away the green courtyards that are inside the blocks. Moreover, ratio of green is 40-50 percent of surface area.	
	The treatment of the water's edge overlooking the Öresund sound has become a defining city amenity.	-In Kronsberg shaded alley gives definition at the edge of the neighbourhood.

4. Result

The focus of the previous analysis of international and national case studies on the urban aspect only because this research deals with this aspect of development only, but it is necessary to study the social and environmental aspects in next researches to reach the comprehensive sustainable development of neighbourhoods through the above has been extracted a set of guidelines for the sustainable development of neighbourhoods as followed in the table 5:

Table 5. Urban aspect guidelines for sustainable neighbourhood development, Developed by authors.

	Guidelines
Urban Aspect	**Consideration of neighbourhood Urban Aspect.** • It's essential to determine the centre of the neighbourhood, The overall size of the neighbourhood should be range from 40-200 acres as well as, 400 meters is a preferable distance in the neighbourhood to most people to reach their needs. • Neighbourhood should have a mix of land uses and housing types. At least three dwelling types are necessary to create architectural diversity. Furthermore, Civic building should be in terminated visit seen to be a land mark building and Squares are often used for civic purposes and the sizes are at least one acre. • Having diversity in block types and architecture ideas which developed by different architecture developer teams. • Transference between different building types should occur at the rear plot lines. The neighbourhood streets should allow pedestrians, cyclists, motorists with safe and comfort and the speed of vehicles inside neighbourhood should be less than 40 Kmh. in addition to, Building Heights is 1:3 of streets width. • Density within 18-29 dwellings unit/acre as well, Priority is given to pedestrians and bicycles. • Make a convenient location of transit stop within 300-400 meter and Transit stops offer real time information on schedules and Headways are at intervals of 6-8 minutes. • Parking ration within 0.5% - 0.8% and creating a compact linear configuration related to a transit spine. • Each dwelling units has access to public green space within 300 meters as well as, Set a green space factor must achieve in neighbourhood.
	Criteria's for sustainable neighbourhood development

5. Conclusion and recommendations

The Research proposed a set of guidelines for sustainable development in addition to improvements the current sustainable urban planning philosophies. By using TND theory's principals represented in paper above. Through this research was realized that sustainability of urban aspect characterize a field of research in continuous development. Specialists are adjusting up-to-date sustainable development solutions for particular apprehension of neighbourhood even though, it is always challenging. The research was set out to recommend a fundamental way out for the neighbourhood development problem in Egypt.

References

[1] Jacobs, A., Klein, J., Koetter, F., Krier, R., & Hill, J. (2003). *TIME-SAVER STANDARDS for URBAN DESIGN.* (U. Des, Ed.) TIME-SAVER Stand

[2] Farr , D. (2007). *Sustainable urbanism.* urban design with natural.

[3] Mccoy , D., Carolina , N., Sanderson , L., & Goins, J. (2000). Division of Highways North Carolina Department of Transportation Traditional Neighborhood Development (Tnd). *Highways*

[4] USGBC. (2012). About USGBC - US Green Building Council-Overview.

[5] ElYamany , S. (2016). Applicability And Implementation Of U.S. Green Building Council Rating System (Leed) In Egypt Case Studies: Leed Certified & Registered Buildings In Egypt. *Procedia Environmental Sciences.* Egpt: ELSEVIER.

[6] NUCC. (2018). *New Urban Community Centre.* Retrieved from: http://ehttp://www.newcities.gov.eg//New_Urban_Community

[7] Fraker , H. (2013). The Hidden Potential of Sustainable Neighborhoods. *Island Press*

Communicating innovations: examples from smart technology retrofits

S J Darby

Energy Programme, Environmental Change Institute, University of Oxford, OUCE, South Parks Road, OXFORD OX3 7LW. E-mail: sarah.darby@eci.ox.ac.uk

Abstract. The importance of effective feedback loops during and after building retrofits is well-established. Good communications between designers, practitioners and building users typically lead to better outcomes for users and can allow designers, investors and builders to learn from experience.

Where 'smart' retrofits are concerned – that is, introduction of ICT for more sophisticated control and new functions such as demand response – it can seem as though the aim is a friction-free transfer of activity and learning from humans to machines. In such a scenario, the complexities of real-life routines and activities are typically minimised or ignored, along with the need for time and effort to put into sense-making and adaptation. The paper argues that this is unhelpful at best. Drawing on the author's involvement in evaluating three contrasting programmes, it illustrates how smart refurbishment is an exercise in which 'smartness' is distributed between technologies and humans; one in which we can expect some emergent (i.e. unpredictable) outcomes. The three case studies, at different scales, are:

- the rollout of smart meters in Great Britain, a large-scale infrastructure refurbishment;
- 'smartening' domestic storage heaters and water heaters, a retrofit exercise in homes;
- 'smartening' electricity demand in a community – small-scale system upgrade.

These programmes show the significance of different types of communication in sociotechnical systems as they incorporate smart technologies. Thematic analysis indicates three ways of categorising communication during these processes: between technologies (connectivity), between technologies and people (controls), and between people (care). As buildings become more closely integrated into energy systems, for example via on-site generation, demand response and heat networks, these categories can be useful when designing, implementing and evaluating smart refurbishment.

1. Retrofits and feedback loops

Buildings do not have the environmental impacts their designers intended; the sheer variety and number of possible human/building and human/technology interactions makes this almost impossible, and new or refurbished buildings typically perform less well than anticipated [3,14]. In order to improve the outcomes from retrofit, a wealth of research now shows the value of developing a 'culture of feedback' that involves designers, practitioners and building users, and more involvement from suppliers in learning how buildings perform in real life [3].

Neither of these pathways seems out of reach, given will and modest resources, yet they have rarely been followed: the structure and practices of the construction industry militate against them [13]. When we live in exceptionally data-rich times and (as explained below), electricity system operators have a direct interest in managing demand, it seems there could be a prospect of improved feedback loops and also of a bringing-together of energy suppliers, at least, with building owners and users. There is a good deal of optimism around the spread and potential of smart (ICT-enabled) technology; yet much of it comes from calculations and marketers rather than from lived experience [4]. This paper argues that we can learn useful lessons from introducing smart technology into the built environment and energy systems and from analysing how this is done. It offers three case studies, involving domestic and small-scale customers with different technologies at different scales.

2. Smart retrofit – the building as part of the grid/network

As electricity comes increasingly from distributed renewable sources, electricity systems need to perform more functions than in the past [10] and they rely on a far greater number of actors in order to perform reliably. Crucially, these actors include electricity users 'beyond the meter' whose demand, storage and small-scale generation must be balanced continually in real time. This has to be achieved not only at the level of a national/ transmission grid but within more local distribution networks. The latter is often more difficult, as new supply or demand technologies such as solar photovoltaics (PV), electric vehicles and heat pumps are likely to be clustered in particular neighbourhoods and this clustering can put considerable strain on local networks that were designed to accommodate demand only up to a certain level and were not designed for two-way flows of electricity.

Traditional systems have offered flexible power supply to meet demand and customers in the more industrialised or post-industrial economies have come to rely on their demands being met as a matter of course: if demand increases, a new generator is simply be brought into action. They are also insulated from the operational realities of generation, typically to paying a flat rate for each unit regardless of how much it costs to provide it at a given time with a specific generating mix. Now customers, along with the supply industry, face a paradigm shift in which they will be called on to provide more flexible demand – demand response (DR) – to match available supply.

A corollary of this is that buildings, traditionally seen simply as 'load', are now being viewed in the electricity industry as potential sites for demand response in order to serve the smooth operation of distribution networks and high-voltage transmission grids. In order to do this to a significant degree, they will need information and communication technology (ICT) networks to link appliances with the electricity network, as in this definition of smart buildings:

> *Smart buildings are flexibly connected and interacting with the energy system, being able to produce, store and/or consume energy efficiently* (Building Performance Institute Europe, 2017)

The BPIE report offers as an example a dozen Belgian houses or varying ages that contain a mix of technologies to allow for load-shifting in order to balance the neighbourhood network: PV, solar thermal capture, heat pumps, and fuel cells or batteries. A monitoring and control system is also part of the smart set-up. Another example includes homes with electric vehicles as devices that can be set to charge at off-peak times and store rooftop solar generation, avoiding the need to export it to the grid.

While it is debateable whether smart buildings will inevitably produce, store or consume energy 'efficiently' [4], the main point is that they have a new dimension: a communications networks to link appliances or subsystems with each other that enables remote access and control. That is, the level of connectedness goes well beyond the use of ICT in smart metering, or remote control of a TV from the sofa. Smart buildings host demand-side assets such as electric water heaters and vehicles, laundry appliance, batteries and other devices. And, of course, they are homes for the occupants themselves, typically people who are not energy or IT professionals. The potential of these buildings for DR will depend on how and when their occupants choose to access the electrically-powered services they want and on how able and willing they are to adjust these services in response to system needs, either by switching manually, setting a programmer or agreeing to pass control to a third party such as a demand aggregator who will trade their flexibility, along with that of many others, into the electricity market.

In short, a building that can provide services to both occupants and system is one that will need occupants to cooperate at some level, even if that is only to the extent of consenting to particular forms of remote control for some devices [6,9]. It is also one that requires effective feedback loops and other forms of communication. Despite a great deal of research literature implying that a smart building offers seamless services and effortless convenience, the much smaller volume of actual evaluations indicates that demand response does in fact require effort and adjustment and it takes time and skill to learn to live in a smart building, with energy and environmental outcomes by no means guaranteed [4, 11].

3. Smart retrofit – communicating the innovation

Effective smart energy retrofits therefore involve training and learning. At a minimum, designers and developers will need to learn the skills needed for effective, while those who are directly involved in operating devices and systems will have to have a basic understanding of how they work. There is also the need to exchange knowledge between specialists and non-specialists, so that building occupants can build some understanding of how their new technology might work for them and, no less importantly, professionals can learn how it is used in different real-life circumstances. An effective smart system therefore requires both human and machine intelligence; communication within and between the human and machine domains. The examples below illustrate a number of issues raised by smart retrofit, at different scales, and how they may be addressed.

3.1. Rolling out smart meters in Great Britain – a large-scale infrastructure refurbishment

Metering of some sort is a standard element in most electricity and gas systems, a link between user and supplier. A traditional or 'dumb' meter simply records cumulative consumption as a basis for billing while a smart meter will be considerably more complex. Smart meters all provide two-way communications between user and supplier, allowing for remote meter reading. They also measure consumption at regular, short intervals in order to allow for time-of-use tariffs that can charge the customer more at times of high demand/ short supply. Beyond that, detailed smart meter functions may be specified in many different ways. They may, for example, record microgeneration, offer remote control of appliances and supply near-real-time usage feedback to customers.

Smart metering therefore does not just involve retrofitting buildings with new meters. It is a process of retrofitting the entire power infrastructure. But if, as argued here, effective smart retrofit is a process of transferring not just technology but the knowledge and practical know-how to operate it, then it is not sufficient simply to replace one meter by another. The British smart meter rollout to homes and small businesses illustrates what can be achieved if a broader view is taken. The rollout, to all domestic and small non-domestic customers, has been undertaken on a voluntary basis and by the end of March 2019, over a quarter of all domestic and over a third of all non-domestic electricity and gas meters (approximately 17m in all) were operating in smart mode or with advanced functionality[1]. Customers are offered not only a meter but an in-home display to show their consumption and costs and act, if they wish, as a tool for budgeting. All smart meter installers are required by the Smart Metering Code of Practice to train in communication skills so that they can explain to customers the purpose of their new meter and how they might benefit from the in-home display, and can offer basic energy advice. They also occupy a useful intermediary role in the process, being able to pass feedback to suppliers on how the rollout is proceeding and how customers are responding [5].

This combination of 'smart meter + in-home display + trained installer' is unusual and probably unique, in national rollouts. While the British smart meter rollout has been marred by a number of technical and infrastructural shortcomings, customer satisfaction levels have been high[2] and there is evidence of sustained reductions in demand of around 3% on average, compared with legacy- metered customers [7, 12]. This is, as far as the research literature shows, the only rollout in the world to combine the three elements of smart meter, customer interface (the in-home display) and guidance before, during and after installation. It is significant and interesting because it demonstrates the value of offering a clear,

[1]https://assets.publishing.service.gov.uk/government/uploads/system/uploads/attachment_data/file/804767/2019_Q1_Smart_Meters_Report.pdf
[2]https://assets.publishing.service.gov.uk/government/uploads/system/uploads/attachment_data/file/650970/Post-install_key_findings_report_FINAL_05102017_PUBLICATION.pdf; https://www.statista.com/topics/4923/smart-meters-in-the-united-kingdom-uk/

accessible in-home display to all customers as part of the metering package, and offering guidance to customers on what meter and display are for and how they can be used to understand and manage energy use. The British experience also points to the risk of 'false economies' if customer service is reduced: for example, if pressure to roll out the meters rapidly means that installers no longer have time for conversation with customers and a chance to answer their questions [7,12].

Smart metering is introduced primarily as a technology to improve data flows in electricity, gas and water systems. During implementation, the focus is normally on connecting up the various physical elements of the system: connectivity. Relatively little attention has been paid to 'control'– the interface between meters and the people who use them – or to 'care' – the person-to-person interactions that can build understanding and keep a socio-technical system functioning effectively. The British smart metering experience to date has some useful lessons to offer on the macro-scale and, as I argue below, control and care also contribute to effective retrofitting of particular technologies, or when adapting electricity systems in order to develop community energy markets.

3.2. Retrofit with smart electric storage heaters and water heaters

Credible financial offers to customers are an important factor in a move towards smarter power networks, to persuade people to take part in providing flexible demand and to keep them doing so. As electric vehicles and heat pumps become more commonplace, we can expect an increase in flexible tariffs. However, residential DR with dynamic tariffs is still in its infancy. In the UK, for example, there is only one 'agile' tariff (from Octopus Energy) at the time of writing, plus the familiar two-band Economy-7 type tariff, typically used by customers with night storage heating. Night storage heating and water heating, charged with cheap off-peak electricity for use later in the day, offers an interesting basis for more ambitious 'smart' thermal storage, in which heaters and water cylinders may be charged up at any time when power is abundant. A 'smart' storage heater is one that is connected via broadband or a modem to the Internet and that can be controlled by the customer (who chooses temperatures and heating times) AND by an actor such as a demand aggregator (who chooses when to charge up the heater or to cease charging, according to grid/network conditions at a given time).

A recent trial of smart storage heating involved roughly 800 customers in three EU countries (Ireland, Germany and Latvia) who made their new smart-enabled heaters available for charging according to grid and network conditions, provided they also met customers' preferences for warmth at different times of day [8]. Customer experience during these trials was a crucial element of the project, as without willing and satisfied customers, there would be no demand response and no business case for smart thermal storage. Accordingly, customer experiences were monitored through surveys, interviews, and focus groups.

From 161 customers in the three countries who completed at least two surveys during the project, and from interviews with 70 of them, five themes emerged as contributory factors to a good customer experience. They were:

- *Comfort*, the single most important consideration when customers were asked about their new and more efficient smart-enabled heaters and their experiences of taking part in fully-flexible demand response. For the most part, they were more comfortable with their new heaters than they had been with the old ones.
- *Cost*. Although cost is often seen as the main consideration when planning or modelling demand response (sometimes the only one), the trial customers, many on low incomes, thought of it as roughly equal with comfort.

The other three themes all involve communicating and are specially interesting with regard to smart retrofit as each relates to a different type of connection or dialogue. They were:

- *Connectivity* (between technologies). This turned out to be the most problematic single operational issue for the project. Although the components of the DR system had been market-ready, it took a lot of effort to get them to connect effectively with each others. Connectivity multiplies the possibilities for system failure, e.g. when software is upgraded or a component manufacturer goes out of business, and both network operators and customers needed reliably flexible heating. As one of the project co-ordinators ruefully acknowledged towards the end of the trials, *"I felt given the amount of hype around this area at the moment, that things would be much more developed and there would be a lot more off-the-shelf solutions available. And that things would fit seamlessly together and just work... (but) it was incredibly difficult."* [8, p.61].

A German installer took a similar view of the complexity of connecting the equipment: *"Initially (I) thought equipment installation (would be the) most effort, but in fact all the rest is more time- consuming. (You have to) ensure that control equipment is working (and the) temperature sensors are integrated. Generally it took one person the same time to do all the programming as the other to do the installations."* [8, p.60]

- *Control* (the link between people and technologies). Customers needed to be able to alter their appliance settings, directly or remotely, to manage comfort and costs. The move to 'smart' entailed a change from analogue to more sophisticated digital controls. This was not a problem for some participants but caused difficulties for others. Although they had been offered professional advice and written information during installation and sometimes afterwards, some householders were not able to demonstrate how they altered their heater settings when interviewed towards the end of the project. Interestingly, a sense of lack of control was correlated with a perceived increase in bills. In Ireland, 12 out of 16 customers who did not know how to change the settings of their heaters (75%) thought their bills had increased. In contrast, of the 86 who reported that they had some control of their settings, only 22 (26%) thought their bills had increased [8, p.39]. While these were perceptions that could not be tested, due to lack of billing data, they clearly relate to customer satisfaction.

- *Care* (thoughtful communication between people). Installers, utility call centre staff and locally-based people such as housing managers or knowledgeable neighbours turned out to be important for achieving affordable comfort and demand response: people who could advise on how best to programme the heaters in normal circumstances and before going away on holiday, or who could troubleshoot if the devices malfunctioned.

The clear need for different types of communication and feedback raised the question: *who* is needed for well-functioning retrofits? The researchers identified a surprisingly long list from their customer surveys and interviews, and from interviews with project personnel [8]. These included:

- Customers – using the heaters, programming and switching;
- Installers for the heaters, water cylinders, sensors and smart meters – installing, explaining, advising, maintaining and repairing;
- Neighbours – advising and passing on practical tips;
- Housing managers – advising, checking welfare, liaising with professionals;
- Supplier call centre staff– advising, referring problems if necessary to other professionals;
- Project coordinators, who played a central role, orchestrating and explaining activity, resolving misunderstandings and troubleshooting;
- Designers and manufacturers of devices, controls, apps and software –acting on feedback to develop products and services further;
- Demand aggregators – recruiting, planning, coordinating and building a market for DR;
- Network operators – evaluating viability and value of DR in their area, giving or withholding permission to carry out DR;
- Grid operators – monitoring system conditions, planning future supply and estimating DR requirements.

The length of this list and the variety of items listed contradict any notion that smart systems offer a rapid, seamless route to uncomplicated, controlled comfort and low-carbon outcomes. But the experience of the smart heating trials did demonstrate how, with the help of three types of communication and a range of actors, it was possible to achieve more flexible demand while maintaining or improving comfort. The communication enabled the retrofit innovation and formed part of it.

3.3. Smart retrofit in a community/neighbourhood setting

Community energy trials offer a good opportunity to observe situated, social learning. This section briefly describes the trial of a business model and technologies that had been designed for groups of customers sharing the same segment of a local electricity low-voltage distribution network with some form of low-carbon generation, e.g. solar PV, micro-hydro, wind turbines or a combined heat and power plant. This particular trial took place between 2015-17 and involved 48 homes in three neighbouring English villages, 14 of which had rooftop solar photovoltaics (PV). It is documented in two papers by Boait and colleagues [1,2]. The business model took into account technological innovations and also the new actors and activities needed for participants to benefit by:

- shifting demand to use or store locally-generated electricity where possible, avoiding the need to export it to the grid at a relatively low price if the 14 PV-equipped households could not use it at the time of generation;
- avoiding import of non-locally-generated electricity at peak times (when, under cost-reflective pricing, the price would be high); and
- reducing their overall consumption.

This model allowed participants to form a co-operative and to use the profits from local electricity generation and management for community benefit. It can be seen as a 'community retrofit' of sorts, involving not only (minor) changes to participants' homes and workplaces but to their electricity-using practices and interactions with the supplier/retailer and network operator.

Retrofitted hardware included storage heaters and batteries, plus technology to schedule loads in such a way that customers could benefit from the time-of-use tariff while respecting their preferences (for example, for timing their laundry or dishwashing, or for indoor temperatures). There was also a time-of-use tariff with a day- ahead adjustment based on the weather forecast, to reflect the amount of locally-generated electricity that could be expected. The trial participants each had a web-based display of their current tariff and usage, and were given regular feedback on the individual and collective savings resulting from their DR. In addition, the project benefited from well-informed, communicative installers and a member of the project team who was available for a day each week for engagement activities, including occasional social events. These proved very effective at developing and maintaining interest, and for gaining useful feedback.

The results of the trial were highly encouraging. For example, during the year of full operation, over 40% of the electricity generated by PV in 14 homes was used by the householders, while almost all the remainder was available to share with other participants. Financial savings varied between £5 and £12.50 per month on average, being highest in the summer when the solar PV was most productive. The 27 (out of 48) customers who were interviewed by phone reported positive experiences, with most claiming some change in their day-to-day electricity-using habits. For example, one participant had put their electric towel rail on a timer plug to avoid peak-time heating; 17 of the 21 interviewed customers who said they had load-shifted during the trial intended to continue doing so, and there were strong indications of the value of an integrated community approach that allowed people to learn in different ways and at their own pace. As one interviewee commented:

> *"When you have the reports that came through and here you have the cheque for your money and your report on your energy use, you could practically see how it all fitted together".* [2, p. 10].

The success of the trial has led to a commercial implementation of the business model in a Welsh community and it is being tested in around 12 locations elsewhere at the time of writing.

4. Summary

Balancing a low-carbon electricity system with the aid of demand response is likely to involve millions of small- scale generation, investment, storage and demand response decisions and actions, often on short timescales. Transforming the built environment also involves millions of decisions and actions, usually on longer timescales. Both can be seen as forms of retrofitting. Both call for competences, accountability and qualified trust, along with reliable technologies and finances. Through qualitative research into smart technology retrofit, we are learning about actors and processes that are relevant to the built environment as a whole, especially as ICT percolates further into energy systems. The three categories of technical connectivity, human-technical interaction in controls, and human-human conversations for support and sensemaking were all shown to be essential ingredients in gaining environmentally and socially positive outcomes from smart technology retrofit.

References

[1] Boait, P.J., Snape, J.R., Darby, S.J., Hamilton, J. and Morris, R.J.R. (2017) Making Legacy Thermal Storage Heating fit for the Smart Grid. *Energy and Buildings* 138, 630-640

[2] Boait, P., Snape, J.R., Hamilton, J., Morris, R. and Darby, S. (2019) The Practice and Potential of Renewable Energy Localisation: Results from a UK Field Trial. *Sustainability 11* doi:10.3390/su11010215

[3] Bordass, B., Leaman, A. and Ruyssevelt, P. (2001) Assessing building performance in use 5: conclusions and implications. *Building Research and Information* 29 (2), 144-157

[4] Darby, S.J. (2018) Smart technology in the home: time for more clarity. *Building Research and Information* 46 (1), 140-147

[5] Darby, S.J. and Liddell, C. (2015) *Communicating 'smartness': smart meter installers in UK homes.* Paper 9-050-15, European Council for an Energy-Efficient Economy summer study, Hyères, June 1-5, 2015. pp 1991-2001

[6] Darby, SJ and McKenna, E (2012) Social implications of residential demand response in cool temperate climates. *Energy Policy* **49**, 759-769

[7] Darby, S.J., Liddell, C., Hills, D. and Drabble, D. (2015) Smart Metering Early Learning Project: synthesis report. For the Department of Energy and Climate Change, London https://www.gov.uk/government/uploads/system/uploads/attachment_data/file/407568/8_Synthesis_FINAL_25feb15.pdf

[8] Darby, S.J., Higginson, S., Topouzi, M., Goodhew, J. and Reiss, S. (2018) *Getting the balance right: can smart thermal storage work for both customers and grids?* Consumer Impact study, RealValue project Horizon 2020. Environmental Change Institute, University of Oxford http://www.realvalueproject.com/images/uploads/documents/RealValue_Consumer_Impact_Report_-_FINAL_%28Compressed_spread%29.pdf

[9] Fell, M.J., Shipworth, D., Huebner, G. M. and Elwell, C.A. (2015) Public acceptability of domestic demand-side response in Great Britain: the role of automation and direct load control. *Energy Research & Social Science* 9, 72-84

[10] FPSA (2017) Future Power System Architecture Project 2: WP2 final report: functional analysis. The Institution of Engineering and Technology with the Energy Systems Catapult, London. https://es.catapult.org.uk/wp-content/uploads/2017/08/FPSA-WP2-Web.pdf

[11] Hargreaves, T., Wilson, C. and Hauxwell-Baldwin, R. (2018) Learning to live in a smart home. *Building Research and Information* 46 (1), 127-139

[12] HoSTC (2017) Evidence check: smart metering of electricity and gas. Sixth report of session 2016-17.House of Commons Science and Technology Committee, London. https://publications.parliament.uk/pa/cm201617/cmselect/cmsctech/161/161.pdf

[13] Killip, G. (2013) Products, practices and processes: exploring the innovation potential for low-carbon housing refurbishment among small and medium-sized enterprises (SME) in the UK construction industry. *Energy Policy* 62, 522-530

[14] Majcen, D., Itard, L.C.M. and Visscher, H. (2013) Theoretical vs actual energy consumption of labelled dwellings in the Netherlands: discrepancies and policy implications. *Energy Policy* 54, 125-136

Acknowledgements
I gratefully acknowledge funding via the Oxford Martin Programme on Integrating Renewable Energy. The Oxford Martin School at the University of Oxford is a centre of pioneering research into global challenges.

IOP Conf. Series: Earth and Environmental Science **329** (2019) 012021 doi:10.1088/1755-1315/329/1/012021

A study on the summer thermal comfort and behaviour patterns in the residential buildings of Kashgar old City, China

T Li[1], X Li[2], X Wang[1], Y Feng[1], Y Ma[1]

[1]School of Architecture, Xi'an University of Architecture and Technology, Yanta Road, Xi'an 710055, China
[2]Welsh School of Architecture, Cardiff University, King Edward VII Avenue, Cardiff, CF10 3NB, United Kingdom

182171912@qq.com

Abstract. Kashgar, located in China's far west, has a typical hot-arid climate. There are a large number of low-rise high-density Islamic housing in Kashgar old City. The thermal comfort in these buildings has not yet been adequately researched. In order to obtain a comprehensive understanding about the real indoor thermal environment and the residents' summer thermal comfort in Kashgar old City, a field investigation of 138 residential buildings was conducted in the summer time, including onsite environmental parameter measurements and simultaneous surveys using subjective thermal comfort questionnaires. Based on a statistical regression analysis of 182 valid questionnaires, a summer thermal comfort model was established. The results show that the measured summer neutral temperature of the local people is 23.4℃, and the upper limit of acceptable summer temperature for 80% of the residents is 26.9℃. The summer thermal tolerance of the residents in the Kashgar old City is lower than expected and also lower than that of the residents in rural Turpan with the same climate features. The survey also shows that the residents in the Kashgar old City have a habit of transfering living spaces over time, and the frequency of passive behaviours is higher than those of active behaviours. The above results are believed to be potentially valuable for the revision of energy conservation policies and the design practice of residential buildings in the hot-arid climate region of China, and also for areas under similar climate conditions in the world.

1. Introduction

Along with the rapid development of the world economy, people's need of quality living environment has been constantly raised. The economic development has led to large amounts of energy and resource consumption. Statistics show the building sector contributes to about 40% of global energy consumption, and China's building sector has accounted for more than 1/4 of the country's total energy consumption [1], among which energy consumption for heating, air conditioning and ventilation accounts for more than 70% of the building sector [2]. As the basis for setting indoor air temperature range, Thermal Comfort Standards play an important role in determining building energy consumption. With global warming and the resulted air temperature rise in summer, the proportion of family using air conditioners and electric fans have increased significantly, as a result more and more attention is paid to the Thermal Comfort Standards. Scholars around the world have conducted a large number of field investigations on thermal comfort [3-4], concluding that there are obvious differences between the actual thermal sensation of people and the predicted results from employing ISO 7730[5] or ASHRAE 55[6] for different regions. These can be attributed to people's behavioral responsibility, living habits and psychological expectations [7].

China is a vast country with varied climates. Residents from different regions have different thermal comfort requirements due to cultural diversity, different living habits and behavioral patterns. In recent years, a lot of thermal comfort investigations have been carried out in different building types in different regions of China, such as Hot and Humid Region, Hot in Summer and Cold in Winter Region, Cold Region, Hot in Summer and Warm in Winter Region [8-11]. However, there are relatively few investigations in the hot-arid climate, which mainly focus on rural dwellings, so there is a lack of research on urban dwellings. For example, Scholars from Xi'an University of Architecture and Technology have been investigating residents' thermal comfort in the rural dwellings of Turpan, and have obtained the thermal neutral temperature and the thermal comfort range of the local residents in summer and winter respectively [12]. However, is the residents' thermal comfort range in Kashgar old City the same as that in Turpan? What are the residents' adaptive behavior patterns in Kashgar? In order to obtain a comprehensive understanding of the questions above, a field investigation was carried out in the residential buildings of Kashgar old City, hoping to get the neutral temperature and the thermal comfort range of the local residents, and to explore the specific characteristics of their adaptive behavior patterns. The aim of the study is to provide a basis for the revision of the Thermal Comfort Standards and the thermal environment design of the residential buildings in hot-arid climate of China, and also for areas under similar climate conditions in the world.

2. Methodology

2.1. Geography and climate

Kashgar lies at N 39° 25′ -39° 35′ and E 75° 56′ -76° 04′ , and is 1260 to 1300m above sea level. The Kashgar old City, located in the Centre of Kashgar, has an area of about 2.5 km^2 and a population of about 221,000(2010), nearly half of the population of Kashgar [13]. It is a multi-ethnic community with Uygur as the main body. Kashgar has a typical hot-arid climate, and is zoned as a part of the Cold Region according to the Standard for Climatic Region-alization for Building and Civil Engineering of China. Its climatic features include long sunshine duration, large annual air temperature range and daily air temperature range, little precipitation and huge evaporation.

2.2. The building features

The residential buildings in Kashgar old City are low-rise and high-density (Figure 1). The streets are narrow and winding, and the buildings grow like cells, forming maze-like blocks with strong Islamic Characteristics (Figure 2). The average building area of each dwelling is about 158 ㎡, while the number of residents is often as many as 7-9. The buildings adopting a courtyard-centered layout, are open to the courtyards but closed to the streets. A large number of semi-open spaces such as eaves and high canopies are developed in the courtyard, and have become important activity venues in summer for the local residents (Figure 3). Most interior spaces have excellent thermal stability, as they are surrounded by 500-700 mm thick walls made by soil or brick-soil (Figure 4).

Figure 1. Low-rise and high-density buildings

Figure 2. The narrow and winding streets

Figure 3. The semi-open spaces

Figure 4. The thick soil wall

2.3. Data collection

The field investigation was carried out from 26th to 29th July 2018, which is normally the hottest period of the year in Kashgar. The sites in Kashgar old City were selected for the survey. The Random Sampling method was used in the field investigation, ensuring the same number of households were selected from each block site. In total 186 residents from 138 dwellings were surveyed, and 182 valid samples were obtained excluding unhealthy individuals and individuals younger than 10 years. Subjective and objective research methods were combined and employed in the survey.

2.3.1. Sample size and the subjects. A summary of the residents' background characteristics is provided in Table 1. For gender composition, there are 52 males (29%) and 130 females (71%). The subjects' ages range from 10 to 85, with an average of 45.5 years. Their average height is 160 cm and their average weight is 64.7 kg. 63.9% of the subjects have been living there for more than 20 years. All the valid samples were in good health. The clothes of individual subjects were recorded in detail. The insulation property of each garment was estimated according to ASHRAE Standard 55-2010 [6], and the clothing insulation values of individual residents were then estimated, varying between 0.3 and 0.6clo, with an average of 0.5clo, which is close to the summer clothe insulation provided by ASHRAE. In order to ensure the objective authenticity of the thermal sensation data from the questionnaire, the subjects were required to stabilize for 20 minutes before filling out the questionnaire, and their metabolic rates are taken as a unified value of 1.1 met, which is the metabolic level in the sedentary state.

Table 1. A summary of the background characteristics of the subjects.

Sample size	Age(years)	Height(cm)	Weight(kg)	Length of residence (years)	Clothing (clo)
Mean	45.50	160	64.70	28.89	0.50
Maximum	85	194	100	83	0.63
Minimum	10	95	20	1	0.37
Standard deviation	18.77	11.59	15.04	20.72	0.10

2.3.2. Subjective investigation. The subjective investigation adopts the method of questionnaire, which mainly covers the following aspects: 1) background information of the residents, including gender, education level, age, height, weight, occupation, length of residence, etc.; 2) residents' subjective sense of the thermal environment, such as thermal sensation, humidity sensation, thermal satisfaction and humidity satisfaction; 3) adaptive behaviour patterns, including clothing habit, living habit and frequency of thermal behaviour. According to ASHRAE thermal sensation scale, the subjective sensation is divided into seven grades (Table 2).

Table 2. The rating scale for the thermal responses of the subjects.

	-3	-2	-1	0	1	2	3
Thermal sensation	Cold	Cool	Slightly cool	Neutral	Slightly warm	Warm	Hot
Humidity sensation	Very dry	Dry	Slightly dry	Neutral	Slightly humid	Humid	Very humid
Thermal satisfaction		Very dissatisfied	Slightly dissatisfied		Slightly satisfied	Very satisfied	
Humidity satisfaction		Very dissatisfied	Slightly dissatisfied		Slightly satisfied	Very satisfied	

2.3.3. Objective investigation. The objective investigation employs two methods, such as measurements of indoor thermal environment synchronized with questionnaire and continuous on-site monitoring of outdoor thermal environment of typical dwellings. Details of the instruments are shown in Table 3. The indoor air temperatures and humidities in the semi-open spaces and interior spaces were measured by hand-held digital recorders (AS817) at the time of the questionnaire survey, and the instruments should be let stand for at least 5 minutes after being taken to the centre of the spaces, to provide reliable readings. The automatic digital recorders (TR72U) were placed in the courtyards which are normally shady and ventilated, and at 2m above the ground. The data loggers were set to record data every 10 minutes over 5 days to obtain outdoor air temperatures and humidities. Through measurements of air speed in most semi-open and interior spaces, it is found that they are all below 0.03m/s during the investigation period, namely the spaces are approximately in a quiet state, therefore impact of air movement can be ignored in the survey.

Table 3. Details of the instruments.

Description	Trade name	Measurement parameter	Range	Accuracy
Hand-held digital recorder	AS 817	Indoor air temperature, Humidity	-10 to 50℃,5-98% RH	±1.5℃, ±4%
Automatic digital recorder	TR 72U	Outdoor air temperature, Humidity	-20 to 70℃,0-100% RH	±0.5℃, ±3%
Hot wire Anemometer	Testo425	Air velocity	0-20 m/s	±0.03m/s, ±5%

2.3.4. Thermal comfort evaluation index. The measured air temperatures (t_a) are used in a thermal comfort analysis. With consideration to the randomness of a single sample, a frequency method is used to group the air temperatures into several temperature ranges at a 0.5℃ interval. Then a linear regression analysis was performed using the mean air temperature of each temperature range and the Actual Mean Vote (AMV) of the subjects as independent and dependent variables, respectively.

3. Results and analysis

3.1. Thermal environment conditions

More than 80% of the indoor measurement points were located in the semi-open spaces, and the rest were located in the interior spaces. A statistic of indoor and outdoor environmental parameters is shown in Table 4. It can be seen that during the investigation period, the average outdoor air temperature was 29.4℃, the maximum outdoor air temperature was 36.1℃, the average indoor air temperature was

28.4℃, the maximum indoor air temperature was 35.8℃. Since the average and the maximum indoor air temperatures are lower than the relevant outdoor air temperatures respectively, it proves the cooling effect from the semi-open spaces.

Table 4. Environmental parameters during the investigation.

	Indoor air temperature (℃)	Outdoor air temperature(℃)	Indoor relative humidity (%)	Outdoor relative humidity (%)
Mean	28.4	29.4	40.7	42.2
Maximum	35.8	36.1	61.6	74
Minimum	23.8	24.7	16.6	19.6
Standard deviation	2.6	2.13	8.53	2.5

3.2. Thermal comfort responses

It is required that the corresponding vote of a comfort temperature range should be between -1 and 1 according to ASHARE. Figure 5 shows the distribution of the residents' thermal sensation votes. It can be seen that the proportion of thermal sensation in the comfort range is about 68%, and the majority of the votes are slightly warm (1), neutral (0) and warm (2), which are 33%, 29% and 22% respectively. It indicates that the residents feel slightly warm in summer in terms of thermal sensation. Figure 6 presents the distribution of the residents' humidity sensation votes. It can be seen that the proportion of humidity sensation in the comfort range is about 94%, and the majority of the votes are neutral (0) and slightly dry (-1), which are 47% and 32%, respectively. This indicates the residents' humidity sensation is moderate in summer. Figure 7 and Figure 8 present the subjects' thermal satisfaction and humidity satisfaction votes. It illustrates that the slightly satisfied vote (1) is the majority for both the thermal satisfaction votes (69%) and the humidity satisfaction votes (53%).

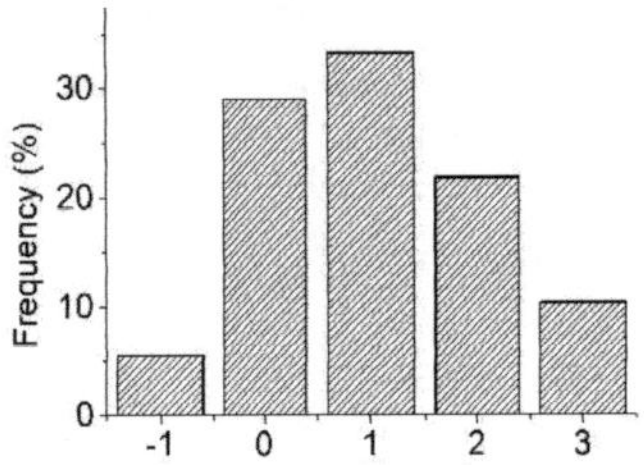

Figure 5. Thermal sensation vote

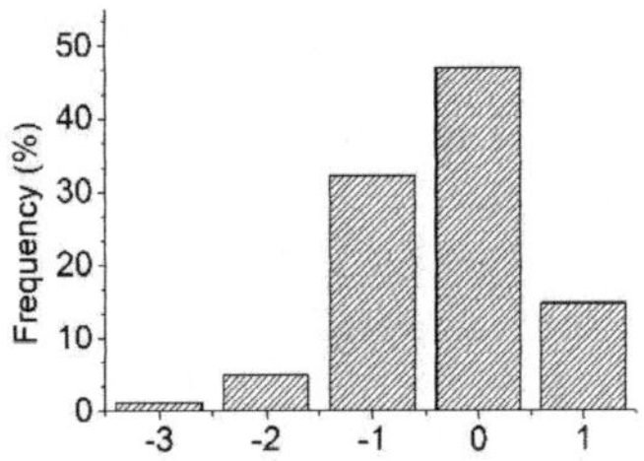

Figure 6. Humidity sensation vote

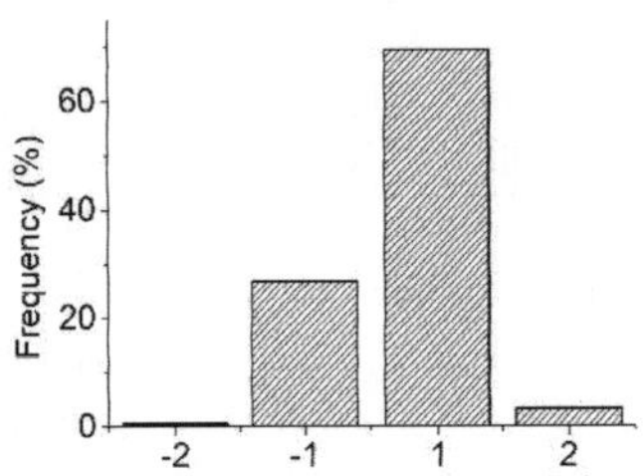

Figure 7. Thermal satisfied vote

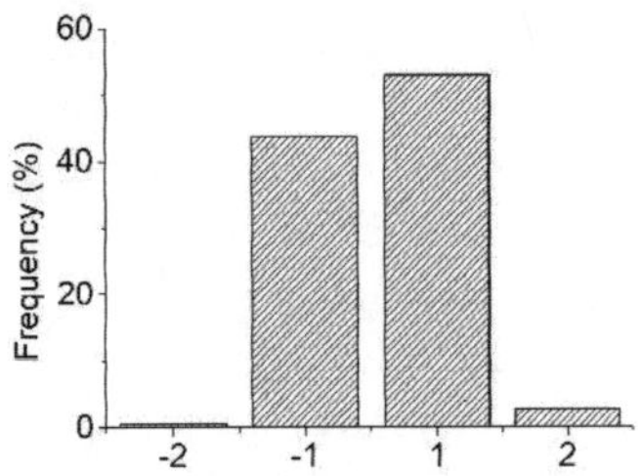

Figure 8. Humidity satisfied vote

3.3. Thermal neutral temperature

The Predicted Mean Vote and the Predicted Percentage of Dissatisfied (PMV-PPD) model were compiled using MATLAB. The measured air temperature, relative humidity, metabolic rate and clothing thermal resistance were brought into the program to calculate the Predicted Mean Vote (PMV) [14]. Actual Mean Vote (AMV) of each temperature range was obtained using the temperature frequency method. AMV and PMV were then linearly regressed with measured air temperature (t_a), as shown in Figure 9. When AMV and PMV values approach 0, the actual and the predicted thermal neutral temperatures fall to 23.4℃ and 25.1℃ respectively. It can be seen that the distribution of AMV is more discrete than that of PMV, implying bigger variations of actual thermal sensation among individuals compared with prediction, mainly impacted by physiological and psychological adaptation. The PMV line is steeper compared with that of the AMV, indicating that the actual thermal sensitivity of the residents is lower than what is predicted. The actual thermal neutral temperature is 1.6 ℃ higher than what's predicted, indicating that the residents' real thermal tolerance is lower than that from prediction.

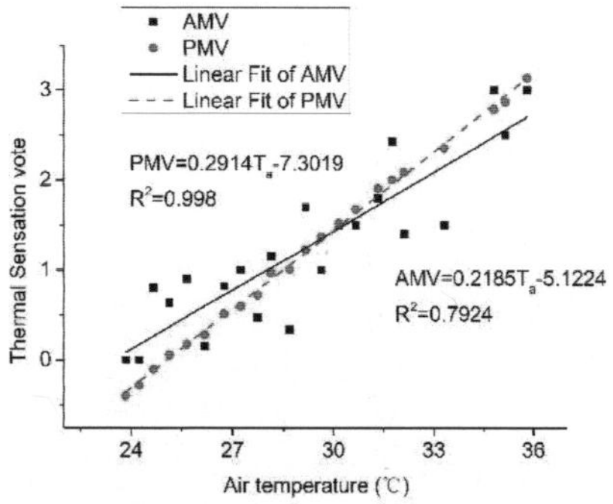

Figure 9. Relationship between PMV(AMV) and air temperature

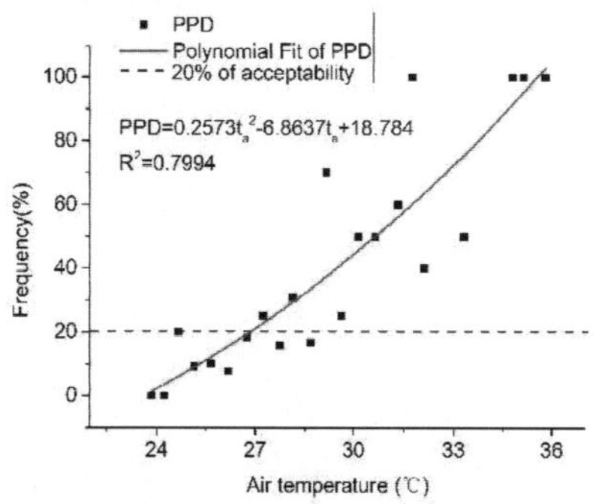

Figure 10. Relationship between PPD and air temperature

3.4. Thermal acceptable temperature

The Predicted Percentage of Dissatisfied (PPD) was obtained by calculating the percentage of the overall votes of -2, -3 and 2, 3 in each temperature range, which had been produced using the temperature frequency method. The Predicted Percentage of Dissatisfied (PPD) and the measured air temperatures (t_a) were then analyzed using polynomial regression, results from which are shown in Figure 10. According to Fanger [15], the thermal sensation vote for a comfort temperature range should be between - 1 and 1, and this temperature range corresponds to the temperature range when the PPD is 20%. Therefore, the intersection point of the PPD curve and the line of the 20% thermal unsatisfactory rate is taken as the acceptable temperature range for 80% of the local residents, thus it can be concluded that the upper limit of the residents' acceptable comfort temperature range is 26.9℃in summer.

3.5. Comparison with other locations in the Cold Region of China

By comparing the findings above with residents' thermal comfort in other locations of the Cold Region of China (Table 5[12,16-17]), it shows that the thermal neutral temperature and the upper limit of the residents' acceptable temperature in Kashgar old City are lower than those in Yinchuan and Weinan, and slightly higher than those in Hanzhong. The actual thermal neutral temperature and the upper limit of the residents' acceptable temperature in Kashgar old City are significantly lower in summer compared with those in Turpan, which has the same climate as that of Kashgar old City. Since the measured daily average air temperatures in Kashgar are lower than those in Turpan, and there are differences in residents' living habits and economic gaps between urban and rural dwellings, the residents in Kashgar old City are less tolerant to summer overheating than those rural residents in Turpan.

Table 5. A comparison of residents' summer comfort in different locations in the Cold Region of China.

Location	Regression equation	Neutral temperature(°C)	Acceptable temperature range(°C)	Mean value and range of indoor air temperature (°C)
Yinchuan[a]	$MTS = 0.357t_a - 9.239$	25.9	28.3/none	28.9 [24.0,35.4]
Weinan	$MTS = 0.178t_a - 4.589$	25.8	29.5/27.0	26.9 [24.0,29.6]
Hanzhong	$MTS = 0.221t_a - 5.091$	23.0	26.6/22.6	26.0 [22.0,20.0]
Turpan	$MTS = 0.225t_a - 7.129$	28.3	31.8/none	35.0 [28.0,41.2]
Kashgar	$MTS = 0.219t_a - 5.124$	23.4	26.9/none	28.4 [23.8,35.8]

[a] From unpublished sources.

3.6. Thermal behaviour pattern

3.6.1. Clothing behaviour. A linear regression analysis of the relationship between clothing insulation and measured air temperature is presented in Figure 11. It shows that with the increase of air temperature, the total amount of clothing tends to decrease, implying that the residents adapt to the climate through increasing or decreasing clothing in summer. However, since very little clothing is needed in summer and there is a large randomness, no significant linear relationship is shown between resident' clothing insulation and measured air temperature in Kashgar old City.

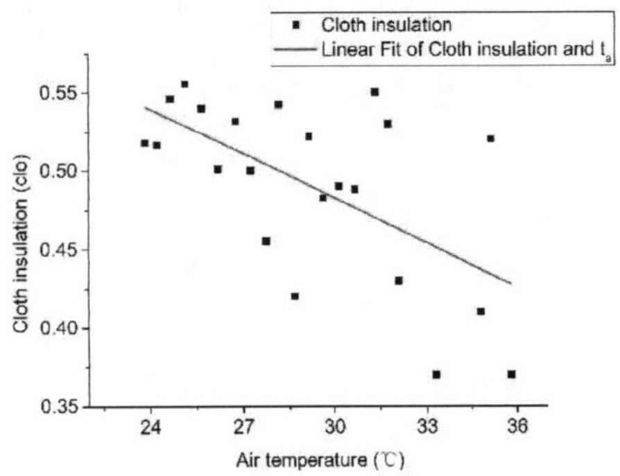

Figure 11. Relationship between clothing insulation and air temperature

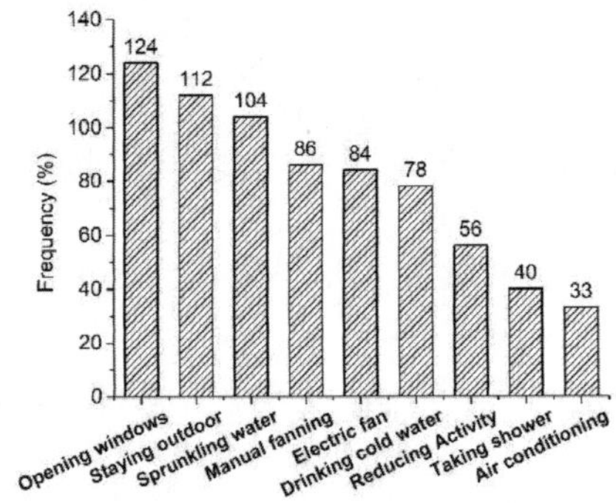

Figure 12. Frequency of thermal adaptation behavior

3.6.2. Living habits. Continuous records of air temperatures and humidities in both semi-open and interior spaces show that the interior spaces possess a good thermal stability, while air temperatures in the semi-open spaces vary greatly with solar radiation on a typical summer day. High temperatures of the day mainly occur from 2pm to 8pm in the semi-open spaces, and the hottest time is around 5pm. According to a survey of the residents' living habits (Table 6), for the time periods of 9am-2pm and 8pm-12am, the residents mainly stay in the semi-open spaces (such as courtyard and Supa) where temperatures are lower than those indoors, while for the period of 2pm-8pm when temperatures in the semi-open spaces are high, the residents would move to the interior spaces (such as bedroom and reception room). It illustrates that the local residents have developed a habit of transferring living spaces over time to adapt to the thermal environment.

Table 6. Daily behaviour pattern of the inhabitants on a typical summer day.

Time	Location		Behaviour
	Interior space	Semi-open space	
12am-9am	Bedroom		Sleep
9am-10am		Supa	Cook, have a breakfast
10am-2pm		Courtyard	Labour
2pm-3pm	Reception room		Cook, have a lunch
3pm-4pm	Bedroom		Take a nap
4pm-8pm	Bedroom		Rest
8pm-9pm		Supa	Cook, have a supper
9pm-12am		Courtyard	Labour

3.6.3. Frequency of Thermal Adaptation Behaviour. Figure 12 summarizes the frequencies of the residents' thermal adaptation behaviours. It can be seen that the residents' adaptation to the hot summer climate includes opening window, staying outdoor, sprinkling water, manual fanning, electric fan, drinking cold water, reducing activity, taking shower, air conditioning. Among them, the frequencies of passive behaviours such as opening window and staying outdoor are higher than those of active behaviours such as electric fan and air conditioning. This indicates that the residents of

Kashgar old City mainly employ passive behaviours for thermal adaption, implying an energy-saving pattern.

4. Conclusions

From the investigation of the residents' summer thermal comfort and adaptive behaviors in the residential buildings of Kashgar old City, the following conclusions can be drawn:

4.1 The local residents feel slightly warm in summer, and the overall thermal satisfaction is good, indicating that the residents have a good psychological adaptation to the local climate.

4.2 The residents' actual thermal neutral temperature is lower than that from prediction, and the slope of the AMV line is lower than that of PMV line, indicating that their thermal tolerance and sensitivity in summer are both lower than those being expected.

4.3 The upper limit of the acceptable temperature range in summer for 80% of the residents in Kashgar old City is close to those in other locations of the Cold Region, but is lower than that in Turpan, indicating that the residents in Kashgar old City have worse thermal adaptability in summer than those living in rural Turpan.

4.4 Among the residents' adaptive behaviors in Kashgar old City, the frequencies of the passive behaviors are much higher than those of the active ones, which can greatly reduce energy consumption from air conditioning and other active cooling devices.

In a word, this study shows that the existing comfort standard in the Cold Region of China (the evaluation standard for indoor thermal environment in civil buildings GB/T 50785-2012) is generally applicable in Kashgar old City. However, differences have been identified between urban and rural residents in regards to their thermal responses to the summer climate, therefore it is recommended that the corresponding thermal comfort standards should be established respectively. The benefit of passive behaviors in terms of energy saving should be considered in the design of thermal environment of local dwellings. It should be noted, there are a couple of limitations with the survey, such as the sample size and the time scale. Further detailed investigation on the resident's thermal comfort in Kashgar old City will be carried out for the other seasons in order to obtain more comprehensive conclusions.

References

[1] S Chen, N Li, J Guan, Y Xie, F Sun, J Ni 2008 *J.Energy Build.* **40** 654-65
[2] F Tu and Q Wang 2004 *J.New Build. Meterial.* **7** 40-2
[3] A Farghal 2011 *Karlsruhe Institute of Technology.* **36** 614-26
[4] M Indraganti 2010 *Appl.Energy.* **87** 866–88
[5] Instruments for Measuring PhysicalQuantities 2005 *ISO 7730–2005*(Geneva)
[6] American Society of Heating,Refrigerating and AirConditioning Engineers 2010 *ANSI/ASHRAE Standard55-2010* (Atlanta)
[7] M A Humphreys 1978 *J.Building Research and Practice Journal of CIB.* **2** 92-105
[8] H Chen, Y Zhang and J Wang 2010 *J.Heating Ventilating. Air Conditioning.* **40** 96-101
[9] J Li, L Yang and J Liu 2008 *J.Heating Ventilating. Air Conditioning.* **38** 20-4
[10] B Cao, Y Zhu and Q Ouyang 2010 *J.Heating Ventilating. Air Conditioning.* **40** 98-100
[11] W Yang, J Guo and S Mei 2006 *J.Taiyuan University of Technology.* **37** 393-96
[12] H Yan, L Yang, W Zheng, W He and D Li 2016 *J.Thermal Biology.* **59** 92-102
[13] H Song and X Wang 2013 *City Planning Review.* **37** 85-9
[14] K Cena and R J Dear 1999 *ASHRAE Trans.* **105** 204-13
[15] P O Fanger 2002 *J.Energy Build.* **34** 533-36
[16] Q Yang, L Yang and J Liu 2011 *J.Heating Ventilating. Air Conditioning.* **25** 49-52
[17] M Zou, L Yang and H Yan 2013 *Building Science Research of Sichuan.* **39** 341-45

Acknowledgments

The work is supported by the Natural Science Foundation of China (Project Nos. 51708439). The authors wish to thank H Li, S Fan and J Zhang for delivering questionnaires, D Abdugupur for translating Uygur into mandarin in the investigation, and all the 186 subjects for giving their comfort votes patiently during.

Agent-Based Modelling of Occupants' Clothing and Activity Behaviour and Their Impact on Thermal Comfort in Buildings

Anxiao Zhang[1], Qiong Huang[2,*], Yiming Du[3], Qi Zhen[4], Qi Zhang[5]

[1] Tianjin University, China, zhanganxiao9012@gmail.com
[2] Tianjin University, China, qhuang@tju.edu.cn
[3] Tianjin University, China, nemo19733@outlook.com
[4] Tianjin University, China, zhenqi0317@163.com
[5] Tianjin University, China, zhangqi_arch@vip.163.com
[*] Corresponding Author

Abstract. This paper proposes a new agent-based approach to model the clothing and activity behaviour of occupants in buildings. An autonomous agent is created to interact with the surrounding thermal environment and make decisions based on its comfort level. The agent- based behaviour model was implemented by MATLAB and was linked with building energy simulation program EnergyPlus via Building Controls Virtual Test Bed (BCVTB) and MLE+. A simulation experiment was conducted to see how an agent adapts to the dynamic thermal changes in different seasons and how it affects the annual thermal comfort. Obvious variation in the predicted annual discomfort time can be observed when using the agent-based method compared with a traditional commonly used method with static parameters. The paper concludes with a discussion of the model's current limitations and possibilities for future development.

1. Introduction

Energy modelling is commonly used during the design phase to estimate future building energy performance and indoor comfort levels. Predictions obtained from common energy estimation software typically deviate from actual energy consumption and comfort levels. This discrepancy can be partially attributed to the misrepresentation of the role that building occupants play in the energy calculation equation. Although occupants might have different and varying comfort-related behaviours over time, current energy estimation tools assume they are constant. For example, clothing adjustment is a very important comfort adjustment available to occupants in office buildings, and has a substantial impact on thermal comfort[1]. A person can also adjust his or her activity level, such as from sitting to walking, to control the body heat generation and resist the outside temperature changes. Furthermore, both clothing and metabolic activity are two of the six variables that affect the calculation of the Predicted Mean Vote (PMV) and are inputs for thermal comfort calculations. However, in energy models, clothing insulation is often assumed to be 0.5/1.0 clo, and metabolic rate is usually assumed 1.2 met (seated) if other information is not available. These simplifications may lead to inaccuracy in the calculation of thermal comfort.

Agent-based modelling (ABM) is a social science method to simulate human behaviour that allows for the observation of aggregate behaviours emerging from the interactions of large numbers of autonomous actors. An agent can be defined as a system that acts and thinks like a human, which operates under autonomous control, perceives its environment, adapts to changes, and is capable of taking on specific goals'[2]. This modelling method provides us with an opportunity to mimic the real- world occupants that can put on or take

IOP Conf. Series: Earth and Environmental Science **329** (2019) 012022 doi:10.1088/1755-1315/329/1/012022

off clothes, or adjust activities according to its comfort level in the built environment. Therefore, this study aims to develop a clothing and activity behaviour simulation engine based on the agent-based technique, while further bridging the gaps between the simulated and actual thermal comfort results.

2. Methodology

2.1. Human behaviour model

The Theory of Planned Behaviour (TPB) developed by Ajzen [3] was adopted in this paper. The theory states that attitude toward behaviour, perceived behaviour control and subjective norms shape an individual behaviour together. TPB is an improvement of the Theory of Reasoned Action (TRA) [4].

Both theories hold that "human social behaviour follows reasonably and often spontaneously from the information or beliefs people possess about the behaviour under consideration". TPB model focuses on the "belief" concept, which can be categorized as follows:

- Behaviour belief: an individual's observable response in a given situation with respect to a given target. As in this case, a person believes that if he or she wears more clothes or increases activity, one will get warmer. Here, the behaviour belief is that to put on more clothes or increase activity intensity is equated with getting warmer.
- Control belief: an individual's perceived ease or difficulty of performing the particular behaviour. For clothing and activity behaviour, the control belief is the person's easiness of wearing or undressing clothes and changing activity status.
- Normative belief: an individual's perception about the particular behaviour, which is influenced by the judgment of significant others. For example, a person might think that it is he or she who should not wear too little, or do heavy load exercise in the office, so as not to affect other people in the office.

The behavioural intention for the Theory of Planned Behaviour can be expressed as a simple mathematical function (Eq. (1)):

$$BI = w_A A + w_{SN} SN + w_{PBC} PBC \qquad (1)$$

BI: Behaviour intention, *A*: Attitudes toward behaviour, namely behaviour belief, *SN*: Subjective norm, namely normative belief, *PBC*: Perceived behaviour control, *w*: empirically derived weight/coefficients.

Once the behaviour intention is determined, the occupant needs to decide the behaviour for the next step. Here the well-known "sense-think-act" loop is adopted to describe the decision-making process: firstly, an agent perceives something (sense), then it processes what it perceives (thinks), and lastly it executes an action (act).

In this case, an agent firstly *senses* its surroundings, which are defined by the clothing and activity values entered by the user and the thermal conditions of the space. The external EnergyPlus simulator provides the environmental thermal parameters to the agent. In order to summarize the holistic effect of environmental factors (including air temperature, humidity, and wind speed) on the human body,
the Predicted Mean Vote (PMV) index of Fanger's model[5] is adopted to calculate the indoor thermal comfort level, which is further identified as the behaviour trigger.

An agent *thinks* itself by following 2 steps. Firstly, the Behaviour Intention (*BI*) is calculated to determine the cost of each behaviour. Secondly, based on the thermal comfort level and costs of behaviours, an agent decides on a set of behaviours to execute. The decision indicates 3 characteristics of the next action: the type of behaviour, the number of behaviour and the incremental changes of the behaviour. The decision process is closely related to an agent's current comfort level and behaviour intention. To be more detailed, if an agent is very uncomfortable (the absolute value of PMV is large), the agent can consider both clothing and activity behaviour to mitigate the discomfort, and the increment can be large. Conversely, the smaller the absolute value of PMV is, the more likely it is to perform only one action with less increment change. On the other hand, the cost determines the likelihood that a type of behaviour will be executed. The larger the cost value of the behaviour is, the more likely it will be executed.

A simple MATLAB code using *rand()* function is used to achieve the above logic:

If rand() < abs(currentPMV)/3, the number of behaviours equals 2 and the increment equals 0.2.

If rand()>abs(currentPMV)/3, the number of behaviours equals 1 and the increment equals 0.1. The execution probability is then determined by the cost of each behaviour: if rand() < (costClo/sum(costs)), the clothing behaviour will be executed. Otherwise, the activity behaviour will be executed.

An agent *acts* itself by communicating with an external simulator EnergyPlus to calculate the behaviour impact on thermal comfort.

2.2. Agent learning
The Planned Behaviour model provides a possibility to fulfill agent learning through the behaviour belief parameter. The effectiveness of a behaviour on thermal comfort is reflected and memorized in the behaviour belief, which is updated every time step. If a behaviour has a positive effect on the agent's comfort by comparing time *t* and *t-1*, the agent will increase its behaviour belief at time *t* to help making decisions at time *t+1*. Otherwise, the agent will decrease its behaviour belief. Therefore, the agent can keep learning the effectiveness of a behaviour and better make decisions based on its experience. A MATLAB code is used to update the behaviour belief as follows (simplified):

If currentPMV > 0:
changePMV = currentPMV – oldPMV, BehavBelief = BehavBelief - (changePMV* rand()/6)

Else: changePMV = currentPMV – oldPMV, BehavBelief = BehavBelief + (changePMV * rand()/6)

The larger the change in PMV is, the more effective the behaviour is, and thus the larger the change in behaviour belief is.

2.3. Simulation coupling
Once the behaviour model is established, an external simulator is needed to calculate the environmental parameters. Here the environmental parameter refers to the PMV value that an agent perceives. In every time step, the PMV data caused by building design and agent properties are passed to the agent-based model (ABM), and then the behaviour model processes the PMV data and executes an action, forwarding the agent properties to the external simulator to get a new PMV value for the next step. Therefore, the interaction between human and the built environment is finally achieved through such iteration.

This paper uses EnergyPlus as the external simulator to implement the environmental calculation, and MATLAB to fulfil the agent-based model. The coupling procedure is achieved though MLE+[6] and Building Controls Virtual Test Bed (BCVTB) architecture[7]. A "*variables.cfg*" document is written to exchange data, including PMV value, clothing, metabolic rate, occupancy status and date/time.

3. Demonstrative example

3.1. Scenario settings
The environment chosen is a typical office room in a lab at Tianjin University (Tianjin, China). The room has a width of 10 m and depth of 8 m with a 30% window to wall ratio. It is equipped with a hot water radiator heating system in winter, while using natural ventilation to cool itself in summer. The heating set point is 22 °C. The total simulation time is 1 year (8760 h) with a simulation time step of 1 hour. The climate is classified as a hot humid continental climate (Dfa) in summer and a cold semi- arid climate (BSk) in winter based on the Köppen climate classification [8].

There are a few observations to be addressed here:

(1) A person cannot wear very little in public places, and does not wear too much as the outdoors to avoid affecting work efficiency. Also, staffs usually don't do very intense exercise in the office. Therefore, meaningful boundaries need to be set for clothing and metabolic rate. With reference to the ASHRAE Fundamentals Handbook[9], the minimum value for clothing is set to be 0.4 (shorts and T- shirts), and the maximum value is 1.2 (business suits). The minimum metabolic heat generation is set to be 1.0 (reading, seated), and the maximum is 2.1 (lifting/packing).

(2) People change clothes with the seasons all year round, so different seasons should have different upper and lower clothing limits. In summer (June, July, August), clothing is set to be bounded between 0.4 and 0.6, while it is set to be bounded between 0.9 and 1.1 in winter (December, January, February) clothing. During transition seasons (March –May, September –November), clothing is set to be bounded between [0.6, 0.9].

Since the month information is not the standard output variable of EnergyPlus, we used the Energy Management System (EMS), a high-level control method to access the month information during simulation. An EnergyPlus Runtime Language (Erl) script is written to get the month data at the simulation time t to determine the clothing boundary.

In addition, people usually wear more clothes when they go to work to cope with the more varied outdoor weather. Therefore, it is possible for people to change their clothes during the day.

(3) People don't change clothes and activities frequently, especially when they feel comfortable, so it is set that if the PMV is within the recommended limits ($-0.5<PMV<+0.5$) [1], the clothing and activity will not change. Besides, as metabolic rates increase above 1.0 met, the evaporation of sweat becomes an increasingly important factor for thermal comfort. The PMV does not fully account for this factor, and this method should not be applied to situations where the time-averaged metabolic rate is above 2.0 met.

(4) Initial values are shown in Table 1. The initial behaviour belief, control belief and normative belief are all set to 0.5 for both clothing and activity. Control belief and normative belief are kept constant during simulation in this paper. The weight coefficients for all beliefs are assumed to be equal.

Table 1: Initial variable values of ABM

Variables	Clothing (clo)	Activity (met)	Behaviour belief (*A*)	Control belief (*PBC*)	Normative belief (*SN*)	Weight coefficients (*w*)
Initial value	Summer: 0.5 Winter: 1.0 Spring & autumn: 0.8	1.2	0.5	0.5	0.5	1

3.2. Results and discussion

3.2.1. Behaviour portrayal. The weekly variation of clothing and activity behaviour in terms of the indoor comfort is shown in Figure 1 to Figure 3. In winter (Figure 1), both clothing and activity vary between days. Moreover, the clothing behaviour mainly occurs in the morning and remains constant during the rest of the day. This is because that the PMV value is the lowest at the beginning of a day—if $PMV < -0.5$, the agent will put on more clothes. With the full opening of the heating system and the warming of the outdoor climate during the day, the PMV increases to the comfort level and there is no need to wear more clothes.

During transition seasons, there are more changes in the clothing and activity behaviour. As shown in Figure 2, each day has a different curve. Taking the first day as an example (April 1), the clothing remains at 0.8 clo during the morning until 14:00, and then dropped to 0.6 clo at 16:00 in the afternoon because of the rising PMV. Similarly, the metabolic rate decreases from 1.2 met to 1.1 met at 14:00 when PMV increases beyond the comfort zone. At 18:00, the metabolic rate again decreases to 1.0.

This action together with the undressing behaviour makes the PMV drop to 0.457 at 18:00, which is within the comfort limits ($PMV<0.5$).

In summer (Figure 3), due to the hot climate and lacking of cooling system, PMV values are out of the comfort zone at the beginning of a day, which leads to the decrease of both clothing and metabolic rate at the beginning. Afterwards, PMV keeps rising and remains at a very high level (>2.5), which is out of the comfort range, and the clothing and activity remain at a lower level since then.

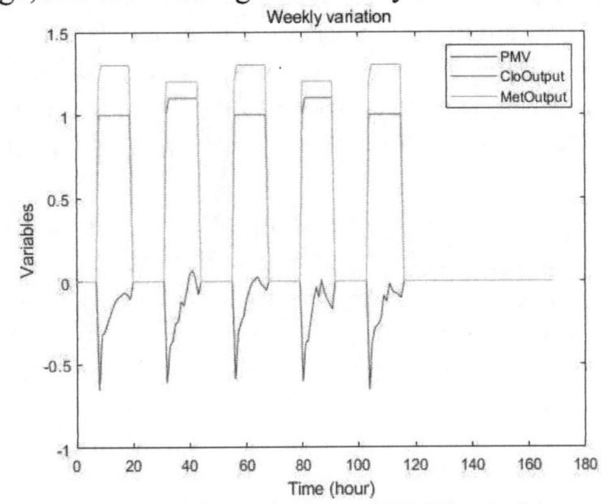

Figure 1: Weekly variation of PMV, clothing and activity level in winter (January 7 to January 13)

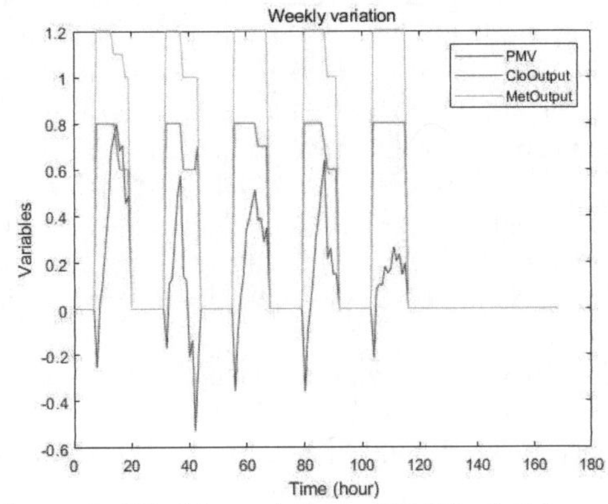

Figure 2: Weekly variation of PMV, clothing and activity level in spring (April 1 to April 7)

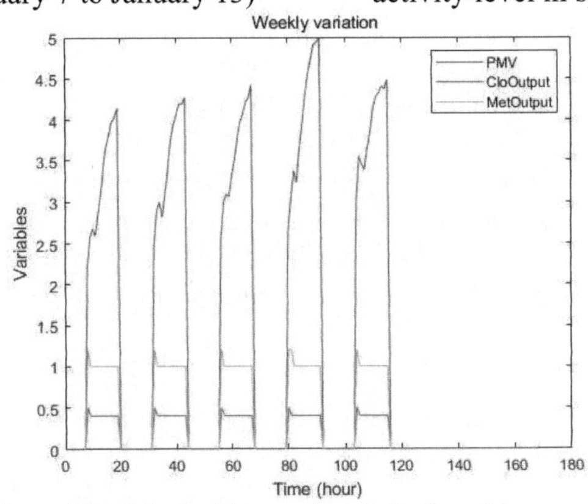

Figure 3: Weekly variation of PMV, clothing and activity level in summer (June 17 to June 23).

Figure 4 presents the annual clothing behaviour variation of the agent. It can be seen that the clothing value decreases regularly from winter to summer. During winter the clothing values are bounded between [0.9, 1.1] clo. Then in the early spring, the clothing values tend towards the upper limit of the transition season (0.9 clo), since the climate is still cold at this time. While approaching summer, the clothing value gets closer to the lower limit (0.6 clo), because the weather is getting hotter in this period. During summer, the clothing values are mostly between 0.4 and 0.5 clo due to the hot weather and high PMVs (Figure 6). Autumn exhibits a similar trend as spring - the closer to the summer it is, the higher the clothing value is, and the closer to the winter it is, the lower the clothing value is.

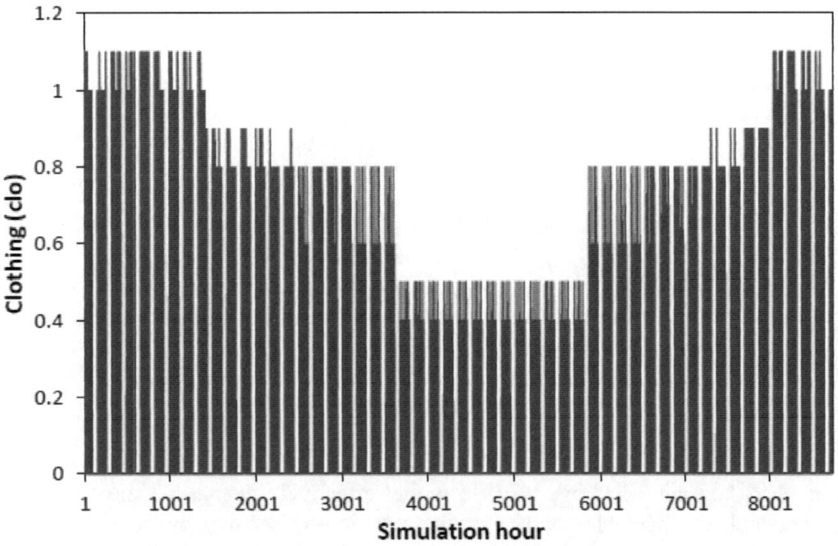

Figure 4: Annual clothing behaviour variations.

The annual activity behaviours are shown in Figure 5. It is observed that the agent's activity level in winter is relatively higher in order to keep the body warm, concentrated between [1.2, 1.3] met. While in summer, the activity level decreases to [1.0, 1.2] met to minimize the body heat production, since the PMV values are very high during summer (Figure 6).

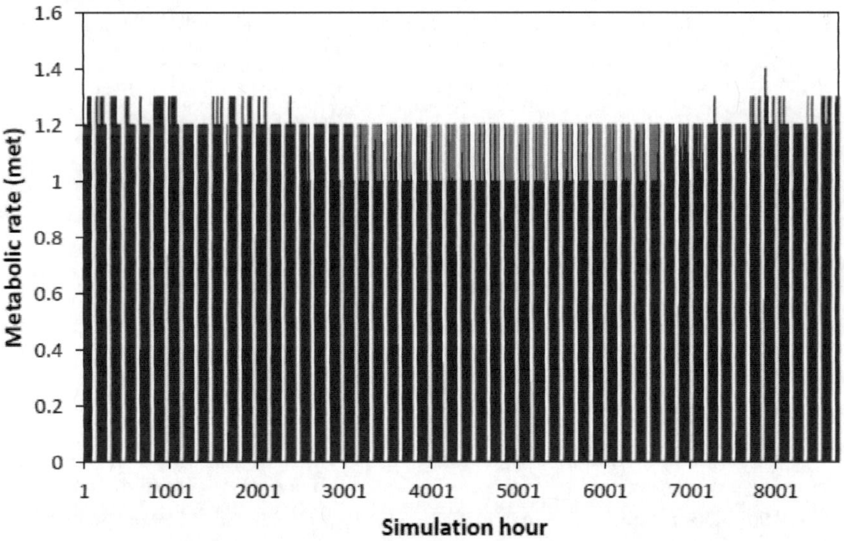

Figure 5 Annual activity behaviour variations.

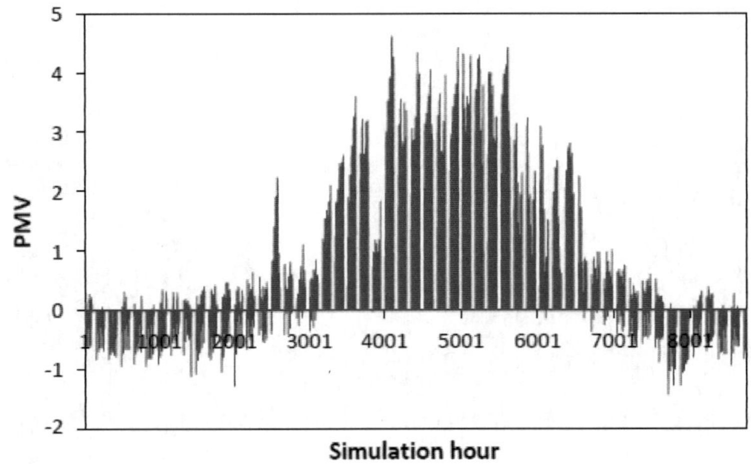

Figure 6 Annual PMV variations.

It can be seen that the established agent-based model effectively depicts people's clothing and activity behaviour. Agent behaviours change logically with the variations of thermal environment, and the simulation results are consistent with our daily observations. Furthermore, Figure 7 and Figure 8 present the behaviour beliefs for clothing and activity in an entire year. Both behaviour beliefs fluctuate significantly throughout the year. Sometimes the agent has higher belief on clothing, and sometimes the "belief" on activity is higher. However, both beliefs are relatively higher in the winter and yet much lower in the summer probably due to the constant high PMV values in summer. In summer days, no matter how the clothing and activity are adjusted, PMV values are stubbornly high (because of the hot climate and lacking of cooling system), causing the agent to lose confidence in these two behaviours. By the time that the hot climate is not so overwhelming in autumn and the effect of clothing and activity starts to manifest itself, both beliefs begin to rise again.

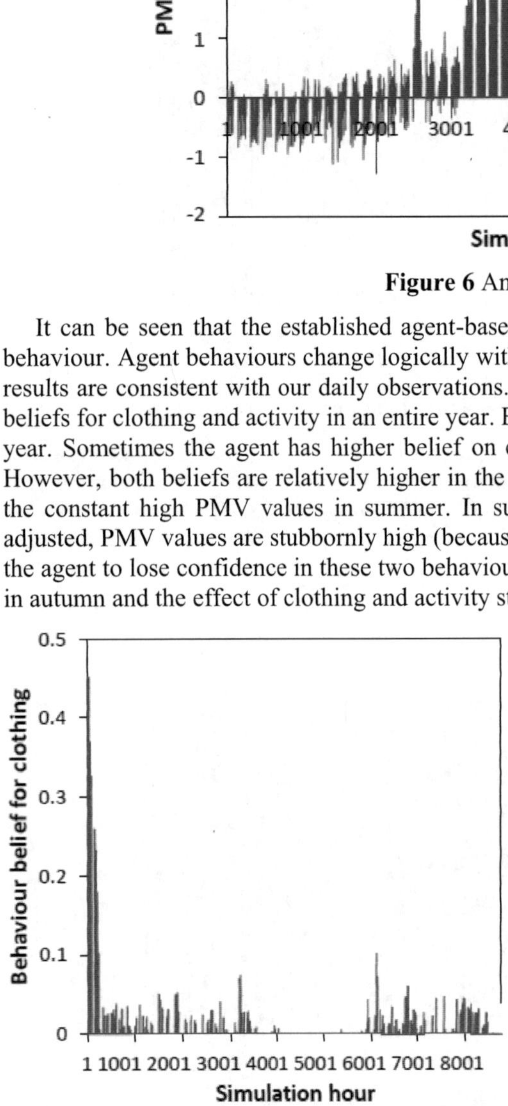

Figure 7 Annual behaviour belief variations for clothing

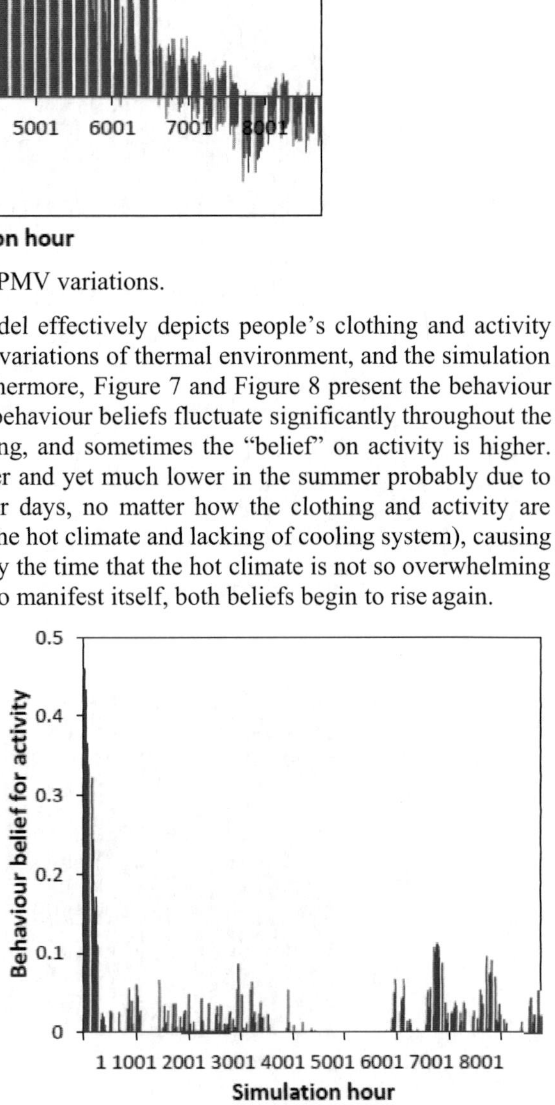

Figure 8 Annual behaviour belief variations for activity

3.2.2. Impact on thermal comfort. Here we compared the comfort results of ABM method with those of the traditional fixed method (Table 2). Adopting summer clothes (0.5 clo) and fixed activity level (1.2 met) in energy simulation leads to a result of 2695.00 uncomfortable hours, while adopting winter clothes (1.0 clo, 1.2

met) results in 2140.00 hours of discomfort. The model using mixed clothes (0.5 clo in summer and 1.0 clo in winter) and fixed activity level has a discomfort time of 1843.00 hours. Lastly, the ABM method results in the shortest discomfort time of 1803.00 hours. Apparently, since the agent can change clothes and the activity level according to the overheated or too cold environment in the agent-based model, its uncomfortable time is the shortest.

Table 2 Time not thermally comfortable in a year in terms of ASHRAE 55-2004

Scenario	Fixed method: summer clothes (0.5 clo) activity level (1.2 met)	Fixed method: winter clothes (1.0 clo) activity level (1.2 met)	Fixed method: summer or winter clothes (0.5 or 1.0 clo) activity level (1.2 met)	ABM method
Time not comfortable	2695.00	2140.00	1843.00	1803.00

4. Conclusion

This paper presents an agent-based method to simulate the occupant's clothing and activity behaviour for thermal comfort in buildings. It can be seen that the agent-based model effectively describes the clothing and activity behaviour of people in response to changes of indoor thermal environment. The ABM allows for intelligent, autonomous agents to adapt to the given environment and make reasonable clothing and activity behaviour decisions based on their comfort level. The energy simulation result using ABM method is quite different from that using the traditional fixed-parameter calculation method, leading to the lowest discomfort hours.

The ABM method to depict occupant's clothing and activity behaviour is still at its early stages and there are many aspects to be improved. Firstly, field study on occupant's behaviour in real situations is needed to validate and improve the ABM method. In addition, this paper does not explore the parameters of control belief and normative belief of the TPB model, which will have a potential effect on the behaviour, such as the difficulty of changing clothes or performing an activity, or other colleagues' attitude towards the agent.

Acknowledgments

This research was funded by the National Key Research and Development Program of China (Grant No.2016YFC0700200) and the Programme of Introducing Talents of Discipline to Universities (Project No.B13011).

References

[1] ASHRAE, A. (2010). Standard 55-2010: Thermal environmental conditions for human occupancy. 2010. American Society of heating, Refridgerating and Airconditioning Engineers: Atlanta.
[2] Russell, S. J., & Norvig, P. (2016). Artificial intelligence: a modern approach. Malaysia; Pearson Education Limited.
[3] Ajzen, I. (1991). The theory of planned behaviour. Organizational behaviour and human decision processes, 50(2), 179-211.
[4] Fishbein, M., & Ajzen, I. (2011). Predicting and changing behaviour: The reasoned action approach. Psychology Press.
[5] Fanger, P. O., & Toftum, J. (2002). Extension of the PMV model to non-air-conditioned buildings in warm climates. Energy and buildings, 34(6), 533-536.
[6] Bernal, W., Behl, M., Nghiem, T. X., & Mangharam, R. (2012, November). MLE+: a tool for integrated design and deployment of energy efficient building controls. In Proceedings of the Fourth ACM Workshop on Embedded Sensing Systems for Energy-Efficiency in Buildings (pp. 123-130). ACM.
[7] Wetter, M. (2011). Co-simulation of building energy and control systems with the Building Controls Virtual Test Bed. Journal of Building Performance Simulation, 4(3), 185-203.
[8] Kottek, M., Grieser, J., Beck, C., Rudolf, B., & Rubel, F. (2006). World map of the Köppen -Geiger climate classification updated. Meteorologische Zeitschrift, 15(3), 259-263.
[9] Handbook, A. F. (2009). American society of heating, refrigerating and air-conditioning engineers. Inc.: Atlanta, GA, USA.

IOP Conf. Series: Earth and Environmental Science **329** (2019) 012023 doi:10.1088/1755-1315/329/1/012023

The relationship between thermal environments and clothing insulation for rural low-income residents in China in winter

Wei Zhao[1, 2], David Chow[1, 3] and Steve Sharples[1, 4]

[1] School of Architecture, University of Liverpool, Liverpool, United Kingdom
[2] Wei.Zhao@liverpool.ac.uk
[3] David.Chow@liverpool.ac.uk
[4] Steve.Sharples@liverpool.ac.uk

Abstract: China is developing rapidly, but there is an obvious imbalance in the development level between urban and rural areas. Differences in economics, living conditions and life style contribute to differences in indoor thermal environments, and may result in differences in the perceptions of thermal comfort between urban and rural areas. Clothing insulation as one of the most important parameters that affect thermal comfort is studied in this research. A field study is conducted in the rural areas of Lankao, China. A total of 384 valid samples were acquired. 54.97% researched houses suffered extreme cold indoor temperatures between 2°C and 6°C. Clothing insulation value of researched population is significantly larger than in other research. Clothing insulation value of female is significantly larger than that of male. It was also found that clothing insulation value correlates positively with age, that clo grow with the increase of age significantly. 7-days running mean outdoor temperature was selected from four temperature indices to predict clothing insulation value. Clo value is negatively correlated with 7-days RM outdoor temperature. But, the slope is quite small in such an extreme cold condition. A case study was conducted to estimate influence of adaptive clothing insulation on heating energy consumption (from Nov. to Mar.). Results show that 863 heating hours are reduced by using adaptive clothing insulation model. The total Heating Degree Hours (HDH) is reduced from 6445 to 1367 by 78.8% which means the same percentage of heating energy conservation. Besides, this case study also concluded that "adaptive clothing insulation model" has higher influence on heating energy consumption in transition season.

1. Introduction

Clothing acts as a layer of insulation between the human body and the ambient environment, affecting the heat exchange between both ends. The resistance of this heat exchange is usually called clothing insulation which is expressed in a unit named "clo". Clothing insulation plays an important role in thermal comfort research. It is one of the six most important parameters that affect the thermal comfort calculation, and is an input of PMV and PPD calculations in many thermal comfort standards such as ASHRAE 55 and ISO 7730. In adaptive thermal comfort research, clothing adjustment is viewed as the most important and efficient adaptive behaviour [1, 2]. For near-sedentary activities with a metabolic rate of 1.2 met, the effect of changing clothing insulation on optimum operative temperature is approximately 6 °C per clo [1]. So, the selection of clothing insulation for thermal comfort calculation during design stage of HVAC system and building affect the energy consumption of the building greatly. Unfortunately, in most standards, only 0.5 clo and 1.0 clo are used, with 0.5 clo for the cooling season and 1.0 clo for the heating season. More accurate research should be conducted, and thus adaptive clothing insulation models should be proposed based on the situation of the research areas.

The relationship between clothing insulation and indoor/outdoor environment has long been a topic of previous researchers. Outdoor temperature is believed to be the most critical factors that affect people`s clothing insulation. However, different indices are used by researchers. Daily mean outdoor dry bulb temperature was used by Morgan and de Dear to develop a linear relationship with clothing insulation [3], while, outdoor air temperature measured at 6 o`clock a.m. was used by De Carli to predict the clothing insulation [4]. As well as the linear models used by these research, other types of model have also been developed to predict clothing insulation. For example, the cubic model used by Nicol et al [5], the logistic model used by Liu et al [6] and a multivariable model used by Schiavon et al [1].

Similar research have been conducted on different populations in different building types in urban areas, such as research on young children in kindergarten [7], students in university [8], workers in workshop [9], people in shopping mall [3] and elderly people in elderly care house [10, 11]. Clothing insulation and its influence on energy consumption in low income rural areas in China is seldom researched. The typical ensemble in winter, the change of clothing insulation with gender, age and indoor/outdoor environments and influence on energy consumption in winter are not clear. These problem are studied in this paper. This research is part of the thermal comfort research in this area. The aims of this research are:

1) To study the difference of clothing insulation between different genders and ages in winter.
2) To study the relationship between clothing insulation and indoor environment and outdoor environment in winter
3) To develop an adaptive prediction model for clothing insulation in winter.
4) To estimate the influence of clothing insulation on energy consumption in winter.

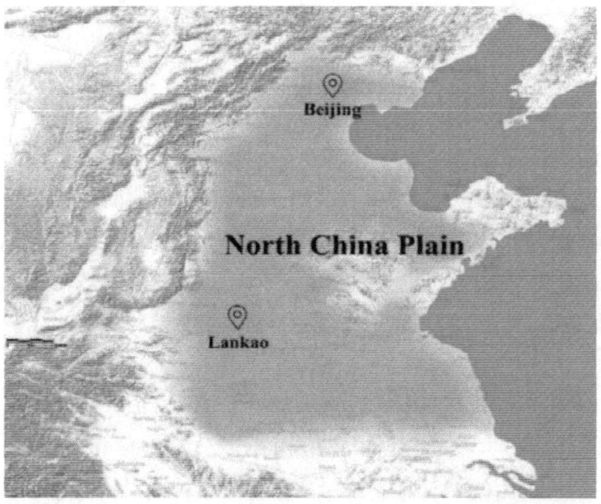

Figure 1. Location of Lankao on North China Plain.

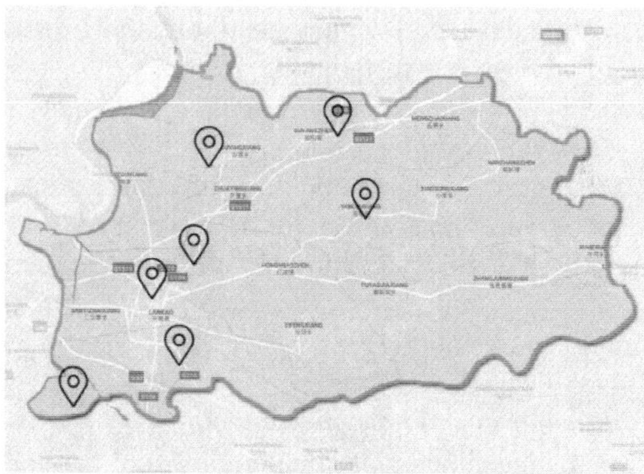

Figure 2. Researched villages in Lankao.

Figure 3. Typical residential buildings in research area.

2. Method

2.1. Sample selection

A field study was conducted in rural regions of Lankao county. Lankao county is located in the "Cold Zone" on the North China Plain. The North China Plain is the most densely populated region in China, and the "Cold Zone" is one of the five "Building Climate Zones in China". The PCDI of rural population in Lankao is only about £1253 per year which is close to but a little smaller than the China average £1543 [12, 13].

Seven villages in Lankao were selected randomly for this field study. Almost all rural residential buildings in research villages are in free-running conditions. Participants were also selected randomly in villages. There is a total of 423 subjects (214 male and 209 female) interviewed in this field study.

2.2. Field measurement

In this study, a transverse field survey method was adopted. The study was conducted in workdays from 21th Dec. 2018 to 23th Feb. 2019, the typical winter season in this region. Visiting time was from 10:00 to 17:00 on every research day. Finally, 423 questionnaires were collected, from which 384 effective questionnaires were selected.

The complete investigation consists of two parts, a questionnaire and a simultaneous environment measurement. The questionnaire contains information on basic information of the participants, their thermal comfort and clothing garments. Indoor parameters were measured during the interview. Usually, the interview took place in the occupant's bedroom or living room. The participants were asked to stay in the room, either seated relaxed or standing relaxed, for at least 15 minutes. Instruments were set either in the centre of the room or in a suitable position close to subject, and were protected from solar radiation and heating devices. Two heights (0.6m and 1.1m) are optional for instruments according to the activity state of interviewed participant.

2.3. Clothing insulation estimate

Clothing garments were noted on the questionnaires according to subjects' description. Participants` ensemble were transformed to clothing insulation value, based on ISO 7730, ISO 9920 and Chinese Standard GBT 50785. Chair insulation was not considered here for 2 reasons. Firstly, most participants were not in a seated position. Secondly, people would not consider chair insulation when choosing garments in the morning.

2.4 Influence on energy consumption

To estimate influence of adaptive clothing insulation on heating energy consumption in winter, two parameters were used and compared. These are heating hours and Heating Degree Hours (HDH). HDH was defined as a measurement to quantify the demand for energy needed to heat a house. It is quite similar to the HDD. The difference is that, HDD uses daily outdoor temperature and base temperature, while HDH use hourly PMV and base PMV for calculation. The calculation is to subtract hourly PMV from Base PMV. If the value is less than or equals zero, that hour has 0 HDH. If the result is positive, the value represent the number of HDH of that hour. Then HDH of a period can be calculated by sum up HDH of each hour. The larger HDH is the more heating energy will be consumed to achieve indoor thermal comfort.

A typical house model was built and simulated in Designbuilder for a case study. Hourly mean PMV-PPD was then calculated based on simulation results. Heating hours and HDH with a base PMV of 0.85 (limit of 80% comfort zone) was calculated and compared.

However, this is a rough estimation of energy consumption. A dynamic simulation and more accurate energy reservation is under process.

3. Results and discussion

3.1. Outdoor climate and indoor environment

This field study was conducted from 21 December 2018 to 23 February 2019, the coldest three months in the research region. Daily outdoor air temperature and relative humidity curves of the period are shown in **Figure 4**. A basic description of outdoor environmental parameters are listed in **Table 1**.

Table 1. Outdoor environmental parameters

	Parameters	Average	MAX	MIN	SD
Outdoor	Air Temperature (°C)	0.67	8.09	-7.34	3.02
	RH (%)	62.63	97.38	29.63	17.83

Distribution of indoor operative temperature measured in surveyed houses is shown in **Figure 5**. Indoor operative temperatures were divided into several bins with a step of 2 °C . The number and frequency of acquired indoor operative temperature drop-in each bin is counted. Indoor operative temperatures between 2.0 °C and 6.0 °C take up more than half, about 54.97%, of all the measured data.

Most families have to suffer extreme low indoor temperature. Building insulation and heating devices are not affordable due to the outdated construction techniques and low income of the local residents. None of researched houses were built with insulation layers, and only 19.2% of the houses were running their heating devices when the interviews took place (Shown in **Figure 6**). Actually, many families do have heating devices, but, none of them keep the heating devices running all day long. Most families only turn on heating devices when they experience extreme cold indoor conditions. The average running time of heating per day is about 2 hours according to local residents` subjective impression.

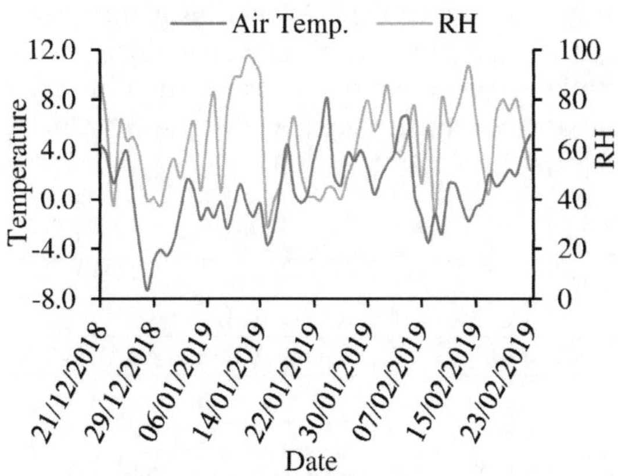

Figure 4. Daily outdoor air temperature and RH

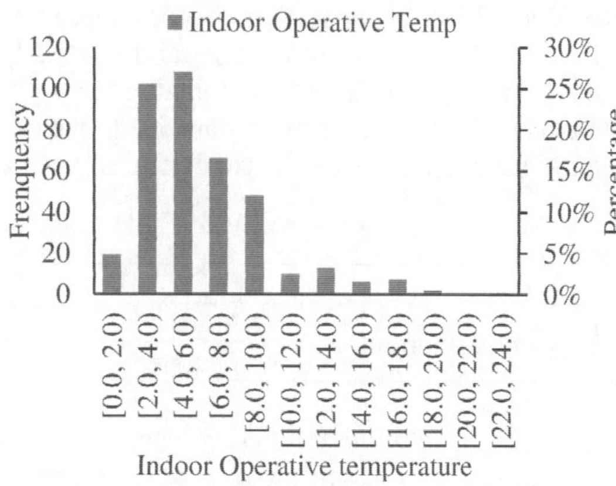

Figure 5. Distribution of indoor operative temperature

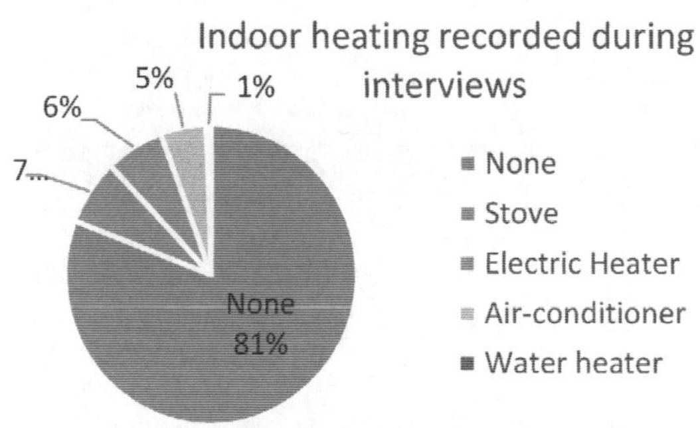

Figure 6. Percentage of heating device in running during interview

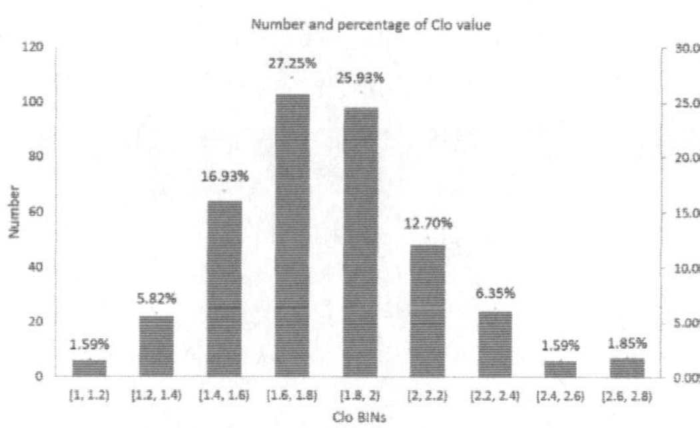

Figure 7. Distribution of clo value.

3.2. Clothing basic

A basic description on clothing insulation value and number of garments of the total sample are shown in **Table 2**. The average, maximum and minimum clothing insulation of the samples are 1.81 clo, 2.78 clo and 1.01 clo, respectively. The standard deviation is 0.30.

The distribution of clothing insulation value is shown in **Figure 7** by Bin method with a step of 0.2 clo. People with clothing insulation value of 1.6 clo to 2.0 clo make up 53.18% of the sample.

Table 2. Basic description on clo value and number of garments

	Average	**MAX**	**MIN**	**SD**
Clo Value	1.81	2.78	1.01	0.30

These results for clothing insulation values are significantly higher than many other previous research [1, 11, 14, 15]. This can be explained by the following reasons: First, both outdoor temperature and indoor temperatures (air temperature and operative temperature) are extremely low. Second, for the low income of rural families, heating devices are unaffordable, and even those who could afford them only use them sparingly, due to the cost of the fuel. Thus, increasing clothing is the main method of increasing thermal comfort for rural residents in cold winters. Other research conducted in regions with similar outdoor weather and indoor operative temperature [10, 16] gave similar levels of clothing insulation values.

3.3. Clothing insulation of different sex

The discrepancies of clothing insulation on sex are listed in **Table 3**. The average clothing insulation value of female is 1.84 clo, larger than male`s 1.78 clo. To test whether females 'clothing insulation value is significantly larger than that of male, a one-tailed T test was conducted with a significant level of $\alpha = 0.05$. The P-value of the one-tail t-test result is 0.026, much smaller than 0.05, implying that the clothing insulation value of the female is significantly larger than that of the male. From the above results, females tend to be more sensitive to the cold environment than males and prefer to wear thicker clothing than males to keep thermal comfort in the research region during winter.

Table 3. Clothing insulation differences on sex.

		Average	Max	Min	SD
Clothing insulation	Male	1.78	2.46	1.01	0.286
	Female	1.84	2.78	1.07	0.315

3.4. Clothing insulation of different ages

In this section, the relationship between clothing insulation value with the age were discussed. The BIN method is again used. "mean clo" for each age group were calculated. The Pearson`s correlation coefficient equals 0.778, means a strong linear relationship between "mean clo" and age. Linear function was developed with a $R^2 = 0.6064$ and P-value < 0.01 (**Figure 8**). This means clo and number of garments show trends of growth with the increase of age. The results agree with previous research that elderly people tend to have lower activity levels and metabolic rates, and prefer to increase clothing insulation level to acquire thermal comfort.

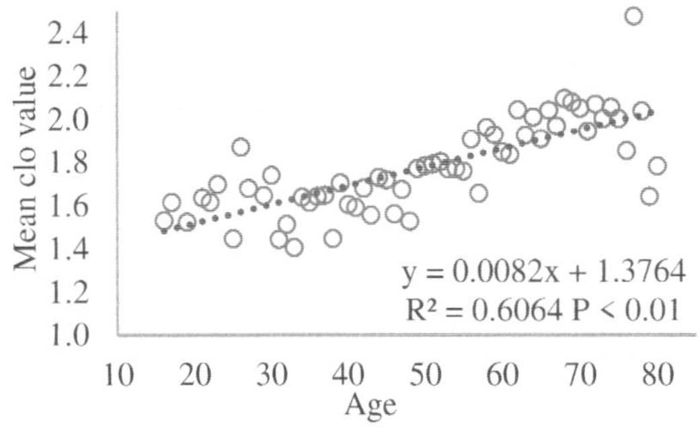

Figure 8. Regression mean clo on age.

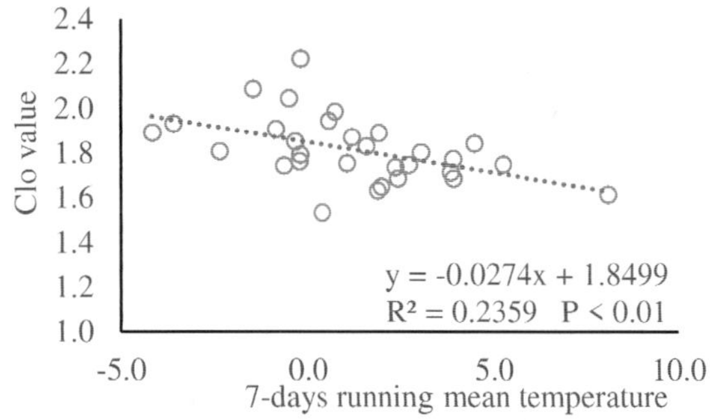

Figure 9. Regression clo on 7-days running mean outdoor temperature

3.5. Clothing insulation and Environment

This part explores the relationship between Clothing insulation with indoor and outdoor environment. As temperature is always the most important factor affecting people's choice of clothing, four temperature indices used most extensively in thermal comfort research were selected to discover their relationship with clothing insulation value. They are: indoor air temperature, indoor operative temperature, outdoor air temperature and the 7-days running mean temperature. For the interior linear relationship between them, only one most correlated index will be selected.

Spearman`s rank correlation coefficient was used to estimate the correlation level of relationship. The Bin method was not used here. The use of bin method will artificially reduce variance and increases the correlation coefficient of determination. This will introduce a loss of information in this case.

Table 4 shows the correlation value (Spearman`s rank correlation coefficient) and significant level. Results of four temperature parameters are similar. All of them have a negative relationship with clothing insulation value. Results show that those two outdoor temperature parameters have closer correlations with clothing insulation value. In addition, the significance level of outdoor parameters are better than that for indoor parameters. The 7-days running mean temperature tend to be the best index to represent the affection of temperature on Clo value than other three indices. Linear regression of Clo against 7-days running mean temperature is shown in **Figure 9**. This function is view as adaptive clothing insulation model, and will be used to predict clothing insulation value in following research.

Table 4. Spearman`s correlation coefficient.

		Indoor		Outdoor	
		Air Temp	Operative Temp	Air Temp	7-days running mean Temp
Clo	Correlation Coefficient	- 0.118	- 0.122	- 0.210	-0.232
	Significance	0.021	0.017	0.000	0.000

Thermal experience (past outdoor temperatures) significantly influences the in clothing insulation value [1, 17]. This can be easily explained by phenomenon in daily life in research area. First, the choice of cloth in the morning is based on not only the forecast temperature of today but more on the thermal experience and weather of past few days. Second, the ensemble for today tend to inherit the ensemble for yesterday. In addition, in this region, people do not usually change clothes. Those low income local residents don`t have much clothes to change. Also, the inconvenience of washing clothes also decreases the possibility of changing.

4. A case study

It is very important to take clothing adjustment into consideration for future design of HVAC system for rural low-income houses. This can be explained from two aspects. Firstly, for rural low-income residents, if they can achieve same level of thermal comfort by adjusting clothing, they would not turn on heating system. Secondly, by adjusting clothing, rural low-income residents can achieve same level of thermal comfort in lower indoor temperature. That means both heating hours and heating setpoint temperature of HVAC system can be reduced. This will surely lower heating energy consumption without reducing occupants` thermal comfort.

In this part, a case study was presented to analysis how clothing adjustment influences occupants` thermal comfort and building`s heating energy consumption. A typical rural low-income house was modelled and simulated in the 3 typical winter months (December, January and February) plus 2 transition months (November and March) adjacent to typical winter months. The "adaptive clothing insulation model" developed above was used to predict clothing adjustment of rural residents in experiment group. A constant 1.0 clo was used in control group for comparison. PMV-PPD index were used to represent thermal comfort. Heating hours and HDH (defined in section 2.4) were used to estimate reduction of heating energy consumption.

This is a rough estimation of heating energy conservation. The relationship between HDH and energy is not clear yet. However, a more accurate research with a dynamic simulation is under process, which will calculate the exact amount of energy saved by using "adaptive clothing insulation model".

4.1 Heating hours

Hourly PMV can be calculated from simulated results. If calculated PMV are beyond the 80% (and 90%) comfort zone limits, the hour is considered as uncomfortable hour. Then, total comfortable hours and uncomfortable hours in each month and the whole period can be summed up. As shown in **Table 7**, if the constant value 1.0 clo is used, only 2 hours out of 3623 hours falls in the 90% comfort zone. However, comfort hours increase to 500 hours when "adaptive clothing insulation model" is used. That means heating hours can be reduced by 498 hours. If the 80% acceptable range is used, number of comfort hours increase from 44 to 907, by changing constant value 1.0 clo to "adaptive clothing insulation model", which means a reduction of 863 heating hours. Heating hours is reduced by 23.8% for 80% comfort zone and 13.7% for 90% comfort zone, by using "adaptive clothing insulation model".

Table 5. Comfort hours

	Calculated with "adaptive clo value prediction model"	calculated with fixed clo value 1.0	Total hours
90% acceptable comfort hours	500	2	3623
80% acceptable comfort hours	907	44	3623

Figure 10 shows the distribution of 80% acceptable comfort hours in each simulated months. Comfort hours in November and March (transition seasons) increase a lot, while remain almost unchanged in December, January and February (typical winter months). That means "adaptive clothing insulation model" can help to reduce heating hours of HVAC system and heating energy consumption in transition seasons significantly. 69.8% and 50.7% heating hours are reduced in November and March respectively.

4.2 Heating Degree Hours

However, heating hour is not able to show all influences clothing adjustment might bring to energy consumption. Heating setpoint also influences energy consumption. Heating setpoint temperature by using "adaptive clothing insulation model" is surely lower than that by using constant value 1.0 clo. Difference in heating setpoint temperature can be mapped to PMV difference. By adjusting clothing, residents can achieve better PMV than using constant clothing insulation, in same indoor environment. In other words, PMV deviate less from comfort zone by using "adaptive clothing insulation model". When PMV deviate from comfort zone, heating is required, energy is consumed. The more PMV deviate from comfort zone, the more energy will be consumed. HDH is based on PMV deviation and is used to

IOP Conf. Series: Earth and Environmental Science **329** (2019) 012023 doi:10.1088/1755-1315/329/1/012023

quantify the energy needed for heating. The more PMV deviating from comfort zone, the larger HDH is, the more heating energy is consumed.

HDH of adaptive clothing insulation model and constant 1.0 clo in each month and whole period was calculated and plotted in **Figure 11**. In November and March, HDH are reduced by 90.8% and 90.7% respectively. The percentage are higher than that in December (74.5%), January (73.7%) and February (77.6%). This result proves that adaptive clothing insulation model has more obvious influence in heating energy consumption in transition season. Total HDH is reduced from 6445 to 1367 by 78.8% which means the same percentage of heating energy conservation.

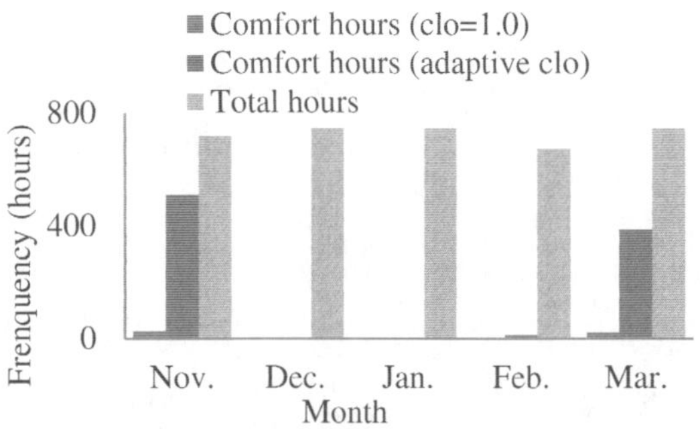

Figure 10. Distribution of 80% acceptable comfortable hours of each month.

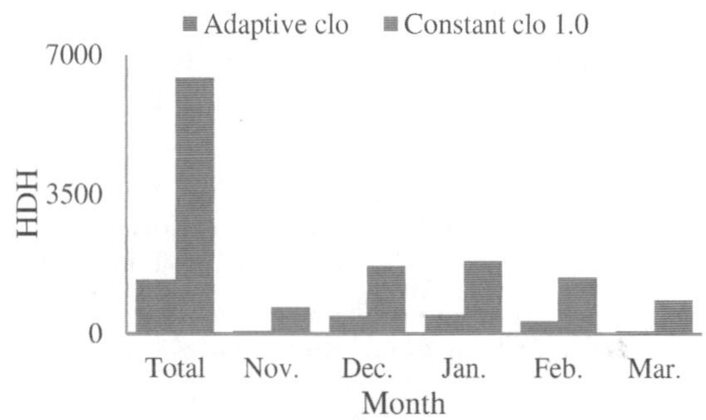

Figure 11. *WT* with different clo value in each month

5. Conclusion

This paper research on clothing insulation and influence of clothing adjustment on heating energy consumption in rural low-income houses in China. Results are based on 382 valid field study questionnaires collected from seven villages in Lankao in winter.

54.97% recorded indoor operative temperature during interview ranges from 2°C to 6°C. Most families have heating devices, including traditional stove, electric heater, air-conditioner and so on. But only 19% of them are running their heating devices when interview took place.

Mean clothing insulation value of researched population is 1.81 clo, significantly larger than that in other previous research, that people have to wear more to maintain their thermal comfort in extremely cold living conditions. Female has a mean clothing insulation of 1.84 clo which is 0.06 clo larger than that of male in research area. It is significantly according to the one-tailed T test. It was also found that clothing insulation value is positively correlate with age, that clothing insulation grow with the increase of age significantly.

The 7-day running mean outdoor temperature was selected from four temperature indices to predict clothing insulation value adaptively. Thermal experience seems to have obvious effect on clothing insulation value. Clothing insulation value is negatively correlated with 7-day running mean outdoor temperature. However, the slope is quite small in such an extreme cold condition.

A case study estimating clothing adjustment's influence on heating energy consumption was conducted. By using adaptive clothing insulation model, heating time is reduced by 863 hours from November to March. Total HDH is reduced from 6445 to 1367 by 78.8% which means the same percentage of heating energy conservation. This case study also concluded that "adaptive clothing insulation model" has higher influence on heating energy consumption in transition season.

A further research on dynamic simulation of exact amount of heating energy consumption using adaptive clothing insulation model is under process.

References

1. Schiavon, S. and K.H. Lee, *Dynamic predictive clothing insulation models based on outdoor air and indoor operative temperatures*. Building and Environment, 2013. **59**: p. 250-260.
2. Newsham, G.R., *Clothing as a thermal comfort moderator and the effect on energy consumption.* Energy and Buildings, 1997. **26**(3): p. 283-291.

3. Morgan, C. and R. de Dear, *Weather, clothing and thermal adaptation to indoor climate.* Climate Research, 2003. **24**(3): p. 267-284.
4. De Carli, M., et al., *People's clothing behaviour according to external weather and indoor environment.* Building and Environment, 2007. **42**(12): p. 3965-3973.
5. Nicol, J.F., et al., *Climatic variations in comfortable temperatures: the Pakistan projects.* Energy and buildings, 1999. **30**(3): p. 261-279.
6. Liu, W., et al., *Indoor clothing insulation and thermal history: a clothing model based on logistic function and running mean outdoor temperature.* Building and Environment, 2018. **135**: p. 142-152.
7. Nam, I., et al., *A study on the thermal comfort and clothing insulation characteristics of preschool children in Korea.* Building and Environment, 2015. **92**: p. 724-733.
8. Cao, B., et al., *Field study of human thermal comfort and thermal adaptability during the summer and winter in Beijing.* Energy and Buildings, 2011. **43**(5): p. 1051-1056.
9. Yang, R. and W. Su, *Thermal insulation of clothing ensembles for Chinese textile workers in the cotton textile workshop.* Case studies in thermal engineering, 2016. **7**: p. 8-13.
10. Jiao, Y., et al., *Thermal comfort and adaptation of the elderly in free-running environments in Shanghai, China.* Building and Environment, 2017. **118**: p. 259-272.
11. Jiao, Y., et al., *The relationship between thermal environments and clothing insulation for elderly individuals in Shanghai, China.* Journal of thermal biology, 2017. **70**: p. 28-36.
12. *Henan Province Bureau of Statistics.* Available from: http://www.ha.stats.gov.cn/sitesources/hntj/page_pc/tjfw/tjgb/szgxgb/article20997a83e9a943bfb63c345c5c176c0d.html.
13. *National Bureau of Statistics.* 2018; Available from: http://www.stats.gov.cn/tjsj/zxfb/201801/t20180118_1574931.html.
14. Huang, L., et al., *Field survey of indoor thermal comfort in rural housing of northern China in heating season.* Journal of Southeast University (English Edition), 2010. **26**(2): p. 169-172.
15. Liu, H., et al., *Seasonal variation of thermal sensations in residential buildings in the Hot Summer and Cold Winter zone of China.* Energy and Buildings, 2017. **140**: p. 9-18.
16. Cheng, B., et al., *Characteristics of thermal comfort conditions in cold rural areas of china: A case study of stone dwellings in a tibetan village.* Buildings, 2018. **8**(4): p. 49.
17. de Carvalho, P.M., M.G. da Silva, and J.E. Ramos, *Influence of weather and indoor climate on clothing of occupants in naturally ventilated school buildings.* Building and environment, 2013. **59**: p. 38-46.

Beyond Energy- efficient Built Environment – Examining the Relationship between the Users' Cultural Values and Energy Consumption.

B K Satish
School of Art, Design and Architecture, University of Plymouth, UK.

Abstract. Recent developments at the global level have forced responsible countries to take decisive actions to mitigate climate change. In the race to reduce energy consumption in the Built Environment, resource efficiency has been underpinned by low embodied energy (low carbon building) and consumption pattern. Recent expectations of energy reduction and energy efficiency have redefined the way buildings are insulated and spaces are utilised. This is particularly critical in the case of Passivhaus design, smart buildings etc, where the users are expected to align themselves to the building specification than the conventional practice of designing and building for the users' requirements. There is a direct correlation between culture informed behaviour and sustainable built environment. Though there is considerable progress in energy efficiency in the built environment, further research is required to understand the individual perception and its implication on energy consumption pattern. Further, research relating ethnic background and their behaviour to energy consumption are scant. In this context, this paper reviews the prevailing research in the three broad areas of Thermal comfort, Ethnicity and Cultural Behaviour and its impact on Energy behaviour. Due to time and scope, this paper will not dwell deep into these topics, rather on the overlapping, interrelated aspects. In the process, it proposes a framework to analyse user energy behaviour while addressing the built environmental issues related to climate change. These topics present many opportunities for productive future research.

Key words: Thermal comfort, Energy Behaviour, Cultural Behaviour

1. Introduction

Climate change and its immediate impacts have had a major consequence in the way we use energy. Meanwhile, technological advancements are leading us to be hypersensitive to the urge to be comfortable is forcing us to attain a higher degree of comfort. P.O. Fanger championed the heat balance calculation of the indoor climate and the users in the 1960s and 70s. His proposed studies, Predicted Mean Vote and Predicted Percentage Dissatisfied (PMV-PPD) indices and PMV model, is still being used as a thermal comfort evaluation tool by international standard organisations such as ISI and ASHRAE Standard 55 [1]. The key shortfall of the heat balance models is that it views occupants as passive recipients of thermal sensation irrespective of different building types and climate zones. On the contrary, later developed adoptive model emphasises more on the occupants and their thermal comfort through the psychological, physiological and behavioural process.

There are considerable innovations while understanding and responding to thermal comfort in buildings and it is the oil crisis in the 1970s, which encouraged the energy-efficient approaches in thermal comfort [2]. The recent developments have shifted the focus of thermal comfort from energy-intensive central systems to locally controllable adoptive comfort [2]. Recently, researchers have highlighted the negative impact of the spaces designed for neutral, uniform and static condition as it is extremely energy intensive to maintain thermal neutral condition [3]. Researchers have argued for the need for a paradigm shift in the notion of comfort, reflecting individuals' preferences and internal ambience, which energises building inhabitants and brings pleasure while saving energy in the process. While introspecting the performance of the occupant, researchers have found out that no best temperature for productivity exists rather it occurs within a wide range of air temperature [3]. One of the crucial limitations of the prevailing studies is that the internal environment is mostly controlled and expressed by air temperature and humidity and building performance are directly related to thermal comfort [3].

While understanding the perception of thermal comfort, one has to acknowledge the distinction between sensation and perception. While sensation detects the stimulus in the ambience, perception refers to the interpretation process of the information [4]. It is this perception which interprets the quality of or changes in the environment to the user [2]. Delzendeh et al in their review paper have identified that hardly 10 % of the research papers related to user behaviour in the 100 papers they have reviewed relating to both social and personal factors [5]. One of the key components of the established Franger's PMV model is the limitation of being descriptive of thermal environmental parameters. For instance, zero on the seven-point ASHRAE scale meant that the environment evokes a neutral, neither warm nor cool, response. However, it does not reflect the perception of the occupants and hence whether they like it or not as the descriptive model assess the acceptability of the environment [2]. Introspecting and changing human behaviour is one of the key paths while addressing the climate change challenge and to reduce energy consumption [6]. Understanding the user behaviour is very important and there is very limited research in this area. This paper focuses on Alliesthesia and elaborates the perception of users beyond thermal neutral conditions. A key driver of the perception and energy behaviour, culture, is explored in the next section. Finally, a cross section of the users, expatriates in the UK, and their energy behaviour is examined.

2. Alliesthesia

Neutral thermal comfort has been the norm of thermal comfort studies for a while and it is only recently researchers have started challenging the notion of thermal neutrality. For instance, Shahzad et al. have in their study of Norwegian and British offices, questioned the reliability of any study that solely relies on neutral thermal sensation [7]. The human brain and physiology are inbuilt to appreciate and adapt to the seasonal variations. People are tempted to go to different climate conditions for holiday and vacations and in the process entice their sensory feelings. In all these cases, they are not necessarily trying to enjoy the extreme climate rather find the variation more stimulating and pleasurable. It is in this context we have to review the current notion of establishing a neutral condition.

To mitigate the climate challenge, we have to find means to reduce energy consumption and in the present climate of the energy-intensive world, we have to be cautious of investing an enormous amount of equipment and energy to produce neutral thermal condition when actually human are by nature are (designed) born to adapt to seasonal climatic variations, which is more pleasurable. Zhang et al have underlined the limitation of the literature pertaining to comfort at segment, body part level and argued that there is limited literature on the sub-segment level of comfort [8]. Ioannou et al have concluded that neutral sensation is not comfortable and variation in the thermal sensation levels could be out of habits and the variation is attributed to alliesthesia [9].

Lichtenbelt et al have demonstrated that the human body will be positively affected by exposure to cardiovascular parameters and challenged the seminal knowledge of comfort zone [10]. It is established that minimal heat is produced while maintaining body temperature and human energy balance is influenced by microclimate [10]. However, it is also proven that mild cold or warm can increase energy metabolism without discomfort. In fact, low or high temperatures in a dynamic thermal environment may be perceived as even pleasant, evoking thermal alliesthesia. These researches underpin the significance of alliesthesia while designing healthy, comfortable and energy-friendly indoor environments [10]. In a Psychology study, Sailer et al have explored the difference between olfactory alliesthesia and olfactory sensory-specific satiety [11]. According to olfactory sensory-specific satiety, there will be a decrease in the pleasantness of an odour as it is smelled and it is the external factor which defines the perception of an individual. However, alliesthesia is defined as "the change in pleasantness that results from a change in the internal state of the subject" [12]. For example, internal state changes due to glucose after eating food results in a decline of the pleasantness of a food stimulus [11].

The Human body responds to the changes through sensors received from the external environment. These sensations are descriptive and hence displeasure or pleasure of a sensation depends on internal signals [4]. These signals from within the body define the external stimulus as pleasant or unpleasant. The early advocate of the concept, Cabanac, coined the term Alliesthesia. Alliesthesia can be defined as the phenomenon whereby 'a given stimulus can be perceived as pleasant or unpleasant according to the inner state of the subject' [4]. The concept of alliesthesia enables us to differentiate thermal pleasure from the established thermal neutrality [2]. Cabanac, in his original study, had focused on thermal, gustatory, and olfactory senses and argued that alliesthesia also exists in bringing signals to the individual's attention [4]. Further, 'a stimulus can feel pleasant or unpleasant depending upon its usefulness as determined by internal signals' Michel Cabanac [4].

Son and Chun have studied the correlation between psychological and physiological measures [13]. Alliesthesia is a compound word; 'Aesthesia' means 'sensation' and 'allios' means 'changed'. Some stimuli may leave either pleasantness or unpleasantness depending on perception. Alliesthesia studies tend to combine physiological (objective) and psychological (subjective) measures. Son and Chun have demonstrated that EEG as a psychological and physiological measurement tool can be used while evaluating individuals' thermal pleasure in thermal alliesthesia [13]. In the case of air movement within an enclosed space, people who feel warm would prefer more air movement and people who feel cold would generally prefer less air movement [14]. Parkinson and Dear have established that contrasting relationships between local and global skin temperatures trends enables positive thermal pleasure [14]. Brager et al have examined the quality of indoor air movement to advance the understanding and use of Alliesthesia [3].

In the race to reduce carbon emission, care has to be taken that there are no adverse effects and negative impact on the comfort of the inhabitants [6]. Traylor et al have demonstrated that one can experience pleasure through change by modulating setpoints [6]. This study suggests a savings of 5-10C by modulating the indoor temperature in cooling applications, whereas in heating applications the variable temperature consumed more energy than the constant set point. It is established that modulating temperature could lead to greater thermal pleasure. Way ahead is to achieve both increased human comfort levels and energy savings. Recent studies demonstrate the significance of users behaviour in energy use [6]. For instance, Hong and Lin have identified a reduction of 50% energy use or increase by 89% due to the behaviour of users compared with the average person [15].

Personal experience and improving comfort condition are paramount while reducing energy reduction. In the thermal sensation vote (TSV) model of seven-points developed by Fanger [16], earlier understanding advocated for maintaining Zero (neutral) to achieve optimum thermal comfort and a PPD below 10% and PMV of between 0.5 and –0.5 [16, 17]. Whereas, research suggests that people tend to prefer a value of 1 as a desired thermal sensation in a cool climate area [18]. Michel Cabanac has established the relation between pleasure and thermal comfort [4] and demonstrated that constant thermal comfort will only avoid thermal discomfort. Building on this, De Dear (2011) has challenged the validity of maintaining the constant indoor temperature. Traylor et al have argued that building energy consumption by changing the human behaviour [6]. Delzendeh et al, through psychological cognitive behavioural methods, have identified the need for understanding socio-personal parameters while analysing the user's behaviour and resultant energy consumption. In their review paper, they have highlighted the lack of impact of occupants' behaviour in building energy performance analysis. One of the key shortfalls is that the impact of occupants is only considered through means of scheduled and fixed patterns of behaviour [5].

IOP Conf. Series: Earth and Environmental Science **329** (2019) 012024 doi:10.1088/1755-1315/329/1/012024

3. Culture informed Energy behaviour

It is established that energy use is influenced by homeowner behaviour and this behaviour varies across the population. The homes of similar build and from the same climate zone can still consume a varying level of energy due to homeowners' behaviour [19]. Human behaviour is quite a complex subject to understand with cognitive biases and behavioural anomalies which have a direct bearing on predicting and changing the behaviour of households. The household's behaviour could be analysed by modelling the formation of behaviours. The individualistic model of behaviour looks at the process of rational decision making focusing on the particular actor. The socially orientated model focuses on action or practices of society. Ofgem and Chatterton have looked at four types of theories to interpret user's energy behaviour [20]; Economic Theories, Psychological Theories, Sociological Theories, and Educational Theories: [20]. When it comes to policy implementation, behaviour plays a crucial role. Frederiks et al. have highlighted the difference between what households admit to what they actually do [21]. For instance, there is an attitude-action gap, a value-action gap, a knowledge-action gap and an intention-action gap. Also, it is likely that households would be aware of the values and hold a positive attitude in a socially desirable way. However, they not necessarily translate into actual behaviour [21]. For instance, people acknowledge that they understand the significance of climate change and the relevance of saving energy. However, research suggests that there is no direct correlation and these concerns not necessarily translate into actions in terms of practical actions to reduce household energy consumption [21].

From the research in behavioural economics and psychology, key suggestions by Frederiks et al. include, Status quo bias and defaults, Satisficing, Loss aversion, Sunk costs, Risk aversion, Normative social influence, Intrinsic and extrinsic rewards, Temporal and spatial discounting, Free-riding and social loafing, Perceived trust, and Availability heuristics. While reviewing the energy conservation behaviour, Delmas et al. have concluded that the energy saving tips and information strategies, the feedback strategies and providing additional pricing information not necessarily lead to additional energy savings [22]. Further, they have argued that providing financial incentives could sometimes be counterproductive. Alloct, in his study of 600,000 households across the United States has demonstrated that non-price interventions can affect consumer behaviour [23].

The underlying determinants of energy-related behaviours and resultant energy use have hardly been examined. This work reiterates the fact that knowledge and attitude are mostly positively related to energy savings. Abrahamse at al. in their review of behaviour studies of the households have concluded that rewards to the households seems to have a positive effect on energy savings and identified the shortfall of relatively little known long-term effects of interventions [24]. Similarly, behaviourally based changes can have a very quick impact on emission reduction, they are highly cost-effective and provides benefits to households directly [25]. The UK government has recognised the significance of behavioural insights to public policy to promote energy-efficient products. They have also demonstrated that carbon emissions could be reduced by awareness of the neighbour's energy use to manage their own emissions [25].

There are many studies reviewing different housing practices for different stakeholders. Chen et al has identified behavioural patterns and their impacts on residential energy use among the low-income population. Further, there is a positive correlation between household size and income to the level of energy use [26]. Based on energy consumption, Barthelmes et al. have classified the occupant behaviour into three lifestyles: low consumer, standard consumer, and high consumer [27]. The occupant's energy behaviour is also underpinned by the values and identity. For instance, people's willingness to be part of the social norm or to see themselves as different from others would be crucial to how they act and respond to societal requirements. There is a correlation between the users' values and behaviour, and occupant's behaviour could be positively influenced by developing new strategies from socially-oriented approaches. Whereas, Individualist approaches to behaviour have its own limitation due to inclusivity, scalability and ethics of an individual's private space [20].

IOP Conf. Series: Earth and Environmental Science **329** (2019) 012024 doi:10.1088/1755-1315/329/1/012024

3.1. User cultural values

Cultural (cul·ture (klchr) n) values are explained by Webster's Dictionary as "'The totality of socially transmitted behaviour patterns, arts, beliefs, institutions, and all other products of human work and thought'. These patterns, traits, and products considered as the expression of a particular community, period, class, or population: Edwardian culture; Japanese culture; the culture of poverty" [28]. Over the years, scholars have articulated the concept of culture in different facets. For instance, Rice has defined culture as "the values, attitudes, beliefs, artefacts and other meaningful symbols that help people interpret, evaluate and communicate as members of society" [29]. In this context, Hofstede argues that it is an important manifestation of culture [30]. Values (system) evolves over a period of time and hence stable and can be defined through research techniques. Hence, the concept of users' culture comparable and measurable.

The terms 'customs', 'culture,' and 'values' are distinctive pieces of a bigger picture, though it is used interchangeably. A 'custom' is an outward sign of the group's cultural values. Whereas, in contrast, a group's values run deep and not always visible. It is the cultural values in the form of customs, people pass down for generations. Tylor's definition of culture is "that complex whole which includes knowledge, belief, art, morals, law, custom, and any other capabilities and habits acquired by man as a member of society". In contrast to Arnold's view, "all folks "have" culture, which they acquire by virtue of membership in some social group – society" [31]. Further, Cultural values are articulated as the predominating attitudes and behaviour that characterize the functioning of a group or organisation.

Culture has various taxonomies and different scholars have explored different facets. Hofstede has reviewed culture as a concept and identified the characteristics of specific societies mapping the data from 50 countries worldwide as Long-term vs. short-term orientation, Power distance, Uncertainty avoidance, Individualism vs. collectivism, and Masculinity vs. femininity [30]. The role of individualism and collectivism plays a key role in the Sustainable Built Environment. Individualism promotes one to take care of self and their immediate family and advocates for a society with loose ties between individuals [30]. On the contrary, Collectivism represents an emphasis on community over the individual and priority for the family requirement than for that of individuals. The key difference is in terms of these community's emphasis on individual values and community values, for instance, Individualism promotes 'I' right from the childhood and Collectivism advocates for 'We'. This will have a significant impact on the built environment as it informs the way people make choices about individual requirement/ spaces. Hofstede's models of understanding culture through the lenses of Power Distance, Individualism, Masculinity and Uncertainty Avoidance, has relevance while understanding the homeowner's behaviours towards the development of Sustainable Built Environment [30].

Significance of culture and its impact is acknowledged in workplaces and resultant business and economic implications. For instance, John et al. have adopted the Hofstede (2001) taxonomies to account for the observed variance of cultural differences in employee commitment [32]. A key challenge of mapping culture to behavioural pattern and sustainable built environment is that the unique aspects of cultures have to be described using a finite set of dimensions. Further with globalisation and movement, it is challenging to compare countries reflecting their relative standing on a set of dimensions [32]. Further, countries are not egalitarian and homogeneous societies and it would be too simplistic to assign one tone for the whole country. Further, consumer-oriented societies have blurred the differences and the culture-based value system is now more driven by affordability and consumption.

4. Expatriates and energy behaviour

A clear and strong relationship has been established between thermal comfort and households. User's cultural background plays a crucial role. Now with more than 15% of people move around for work and settle, the way expatriates adopt themselves to the new place is important [33]. It is even more central to the discussion in the way the expatriates adopt, internalise the prevailing housing typology. Adjusting to the local housing is a complex phenomenon as it had to meet the climatic condition, cultural and social conditions and bridge the gap between two typologies and alleviate the homesickness of the expatriates.

IOP Conf. Series: Earth and Environmental Science **329** (2019) 012024 doi:10.1088/1755-1315/329/1/012024

There are about 60 million expatriates globally and the annual growth rate is growing exponentially [33]. The size of the expatriates is nearly equal to the population size of the UK. The behaviour pattern of the Expats is complex and constantly prodded by culture, ethnicity they belong to and the place and culture they have moved. According to UN statistics, Expatriates can be ranked as the fifth most populous nation on the planet with more than 235 million [34]. Nearly 15% of the UK population is expats [34]. With the reconfigured political system, more EU Expatriates from Bulgaria, Latvia, Romania, Poland and Lithuania have entered the UK in the last decade [34]. Review of literature suggests that immigrants' demonstrate similar or at times, even stronger concern for the environment compared to that of native-born individuals [35]. However, researchers have identified a lack of concern and lower levels of environmental awareness among immigrants compared to native-born individuals [36]. These studies have examined the concerns and attitude of ethnic individuals, however, there is limited research to prove how these attitudes translate into pro-environmental behaviour [35]. On the contrary, researchers have demonstrated that environmental concern not necessarily reflects environmentally responsible behaviour, known as the attitude-behaviour gap [35].

Values act as a guiding principle and enable people to develop their beliefs and attitudes about their physical environment [37]. According to the cultural cognition thesis, cultural values operate as a type of heuristic in the rational processing of information related to public policy issues. At the centre of the cultural cognition thesis, Douglas has articulated framework for classifying cultural values, which defines cultures based on their characteristics; hierarchy-egalitarianism and individualism-communitarianism [38]. The cultural cognition thesis asserts that people tend to form perceptions of societal risks based on the values characteristic of the groups with which they identify [39].

The hierarchical worldview advocates a hierarchy in the system and believes in distributing the rights, duties and goods based on the established social characteristics such as wealth, gender or ethnicity. Contrarily, equal distribution of the rights, duties and goods equally in the society is the egalitarian worldview. Further, individualistic worldview advocates that individuals should secure their own well-being and succeed without any collective assistance or interference. Whereas, societal interests are considered before individuals by the 'communitarian' or collectivistic worldview and consider society as responsible for securing the conditions necessary for individuals to thrive [35]. Environmental values: Pro-environmental perceptions (environmentalism) characterise people who show environmental responsibility.

Environmentalism can be viewed as a more specific, context-oriented value that is key to pro-environmental behaviour and includes a set of perceptions about the relationship between human beings and their environment [35].

5. Conclusions

To mitigate the climate challenge, we have to find means to reduce energy consumption and in the present climate of the energy-intensive world, we have to be cautious of investing an enormous amount of equipment and energy to produce neutral thermal condition when actually human are by nature are (designed) born to adapt to seasonal climatic variations, which is more pleasurable.

The user's cultural values and energy behaviour are explored in depth and debated. However, further work is required as to how homeowner action has either contributed to the growth of household energy use over time or offset it. Further, catalogue the changes in homeowner's behaviour over time and examine whether such changes would lead to more or less energy consumption.

The relation between the environmental concern and behaviour is well researched and the relationship between environmental concerns and pro-environmental intentions and behaviour is debated both for and against. A key concern of the individual's behaviour and values is that despite individual's high concern for the environment, there is a tendency to depend on the government and big business. Further, the cost of complying with environmental restrictions would influence the energy behaviour along with the price, quality, and convenience of the product.

IOP Conf. Series: Earth and Environmental Science **329** (2019) 012024 doi:10.1088/1755-1315/329/1/012024

6. References

[1] ASHRAE, *Standard 55 - Thermal environmental conditions for human occupancy*. 2017, accessed on 27 Oct 2018: https://www.ashrae.org/.

[2] De Dear, R., *Revisiting an old hypothesis of human thermal perception: alliesthesia.* Building Research & Information, 2011. **39**(2): p. 108-117.

[3] Brager, G., H. Zhang, and E. Arens, *Evolving opportunities for providing thermal comfort.* Building Research & Information, 2015. **43**(3): p. 274-287.

[4] Cabanac, M., *Physiological Role of Pleasure.* Science, 1971. **173**(4002): p. 1103-1107.

[5] Delzendeh, E., et al., *The impact of occupants' behaviours on building energy analysis: A research review.* Renewable and Sustainable Energy Reviews, 2017. **80**(C): p. 1061-1071.

[6] Traylor, C., W. Zhao, and Y.X. Tao, *Utilizing modulating-temperature setpoints to save energy and maintain alliesthesia-based comfort.* Building Research & Information, 2019. **47**(2): p. 190-201.

[7] Shahzad, S., et al., *Does a neutral thermal sensation determine thermal comfort?* Building Services Engineering Research & Technology, 2018. **39**(2).

[8] Zhang, H., E. Arens, and Y. Zhai, *A review of the corrective power of personal comfort systems in non-neutral ambient environments.* Building and Environment, 2015. **91**(C): p. 15-41.

[9] Ioannou, A., L. Itard, and T. Agarwal, *In-situ real time measurements of thermal comfort and comparison with the adaptive comfort theory in Dutch residential dwellings.* Energy & Buildings, 2018. **170**: p. 229-241.

[10] Van Marken Lichtenbelt, W., et al., *Healthy excursions outside the thermal comfort zone.* Building Research & Information, 2017. **45**(7): p. 819-827.

[11] Sailer, U., C. Triscoli, and I. Croy, *Still Eating Despite Decreased Olfactory Pleasure—The Influence of Odor Liking and Wanting on Food Intake.* Chemical Senses, 2016. **41**(6): p. 497-504.

[12] Rolls, E.T. and J.H. Rolls, *Olfactory Sensory-Specific Satiety in Humans.* Physiology & Behavior, 1997. **61**(3): p. 461-473.

[13] Son, Y.J. and C. Chun, *Research on electroencephalogram to measure thermal pleasure in thermal alliesthesia in temperature step-change environment.* Indoor Air, 2018. **28**(6): p. 916-923.

[14] Parkinson, T. and R. De Dear, *Thermal pleasure in built environments: spatial alliesthesia from air movement.* Building Research & Information, 2017. **45**(3): p. 320-335.

[15] Hong, T. and H.-W. Lin, *Occupant Behavior: Impact on Energy Use of Private Offices.* Paper published at ASim 2012 - 1st Asia conference of International Building Performance Simulation Association., 2013: p. 190-201.

[16] Fanger, P.O., *Thermal comfort : analysis and applications in environmental engineering*. 1972, New York: New York : McGraw-Hill.

[17] ASHRAE, *ASHRAE Handbook - Fundamentals.* 2009: American Society of Heating, Refrigerating and Air-Conditioning Engineers, Inc. Atlanta, GA.

[18] Nikolopoulou, M. and K. Steemers, *Thermal comfort and psychological adaptation as a guide for designing urban spaces.* Energy & Buildings, 2003. **35**(1): p. 95-101.

[19] Barkenbus, J., *Indoor Thermal Comfort: The Behavioral Component.* Sustainability, 2013. **5**: p. 1680- 1699.

[20] Chatterton, T., *An introduction to thinking about 'energy behaviour': A multi-model approach.* 2011, http://eprints.uwe.ac.uk/17873: Department of Energy and Climate Change, Bristol.

[21] Frederiks, E.R., K. Stenner, and E.V. Hobman, *Household energy use: Applying behavioural economics to understand consumer decision-making and behaviour.* Renewable and Sustainable Energy Reviews, 2015. **41**: p. 1385-1394.

[22] Delmas, M.A., M. Fischlein, and O.I. Asensio, *Information strategies and energy conservation behavior: A meta-analysis of experimental studies from 1975 to 2012.* Energy Policy, 2013. **61**: p. 729-739.

[23] Allcott, H., *Social norms and energy conservation.* Journal of Public Economics, 2011. **95**: p. 1082- 1095.
[24] Abrahamse, W., et al., *A review of intervention studies aimed at household energy conservation.* Journal of Environmental Psychology, 2005. **25**: p. 273-291.
[25] CabinetOffice, *Behaviour Change and Energy Use*, T.C.O.c.B.I. Team, Editor. 2011: London.
[26] Chen, C.-f., X. Xu, and J.K. Day, *Thermal comfort or money saving? Exploring intentions to conserve energy among low-income households in the United States.* Energy Research & Social Science, 2017. **26**: p. 61-71.
[27] Barthelmes, V.M., C. Becchio, and S.P. Corgnati, *Occupant behavior lifestyles in a residential nearly zero energy building: Effect on energy use and thermal comfort.* Science and Technology for the Built Environment, 2016. **22**: p. 960-975.
[28] Gates, E., *Webster's New World Dictionary.* English Today, 1989. **5**(2): p. 52-54.
[29] Wijnen, G., A. Kemperman, and I. Janssen, *Shopping behaviour and attribute evaluation of expatriates - a cross-cultural study.* European Real Estate Society (ERES), 2011. **70**.
[30] Hofstede, G.H., *Culture's consequences: comparing values, behaviors, institutions, and organizations across nations.* 2nd ed. ed. 2001, Thousand Oaks, Calif: Thousand Oaks, Calif : Sage Publications.
[31] Bennett, T., *Cultural Studies and the Culture Concept.* Cultural Studies, 2015. **29**(4): p. 1-23.
[32] Meyer, J.P., et al., *Affective, Normative, and Continuance Commitment Levels across Cultures: A Meta-Analysis.* Journal of Vocational Behavior, 2012. **80**(2): p. 225-245.
[33] @ExpatSurvey, *How many expats are there? - The Expat Survey.com, World's largest global research programme into expatriates.* 2015. **Accessed on 15 Oct 2018**(Paragon Relocation, https://paragonrelocation.com).
[34] Organisation for Economic Co-operation, D., *Greening household behaviour: overview from the 2011 survey.* Revised edition. ed. Organisation for Economic Co-operation and Development. 2014.
[35] Segev, S., *Modelling household conservation behaviour among ethnic consumers: the path from values to behaviours.* International Journal of Consumer Studies, 2015. **39**: p. 193-202.
[36] Johnson, C.Y., J.M. Bowker, and H.K. Cordell, *Ethnic Variation in Environmental Belief and Behavior: An Examination of the New Ecological Paradigm in a Social Psychological Context.* Environment and Behavior, 2004. **36**(2): p. 157-186.
[37] Schwartz, S.H., *Universals in the content and structure of values: theoretical advances and empirical tests in 20 countries.* Advances in Experimental Social Psychology, 1992. **25**: p. 1-65.
[38] Douglas, M. and A.B. Wildavsky, *Risk and Culture: An Essay on the Selection of Technical and Environmental Dangers*. 4th ed. ed. 1982, New York: London: University of California Press, Berkeley & Los Angeles, CA.
[39] Kahan, D.M., et al., *The polarizing impact of science literacy and numeracy on perceived climate change risks.* Nature Climate Change, 2012. **2**: p. 732 - 734.

Potential shift of Integrated Design (ID) through BIM in Sustainable Building Renovation

A Kamari and PH Kirkegaard
Department of Engineering, Aarhus University, Aarhus, Denmark

E-mail: ak@eng.au.dk

Abstract. Future building renovation concerns more holistic perspectives related to both integrations through the involvement of key stakeholders and fulfillment of sustainability in connection to a broader range of objectives and criteria facilitated by the renovation scenarios. This entails re-thinking, reviewing, and application of novel integrating methods and approaches through an integrated design (ID) schema, towards establishing a richer early design stage to deal with the multifaceted challenges in building renovation context. Through its technological capabilities (modeling techniques), shifting into Building Information Modelling (BIM) as a methodology for integration can be useful bringing more efficiency in (particularly the early design stage of) renovation projects by creating and adding values through the five strands of *product, people, process, policy*, and *technology*. On the basis of this underlying hypothesis, the aim of this paper is firstly to review the characteristics of ID, BIM, and their interactions on each other, and secondly to explore and address the potential shift of integrated design through BIM (ID+BIM) for sustainable building renovation. This contribution is relevant to building design researches and practitioners, who can use the resultant to better understanding and navigating the integrated design through BIM, and its potential shift in sustainable building renovation with multiple stakeholders.

Keywords: Sustainable Building Renovation; Integrated Design (ID); Building Information Modeling (BIM); Integrated Design through BIM.

1. Introduction

The term 'holistic renovation scenario' in this paper is adapted from [1]. It serves to underline a holistic approach where various renovation objectives (i.e. energy consumption, indoor thermal comfort etc.) linking to the sustainability in its full sense [2] are achieved in a balanced way, through the development of a renovation scenario. A 'renovation scenario' refers to a set of design changes that will be made to a building, e.g. replacing all north-facing windows with triple glazing, adding a new cladding system (i.e. composite panels), and HVAC equipment. Identification of renovation objectives and criteria, an indication of the renovation approaches [3], and necessarily how to deal with the complexity of decision-making within many involved stakeholders, are the key points and essential factors for the development of holistic renovation scenarios.

The new vision and tightened sustainability demand for buildings [4] have led to an increased demand for sustainable development of the renovation projects facilitated by renovation scenarios. However, investigations have documented that the efficiency of the building renovation industry still develops slowly and is low compared to other contexts [5]. New methods to support the processes within the conventional renovation process are therefore needed. This entails re-thinking, reviewing, and application of novel integrating methods and approaches through an integrated design schema, towards establishing a richer early design stage to deal with the multifaceted challenges in the sustainable renovation context. One suggestion can be the use of integrated design (ID) through the application of BIM, that exists and being practiced by relevant practitioners in the building design context.

Besides many advantages of performing renovation design within an integrated design schema due to encompassing the sustainability tendencies, shifting into Building Information Modelling (BIM) as a methodology through its technological capabilities (modelling techniques) for integration can be effective in bringing richer early design stage for renovation projects by creating and adding values through various perspectives. Based on this underlying hypothesis, the aim of this paper is firstly to review the characteristics of existing integrated design (ID) methodologies, BIM, and their interactions on each other, and secondly to remark the potential of how shifting into the ID+BIM -that can also be supported via application of BIM-based Decision Support Systems (DSS)- can enhance the renovation context to meet the sustainable building renovation objectives and criteria in a broader sense. This is done through establishing and evaluating crucial seminal strands in the ID+BIM context in this paper.

The paper presents the results of a literature review towards establishing ID+BIM's seminal strands, which is substantiated through conducting seven unstructured interviews of practitioners in renovation context on evaluating their importance on the improvement of the sustainable renovation context. The paper in section 1.1 provides the common existing barriers in renovation context. Sections 2.1 and 2.2 review the concept of BIM and ID via investigation of the existing critical literature in these areas, and then section 2.3 presents a brief discussion on how these topics may interact together towards establishing some seminal strands in the ID+BIM context. Thereafter, on the basis of the discussed theoretical frameworks as well as the application of established strands as an analytical lens, in section 3 the paper explores the potential of how shifting into the ID+BIM can enhance the renovation context to attain the sustainable building renovation. This is carried out through visiting and conducting seven unstructured interviews of real practitioners from four large scale architecture and construction consultancy companies located in Aarhus, Denmark. Finally, section 4 discusses the conclusion and further work. This contribution is relevant to building design researches and practitioners, who can use the resultant to better understanding and navigating the integrated design through BIM, and its potential shift in sustainable building renovation with multiple stakeholders.

1.1. Existing barriers in the renovation context

Experience from projects and research carried out over recent decades has identified numerous barriers that hinder the uptake of a holistic building renovation. BPIE [6] reported existing barriers in this context through categories such as 'financial', 'institutional & administrative', 'awareness, advice & skills' and 'separation of expenditure and benefit'. Further, a list of five main constraints that building renovation projects face, from pre-retrofit to post-retrofit stages, were explored by Cattano et al. [7], including identification of pre-existing hidden conditions late in the design process, interactions between building systems, lack of experience with the methods and materials required to deliver successful sustainable renovations, and poor measurements.

Galiotto et al. [8] discuss retrofitting barriers that occur due to politico-economic barriers (which need to be addressed by policymakers and market developers), technical barriers (which need to be addressed by architects and engineers), and behavioral barriers (which are the direct impact of building owners and occupants). In relation to the last-mentioned barrier, the authors [8] emphasize the role of occupants' behavior in building energy consumption, and reasons for behavioral barriers are argued, e.g. limited knowledge about the building renovation process [9] and its benefits among stakeholders, or lack of guidance from the government and responsible institutions etc.

The building occupants indirectly influence the pattern of energy demands due to the changes over time in occupancy schedules and usage patterns [10]. In connection to this, Booth and Choudhary [11] categorize the most common barriers in a building renovation as the pre-bound effect which is known as the divergence between modelled and actual energy consumption for the pre-retrofit, and rebound effect in which the post-retrofit energy consumption is higher than predicted, due to either technical issues such as incorrect design options, failures and mistakes during construction works, pre-existing conditions (solar exposure conditions and orientation, historical and cultural heritage interest, local, many others) or more importantly changes in occupant behavior. The essence of the pre-bound and rebound effects lead to a substantial disparity between the predicted and actual energy savings. The authors in [12] consider that the removal of these barriers may reduce renovation costs and yield buildings that consume less energy and resources. Nevertheless, Yu et al. [13] state that understanding of building occupants' behavior in a renovation field is not addressed adequately through the renovation process since the general focus in this field is still on technical goals i.e., buildings' energy efficiency improvements. It seems essential that the design approach should integrate the effects of a building's technical aspects with the users' behavior representation, giving them the same importance [13,14]. From another perspective, Acre et al. [15] discuss that post-occupancy evaluation of renovated buildings, which is often used to assess the impact of energy renovation, fails to examine the social context correctly because many of the energy efficiency measures and technical issues in energy renovation remain abstract to the occupants.

2. Theoretical concepts

The literature review is divided into three sub-sections. The first two parts present BIM and integrated design, which are currently applied in many existing consultancy companies. The third part provides a brief description regarding the BIM and integrated design and how these topics may interact together.

2.1. Building Information Modeling

Building Information Modelling (BIM) is a method that focuses on generating and administrating building data [16,17]. In the year 2003, the Danish government launched a new political plan of action [18]. The goal was to increase the focus on how to optimize the AECO industry. The political initiative resulted in a project called "The Digital Construction" to improve the interdisciplinary use of digital information within the industry. The use of BIM created the foundation of the project, and it is the main method for handling and generating digital data. As a result, the Danish government demanded that all governmental clients should make use of digital applications in their building projects, starting from the beginning of 2007. In the other hand, the economic potential of using digital data in the AEC industry is significant. A report from 2009, argues that digitalization can generate an economic potential of 17 billion DKK / year in Denmark [19]. Another research study from the National Institute of Building Sciences in the United States argues that digitalization in the AECO industry may result in savings in the range of 15.8 billion USD / year in the USA [20].

According to Krygiel et al. [21], BIM is an emerging methodology in the AECO industry since the intelligent digital three-dimensional model-based process of BIM can be used to plan, design, structure, manage, and analysis buildings. The BIM-model represents the foundation for the BIM-process. It functions as a database for a project and may contain information about the building geometry, spatial relations, building location, properties, quantities of building components etc. These can be extracted and used in applications for, i.e. energy simulations, acoustic conditions, facility management. These interactions between the BIM-model and other applications can be used throughout the whole lifecycle of the building, from the cradle to the grave, and all parties in the project can participate [22]. Succar [23] addresses BIM is a set of *technologies*, *processes*, and *policies* enabling multiple stakeholders to design, construct and operate a facility collaboratively. The communication between the BIM-model and other applications useable in a building construction project is the central concept of the BIM-process [20]. The digital interoperability makes the building construction process transparent for all the participants. This makes it possible to give continuous feedback on results in relation to energy, construction, maintenance etc., or to explore the relative effect of different design alternatives. The BIM-process depends on the ability to share information between applications, organizations, IT systems, and databases. Open file standards that consider the benefits of all applications in the process and not one software vendor may be the solution for achieving an optimal BIM-process [24]. This issue is compiled by the Industry Foundation Classes - IFC, that are the open and neutral data format for openBIM [25]. The IFC specification is developed and maintained by buildingSMART International as its "Data standard". IFC has been being released since 2000 through various versions, from IFC1 to IFC4. Nowadays, IFC5 is currently in the early planning phase, and it is expected to include full support for various infrastructure domains and more parametric capabilities [26].

2.2. Integrated design schema

Integrated design is a method that has been evolved mostly during the last two decades, due to the new sustainability tendencies in the AECO industry. As such, the growing focus on sustainability is one of the areas that make integrated design highly relevant [27]. This method utilizes the strength of multidisciplinary design consultancies to optimize the building design and reduce its energy consumption [28]. Previous experiences of using the integrated design method in building construction projects show a relative potential reduction of energy consumption by 60 to 70 percent compared to a traditional building project [29]. Figure 1 demonstrates the difference between conventional/traditional design process versus an integrated design schema.

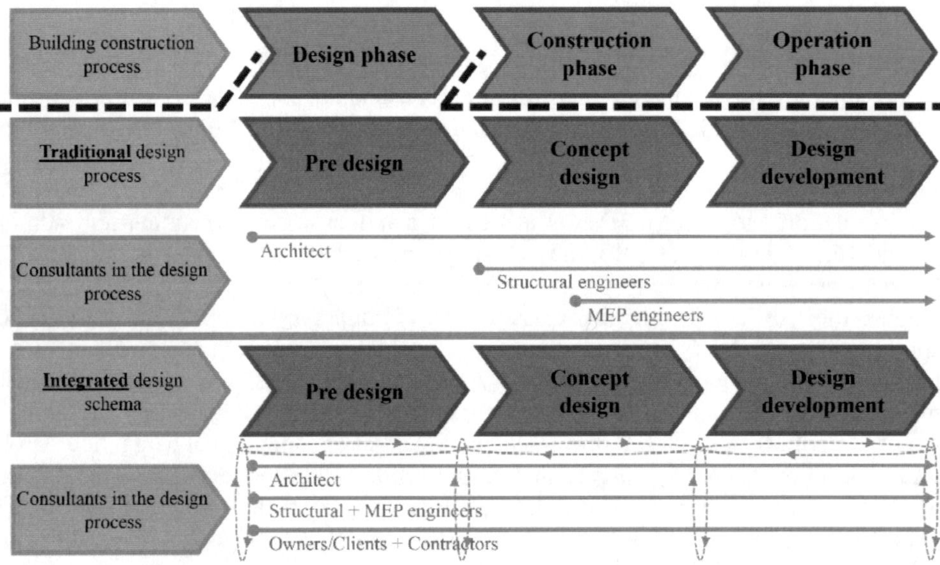

Figure 1. Traditional design process versus integrated design (ID) schema in the AECO industry

In an integrated design schema, iterations are performed in each sub-phase, and each iteration results in different design alternatives, which are discussed in an evaluation round [30]. By the evaluation round, one of the design alternatives is selected for further development based on the design goals established. Thus, the evaluation creates a milestone on the linear backbone of the design process and forms a transition between two design sub-phases. These milestones ensure a constant design progression towards the final goal (with reference to [31]). In essence, some characteristics may describe the integrated design process (according to [31]):

- Inter-disciplinary work from the beginning of the design process between all consultants
- Test of various design alternatives by use of simulation tools
- Development of well-formulated performance targets and design strategies
- A more active role of the client
- The architect as teamwork leader instead of the sole form designer

Integrated design can be understood as a collaborative method for designing buildings that emphasizes the development of holistic design [32]. The implication is clear: integrated design is holistic in that it involves all stakeholders from the earliest stages, each having input into what goes into making the decisions that will lead to the completed project. It is holistic in that it takes every team member's point of view into consideration [33]. The financial advantages of integrated design have been thoroughly researched and documented through large EU-financed projects such as Task 23 [31].

Concerning who benefits from integrated design, there is a perception that all sacrifice for the owner - that the owner has the most to gain, then the contractor, and lastly-if at all- the design team. In fact, given time, trust, transparency, and not a little effort on everybody's part, all benefit. The more who participate, the more who benefit- and the more benefits there are. With integrated design, the focus is on the owner - the owner's needs - and the end result, the completed building, optimized for a higher value and reduced waste.

Figure 2 illustrates the ability of an integrated design schema to significantly impact projects in the early design stages along with the fact that the cost of changes is the largest in the late design phases. This highlights the importance of integrating technical knowledge in the early design stages. Deutsch [32] states that because the stakeholders are together from the earliest project stages, decisions can be made from siting and orientation of the building to the specification of green components in a proactive and coordinated manner. The integrated design has its origin in the development of new buildings; however there are recent efforts to adapting and applying it for the renovation as well [33].

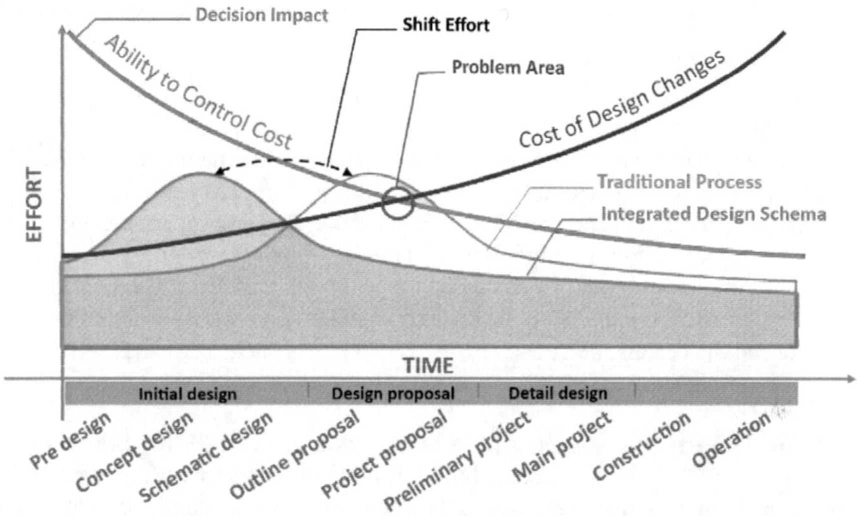

Figure 2. Integrated design effort - the blue curve indicates the cost of design changes. The green curve indicates the possibility of influencing the project (adapted from [34]).

2.3. Integrated design through Building Information Modelling (BIM)

On the basis of the precedent subsections, and towards providing a comprehensive categorial scheme to existing integrated design and BIM (ID+BIM) that encompasses its multifaceted perspectives, when we inspected our collection we observed that the ID+BIM's descriptions are not mutually exclusive. They overlap and intertwine. When analyzed, as through a prism, the content of the descriptions forms five strands. One of these strands pertains essentially to the *people* as involved stakeholders in the process. Another strand pertains to the real design solutions and/or digital prototype (with reference to BIM-model), which we refer as the *product*. Third and fourth strands pertain to the role of *process* and *policy* on the *people* and upon the developing *product*. And the fifth strand pertains to the used *technologies*. Hereafter, we shall refer to these strands as the four P's + one T (4Ps+T) of ID+BIM (see Figure 3), i.e. (1) people, (2) product, (3) process, (4) policy, (5) technology.

- *People*: the collaborative schema between involved stakeholders including architect/ engineers/ contractors/ clients/ owners etc. that contributes to increasing the level of awareness and informed decision-making.
- *Product*: refers to the real design solutions and/or digital prototype of a project which contributes to more sustainable buildings.
- *Process*: encompasses an integrated process including iterative workflows for generating and leveraging building data to design and construct the building.
- *Policy*: the streamlined steps related to contractual, regulatory, and preparatory.
- *Technology*: the requirements concerning software, hardware, and computer network

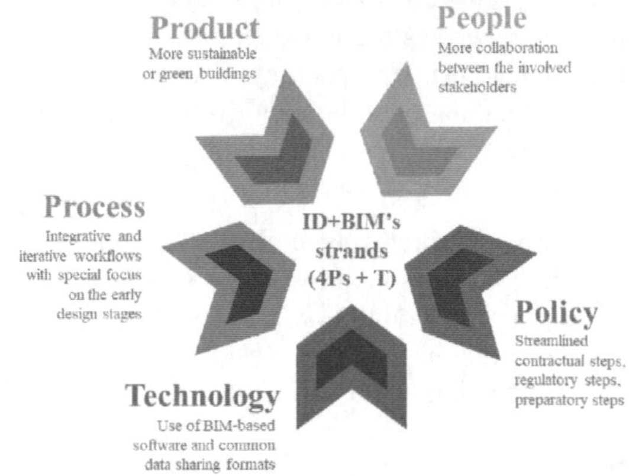

Figure 3. The ID+BIM strands/categorial scheme

ID+BIM can be understood as a collaborative method for designing buildings that emphasizes the development of holistic design. It strives to be holistic in that it involves stakeholders (architects, engineers, contractors, and clients) from the earliest stages, each having input into what goes into making the decisions that will lead to the completed project. It, therefore, strives to take every team member's point of view into consideration, and it is holistic in that these decisions are made with all the information shared at one time, up front and not in the more traditional linear fashion, each entity maintaining and controlling the distribution of its own locus of information (adapted from [32]). BIM technology allows the integrated design to flourish, encourages - and provides a vessel and conduit for - the sharing of information between the design and construction team [32]. As such, BIM enables integrated design and therefore makes it possible.

3. Potential shift of integrated design and BIM for sustainable building renovation

As discussed in section 1.1, the practice of building renovation faces various types of challenges and barriers, where economic, human behavior, and technical barriers seem to be dominating. There are several reasons for these barriers including variety of the involved stakeholders and their specific interests in the process [33], a wide range of available renovation approaches (e.g. insulation approaches, window replacement, HVAC systems etc. - see [3]), and broad number of objectives and criteria [or sub-criteria] (e.g. regarding energy efficiency, spatial quality, investment cost etc.) that needs to be met [2]. In this section, we use the ID+BIM' strands as an analytical lens to explore and address the potential shift of ID+BIM for the improvement of sustainable building renovation context. To this end, we conducted seven unstructured interviews of real practitioners from four large scale architecture and construction consultancy companies, who are also involved in the ReVALUE (see revalue.dk) research project and have offices/branches located in Aarhus, Denmark. Three of the interviewees were BIM coordinators in their companies, two engineers, one experienced architect, and one owner. They have initially discussed their relevant knowledge about either of ID and BIM. Then, we questioned them about the potential of using these approaches in their work environment, focusing on their recent experiences in renovation projects. In the context of this study, the focus was on a scenario in which the design specification and overall plan form the foundation for a turnkey contract tender and associated competition process.

Looking into the renovation context from the 4P+T strands at the time of interviews was effective in terms of bringing focus into the discussions with the relevant practitioners, and it significantly helped in exploring the potential shift of the ID+BIM for the sustainable building renovation in more detail. Table 1 sums up these insights.

Table 1. The ID+BIM's potential shift in sustainable building renovation as explored through the analytical application of 4P+T seminal strands

Focus	Implications of ID+BIM in sustainable building design
People	- improving the level of awareness, communication, and cohesion between the entire involved stakeholders - establishing a more effective collaborative and interdisciplinary work culture, especially in Denmark that has a well-documented culture of trust - moderating disputes and reduces adversarial relations between design team including architect and engineers, as well as owners and contractors
Product	- encompassing the overall project goals including functionality, accountability, feasibility, which leads to attaining a more sustainable product - strengthening the possibility that the owners' project goals will be attained - enhancing the final constructed project quality because of shared goals, particularly with reference to investment cost and clients' budget - consolidating building elements toward a sustainable and synthetic whole
Process	- improving the program and project time frame by reducing schedule waste and time overruns - contributing to shorter and informed decision-making on early design stages - streamlining data exchange and share of information between design stages and project phases (i.e. design, construction, operation and maintenance)
Policy	- increasing the likelihood that the sustainability certification systems such as DGNB-DK will be used in the design process with the focus on earlier stages - creating a secured job site within a further and better understanding of the real renovation practice and workflows by all - increasing the chance of winning the project competitions - streamlining the implementation and development of the technical standards based on the consensus of different parties
Technology	- reducing the project costs through time frame optimization and systems coordination, decreasing modifications made in the context, loss of data, and reducing the need for rework during costliest and most prone to error - improving the productivity of the design team, giving the accurate and right data, and while discussing with non-experts, i.e. owners/clients

Figure 4 demonstrates the relative graphical positioning of the ID+BIM strands as explored through the analytical application of 4P+T seminal strands. The ID+BIM is enabling strategies to achieve sustainable building design through early intervention and the bottom-up consideration by stakeholders in how the design program, materials and systems, and components and products in a renovation project influence each other. Instead of engaging sustainability specialist working in isolation, ID+BIM implies a holistic, collaborative approach comprising of the insights and experiences of all teammates. ID+BIM has the potential to aid project stakeholders to provide the required geometrical and analytical data of an existing building 3D model to select and evaluate the proper type of renovation actions during the early design stage and to make decisions that have a significant impact on the life cycle of projects.

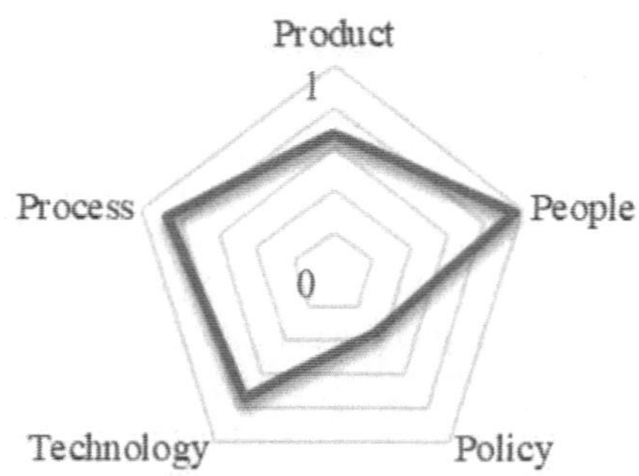

Figure 4. The relative graphical positioning of the explored ID+BIM strands for sustainable building renovation

This can significantly be supported by development and application of BIM-based hybrid Decision Support Systems (DSSs) to rapidly generate various scenarios and evaluate them against the essential sustainability Key Performance Indicators (KPIs), i.e. energy consumption, day-lighting parameters, indoor comfort [i.e. 35-36]. Ultimately, application of BIM-based DSSs within ID+BIM will bring an enormous potential to aid the design team together with the other stakeholders (motivating them) to comprehend and develop holistic renovation scenarios and to make informed decisions in a shorter period with focus on the early design stages, and thus significant impacts on considering/addressing the trade-offs between architecture and engineering principles for the sustainable building renovation.

4. Conclusion

This paper reviewed the characteristics of integrated design (ID) approach, BIM, and their interactions on each other, which are applied in many consultancy companies. The study established the ID+BIM's 4P+T strands, and then used it as a tentative analytical lens to explore the potential of how shifting into the ID+BIM can enhance the renovation context. The study considers that ID+BIM can significantly influence in enhancing the current renovation *process*, by promoting the iterative decision-making and bring focus to the earlier stages of the renovation design process. This improves the level of awareness, communication, and cohesion between the entire involved stakeholders and their relevant demands the renovation of existing buildings. Likewise, it offers more productive teamwork and process from the used technology perspective. ID+BIM shifts the focus on the early design stages as well as promotes the iterative decision-making process, which leads to achieving more sustainable solutions for the renovation purpose. That will require changes in the culture and existing policies in the contemporary renovation practice. Accordingly, it can be concluded that ID+BIM serves at coping with the complexity within the renovation context, and as a result of this, it contributes in the time-saving, improvement of the accuracy and quality of the final decision, as well as encouraging stakeholders to accommodate holistic renovation scenarios in the early design stages of the renovation projects.

As the basis for our proposed 4P+T model as a tentative analytical lens, it presumably does not capture the full complexity of a sustainable renovation design process given its highly complex value profile and many heterogeneous stakeholders. Still, we argue that the seminal model offered may serve as a useful, analytical perspective and reflection method to facilitate the understanding of the explained complexity. The research in future concerns the exploration of ID+BIM methodology in real cases related to the several existing obstacles which may hinder the application of it or reduce its advantages.

Acknowledgments

The authors of the paper would like to show their gratitude to the Danish Innovation Foundation for financial support through the ReVALUE research project.

References

[1] Kamari A, Corrao R, Petersen S and Kirkegaard P H 2018 Tectonic Sustainable Building Design for the development of renovation scenarios—Analysis of ten European renovation research projects *Proc. Int. Conf. on Seismic and Energy Renovation for Sustainable Cities* (SER4SE 2018) vol 1 (Catania: Italy) pp 645–656

[2] Kamari A, Corrao R and Kirkegaard P H 2017 Sustainability focused Decision-making in Building Renovation *Int. J. of Sustainable Built Environment* **6** pp 330-350

[3] Kamari A, Schultz C P and Kirkegaard P H 2018 NovaDM: Towards a formal, unified Renovation Domain Model for the generation of holistic renovation scenarios *Proc. Int. Conf. on 12th European Conference on Product & Process Modelling* (ECPPM 2018) vol1 (Copenhagen: Denmark) pp 197-205

[4] Keeler M and Vaidya P 2016 *Fundamentals of integrated design for sustainable building* (Hoboken: Wiley & Sons)

[5] Kristensen E L 2007 Byggeriets produktivitet – en analyse fra Byggeriets Evaluerings Center Retrieved from http://www.byggeevaluering.dk/media/5364/produktivitet_feb_07.pdf

[6] BPIE - Buildings Performance Institute Europe 2011 Europe's buildings under the microscope. Retrieved from http://bpie.eu/wp-content/uploads
[7] Cattano C, Valdes-Vasquez R, Plumblee J and Klotz L 2013 Potential solutions to common barriers experienced during the delivery of building renovations for improved energy performance: literature review and case study *J. of Architectural Eng.* **19** pp 164-167
[8] Galiotto N, Heiselberg P and Knudstrup M 2015 Integrated renovation process: overcoming barriers to sustainable renovation *J. of Architectural Eng.* **22** pp 1-12
[9] Boeri A, Antonin E, Gaspari J and Longo D 2014 *Energy design strategies for retrofitting: methodology, technologies, renovation options and applications* (Southampton: WIT Press)
[10] Masoso O T and Grobler L J 2010 The dark side of occupants' behaviour on building energy use *Energy and Build.* **42** pp 173-177
[11] Booth A T and Choudhary R 2013 Decision making under uncertainty in the retrofit analysis of the UK housing stock: Implications for the Green Deal *Energy and Build.* **64** pp 292-308
[12] Yu Z, Fung B C M, Haghighat F, Yoshino H and Morofsky E 2011 A systematic procedure to study the influence of occupant behaviour on building energy consumption *Energy and Build.* **43** pp 1409-1417
[13] Degan G, Rode C, Vettorato D and Castagna M 2015 Holistic method for energy renovation of buildings: focus on users involvement *Proc. Int. Conf. on Building Simulation Application*, (2nd IBPSA-Italy Conference) British Sociological Association UK Retrieved from http://www.eurac.edu/en/research/technologies/renewableenergy/publications/Documents/EURAC-RenEne_GDegan-CRode-DVettorato-MCastagna_IBPSA-Italy2015.pdf
[14] Tweed C 2013 Socio-technical issues in dwelling renovation *Build. Research and Info.* **41** pp 551-562
[15] Acre F and Wyckmans A 2015 Dwelling renovation and spatial quality: The impact of the dwelling renovation on spatial quality determinants *Int. J. of Sustainable Built Environment* **4** pp 12-41
[16] Eastman C, Teicholz P, Sacks R and Liston K 2011 *BIM Handbook: A Guide to Building Information Modeling for Owners Managers Designers Engineers and Contractors* (Hoboken: Wiley & Sons)
[17] Kensek M and Noble D E 2014 *Building Information Modeling: BIM in current and future practice* (Hoboken: Wiley & Sons)
[18] Erhvervs og Byggestyrelsen 2003 Staten som bygherre Vækst og effektivisering i byggeriet Retrieved from https://bibliotek.dk/da/work/870970-basis:24982483
[19] COWI 2009 Digital forvaltning af bygninger fra vugge til grav Retrieved from http://byggeold.byggecentrum.dk/fileadmin/templates/byggecentrum/images/Efteruddannelse/Kursusmateriale/IKT_koordinator_og_leder/Byggeri_fra_vugge_til_grav_COWI-2009.pdf
[20] National Institute of Building Science 2007 National Building Information Modeling Standard™ (NBIMS) Version 1 - Part 1 Retrieved from https://buildinginformationmanagement.files.wordpress.com/2011/06/nbimsv1_p1.pdf
[21] Krygiel E and Nies B 2008 *Green BIM: Successful sustainable design with building information modeling* (Indianapolis: Wiley & Sons)
[22] Garber R 2014 *BIM Design: Realising the creative potential of Building Information Modeling* (Chichester: Wiley & Sons)
[23] Succar B 2009 Building information modelling framework: A research and delivery foundation for industry stakeholders *Automation in Construction* **18** pp 357- 375.
[24] BuildingSMART 2006 Open BIM Retrieved from https://www.buildingsmart.org/standards/technical-vision/
[25] BuildingSMART 2008 Industry Foundation Classes Retrieved from http://www.buildingsmart-42 tech.org/specifications/ifc-overview
[26] BuildingSMART 2008 Industry Foundation Classes Retrieved from http://www.buildingsmart-tech.org/specifications/ifc-releases/summary

[27] 7group and Reed B 2009 *The Integrative Design Guide to Green Building* (Hoboken: Wiley & Sons)
[28] Moe K 2009 *Integrated Design in Contemporary Architecture* (New York: Princeton Architectural Press)
[29] Jørgensen P F, Andresen I and Bramslev K 2009 Integrert energidesign IED - Grønn byggallianse Retrieved from https://byggalliansen.no/wp-content/uploads/2018/11/Integrert_energidesign.pdf
[30] Hansen H T R and Knudstrup M A 2005 The Integrated Design Process (IDP): a more holistic approach to sustainable architecture. *Proc. Int. Conf. on World Sustainable Building Conference 2005* (Tokyo: Japan) pp 894-901
[31] Löhnert G, Dalkowski A and Sutter W 2003 Integrated Design Process: a guideline for sustainable and solar-optimised building design Retrieved from http://task23.iea-shc.org/data/sites/1/publications/IDPGuide_internal.pdf
[32] Deutsch R 2011 *BIM and Integrated Design* (Hoboken: Wiley & Sons)
[33] Kamari A, Jensen S R, Corrao R and Kirkegaard P H 2018 A Holistic Multi-methodology for Sustainable Renovation *Int. J. of Strategic Property Manag.* **23** pp 50–64.
[34] Landgren M, Jakobsen S S, Wohlenberg B and Jensen L M B (2018) Integrated design processes – a mapping of guidelines with Danish conventional 'silo' design practice as the reference point *Archi. Eng. and Design Management*, https://doi.org/10.1080/17452007.2018.1552113
[35] Kamari A, Christensen M L, Jensen S, Petersen S and Kirkegaard P H 2018 A hybrid decision support system for generation of holistic renovation scenarios – cases of energy consumption, investment cost, and thermal indoor comfort *Sustainability* **10** pp 1255-1277.
[36] Kamari A, Laustsen C, Petersen S and Kirkegaard P H 2018 A BIM-based decision support system for the evaluation of holistic renovation scenarios *J. of Inf. Tech. in Cons.* **23** pp 354-380

IOP Conf. Series: Earth and Environmental Science **329** (2019) 012026 doi:10.1088/1755-1315/329/1/012026

Residential building energy performance evaluation for different climate zones.

S M Albogami[1,2] and R Boukhanouf[1]

[1]Department of Architecture and Built Environment, University of Nottingham, Nottingham, UK
[2]Corresponding Author: Sultan.Albogami@nottingham.ac.uk

Abstract. The building sector in the Kingdom of Saudi Arabia accounts for a large proportion of energy consumption. This paper investigates energy consumption in residential type A villa buildings constructed by the Ministry of Housing. The study specifically investigates energy consumption of a case building in three climate zones of the country identified in the KSA Building Code (SBC-602E). The software package IES-ve was used to simulate power consumption of the case building taking into account the climate characteristics of each zone. In total 6 cities were considered from the three zones to evaluate and compare annual and monthly power consumption loads of the building. It was found that there is some distinction of energy consumption for heating and cooling in the three zones (e.g., peak monthly power energy consumption for cooling varied from 6.371 MWh in Riyadh (zone1) to 1.663 MWh in Abha (zone3) and no heating requirement for Jaddah (zone 1) while Turaif (zone 3) heating consumption is 23.835 MWh) however, further work is required to identify clearer the climate specifications within each zone to address the excess energy consumption of this type of buildings.

1. Introduction

Increasing energy consumption in buildings and resulting pollutants emissions has led many countries to introduce tighter construction regulations and best practice codes. For example, in the Kingdom of Saudi Arabia (KSA) the building sector has expanded very fast from early 1980s to address in part housing shortage for the growing population. Figure 1(a) shows that the building sector is a major energy consumer in KSA with an annual growth rate of about 10% and a share of 79% of the total electricity used in KSA. Furthermore, Figure 1(b) illustrates that the residential sector alone accounts for the consumption of more than half of the total electricity generated in KSA [1, 2]. This energy consumption trend of the building sector is expected to be maintained in the foreseeable future [1]. Therefore, to alleviate the economic and environmental impacts of the sector, the Saudi government initiated an energy efficiency strategy that attempts to reduce demand for fuel and raw materials in the construction sector [1]. For example, a National Energy Efficiency Program (NEEP) has been launched at King Abdulaziz City for Science and Technology (KACST) to support research projects that rationalise energy use as part of the government's policy to reduce carbon emissions by 130 million tonnes by 2030 as inscribed in Saudi Arabia's 2030 vision. Among the promoted solutions include adoption of energy efficiency and renewable energy technologies [3] through, for example, the Saudi Energy Efficiency Centre (SEEC) and a Mandatory Energy Efficiency Plan for government organisations [4].

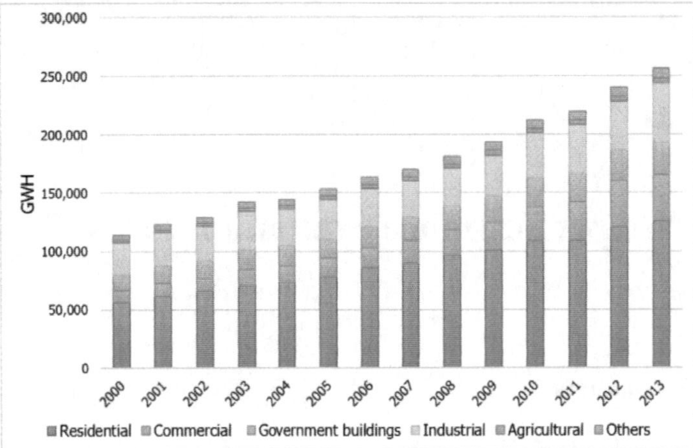

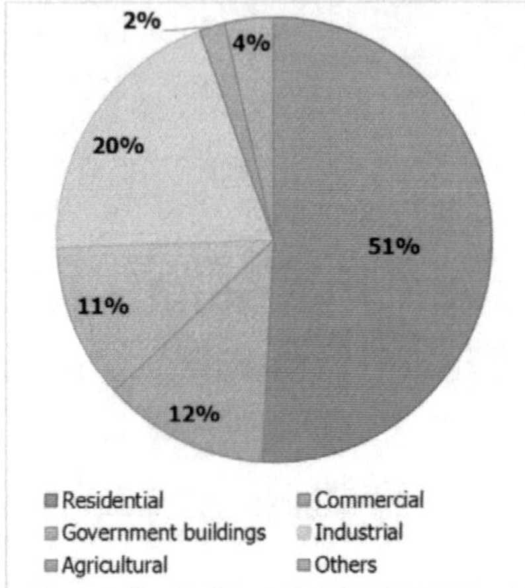

Figure 1: Building sector electricity consumption in KSA (a) Trend of energy consumption (b) Average breakdown [1]

Low energy prices, improved living standards and access to air conditioning systems created a strong correlation between power demand profile and ambient temperature. As illustrated in Figure 2, electricity demand nearly doubles in hot summer months [2]. It is estimated that over 50% of total electricity use in buildings is for space air conditioning in buildings [2], creating a compelling case for improving energy use in the domestic sector through the introduction of energy conservation measures and efficient building designs that do not compromise thermal comfort of buildings' occupants.

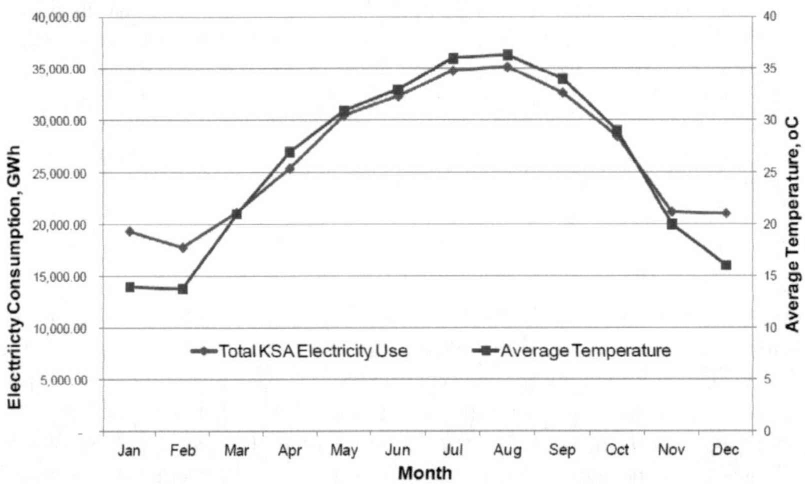

Figure 2: Monthly total KSA electricity consumption and average ambient temperature during 2014 [2].

KSA is located 23.88° N and 45.08° E and has an area of approximately 2.15 million km^2 and coast line of 3800km, creating distinctive climatic regions [5]. However, for a long time, the prevailing climate of KSA was classified as a single zone of arid desert, disguising the climatic differences between continental desert (hot and dry), costal line (hot and humid) and high altitude regions (hot/moderate in summer and cold in winter) [6]. This diverse climate has been reflected in the Saudi Buildings Energy Conservation Code 602E which divides the country into three major climatic zones as can be seen in Figure 3. However, the classification still does not consider the apparent disparities of climatic characteristics of Riyadh and Jeddah with hot dry and hot humid climates respectively. This was pointed out by Said et al.[7] in the study on climatic zones for building design purposes which proposed six climatic zones.

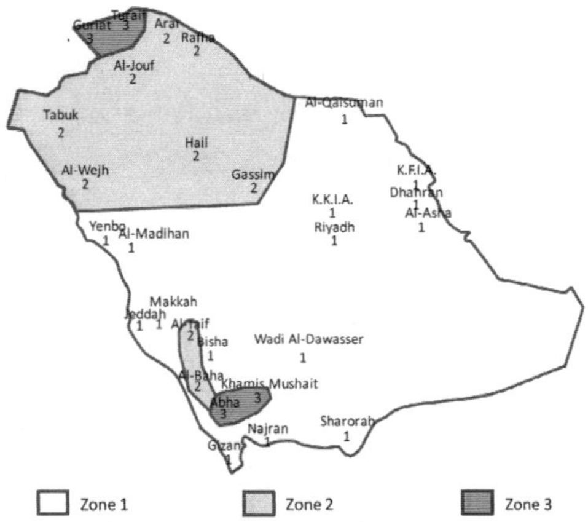

Figure 3: KSA Building Code (SBC-602E) climatic zones [8]

The main objective of this paper is to investigate the thermal performance and associated energy consumption of a typical individual residential villa building type in KSA three climatic zones. This is carried out with the aid of computer simulation using IES-ve software package, data of construction materials and building envelope extracted from the project document and energy bills of residents. Areas of improvement of the building envelope to conserve energy is highlighted.

2. Selection of cities in the three climates zones

To cover the wider climatic diversity of KSA, six cities located in the three climate zones identified in the Saudi Buildings Energy Conservation Code 602E and the study conducted by Said et al. [10] were selected. This includes Riyadh (which is a hot and dry with a desert subzone) and Jeddah (which is a hot and dry maritime subzone) in zone 1, Qassim (which is a hot and dry with a desert subzone) and Alwajh (which is a hot and dry maritime subzone) in zone 2, and Turaif (that a cold and dry with a desert subzone) and Abha (which is a subtropical with a Mediterranean subzone and a mountainous subtype) in zone 3. Weather data for these cities was taken from meteorological stations of the General Authority for Meteorology and Environmental Protection.

3. Selection of residential building type

The residential sector in KSA is varied and predominantly composed of flat apartments and individual villas. In this work type A villa building constructed by the Ministry of Housing has been selected as the base case study. The Ministry of Housing records show that 12,161 buildings of this type were constructed in 38 different regions so far. These buildings were designed to take into account local architectural and construction practices with the construction phase be implemented by an approved contractor and under competent supervision. A layout plan and photo of the building front façade are shown in Figure 4. It consists of 3 bedrooms, a guest room, kitchen, living room, dining room and four toilets with a total floor area of 245 m^2. The building walls were constructed from insulated hollow concrete blocks with windows and doors to wall area ratio of 6.29%. Rooms are air conditioned using window mounted units.

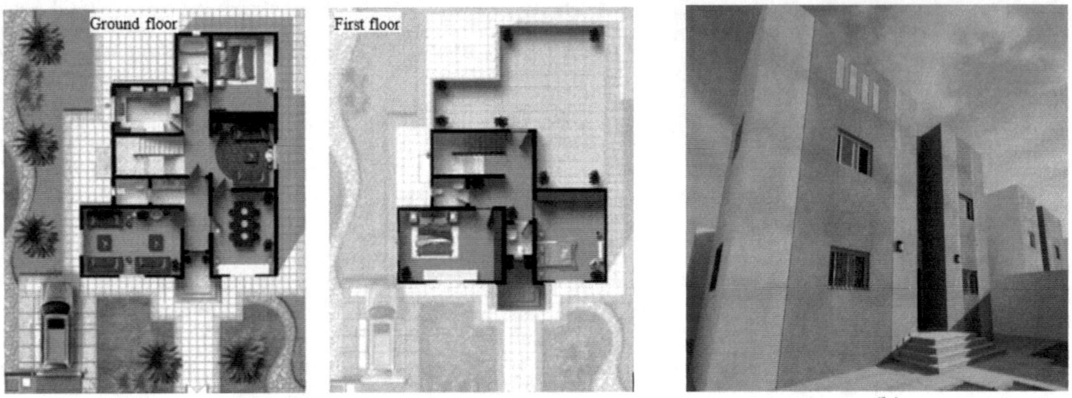

(a) (b)

Figure 4: (a) Type A villa layout plan. (b) Front elevation.

The main design parameters of the building envelope as specified in the project documentation are summarised in Table 1.

Table 1: Envelope elements U-Values.

Envelope Elements	U-Value (W/m^2K)	Envelope Elements	U-Value (W/m^2K)	Envelope Elements	U-Value (W/m^2K)
Exterior Wall	0.6993	**Internal Floor**	2.3417	**Interior Door**	1.8661
Interior Wall	1.3858	**Slab on-grade**	2.3876	**Windows**	2.2310
Roof	0.4510	**Exterior door**	Steel with Lexan (Polycarbonate) 8mm thickness		

Other design parameters include indoor design temperature of 24° C in summer and 20° C in winter and air infiltration rate of 1 ACH.

4. Building Energy Performance simulation

The building plan profile and energy performance evaluation was carried out using IES-ve software. Input parameters from the location weather database, the building energy model (envelope u-values, infiltration rates, indoor design temperatures and approximate occupancy schedule) to estimate annual energy consumption of the whole building. Operation parameters of the HVAC plant were also considered in IES-ve using specification data from commonly installed air conditioning equipment. To establish good confidence of the simulation results, a 3-year average electricity consumption data from energy bills of a case building in Qassim city (zone 1) was used to validate the model. Figure 5(a) shows a 3-D plan and sun path of the simulated building and Figure 5(b) shows a flow chart of IES-ve simulation block parameters of the buildings.

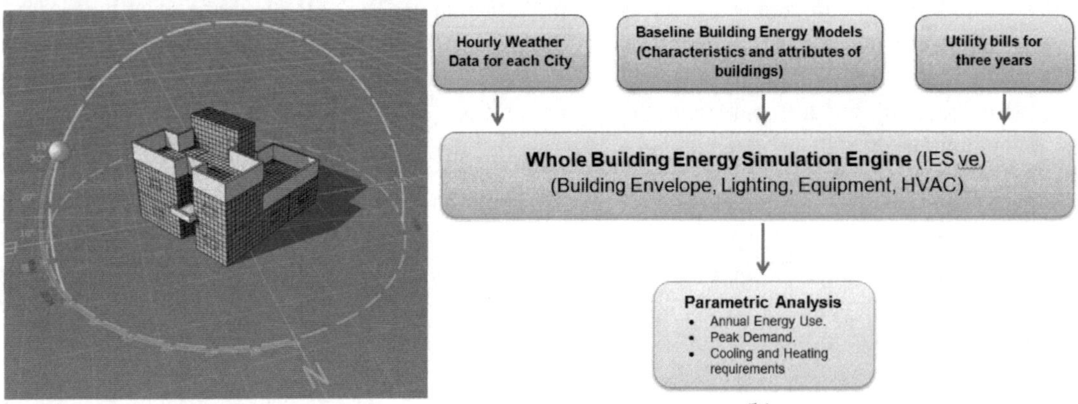

(a) (b)

Figure 5: (a) 3D plan of the building, (b) IES-ve simulation flow chart.

In this work the availability of reliable long-term climate data was particularly helpful in undertaking a comprehensive a comparative analysis of energy consumption in the three climate zones described in the building code.

5. Energy performance results and analysis

The power consumption simulation for the typical A type villa building in six cities located within the three climate zones was performed. The energy performance data assesses the building heating, cooling and total power consumption presented as monthly load profiles, as shown in Figure 6 (a), (b) and (c) respectively.

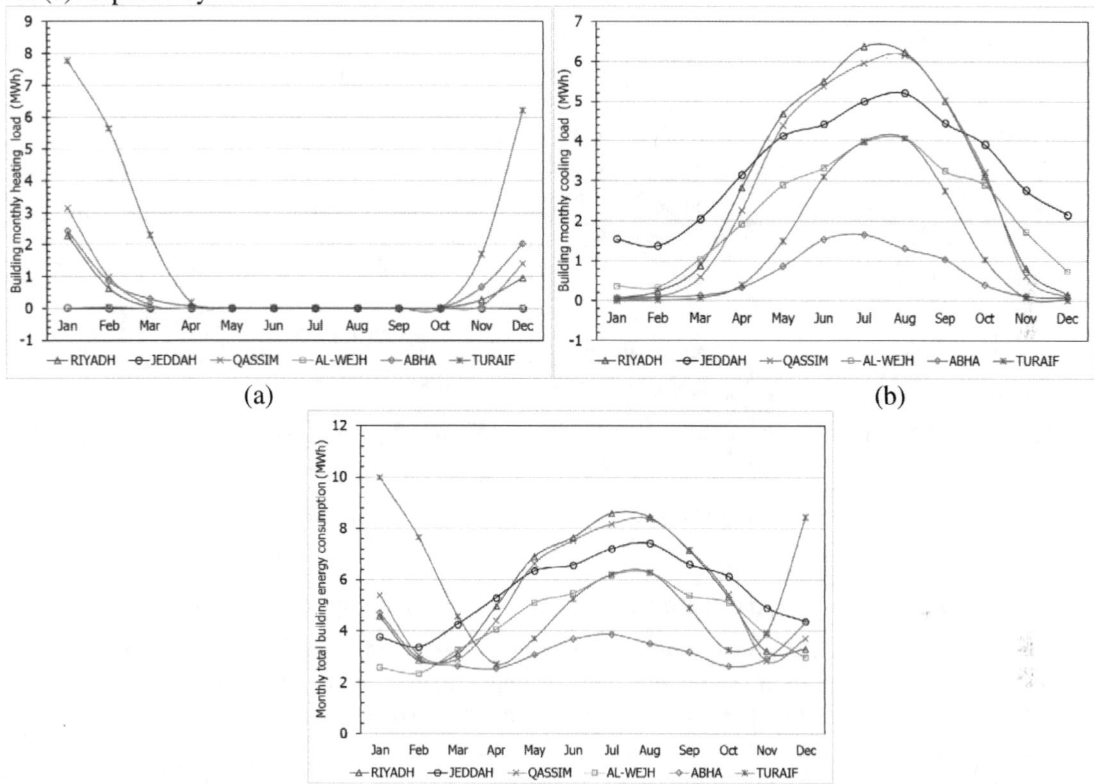

Figure 6: Building monthly power consumption load profile, (a) heating (b) cooling and (c) total.

Figure 6 (a) vindicates in part the climate zone classification of buildings in KSA. For example, Turaif (zone 3) requires the largest power consumption for heating during the heating season (October to April). Whereas the coastal city of Jeddah (zone1) has no heating load at all for the whole of the year, while Alwajh (zone 2) recorded only a low monthly heating load of about 0.0223 MWh in February.

Similar analysis was conducted for the power consumption for cooling loads, where Figure 6(b) shows monthly consumption profiles for the six cities. As expected, Riyadh (zone 1) and Qassim (zone 2) with prevailing hot and dry climate have the largest power consumption for cooling with peak demand in July of 6.371 and 5.954 MWh respectively. Power consumption for cooling in Abha and Turaif (both in zone 3) is lowest and recording 79% and 37% lower than Riyadh.

Figure 6(c) shows the aggregate monthly power consumption for space heating, cooling, and other electric appliances (lighting, fans/pumps, and Domestic Hot Water (DHW)) of the considered type of residential building. It is markedly clear the building located in climate zone 1 (Riyadh and Jeddah) consumes more power for cooling. It is also worth noting the large swing in demand for heating and cooling in Turaif (Zone 3) due to large variation in ambient temperature between the heating and cooling season with peak heating load in January exceeds that of cooling load in summer. In contrast, the building located in Abha (zone 3) exhibits particularly a flat annual energy load profile due mainly to the smaller seasonal temperature variations associated with mild climates.

The total power consumption of the building was further broken down into weather sensitive (i.e., HVAC consumption such as heating and cooling) and non-weather sensitive (i.e., lighting, DHW, and other equipment). The simulation input parameters for non-weather related energy consumption and occupancy pattern is assumed to be the same for all six locations. Figure 7 presents the proportion of energy consumption by category, location of the building and its zone. It can be seen that weather related energy consumption (i.e., heating and cooling) forms the largest part; 54 and 61% in zone 1; 51% in zone 2; and 35 and 61% in zone 3.

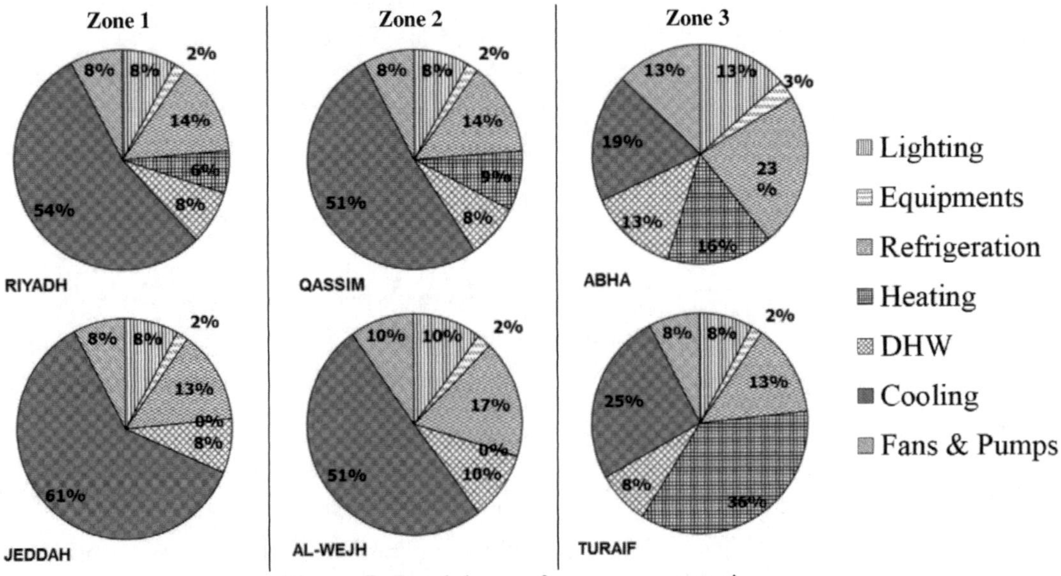

Figure 7: Breakdown of power consumption

Further analysis of annual power consumption data is presented in Figure 8. Figure 8 shows that the total energy consumption in the three zones is comparatively similar in Riyadh and Jeddah (both zone 1), Qassim (zone 2) and Turaif (Zone 3) amounting to about 66 MWh/year. While Alwajh a (zone 2) and Abha (zone 1) recorded the lowest power consumption of 52 and 40MWh/year. It is also shown that HVAC (heating and cooling) account for a minimum of about 14 MWh in Abha to a maximum of about 41 MWh in Turaif. The power consumption for heating and cooling varies significantly between the three zones (cooling: 7.641MWh in Abha and 40.260MWh in Jeddah; heating: 0 MWh in Jeddah and 23.835 MWh in Turaif).

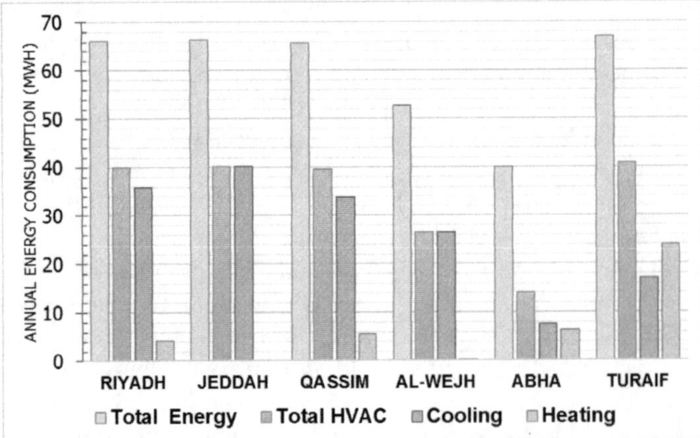

Figure 8: Annual power energy consumption.

6. Discussion and conclusions

There is a growing concern about energy use in Saudi Arabia. With rapid building development programmes and improvements of living conditions, the building sector will continue to be a major energy end consumer. This study examines annual power consumption of a type A villa building in six major cities in three different climate zones as identified in KSA Building Code (SBC-602E) - namely zone 1: Riyadh (hot and dry- desert subzone), and Jeddah (hot and dry - maritime subzone); zone: Qassim (hot and dry - desert subzone) and Alwajh (hot and dry- maritime subzone); and zone: Abha (subtropical- Mediterranean subzone and mountainous subtype) and Turaif (cold and dry- desert subzone). The annual power consumption in the case building was analysed for the three zones using the building envelope design parameters for villa type A that built according to the specifications of ministry of housing, along with HVAC equipment specification and occupancy schedule. The computed results from IES-ve were also validated using electric consumption bills. It was found that there is a wide variation in weather related energy consumption. For example, the annual power consumption for heating ranged from 0 MWh in Jeddah (zone 1) to a peak of 8 MWh in January in Turaif (zone 3), which is characterised by cold winter climate. On the other hand, Jaddah (zone 1 with hot, humid climate) had the largest annual power consumption for cooling- 40.260 MWh, while Riyadh (zone 1 with hot, dry climate) had the largest cooling peak of 6.371 MWh in July. This is about 74% bigger than that power consumption for cooling in Abha (zone 3), reflecting the extent of the climate differences in the two zones. The total annual power consumption in the three zones also varied greatly.

The study shows that current KSA Building Code (SBC-602E) differentiates to a certain extent between the construction specifications of the three zones based on climatic conditions. It was also shown that maybe further zones such as coastal lines (e.g., Jeddah, Alwajh, etc) be classified as a separate zone due to humid climate and slightly lower temperature variation that do not need heating in winter. Further work of this study will focus on reducing weather related heating and cooling loads through using lower u-values envelope elements, shading and other passive construction techniques.

References

[1] SEC, "Electrical data 2000-2013," Saudi Electricity Company, KSA, Riyadh2014, Available: http://www.se.com.sa, Accessed on: 28th March 2019.
[2] M. Krarti, K. Dubey, and N. Howarth, "Evaluation of building energy efficiency investment options for the Kingdom of Saudi Arabia," *Energy,* vol. 134, pp. 595-610, 2017.
[3] J. Mitchell and B. Mitchell, "Paris Mismatches: The Impact of the COP21 Climate Change Negotiations on the Oil and Gas Industries," *Chatham House Energy, Environment and Resources Department Research Paper,* 2016.
[4] M. Asif, "Growth and sustainability trends in the buildings sector in the GCC region with particular reference to the KSA and UAE," *Renewable and Sustainable Energy Reviews,* vol. 55, pp. 1267-1273, 2016.
[5] SGS, *Kingdom of Saudi Arabia: Facts and figures*, 2 ed. Jeddah: The Saudi Geological Survey, 2017, p. 170.
[6] B. Y. M. Ahmed, "Climatic classification of Saudi Arabia: An application of factor--Cluster analysis," *GeoJournal,* vol. 41, no. 1, pp. 69-84, 1997.
[7] S. Said, M. Habib, and M. Iqbal, "Database for building energy prediction in Saudi Arabia," *Energy conversion and management,* vol. 44, no. 1, pp. 191-201, 2003.
[8] *Saudi Energy Conservation Code, Low-Rise (Residential) Buildings, SBC 602 E*, 2018.

A simulation-based procedure for the holistic resilience testing of building performance

J A Clarke

Energy Systems Research Unit, Department of Mechanical & Aerospace Engineering, University of Strathclyde, 75 Montrose Street, Glasgow G1 1XJ.

joe@esru.strath.ac.uk

Abstract. The design of an energy efficient urban environment is a formidable task because the underlying processes are complex, coupled and time varying. Competing objectives from a variety of stakeholders relating to occupant wellbeing, fuel poverty alleviation, improved air quality, low energy use, emissions reduction, and legislative compliance then act to confound the attainment of operational quality and resilience of particular design solutions. This gives rise to a persistent and growing gap between design intent (in both new build and retrofit contexts) and the operational reality. Unfortunately, there are no current standards for building performance appraisal and operational resilience testing so that such assessments, where undertaken, are *ad hoc* and implemented as an arbitrary function of the particular project context. The result is that it is impossible to compare alternative proposals in terms of the performance that is likely to be delivered in practice. This paper reports recent developments in automating the application of building performance simulation as a means to address this deficiency.

1. Introduction

The drive towards a more sustainable built environment gives rise to challenges for construction industry practitioners. From an energy viewpoint, these challenges stem from the need to reduce energy consumption, integrate sustainable energy supplies and mitigate environmental impacts, all while meeting expectations for human wellbeing and economic growth. The design of an energy efficient building or community is a formidable task because the underlying physical processes are complex: coupled heat and mass transfers, possibly bidirectional electrical power flows, radiation exchanges, distributed control actions, uncertain occupant behaviours and equipment failures, all of which vary spatially and temporally.

The design task is then confounded by myriad policy agendas aimed at stimulating a clean energy transition through accelerated renewables deployment and the electrification of building heating and transport. Such policy goals usually give rise to competing aspirations relating to occupant satisfaction, fuel poverty alleviation, local air quality improvement, energy demand reduction, emissions abatement, energy supply resilience, and more stringent legislative compliance. Because of this socio-technical complexity, the gap between design intent and the operational reality is both persistent and growing [1] [2]. This paper reports on a simulation-based procedure that provides a means to address this gap.

In relation to reducing the energy demand associated with the built environment, there are many competing technology options including, but not limited to, those listed in Table 1. Such options are typically unable to provide an effective solution on their own and in many cases may be mutually exclusive. The need then is to identify fit-for-purpose blends as a function of the requirements and constraints of a particular scheme. This is an obvious role for *Energy Systems Simulation* (ESS), a term used hereinafter to refer to building performance simulation when applied to buildings and their related energy supply systems within the urban environment at whatever scale. ESS probably provides the most effective mechanism to identify appropriate solutions within the time and resource constraints of design practice.

Table 1. Example energy demand reduction options.

Passive features	Heat-related	Electricity-related
▪ daylight utilisation	▪ heat pumps	▪ integrated photovoltaics
▪ adaptive facades	▪ solar thermal	▪ smart meters
▪ solar ventilation pre-heat	▪ electricity-to-heat	▪ smart grids
▪ switchable glazing	▪ condensing boilers	▪ urban wind power
▪ selective films	▪ smart heating control	▪ fuel cells and hydrogen
▪ advanced insulation	▪ heat recovery	▪ electric vehicles
▪ desiccant cooling	▪ biomass/biofuel heating	▪ demand management
▪ evaporative cooling	▪ culvert heating/cooling	▪ demand response
▪ moveable devices	▪ district heating/cooling	▪ smart water heating
▪ breathable walls	▪ energy storage	▪ combined heat/ power
▪ phase change materials	▪ tri-generation	▪ ducted wind turbines

Less obvious, but no less powerful, is the use of ESS to ensure the best temporal match between energy demand and energy supply as derived in the latter case from appropriate blends of conventional (fossil fuel and nuclear) and renewable (national and urban scale) energy technologies assisted by energy storage and/ or utility grid interaction.

The conflicts that underlie policy and business agendas that seek to electrify building heating and transport, to encourage net zero energy in the housing stock, to establish community energy schemes via smart control, or encourage urban renewables, can best be resolved by utilising this dynamic matching capability to design hybrid schemes. Examples of such conflicts include the tensions between cost reduction, occupant wellbeing, fuel poverty, indoor/ outdoor air quality, real-time utility transactions and so on. The merits of an ESS approach that enables whole system, multi-variate performance appraisal under realistic operational scenarios are summarised in Table 2. It is likely that no other approach can accommodate the underlying physics while addressing all relevant performance aspects and satisfying the time and cost constraints of business – or as succinctly summarised by Hong *et al* [3]: "Building simulation serves not only to reveal the interactions between the building and its occupants, HVAC systems, and the outdoor climate, but also to make possible environmentally-friendly design options".

Table 2. Merits of the ESS approach.

Can address all relevant issues:	Can encapsulate all relevant processes:
▪ technical feasibility;	▪ building physics;
▪ thermal and adaptive comfort;	▪ thermo-fluids;
▪ indoor and outdoor air quality;	▪ heat and mass transfer;
▪ life cycle operation;	▪ radiant heat exchange;
▪ energy and carbon economics;	▪ plant and systems processes;
▪ environmental emissions;	▪ electrical power flows;
▪ controllability assurance;	▪ micro-climate;
▪ operational resilience.	▪ renewables stochasticity;
▪ wellbeing and productivity;	▪ control system response
▪ uncertainty and risk;	▪ innovative materials and systems;
▪ increasingly stringent legislation	▪ stochastic events.

The projects reported in this paper seek to standardise performance assessments in relation to the resilience testing of new designs and retrofits. Currently, there is no standard way to assess building performance in a manner that addresses lifetime resilience. These projects aim to create industry standard performance assessment procedures that require no specialist knowledge of any particular modelling tool on the part of the user.

The overall message of the paper is that clean energy transition aspirations have fostered conflicting technology solutions that are not generally applicable. Holistic performance simulation is one way to select appropriate solutions for specific contexts, ensure operational resilience and reduce the gap between design intent and performance in use.

2. EES theoretical basis

To deliver such functionality, ESS requires high resolution models, implying a profound departure from traditional approaches to design appraisal [4] [5]. All energy systems – heat pumps, buildings, district heating, photovoltaic panels *etc.* – have 4 characteristics that need to be preserved where the aim is to enable an emulation of reality. First, they are intrinsically dynamic due to the spatially distributed heat, mass and momentum exchanges that vary with time and at different rates. Second, defining data (e.g. material thermo-physical properties and plant efficiencies) are non-linear in that they vary as a function of the evolving variables of state (e.g. temperature and moisture content). Third, the system overall is systemic in that component parts impact variously on capital & running cost, thermal & visual comfort, emissions & air quality, electricity network interactions & power quality, and controllability & resilience. Fourth, some principal influences are stochastic in that they derive from uncertain processes associated with occupant behaviour, weather and equipment failure.

Where one or more of these characteristics are overly simplified, the resulting ESS tool will reduce to a mere calculator that, seemingly by design, is misaligned with the reality. Within high integrity ESS, there is no justification for imposing actions such as, for example:

- representing constructions dynamically but plant components in steady-state;
- representing infiltration and ventilation via imposed air change rates when flows depend on time varying pressure and temperature differences;
- prescribing equipment efficiencies when these depend on the as yet unknown values of state variables (such as the dependency of a heat pump's coefficient of performance on evaporator and condenser temperatures, which may be highly variable); and
- decoupling system parts as a computational convenience.

As summarised by the collage of Figure 1, the ESS numerical approach conserves energy, mass and momentum within connected finite volumes corresponding to a discretised equivalent of some energy system at whatever scale and when subjected to heat transfer mechanisms, sources of the conserved properties, and control actions.

Even with such a high resolution approach to performance simulation, it is hubris to suggest that it is possible to predict the future state of a system. Instead, the aim is to emulate reality as a means to ensure operational resilience. This is achieved by ensuring that the causal relationships between different aspects of a system are explicitly represented. For example, to test the acceptable performance and resilience of a community energy scheme utilising building-integrated photovoltaic (PV) components and interactions with the low voltage electricity network (LVN), several couplings would need to be represented, such as:

- increased daylight penetration to indoor spaces (needs daylight modelling);
- resulting in artificial light dimming (needs photocell modelling);
- giving rise to a reduced electrical load (needs power flow modelling);
- requiring the power flow from a local PV array to be exported (needs PV cell modelling);
- perhaps resulting in a power quality impact on the LVN (needs network modelling);
- requiring an export refusal or tariff adjustment (needs utility control modelling); and
- necessitating an alternative dump load for the PV power (needs storage modelling).

Such a simulation would result in an entirely different outcome than one that merely determined the output power from the PV array for a given irradiance with no consideration of system interactions; and would lead naturally to a robust practical scheme that is more likely to perform as required when established.

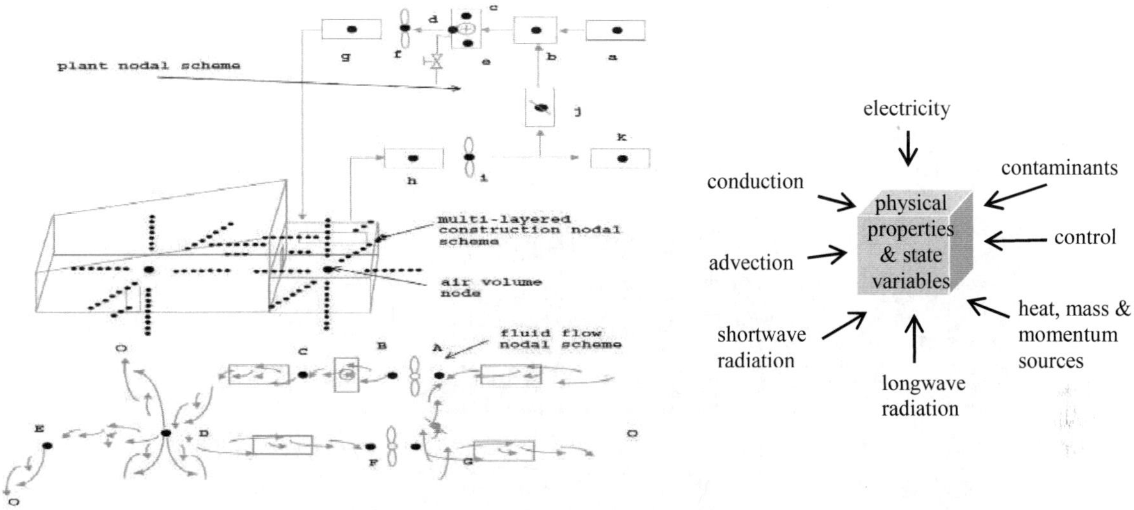

Figure 1. The ESS numerical approach.

3. High resolution modelling

The use of ESS for future reality emulation requires a high resolution input model [6]. Figure 2 shows such a model for a dwelling.

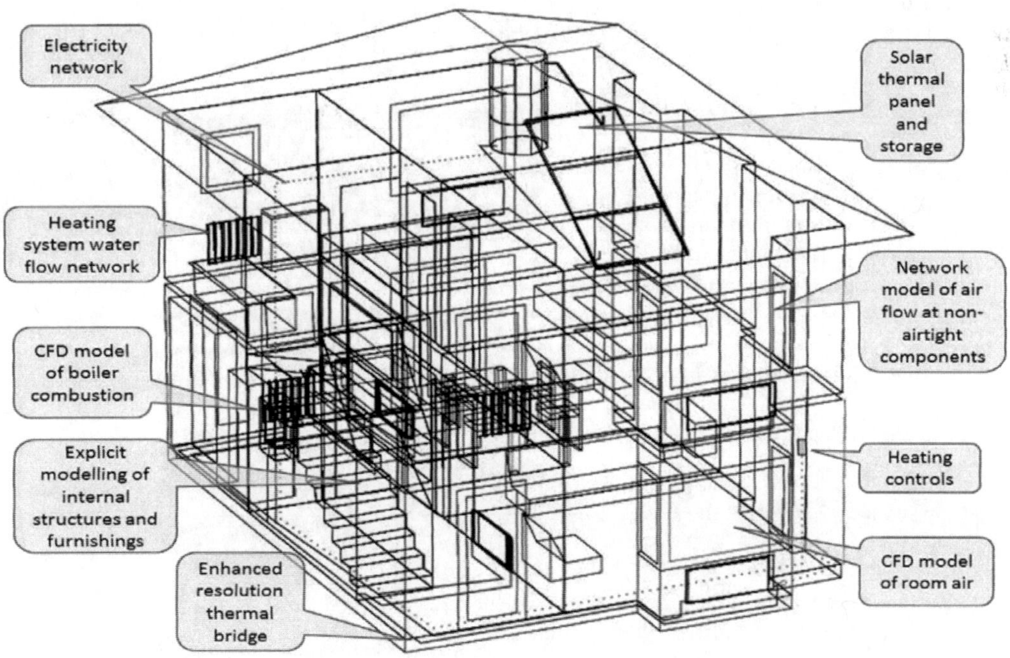

Figure 2. A high resolution house model.

This model comprises all relevant physical entities – 3D form, construction material attribution, internal fittings & furnishings, heating system & hot water components, low voltage electrical network, solar thermal panel (could be PV), and control system components. To such a model must be added (automatically where possible) virtual constructs relating to network air & electricity flow, room air movement, gas boiler combustion-related emissions and occupant behaviour. Such a model is capable of providing a range of relevant outputs addressing the different aspects of overall performance, such as the snapshot examples depicted in the collage of Figure 3.

Similar high resolution models can be created for the non-domestic case where data defining form, fabric and contents are augmented by descriptions related to occupant presence, control systems, HVAC equipment, renewable energy components, artificial lighting and electrical networks. In this case, outputs can, additionally, include the spatial and temporal distribution of thermal comfort, discomfort glare and indoor air quality (as evaluated against any relevant standard) as depicted in Figure 4.

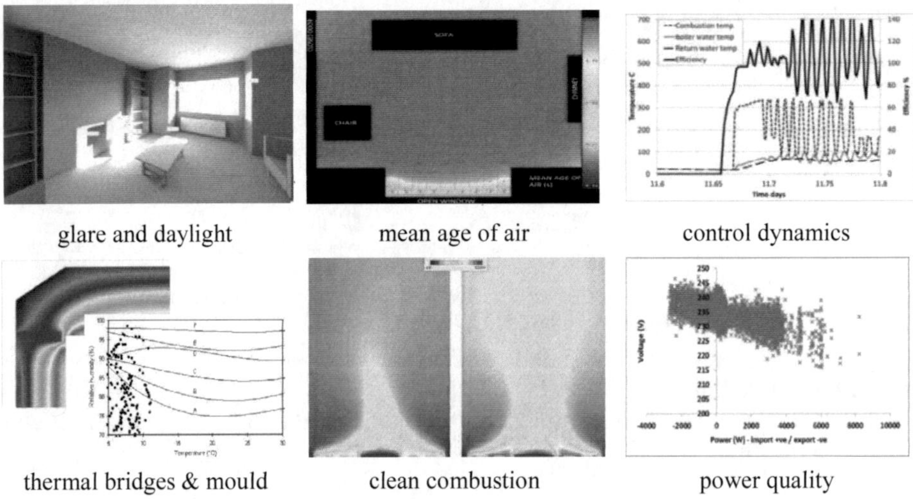

Figure 3. Some possible outputs from a high resolution dwelling model.

Thermal comfort – ISO 7730
- PMV
- draught risk
- vertical air temperature stratification
- floor temperature
- radiant asymmetry

Visual comfort – BS EN 12464-1
- glare related to daylight
- unified glare rating, UGR ≤ 19

Air quality – BS EN 15251
- CO_2 concentration

Figure 4. Wellness outputs from a high resolution office model.

Two features of the high resolution modelling approach are emphasised here. First, the approach is holistic and therefore enables performance to be represented in terms of a range of relevant criteria. Second, it is possible to semi-automate the generation of models for large building stocks based on design parameter diversification applied to 'seed' models corresponding to particular archetypes [7]. The former feature provides a means to make trade-offs explicit (e.g. between energy reduction and occupant wellbeing); the latter feature reduces the model creation burden and helps harmonise the simulation process because the same model can be utilised by different users and for different applications.

4. Resilience testing

While the analysis capability of EES covers an extensive range of performance aspects, the technology is not particularly easy to apply in practice. One approach to alleviate this dilemma is to adapt the manner in which simulation is accessed by providing cloud-based services offering industry standard performance assessments. Imagine a future in which simulation services exist that:

- operate on the basis of design proposals delivered as high integrity BIM models [8] (extensions to current data schemas will be required to cover all descriptive and performance domains covered by ESS in future);
- require no user involvement in the simulation process;
- automatically initiate standard performance assessments;
- evaluate the performance outcomes against agreed criteria;
- judge overall acceptability using models of building users/ operators (as opposed to simulation tool users as now); and
- facilitate the unambiguous comparison of alternative proposals.

This is the approach adopted in the Horizon 2020 Hit2Gap [9] and Construction Scotland Innovation Centre (CSIC) Resilience Testing [10] projects; the former nearing completion, the latter recently commenced. Figure 5 summarises the concept.

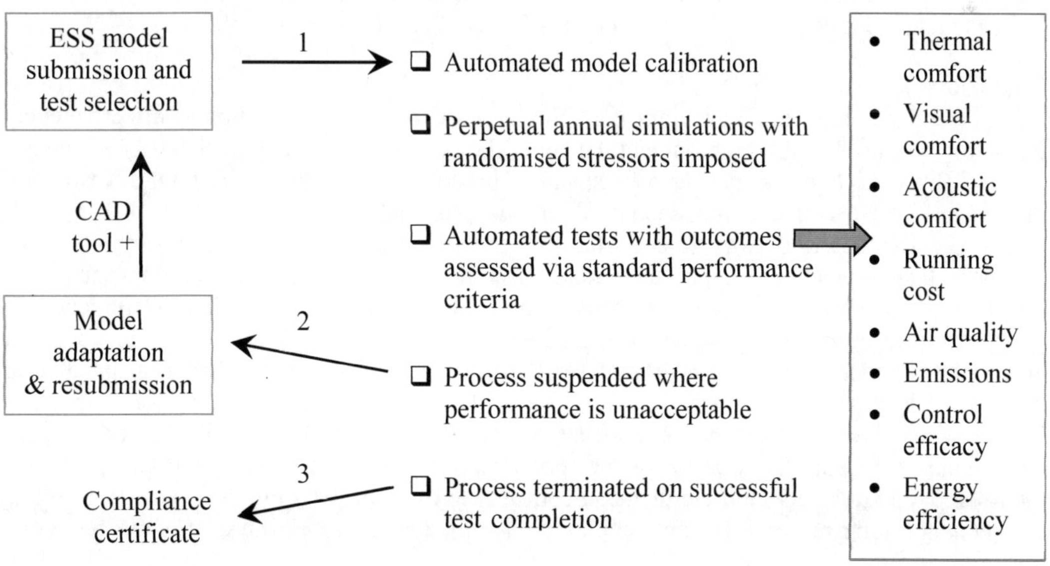

Figure 5. A cloud-based resilience testing service powered by an approved ESS.

The aim is to standardise and automate performance assessments corresponding to different design targets – low energy dwellings, estate facility management, community energy schemes, critical environments in the health service *etc.* – and levels of test stringency, e.g. current regulations, Passivhaus standard [11] or net-zero energy in the case of housing.

Within the Hit2Gap project, automated simulation services are targeted on the energy and wellbeing aspects of facility management within large estates. Within the CSIC project, the focus is building design and refurbishment, with several companies led by CIBSE and RIAS collaborating to establish performance assessment procedures, assessment criteria and resilience impacting events for different building types. The outcomes are encapsulated within a 'Resilience Testing Environment' prototype (Figure 5), which is being field trialled within live projects at different scale.

Such emulations of reality, have significant benefits. Assuming that they include a realistic representation of occupant behaviour (e.g. [12]), they relate more closely to reality and not to an inappropriately simplified, partial representation. They do not require the simulation tool user to define performance assessments and interpret outcomes, thus standardising the assessment process and facilitating the inter-comparison of alternative proposals in terms of whether or not they passed the test. Such attributes leave the design team free to innovate, supported by a computational environment that assures acceptable performance under a range of conditions likely to be encountered in practice, and in terms of performance criteria that can be readily understood by different stakeholders. The approach also provides a mechanism for building regulation compliance that overcomes the constraints of the present rudimentary methods that fail to respect the thermodynamic integrity of contemporary and future energy systems. (It is emphasised here that simulation models may still be configured against standard prescriptions in order to obtain outputs from simulation tools for legislative compliance purposes, to size system components for peak demand, or to obtain deeper performance insight – i.e. the traditional application of simulation techniques.)

At the present time, the simulation environment depicted in Figure 5, when powered by the ESP-r system [13], has been connected to the BEMServer platform as established within the Hit2Gap project [9] and is being field trialled by application to two pilot buildings. Within this implementation, pre-constructed, high resolution models are automatically calibrated using the Calibro application [14] prior to being subjected to requested performance assessments that are entirely automated. At the same time, an industry consortium led by CIBSE has been established within the CSIC project [10] to harmonise approaches to performance assessment in support of building design and identify the 'gremlins' that might be randomly applied during resilience tests to ensure the robustness of outcomes.

5. Conclusions

Explicit performance simulation provides powerful support for the identification of effective energy-related actions within the built environment. Hitherto, the potential of the approach has been hampered by the lack of standards relating to the assessment of proposals in a manner that ensures resilience in operation and facilitates direct comparison of alternative proposals.

This paper has reported two projects that are researching the concept of a simulation-based environments that automatically apply standardised performance assessments to proposals and provides a means to issue compliance certificates corresponding to increasing levels of resilience stringency. This is an approach which, it is hoped, will bring benefit to design practitioners, policy makers, estate managers and researchers because it harmonises the use of performance simulation in practice as a means to explore dynamic behaviour over a proposal's lifetime in order to reduce the performance gap.

The use of automatic performance assessment provides access to sophisticated simulation scenarios with minimum user input required because the approach incorporates assessment knowledge alongside simulation tool control. Service users then can concentrate on improving building performance rather than expending effort and resources on complex *ad hoc* model calibration, simulation process control and results analysis.

6. References

[1] Menezes C, Cripps A, Bouchlaghem D and Buswell R (2012). 'Predicted vs. actual energy performance of non-domestic buildings: using post-occupancy evaluation data to reduce the performance gap', *Applied Energy*, 97, pp. 355-364.

[2] De Wilde P (2014). 'The gap between predicted and measured energy performance of buildings:

A framework for investigation', *Automation in Construction*, 41, pp. 40-49.
[3] Hong T, Chou S K and Bong T Y (2000). 'Building simulation: an overview of developments and information sources', Building and Environment, 35(1), pp. 347-361.
[4] Clarke J A (2015). 'A vision for building performance simulation: a position paper prepared on behalf of the IBPSA Board', Building Performance Simulation, 8(2), pp. 39-43.
[5] Clarke J A and Hensen J L M (2015). 'Integrated building performance simulation: Progress, prospects and requirements', Building and Environment, 91, pp. 294-306.
[6] Clarke J A, Cockroft J, Cowie A, Hand J, Kelly N, Samuel A and Tang D (2015). 'On the high-resolution simulation of building/ plant systems', *Proc. Building Simulation 2015*, Hyderabad, India.
[7] Clarke J A, Ghauri S, Johnstone C, Kim J and Tuohy P G (2008). 'The EDEM Methodology for Housing Upgrade Analysis, Carbon and Energy Labelling and National Policy Development', *Proc. eSim 2008*, Quebec.
[8] Eastman C, Tiecholz P, Sacks R and Liston K (2011). *BIM Handbook: A Guide to Building Information Modeling for Owners, Managers, Designers, Engineers and Contractors (2nd ed.)*, Hoboken, New Jersey: John Wiley, pp.36–37.
[9] Clarke J A, Costola D, Kelly N J and Monari F (2017). 'A big data approach to the application of building performance simulation to improve the operational performance of large estates', *Proc. Building Simulation '17*, San Francisco.
[10] Construction Scotland Innovation Centre (2019). 'Development of standardised approaches and a prototype software tool for resilience testing of building performance', Research project awarded 25.02.2019.
[11] Passivhaus (2019). http://www.passivhaustrust.org.uk (viewed 8 April 2019).
[12] Hong T, Sun H, Chen Y, Taylor-Lange S C and Yan D (2016). 'An occupant behavior modeling tool for co-simulation', *Energy and Buildings,* 117, pp. 272-281.
[13] ESP-r (2019). https://www.strath.ac.uk/research/energysystemsresearchunit/applications/esp-r/ (viewed 8 April 2019).
[14] Monari F (2016). 'Sensitivity analysis and bayesian calibration of building energy models', *PhD Thesis*, Glasgow: University of Strathclyde.

Acknowledgments

The author wishes to thank the European Commission and Construction Scotland Innovation Centre for their research support, and his many colleagues within and outwith ESRU for their valuable inputs and unwavering support in the evolution of advanced simulation software for the construction industry.

IOP Conf. Series: Earth and Environmental Science **329** (2019) 012028 doi:10.1088/1755-1315/329/1/012028

Energy and Environmental performance of a green certified office building in the hot dry climate of India

Rajat Gupta*, Matt Gregg and Rupesh Panchal

Low Carbon Building Research Group, Oxford Institute for Sustainable Development, Oxford Brookes University, Oxford, UK

*Corresponding Author: rgupta@brookes.ac.uk

Abstract. This paper applies a customised post-occupancy building performance evaluation (BPE) approach to evaluate the actual performance of a Platinum-certified green office building in the extreme hot dry climate of India from both technical and occupants' perspectives. The in-use energy and environmental performance of the office building was examined using energy data on consumption and generation, monitoring of environmental conditions (temperature, relative humidity, CO_2), along with occupant satisfaction survey. Total annual energy use of the building was measured as 100 kWh/m^2/year with photovoltaic generation contributing 18kWh/m^2/year. Although total annual energy use was found to be 42% less than the typical office benchmark, about 6% more energy was used than predicted. Indoor air quality was perceived to still but fresh and neither too dry nor too humid. Despite this, some spaces had relative humidity measurement over 70% for more than 50% of occupied hours. Daylight factor was consistently below 2% in many areas due to internal partitions; however, BUS survey (105 responses) results showed that 71% of occupants were satisfied with lighting quality, overall comfort and other variables. Such differences between measured and perceived indoor environment indicate the role of occupant adaptation in building performance.

1. Introduction

Due to rapid growth [1] and global concern for environmental impact, India is an important focus for reducing energy consumption in buildings and green building certification councils have seen this as an opportunity. The Indian Green Building Council (IGBC) claims that India is the second country in the world with the largest registered green building footprint with over 5,142 projects registered for green building ratings, of which 1,645 certified and fully functional (as of February 2019) [2]. Research, however, continually demonstrates that green building rating and certification systems do not always ensure greater energy performance [3, 4], occupant satisfaction [5, 6] or better indoor environmental quality (IEQ) over conventional buildings [7].

The energy performance gap (EPG) has been demonstrated in international research. Research has shown LEED (Leadership in Energy and Environmental Design) certified building using twice the energy beyond model predictions in the USA [8]. In the UK this has been reported as up to five [9] and nine [10] times the energy predicted for exceptionally energy efficient designed buildings. Also, in India [4], LEED certified buildings are not performing as expected with respect to certification, i.e. the LEED Silver and Gold facilities of the same type are performing better than one certified as LEED Platinum. The EPG may not necessarily indicate poorly designed or failed green buildings but simply indicate a lack of rigour in the prediction / certification process. One such study demonstrates the way models can create this dilemma by underestimating occupant numbers and behaviour in office buildings [9]. Another significant problem is attributed to how the prediction side of some building regulation only predicts heating and cooling load i.e. *regulated energy*, while the measurement side includes total energy consumption [11]. Nonetheless, the frequency of the performance gap internationally indicates the necessity to demonstrate claims of efficiency, sustainability and comfort through evaluation in India.

Building Performance Evaluation (BPE) can be used to identify, quantify and resolve the gap between 'as-designed' and 'built / in-use' performance through a systematic collection and analysis of qualitative and quantitative information related to fabric performance, energy performance, environmental conditions and occupant behaviour. BPE can involve feedback and evaluation reviews at every phase of the building delivery from strategic planning to occupancy, adaptive reuse and recycling [12]. This paper seeks to apply a customised post-occupancy BPE approach developed in [13] (I-BPE) for Indian green buildings, to evaluate the actual performance of an IGBC Platinum-rated office building in the hot and dry climate of India, from both technical and occupants' perspectives. Following the literature review of relevant studies to-date in India (following section), the paper then introduces the case study building and the methods used to evaluate it. The results are then presented followed by a discussion on future application and key conclusions.

2. Review of building performance evaluation in India

There is a growing body of published research on assessing actual building performance in India. Most of these studies are focussed on field studies in thermal comfort (FSTC) utilising thermal comfort questionnaires, interviews, temperature and relative humidity (RH) logging, and spot measurement tools to evaluate environmental parameters of buildings [14]. Interior environment assessment in the literature include spot measurements at the time of survey [15, 16], thermal comfort questionnaires (including long-term/seasonal outlook and/or thermal sensation and preference votes) [15-18] and long-term logging/monitoring of temperature, relative humidity (RH) and other parameters [17, 19]. In addition, several studies, of naturally ventilated domestic and non-domestic, concluded that occupants are comfortable at temperatures greater than comfort ranges recommended by ASHRAE 55, ISO-7730 standards, and the National Building Code of India [16, 20, 21].

Generally, POE/BPE studies differ from FSTC in that they include the addition of a long range questionnaire on such variables as work area satisfaction, lighting, productivity, and health [15]; a review of project information, interviews with key stakeholders [18]; design and system installation review, monitoring plan walkthrough, monthly energy bill collection for one year combined with seasonal energy monitoring, data logging of electricity distribution, spot measurements of lighting, temperature, RH, and envelope temperature [22]; and aggregated, sub-metered and appliance energy consumption monitoring [23]. The largest gap in BPE methods relate to those generally applied before post-occupancy, e.g. evaluation of systems installation, commissioning, and fabric performance. In summary, most studies are focused on thermal comfort and less on energy consumption, with little cross-over between the two subjects.

3. Case study and research methods

A previously developed BPE methodology [13] developed for the Indian context (I-BPE), as part of a Newton Fund UK-India research project, was tested on a green office building as part of a postgraduate dissertation (by one co-author), with the intent to provide feedback on the relevance and effectiveness of the I-BPE methods as a research tool in the Indian context. A key aim of the case study is to better understand the challenges in applying the methods and tools of the I-BPE methodology, and how these might be improved to continue BPE studies in India.

The case study building, the operational headquarters of a company located in Rajasthan, India, was originally constructed in 1966 and has been retrofitted many times over its life. It was finally renovated to receive an IGBC Platinum rating in the category of *existing building - operation and maintenance* in 2017. The building is comprised of five levels with a gross floor area of 9,867 m^2 and a built-up area of 7,800 m^2; however, much of the second and third floors are currently unoccupied. The building is occupied from Monday to Friday from 9:30 – 18:30. Occupancy varies every day with 80 permanent occupants, 15-25 part-time occupants, 20-30 daily visitors.

The building is constructed of reinforced cement concrete (RCC) and masonry to create a heavy mass structure. To its benefit, the building is oriented long-ways along the north and south and is designed around a central atrium which allows daylight into the centre of the building. To each side of the central atrium are double loaded corridors with open office space in the centre. The central atrium was originally open to the sky but was covered to restrict bird intrusion. The covering is translucent, but it has reduced the level of daylight in the space. The building uses a mix of cassette ceiling air conditioning (AC) units and a variable refrigerant flow (VRF) system for cooling; the offices on the fourth floor are cooled with split ACs, and the auditorium has separate chiller plant for cooling. In addition, tower air conditioners are in place as back-up cooling units. The auditorium also has a separate chiller plant for cooling. The stairwells and the lift lobbies & corridors of the fourth floor are naturally ventilated. The remainder of the building is mechanically ventilated throughout the entire year, and all windows are fixed. The site uses grid electricity and a 100kWp photovoltaic (PV) array; there are also onsite diesel generators used as a backup during power outages. The building also performed excellent in 'site and facility management' and 'water efficiency' categories achieving targets such as waste collection and disposal, eco-friendly commuting and landscaping practices, heat island reduction (roof and non-roof) and outdoor lighting pollution reduction.

3.1. *BPE field study methods*

The field study was carried out for one month in the monsoon season (5 July – 6 August 2018). The primary components of the study included design and construction audit, energy audit, environmental audit and occupant survey. Table 1 shows the I-BPE recommend study elements divided in four levels of increased complexity. For each level, the table indicates the action taken and/or tools used.

Table 1. Adaptation of the I-BPE methodology for this study *(NP = not performed)*

	Level 1	Level 2	Level 3	Level 4
Review of design intent	Collection and review of design docs.: *Plans and IGBC credit doc.*	Review of services and energy systems: NP	Interviews with key stakeholders: *facility manager (FM)*	Walkthrough with key stakeholders: *FM*
Technical building survey	Inspection of build quality and services: *photographic survey*	Controls interface survey: NP	Review of installation and commissioning of services: NP	Thermography of building fabric: NP
Energy assessment	From BMS: *Electricity use: Jan 2017 – Jun 2018 monthly & incomplete daily data*	Energy monitoring: NP	Sub-metering: From BMS: *Diesel generator (Jan – Dec 17) & PV (Apr 17 – Mar 18) monthly & incomplete daily data*	Plug load monitoring of individual appliances: NP
Environ. monitoring	Temperature, and RH spot readings: *tool: WatchDog A160*	Temperature and RH monitoring: *5-min. 4-week duration; tools: Hobo UX-100, I-button*	Additional spot read/ logging (e.g. CO_2, lux, wind speed): CO_2 spot reading/logging: *tool: WatchDog A160, Tinytag* Light reading: *Lux meter*	Additional pollutants: *PM2.5 & VOC:3-hr. ~4-week duration; tool: Foobot*
Occupant feedback	Occupant satisfaction survey: *BUS (78 of 105 returned)*	Semi-structured interview: *10 interviews*	Thermal sensation and preference survey: *self-completed diary (~15 returned; incomplete)*	Focus group: NP

4. Results

4.1. Review of design intent
According to BUS[1] [24] results and FM interview, the exterior and interiors of the building are considered well designed and facilities are overall satisfactory; however, lack of meeting rooms and storage space, open plan office noise, and lack of control over temperature were noted as issues.
Variation in use between 'as-built' and 'in-use' include: most the spaces on second and third floors of the buildings are unoccupied and are being used to store unwanted things; balconies are used to house split AC units to cool the newly designed occupancy of the fourth floor; and improper wiring, unnecessary spotlights, broken window shade controls, water leakage in the false ceiling, lack of storage at the workspace were noted at the time of walkthrough survey. Overall, the design review process proved to be challenging as much of the core team responsible for the coordination of the certification process had left or retired and even the plans received from the FM consultants did not represent the current state of in-use.

4.2. Energy assessment
The building received two *Energy Efficiency* credits for *Minimum Energy Performance* (mandatory minimum) for projecting a 10% reduction in total energy consumption beyond the design baseline of 817,000 kWh. The PV system's capacity goes beyond the 7.5% that was required to gain the maximum credits for on-site renewables. Though grid connected, all PV electricity generated is consumed by the building. The building also gained two credits for purchasing 273.5 MW of off-site renewable energy.

The following energy analysis considers energy data for the year covering 1 April 2017 – 31 March 2018 as this is the extent of data available to the researcher for PV generation. Though the data for grid and diesel consumption span from January 2017 – June 2018, there is a gap in the data wherein the monthly totals for March – May 2018 are missing. March 2018 is therefore proportionally estimated from March 2017. Overall, the building is not performing as well as expected; it is also consuming more energy than it did in 2015 (pre-certification). Based on current consumption, in relation to the design baseline, the building would not qualify for the *Minimum Energy Performance* credits to achieve certification. The total EPG is +6%; however, the renewable system is performing better than predicted: EPG = -21%. Overall, this lowers the net EPG to +3%. Table 2 shows the results for the as-designed and as-built energy consumption and generation.

Table 2. Annual energy data from 1 April 2017 – 31 March 2018.

	Pre-certification (2015)	As-designed	In-use
Total energy use (kWh/yr)	734,424	735,125	779,789
Renewable generation (kWh/yr)	115,760	115,856[a]	140,671
Net energy use (kWh/yr)	618,664	619,269	639,118
Net energy use/m^2 (kWh/m^2/yr)[b]	79.3	79.4	81.9

[a] Given as percentage of total energy use (15.76%) based on historic performance.
[b] Energy Performance Index.

Figure 1 shows the case study building's energy performance index (EPI) in relation to relevant Indian benchmarks [25, 26]. The building is performing better than other green buildings and the Indian Energy Conservation Building Code (ECBC) office building benchmark for the hot & dry climate.

[1] The Building Use Studies (BUS) methodology obtains feedback data on building performance through a self-completion occupant questionnaire; the results can be compared against a national benchmark database. The questionnaire prompts the respondents to comment on the building's image and layout, comfort, and daily use of the building features.

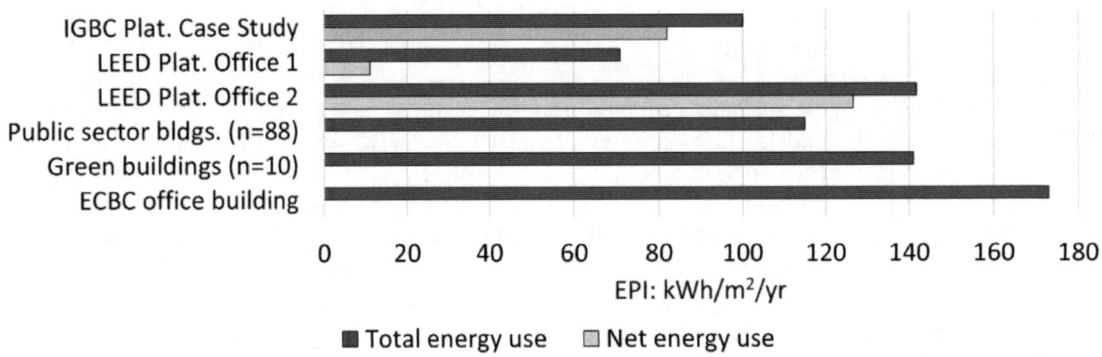

Figure 1. Annual energy data for building (1 April 2017 – 31 March 2018) with benchmarks.

There is a strong correlation (r = 0.68) between total energy consumption and total degree days [27]. Total energy consumption proportionally tracks degree days from April – October 2017. However, despite the low degree days for the months of November 17 – February 18, more energy is consumed in these months than would be expected. Part of this is attributed to the continuous ventilation and fixed windows in the building. The reported lack of handover provided to the current FM onsite could also contribute to higher than expected energy consumption, especially during the winter months. During the previous change in FM, no handover was provided to the current FM. This resulted in lack of awareness of the extent of BMS monitoring, including air handling unit temperature monitoring.

4.3. Indoor environment

The acceptable operative temperature range for offices in summer is 24.5°C +/- 2.5°C according to the ISHRAE standard [28]. Figure 2 shows the temperature measurements for four spaces during the monitoring period. As operative temperature was not observed, dry bulb temperature is used here as a proxy. Based on the temperature data alone, the comfort temperatures in the spaces of the building are well-managed though not consistent from floor to floor. The 4th floor office has the most occupied hours (80%) within 0.5 degrees of the target temperature. The ground floor is well within acceptable range but could experiment with scaling back on the cooling a little to reduce energy consumption. RH is not as well managed as temperature. The 4th floor office also has the smallest percentage of hours (1%) over 70% RH; however, the 1st floor open office has 60% of occupied hours over 70% RH. This should be reviewed and monitored as it could increase formaldehyde and microbial concentrations in the air, leading to health problems or allergic reactions.

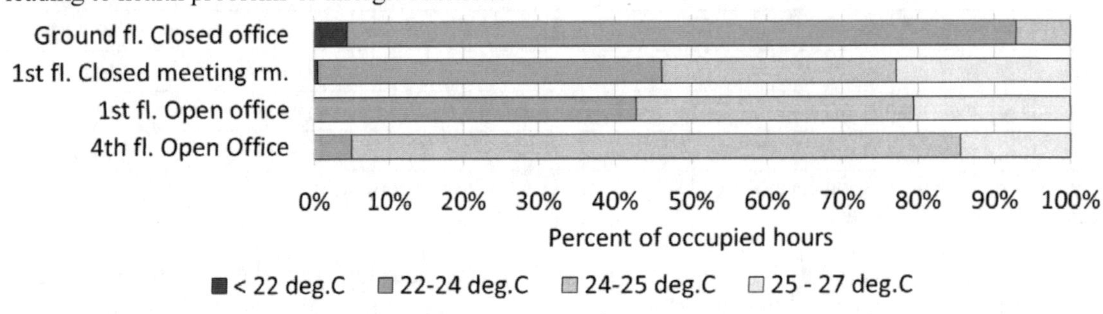

Figure 2. Temperatures (9 July – 3 August 2018).

Figure 3 shows the indoor air quality (IAQ) measurements taken in four offices during the period 10 July – 6 August 2018. Each office was measured for six days at a time, sequentially with the same device. *Note: 1FOC = first floor open centre office, 4FOB = fourth floor open back office, CF = closed front office.* ISHRAE (2016) thresholds are shown in the graphs for PM2.5 and CO_2. For these graphs the lower band indicates the Class A threshold: aspirational and the upper band indicates the Class C threshold: marginally acceptable. Though mean PM2.5 concentrations in the offices are below the ISHRAE Class A threshold, two offices have maximum values which have reached unacceptable levels. A total of 13% of occupied hours combined in these spaces had CO_2 concentrations between the *Class B: acceptable* and *Class C: marginally acceptable* levels. 3% of occupied hours were beyond *marginally acceptable*. Most of these hours were experienced at the end of the workday in the fourth-floor open office. This corresponds with the BUS findings where occupants rated the air quality as fresh; better than the BUS mean.

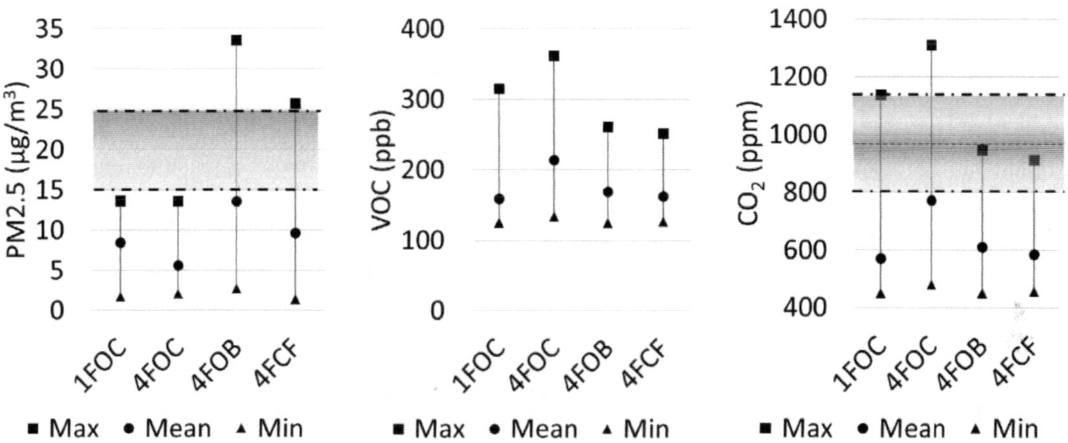

Figure 3. Office IAQ measurements and ISHRAE standard thresholds.

4.4. Occupant survey

According to BUS survey results, the design, the building's image to visitors, whether the building meets the user's needs and effectiveness of the use of the space were considered satisfactory. Overall lighting is considered satisfactory. In addition, the respondents considered there to be enough natural light on the scale toward 'too much'; however, there is little to no issue with glare in the building. Measurements showed higher than required levels of daylight with exception of ground floor and some deeper, central areas of open floor plan offices. In these areas the daylight factor was consistently below 2% due to internal partitions. Control over lighting is considered good, though no better than the BUS benchmark mean. Overall comfort is rated moderately high for both summer and winter. In both seasons the temperature is perceived to be on the cooler side, cooler in winter, and the air is considered neither too dry nor too humid, somewhat still, and fresh & odourless. Though comfort and temperatures are considered satisfactory, the respondents feel that they have little control over cooling or ventilation. Productivity at work is perceived to have increased because of the environmental conditions in the building and occupants feel healthier when in the building.

5. Discussion

Though there is generally a lack of procedure to ensure that the design intent of green buildings is realized, this certification process was special in that a year of pre-certification energy data were available with which to compare projection calculations. This was because the building was certified as an existing building. However, no performance evaluation follow-up was officially performed after the certification process. Though the building is performing worse than it did in 2015 there could be external environmental factors, occupancy shifts, etc. that contributed to this. This is a clear case of a Platinum rating that was not dependent on energy performance but a significant list of other green building features to qualify for certification. This raises two questions: (1) Is a 10% reduction on baseline sufficient to earn a platinum rating knowing that there is often an EPG? (2) As the building in-use is only performing at a 5% reduction on the baseline can it still be considered a Platinum rated building? Recommendations to improve performance include low cost options such as awareness campaigns

IOP Conf. Series: Earth and Environmental Science **329** (2019) 012028 doi:10.1088/1755-1315/329/1/012028

addressing energy-related use of the building among occupants and high cost options such as, consider operable windows and or trickle vents to reduce the need for mechanical ventilation in the winter months, mixed-mode ventilation and or night-purging. It was also recommended that the BMS system be commissioned on regular basis and that meetings are held to discuss the findings of the BMS system amongst facility and building management to improve performance.

The learnings from this study will be used to improve the I-BPE process. When assessing an existing building after much time has passed, lack of access to design intent data can occur. Knowing beforehand that these issues exist would require more time to survey the building and systems in order to develop an idea of the difference between as-designed and in-use. Another lesson learned is that monitoring devices that are dependent on Wi-Fi connection or outlet connection can suffer interruptions during power outages which happen often. This suggests the need to consider, as an option, battery driven logging equipment with built-in storage accessible by Bluetooth. Coordination with management in the building helped increase the return rate of BUS responses. Selling the concept to the building owner/management is essential to success. Delivering recommendations for improvement can help in this regard.

6. Conclusion

This study shows the process of testing the I-BPE methodology on a Platinum-certified green office building in the hot and dry climate of India. The field study was carried out for 30 days which included spot measurements, data monitoring, walkthroughs and occupant surveys. The field study offers a template for replication of BPE and benchmarking data for green residential buildings in India. The next step in the Learn-BPE project involves testing the I-BPE approach on several other case studies implemented by students using a programme developed for this purpose. The I-BPE case studies intend to demonstrate actual performance of certified green buildings in India, publish the data, and continually provide a testing platform for refinement of the I-BPE framework for application in India. Finally, the I-BPE case studies are also intended to build trust in the industry by strengthening the relationship between the industry professionals and researchers in academia.

Acknowledgements

The authors gratefully acknowledge the support of the Newton Fund through the Royal Academy of Engineering (RAEng) Industry Academia Partnership Programme for financially supporting the Learn-BPE project [Grant number IAPPI\74].

References

[1] IEA *India energy outlook* Paris: International Energy Agency; 2015.
[2] IGBC. Green building movement in india 2019 [Available from: https://igbc.in/igbc/redirectHtml.htm?redVal=showAboutusnosign.
[3] Sawyer L, de Wilde P and Turpin-Brooks S Energy performance and occupancy satisfaction: A comparison of two closely related buildings. *Facilities*. 2008 **26** 542-51.
[4] Sabapathy A, Ragavan SK, Vijendra M and Nataraja AG Energy efficiency benchmarks and the performance of leed rated buildings for information technology facilities in bangalore, india. *Energ. Buildings*. 2010 **42** 2206-12.
[5] Alborz N and Berardi U A post occupancy evaluation framework for leed certified us higher education residence halls. *Procedia Engineer.* 2015 **118** 19-27.
[6] Gupta R and Gregg M Empirical evaluation of the energy and environmental performance of a sustainably-designed but under-utilised institutional building in the uk. *Energ. Buildings*. 2016 **128** 68-80.
[7] Tham KW, Wargocki P and Tan YF Indoor environmental quality, occupant perception, prevalence of sick building syndrome symptoms, and sick leave in a green mark platinum rated versus a non-green mark-rated building: A case study. *Sci. Technol. Built. En.* 2015 **21** 35-44.

[8] Chen Q, Kleinman L and Dial A Energy performance of campus leed® buildings: Implications for green building and energy policy. *J. Green Build.* 2015 **10** 137-60.
[9] Menezes AC, Cripps A, Bouchlaghem D and Buswell R Predicted vs. Actual energy performance of non-domestic buildings: Using post-occupancy evaluation data to reduce the performance gap. *Appl. Energ.* 2012 **97** 355-64.
[10] Palmer J and Armitage P *Early findings from non-domestic projects*: Innovate UK; 2014.
[11] De Wilde P The gap between predicted and measured energy performance of buildings: A framework for investigation. *Automation in Construction*. 2014 **41** 40-9.
[12] Preiser WF The evolution of post-occupancy evaluation: Toward building performance and universal design evaluation. *Learning From Our Buildings A State-of-the Practice Summary of Post-Occupancy Evaluation*. 2001.
[13] Gupta R, Gregg M, Manu S, Vaidya P and Dixit M Customized performance evaluation approach for indian green buildings. *Build. Res. Inf.* 2019 **47** 56-74.
[14] Nicol F and Roaf S Post-occupancy evaluation and field studies of thermal comfort. *Build. Res. Inf.* 2005 **33** 338-46.
[15] Manu S, Shukla Y, Rawal R, Thomas LE and de Dear R Field studies of thermal comfort across multiple climate zones for the subcontinent: India model for adaptive comfort (imac). *Build. Environ.* 2016 **98** 55-70.
[16] Kumar S, Singh MK, Loftness V, Mathur J and Mathur S Thermal comfort assessment and characteristics of occupant's behaviour in naturally ventilated buildings in composite climate of india. *Energy Sustain. Dev.* 2016 **33** 108-21.
[17] Dhaka S and Mathur J Quantification of thermal adaptation in air-conditioned buildings of composite climate, india. *Build. Environ.* 2017 **112** 296-307.
[18] Thomas L Combating overheating: Mixed-mode conditioning for workplace comfort. *Build. Res. Inf.* 2017 **45** 176-94.
[19] Ford B, Patel N, Zaveri P and Hewitt M Cooling without air conditioning: The torrent research centre, ahmedabad, india. *Renewable energy*. 1998 **15** 177-82.
[20] Sharma M and Ali S Tropical summer index—a study of thermal comfort of indian subjects. *Build. Environ.* 1986 **21** 11-24.
[21] Indraganti M Thermal comfort in naturally ventilated apartments in summer: Findings from a field study in hyderabad, india. *Appl. Energ.* 2010 **87** 866-83.
[22] Bhanware P, Jaboyedoff P, Chetia S, Maithel S and Reddy B *Case study of an energy efficient commercial building: Validating deisgn intent and energy simulation results with monitored performance data*. Inspire 27 November - 1 December 2017 Jaipur, Rajasthan, India: Alliance for an Energy Efficient Economy; 2017. p. 34-43.
[23] Batra N, Gulati M, Singh A and Srivastava MB, editors. *It's different: Insights into home energy consumption in india*. Proceedings of the 5th ACM Workshop on Embedded Systems For Energy-Efficient Buildings; 2013: ACM.
[24] ARUP. Bus methodology 2014 [Available from: http://www.busmethodology.org.uk/.
[25] Kumar S, Kapoor R, Rawal R, Seth S and Walia A Developing an energy conservation building code implementation strategy in india. *Proceedings of the 2010 ACEEE Summer Study on Energy Efficiency in Buildings*. 2010 **8** 209-24.
[26] BEE. Energy benchmarks for commercial buildings. In: Efficiency BoE, editor. New Delhi: Bureau of Energy Efficiency; nd. p. 2.
[27] Degree days.Net - custom degree day data [Internet]. BizEE. 2019 [cited Feb 19, 2019]. Available from: https://www.degreedays.net/.
[28] ISHRAE. Indoor environmental quality standard. Indoor Environmental Quality: Indian Society of Heating, Refrigerating and Air Conditioning Engineers; 2016. p. 59.

IOP Conf. Series: Earth and Environmental Science **329** (2019) 012029 doi:10.1088/1755-1315/329/1/012029

Study of the Uncertainties Influencing the Energy Performance Assessments of Two Very Different Urban Houses in Wallonia

S Monfils[1], J-M Hauglustaine[1]

[1]Energy and Sustainable Development, Department of Environmental Sciences and Management, University of Liege, 185 Avenue de Longwy, 6700 Arlon, Belgium
Corresponding Author email: Stephane.monfils@uliege.be

Abstract. Implemented by the European EPB directives, the Energy Performance Certification has been designed by authorities to "provide clear information about the energy performance of a building". Studies led in Wallonia show however that official results are often too distant from reality and difficult to understand for the lay person. This paper focuses on the comparison between two certified old urban brick single-family houses in the Walloon residential stock: a highly efficiently renovated "exemplary building", described in detail, versus an inefficient and poorly described second house. After presentation of the case studies, this paper will investigate the uncertainties that surround the description of the "energy systems" and the standardized calculation method. This research explores the possibilities of elevating the confidence in the certification results by adding information on the description of the envelope and systems, by re-integrating the human factor in the equations (through a questionnaire collecting by interview the dwellers' energy-related behaviours and practices), and by using dynamic simulations. The comparison of the results to the official standardised results indicates that additional data on the household's composition, practices and energy-related behaviours are as much needed as a more accurate description of the energy system, which is penalized by default values and a severe standardisation of the calculation method.

1. Introduction

The EPC (Energy Performance Certificate) has been designed by authorities to "provide clear information about the energy performance of a building", in order to influence the real-estate market by bringing the general public to integrate energy into their dwelling decision-making processes, promote energy performance improvements and help build up comprehensive benchmarking databases, fundamental for shaping strategies on a local or regional level.

The certification procedure has been imposed by the 2002/91/CE [1] and 2010/31/UE [2] European Directives, and the definition of the calculation method was left to the member states (or regions, in the case of Belgium). The Walloon regulatory calculation method is described in [3] and had to be adapted in its Annex D [4] for the assessment of existing residential buildings. The EPC regulatory calculation method is standardized in the honourable and understandable objective to ease the comparison between buildings. This leads to the use of a tight and strict protocol in the certification procedure that regulates the acceptable proofs of performance, and delivers many default values to fill the holes.

The general picture, nearly ten years later, shows results that are often distant from reality. The Table 1 hereunder gives an overview of average gaps between EPCs and actual energy performance of buildings, evaluated by [11].

IOP Conf. Series: Earth and Environmental Science **329** (2019) 012029 doi:10.1088/1755-1315/329/1/012029

Table 1. Conversion of consumption per EPB scale level.

EPB scale level	Consumption ratio : real / theoretical
A++ or A	100%
B	82.1%
C	75.1%
D	66.6%
E	58.2%
F	54.3%
G	42.5%

It seems also that EPC results are too difficult to understand by the lay person. By expressing performance in kWh of "primary energy", it bypasses the fact that consumers' only knowledge of their energy consumption are the final energy bills, expressed in money terms. It is believed that the input data used in the quasi-steady state calculation method used in Wallonia could be used to display more understandable complementary results. Additional data on the household's composition, their practices and behaviours related to energy consumption, as well as a more accurate search in the description of the energy system, could help close the gap between real and theoretical consumptions, allowing future owners to better understand and appropriate the EPC results, foresee a rough monthly energy bill, and make better decisions for their real-estate renovation ambitions.

2. Uncertainties

The uncertainties that appear in the Walloon EPC procedure can be distinguished between the ones related to the procedure (the protocol that brings simplifications and shortcuts by design) and those related to the standardization of the calculation method. Only the second kind will be of interest in this study, resulting from the use, in Wallonia, of a standardized consumption calculation method (asset rating). Standardization usually means that the targeted parameters cannot be replaced by other values, even if they are known with precision. This is the case of several default values used in the Walloon calculation method, such as the definition of the set temperature (constant all year long at 18°C) or the heated volume (equal to 100% of the volume, all year long). The standardization aims to take the human factor out of the equations, by using objective parameters and statistical values as a replacement (the protected volume (V_p) is used with this objective in the Walloon method). It may objectify the assessments, but denies all known influences on energy consumption from the building state and condition, its age and typologies characteristics, its dwellers and their characteristics.

A list of accepted sources of accurate data in the dwelling description ("acceptable proofs") has been developed to impose a rigid EPC assessment method, leaving few liberties to the assessors in the process. Default values also dominate the methodology, when no acceptable proof of a more accurate value is available. This research proves that, of all those default values, those describing the envelope are most influential on the results, as default values can rule on the presence of insulation, depending on the age of the construction or the renovation, impose thicknesses and thermal properties of all materials, including the insulation layers.

3. Methodology

This research is based on the manipulation of the certification method for two case studies, in order to try and bring the results closer to their real consumption results. This implies the need to:

- Describe, as accurately as possible, those two houses and their energy systems;
- Interview their owners in order to gain additional data on the house, their data of real energy consumption, their habits and behaviours that are believed to be needed in order to de-standardize the calculation method;

- Re-write some parts of the calculation method, and integrate this "new method" (tested several times in order to insure the absence of errors) in an Excel spreadsheet in order to obtain the new de-standardized results;

Those two case studies are extracted from a larger group of 16 houses studied in [6]. The goal of that research was to explore the qualitative requirements to close the gap between real and theoretical EPC consumptions. The objective was not statistical: where individual buildings are concerned, EPC values are bound to vary from actual consumption values, depending on a wide range of factors, relating to the buildings "energy systems" themselves, weather patterns and behaviours, which are among the standardized uncertainties studied in these example. Although the EPC, as a regulatory scheme, might not seem the right tool to evaluate actual performance, one of the constraints of the research is to prove the potentialities of the EPC scheme without the need for too deep a change in the calculation method. As said above, it is believed that the input data used in the quasi-steady state calculation method used in Wallonia could be used to display more understandable complementary results.

3.1. *Case studies*

Both houses studied in this paper date from the 1950's, and are located in the suburbs of Liege, in Wallonia, Belgium. Their main characteristics are displayed in Table 2 hereunder.

Table 2. Overview of the main energy characteristics of both case studies.

	Case study 1	**Case study 2**
Number of inhabitants *(from interview)*	2 young adults + 1 child	2 seniors
Heated floor area A_{ch} [m²] *(from EPC)*	261	153
Protected volume V_p [m³] *(from EPC)*	812	476
Total transmission loss area A_T [m²] *(from EPC)*	584	350
Global envelope thermal transmittance U_m [W/m²K]	0.17 *(obtained from file of acceptable proofs)*	1.86 *(mainly obtained by default values)*
Global coefficient for heat losses by transmission $H_{T,heat}$ [W/K]	99.56	652.73
In/exfiltration rate at 50Pa v_{50} [m³/h.m²]	1.6 *(measured)*	12 *(default value)*
Global heating installation efficiency [%]	77 *(acceptable proof)*	67 *(default value)*
Global coefficient for heat losses by air renewal $H_{T,heat}$ [W/K] *(from EPC)*	49.79	152.54
Theoretical annual consumption for heating [kWh/year] *(from EPC)*	3,906 [wood pellets]	65,637 [Natural Gas]
Heated spaces during daytime *(obtained from interview with owners)*	Living room, dining room, kitchen, bedrooms, "other" ($f_{Vp,hs}$ = 78.8%)	Living room, dining room, kitchen, bedrooms, circulations ($f_{Vp,hs}$ = 64.9%)
T_{set} in heated rooms during daytime *(obtained from interview with owners)*	20°C downstairs, 18°C upstairs (average 19.3°C)	22.6°C downstairs, 20°C upstairs, 24°C in the bathroom when in use (average 21.6°C)
Heated rooms during night time *(obtained from interview with owners)*	Main bedroom ($f_{Vp,hs}$ = 14.8%)	Living room, dining room, kitchen, bedrooms, circulations ($f_{Vp,hs}$ = 64.9%)
T_{set} in heated rooms during night time *(obtained from interview with owners)*	16°C	18°C

The first house has been renovated in 2016-2017 to reach a highly efficient standard (it is now labelled A+ on the official certification scale). The whole existing envelope has been covered with thick layers of thermal insulation that allowed to reach better thermal transmittance than those asked today for new buildings in Wallonia. It has been equipped with highly efficient systems, using wood pellets for its production of heat and Domestic Hot Water (DHW), and a completely mechanical ventilation system with heat recovery. There was, for this house, a complete file of acceptable proofs in order to use the most accurate values everywhere possible, from the U-values of the different walls, to the in/exfiltration rate and systems efficiencies.

The second house shows far less impressive characteristics, its envelope seems to have stayed in its original condition (without any insulation layers). It is equipped with a natural gas boiler for the production of heat, an electric boiler for DHW production, and no ventilation system whatsoever. The owner of this house could not mention any presence of insulation in the house, nor could she give any acceptable proof to feed the description. No insulation was visible, suspected, or found in the walls of the house (without dismantling or destructive tests), so that default values govern most of its description in the software.

This paper will focus entirely on the heating consumptions. The owners of the first case study have followed closely the settings and parametrization of the systems in order to maximize their efficiencies and minimize their consumptions. Besides showing validation that there is an element of fun, interest and intellectual stimulation and real energy saving potential in a technology that can be appropriated by the owner [8], this behaviour allowed them to separate their wood pellets consumptions between DHW and heating end-uses, and detail their solar production. In the second case study, heating is the only end-use consuming natural gas; all other equipment, including cooking appliances and DHW installations, use electricity.

The revaluation of energy consumptions related to heating will mainly focus on the revaluation of net heat demands, through redefinitions of set temperatures and heated volumes, and revaluation of internal loads and heat losses.

3.2. De-standardization

The consumption of energy for space heating is calculated using a single zone, quasi-steady-state, monthly calculation method. This imposes to keep a monthly approach on the evaluation of heating demands and consumptions, implying the use of average monthly values to define each parameter of the calculation, such as the climate, set temperature, heat losses by transmission and air renewal, internal and solar gains… It is the uncertainty on these monthly values that is questioned in this research. While leaning on the existing calculation method, there is a clear need to inquire about the heating patterns and energy consuming habits of the households, in order to allow a revaluation of the net heat demand. It appears possible, therefore, to discretize inputs and so replace a big uncertainty on monthly averages inputs by a string of smaller uncertainties on periods of time that are better described in occupancy and heating. This method remains subjected to averages and default values, on which the uncertainty should be lowered, either by a better description of the building, of the household, or their energy consuming habits.

In order to gain additional data on inhabitants' behaviour and energy consumption habits, a questionnaire has been built, submitted by the author to the respondents during in-depth, open-ended interview in which questions are worked into a conversational flow. It also permitted witnessing some habits and behaviours that were not considered or not verbalized by respondents. Apart from a selection of additional data on the building and its users (socio-demographic variables), the questionnaire mainly asks the owners to specify their heating habits (days and hours of heating, set temperatures, presence of temperature controls…) for the six main spaces listed above; their ventilation habits in four of them (living room, kitchen, bedrooms and bathrooms); a list of "rational use of energy behaviours"; a list of appliances and equipment to re-evaluate the internal gains; and, of course, their real consumption data (with the corresponding periods, in order to adjust the climatic data accordingly).

The discretizing process consists in splitting monthly lengths into sub-periods of time, depending on the owners' answers to the questionnaire. It appeared that each day could be divided into eight periods: "sleep", "wake up and get ready", "morning", "noon", "afternoon (school schedule)", "afternoon (rest)", "evening", and "ready for bed and turn off". The second and last one of them are mainly devoted to the periods of time the bathrooms are used (and often heated more than usual).

The definitive calculation method keeps the single-zone model and its important parameters, such as the protected volume (V_p), heated floor area (A_{ch}), total heat loss area (A_T), coefficient of heat losses by transmission ($H_{T,heat}$), etc. Contrary to the regulatory calculation method hypotheses that consider the whole protected volume as heated, all year round, at the same set temperature of 18°C, this method needs recalculating the global temperature of the building for each of those periods, depending on heating patterns that have been created based on the owners' interviews. It uses the discretization only to reassess the global internal temperature that translates the influence of the proportion of V_p that is directly heated and the comfort temperatures demanded in those parts. As a consequence, Net Heat Demands (NHDs) have to be revaluated for the whole dwelling, heated and unheated spaces alike.

For that purpose, the mono-zone model needs a hypothesis on the temperature of unheated (or indirectly heated) spaces in order to estimate the global average internal temperature of the dwelling. At first, this temperature in unheated spaces is defined thus in this research:

$$T_{uhs,m} = (T_{set,hs} + \Delta T_{tset}) - \left[\left(\frac{((T_{set,hs} + \Delta T_{tset}) - \theta_{e,m})}{2} \right) \times (1 - f_{Vp,hs}) \times f_{\Delta T,uhs} \times f_{pct} \right] \quad (1)$$

where:

- $T_{uhs,m}$ = monthly average temperature in unheated spaces during heating periods [°C];
- $T_{set,hs}$ = set temperature in heated spaces [°C];
- ΔT_{tset} = positive or negative increment in set temperature according to sensitivity to cold [°C] (data extracted from the questionnaire);
- $\theta_{e,m}$ = monthly average exterior temperature [°C];
- $f_{Vp,hs}$ = ratio of the heated spaces' volumes on the total protected volume, during the heating period considered.
- $f_{\Delta T,uhs}$ = empirical factor (0.8 to 1.2), depends on the tendency to close doors between them (data extracted from the questionnaire).
- f_{pct} = multiplicative factor considering the quality of the envelope as influence on temperature homogenization, equal to the sum of transmission and air change heat loss coefficients, divided by the protected volume [W/m³.K].

The average periodic temperature $T_{set,i}$, and the resulting global monthly internal temperature $T_{set,m}$ can then be defined thusly:

$$T_{set,i} = T_{set,hs} \times f_{Vp,hs} + T_{uhs} \times (1 - f_{Vp,hs}) \quad (2)$$

$$T_{set,m} = \sum_{i=1}^{i=8} \frac{(T_{set,i} * t_{m,heat,i})}{t_{m,heat}} \quad (3)$$

The software developed for the regulatory EPC assessment of existing dwellings, PACE (for "Procédure d'Avis et de Certification Energétique" – "Energy Certification and Audit Procedure"), does not allow alterations on the calculation method. The software used in this research, therefore, is an Excel spreadsheet in which the method could be implemented and modified. A series of input parameters were gathered from the PACE software then directly introduced in the spreadsheet, such as the average thermal transmission coefficient (U_m), the solar gains; or the efficiencies for the technical systems. Tests have been run, between the PACE software and the spreadsheet to ensure the correspondence of results, before implementing any other modification made necessary by the objectives of this research and integrate the (coded) answers to the questionnaire.

4. Results

These results are very convincingly close to reality, as the revaluations made under real climatic data hypotheses only show a 0.5% gap with the real consumption data given by the owners. Revaluations under the average climatic data are overestimated by 52.5%, which emphasizes the need to use adequate climatic data in order to reach accurate results. The average climate used in the EPB regulatory method is based on the data registered in Uccle (near Brussels) during the last 50 years.

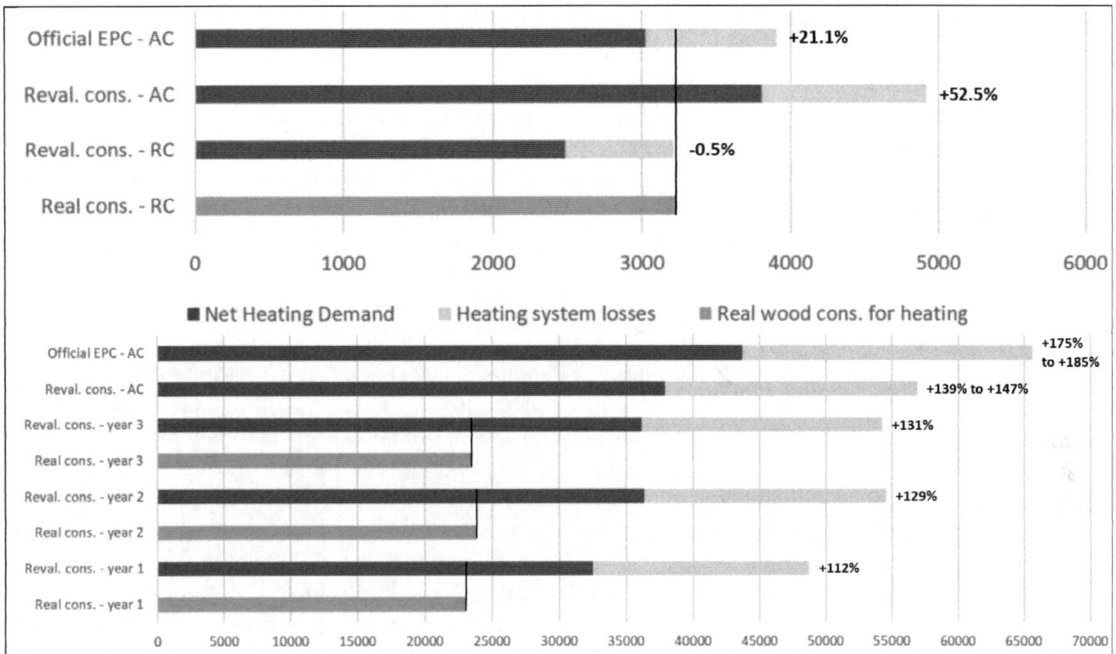

Figure 1. Results in final energy consumptions (kWh/year for heating) for (a) the first case study and (b) the second case study. "Reval." = Revaluations; RC = Real Climate (obtained from local weather stations); AC = Average Climate (Belgian average used in regulatory EPC calculations).

In the second case study, there is a discrepancy between the revaluations and the real consumption data that cannot be wiped away by the de-standardization proposed here. The theoretical EPC results (65,637kWh of natural gas final consumption) were estimated 2.8 times higher than the average real consumption data (23,427kWh). Revaluations lowered these theoretical consumptions, which remained as high as twice the real data. It is believed that the parametrization of the calculation method cannot be the only questionable determiner of this discrepancy.

The ratio between natural gas consumptions revaluations and real data shows a wider range of variation when using the real climatic data, which means that the real consumptions data vary less, from one year to another (standard deviation of 421kWh), than the evaluations that were made using the real climatic data (standard deviation of 3,290kWh). This would incriminate the characterization of the energy system in this case: should its envelope and heating installation be more accurately described, revaluations results would better follow the dynamics of real data, and be less influenced by climate. Looking closely, the first year's revaluations are "particularly low" (+112%), whereas the second and third year are both overestimated by 130%. A behavioural change in habits at that time, or in the energy system, might bring a better explanation; unfortunately, this kind of dynamics is not possible in a "one-shot" interview and description of the heating habits.

To try and bring those results closer to real consumption data, several additional results were sought:

First, the presence of an un-witnessed presence of thermal insulation could partially explain the gap. A hypothetical insulation layer of 6cm of mineral wool in the roof brings the U_m to 1.49W/m^{2}K, and lowers the theoretical consumptions by an average 12,600kWh under average climate, 11,527kWh under real climate (10,765 to 12,327kWh, depending on the year). The standard deviation mentioned above slightly decreases to 3,158kWh, but this improvement does not allow to better explain the consumption dynamics on the period covered, nor does it allow to close the "prebound" gap. At their lowest, these consumption revaluations are still on average 72.6% higher than the real consumption data. The hypothetical insulation layer in the roof could be thicker than 6cm, but this hypothesis has a limit, both in credibility and on the total heat losses by transmission, as 480W/K of them are attributed to the other uninsulated walls and windows.

Then in addition to a permeability test (in order to get a more accurate infiltration air rate, which proved to be around 9 $m^3/h.m^2$ under a 50 PA pressure difference), a monitoring campaign of this case study's internal temperatures has been realized between February 21st, 2018 and March 4th, 2018, a cold period in Wallonia. Sensors have been placed in exterior (protected from wind and sun influences, to monitor $\theta_{e,m}$ in Eq. 1); in heated spaces (living room, where the thermostat is placed, kitchen, bathroom and two bedrooms, including the main one) and in unheated spaces inside the Vp (those which temperature is revaluated by Eq 1: the garage and veranda). This monitoring campaign allowed to draw the following conclusions:

- The heating pattern was no so far off the reality. The bathroom temperature often rises above 27°C, which is more than the model hypothesis of 24°C but can be explained by the running of hot water in such a small space. During [10AM – 10PM] "daytime" periods, the monitored average temperature is 22.08°C in daytime spaces (lower than the 22.6°C used as input in the model), and 20.82°C in night-time spaces (higher than the 20°C used as hypothesis). This is an example of compensation, where overestimated consumptions for one end-use offset another end-use's consumption underestimation. Furthermore, the heating system needs more than two hours to raise the living room temperature, which never fell under 18°C, above 22°C. This sheds another light on the calculation method, which considers the heated share of the volume to reach the set temperature within seconds. Old, uninsulated, thermally inert and globally inefficient homes need time for the installation to heat the dwelling up, whereas efficient buildings reach comfortable temperature quicker, which might partly explain their more accurate results so far. The adjustment of the heating pattern to these monitored data results saw a reduction of the total final energy consumption of around 2 to 3%, which is mainly attributed to the first "start of the heating system" period.
- There does not seem to be any indication of a night time heating, despite the declared setting. The thermal inertia of the house, and the high temperature to which it is constantly heated daily, allow a slow reduction of the internal temperature during the cold nights of the campaigns without resorting to active heating. The choice was made to keep the night time settings in the calculation method, hoping that it would come to the same conclusions by itself.
- The veranda is central in the definition of the V_p. Its temperature curve shows that it should probably be excluded from it, which would be quite in accordance with the owner's use of the space (they only use it in the summer, never when the heating system is on) but had to be included when considering the certification protocol. The garage would then become the only unheated space in the V_p, and the Eq 1 then becomes accurate enough to define its theoretical temperature. However, the exclusion of the veranda from the Vp means some important changes in simulations parameters (V_p, A_{ch}, A_T, U_m, $H_{T,heat}$, $f_{Vp,hs}$). The reduction in solar gains is also important, from 14,798 annual kWh in the initial EPC, to 4,640kWh/year after the exclusion of the veranda's windows from the V_p envelope. The heat gains were probably overestimated before this change, as the owners keep the access to the veranda shut during winter times and do not use the solar gains that are captured by the veranda's glazed areas but for the small share that reaches the living room. Now that the veranda is out, the calculation method cancels any solar gains that could be considered coming through the 3.6m^2 of window that separate the living room from the veranda. This scenario delivers more surprising revaluated results, thanks to the diverse consequences on the results that this scenario brings: there is a compensation between effects, as the revaluated results of final energy consumption do not change much on average.

IOP Conf. Series: Earth and Environmental Science **329** (2019) 012029 doi:10.1088/1755-1315/329/1/012029

5. Discussion

Multi-zone dynamic calculations could perhaps render more precise (and closer) results, but this study, in order to evaluate the potential of the existing protocol and assessment method, is purposefully based on the existing steady-state regulatory calculation method. "All models are wrong, but some are useful" [9]. Dynamic simulations are also expected to ask for a more precise input data collection, in order to deliver a precision in results that is not really sought in the EPC procedure. There are, however, two major parameters of the dynamic method that could perhaps be wisely integrated into the EPC calculation method: the available power of the systems, and the thermal inertia.

The collection of accurate acceptable data is not the assessor's job, but the applicant's, which implies that all assessors should start their EPC with the same information, use the same default values when there are no information, and therefore reach similar results. When no such data exists, or is available, the method proposes many default values. They allow a simplification of the protocol, so that the certification procedure can remain affordable for the owner, and worthwhile for the assessor. It is an admitted principle in EPB procedures in Wallonia, that the use of default values should be disadvantageous for the results. "Choosing more realistic default values can help bridging the prediction gap. However, this would oppose itself to the role of conservative default values, namely to admonish building teams to perform better and to verify their results, by rewarding these efforts through better energy labels based on measured values." Additionally, "more positive default values could even result in some kind of impunity for those buildings that really do perform badly." [7]

Last but not least, the prediction of energy bills is, by essence, uncertain, and in that regard, behaviour can be the main determiner of annual variations. As the refinement of calculation parameters progresses, the influence of the remaining pool of unknown (or default) parameters increases. Exact correspondence between real consumption data and theoretical revaluations is limited by the high number of uncertainty parameters in the method to control. These revaluations only considered the "habitual consumption resulting from routine conscious and unconscious management", one of the three dimensions of consumption patterning identified by [10] (the other two being the "structural consumption that occurs when the building is unoccupied", and the "daily variation consumption resulting from unusual events").

6. Conclusions

This research has been devoted to explain past energy consumptions by increasing the accuracy of their determiners in the calculation method. The objective is clearly not statistical, given the low numbers of case studies in this paper (2) or in the thesis [6] from which they are extracted (16). With the necessary understanding that the EPC calculation method needs to remain standardized in order to allow the comparison of buildings, this research mainly aimed to answer to the question "could it also give more accurate complementary predictions in energy consumption, with a number of additional data?"

It seems possible to predict with an interesting accuracy margin, the final energy consumptions of a household which profile is known, in a house which is accurately described. It is quite clear, when comparing those two case studies' results, that the calculation method is more accurate for new buildings than existing ones; or rather, for insulated buildings which owners can prove their information, than for inefficient or undocumented buildings. The EPC calculation method was designed on the basis of the Walloon EPB regulatory calculation method, which was initially developed for new and efficient buildings. Its standardization, considering 100% of the V_p heated at 18°C, is closer to the reality of the first case study.

This research (along with other case studies presented in [6]) permitted to highlight the inaccuracy of the envelope description as most likely source of that remaining gap. The next challenge, therefore, concerns the possibilities to refine the set of default values, without increasing the cost of data collection, bearing in mind that building documentation is not available for the majority of the building stock in need of renovation. A more thorough analysis of the existing Walloon dwelling stock could lead to a matrix of more accurate default values based on typologies, to replace the actual list of values that characterize the whole building stock.

Although the standardized view on the stock's energy performance is much needed, this research stresses the importance of an increased accuracy in the energy system's description, and the need for nuance in the results given by energy audits. One way to achieve that could be based on statistics and added information, in order to deliver more accurate results in final energy consumption, and the integration in the calculation method, of behavioural determiners of energy consumption.

References

[1] European Parliament and Council, 2002. Directive 2002/91/CE Approved the 16th of December 2002, about Energy Performance of Buildings. 2002. Official Journal of the European Communities 4.1.2003 L 1/65 Available online: http://eur-lex.europa.eu/eli/dir/2002/91/oj (accessed on November 9th, 2016)

[2] European Parliament and Council, 2010. Directive 2010/31/UE Approved the 19th of May 2010, on energy performance of buildings (recast). 2010. Official Journal of the European Union 18.6.2010 L 153/13 Available online: http://data.europa.eu/eli/dir/2010/31/oj (accessed on April 10th, 2019)

[3] Service Public de Wallonie. 2014. Arrêté du Gouvernement Wallon du 30 Juillet 2014 Portant Exécution du Décret du 28 Novembre 2013 Relatif à la Performance Énergétique des Bâtiments; Moniteur Belge: Namur, Belgium; pp. 56172–56294.

[4] Service Public de Wallonie, 2014. Annexe D: Méthode de Détermination de la Consommation Spécifique des Bâtiments Résidentiels dans le Cadre de la Certification PEB; moniteur Belge: Namur, Belgium; pp. 61636–61767.

[5] M. Delghust, W. Roelens, T. Tanghe, Y. De Weerdt, A. Janssens, 2015. Regulatory energy calculations versus real energy use in high-performance houses, Building Research & Information, 43:6, 675-690

[6] Monfils, S., 2018. The human factor in the energy performance assessments for renovation strategies of existing houses in Wallonia, PhD thesis, University of Liege, Belgium

[7] M. Deurinck, D. Saelens, S. Roels, 2012. Assessment of the physical part of the temperature takeback for residential retrofits, Elsevier Energy and Buildings 52: 112-121

[8] N. Kalampalikis, M. W. Bauer, T. Apostolidis, 2013. International review of social psychology, Presses universitaires de Grenoble, France, 232 p.

[9] G. E.P. Box, 1979. Robustness in the strategy of scientific model building, p.202 of Robustness in Statistics, R.L. Launer and G.N. Wilkinson, Editors.

[10] M. J. III Bernard, J. R. McBride, D. J. Desmond, N. E. Collins, 1988. Events-The third variable in daily household energy consumption. Proc. Am. Cour/c. Energy Effie. Eeon., pp. 11.11-14. Washington, DC: ACEEE Press

[11] Secrétariat général du Service public de Wallonie, Département du Développement durable, 2010. Etude « Alliance Emploi Environnement » : Proposition de modalités opérationnelles et affectation des moyens budgétaires aux mesures envisagées par le groupe de travail financement, Namur, Belgium, p. 2 and 8.

IOP Conf. Series: Earth and Environmental Science **329** (2019) 012030 doi:10.1088/1755-1315/329/1/012030

Evaluating Incidental Thermal Performance Improvements of a Historic Timber-Framed Building in Central Hereford

Christopher J Whitman[1]
[1]Welsh School of Architecture, Cardiff University, Cardiff, Wales, CF10 3NB, UK.

WhitmanCJ@Cardiff.ac.uk

Abstract. As we strive to decarbonise our historic built environment, a balance must be struck between technical, aesthetic, philosophical and economic concerns. Compromises must be made and opportunities seized. At the same time, evaluation allows us to assess decisions previously made and reflect on their outcomes. This paper presents the evaluation of the refurbishment of The Old Mayor's Parlour, Church Street, Hereford, a historic timber-framed building now used as a gallery and exhibition space. The conservation work undertaken was not specifically envisioned as an energy retrofit, however the necessary replacement of failing concrete block infill, the legacy of a 1970s renovation, allowed improvements to be made to the thermal performance of the external envelope. Environmental monitoring and digital simulation have been used to assess the impact of these interventions. In situ U-value measurements show the success of the replacement infill panels and associated internal lining, although digital energy simulations suggest a limited improvement to the building's overall energy efficiency. At the same time thermography suggests a potential threat of increased condensation risk to the uninsulated ornate 17th century plaster ceiling. The results of this paper show the risk of unintended consequences and the challenges faced by sustainable building conservation.

1. Introduction

The retrofit of our existing building stock has been identified as a key factor in achieving international goals for mitigating climate change [1]. This has been reflected in policy by both the EU [2] and the UK governments [3]. Although in the UK historic buildings and those of a traditional construction are not required to comply with fully comply with the energy efficiency requirements of the building regulations [4, 5], they must still aim to "improve energy efficiency as far as is reasonably practicable" [6]. At the same time, building owners and occupants wish to improve the thermal comfort of their properties and reduce heating bills. As such, both the extent and detail of any retrofit remains at the discretion of the building owner. Whilst it is hoped that they will seek advice from qualified professionals, the lack of knowledge in the construction industry with regard to energy retrofit in general [7], and especially related to historic and traditional buildings [8], combined with a reduction in historic environment specialist within local authorities [9], means that too often they do not. By undertaking the evaluation of those buildings where retrofits have been undertaken, we can assess their success and apply those lessons learnt to future projects. Research into the energy retrofit of historic buildings in the UK has to date focused on solid masonry construction [10-12], with little research covering the 68,000 timber-framed buildings that form an integral part of the UK and specifically England's cultural identity[13]. This paper therefore begins to explore this previously under-researched area with the evaluation of one such building.

2. Case Study

2.1. Introduction.

The Old Mayor's Parlour (OMP), 24 Church Street, Hereford is a gallery space owned by the Church Street Charitable Trust, together with the adjacent property, 25 Church Street (Figure 1). This building was selected as a case study as part of a larger research project [13] due to the use of a replacement infill panel detail suggested by Historic England in their "Practical Building Conservation" series [14] a popular reference source for conservation professionals.

Figure 1. Nos.24 and 25 Church Street, Hereford. Source: (Author's own, 2016)

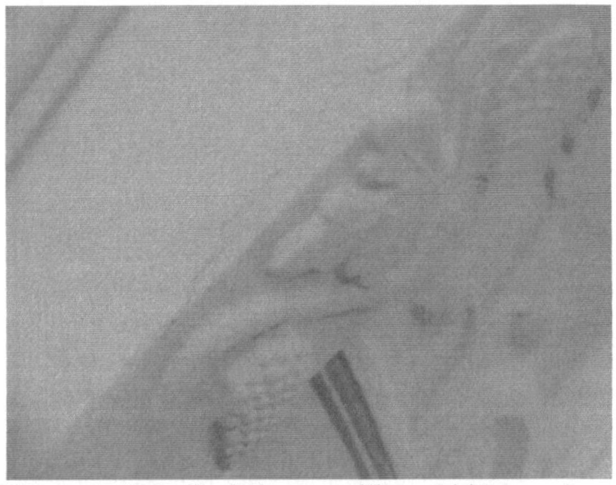

Figure 2. Detail of plaster ceiling, Old Mayor's Parlour. Source: (Author's own, 2016)

The building is divided into three separate entities; The OMP on the first floor of 24 Church Street (the right-hand gable in Figure 1), accessed via a staircase within 23 Church Street; "Rocket" café on the ground floor of no.24; and "Layers" women's clothes store at no.25, with a sales area on the ground floor and storage and office on the first. The interconnecting door between the OMP and the Layers' storage area was locked shut at all times and was sealed during pressure testing.

2.2. History

According to the designation description, the buildings date from the early 17th century [15], however, other sources state that the buildings origins are 14th century [16, 17]. The original description by The Royal Commission on Historical Monuments records the building as probably built early in the 16th century but with a stone-built cellar under the north part of the building, containing 15th century doorways [18]. The east façade onto Church Street is timber-framed at first floor with underbuilding and 20th century shopfronts. It is claimed that it once was used by the *Custos Rotulorum*, the keeper of the rolls, and the Vicars Choral (the men of the Cathedral Choir) [17]. Its most notable feature is the ornate early 17th century plaster ceiling to the gallery space (Figure **2**) and a fresco of what is believed to be Hereford Castle [17]. The building was saved from demolition in 1969 with help from Ivor Bulmer and the Ancient Monuments Society and had recently been refurbished [16, 19].

2.3. Retrofit

The first phase of the most recent refurbishment work consisted of internally lining the east first floor façade of the OMP with polyisocyanurate (PIR) insulation and wood wool boards, finished in gypsum plaster. During this first phase of work it was discovered that the infill to the timber-frame was very loose modern concrete blockwork, rendered with a cement lime render externally and gypsum plaster internally [19]. It was therefore decided that this would require replacement during a second phase of work. When this took place, the aforementioned concrete block infill was removed from the timber-framed east façade of both the OMP (no.24) and no. 25 and replaced with wood-fibre insulation based on the previously mentioned detail published by Historic England [14] (Figure 3).

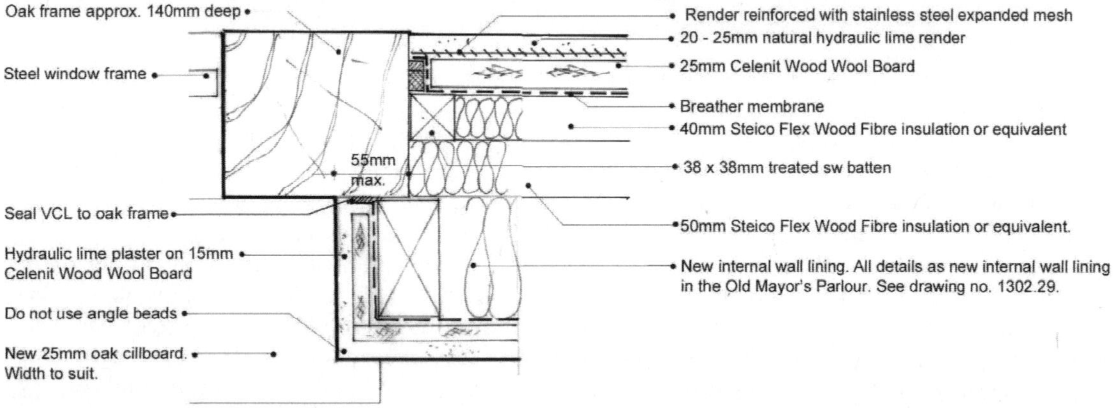

Figure 3. OMP replacement panel infill detail with internal lining. Source: (Project Architect, 2014)

At the same time an internal lining to the first floor of no.25 was inserted using the same detail as that used for the OMP but with mineral wool insulation in the place of PIR [19]. Whilst installing the replacement panel infill detail, the conservation contractor found the wood fibre insulation difficult to use due to its friable nature and the challenge of accurately cutting a board material to fit the irregular timber-frame [20]. In addition to these technical issues, in the opinion of the author, the regularity of the finish created by the board substrate lacks the character of panels with a wattle or oak lath background.

3. Methodology

In order to evaluate the performance of the retrofit actions, the following monitoring was undertaken; in situ U-value measurements, thermography, pressure testing and monitoring of internal hygrothermal comfort conditions. Measured data was then used to perform digital energy demand simulation. There follows a brief description of the methodologies employed. Where possible the relevant British and International standards were followed and best practice guidelines consulted in order to maximise the validity of the data collected.

3.1. In Situ U-value.

Two infill panels on the retrofitted east facade were monitored following BS ISO 9869-1:2014 [21], one in the OMP and the other in the first floor office of no.25. Huxeflux HFP01 heat flow plates were held by pressure against the wall surface with building props and flexible plastic clips. Petroleum jelly and plastic film was used to ensure a continuous contact between plates and wall. The outputs from the plates were connected directly to an Eltek® Squirrel® data logger with the voltage recorded at 5 minute intervals. The internal and external ambient dry bulb air temperatures (˚C) were measured with thermistors also wired directly back to the datalogger, with readings at the same frequency. The external temperature thermistor was protected from direct solar radiation by a ventilated, plastic and foil cover. As per the BS, only data collected one hour after the surface had passed into shade was utilised. The in situ U-value monitoring was undertaken between 18/02/2016 and 10/03/2016, with a measurement period of 21 consecutive days.

3.1.1. Assumptions and limitations. The panels can be classified as "quasi-homogenous" according to the BS's definition [21], however overall the facade is heterogeneous with the timber-frame forming a thermal bridge. The BS states that heat flux sensors "shall not be installed in the vicinity of thermal bridges, cracks and similar sources of error" [21]. No distance is specified, however, the surrounding timber-frame constitutes such a source of error. This is a problem common to all timber-frame infill panels in general, therefore direct comparison between panels can be made, however care should be taken in comparing these results with those of solid wall constructions [10, 22, 23], without timber-frames and the associated errors.

3.2. Pressure Testing.

The air permeability index ($m^3/hr/m^2$) and air change rate (/hr) were measured following BS EN ISO 9972:2015 [24]. All intentional openings in the building envelope were sealed and all doors and windows closed. A Minneapolis Blower Door, was used to depressurise the building, with building pressure and the fan pressure measured using magnehelic analogue pressure gauges. The gauges were zeroed prior to commencing depressurisation. Measurements were taken at regular intervals both up until a >50Pa pressure difference had been achieved and as pressure returned to normal.

Pressure testing took place on 11/03/2016. Due to the compartmentalised configuration of the property, only the OMP, located on the first floor, was pressure tested. The Minneapolis® blower door was inserted in the doorway between the semi-external staircase and the gallery. The tube measuring external air pressure was extended down and out onto the street. It cannot be assumed that the dividing floor been gallery and the cafe below has been designed to act as an air barrier. Additional leakage through this element must therefore be considered when reviewing the results.

3.3. Thermography

Given that no British Standard exists for infra-red thermography of buildings, the Building Research Establishment (BRE) [25] and Historic Environment Scotland's (HES) [26] guides were consulted in the creation of this methodology. The BRE guide recommends a minimum temperature difference between inside and outside of 10°C or 5°C if the building is mechanically pressurised or depressurized [25]. It is recommended that this temperature difference is maintained for a 24 hour period prior to the commencement of the monitoring, with a variation of external temperature of >±2°C. This in reality is difficult to achieve. The HES guidance is less stringent, only stating that a "significant temperature difference" is achieved, with a recommendation of a minimum difference of 5°C [26]. The temperature differences achieved are stated with the results. In order to minimise the effects of direct solar radiation and to maximise the temperature difference, thermography was undertaken just prior to sunrise on 11/03/2016, at 06:00 for the external and 06:30 for the internal. A FLIR® B250 thermal imaging camera was used for both. Using the Minneapolis Blower Door the OMP was pressurised during the thermography of external surfaces, and depressurised for that of the internal surfaces.

3.3.1. Assumptions and limitations. Care must be taken with variations in surface emissivity, thermal influences from building services and genuine variations in temperature which must be expected due to the physics of heat flow, such as at the corner of walls, and the junction between floors and ceilings [25, 26]. Care must also be taken over reflective surfaces such as glazing, as measurements will be of the objects reflected and not the surface itself [26].

3.4. Internal Hygrothermal Comfort.

According to ISO EN 7726: 2001 [27] sensors for monitoring hygrothermal comfort should be mounted at 0.1m, 1.1m and 1.7m above finished floor level in the centre of the space or close to the typical location of the occupant(s). This is however impractical for long-term measurements within an occupied space, where sensors would be intrusive and subject to possible interference. Instead, the sensors were located in discrete locations, selected to minimise both the influence of localised microclimates and inadvertent disturbance by the occupants. Whilst not ideal, this methodology enabled monitoring to occur over a longer period of time undisturbed. For the first three week period, between 18/02/2016 – 10/03/2016, TinyTag Ultra 2 TGU-4500 sensors were used. Due to concerns from the building's trustees over their visual intrusion during exhibitions, subsequent monitoring was continued with Maxim Hygrochron iButton DS1923 Sensors. With a similar accuracy but a reduced resolution (0.5°C and 0.6%), at less than 20mm diameter they are unobtrusive. These were left in place until July 2017 with the intention of obtaining at least one calendar year of measurements. Unfortunately, the capacity of the sensors memory was exceeded on the 03/06/2016 having recorded

only three months of data. Two locations within the OMP were monitored one to the east end and one to the west. In no. 25 Church Street a further two locations were monitored, one within the WC and the other in the office, both of which are located on the first floor to the east of the property.
The measurements recorded were then used to assess the hygrothermal comfort conditions by plotting the results on a bioclimatic chart, as developed by Givoni [28] who defines the comfort zone between 17°C and 27°C, with a range of relative humidity between 20% and 70%.

3.5. Digital Energy Demand Simulation.
Dynamic digital energy simulation was carried out using DesignBuilder Version 4.2.0.54. The weather file was created using Meteonorm version 6.1 using the time period 1996-2005. Each retrofit action was simulated separately, in addition to the hypothetical actions of insulating the roof and installing secondary glazing. In addition, scenarios combining retrofit actions were also simulated. As the OMP is located on the first floor between adjacent buildings, these and the ground floor were modelled as adiabatic volumes.

4. Results and Analysis

4.1. In situ U-Value monitoring.
The measured U-values are presented in Table 1. When the standard deviation is considered, these are similar those calculated according to BS EN ISO 6946:2007. It is to be expected that the measured and calculated U-values are similar as, unlike measurements of historic or non-conventional constructions, these infill panels are of standard layers of know materials. These U-values are well within the standards for new thermal elements for existing buildings as defined by the Building Regulations [5].

Table 1. OMP and No.25 Church Street. Measured and calculated U-values.

	Measured U-value (W/m²K)	Standard deviation (W/m²K)	Calculated U-value (W/m²K)
Old Mayor's Parlour	0.11	±0.04	0.13
No 25 Church Street	0.11	±0.03	0.17

4.2. Pressure Testing.
Based on the readings measured[1] the calculated air permeability index was 17.6 $m^3/h/m^2$ and air change rates were 22.5 /hr@50 Pa and 1.12 /hr unpressurised. As such the building does not achieve the 10 $m^3/h/m^2$ air permeability index required by building regulations for new-build [29] and is higher than the average air change rate for pre-1900 UK buildings of 12.3 ac/hr@50 Pa [30]. It is however similar to other historic timber-framed buildings measured by the authors [13]. The weakest areas are most probably the windows and the floor separating the OMP from the café below.

4.3. Thermography.
The recommended temperature difference of >10°C was achieved for both the external and internal thermography as presented in Table 2.

Table 2. Temperature difference for thermography at The Old Mayor's Parlour, 11/03/2016

	External Temp. (°C)	Internal Temp. (°C)	Temp. difference (°C)
External Thermography	2.4	20.4	18
Internal Thermography	0.4	22.9	22.5

[1] (R^2) of the best-fit line of the ln(flow) against ln(pressure difference) = 0.9781 showing a good degree of accuracy.

Figure 4 shows that the new infill panels are performing better than the surrounding timber-frame. The weak points of the façade are the windows and some joints between the timbers of the frame, especially around the first floor window.

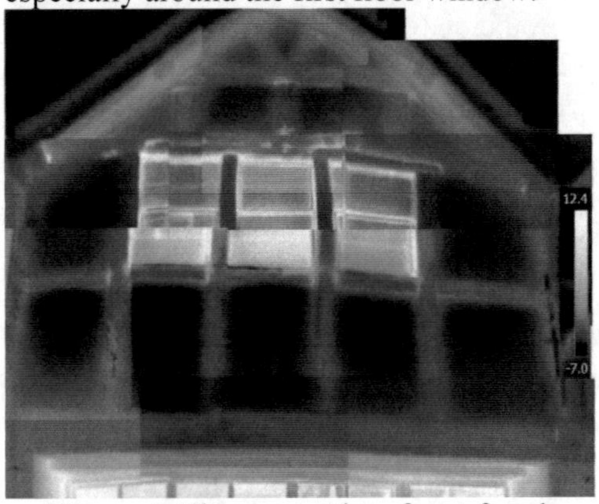

Figure 4. Thermography of east façade.

Figure 5. Internal thermography of east façade.

Figure 5 shows the efficacy of the internal lining, with no noticeable cold bridging or temperature difference across the internal surface. The upper portion of the wall with no internal lining is however cooler, with marked cold spots in the recesses of the historic plasterwork. Whilst the physics of heat flow dictates that recesses and junctions will naturally be cooler, this image still raises concern over the decision not to internally line the upper portion or introduce roof insulation. The rational of minimising intrusion and damage to the historic fabric is understandable but potentially, over time, may lead to the concentration of condensation and accelerated decay. This issue is explored further below.

4.4. Hygrothermal monitoring.

It was known that the OMP would be unoccupied and unheated during the first measurement period. It was however expected that the office and WC of No. 25 Church Street, both in daily use, would have some degree of heating. It was therefore surprising to see from the results that neither of these spaces were heated, with temperatures in the WC closely following those in the OMP at around 10°C. Temperatures in the office were a little higher, occasionally exceeding 15°C, probably due to heat rising from the shop below but fail to achieve levels of comfort. Whilst the author has encountered similar situations in a developing country or fuel poor households, it is perhaps unexpected in a commercial premises in Hereford. The results for all spaces show that comfort conditions were not achieved during office hours. Although this lack of heating reduced energy consumption and associated costs, it may create unintended consequences such as increased condensation and accelerated deterioration of interior finishes and building fabric.

The second stage of monitoring 10/03/16 – 04/06/2016 showed that once in use for exhibitions, the temperatures in the OMP did exceed 17°C for parts of the working day (Figure 6). However, the office of No.25 did not begin to enter the thermal zone until the external temperatures rose in early May.

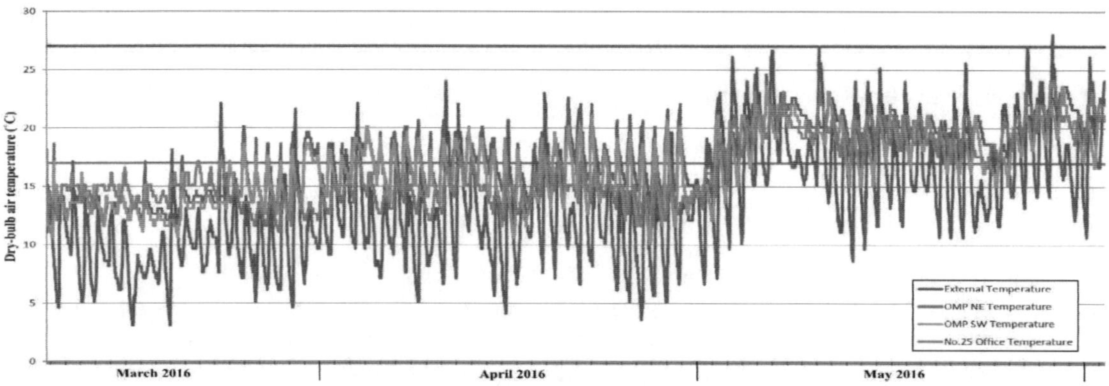

Figure 6. Dry-bulb air temperatures 10/03/2016 to 04/06/2016. Source: (Author's own, 2017).

During opening hours (09:00-17:00) over the four and a half months of monitoring, hygrothermal comfort was achieved 50% of the time in the east end of the OMP, 55% of the time at the west end but only 28% of the time in no.25. The difference between the east and west end of the OMP may be due to the door to the semi-external staircase being located towards the east, although the electric heater is also located at this end of the room.

As noted previously, the lack of insulation to the roof raised some concern over the possible concentration of condensation on the cold surface of the 17th century decorative plaster ceiling. This would most likely under conditions of high vapour pressure, coupled with a large diurnal temperature variation. The highest vapour pressure recorded in the OMP was 1557Pa, occurring at 15:30 on 12/05/2016 at a dry-bulb temperature of 23°C. Tracing this vapour pressure across on the psychrometric chart, condensation would occur at a dew point temperature of 13.5°C. Within the following 24 hours, the minimum internal dry-bulb temperature was 18.6°C and the minimum external temperature 11.1°C. As such, it is possible that the internal surface temperature of the ceiling could have dropped, in places, below the dew point temperature and condensation may have occurred. Equally, on the day with the largest diurnal temperature oscillation, the 28/04/2016, a maximum vapour pressure of 913Pa was recorded at a temperature of 17°C. The internal dry-bulb temperature then dropped to 11.6°C with a minimum external dry-bulb temperature of 5.6°C. Given that the dew point temperature for the maximum vapour pressure measured (913Pa) would be 5.7°C, there again exists a small possibility that conditions for condensation may have occurred. In order to prove this, further monitoring of surface temperatures of the plaster ceiling would be required.

4.5. Digital Energy Demand Simulation.

The results showed that the thermal upgrading of the walls that was undertaken led to a reduction in energy demand of 12%. It is interesting to note that, assuming secondary glazing would not only improve the thermal performance of the windows but also lead to increased airtightness, this alone could potentially achieve a reduction of 15%. The introduction of 200mm wood fibre insulation to the roof could potentially have achieved a 17% reduction alone, or 42% when combined with the walls, and 58% combined with walls and secondary glazing. This calls into question if the thermal upgrading of the walls was actually the most efficient retrofit action to undertake. However, given that this project was not envisioned primarily as an energy retrofit, any reduction in energy demand should be seen as an added benefit and must be weighed against the other complex decisions of traditional building conservation.

5. Conclusions

The incidental thermal performance improvements achieved by the necessary replacement of failed 20th century panel infill materials have shown to be successful at the level of the building component, with U-values of the new panels achieving Building Regulations, however at a building scale the positive impact on energy demand reduction (12%) and hygrothermal comfort levels is limited (achieved only an average of 53% of occupied hours). At the same time there exists the possibility that the work undertaken may have a negative impact through the concentration of condensation on the 17th century plastered ceiling. In order to verify if condensation is occurring, further detailed monitoring of surface temperatures and hygrothermal conditions is required. The results presented here underline the need for wherever possible a holistic approach to energy retrofit, however this must be balanced by the complexities of sustainable building conservation.

References

[1] IPCC 2018 Global Warming of 1.5 ºC. (The Intergovernmental Panel on Climate Change)
[2] OJEU 2010 Directive 2010/31/EU of the European Parliament and of the Council on the energy performance of buildings.: (Official Journal of the European Union) pp 1-56
[3] DECC 2014 UK National Energy Efficiency Action Plan (London)
[4] Scottish Government 2017 Technical Handbook 2017 Domestic - Energy. (Online)
[5] HM Government 2016 Approved Document L2B; Conservation of fuel and power in existing buildings other than dwellings. (NBS)
[6] HM Government 2016 Approved Document L1B; Conservation of fuel and power in existing dwellings. (NBS)
[7] Bonfield P 2016 Each Home Counts. (Department for Buisness, Energy and Industrial Strategy & Department for Communities and Local Government)
[8] STBA 2012 Performance and Energy Efficiency of Traditional Buildings: Gap Analysis Study. STBA (Sustainable Traditional Buildings Alliance)
[9] Historic England, Association of Local Government Archaeological Officers and Institute of Historic Building Conservation 2017 The Ninth report on Local Authority Staff Resources.
[10] Baker P and Rhee-Duverne S 2015 A Retrofit of a Victorian Terrace House in New Bolsover a Whole House Thermal Performance Assessment. (Historic England)
[11] Currie J, Williamson J B and Stinson J 2013 Technical Paper 19: Monitoring thermal upgrades to ten traditional properties. (Glasgow: Historic Environment Scotland,.)
[12] Gandhi K, Jiang S and Tweed C 2012 Field Testing of Existing Stone Wall in North Wales Climate. In: *SusRef: Sustainable Refurbishment of Building Facades and External Walls*: (Cardiff University)
[13] Whitman C 2017 The distribution of historic timber-framed buildings in the UK and the impacts of their low energy retrofit. (Cardiff University)
[14] McCaig I and Ridout B 2012 *English Heritage practical building conservation- Timber* (London; Farnham, Surrey; Burlington, VT: English Heritage ; Ashgate)
[15] Historic England 2014 The National Heritage List for England. ed Historic England.
[16] BBC 2014 Old Mayor's Parlour in Hereford renovation work begins. (online)
[17] The Church Street Charitable Trust 2016 Old Mayor's Parlour Information Pack.
[18] Royal Commission on Historical Monuments 1931 *An Inventory of the historical monuments in Herefordshire Vol. 1, Vol. 1* (London: H.M.S.O.)
[19] Demaus J 2015 The Old Mayor's Parlour & Gallery 25. ed C Whitman
[20] Richards C 2014 Re: PhD in retrofitting of historic timber framed buildings. Private communication with C Whitman
[21] British Standards Institution 2014 BS ISO 9869-1:2014 Thermal insulation- Building elements- in situ measurement of thermal resistance and thermal transmittance Part 1: Heat flow meter method. (British Standards Institution)
[22] Baker P 2011 U-Values and Traditional Buildings: In situ measurements and their comparison to calculated values. ed Historic Environment Scotland.
[23] Rye C, Scott C and Hubbard D 2012 THE SPAB RESEARCH REPORT 1. U-Value Report.: (Society for the Protection of Ancient Buildings)

[24] British Standards Institution 2015 BS EN ISO 9972:2015 Thermal performance of buildings - determination of air permeability of buildings - fan pressurization method. (British Standards Institution)
[25] Hart J M 1991 Practical guide to infra-red thermography for building surveys. (Bracknell, UK: Building Research Establishment)
[26] Young M 2015 Thermal Imaging in the Historic Environment. In: *Short Guide*: (Historic Environment Scotland)
[27] ISO 1998 ISO EN 7726: 2001 Ergonomics of the thermal environment—instruments for measuring physical quantities.
[28] Givoni B 1998 *Climate considerations in building and urban design* (New York: Van Nostrand Reinhold)
[29] HM Government 2016 The Building Regulations. Approved Document Part L1A conservation of Heat and Power. New Dwellings. (NBS)
[30] Hubbard D 2011 Ventilation, Infiltration and Air Permeability of Traditional UK Dwellings *Journal of Architectural Conservation* **17** 59-73

Adaptive comfort assessment for different thermal insulations for building envelope against the effects of global warming in the mid-western Brazil

Emeli Lalesca Aparecida da Guarda[1], Elaise Gabriel[2], Renata Mansuelo Alves Domingos[3], Luciane Cleonice Durante[1], Ivan Julio Apolonio Callejas[1], João Carlos Machado Sanches[4], Karyna de Andrade Carvalho Rosseti[4]

[1]Universidade Federal de Mato Grosso, MT, Brazil
[2]Universidade Federal de Santa Maria, RS, Brazil
[3]Universidade Federal de Santa Catarina, SC, Brazil
[4]Universidade do Estado de Mato Grosso, MT, Brazil
[5]Corresponding Author Email: emeliguarda@gmail.com

Abstract. The consequences of global warming have gained worldwide importance, also in the scope of built environment when related with the thermal comfort conditions of users. In this way, the objective of this research is to analyse the effects of global warming on the hours of thermal comfort of dwellings, with different thermal insulators in the external walls, considering the emission scenario A2 of the Fourth Assessment Report (AR4) of the IPCC. The methodological procedures consist of four stages: preparation of climate files for Future Scenarios (2020, 2050 and 2080) from the base climate file (base scenario); definition of a dwelling for study object (Tbase); preparation of thermal insulating intervention proposals, such as: EPS (T1), rock wool (T2) and glass wool (T3) and evaluation of thermal comfort conditions using the adaptive thermal comfort method. The results indicate that the air temperature increases by 21.5% and the relative air humidity reduces by 22% until the 2080 scenario if compared to the base scenario. From these new conditions, the percentages of Tbase comfort hours suffer a reduction of 63.61%, increasing the hours of heat discomfort by 98.48%, in the scenario of 2080 in relation to the base scenario. With the adoption of insulation, the T1 typology presented comfort hours of 58.4%, reducing to 17.9%. However, typologies T2 and T3 presented similar behaviour, with 27.2% of hours in comfort and 67.4% of hours in heat discomfort, in the 2080 scenario in relation to the baseline scenario. Therefore, it can be concluded that typologies T2 and T3 presented greater resilience to the effects of global warming, but it is necessary to incorporate constructive interventions to absorb the impacts of climate changes and provide better conditions of thermal comfort.

Keywords: Climate changes. Built Environment. Thermal comfort.

1. Introduction

Scientific research evidences a progressive increase of terrestrial temperature as one of the present transformations of the planet. Different studies, such as Duursma [1], point out that between the end of the 19th century and the beginning of the 20th century, the temperatures showed quite invariable behaviour. Marengo and Soares [2] estimated an addition of 0.85°C, every 100 years, in the Amazon region. Casagrande and Alvarez [3] points out that since the 1920s, there has been an increasing trend of temperatures, with an addition of 0.8 ° C in the year 2000, compared to the averages of the period 1961-1990.

Knowledge about climate change, motivated by anthropogenic actions and based on greenhouse gas (GHG) emissions, has been developed by the Intergovernmental Panel on Climate Change (IPCC). The IPCC has published scientific reports, on which scientists from a wide range of fields base their research on the behaviour of terrestrial ecosystems. In 2007, IPCC released the results of its Fourth Climate Change Assessment Report (AR4). The results are based on three working groups, which elaborate models of GHG emissions and their respective impacts on global warming. The scenarios are divided into four families, called A1, A2, B1 and B2, where 'A' and 'B' mean low (pessimistic) and high (optimistic) commitment to sustainable development, and '1' and '2 ', regional integration or fragmentation, respectively [4]. The GHG emissions scenarios developed by the IPCC are future conditions with similar demographic, social, economic and technological histories. This report highlights that most of the temperature increase events observed in the last 50 years were caused by anthropogenic actions, and alert for the average increase in global temperatures of 1.1 ° C to 5.4 ° C, reaching 6.4 ° C by the year 2100, if the population and economy continue to grow rapidly, with heavy consumption of fossil fuels [4].

It is known that the thermal energy performance of buildings depends on the climate they are exposed [5] and new climatic conditions influence and impose new impacts on buildings and user thermal comfort [6]. Considering that the useful life of a building is 50 to 75 years [7], it is inferred that the performance of the current buildings will be affected by the expected climate changes for the coming decades. Nico-Rodrigues et al. [8], suggest that adopting guidelines that consider the relationships between climate and humans is essential to define dwellings with adequate thermal comfort. Therefore, it is necessary to update the climatic data from the survey by IPCC emission scenarios and verify the current performance of buildings in the future environmental conditions. Thus, researches that address climate change have acted in the simulations of thermal and energetic performance of buildings through the effects of heating, obtaining as a result conditions of comfort of the internal environments, consumption of electric energy [9,10] and passive design strategies to provide better housing conditions [11,12,13].

In this way, the mitigation of climate change effects has been a guideline of public policies in many countries and, researches to adapt the buildings to the new climatic scenarios has been developed in several areas of construction field. Therefore, this work has as general objective to analyse the effects of global warming on the hours of comfort and thermal discomfort of the dwellings with different thermal insulation, for mid-western Brazil.

2. Methodology

2.1. Elaboration of future climate files

The survey of published researches, on scientific platforms on studies of future climate projections, made possible the knowledge of the methodology indicated by the IPCC for the elaboration of climatic archives, with influence of global warming [9, 10, 11, 12, 13 14]. By means of these literature the morphing methodology published by Belcher, Haker and Powell [15] for the elaboration of the future climate archives was chosen. This methodology modifies a set of historical climatic variables (1961-1990) of 8,760 hours per year, without the influence of urbanization and incorporates the effects of global warming on the climatic archives, thus obtaining projections of future climatic data. The process used by this method evidenced the complexity in developing future climate scenarios with the vision of application in simulation software for buildings. In order to facilitate and consolidate the aforementioned algorithmic operations, the Sustainable Energy Research Group (SERG) from the University of Southampton in the United Kingdom has developed the "Climate Change World Weather Generator" (CCWorldWeatherGen) tool incorporated into Excel Software. This tool uses the emission scenario A2 from the IPCC Fourth Report (AR4), for time-slice 2020 (period 2011-2040), 2050 (period 2041-2070) and 2080 (period 2071-2100).

2.2. Definition of the study object

A standard single-family housing (Tbase), located in the city of Cuiabá, Brazil, with a total area of 39.18m², containing the Living Room/Kitchen (17.44m²), Bedroom 1 (7,78m²), Bedroom 2 (7,57m²) and Bathroom (1.75m²) as shown in Figure 1. The house has a gable roof with eaves of 30 cm. The ceiling is 3,00 m high.

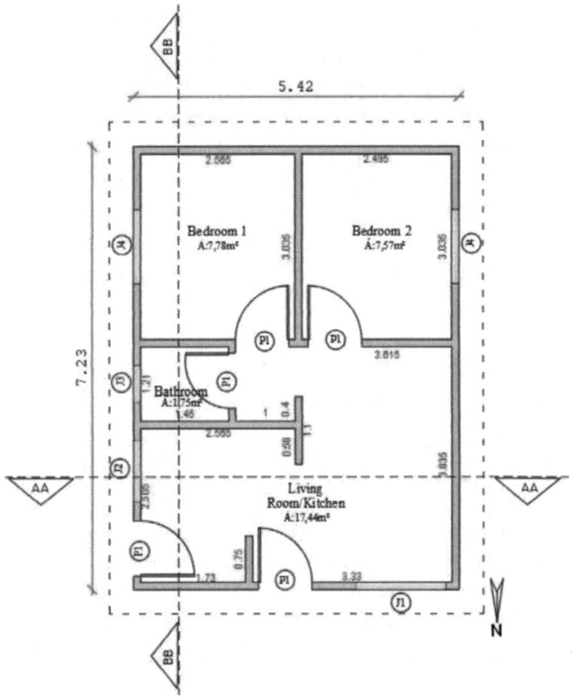

Figure 1: Study object (Tbase)

The Bedrooms and the Living room window frames are metallic, Venetian type with glass, with four sliding panels (two fixed and two mobile). Their dimensions are 1.50 x 1.10 m (living room), 1.20 x 1.10 m (Bedrooms 1 and 2) and 1.20 x 1.10 m (Kitchen). The external doors are made of metal sheet and the internal doors are made of wood. The vertical and horizontal closures were defined by the Brazilian standard NBR 15.220 [16], with the thermal properties of building materials expressed in Thermal Resistance (R), Thermal Transmittance (U), Thermal Capacity (C) shown in Table 1. The thermal resistance of the air chamber is 0.21 m²K/W, high emissivity, thickness greater than 5 cm.

Table 1: Study object thermophysical properties (Tbase)

	Envelope	Composition	Thickness (cm)	R (m²K/W)	U (W/m²K)	C (J/m²K)
Tbase	External walls 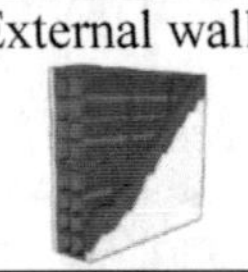	External mortar	2.50	0.2991	3.34	166.87
		Ceramic brick	9.00			
		Internal mortar	2.50			
	Roof 	Ceramic roof tile	2.00	0.4795	2.08	41.92
		PVC liner	1.00			

Proposals for constructive interventions were inserted in Tbase, with the purpose of analysing the influence of thermal insulation on thermal comfort. Without modifying the Tbase cover materials, thermal insulation was inserted inside the outer walls, with Expanded Polystyrene (EPS) composing T1, Rock wool in typology T2 and glass wool in typology T3, all of them with thickness of 2.50cm.

2.3. Adaptive comfort conditions according to De Dear and Brager methodology (1998)

Standard 55 (ASHRAE, 2013) presents an index based on the studies of adaptive thermal comfort proposed by De Dear and Brager [17]. The method can be applied to naturally ventilated buildings, considering occupancy and internal sources of heat to better represent the users comfort condition in the buildings. The comfort levels are established by the ideal indoor operative temperature or the monthly neutral temperature (Tn, in °C) related to the monthly averages outdoor temperature (TEmed., in °C), according to Equation 01. It is emphasized that the equation is valid for TEmed between 10.0°C and 33.5°C.

$$Tn = 17.8 + 0.31 \times TEmed \qquad \text{Equation 01}$$

After determining the neutral temperature, ASHRAE establishes a comfort temperature range for 80% or 90% of satisfied users. The bands are determined by upper and lower limits given by Tn+Tolerance and Tn-Tolerance, respectively and the indoor operative temperature (Top), representing the interval for 80% of satisfied users of Tn-3.5 and Tn + 3.5 and to 90% of satisfied users of Tn-2.5 and Tn+2.5. With the monthly ideal indoor operative temperatures, it is possible to quantify the hours of discomfort by heat or cold and thermal comfort inside the building, for the current scenario and future climate projections. Therefore, this study considered the range of 90% of satisfied users, as well as Alves [16], and Tateoka and Duarte [18]. In addition to these studies, Sánchez-García et. al [19] used this methodology to investigate adaptive comfort in Southern Spain.

3. Results

3.1. Air temperature and relative humidity

For Cuiabá, the results of simulations showed that future climatic effects caused an increase in monthly temperature averages and a decrease in relative humidity, corroborating the studies from Triana, Lamberts and Sassi [20]. The average annual temperature increased by 21.50% up to the 2080 scenario, when compared to the Base Scenario. The average annual temperature in the current scenario is 26.73 ° C, going to 28.24 ° C in the 2020 scenario, to 29.90 ° C in the 2050 scenario and to 32.48 ° C in the 2080 scenario (Figure 2-a). Relative annual air humidity decreased by 22% from the Base Scenario for the 2080 scenario. The annual average base is 69.08%, rising to 64.75% in 2020, 60.41% in 2050 and 53.67% in 2080 (Figure 2-a and b).

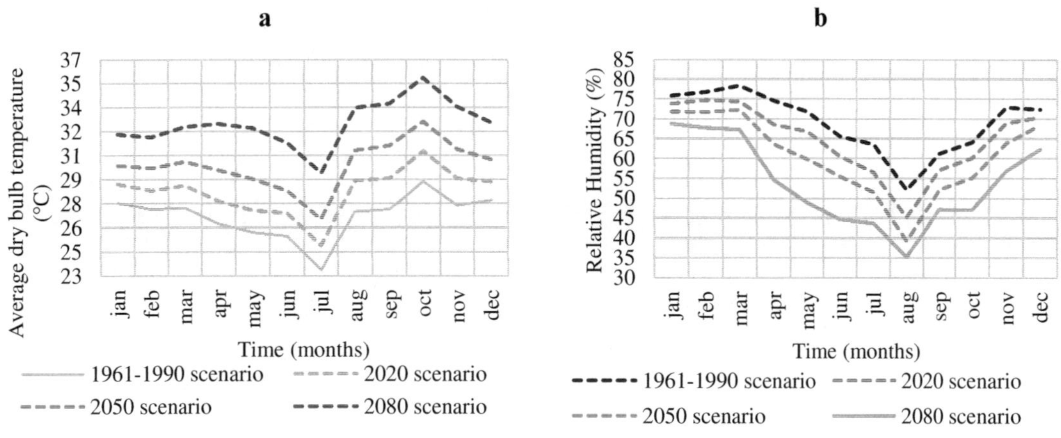

Figure 2: a) Dry bulb temperature (°C) and b) Relative humidity (%)

The increase in mean annual dry bulb temperatures was estimated at 1.51°C in the 2020 scenario, 3.17°C in the 2050 scenario and 5.75°C in the 2080 scenario. Annual average relative humidity decreased by 4.33% in the 2020 scenario, 8.67% in the 2050 scenario and 15.40% in the 2080 scenario. It has been found that the increase of dry bulb temperature is more significant between the 2050 and 2080 scenarios, as well as the reduction of relative humidity.

3.2. De Dear and Brager scenarios: Base (1961-1990) and Future (2020, 2050 e 2080)

The annual percentages of satisfied and unsatisfied hours by cold and heat were quantified for the four typologies of dwellings and for the four study scenarios. The base typology (Tbase) presented increased hours of heat discomfort and reduced hours of cold discomfort, as well as the hours of comfort for the 2080 scenario. In this way, Tbase showed a variation of + 35.15%, + 65.88% and + 98.48% of the hours of heat discomfort, -21.22%, -37.95% and -59.46% of the hours of cold discomfort and -22.71%, -42.99% and -63.61% of the comfort hours, respectively, in the scenarios of 2020, 2050 and 2080, demonstrating the need to adapt the dwellings for mitigation effects of global warming (Figure 3).

By adopting the thermal insulation and comparing with the base case, the dwellings presented higher satisfaction hours of comfort in the scenario of 2080 and, consequently, they obtained greater resilience to the increase of the outdoor temperatures. In the base scenario, the typologies with thermal insulation in the external walls, such as T2 and T3, presented higher satisfaction hours of comfort, 5,242 (59.8%) and 5,242 (59.8%), reducing both 2,383 (27.2%) hours in the 2080 scenario. It should be noted that the rock wool and glass wool insulators did not show differences in the percentages of comfort hours, as well as in the hours of heat and cold discomfort, presenting similar results for the four scenarios analysed. The typology with thermal insulation of EPS (T1) presented 5,151 (58.4%) of satisfied hours for comfort, reducing to 2,179 (24.9%) in the 2080 scenario.

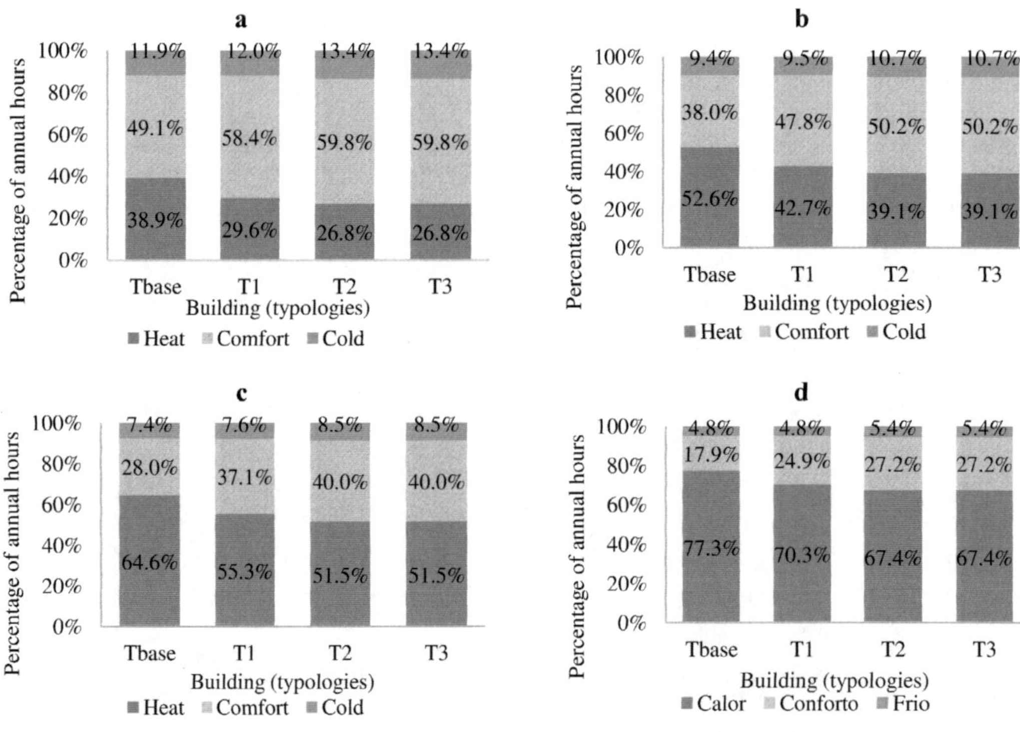

Figure 3: Hours of comfort and discomfort for dwellings: a) in the Base Scenario, b) 2020, c) 2050 and d) 2080.

By means of the neutral temperatures obtained from the monthly averages of Te,med. for the Base Scenario and the scenarios of 2020, 2050 and 2080, it was possible to verify the thermal comfort temperature ranges, considering the satisfaction band of 90% of the users. In the Base Scenario, the adaptive comfort range ranged from 22.54°C to 29.26°C. Thus, the thermal comfort conditions of Tbase were above the tolerance range for 90% of users, presenting discomfort in 16.6% of the year. It is noteworthy that Tbase average indoor temperature was higher than 29.2°C in the months of October and December. The other typologies accompany the behaviour of the comfort range, remaining between the upper and lower limits (Figure 4-a).

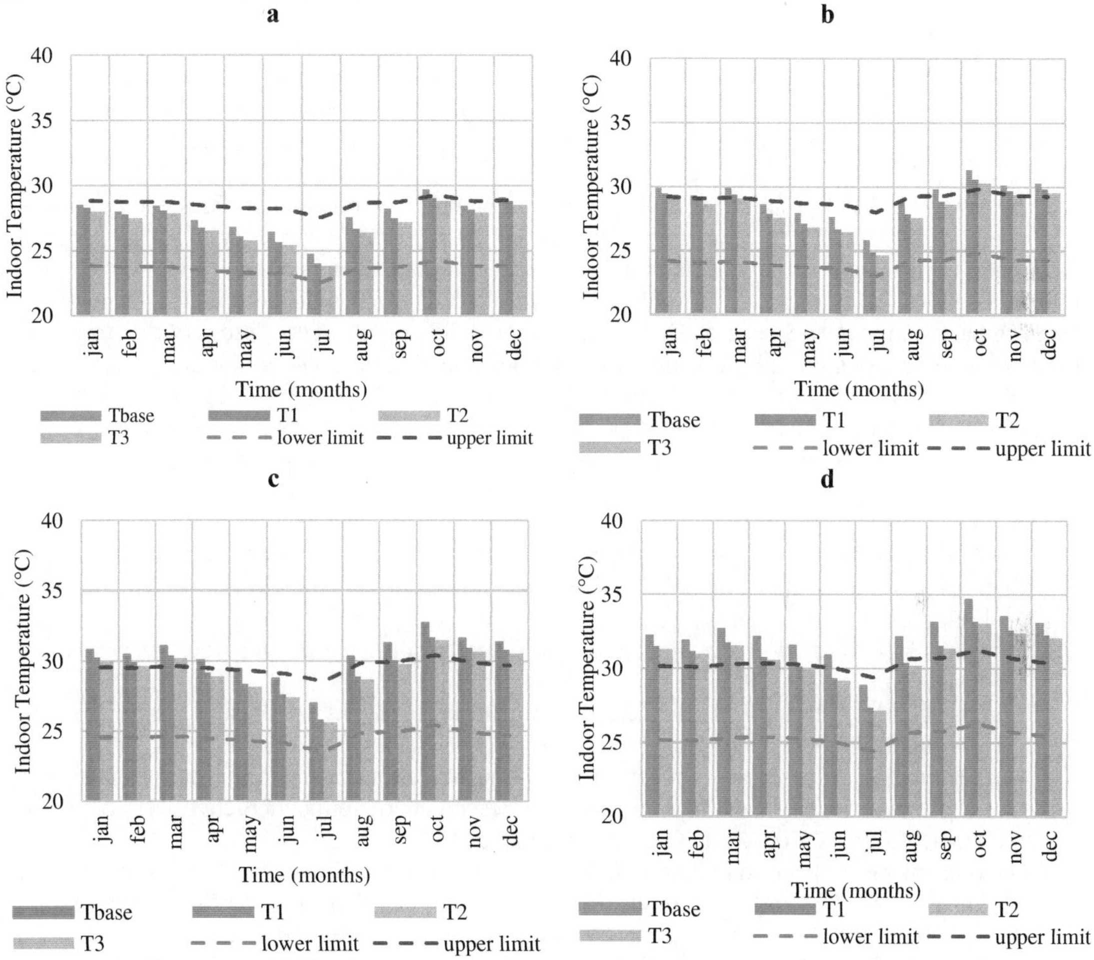

Figure 4: Building comfortable temperatures ranges: a) in the Base Scenario, b) 2020, c) 2050 e d) 2080.

In the 2020 scenario the comfort range increased by + 0.58°C compared to the current scenario, changing from 23.01°C to 29.84°C. Tbase presented conditions of thermal discomfort in 58.3% of the year, increasing by + 40% in relation to the Base Scenario. With the increase of TE,med, T1 presented conditions of thermal discomfort in 41.66% of the year, being more representative in the month of October with Ti,med of 30.55 ° C. The other typologies (T2 and T3) presented thermal discomfort in 25% of the year (Figure 4-b). In 2050, the comfort range increased by +1.15°C compared to the Base Scenario, rising to 23.51 ° C at 30.41 ° C. Tbase's annual thermal discomfort increased by + 78% from the Base Scenario to 91.6% of the year. This increase is also verified in the other typologies, presenting discomfort in 58.3% in T1 and 50% in T2 and T3 in the scenario. It should be noted that Tbase provided thermal comfort conditions only in June and July (Figure 4-c). Finally, in the scenario 2080 (2071-2100), the comfort range increased by + 1.50°C, when compared to the Base Scenario, from 24.41°C to 30.76°C. It was verified that in this scenario, Tbase presented thermal discomfort during all year, except for the month of July. The other typologies presented thermal discomfort of 66.6% of the year, providing thermal comfort only in the months of May, June, July and August (Figure 4-d). This increase in the thermal discomfort of the typologies can be justified by the significant increase of Te,med over +5.5°C, directly affecting the comfort conditions inside the dwellings, resulting in Ti,med above 30.0°C in the scenario.

4. Conclusions

The effects of climate change due to global warming have affected habitability and thermal comfort. In this way, the climatic data of the building implantation is an important parameter to evaluate its behaviour, also, the climatic forecast is significant for habitability planning and conditions of thermal comfort.

By means of the De Dear and Brager [17] model for naturally ventilated buildings, the behaviour of comfort and discomfort hours were similar, presenting an increase in the discomfort from heat and reduction of the discomfort from cold, as well as the reduction of the hours of comfort until the scenario of 2080. The hours of heat discomfort increased and cold discomfort reduced until the scene of 2080, when compared to the Base Scenario. The Tbase typology, did not obtain conditions of thermal comfort in the scenario of 2080, due to the increase of outdoor temperatures. The typologies T1, T2 and T3 obtained better conditions of thermal comfort in the scenario of 2080, with 24% of hours of comfort. It is concluded that the adaptation of the typologies by means of constructive interventions becomes important to achieve thermal resilience and provide comfort conditions until the scenario of 2080.

References

[1] Sze S M 1969 *Physics of Semiconductor Devices* (New York: Wiley–Interscience)

[2] Dorman L I 1975 *Variations of Galactic Cosmic Rays* (Moscow: Moscow State University Press) p 103

[1] Duursma, E. River flow and temperature profile trends: consequences for water resources. N.V Company. Amsterdam, 2002.

[2] Marengo, J. A; Soares, W. R. Impacto das mudanças climáticas no Brasil e possiveis cenários climáticos: síntese do terceiro relatório do IPCC de 2001. CPTECINPE, p. 29, 2005.

[3] Casagrande, B. Alvarez, C. Preparação de arquivos climáticos futuros para avaliação dos impactos das mudanças climáticas no desempenho termoenergético de edificações. Ambiente Construído, Porto Alegre, v.13, n.4, p. 173-187, 2013.

[4] Intergovernmental Panel on Climate Change (IPCC). Climate Change 2007: Synthesis Report. Contribution of Working Groups I, II and III to the Fourth Assessment Report of the Intergovernmental Panel on Climate Change. Geneva, Switzerland, 2007.

[5] De Wilde P; Coley D. The implications of a changing climate for buildings. Build Environment, v.55, n.1-7, 2012.

[6] Alves, C. A. Resiliência das Edificações às mudanças climáticas na região metropolitana de São Paulo – Estudo de caso: Desempenho térmico de edifícios residenciais para idosos. Dissertação (Mestrado). Faculdade de Arquitetura e Urbanismo, Universidade de São Paulo. São Paulo, 2014

[7] Associação Brasileira de Normas Técnicas (ABNT). NBR 15.575: Edificações habitacionais - Desempenho. Rio de Janeiro, 2013.

[8] Nico- Rodigues, E. A; Alvarez, C. E; Santo, A. D; Piderit, M. B. Quando a janela define a condição de desempenho térmico em ambientes ventilados naturalmente: caso específico das edificações multifamiliares em Vitória, ES. Associação Nacional de Tecnologia do Ambiente Construído (ANTAC), v, 15, p. 7-23, 2015.

[9] Wang, L; Liu, X; Brown H. Prediction of the impacts of climate change on energy consumption for a medium-size office building with two climate models. Energy and Buildings, 2017.

[10] Song, X; YE C. Climate Change Adaptation Pathways for Residential Buildings in Southern China. Energy Procedia, v. 105, p. 3062–7, 2017.

[11] Triana, M. A; Lamberts, R. Sassi, P. Desempenho de Habitações de interesse social frente às mudanças climáticas. In: Encontro Nacional de Tecnologia do Ambiente Construido. São Paulo. Anais.... Porto Alegre, 2016.

[12] Invidiata, A; Ghisi, E. Impact of climate change on heating and cooling energy demand in houses in Brasil. Energy and Building 130, p. 20-32, 2016.

[13] Sabunas A, Kanapickas A. Estimation of climate change impact on energy consumption in a residential building in Kaunas, Lithuania, using HEED Software. Energy Procedia, vol. 128, p. 92–9, 2017.

[14] Casagrande, B. Alvarez, C. Preparação de arquivos climáticos futuros para avaliação dos impactos das mudanças climáticas no desempenho termoenergético de edificações. Ambiente Construído, Porto Alegre, v.13, n.4, p. 173-187, 2013.

[15] Belcher, S. E; Hacker, J. N; Powell, D. S. Constructing desing weather data for future climates. Building Services Engineering Research and Technology, v. 26, p. 49-61. 2005.

[16] Associação Brasileira de Normas Técnicas (ABNT). NBR 15.220: Desempenho térmico de edificações - Método de cálculo da transmitância térmica, da capacidade térmica, do atraso térmico e do fator solar de elementos e componentes de edificações. Rio de Janeiro, 2005.

[17] De Dear, R; Brager, G; Cooper, D. Developing an adaptive model of thermal comfort and preference – Final Report on ASHRAE RP 884. American Society of Heating Refrigerating and Air Conditioning Engineers. Atlanta, 1997.

[18] Tateoka, S. S; Duarte, D. H. Adaptação às mudanças climáticas: desafios para requalificação de edifícios existentes em São Paulo. In: Encontro Nacional de Conforto no Ambiental Construído. Balneário Camboriú, 2017.

[19] Sánchez-García, D; Rubio-Bellido, C; Pullido-Arcas, J; Guevara-García, J; Canivell, J. Adaptative comfort models applied to existing dwellings in Mediterranean climate considering global warming. Sustainability, v.10, p. 3507, 2018.

[20] Triana, M. A; Lamberts, R; Sassi, P. Should we consider climate change for Brazilian social housing? Assessment of energy efficiency adaptation measures. Energy and Buildings, v. 158, p. 1379-1392. 2018.

IOP Conf. Series: Earth and Environmental Science **329** (2019) 012031 doi:10.1088/1755-1315/329/1/012031

Comparison between environmentally designed houses and standard houses regarding thermal behavior

M Nessim[1,3], T Abdin[2] and Z Mostafa[2]
[1] Housing and building Research Center, 87, Tahrir Street, Dokki, Egypt.
[2] Arab Academy for Science and Technology, Alexandria Desert Road, Egypt
[3] Corresponding Author email: mariannessim7@gmail.com

Abstract. Based on the literature review and environmental studies, it is well known that the fundamental design standards focus on the building orientation, materials, form, window wall ratio, and passive tools for cooling in hot climates. This was the reason behind this paper, to verify and prove that if a building is designed according to these standards then it could achieve thermal comfort and energy efficiency. So, a comparison was held between four case studies, two of them were designed according to those standards and the other two were not. Hygrometer data loggers that measure ambient air temperature and relative humidity were placed in living rooms and on roofs in the four villas for one month in order to compare and analyse the measurements. In the coming paragraphs, each case study is explained and measurements and analysis are presented in order to reach the conclusion of this paper study. Fixed parameters were set for comparison between the four cases, those parameters were as follows: Orientation- wall thickness- building material- WWR (window wall ratio) - Number of Floors- Type of roof – Surroundings. Data was collected based on these parameters to be able to compare between the results. After comparing those parameters in the four villas, it was concluded the great influence of the surrounding buildings on each villa on its internal temperature. And so therefore a simulation has been made to measure the effect of the surrounding buildings.

1. Introduction

According to the annual report issued by the Ministry of Electricity in Egypt (2016-2017), the greater source of energy is thermal (92%) which depends on burning fossil fuel that of course produces tons of CO_2 contributing in the greenhouse effect and raising the temperature of the planet globally.

Also the greater sector of consumer is the residential sector 50.5% [1].Since Egypt is committed to the "Paris Agreement" held in 2015, Egypt should improve the climate change performance according to the "Climate change performance index" that is issued annually where Egypt holds the 24th in the ranking over the world. The ranking results are defined by a country's aggregated performance regarding 14 indicators within the four categories "GHG Emissions", "Renewable Energy" and "Energy Use", as well as on "Climate Policy", in a globally unique policy section of the index [2]. Based on the above mentioned figures, serious action should be taken via Architects to help decrease these percentages, since energy is used mostly in houses for cooling purposes beside other domestic purposes. This consequently, derives Architects towards designing and building houses using Environmental design standards.

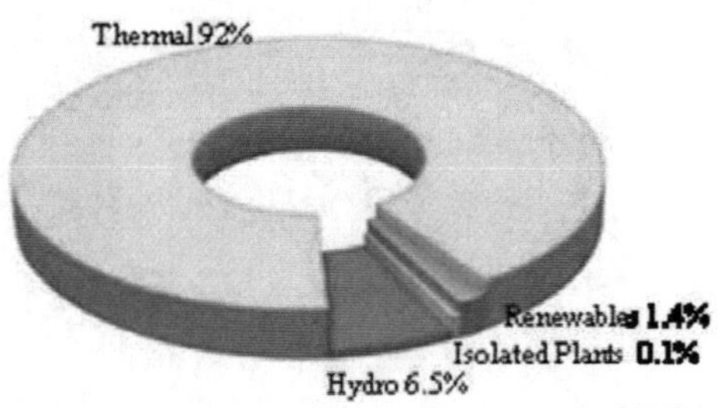

Figure 1: Types of Energy Generated (GWh)

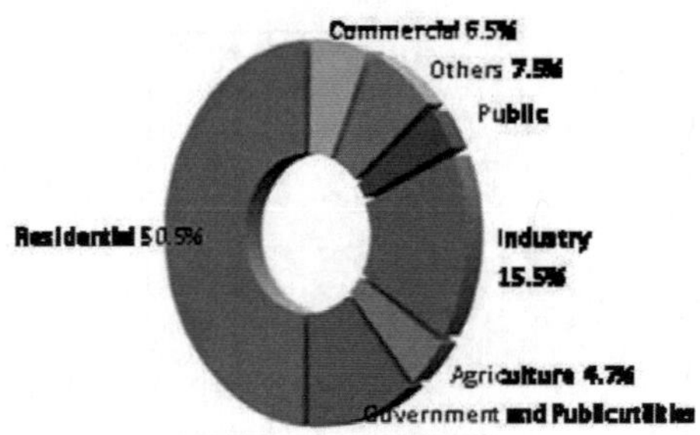

Figure 2: Energy Distribution according to purpose (%)

2. Case Studies

For each of the following case studies, two hygrometer data loggers were placed, one was placed in the centre of each zone (reception area) and another outdoor. The parameters measured in this research were indoor and outdoor temperature. The devices used were eight Perfect Prime TH0160, USB temperature and humidity data loggers. The runtime for the measurements was 20 days.

2.1. Villa A:

Location	: El Thawra El Khadra next to El Sheikh Zayed- 6th of October City		
Orientation	: South West/North West	Wall thickness	: 45cm
Building materials	: Lime Stone		
WWR for the living room	: SW orientation =20%	SE orientation =14%	
	NW orientation=12%	NE orientation= 40%	
Height of floor	: 5m	Roofs: Cross Vaults	
Number of floors	: Basement +2 floors	Surface area to volume ratio: 0.2	
Neighbours	: No neighbours or trees surrounding the villa from all directions		

The Villa is located near El Sheikh Zayed which is considered hot arid climate. The architect designed his Villa to be an icon for Vernacular Architecture. There is a wind catcher on the ground floor that is connected to the basement where the air circulates and drops more of its temperature. They are tall towers on the building's roof tops with several vents to capture wind for cooling the interior spaces. Wind catcher openings are usually positioned facing desired wind directions. Wind catcher functions based on pressure difference between leeward and windward which is positive in inlet opening and negative outlet vent. Venturi effect causes air to come down through the wind catcher channel at a higher velocity and lower pressure [3]. The Villa contains air shafts that are embedded in the walls; these shafts pull the cool air from the basement to the upper two floors through small openings in the walls of the rooms, this help deliver cool air to most of the rooms of the villa.

There is also an inner court that also helps in enhancing the natural ventilation and day lighting. Courtyard buildings have been always recommended as a passive architectural technique in desert environments in order to maintain indoor thermal comfort. A previous study, however, showed that in desert environments, the energy performance of two-storey residential courtyard buildings proved less efficient than other solid forms, even when attached to neighbouring buildings from three sides in a compact urban fabric. Their performance was relatively better in mild desert climates than in extreme hot ones. Another previous study showed that in Cairo, significant saving were only achieved at large building depth (18m-20m) and low height proportions (1:0.25 to 1:1) while a significant increase in consumption occurred especially at small building depth and higher height proportions in most cases [4]. The structure of the villa is bearing walls structure. Wall thickness is 45 cm of lime stone that delay heat from entering at day till it reaches the night hours. All roofs are in the form of Vaults or Cross Vaults or Domes which also help cooling the spaces underneath it. The hygrometer data logger was placed in the living room in the ground floor where its roof is cross vault so there is no other floor above this part of the ground floor.

IOP Conf. Series: Earth and Environmental Science **329** (2019) 012031 doi:10.1088/1755-1315/329/1/012031

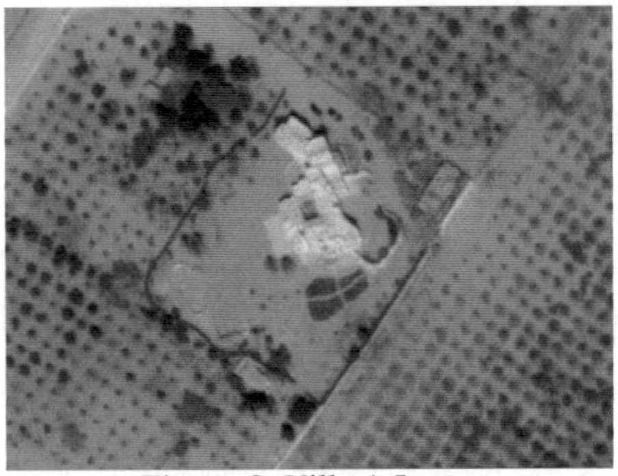

Figure 3: Villa A Layout

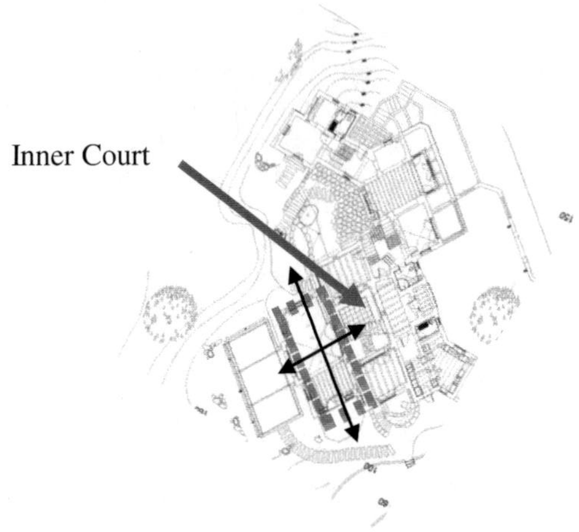

Figure 4: Villa A Ground Floor Plan

Figure 5: Villa A Section

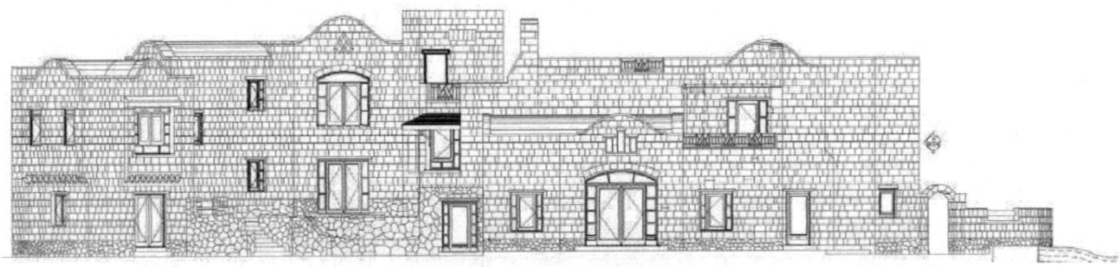

Figure 6: Villa A Elevation

Figure 7: Wind Catcher

Figure 8: Cross vaults roof

Figure 9: Air shafts

2.2. Villa B:

Location	: El Sheikh Zayed- 6th of October City		
Orientation	: South West/North West		
Wall thickness	: 25cm	Building materials	: Bricks
WWR for the living room	: SW orientation =40%	Height of floor	: 3m
Roof type	: Flat slab	Number of floors	: Basement + 2
Surface area to volume ratio	: 0.16		
Neighbours	: Neighbours are surrounding the villa from three directions and all have the same number of floors.		

The Villa is in a compound in El Sheikh Zayed and surrounded by neighbours from 3 sides. It is not designed based on any environmental design standards or used any passive cooling tools. It is a standard basic prototype of a villa composed of a basement and two floors, the basement is connected with the other floors through external staircase. The structure of the villa is column and beam structure with wall thickness 25cm.The hygrometer data logger was placed in the reception in the ground floor.

Figure 10: Villa B Layout

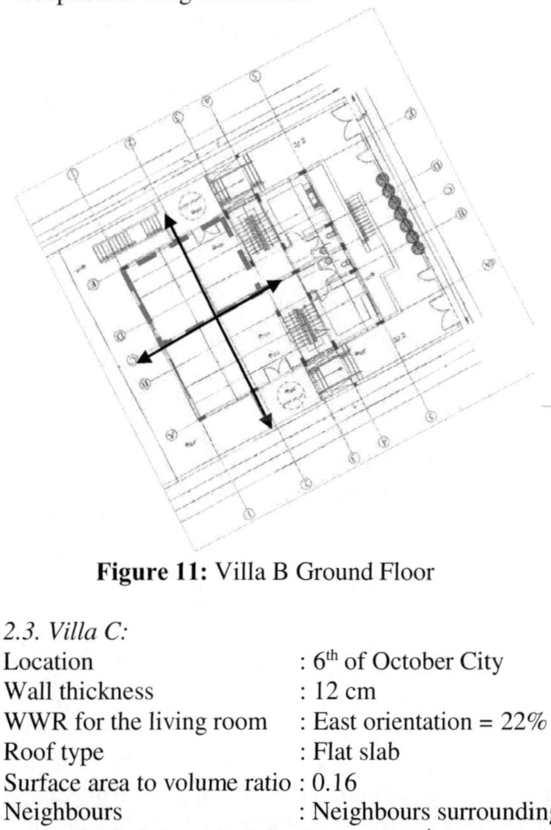

Figure 11: Villa B Ground Floor

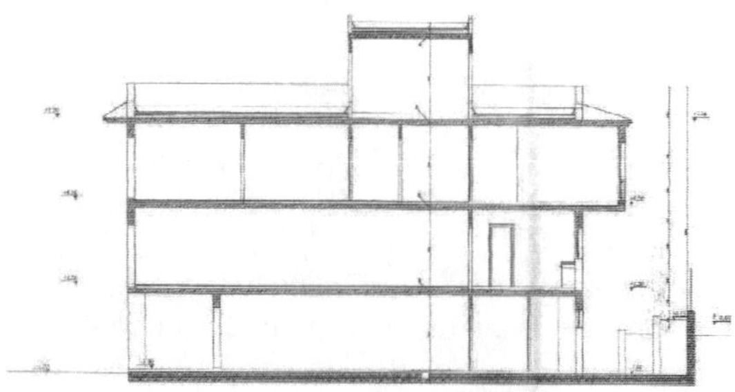

Figure 12: Villa B Section

2.3. Villa C:

Location	: 6th of October City	Orientation	: East
Wall thickness	: 12 cm	Building materials	: Bricks
WWR for the living room	: East orientation = 22%	Height of floor	: 3m
Roof type	: Flat slab	Number of floors	: 2 floors
Surface area to volume ratio	: 0.16		
Neighbours	: Neighbours surrounding the villa from three directions		

The Villa is in one of the districts in 6th October and surrounded by neighbours from 3 sides. It is also not designed based on any environmental design standards or used any passive cooling tools.

IOP Conf. Series: Earth and Environmental Science **329** (2019) 012031 doi:10.1088/1755-1315/329/1/012031

It is a standard basic prototype of a villa composed two floors and a basement. The structure of the villa is column and beam structure with wall thickness 12 cm. The hygrometer data logger was placed in the reception in the ground floor. All the surrounding buildings are 5 stories but the villa is only two floors high which cast shadow on the villa, there is also a fence on the eastern side between the villa and its neighbour which funnels the air through this zone.

Figure 13: Villa C Layout

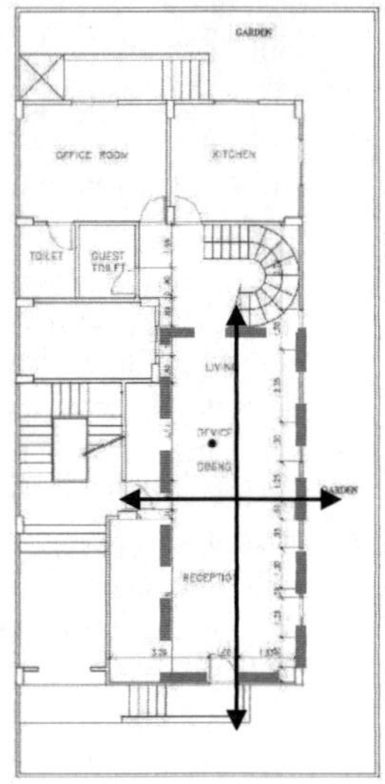

Figure 14: Villa C Ground Floor

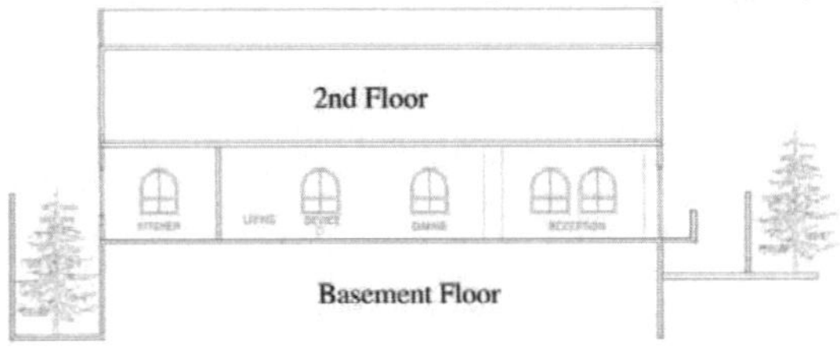

Figure 15: Villa C Section

2.4. Villa D:

Location	: 6th of October City
Orientation	: North East/ South East
Wall thickness	: 30 cm
Building materials	: Double wall from bricks with insulation in between
WWR for the living room	: North East orientation = 11% South East orientation = 34% (Shaded)
Height of floor	: 3m Roof type : Flat slab
Number of floors	: Basement + 3 Surface area to volume ratio: 0.1
Neighbours	: Neighbours surrounding the villa from three directions

The Villa was designed using passive cooling design strategies. It consists of three floors beside the basement. The wall thickness is 30cm and built of double wall with inner insulation. The Villa contains a wind catcher and is oriented to the North West and rises above the roof by 1.5m. It also contains a solar chimney that is painted black and works along with the wind catcher to enhance the natural ventilation in the villa. The wind catcher is connected to the solar chimney through tubes buried under the basement floor. The temperature of the air flowing in the solar chimney is higher due to the black paint of the chimney, this air escapes through the vents of the solar chimney and is replaced by cooler air, which comes from pulling the air coming through the wind watcher and cooled through the buried tubes in the basement floor. This air cools the rooms of the villa without having to use any mechanical cooling. There is also a skylight centred in the villa that provides natural lighting to the living spaces. In past, usually one of these items was used in architectural designs to help natural ventilation but with this combination (wind catcher and solar chimney) we can achieve more efficient ventilation system. Meir explains that the air trap operates with the change of air temperature and the difference of weight between the inside and outside of the trap. The difference of weight of the air impels a suction process which causes the air to flow either to the bottom or to the top [5].

Figure 16: Villa D Layout

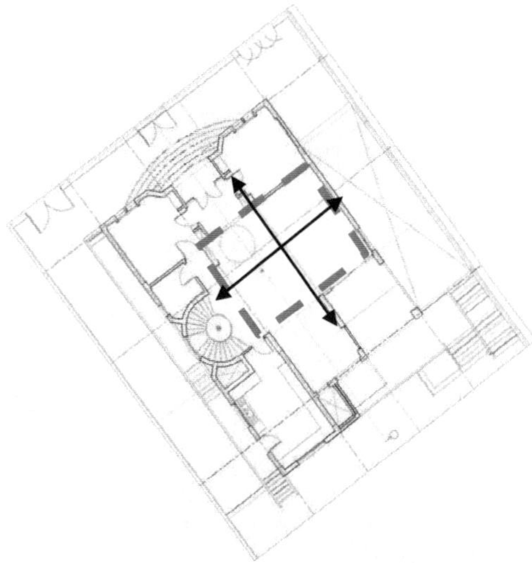

Figure 17: Villa D Ground Floor

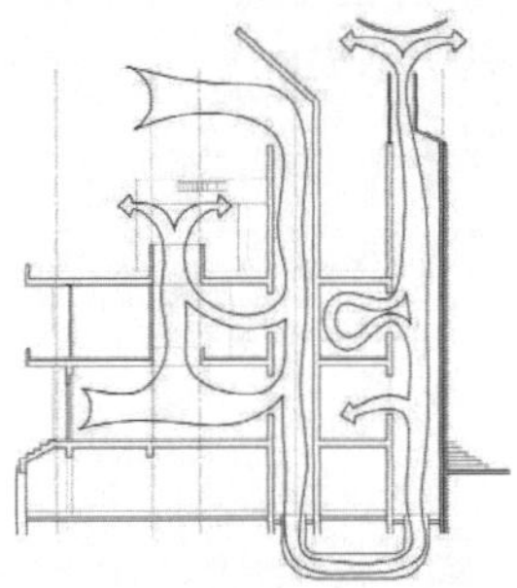

Figure 18: Section showing the wind catcher and solar chimney

Figure 19: Solar Chimney

Figure 20: Wind catcher

3. Analysis

Four Hygrometer data loggers were placed for 15 days (from 20th of October to 5th of November 2018) on the top of roofs in each case to record the ambient air temperature and relative humidity, and accordingly it was divided into two climatic zones: Zayed and 6th of October. Villa A and B are in Zayed zone and Villa C and D are in 6th of October zone. Four other Hygrometer data loggers were placed at the reception in the ground floor in each Villa to record the indoor temperature degrees at the same time duration. Readings were taken and presented in the form of charts.

IOP Conf. Series: Earth and Environmental Science **329** (2019) 012031 doi:10.1088/1755-1315/329/1/012031

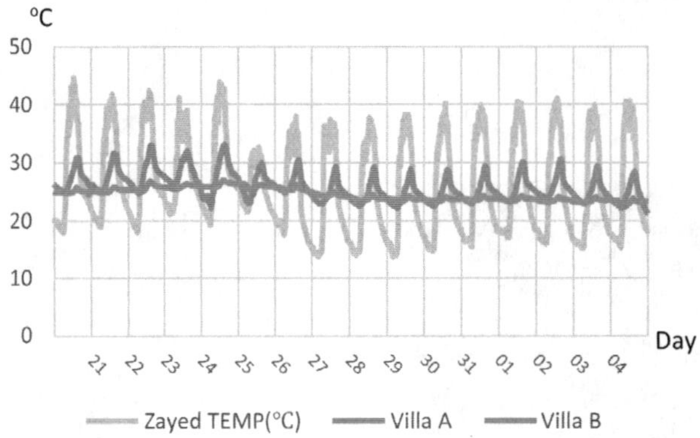

Figure 21: Chart analysis for villa A and B over two weeks

In Figure 21, the chart shows the outdoor ambient air temperature in Zayed zone throughout the day and night for 2 weeks and it also shows the indoor air temperature in Villa A and Villa B for the same duration. Form the Chart, it shows that although villa A is design according to environmental design standards as explained before, its indoor air temperature recorded is higher than Villa B which is not designed according to any environmental design standards.

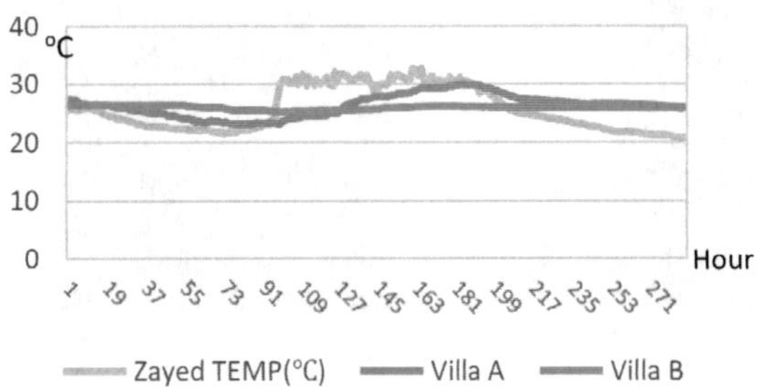

Figure 22: Chart analysis for villa A and B over 24 hours

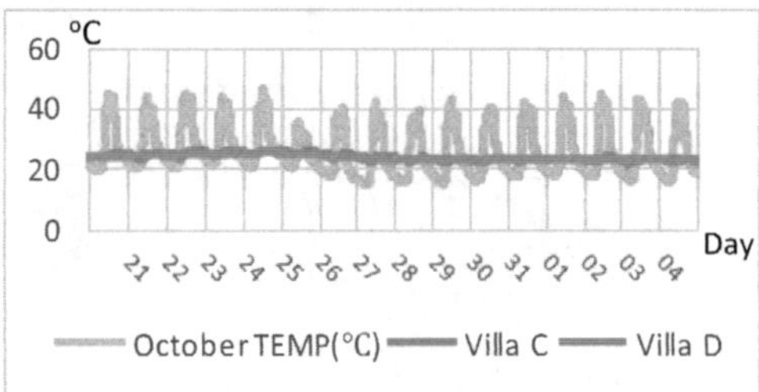

Figure 23: Chart analysis for villa C and D over two weeks

In Figure 22, the chart shows the outdoor ambient air temperature in Zayed zone for 24 hours and it also shows the indoor air temperature in Villa A and Villa B. Form the Chart, it shows that the indoor air temperature recorded in Villa A has reached 30°C while Villa B has reached 26°C.

In Figure 23, the chart shows the outdoor ambient air temperature in 6th October zone throughout the day and night for two weeks and it also shows the indoor air temperature in Villa C and Villa D. From the Chart, it shows that villa D reached 26°C and villa C reached 27°C, which is so close to each other although villa D is designed according to environmental design standards as explained before, but Villa C is not designed according to any environmental design standards, this is because villa C is surrounded by tall and close buildings which cast shadows on the villa and protect it from heat.

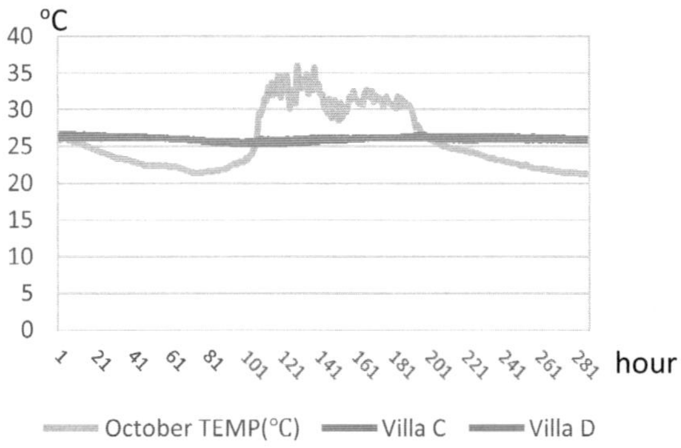

Figure 24: Chart analysis for villa C and D over 24 hours

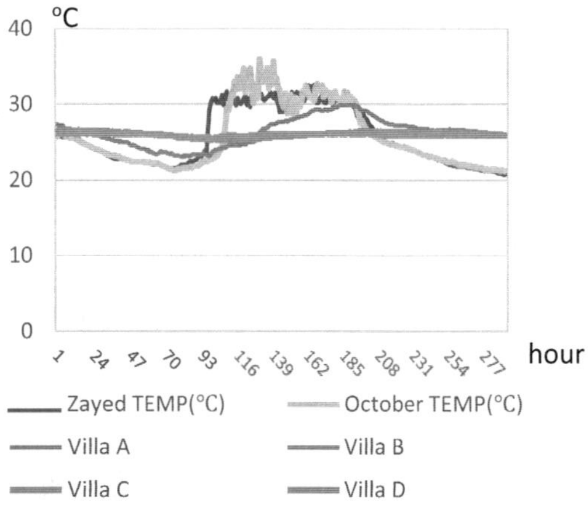

Figure 25: Chart analysis for the four houses over 24 hours

In Figure 24, the chart shows the outdoor ambient air temperature in 6th October zone for 24 hours and it also shows the indoor air temperature in Villa C and Villa D. Form the Chart, it shows that the indoor air temperature recorded in Villa D is almost the same as Villa C.

Figure 25 compares between the four cases throughout 24 hours, it shows the ambient outdoor air temperature in Zayed zone and 6th of October zone, and it also shows the indoor air temperature recorded in the four villas, where Villa A came out to be the highest case in the four cases.

This result was unexpected since Villa A was designed based on environmental design strategies for hot arid regions, but from comparing the four cases based on the parameters previously mentioned (Orientation- wall thickness- building material- WWR (window wall ratio) - Number of Floors- Type of roof –Surroundings), the following could be concluded:

- The surroundings had a great effect on the indoor air temperature, since Villa A was the only case without any surrounding, neither buildings nor landscape, which means that shading the roof and facades had a major role in decreasing the indoor air temperature.
- The WWR especially for SE and SW orientation also had a great effect since villa A had 20% WWR in SW orientation which is considered a high percentage.
- Surface area to volume ratio, where Villa A had no upper floors above the living area and so the ratio was 0.2 which affected negatively on the indoor temperature since it exposed larger areas to direct sunlight radiation.

And so, to verify this conclusion, a simulation has been made for Villa A, using Design Builder, where surrounding buildings were added to see its effect on the internal ambient temperature.

In the simulation, a model for the living space was built with abstract forms, and the weather file of the program has been used for the same duration that the Hygrometer data loggers recorded the ambient temperature. The results were as follows:

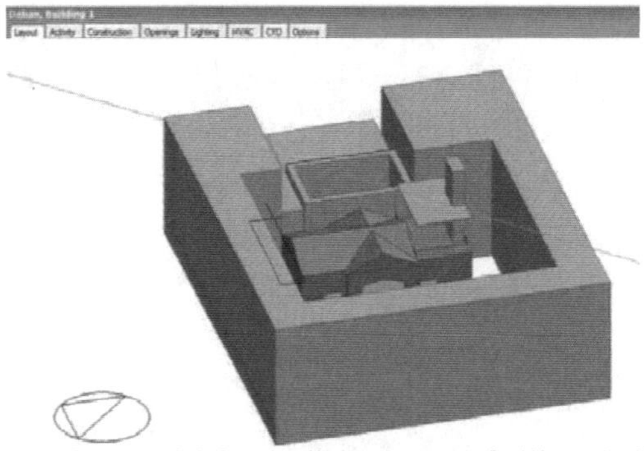

Figure 26: Model for the living space of villa A built using design builder

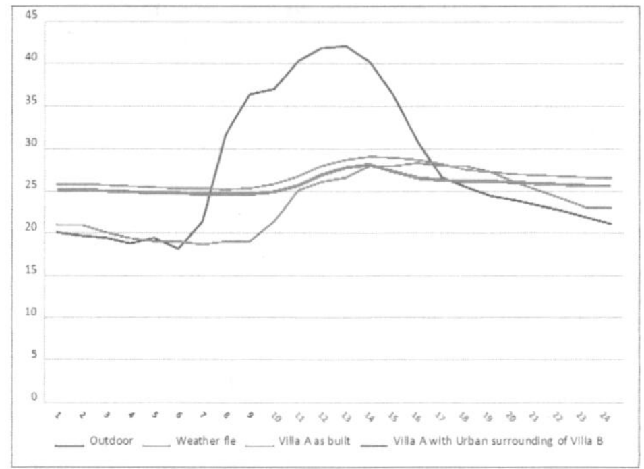

Figure 27: Simulation chart for villa A after adding surrounding context

Figure 27 shows that just by adding a surrounding context, the maximum indoor ambient temperature dropped to 28°C after being 30 ° C, and the minimum indoor ambient temperature is almost stable at 25°C after being fluctuating between 23°C and 27°C.

4. Conclusion

Villa A and D didn't have any air conditioning system, but villa B and C had air conditioning system but were turned off during the monitoring since the weather was pleasant at that time of the year and because they were surrounded by buildings same height or higher that cast shade on them, so there was no need for the ACs. The Four villas were occupied by senior residents. Fixed parameters were set for comparison between the four case studies. Those parameters were orientation, wall thickness, building material, WWR (window wall ratio), number of floors, type of roof and surroundings. The paper concluded the great effect of the surroundings in a hot arid climate where buildings should be arranged in compact arrangement. The simulation proved the importance of the surrounding urban context and its effect on decreasing internal ambient air temperature where it dropped two degrees Celsius. Therefore, it is expected that by decreasing the WWR for the SW oriented wall, and also by decreasing the surface area to volume ratio of the villa, the internal ambient temperature will drop further more degrees and could become less than the other cases. This conclusion could be helpful when designing new settlements in hot arid regions.

Acknowledgments

This study wouldn't have been possible without the support and collaboration of: Architect Ramy EL Dahan, Prof. Dr. Azza Sirry, Prof. Dr. Ahmed Abdin and Engineer Mostafa Hassan, and providing all the data needed.

References

[1] *Egyptian Electricity Holding Company Annual Report* (2016-2017). Egypt. Retrieved from http://www.moee.gov.eg/english_new/EEHC_Rep/2016-2017en.pdf

[2] Burck J, Hagen U, Marten F, Höhne N and Bals C (2019). *Climate change performance index.* Germanwatch, New Climate Institute & Climate Action Network. Retrieved from https://www.climate-change-performance-index.org/sites/default/files/documents/ccpi2019_results.pdf

[3] Hossein M, Ghadiri N, Lukman N, Ibrahim R. Aayani (2011) The Effect of Wind catcher Geometry on the Indoor Thermal Behavior, *45th Annual Conference of the Australian and New Zealand Architectural Science Association.*

[4] El Deeb K, Sherif A, El Zafarany A (2014) Effect of Courtyard Height and Proportions on Energy Performance of Multi-Storey Air-Conditioned Desert Buildings, *30th International Plea Conference.*

[5] Mahdavinejad M and Khazforoosh S (2014) Combination of Wind Catcher and Chimney for More Energy Efficient Architectural Buildings, *Sw;tainable Energy,* 2(1), 35-38.

IOP Conf. Series: Earth and Environmental Science **329** (2019) 012032 doi:10.1088/1755-1315/329/1/012032

Seasonal thermal adaptability of shopping arcades in hot and dry climates. The case of Nicosia's historic centre

E Malaktou[1, 2] **and M Philokyprou**[1]

[1] Department of Architecture, University of Cyprus, P.O. BOX 20537, 1678, Nicosia, Cyprus

[2]meleni@ucy.ac.cy

Abstract. 'Traditional' semi-open retail environments have recently received increasing scientific attention due to their 'rediscovery' as passive and low-energy architectural elements, which could promote sustainability in the urban realm. This is of particular importance when considering that fully-closed contemporary shopping centres constitute high-energy consuming spaces during the cooling period, in hot climates. The present study deals with the thermal behaviour of shopping arcades, a 'traditional' bioclimatic urban feature, in the hot and dry climate of Cyprus, by means of on-site air temperature recordings. The aim of this research is the bi-seasonal thermal performance assessment of this semi-open space, so as to inform energy retrofitting practices and promote bioclimatism in contemporary designs. The research shows that arcades were designed mainly to address the hot climatic conditions of the island. The sample tested achieved a significant daytime cool island up to 4.4K, during the cooling period. The best performance was achieved in arcades which presented minimum openings to envelope areas. In contrast, a daytime air temperature reduction which reached 2.1K, was observed in arcades during the heating period, indicating a negative thermal effect. Overall, the positive thermal contribution of arcades during the cooling period proved stronger than the negative thermal effect during the heating period.

1. Introduction

The predominance of fully-closed, conditioned and energy-intensive shopping centres, has recently renewed interest on 'traditional' passive and low-energy, semi-open retail prototypes, such as shopping arcades and covered street markets. Shopping arcades – i.e. covered passages, aligned with shops on either side, which were designed to introduce natural daylight, sunlight and air in the underdeveloped interiors of urban blocks – formed focal points for social interaction, entertainment and retail activity, mainly from the 19th until the mid-20th centuries, which marked the decline and abandonment of this building type at an international level. By providing moderated control over sun, wind and light, these semi-open spaces achieved favourable microclimatic conditions relying completely on passive means.

In the last decades, a number of studies redrew attention on arcades. These studies mainly dealt with the architectural history, social history, typology and conservation of arcades, rather than with thermal issues [1, 2, 3, 4]. Interestingly, the study of Andre Potvin [5] on the microclimatic environment of several – mainly glass-covered – arcades in London and Cardiff, addresses this subject. His research demonstrated that such spaces offer significant thermal advantages, particularly during the cold days of the heating period, compared to open streets, while they do not overheat during the cooling period despite their lightweight glass roofs. Other studies [6, 7] carried out on the thermal performance of semi-open spaces, i.e. covered passages, in the context of the Mediterranean vernacular architecture, illustrated a cooling effect of such spaces over the external environment during the cooling period.

Overall, the microclimatic response of semi-open arcades still remains unexplored to a great extent and their thermal contribution and seasonal thermal response requires systematic documentation. In particular, the thermal performance of arcades with high-mass roofs – which are common in the hot climates of the Mediterranean region – remains poorly investigated up to now. This study intends to deepen the knowledge on the seasonal thermal behaviour of shopping arcades in the Mediterranean context. In order to investigate the thermal behaviour of arcades, a representative sample was selected in the historic centre of Nicosia, for seasonal microclimatic recordings. Seasonal monitoring was undertaken in order to capture the thermal benefits and limitations of arcades over the external environment during the cooling and heating period, and to investigate the impact of arcade geometry, orientation and surrounding built environment on microclimate dynamics. The study of the thermal behaviour of arcades in Cyprus, would inform and encourage energy retrofitting practices and promote energy efficiency policies regarding contemporary design applications.

2. Methodology

The research methodology was based on on-site air temperature measurements for a sample of 14 arcades (Arc_1 to Arc_14) in the historic centre of Nicosia, which constitutes a typical Mediterranean historic centre. The case study arcades have been selected over a sample of more than 50 arcades that were identified in the designated area, based on their representativeness, historical, social and architectural significance and accessibility. Case study arcades distribute around a small area, in the Southwest part of the historic centre, which shares common urban design and microclimatic characteristics. One data-logger was positioned in each case study arcade, except in arcades Arc_2, Arc_3 and Arc_7, which were recorded in two different spots (A and B) (Table 1). Data-loggers were positioned above average human height to ensure equipment safety, under shade for protection from incident solar radiation and, where possible, in the middle of the space. The measurements were undertaken for a full-year. In order to examine the seasonal thermal variations of the arcade type, representative monthly data of the cooling (August) and heating (January) period were further analysed. Continuous measurements of air temperature were carried out using HOBO UX100-003 data-loggers. Complementary data of outdoor climatic conditions was obtained by a VantagePro weather station installed near the examined sites, at a height above the surrounding built environment. The analysis of results included the presentation of: (a) mean minimum and maximum air temperatures and mean air temperature fluctuations for the cooling and heating period, in order to estimate the level of seasonal thermal response of each arcade, over the outdoor environment and compared to the rest of the sample, and (b) median values of mean hourly air temperature differences between arcades and ambient conditions so as to assess the diurnal thermal pattern of arcades and especially to identify the magnitude and duration of seasonal cooling and warming effects within a 24-hour period. This research is now being followed by a more detailed investigation of 4 arcades (taking into consideration other meteorological parameters such as wind speed, globe temperature and relative humidity, as well as use which allowed a quantitative assessment of thermal comfort) in the framework of an ongoing PhD thesis.

3. Description of examined arcades

The examined arcades, built between early 20th century and 1970, are characterised by common materials, local techniques and various architectural styles. Depending on the period of construction, arcades were influenced by traditional, Neoclassical, British colonial, Art Deco and/or Modernist prototypes. The main materials used for their construction are thermally massive, i.e. local stone or concrete, in combination with local stone or brick. Case studies are built in a dense and compact urban fabric that limits openness to the sky. They offer a rich variety in terms of geometry, typology, orientation and level of exposure of the building envelope, though it can be noted that specific design tendencies are clearly identifiable (Table 1). More specifically, since they are directly related to the commercial routes of the historic centre which are mainly north-south oriented, the majority of the examined sites have an east-west orientation, which is considered to be less favourable in terms of solar gains and solar control potential. In terms of typology, most of the examined arcades feature a linear straight passage which prevails in the Nicosia paradigm, while a limited number represent more complex forms, such L-shaped and U-shaped layouts. The particularity of Nicosia's arcades, in the way they have been developed in landlocked sites, is also examined through the analysis of Arc_4, Arc_5 and Arc_7. Interestingly, these arcades, due to reduced accessibility, are characterised by limited openings to the outdoor environment. The examined arcades have a moderate length (5.5–30.0m). They have high-mass roofs which predominate in the historic urban fabric, with the exception of a glass-covered arcade (Arc_12), which is also surveyed for comparison reasons. Their building envelope mainly has limited opening areas with direct exposure to the outdoors, due to the compact build form and high-mass roofs, which probably allows for enhanced protection from direct solar gains. The majority of arcades with the exception of Arc_10 and Arc_12 remain vacant or with limited shops in use and are empty of people apart from occasional passer-by. For this reason, the possible gains from people or shops is not considered to be a significant factor affecting thermal performance for most of the cases. Special conditions, however, were recorded in Arc_2B, Arc_9, Arc_10 and Arc_12. Arc_2B, Arc_9 and Arc_12 are exposed to air-conditioning units that are dumping warm air; Arc_10 benefits from the operation of fans; Arc_2B and Arc_12 are semi-climatised and Arc_10 is fully climatised during the heating period.

Table 1. Sample images and architectural plans

4. Results

4.1. Thermal environment analysis

4.1.1. Air temperature monitoring results during the cooling period. The analysis of air temperature data during the cooling period, shows that arcades present lower mean maximum air temperatures (mT_{max_arc}) than the corresponding values of the outdoor environment (mT_{max_out}) (Table 2). Particularly, mT_{max_out} reaches 37.0s°C, while mT_{max_arc} ranges between 33.0°C and 35.9°C for the sample tested. Mean temperature differentials are between 1.1K and 4.0K (Table 2). The positive cooling effect of arcades is related to the combined effect of the compact built environment, the high-mass roofs, walls and floors and the deep plan layout, which reduce solar gains. It is interesting to mention that the arcades oriented along the least desirable east-west axis, or within ±30° of due east-west (e.g. Arc_6, Arc_8, Arc_12, Arc_13), do not generally elevate temperatures, most probably due to the effective shading of the surrounding compact built environment. It is noted that mT_{max_arc} tends to be lower: (a) in the arcades that have been developed in landlocked sites, i.e. Arc_4, Arc_5, Arc_7A and Arc_7B, due to the limited direct openings of their building envelope to the outdoor environment and their favourable orientation in terms of solar control which moderates external environmental extremes, (b) in Arc_10, that benefits from the ceiling fans in operation which remove excessive heat and (c) in Arc_3B that benefits from an indirect cooling effect of the cooling units that operate in the adjacent shops. Particularly, the mT_{max_arc} ranges between 33.0–34.1°C in Arc_4, Arc_5 and Arc_7, equals to 33.5°C in Arc_10 and to 33.8°C in Arc_3B, while for the rest of the sample ranges between 34.2–35.9°C. Interestingly, Arc_11 presents the highest mean maximum temperature (35.9°C), due to the fact that its roof is ineffectively insulated and is directly heated by the sun, in contrast to the other arcades that are thermally protected by the upper floors of the buildings. Arc_2B which is negatively affected by the air-conditioning units that are dumping warm air directly into the arcade, presents the second highest mean maximum temperature (35.8°C) despite its favourable north-south orientation in terms of solar protection.

In contrast, a negative warming effect is observed in all examined arcades when minimum temperatures occur. More specifically, the mT_{min_out} is 24.4°C and mT_{min_arc} ranges between 26.5°C and 28.9°C. This means that there is a temperature differential between 2.1K and 4.5K (Table 2). This negative warming effect can be attributed to the low sky view factor which reduces heat loss through net long-wave radiation to the sky, the thermally massive materials which emit the absorbed heat during daytime and to the reduction of the wind speed in the build environment. High night-time temperatures then, confirm the presence of an urban heat island effect that is intensified due to the sheltered design of arcades.

In terms of mean air temperature fluctuations (mT_{fluct}), a 'damping' effect is evident in arcades compared to the external environment, indicating the microclimatic advantages of the arcades during the cooling period. In particular, mean outdoor temperature fluctuations (mT_{fluct_out}) equals to 12.6°C, while the corresponding values of arcades lay between 4.5°C to 8.5°C. Mean air temperature fluctuation differentials are estimated between 4.1K and 8.1K. This 'damping' effect is mainly attributed to the thermally massive materials and the compact build form of both the arcades and the site which allows effective shading and absorption of solar heat. Thermal stability is much stronger in Ac_4 and Arc_5 due to the combined effect of additional parameters. In particular, Arc_4 has minimum opening areas to the outdoor environment and is protected by solar gains due to its north orientation; Arc_5 is effectively protected and shaded by the adjoining deep atrium.

4.1.2. Air temperature monitoring results during the heating period. The monitoring results of the entire heating period indicate a negative cooling effect, since the majority of arcades present lower mean maximum temperatures (mT_{max_arc}) than the mean maximum temperatures of the outdoor environment (mT_{max_out}). Arc_2B, Arc_12 and Arc_14 – which present slightly higher mean maximum temperatures compared to the corresponding values of the outdoor environment – are exceptions. Arc_10 is not included in this analysis because is fully climatised. The mT_{max_arc} in Arc_2B equals to 15.1°C, in Arc_12 and Arc_14 corresponds to 14.3°C, while mT_{max_out} reaches 14.2°C. The temperature increase in the cases of Arc_2B and Arc_12, is mainly attributed to the heat gains by conduction and convection from shops on either side of the arcades and/or by the fact that these spaces are semi-climatised. As for Arc_14, its south orientation combined with its short length, maximise solar gain potential, which in turn improves overall thermal behaviour during the heating period. The mT_{max_arc} for the rest of the samples, ranges

Table 2. Summary of recorded air temperature data during the cooling and heating period

Arc	Plan	Le.[a] (m)	Or.[b] (°)	DE[c]	P[d]	mT_{max} (°C)	mT_{min} (°C)	mT_{r} (°C)	T_{max} (°C)	T_{min} (°C)	T_{max_out}- T_{max_arc} (°C)	T_{min_out}- T_{min_arc} (°C)
Out.					C	**37.0**	**24.4**	**12.6**	**39.9**	**22.7**		
					H	14.2	5.3	9.0	18.2	0.8		
1		8.9	202	5.5	C	**34.2 (-2.8)**	**27.4 (+3.0)**	**6.8**	**36.0**	**25.7**	**1.7 – 4.0**	**-3.6 – -2.0**
					H	13.9 (-0.3)	8.6 (+3.3)	5.3	16.7	4.5	-2.3 – 2.3	-4.3 – -2.1
2A	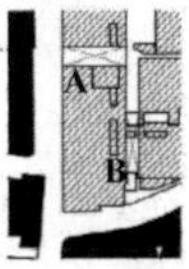	19.7	90	8.7	C	**35.6 (-1.4)**	**27.8 (+3.4)**	**7.8**	**37.6**	**25.7**	**0.3 – 2.4**	**-4.6 – -2.8**
					H	13.5 (-0.7)	8.3 (+3.0)	5.2	16.3	4.1	-2.7 – 3.5	-4.0 – -2.4
2B		24.1	180	10.3	C	**35.8 (-1.2)**	**28.1 (+3.7)**	**7.7**	**38.1**	**26.0**	**0.0 – 2.2**	**-5.0 – -3.0**
					H	15.1 (+0.9)	9.7 (+4.4)	5.4	17.3	4.8	-3.9 – 1.3	-5.8 – -3.3
3A		16.5	158	7.8	C	**34.7 (-2.3)**	**26.5 (+2.1)**	**8.2**	**37.2**	**24.4**	**1.2 – 3.1**	**-2.9 – -1.5**
					H	13.8 (-0.4)	8.1 (+2.8)	5.7	17.0	3.0	-1.5 – 1.8	-4.6 – -1.5
3B		22.5	67	10.3	C	**33.8 (-3.2)**	**28.3 (+3.9)**	**5.5**	**35.5**	**26.4**	**2.0 – 4.7**	**-4.9 – -2.6**
					H	14.2 (0.0)	8.6 (+3.3)	5.6	16.6	4.4	-2.2 – 1.9	-4.2 – -2.1
4		21.9	32	25.5	C	**33.0 (-4.0)**	**28.1 (+3.7)**	**4.9**	**35.1**	**26.4**	**2.8 – 5.4**	**-4.5 – -2.8**
					H	12.6 (-1.6)	9.7 (+4.4)	2.9	14.9	6.1	-2.7 – 3.4	-5.6 – -2.6
5		12.2	117	9.8	C	**33.1 (-3.9)**	**28.5 (+4.1)**	**4.5**	**34.8**	**26.8**	**3.0 – 5.2**	**-5.0 – -3.1**
					H	12.8 (-1.4)	9.5 (+4.2)	3.3	15.4	5.8	-2.9 – 4.3	-5.7 – -2.5
6		18.3	117	11.6	C	**35.2 (-1.8)**	**28.9 (+4.5)**	**6.2**	**37.9**	**27.5**	**0.0 – 2.7**	**-5.8 – -2.8**
					H	13.4 (-0.8)	9.7 (+4.4)	3.7	16.1	5.9	-2.8 – 2.7	-5.8 – -2.7
7A		28.2	107	23.1	C	**33.3 (-3.7)**	**27.4 (+3.0)**	**5.9**	**35.6**	**25.8**	**2.8 – 4.8**	**-4.0 – -2.1**
					H	13.0 (-1.2)	8.6 (+3.3)	4.4	15.8	4.3	-2.0 – 4.2	-4.6 – -2.2
7B		17.1	197	21.9	C	**34.1 (-2.9)**	**27.5 (+3.1)**	**6.6**	**35.7**	**26.0**	**1.1 – 5.0**	**-4.4 – -1.6**
					H	13.6 (-0.6)	8.9 (+3.6)	4.7	16.2	4.6	-2.3 – 3.1	-4.6 – -2.3
8		14.4	106	9.3	C	**35.2 (-1.8)**	**27.0 (+2.6)**	**8.2**	**38.3**	**25.3**	**0.9 – 2.2**	**-3.3 – -1.8**
					H	13.2 (-1.0)	7.6 (+2.3)	5.6	16.4	3.0	-1.3 – 3.2	-3.2 – -1.1
9		28.1	82	8.7	C	**35.6 (-1.4)**	**27.7 (+3.3)**	**7.9**	**37.7**	**26.3**	**0.6 – 2.5**	**-3.8 – -2.7**
					H	13.9 (-0.3)	8.4 (+3.1)	5.5	16.7	4.5	-2.0 – 2.1	-4.2 – -1.8
10		29.5	103	8.1	C	**33.5 (-3.5)**	**28.6 (+4.2)**	**4.9**	**35.9**	**26.8**	**2.1 – 4.6**	**-4.8 – -3.3**
					H	mechanical heating						
11		30.0	295	31.0	C	**35.9 (-1.1)**	**28.9 (+4.5)**	**7.0**	**38.8**	**27.1**	**0.4 – 1.7**	**-6.1 – -3.0**
					H	13.3 (-0.9)	8.4 (+3.1)	4.9	16.5	4.3	-2.5 – 2.4	-4.2 – -1.0
12	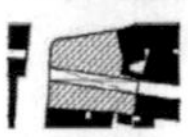	27.7	99	13.3	C	**35.2 (-1.8)**	**28.6 (+4.2)**	**6.7**	**37.5**	**27.0**	**0.5 – 2.7**	**-4.7 – -3.3**
					H	14.3 (+0.1)	9.5 (+4.2)	4.8	16.6	5.4	-2.6 – 2.8	-5.1 – -2.9
13	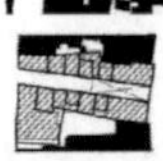	18.4	98	8.5	C	**34.8 (-2.2)**	**27.4 (+3.0)**	**7.5**	**37.1**	**25.8**	**1.4 – 2.9**	**-3.5 – -2.0**
					H	12.4 (-1.8)	7.5 (+2.2)	4.9	15.4	3.1	-1.5 – 4.4	-3.3 – -1.0
14		11.5	158	6.6	C	**35.5 (-1.5)**	**27.0 (+2.6)**	**8.5**	**37.7**	**25.3**	**0.1 – 2.6**	**-3.5 – -1.9**
					H	14.3 (+0.1)	8.0 (+2.7)	6.3	17.3	2.7	-2.1 – 1.6	-3.9 – -1.5

[a]Length
[b]Orientation, Degrees from North
[c]Degree of Enclosure; DoE = $\Sigma A_{total}/\Sigma A_{open}$ ΣA_{total}, Sum of all surface areas; ΣA_{open}, Sum of opening areas
[d]Monitoring Period, C, Cooling Period; H, Heating Period
(shaded) Shopping arcades developed in landlocked sites

between 12.4°C and 13.9°C (Table 2). Differences in mean maximum air temperature between arcades and the outdoor environment vary from 0.3K to 1.8K. These unfavourable microclimatic conditions can be explained by the high-mass roofs, the compact surrounding urban fabric and the deep plan of arcades which enhance shading and minimise the potential for direct solar access and solar passive heating.

On the other hand, arcades demonstrate a positive thermal effect when minimum temperature occurs, as all examined arcades present higher mean minimum temperatures (mT_{min_arc}) than the outdoor environment (mT_{min_out}). More specifically, mT_{min_out} reaches 5.3°C, while mT_{min_arc} ranges between 7.5°C and 9.7°C (Table 2). Differences in mean minimum air temperature between arcades and the outdoor environment vary from 2.2K to 4.4K. This positive thermal effect can be explained by the compact build form, which reduces the thermal heat losses, the slow release of heat of thermal mass which warms up the space and the effective protection from cold winds.

Table 2 indicates that the mean air temperature fluctuation of arcades (mT_{fluct_arc}) is significantly lower compared to the air temperature fluctuation of the external environment (mT_{fluct_out}). The mT_{fluct_arc} varies from 2.9°C to 6.3°C, while mT_{fluct_out} equals to 8.9°C. This 'damping' effect ranges from 2.6K to 6.0K.

4.1.3. Thermal behaviour assessment during the cooling period. The results of median values of mean hourly air temperature differences between each case study arcade and the outdoor environment during the cooling period are shown in Table 3. It is demonstrated that all case study arcades present a daytime cool island, i.e. the arcade is cooler than the outdoor environment. More specifically, the cooling effect takes place during morning to noon hours, while in some cases it lasts until late afternoon hours. The daytime cool island occurs between five to nine consecutive hours. The cooling effect becomes highest during noon or early afternoon, i.e. between 12:00h and 15:00h, at the hottest time of the day. At this critical time of the day, in terms of thermal stress or comfort, the arcade achieves a significant mitigation of outdoor temperature extremes. The highest median air temperature differences of the sample ranges between 1.0K and 4.4K. It is noted that the cooling effect in the arcades developed in landlocked sites (Arc_4, Arc_5, Arc_7A, Arc_7B) is stronger (3.3–4.4K) due to the enhanced protection from the outdoor climatic extremes. Arc_10 also presents a significant outdoor temperature reduction (3.7K) due to the effective contribution of ceiling fans. The highest median value of mean hourly air temperature differences for the rest of the sample is reduced between 1.0K and 3.1K.

During the late afternoon, night-time and early morning hours, the arcades under investigation present a heat island, i.e. the arcade is hotter than the outdoor environment (Table 3). This unfavourable warming effect reaches a peak either during late night, midnight and after midnight hours, i.e. 22:00h, 24:00h, 01:00h and 03:00h, or during early morning time, i.e. 05:00h and 06:00h, when ambient temperatures are at a minimum. It lasts between 15 and 19 hours. The highest median values of mean hourly air temperature differences registered for the sample was found to be between 2.3K and 5.1K. It is noted that the highest intensity of the heat island effect occurs when human activity is rather limited and outdoor temperatures are relatively low and thus less critical in terms of thermal comfort.

A further quantitative analysis of the cooling and warming effects of various arcades shows that the night-time heat island effect is stronger and lasts longer compared to the daytime cool island effect, with the exception of arcades developed in landlocked sites, i.e. Arc_4, Arc_7A and Arc_7B (Table 3).

4.1.4. Thermal behaviour assessment during the heating period. The results of median values of mean hourly air temperature differences between each case study arcade and the outdoor environment during the heating period are shown in Table 3. As shown in the Table, the majority of arcades tested are cooler compared to the outdoor environment during morning, noon or even early afternoon hours. Only Arc_2B, Arc_3B and Arc_14, remain warmer than the outdoor environment throughout the day. When a negative daytime cooling effect exists, it lasts from as little as two hours, to as long as eight hours. The highest median value of air temperature differences is registered at late morning or noon, i.e. between 11:00h and 12:00h. Although solar availability is maximum at this time, it remains unexploited to a large extent in the case of arcades. The highest median hourly air temperature differences of the sample ranges between 0.3K and 2.1K.

Table 3. Summary of median values of mean hourly air temperature differences between the arcades and the outdoor environment during the cooling and heating period

Arc	Plan	Le.[a] (m)	Or.[b] (°)	DE[c]	P[d]	$T_{out}>T_{arc}$ Cooling Effect Hours[e] (h)	$T_{out}-T_{arc}$[e] (°C)	Max[e] (h)	$T_{out}<T_{arc}$ Warming Effect Hours[e] (h)	$T_{arc}-T_{out}$[e] (°C)	Max[e] (h)
1		8.9	202	5.5	C	**09:00–18:00 (9)**	**0.1–2.9**	**13:00**	**18:00–09:00 (15)**	**0.6–3.1**	**06:00**
					H	10:00–13:00 (3)	0.4–0.9	11:00	13:00–10:00 (21)	0.7–3.2	04:00
2A	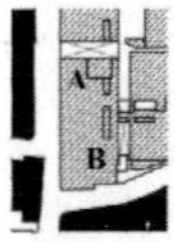	19.7	90	8.7	C	**11:00–16:00 (5)**	**0.1–1.4**	**13:00–14:00**	**16:00–11:00 (19)**	**0.2–3.9**	**22:00**
					H	11:00–15:00 (4)	0.1–0.6	11:00	15:00–11:00 (20)	0.1–3.0	01:00
2B		24.1	180	10.3	C	**11:00–16:00 (5)**	**0.1–1.0**	**14:00**	**16:00–11:00 (19)**	**0.3–3.8**	**06:00**
					H	0-hours	–	–	24-hours	1.0–4.9	02:00
3A		16.5	158	7.8	C	**09:00–18:00 (7)**	**0.1–2.3**	**14:00**	**18:00–09:00 (17)**	**0.5–2.3**	**06:00**
					H	11:00–14:00 (3)	0.1–0.3	12:00	15:00–10:00 (21)	0.2–3.1	05:00/07:00
3B		22.5	67	10.3	C	**10:00–18:00 (8)**	**0.1–3.1**	**13:00–14:00**	**18:00–10:00 (16)**	**0.8–3.8**	**05:00–06:00**
					H	0-hours	–	–	24-hours	0.1–3.3	06:00
4		21.9	32	25.5	C	**10:00–18:00 (9)**	**1.3–4.4**	**13:00–15:00**	**18:00–10:00 (15)**	**0.7–3.8**	**06:00**
					H	10:00–16:00 (6)	0.4–1.6	12:00	16:00–10:00 (18)	0.4–4.2	06:00–07:00
5		12.2	117	9.8	C	**10:00–18:00 (8)**	**1.3–3.8**	**14:00**	**18:00–10:00 (16)**	**0.3–4.1**	**06:00**
					H	10:00–16:00 (6)	0.5–1.7	12:00	16:00–10:00 (18)	0.4–4.2	06:00
6		18.3	117	11.6	C	**11:00–17:00 (6)**	**0.6–1.7**	**14:00**	**17:00–11:00 (18)**	**0.6–4.7**	**06:00**
					H	10:00–16:00 (6)	0.3–1.0	12:00	16:00–10:00 (18)	0.7–4.4	02:00
7A		28.2	107	23.1	C	**09:00–18:00 (9)**	**0.2–3.9**	**14:00**	**18:00–09:00 (15)**	**0.5–3.1**	**06:00**
					H	10:00–16:00 (6)	0.2–1.2	12:00	16:00–10:00 (18)	0.5–3.3	02:00/05:00
7B		17.1	197	21.9	C	**10:00–18:00 (8)**	**0.9–3.3**	**14:00**	**18:00–10:00 (16)**	**0.3–3.2**	**06:00**
					H	10:00–14:00 (4)	0.1–0.6	12:00	14:00–10:00 (20)	0.2–3.4	04:00–07:00
8		14.4	106	9.3	C	**10:00–17:00 (7)**	**0.3–1.9**	**14:00**	**18:00–09:00 (17)**	**0.3–3.0**	**06:00**
					H	10:00–16:00 (6)	0.2–1.3	12:00	16:00–10:00 (18)	0.4–2.5	07:00
9		28.1	82	8.7	C	**10:00–15:00 (5)**	**0.4–1.6**	**13:00–14:00**	**15:00–10:00 (19)**	**0.1–3.5**	**06:00**
					H	10:00–13:00 (3)	0.1–0.5	11:00	13:00–10:00 (21)	0.1–3.0	02:00/06:00–07:00
10		29.5	103	8.1	C	**10:00–18:00 (8)**	**0.1–3.7**	**14:00**	**18:00–10:00 (16)**	**0.6–4.3**	**06:00**
					H	mechanical heating					
11		30.0	295	31.0	C	**10:00–16:00 (6)**	**0.4–1.8**	**12:00**	**16:00–10:00 (18)**	**0.4–5.1**	**22:00–24:00**
					H	10:00–15:00 (5)	0.4–2.1	11:00	15:00–10:00 (19)	0.3–3.1	02:00
12		27.7	99	13.3	C	**11:00–17:00 (6)**	**0.1–1.5**	**13:00–14:00**	**17:00–11:00 (18)**	**0.8–4.2**	**06:00**
					H	11:00–13:00 (2)	0.3	11:00–12:00	13:00–11:00 (22)	0.1–4.2	02:00–03:00
13		18.4	98	8.5	C	**09:00–17:00 (8)**	**0.1–2.4**	**13:00**	**17:00–09:00 (16)**	**1.0–3.1**	**22:00/01:000 3:00/05:00**
					H	09:00–17:00 (8)	0.5–2.0	11:00–12:00	17:00–09:00 (16)	0.5–2.2	02:00–03:00
14		11.5	158	6.6	C	**10:00–16:00 (6)**	**0.2–1.1**	**14:00**	**16:00–10:00 (18)**	**0.1–2.8**	**06:00**
					H	0-hours	–	–	24-hours	0.1–3.1	01:00

[a]Length
[b]Orientation, Degrees from North
[c]Degree of Enclosure; DoE = $\Sigma A_{total}/\Sigma A_{open}$ ΣA_{total}, Sum of all surface areas; ΣA_{open}, Sum of opening areas
[d] Monitoring Period, C, Cooling Period; H, Heating Period
[e]Based on median values
(shaded) Shopping arcades developed in landlocked sites

Interestingly, a reverse phenomenon is observed from early afternoon, or in some cases late afternoon to morning hours, where arcades are warmer than the outdoor environment. This positive warming effect is much appreciated during the heating season, especially at these coldest hours of the day. The heat island occurs between 16 and 22 consecutive hours. However, by the time that this beneficial warming effect is at a maximum, the human activity within the spaces tested is significantly declining. In particular, the heat island intensity becomes highest a few hours after midnight, i.e. at 01:00h, 02:00h, 03:00h and 04:00h, or early in the morning, i.e. at 05:00h, 06:00h and 07:00h. The highest median value of mean hourly air temperature differences registered, was found to range between 2.2K and 4.9K. A comparative analysis shows that the heat island effect is stronger and lasts longer compared to the daytime cool island.

5. Discussion and Conclusions

Despite the restricted consideration of other than air temperature parameters in the framework of this research, which could allow a quantitative assessment of thermal comfort, some interesting outcomes were derived. Specifically, research results reveal that arcades in Nicosia were clearly designed to mainly address hot summer climatic extremes, rather than cold winter conditions, demonstrating a medium seasonal response. The hot climatic context of the island justifies this research outcome. The study revealed a positive daytime cool island in arcades during the cooling period at the hottest time of the day, which enables a significant moderation of thermal stress. Arcades which are characterised by limited opening areas to the outdoors proved to perform best during the cooling period. In addition, it is concluded that design features such as the high-mass roofs, deep shady atriums and the non-direct solar exposure of the roof – i.e. the coverage of the roof by upper floors – as well as the compact urban fabric, offer adaptability to the hot climate of the island. The operation of ceiling fans has also proved to be effective in moderating temperature extremes and fluctuations, while the negative impact of air-conditioning units has been also documented. During the heating season, it has been documented that the daytime bioclimatic potential of such urban and architectural elements is rather limited, mostly due to the shading effect. Thermal improvements need to be considered to tackle winter cold. However, it is noted that internal heat gains from shops on either side of arcade proved to be a significant source of heating. Overall, thermal differentials increase from the heating to the cooling periods. In particular, maximum temperature differentials range between 0.3–1.8K, and 1.1–4.0K during the heating and cooling season respectively. In line with this, the highest median value of mean hourly air temperature differences (i.e. when the arcade is cooler than the outdoor environment) ranges between 0.3–2.1K and 1.0–4.4K during the heating and cooling seasons respectively. This means that the cooling effect of arcades is stronger during the cooling period, where this effect is considered to be positive. These results document that the positive thermal contribution of arcades during the cooling period is stronger than their negative thermal effect during the heating period. This rather strikingly illustrates the positive bioclimatic effect of arcades.

References

[1] J. F. Geist, Arcades: the history of a building type, The MIT Press, 1989.
[2] M. MacKeith, The history and conservation of shopping arcades, Mansell Pub, 1986.
[3] M. MacKeith, Shopping arcades: a gazetteer of extant British arcades, 1817-1939, Mansell Publishing, 1985.
[4] B. Plevoets and K. Van Cleempoel, "Assessing authenticity of nineteenth-century shopping passages," *Journal of Cultural Heritage Management and Sustainable Development*, pp. 135-156, 2011.
[5] A. Potvin, "The arcade environment," *Architectural Research Quarterly*, vol. 2, pp. 64-79, 1997.
[6] E. Andreou and K. Axarli, "Investigation of urban canyon microclimate in traditional and contemporary environment. Experimental investigation and parametric analysis," Renewable Energy, vol. 43, pp. 354-363, 2012.
[7] M. Sinou and K. Steemers, "Urban space - thermal analysis of case studies," in *WIT Transactions on Ecology and the Environment*, 2003.

IOP Conf. Series: Earth and Environmental Science **329** (2019) 012033 doi:10.1088/1755-1315/329/1/012033

The Intuitive Bioclimatism and Embedded Sustainability of Cypriot Vernacular Farmhouses, as Principles for their Strategic Restoration and Reuse.

A Sivitanidou[1 2] and M Nikolopoulou[1]

[1] Kent School of Architecture and Planning, University of Kent, Canterbury, UK.
[2] School of Architecture, Engineering, Land and Environmental Sciences, Neapolis University Pafos, Cyprus

a.sivitanidou@nup.ac.cy

Abstract. This research aims to investigate the bioclimatic design, sustainability and environmental behaviour of Cypriot vernacular farmhouses, as part of small-scale family-owned farmsteads. These farmhouses, located in village outskirts, are still in operation, greatly supporting the primary sector of Cypriot economy. Current uses have conduced arbitrary modifications in their morphology, layout, construction and usage, which in the past had contributed positively to thermal comfort and household autonomy. In the present, these modifications caused farmhouses to become climatically inept, devoid of Bioclimatism and Sustainability. This research employs multiple case studies with a fully-integrated mixed-methods design. Findings are drawn via triangulation of qualitative and quantitative data from ethnographic participant observation, post-occupancy evaluation survey, interviews, in-situ documentation and environmental monitoring. Initial findings show that there is a tangible and intangible relationship between the dwelling, its inhabitants and the environment, strengthened in time due to tradition, accumulated knowledge and experience, and born out of necessity, scarcity and practicality. It is an interdependent, dynamic and adaptive reciprocity, induced by the complex interplay of socio-cultural, economic, technological, aesthetic and environmental factors. The end result was for bioclimatic design to be applied intuitively and sustainability to be embedded in the form, configuration, construction and operational modes of farmhouses. As Cyprus attempts to conform to its EU obligations regarding the sustainable development of rural environments and make the transition towards nearly-zero energy buildings, the Cypriot vernacular farmhouse can offer valuable lessons in building design and performance, whilst provide clear guidance for its strategic restoration and reuse.

1. Introducing the Cypriot Vernacular Rural Dwelling and Farmhouse

In general, the Cypriot vernacular rural dwelling, its granary, storerooms, stables, inner courtyard and orchard functioned as a small farm. It evolved from a *monochoro* (single room) to a *makrinari* (increased longitudinal space), which gradually became the archetypal shelter of Cyprus. The need for larger and separate spaces led to the development of the *dichoro* (double-space room divided by arches or posts) and the *anoi* (upper floor). The necessity to safely store items and agricultural produce added the *sospito* (inner storage room and ground-level cellar) and the *sente* (loft for crop storage). Encouraged by mild winters yet mindful of harsh summer conditions, peasants devised courtyards and transitional spaces to create outdoor working spaces throughout the year, such as the *hliakos* (semi-open space attached to the south building façade) and the *portico* (covered street access to courtyard). These elements were widely employed throughout the island, but their size, orientation and spatial arrangement created different house forms that adapted to terrain and climate variations, inhabitant activities, household needs and socio-cultural habits associated with family and communal life [1].

Farmers' main concerns were security, longevity and survival; thus, they laid their villages cascading down mountain slopes or perched on hilltops to avoid floods and disease-ridden stagnant waters; seek protection from harsh cold winds whilst exploit cool breezes and solar radiation; and free up arable land, to be cultivated and used as pastures for their domesticated animals. Rural dwellings were tightly-clustered to form neighbourhoods around church squares and water fountains, radiating from village cores and surrounded by fertile lands, apportioned to families. Inhabitants gained communal support and established trading relationships in order to amplify production and collectively manage the community's services and resources. Each community sustained a certain autonomy out of necessity, since until the 19th century no comprehensive route network serviced the island [2]. Under heavy taxations and demand for extensive exports during the British Rule, farmers were forced to maintain high productivity, though agriculture was far from being self-sufficient and ploughing was still primitive [3]. To survive, new families built farms in the sparsely-populated village fringes. Unlike village dwellings, these farmsteads were bigger in size, consisting of the farmhouse, numerous outbuildings and larger open spaces to process and store increased amounts of agricultural produce, orchards, vegetable gardens and grazing grounds for animal herds.

2. Methodology

This research focuses on vernacular farmhouses, as part of small-scale family-owned farmsteads, instead of dwellings in densely-built rural settings. These farmhouses, as an important building type, exhibited similar but also additional building forms, spatial configurations, usages and bioclimatic strategies. Furthermore, they are still in operation, greatly supporting the primary sector of Cypriot economy. Current uses have conduced arbitrary modifications in their morphology, layout, construction and usage, which in the past had contributed positively to thermal comfort and sustainable management. In the present, unregulated modifications have caused farmhouses to become climatically inept, devoid of Bioclimatism and Sustainability. As Cyprus attempts to conform to its EU obligations regarding the sustainable development of rural environments and make the transition towards nearly-zero carbon buildings[1], the Cypriot vernacular farmhouse can provide guidelines for its strategic restoration and reuse, and in general, offer valuable lessons in bioclimatic design and sustainable performance.

Previous studies have identified the form, materiality, construction and bioclimatic strategies of traditional buildings in compact built settings [4-6]. As they examine factors separately, it is difficult to consider a potentially complex interplay of several factors related to building form, organization and usage. They also provide minimal contextualization, as they take a section across settlements to study convergent typologies and characteristics [7]. In their majority, they have produced quantifiable evidence via in-situ documentation, environmental monitoring and computer simulation, but few were supported by qualitative data [8]; and rarely did they explicitly attributed inhabitant's values, perspectives and understandings, habits and norms (socio-cultural intangible aspects), as causes to form creation, placement, layout and operational mode of vernacular dwellings. Thus, this research seeks to identify the socio-cultural, economic, technological, aesthetic and environmental factors that influence form creation, placement, orientation, spatial configuration and usage. Focusing on the relationship *"dwelling – inhabitant – environment"*, it examines how these parameters affect or relate to each other; and finally, it investigates the environmental behaviour, passive heating, cooling and ventilation strategies, as well as the sustainable operational modes of farmhouses.

In order to construct a holistic and comprehensive account of the vernacular farmhouse, this study employs multiple case studies with a fully-integrated mixed research design (QUAL+QUAN). Qualitative and quantitative methods occur concurrently in an interactive manner at all stages, so that each affects the formation of the other and leads to multi-strand directions of inquiry [9]. Data triangulation enhanced the ability of case studies to generate theory, provided additional validity to assertions made by either the researcher or the case-study participants, and maintained researcher's objectivity as cases are allowed to speak for themselves [10]. The objective of field studies is to construct life narratives for each farmhouse, informed by: inhabitants' meanings ascribed to their dwelling as a family hearth, a shelter, and an economic production unit; inhabitants' experiences, actions, uses and adaptive opportunities obtained via interviews and observed in the field, as they go about their everyday

[1] European Directive 2010/31/EE was incorporated into Cypriot legislation [Law 2012 N.210(I) /2012-5A], which stipulates that after 31/12/2020 all new and renovated buildings must be Nearly Zero Energy Buildings.

life conducting daily chores and agricultural work, and seek ways to create comfortable, liveable and functional homes. Data were collected through participant-ethnographic observation, semi-structured interviews and post-occupancy evaluation (POE) survey, and in-situ documentation with field notes, architectural drawings and audio-visual media. To assess building performance, environmental monitoring provided data for 202 days, split in five periods[2]. 12 data loggers, located in indoor and outdoor spaces most frequently used by inhabitants, recorded air temperature and relative humidity every 15 minutes. Through thematic analysis of the qualitative material [11], placement, orientation, configuration, use of spaces, construction techniques, sustainable operation and management emerge as some of the main themes. Findings from quantitative data analysis were employed to triangulate and reinforce emerging main qualitative themes.

3. Results and Discussion

3.1. Case Study (Life) Narratives

The MK, GL and EL farmsteads, built in 1949, 1935 and 1925, are located in the outskirts of a village, on a natural terrace meandering along a stream between two foothills. This region boasted most farmsteads because of its well-irrigated rich soils, accessible water bodies and natural resources, which supplied a constant source of food, clean water, fuels and supplies to be transformed into everyday tools and building materials. In the past, the inhabitants of the three case studies engaged primarily in agriculture and animal husbandry; however, in order to supplement household income, they increased functionality and diversified production by establishing light industrial facilities, arts and crafts workshops and mastering other occupations (Table 1). In the present, the MK and GL farms are still in operation though at a reduced rate. As original owners grew older, they abandoned most of their activities; and their descendants chose other occupations and migrated to near-by cities, returning to their ancestral homes to harvest, process and sell agricultural products. EL farmhouse was transformed into a vacation home, its flour mill and kiln abandoned. In all cases, changes occurred gradually as families increased and required new living spaces, either housed within or built attached to existing building cells. These modifications altered original building forms to a varying degree, created new spatial arrangements and required new apertures; however, they were based on the existing bioclimatic design, employing the same passive cooling and heating strategies. In the last decade, all three farmhouses were retrofitted with thermosiphon solar water heaters and mechanical heating and cooling systems (HVAC units, underfloor heating for MK house and electric heaters in GL and EL houses). EL house installed photovoltaic panels for its demands, the surplus sold to the electricity company.

Table 1. Case study demographics, key characteristics, occupations and activities.

C. S. Codes	No# - Gender	Age Range	Relation	Habitation Status	Owners' Occupations – Activities (Typical and representative of the region)
MK	1F- 1M	82-86	O	P	Farmer, stock breeder, agricultural produce
	1F -1M	64-67	D	Oc	processing, stone mason, carpenter, basket weaver, craftsman
GL	1F- 1M	85-89	O	P	Farmer, stock breeder, agricultural produce
	1F	47	U	P	processing, lorry driver, tailor, textile and fabric production at the loom
EL	1F -1M	92-93	O	P	Farmer, stock breeder, agricultural produce
	2F	50-68	D	Oc	processing, bus driver, flour mill, stone quarry and lime kiln operator

F= Female, M= Male, O= Owners, D= Descendants, U= Unrelated. P= Permanent, Oc= Occasional

3.2. Placement – Orientation – Configuration – Use of Spaces

In all case studies, the original farmhouse, granary and stables were located on a higher plateau within the site, further away from the street and bordering properties, in order to have an even surface onto which to build, direct rainwater run-off into the orchard, maintain dry, well-ventilated and southerly-

[2] P1: 25.12.15-11.01.16, P2: 01.05-30.05.16, P3: 11.07-21.08.16, P4: 02.09-27.10.16 and P5: 12.11-06.01.17

orientated living/working spaces and stables, and protect storerooms from damp, mould and flooding (Interviewees 3, 6 and 10). The predominant building form for a peasant's house was the *makrinari*; it was preferred not only because it was feasible, in terms of economy, practicality and availability of locally-sourced materials, but also because its shape best exploited the microclimatic conditions of the region. Based on localized wind patterns, the elongated single-spaced building and its immediate principal courtyard were orientated southeast, in order to take advantage of summer winds, sea and land breezes for passive cooling and natural ventilation, allow daylight to infiltrate the interiors and utilize passive solar heating especially in the morning during the cold months (Interviewees 2, 5 and Figure 1).

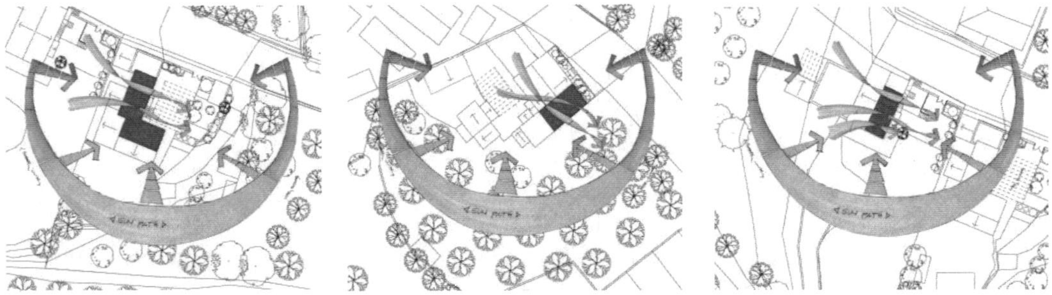

Figure 1. Placement and orientation of MK, GL and EL Farmhouses, courtyards and threshing floors, based on solar geometry and predominant summer winds. Red indicates the original *makrinari(a).*

In MK and EL farmhouses, the *makrinari* was duplicated in height so that the *katoi* functioned as a ground floor storeroom, kitchen and working area, whereas the *anoi* served as upper floor sleeping quarters. All three life narratives support that whenever inhabitants acquired economic means, they gradually added specific spaces primarily to satisfy the demands of agricultural work, e.g. a *sente* to store grains, vetch, oats and nuts, a barn to store fodder and straw, stables and a threshing floor equipped with processing equipment (i.e. grape and olive press, cisterns, distillery). Interviews revealed that traditional farmers prioritized their needs: firstly to amplify productivity and adequately store agricultural produce, and secondly to upgrade living spaces. The farming complex was enlarged not for the improvement of comfort or aesthetics, but instead, it sought to better adapt and satisfy specific needs related to the demands and necessities of agricultural life. For the Cypriot peasant, the *house* was primarily a shelter to sleep in during the cold winter nights and an enclosure to ensure family privacy, away for prying eyes; because in reality, daily life and work happened outdoors throughout the year, in open-air spaces, on flat roofs and surrounding fields. Therefore, the *house* was small enough to just fit the family, and minimal in its amenities and fixtures. It was more important to acquire and cultivate as much land as possible because it was the sufficiency of agricultural produce that provided the livelihood of peasant families (Interviewees 1, 6, 10 and Figures 1 and 2).

Studies of vernacular dwellings attest to a predominant spatial configuration: outbuildings juxtaposed in a linear fashion on each side of the farmhouse and around the courtyard. Secondary structures and open-air working spaces, perimeter walls, corrals, gardens and orchards radiated out from this centre [1]. Their siting, orientation, organization, materiality and construction techniques, heating, cooling and ventilation approaches were carefully selected for each geographic location, based on solar geometry, wind patterns, microclimate, topographical restrictions, family needs, societal structure, practical and functional requirements associated with running a farm, e.g. crop and livestock quantity and diversity, light industrial, arts and crafts activities [12]. The case studies provide evidence of the logic and intention behind these decisions. The strong bond with the land, as a principle for building form and spatial configuration, was manifested in the harmonious integration of the farming complex with its surrounding environment. The inhabitants' efforts to organise their life and work according to agricultural demands created rooms for specific use, i.e. dark, dry and cool storerooms for the preservation of foodstuff and agricultural produce, but also multi-purposed spaces to be used differently based on the agrarian calendar (interviewee 4). The need to live and work closer to nature and in accordance to the climate, manifested in open-air working and living spaces such as the courtyard, divided the farming compound into thermal zones and allowed for seasonal use of indoor and outdoor spaces: summer (cool and breezy) and winter (sun-bathed or protected) living and working quarters, sheltered stables, airy and shaded animal corrals (interviewee 9). As times changed, the family increased and required new living spaces, either housed within or built attached to existing building cells. These modifications altered original building forms to a varied degree, created new spatial arrangements and required new apertures; however they were designed in relation to the courtyard in order to contribute to natural cross ventilation, solar heating, passive cooling and night purging, and increase the existing seasonal use of spaces (interviewees 2, 9 and Figures 1 and 2).

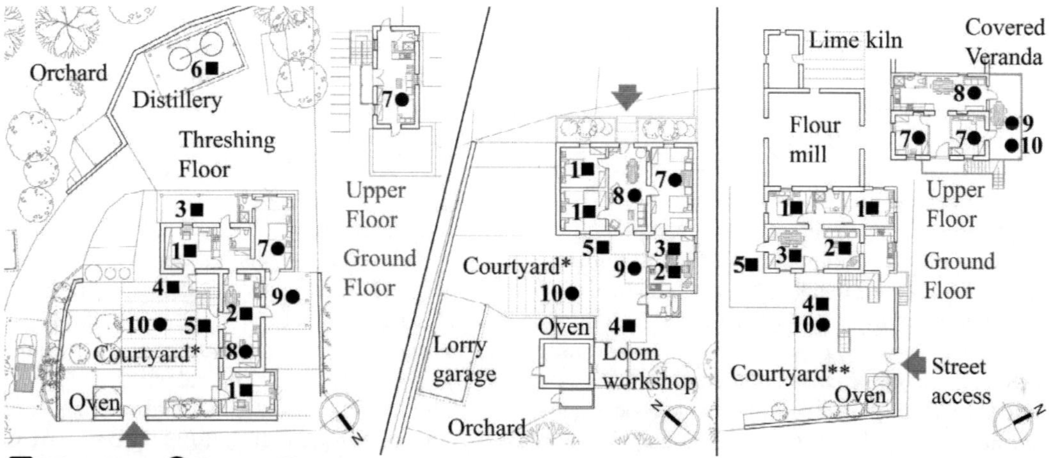

■ Winter Use ● Summer Use

1: Sleeping, 2: Living - cold and rainy days, 3: Working/ living - sunny days, 4: Outdoor working – morning, 5: Outdoor working – afternoon, 6: Outdoor living – evening (stills operated), 7: Sleeping, 8: Living - noontime/late evening, 9: Outdoor working – morning, 10: Outdoor working/eating/relaxing - late afternoon and evening.

Figure 2. Seasonal spatial use in MK, GL and EL Farmhouses, based on location of fireplaces, openings, sunspots, shaded/airy outdoor areas, orientation, cross ventilation and sunlight penetration.

Field observations and inhabitant's accounts, provided evidence that the courtyard was a necessity for Cypriot peasants. Fundamentally, it functioned as a transitional space between the public community realm and the private sphere of family life. It also had a socio-cultural character: during family gatherings, coinciding with religious celebrations and agrarian cycles, extended kin and friends feasted under the shade of climbing vines in the courtyard (Interviewee 3). In terms of environmental ambience and comfort, the courtyard enabled inhabitants to conduct their chores in the open air, surrounded by nature, protected from the scorching sun but in broad daylight and in the breeze, away from the airless, smelly, gloomy and crumped indoors. During temperate weather (lasting most of the year), residents cooked and dined daily under its shade; gathered as a family; relaxed after a hard day's work; engaged in arts and crafts; took an afternoon nap, and slept during breezy nights. In winter and autumn, inhabitants preferred to sit in different spots in the courtyard (depending on the hour) to soak up the sun (interviewee 5). The courtyard, as the main living and working area, was designed with practicality, multiple functionality and flexibility to accommodate seasonal and daily household and agrarian activities. The numerous installations and equipment needed to process agricultural produce and carry out chores (e.g. fire pit, water basin and oven) were placed around its perimeter to facilitate egress, and grouped to subdivide the space into smaller sections. The olive and grape press, distillery and smoke house were similarly placed around the threshing floor, the secondary multi-purpose working area (interviewee 1). All three farmhouses were sustainably managed. Inhabitants preserved vegetables, plants, herbs and fruits for long-term consumption; processed harvested crops (grains, olives, fruits, nuts, legumes, beans etc.); and prepare dairy products and culinary delicacies (e.g. cured pork, sausages, pies and pastries). In the courtyard and the threshing floor, various agricultural produce were processed, stored, sold or traded for other necessities and precious commodities (interviewees 2, 3).

The MK farmhouse, barn and stables were organized in a Π-shape around the courtyard, so that the latter served as the main movement distributor (spaces did not communicate internally and upper floors was accessed via an open-air staircase). The building acted as a buffer against winter winds, whereas in the summer it acted as a protective barrier, stopping any by-products from agricultural activities at the threshing floor from entering indoor living spaces, e.g. winnowing of grains when winds blew from the west (interviewee 1). The courtyard exploited diurnal temperature fluctuations to become a thermal comfort regulator and microclimatic modifier, employed by indigenous builders based on amassed experience [13]. In all houses, the courtyard acted as a light well that trapped sunrays in winter when solar radiation was desirable. Its light-coloured stone paving acted as a heat sink, storing solar heat and radiating it at night. In the summer, climbing vines and trees shaded the courtyard's floor, walls and openings, thus reducing solar glare and overheating. Temperature difference underneath the vegetated canopy created a descending flow of cool air, maintaining comfortable air temperatures until the following day. The courtyard as a coolth sink, captured air movement: orientated towards summer winds, it acted as a breezeway, increasing wind speed and cooling down walls via convection. Surrounding buildings featured openings on opposite walls, so that the courtyard induced cross ventilation, fresh-air intake and passive cooling of indoor spaces (interviews, field observations).

3.3. Construction techniques and thermal inertia of building envelope

Due to scarcity of resources and transportation difficulties, peasants found materials in the vicinity of the building site. Builders prolonged the life and use of construction materials because they had knowledge for their innate properties and behaviours in response to climatic conditions, weathering and structural laws [14]. Hence, they chose specific locally-sourced and climatically- suitable materials to create air-tight and waterproofed shelters and provide certain levels of thermal comfort to inhabitants. In these farmhouses, building thermal inertia was secured by solid stone walls (thermal mass) and multi-layered roofs (high insulation). The load-bearing walls were built with coursed ashlar masonry, because sandstones and limestones could be quarried from near-by locations [15], cut into blocks and set in alternating courses, to erect an inner and an outer wall, with an overall 50-60 cm thickness (Figure 3). 'Through stones' (placed perpendicularly on both walls) were used to prevent wall delamination and crumbling. The space in-between was filled with a mixture of stone chippings, clay, straw and vinegar, in order to serve as an aggregate reinforcer to the composite double wall. Mortar, made with crushed stone powder, fly ash and clay, increased wall hydrophobicity and resistance against damp accumulation. Chamfered corner stone quoins increased aesthetics and strengthened wall junctions against earthquakes. The original buildings featured roofs composed of many layers. The width of *makrinaria* was determined by the length of tree trunks used as beams; oak, poplar and cypress were preferred, as they required minimum curing, weatherproofing and maintenance due to their high resin content. Reed wattles and dry rushes were laid on top to trap air and provide thermal insulation. Layers of reddish clay, known for its plasticity, uniformed shrinkage, fineness of grain, cohesion and hardness sealed the roof. To fortify its imperviousness and weatherproofing against possible cracking during large diurnal temperature differences, every year inhabitants added layers of packed earth, so that the rain would wash the mud into forming breaks (interviewee 3).

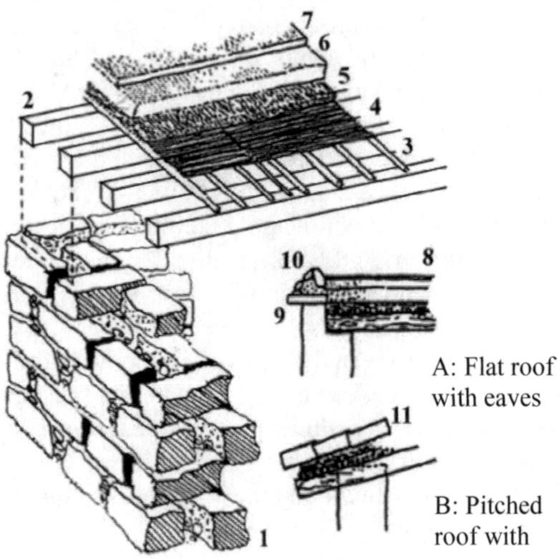

1: double stone walls with 'through stones', 2: timber beams, 3: pole rafters, 4: reed wattles, 5: rushes, shrubs and straw matting, 6: clay, 7: packed earth, 8: reddish clay or lime plaster, 9: stone wall plate (eave), 10: stone kerb, 11: terra cotta roof tiles or stone plates.

Figure 3. MK Farmhouse roof system and wall construction with 'through' stones and corner quoins.

The adaptive thermal comfort model [16] was used to determine indoor optimum comfort temperature (Topt comf) in a linear relationship with the exponentially weighted running mean of the daily mean outdoor air temperature (Trm), taken from MK Courtyard. Based on BS 16798-1:2019, the formula for Category III: Existing naturally-ventilated buildings with moderate expectations for thermal comfort was selected. Environmental monitoring and POE surveys showed that inhabitants were satisfied with the deliberate passive cooling, ventilation strategies and adaptive opportunities their houses provided, which combined with the thermal inertia of the building envelope, maintained indoor temperatures within the thermal comfort range for most of the year (Figure 4: Spring period). The EL living room temperature had the steadiest moving average, indicating smaller diurnal fluctuations and day-to-day changes, mainly because the farmhouse was uninhabited and no openings were operated during this monitoring period[3]. In the summer, indoor evening temperatures in all case studies remained within the comfort range. Day temperatures exceeded comfort values in MK and GL Farmhouses, except in EL residence, which was again largely unoccupied. Thus, the EL farmhouse was used to determine the dynamic effects of masonry-wall thermal mass on indoor temperatures, indicating a time lag of approximately four hours; and confirmed that the multi-layered roofing system offered protection against temperature extremes and intense insolation, keeping indoor temperatures at comfortable levels even during hotter months and reducing winter conventional heating loads, which is in accordance to recent studies [7, 17]. Field observations and POE surveys attested that due to the nature of their daily chores, inhabitants lived outdoors, preferring to work in shaded corners and breezeways in the courtyard. In the summer, they opened all apertures to induce air currents from and to internal and external spaces and dissipate heat build-up, in preparation for evening indoor living, instead of turning on mechanical cooling. They operated mechanical heating and fireplaces during winter, as the building envelope could not maintain satisfactory thermal comfort levels.

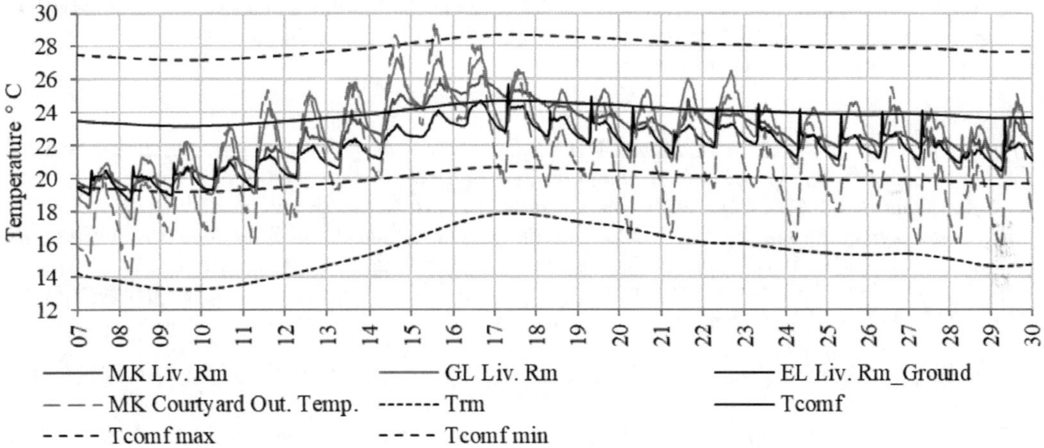

Figure 4. Living room indoor versus outdoor temperatures, Trm and Tcomf, during 07-30.05.2016.

3.4. *Sustainable Management*

In terms of waste management, they recycled and reused everything, from broken items transformed into tools, to multi-purposed apparatuses and riggings. At the garbage site, fruit and vegetable peels and grape pomace were mixed with fodder as livestock feed, or decomposed and mixed with manure to make a richer fertilizer. Women collected water from wells and streams, and harvested rainwater from tiled roofs to store in reservoirs (interviewees 5, 9). Mustard greens, pumpkin vines, sunflowers, mulberry, poplar and eucalyptus trees were planted near outhouses and stagnant waters to decontaminate the soil as biological hydraulic containment systems, contaminant root up-takers, stimulators and hyperaccumulators. Wild grasses and clover thrived in the orchard as stabilizers and as fresh food for animals. Farmers, equipped with an in-depth knowledge of botany, cultivated herbs as medicines, food flavouring agents, preservatives and natural repellents (interviewees 4 and 11). Nut and fruit trees and

[3] EL living room: mean Diurnal Temperature Difference (DTD) = 2.2 °C, StD= 0.84 °C, Diurnal Temp. Range (DTR) = 18.6-25.8 °C (abs. max-abs. min. temp.). Outdoor Temp.: mean DTD = 6.5 °C and DTR = 14-29.2 °C.

sycamores for in-house sericulture were planted in the orchard. Rotational crop management of vegetables increased crop yield, replenished nutrients, improved soil structure, reduced erosion and pollution, controlled pest infestation and diseases, and reduced stress of weeds (interviewees 3 and 6).

4. Conclusions

The adaptive building form and spatial configuration of Cypriot vernacular farmhouses and their compounds evolved by correlating microclimatic and environmental characteristics, socio-cultural needs related to family and communal life, material scarcity and construction limitations, practical and functional farming requirements. Each indoor and outdoor space acquired multiple functionality and usage, certain materiality, design, placement and orientation in order to: fortify each other to optimize building performance, quality and comfort; offer thermal diversity and adaptive opportunities to occupants; and allow for a more feasible management. Economic viability necessitated these practices; but it was the strong relationship between the inhabitants, their dwelling and the environment, ingrained by tradition, accumulated knowledge and tried experience that led to the intuitive application of passive heating and cooling strategies and sustainable operational modes. The three case studies, as examples of modified and current use, support that if similar vernacular farmsteads are restored and reused based on their embedded bioclimatic design, they can reduce heating and cooling demands on mechanical heating and cooling systems. Solar water heaters and renewable energy systems can be installed either on their roofs (as they are properly oriented) or in their larger plots, to satisfy electricity requirements and provide a constant supply of hot water. Since most of them have wells and harvest rainwater, they can operate small water-treatment facilities for irrigation and animal watering. They can also increase their productivity and profits, by utilizing their existing resource and waste management and seasonal crop rotation practices. As costs for new construction are increasing, due to the fast transition towards zero carbon buildings, these retrofitted and upgraded vernacular buildings and sustainable farms offer more viable solutions, especially to low-income working families, and provide employment alternatives in the primary sector of the Cypriot economy.

References

[1] Ionas I 2003 *La Maison Rurale de Chypre XVIIIe- XXe siècle* (Nicosia: Cyprus Research Centre).
[2] Gaudry A 1855 *Recherches Scientifiques en Orient* (Paris: Ministry of Public Works).
[3] Richter H A 2007 *A Concise History of Modern Cyprus* (Ruhpokling: Franz Philipp Rutzen).
[4] Philokyprou M and Michael A 2014 *The Bioclimatic Dimension of Traditional Architecture of Cyprus* (Nicosia: Republic of Cyprus and European Regional Development Fund).
[5] Papacharalambous G Ch 2001 *The Cypriot Residence* (Nicosia: Cyprus Research Centre).
[6] Theodosiou A and Pitta A 1996 *Settlements of Akamas: Architecture* (Nicosia: Leventis Foundation).
[7] Philokyprou M, Michael E, Malaktou E and Savvides A 2017 *J. Building and Environment* **111** pp 91-109.
[8] Dincyurek O, Mallick F H and Numan I 2003 *J. Building and Environment* **38** pp 1463-73.
[9] Tashakkori A and Teddlie C 2009 *Foundations of mixed methods research: Integrating quantitative and qualitative techniques in the social and behavioural sciences* (Thousand Oaks: Sage).
[10] Stake R E 2005 *The Sage Handbook of Qualitative Research* ed N Denzin and Y Lincoln (Thousand Oaks: Sage) pp 443-466.
[11] Miles M B, Huberman A and Saldana J 2014 *Qualitative Data Analysis* (Thousand Oaks: Sage).
[12] Oktay D 1999 Sustainability of Housing Environments: Assessments in Cypriot Settlements *The Power of Imagination* ed T Mann (Orlando: EDRA Publications) pp 147 – 158.
[13] Serghides D 2010 *J. Open Construction and Building Technology* **4** pp 29-38.
[14] Papadouris G 1990 *J. Archaeologia Cypria* **2** pp 123- 135.
[15] Christou P and Elliotis M 2016 *Construction and Retrofit Methods of Stone Masonry Structures in Cyprus* (Sharjah: Bentham Open).
[16] Humphreys M, Nicol F and Roaf S 2016 *Adaptive Thermal Comfort: Foundations and Analysis* (London: Routledge).
[17] Heracleous C, Ioannou I, Philokyprou M and Michael E 2017 *Proc. Int. PLEA Conf. on Design to Thrive (Edinburgh)* vol 3 ed L Brotas *et al* (Edinburgh: NCEUB 2017) pp 5030-37.

IOP Conf. Series: Earth and Environmental Science **329** (2019) 012034 doi:10.1088/1755-1315/329/1/012034

Renovation activities by home-owners. Analysis of energy and non-energy renovation activities and possible connections.

B Gepts[1,2], E Nuyts[1,3] and G Verbeeck[1]

[1] Hasselt University, Faculty of Architecture and Arts, Agoralaan Building E, B-3590 Diepenbeek, Belgium
[2] Essencia, marketing agency specialized in construction, Beverlosesteenweg 114, B–3583 Paal, Belgium
[3] PXL University College, Elfde-Liniestraat 24, B-3500 Hasselt, Belgium

bieke.gepts@uhasselt.be

Abstract. In order to achieve a sustainable built environment by 2050, the European building stock needs to be retrofitted. Also in Belgium, with 75% of the residential buildings built before 1980, deep energy renovation is stimulated. The Flemish government premises long term goals for the energy performance of the existing building stock in the Renovation Pact (2014). To be able to monitor the progress and effectiveness of the Renovation Pact, knowledge of the actual renovation rate is indispensable. Since years, a renovation rate of 0.7% to 1% is communicated as an argument that renovation should be stimulated. However, there is little empirical evidence for this figure and how it reflects the actual energy renovation activity. Furthermore, non-energy renovation activities are considered as possible gate for home owners towards energy renovation, but little is known about these renovation activities. Since 2013, a yearly survey is done by Essencia Marketing, a marketing agency specialized in the construction sector, on the renovation activity of Belgian households. A representative sample of 1000 households indicates their renovation activity in the previous year through an online questionnaire on structural works (façade, wall, roof, etc.), technical works (HVAC, electricity, etc.) and energetic works (insulation, windows, etc.). For the period 2012-2018, the results show that each year almost 2,1% undertakes a deep energy renovation activity (excluding decorative works). To better estimate the energy renovation activity in Belgium, the paper will firstly discuss the different types of renovation works home-owners execute, and determine the share of energy related renovations. This will be confronted with official statistics on (energy) renovation. Then the non-energy renovation activity will be discussed to determine the share of renovations that could serve as a gate to promote deep energy renovation. This could inform policy makers on policy actions to stimulate combined renovations.

1. Introduction

In the context of achieving a sustainable built environment by 2050 the regional governments in Belgium spelled out measures and regulations to retrofit the existing building stock. To monitor the progress and to assess the need for adjustments of the policy actions for energy renovation, detailed information about the renovation activity is indispensable. Since years, a renovation rate for Flanders of 0,7 to 1% is communicated in the Renovation Pact [1] and used as an argument that renovation should be stimulated. However, the empirical basis for this figure is not clear and it is assumed that it only represents deep energy renovations with building permit. It is highly probable that the actual renovation rate is higher, since many renovation measures in Belgium, especially energy-related such as roof insulation, replacement of windows or heating system, installation of solar panels, can be executed without building permit and are thus poorly documented.

Therefore, an in-depth analysis of the entire renovation activity is necessary. At first to clarify the mentioned 1% and in a next step to enhance the knowledge about the type of renovation and the combination of renovation works that take place. This could inform policy makers on the progress of their national action plans to meet up the Energy Performance of Buildings Directive (EPBD) [1] requirements, which urge member states to lower the CO2 emissions of the new and existing building stock. The relation between energy related renovation works and CO2 emission reduction is already provided in other literature [1].

2. Methods and means

In Belgium, where only deep renovations need a building permit, most renovation activity remains under the radar. As a marketing agency specialized in construction, Essencia Marketing explores yearly the renovation market by means of an online survey research among home-owners [4]. Since 2013 the renovation activities of a sample of the Belgian population, representative by age, region and education is examined. At the beginning of a year, respondents are invited to participate in an online survey. Besides general questions on their house, they are surveyed on eventual renovation activities they executed in the previous year. Tenants and owners of an apartment are screened out. Also the home-owners of a house built less than five years ago are left out. With a yearly sample between 500 and 1000 respondents since 2013, a database of 5449 respondents is obtained, considering the renovation activity between 2012 and 2018. Although through the years the questionnaire slightly changed, the methodology and used definitions remained unchanged. The analysis of the data is performed in JMP Pro 14.1 by means of consumer research method.

2.1. Characteristics of the Belgian home-owner

Most of the home-owners live in Flanders and 74% is older than 40. 47% owns a house which is built before 1970, so before the oil crises that induced the public attention for energy efficiency. 63% bought an existing house, whereas 37% was the first inhabitant of the (new built) house (Figure 1).

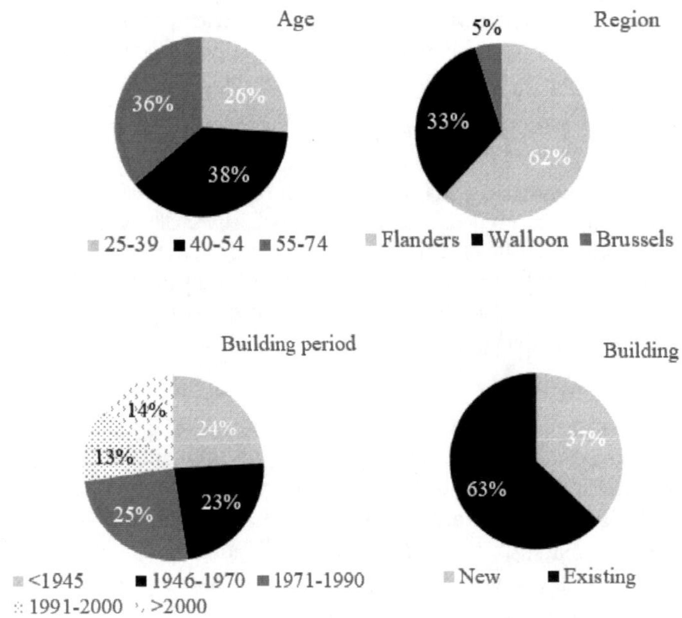

Figure 1. Characteristics of the Belgian home-owner.

2.2. Renovation activities considered

The considered type of renovation activity is limited to structural work (facades, walls, roofs, etc.), energy related work (insulation, heating, doors and windows, etc.) and other work (kitchen, bathroom, etc.). Decorative work (paint, furniture, carpeting, etc.) is not included. In this paper only the renovation activity of home-owners is under study. The activities of tenants or apartment owners are left out. An overview of the studied renovation work and their classification is given (Table 1).

Table 1. List of the renovation works and their classification.

Renovation work	Classification
Ventilation	Energy related
Heating elements (radiators)	Energy related
Doors, windows and glass	Energy related
Insulation (roof, wall and floor)	Energy related
Heating installation	Energy related
Heating pump	Energy related
Solar panels	Energy related
Solar water heater	Energy related
Roof covering	Structural
Wall (Exterior and interior)	Structural
Extension of the house	Structural
Porch	Structural
Garage	Structural
Garage door	Other
Works on electricity net (cables, switch,…)	Other
Roof window or dome	Other
Bathroom	Other
Kitchen	Other

3. Results

At first, the type of renovations and their evolution from 2012 until 2018 is presented. An analysis of their characteristics is discussed. A yearly renovation rate for the sample and its evolution is obtained. Secondly, an estimation of the overall renovation rate for Flanders is performed by relating the obtained results of the Belgian home-owners to the Flemish population including tenants, apartment owners and home-owners of whom the house is older than five years. Finally combinations of renovations are discussed.

3.1. General overview of renovation activity

Overall, the results of the online survey of Essencia marketing show that on average yearly 37% of home-owners executes at least one of the listed renovation works above, but with a tendency towards a decreasing percentage in the period 2012-2018. With an average of 2.4 renovation works that renovators perform in the same year, it is obvious that several renovation works are combined. Between renovation types, energy-related renovations are most popular followed by other types of work and structural work (Figure 2). Every year on average 24.6% of home-owners performs at least one energy-related renovation. Other works are performed yearly by 20.2% of the home-owners and structural works by 13%. A more detailed insight per year shows a declining trend in the energy related renovations for the period 2012-2018, although still most popular. The number of structural renovators fluctuates in time, but shows a steady trend line. The other renovation works suffer from a decline, but not as severe as the energy related renovations.

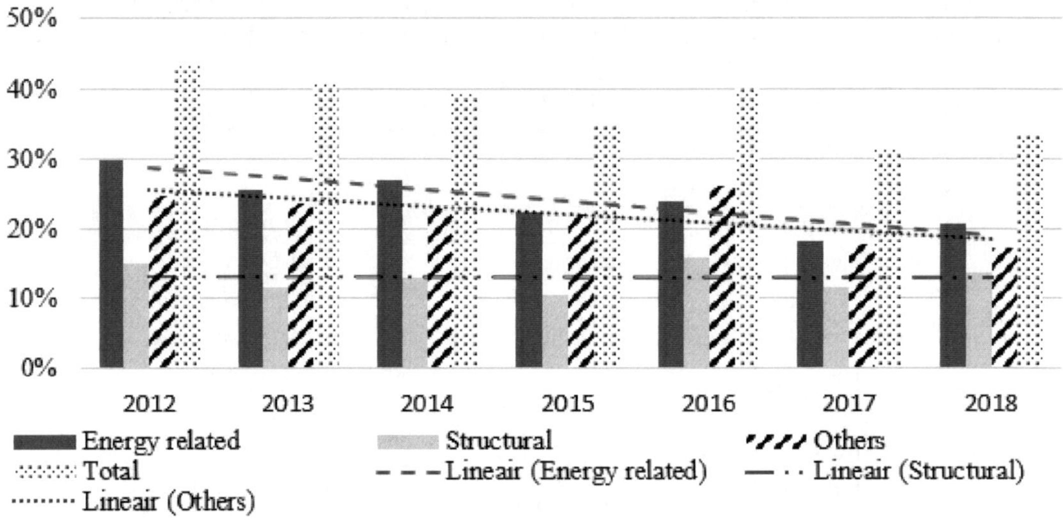

Figure 2. Overview of the renovation activity (in % of home-owners).

With regard to the presence of a building permit, yearly an average of 13.1% of all renovators applies for a building permit (period 2012-2018). Due to an adaptation of the questionnaire since 2016, a subdivision by type of work is only available until 2015. Before 2016, the application of a building permit was questioned for each renovation work separately, whereas since 2016 the application is questioned for the renovation in general. An average of 17.4% who performed an energy-related renovation received a permit. For structural works the percentage with building permit is expectedly higher (25.6%). While for other works the percentage with building permit remains low at an average of 9.9%.

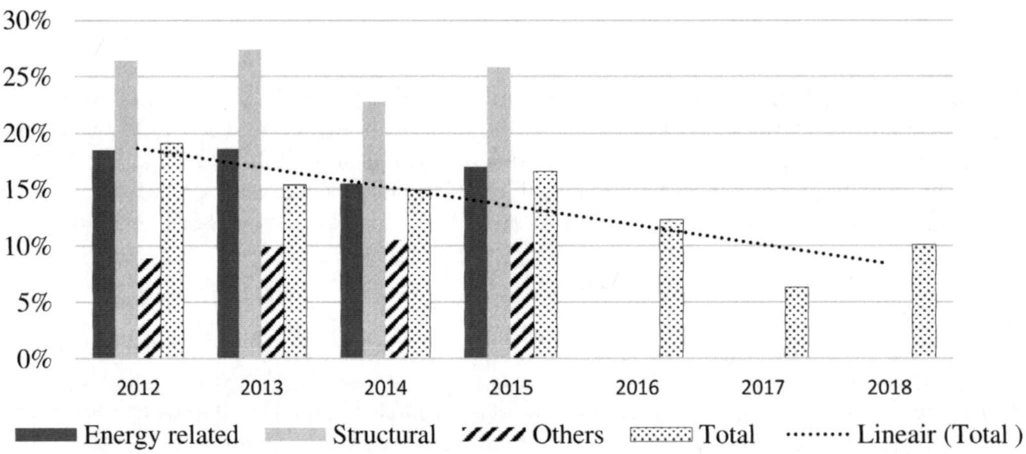

Figure 3. Overview of the application for a building permit
(in % of renovators per type of renovation work).

Furthermore, an analysis of the renovation activity by building period shows more activity in houses built before 2000 (Figure 4). 40.5% of houses built before 1970 underwent a renovation activity compared with 27% of houses built after 2000. In addition, energy-related and structural works are performed more in houses built before 1990.

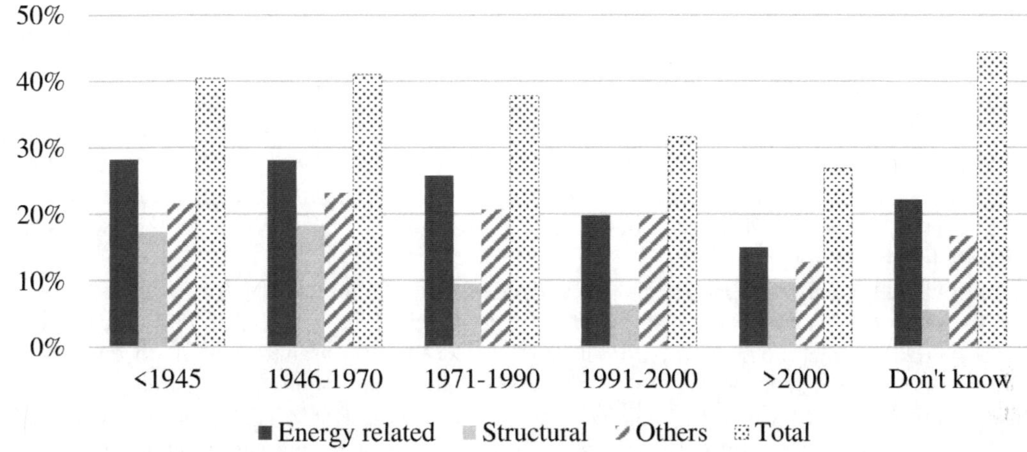

Figure 4. Overview of the renovation activity per type of renovation per building period (in % of renovators per building period).

3.2. Calculation of renovation rate

To compare the renovation rate of 37% of Belgian home-owners with the mentioned renovation rate in the Renovation Pact, further calculations are necessary. Firstly, to reconstruct the 1% renovation rate an analysis of the delivered building permits for renovation is performed. Secondly, a comparison is made with the Essencia Marketing survey. Finally, the reliability is explored by confronting the findings with other sources.

Statbel, the Belgian statistical office, reports on the delivered building permits for renovation [5]. In time, the regulation of building permits for renovation is severely adapted by making the application for a building permit no longer compulsory for small renovation works. Only deep renovation works (e.g. structural adjustments or expansion of the living area with more than 40m^2) need to be accompanied by a permit. A comparison of the delivered building permits with the existing building stock shows an average renovation rate of 0.7% for buildings and 0.6% for living units in the period 2012-2018 in Flanders (Table 2).

Table 2. Overview of the building permits and building stock of Statbel for the period 2012-2018 (Flanders) [5].

	Building permits		Building stock		Renovation Rate		Building stock >5Y		Renovation Rate >5Y	
	Buildings	Dwellings	Buildings	Dwellings	Buildings	Dwellings	Buildings	Dwellings	Buildings	Dwellings
2012	15 846	17 593	2184 307	2836 293	0.7%	0.6%	2095 351	2653 482	0.8%	0.7%
2013	16 258	18 171	2197 301	2866 649	0.7%	0.6%	2110 203	2681 973	0.8%	0.7%
2014	16 435	18 487	2209 107	2894 414	0.7%	0.6%	2126 489	2712 703	0.8%	0.7%
2015	16 549	18 161	2220 608	2920 045	0.7%	0.6%	2142 761	2746 014	0.8%	0.7%
2016	15 760	17 856	2231 800	2944 668	0.7%	0.6%	2156 798	2775 682	0.7%	0.6%
2017	12 752	14 043	2244 288	2972 921	0.6%	0.5%	2170 299	2805 902	0.6%	0.5%
2018	16 526	16 526	2257 019	3006 601	0.7%	0.5%	2184 307	2836 293	0.8%	0.6%
Average	15 732	17 262	2220 633	2920 227	0.7%	0.6%	2140 887	2744 578	0.7%	0.6%

Until January 2018, a home-owner of a building older than five years could profit a reduced VAT rate of 6% instead of 21%. Therefore, also a calculation of the renovation rate is also made based on building permits and buildings older than five years. However, still a renovation rate of 0.8% for buildings and 0.7% for dwellings is obtained.

When the renovation activity in the Essencia Marketing survey for the same period is analysed, a yearly rate of 37% of the Belgian home-owners is obtained (Figure 2). To compare the achieved renovation rate with the pronounced number of 0.7% to 1%, the Essencia Marketing survey need to be transposed from 1) Belgian home-owners to the Flemish population of owners and tenants, 2) home-owners to owners of dwellings, 3) houses older than five years to all houses and 4) all renovation to deep energy renovation.

1) Instructed by the Flemish government, a face-to-face survey among 10 000 households is performed in 2012-2013 about the living conditions and quality in Flanders, further referred to as GWO2013 [6]. 70% of the households owns their dwelling. However, the online survey of Essencia Marketing results in an owners rate of 83% probably caused by an underrepresentation of the social weaker households. Due to the larger sample and the higher degree of representativeness in GWO2013, the owners rate of 70% is used for conversion.

2) In addition, the type of dwelling needs to be added to transpose home-owners to dwelling owners. In the Essencia Marketing survey 14% of the owners were screened out because of their type of dwelling (apartment), so a correction factor of 86% is added. Hereby, the assumption of no energy-related renovation for owners of apartments is made, this could cause a small underestimation.

3). Based on the results of Table 2 a rounded average correction of 85% is necessary to expand from houses older than five years to the entire housing stock (comparison of renovation rates 0.7/0.8 (87.5%) - 0.6/0.7 (85.7%) or 0.5/0.6 (83.3%)).

4) Where the Essencia Marketing survey takes into account all renovation works, the mentioned renovation rate in the Renovation Pact is focused on deep energy renovation. Due to the absence of more specific information, it is assumed that renovations with building permit can be seen as deep (energy) renovation. Therefore, only renovators with building permit are withheld.

A recalculation of the obtained 37% home-owners renovators (Table 2) based on the explained assumptions for all renovation and for energy-related renovations is shown in table 3. The Essencia Marketing survey reports for the Flemish population a yearly rate of 2.5% for overall renovation activity and 2.1% for deep energy renovation. It is obvious that the detailed calculation of the Essencia Marketing survey surpasses the assumed 1%. Furthermore, the Essencia Marketing survey results in a yearly energy renovation rate of 13%, taken into account energy renovations with building permit as well energy renovations without building permit. This could give new insights on the need for adjustments of the policy actions for energy renovation.

To evaluate the reliability of these findings they need to be confronted with other sources, for example the Rational Use of Energy survey (RUE) [7]. Commissioned by the Flemish government, every two year a representative sample of Flemish households is surveyed considering their energy-related behaviour and energy-related investments. Specific attention is payed to the policy measures for energy renovations. Since, the RUE survey analyses the policy measure separately, the comparison will be performed on that level.

Table 3. Calculation of the correction factor for transposing Essencia Marketing results.

All renovations		Correction	Source	Assumption
Flemish population	100%			
Only owners	70%	70%	GWO2013	Tenants don't perform deep renovations
Only home-owners	60%	86%	Essencia	Apartment owners don't perform deep renovations
Only houses > 5 years	51%	85%	StatBel	Correction is remained form buildings and dwellings
% of renovators	19%	37%	Essencia	
With building permit	2.5%	13%	Essencia	

Energy renovations		Correction	Source	Assumption
Flemish population	100%			
Only owners	70%	70%	GWO2013	Tenants don't perform deep renovations
Only home-owners	60%	86%	Essencia	Apartment owners don't perform deep renovations
Only houses > 5 years	51%	85%	StatBel	Correction is remained form buildings and dwellings
% of renovators	13%	25%	Essencia	
With building permit	2.1%	17%	Essencia	

The RUE survey 2017 of the Flemish government indicates that 25% of a representative sample of the Flemish households installed roof insulation during the last ten years, of which 29% during the last two years. This means that on average 3.6% per year (25% * 29%/2) installed roof insulation during the last two years. Based on the Essencia marketing survey yearly an average of 5.5% of the Flemish home-owners installed roof insulation in the period 2016-2017. For a population with 60% home-owners this leads to an overall percentage of 3.3% (5.5% * 60%) of the Flemish population. Again, given the assumptions made to calculate the present estimation, both estimates are quite near one another, confirming the validation of our estimation.

Although further and more profound research is necessary to underpin the results of the survey through confrontation with other sources, it is clear that the 1% renovation rate of the Renovation Pact is rather a reflection of deep energy-related renovation with building permit than of the actual renovation rate. Furthermore, deep energy-related renovation with building permit is an underestimation of all energy-related renovation. The more detailed information about the yearly renovation activity provides more insights on the feasibility of the sustainability goals towards 2050.

3.3. Type and combination of renovation works

3.3.1. Type of renovation works. The top five of most pursued renovation activity is in order of appearance 1) works on the electricity net, 2) doors, windows and glass, 3) roof insulation, 4) bathroom and 5) the heating system. The highest rate for works on the electricity is not surprisingly since this work is often part of another renovation work (e.g. installation of solar panels mostly require extra electricity work, works on the interior wall are mostly combined with the installation of extra switches or sockets, …). We notice that the top five contains three energy efficiency measures. Although along the years the shares of the works have changed, the composition of the top five remains mostly the same. In recent years roof insulation is sometimes replaced by works at the interior wall.

3.3.2. Combination of renovation works. Figure 5 shows the interrelational dependence of renovation types. 37% (15% + 8% + 14%) of the renovators combines an energy-related renovation with structural or other works. However, 33% (4% + 21% + 8%) renovates without any energy efficiency measures. For 12% (4% + 8%) it involves structural work, meaning renovating walls, roof or expanding the living space. Obviously, these types of work are most likely to be accompanied by energy related work. Further research will be necessary to determine if these structural renovation works could serve as a gate to promote deep energy renovation.

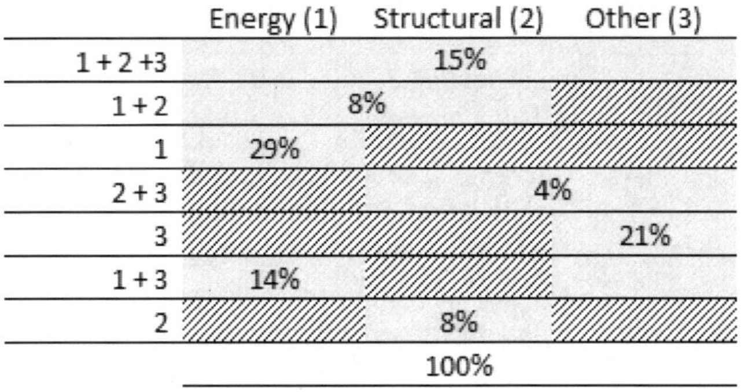

Figure 5. Combination of renovation works

4. Conclusion and discussion

Based on a yearly online survey among a representative sample of Belgian households the renovation activity of Belgian home-owners is analyzed. Yearly, an average of 37% of home-owners renovates or rebuilds something in their house. Energy related renovation works are most executed. When the results are recalculated – as good as possible with the available data- to the population of Flanders, a renovation rate of 19% is obtained or 2.1% of deep energy related renovations. This result surpasses the 1% renovation rate, which is often referred to in the communication on the need for a higher renovation activity to achieve the European goals.

However, some limitations have to be mentioned. The Essencia marketing database indicates an underestimation of the lowest socio-economic households and it can be expected that due to budget limitations, renovation activities are lower among these households compared to higher income groups. Furthermore assumptions had to be made since the survey only questions home-owners, and excludes tenants and apartment owners. Also, the database under study was an online research among consumers. Although home-owners in general are aware of the type of renovation works executed, differences in interpretations cannot be avoided; e.g. a porch is seen as a structural work, although it can also be seen as replacing windows and doors which is an energy related measure. Finally, the size of the used database (5.449 respondents) permits a profound calculation of the evolution of the renovation activity. However, it makes it impossible to complement with results on energy use before and after the renovation. But as shown by [3], it can be expected that explication of energy efficient measures in general also lead to lower energy use.]

References

[1] Renovation Pact n.d. *Bis-Conceptnota aan de Vlaamse Regering, betreft: Renovation Pact*
[2] European Parliament, Council of the European Union Directive 2010/31/EU of the European Parliament and of the Council of 19 May 2010 on the energy performance of buildings
[3] Aydin, E., & Brounen, D. (2019). The impact of policy on residential energy consumption. *Energy*, *169*, 115-129.
[4] Essencia Marketing 2018 Analyse van de Belgische renovatiemarkt
[5] Economie–Statbel FOD n.d. Algemene directie Statistiek en Economische informatie https://statbel.fgov.be/nl/themas/bouwen-wonen/gebouwenpark#figures
[6] Winters S, Ceulemans W, Heylen K, Pannecoucke I, Vanderstraeten L, Van den Broeck K, De Decker P, Ryckewaert M and Verbeeck G 2015 *Wonen in Vlaanderen anno 2013. De bevindingen uit het Grote Woononderzoek 2013 gebundeld* Steunpunt Wonen, Leuven
[7] Vlaams Energie Agentschap 2017 Beknopte resultaten van de REG enquête 2017

IOP Conf. Series: Earth and Environmental Science **329** (2019) 012035 doi:10.1088/1755-1315/329/1/012035

The energy performance of dwellings of non-profit housing associations in the Netherlands 2017 - 2018

H S van der Bent[1,3], H J Visscher[1], A Meijer[1] and N Mouter[2]

[1] Faculty of Architecture and the Built Environment, Delft University of Technology, Julianalaan 135, 2628BL, Delft, The Netherlands
[2] Faculty of Technology Policy and Management, Delft University of Technology, Jaffalaan 5, 2628BX, Delft, The Netherlands
[3] Corresponding author: h.s.vanderbent@tudelft.nl

Abstract. Following regulation of the European Union, objectives were formulated to reduce energy consumption of the built environment in the Netherlands. For the stock of Dutch non-profit housing associations it was agreed to improve the average energy performance to an average energy index of 1.40 in 2020. This research assesses and gives insights in the progress to this objective for over 2.0 million dwellings of over 250 Dutch non-profit housing associations in 2017 and 2018. The assessment consists of an analysis of applied renovation measures, changes of the stock like new construction and demolishing, and clarifying characteristics of housing associations. It is concluded that large urban housing associations with adequate financial positions drive the improvement of the average sectoral energy performance. The improvement happens for a large part within the existing stock, mostly with traditional improvements like improved heating installations and improved insulation. Innovative solutions like: photovoltaic solar systems, combined heat and power systems, biomass systems, heat pumps and external heating, are responsible for a relative small part of the improvement within renovations. New construction and demolishing are also responsible for a relative small part of the annual improvement, but there is potential to improve this.

1. Introduction

The European Union agreed in 2008 on goals to reduce the impact of human activities on the climate by 2020 [1]. In the Netherlands these goals were translated in several sectoral covenants. Non-profit housing associations in the Netherlands have a large impact on the built environment because they manage about one third of the Dutch housing stock, totalling 2.4 million dwellings. For the Dutch non-profit housing sector, agreements were made in 2008 [2] and in 2012 [3], which lead up to the Dutch Energy Agreement in 2013 [4]. Amongst others it was agreed to improve the average energy performance of dwellings of non-profit housing associations to an average energy index (EI) of 1.25 in 2020. The energy index is measured with Dutch legislation on the energy performance [5] which is the Dutch translation of the European Energy Performance of Buildings Directive [6], [7]. In 2015 the determination method of the energy index changed [8] and also the related goal changed to an average EI of 1.40 for housing associations in 2020 [9].

Several sources have discussed the progress of the improvement of the energy performance of the Dutch non-profit housing sector. In the article of Filippidou et al [10] it is concluded that, based on the progress made from 2010 to 2014, the goal of an EI = 1.25 will not be met in 2020. Filippidou et al [11] concluded in another research that housing associations in the period up to 2014 generally did not carry out major renovations, but many smaller investment projects, mostly as part of planned maintenance work. The most frequently applied changes were changes in heating systems, hot water installations and glazing. Other researchers [12], [13] have discussed whether energy performance measures need to be implemented in small steps or whether large-scale renovations are needed, or whether demolition and new construction are better alternatives.

With time progressing, there is a need for more updated insights of the progress of the energy performance of non-profit housing associations towards the goal in 2020. Three elements of the progress are not well understood. First, the effects of changes of the stock, like new construction and demolishing, were not examined in previous researches or papers. Second, the effects of taken renovation measures in the period after the changed determination method in 2015 are unclear. Third, it is not well understood which characteristics of housing associations, for example the size, financial position, or location, can explain the improvement of the average energy performance of the non-profit housing sector.

The following research question was formulated to address this knowledge gap: How does the energy performance of dwellings of non-profit housing associations in The Netherlands improves between 2017 and 2018 and how does this relate to the goal in 2020? Subsidiary questions:

- What is the improvement of the energy index between 2017 and 2018?
- How do changes of the stock and renovation measures taken within the stock explain the improvement of the energy index?
- How do characteristics of housing associations explain the improvement of the energy index?
- How do the insights of the progress of the energy index between 2017 and 2018 relate to the goal of an energy index of 1.40 in 2020?

2. Data and method

2.1. Data

In Europe each country has its own data and method to monitor the building stock. In the Netherlands Aedes, the Dutch umbrella organization of housing associations, started a monitoring system of dwellings called *Sociale Huursector Audit en Evaluatie van Resultaten Energiebesparing* (Social Rental Sector Audit and Evaluation of Energy Saving Results) abbreviated SHAERE [10]. The SHAERE database contains the energy performance characteristics of dwellings of non-profit housing associations. After the change of the calculation method in 2015 [8] the structure of the SHAERE database changed. Data collection in this new framework started in 2017. In the database are 2.03 million dwellings of 246 Dutch housing associations for 2017 and 2.07 million dwellings of 253 Dutch housing associations for 2018. Of every dwelling, the energy index under the new calculation method is available, and also 30 clarifying indicators, such as the type and size of the dwelling, types of installations and the degree of insulation of the windows and the envelope. This database is the main source of data for this paper.

Secondly, a dataset of Aedes with descriptive parameters of individual housing associations was used. Of every housing association in the Netherlands the following parameters were available: the size (classification of the number of dwellings), location (province), the financial position (weak, average or above average), the degree of urbanity (rural to urban) and the average energy performance (high, average, low).

2.2. Method

The average energy index of the Dutch non-profit housing sector is calculated as the mean of the available dwellings in the SHAERE database for 2017 and 2018. To explain the improvement of the energy index, two insights are given. First, the improvement of the energy index from 2017 to 2018 is explained by analysing changes of the sample, changes of the stock and changes within the stock. This was done by explaining the relative contribution of changes of the sample, changes of the stock and changes within the stock, to the improvement of the energy index between 2017 and 2018. Second, the improvement of the energy index from 2017 and 2018 is explained with the characteristics of housing associations. This was done by coupling the progress of the average energy performance of each housing association to descriptive parameters of the same housing association as discussed in the previous paragraph. The relative contribution to the improvement of the energy index between 2017 and 2018 of these descriptive parameters was thereafter analysed.

3. Results

The results of the analysis are presented in four paragraphs. First, the progress of the average energy performance is explained. Second, changes of the sample, changes of the stock and changes within the stock are discussed. Third the progress of the energy index of groups of housing associations with the same characteristics are explained. Fourth, the chapter finishes with a paragraph which relates the progress made between 2017 and 2018 to the agreed goal of EI=1.40 in 2020.

3.1. What is the improvement of the energy index between 2017 and 2018?

Figure 1 shows the progress of the average energy index of dwellings of the non-profit housing sector. Filippidou et al. [10] showed that the progress of the energy index between 2010 and 2015 decreased from 1.80 in 2010 to 1.61 in 2015. A linear extrapolation would exceed the target of EI = 1.25 in 2020. In 2015 the calculation method of the energy index changed. Also, the objective changed to an average energy-index of 1.40 for housing associations in 2020 [9]. Aedes collected data for 2016 in an old database format and data of 2017 and 2018 was collected within this research in a new database structure. A new trend line shows that a linear extrapolation of the average energy index after the change of the calculation method also exceeds the new target of now EI = 1.40 in 2020.

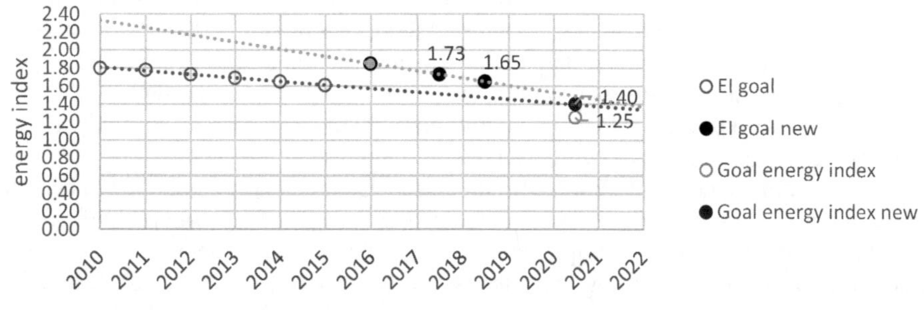

Figure 1: Progress of the average energy index of dwellings of the non-profit housing sector

3.2. How do changes of and in the stock explain the improvement of the energy index?

As shown in Figure 1 the improvement of the energy index between 2017 and 2018 improved from 1.73 to 1.65. The difference of 0.082 can be explained by changes of the sample, changes of the stock and changes within the stock. Table 1 shows that the changes of the sample were found to be responsible for an increase of the average energy index by 2.0%. Changes of the stock were found to be responsible for 14.3 % of the improvement of the energy index and the other 87.7% of the improvement was found to be caused by changes within the stock. This is explained in the following paragraphs.

Table 1. Contribution to the improvement of the energy index 2017-2018 by changes of the sample, changes of the stock and changes within the stock.

Progress EI	Δ EI	%
Changes of the sample	+0.002	+ 2.0 %
Changes of the stock	-0.012	-14.3 %
Changes within the stock	-0.072	-87.7 %
Total change	-0.082	-100.0 %

3.2.1. Changes of the sample. The change of the sample of housing associations in the SHAERE database of 2017 and 2018 has an effect on the average energy index between 2017 and 2018. 11 housing associations did participate in 2017 but not in 2018, and 18 new housing associations delivered data in 2018. The effect on the change in energy index between 2017 and 2018 is EI= 0.002, which is positive. It means that only because of the change of the sample the average EI in 2018 is 0.002 higher than in 2017. This is 2.0% of the total change between 2017 and 2018.

3.2.2. Change of the stock. -14.3% of the improvement of the energy index is explained by changes of the stock. This consists of new built, purchased, sold and demolished dwellings or administrative corrections as shown in Table 2. The percentages do not add up to the total, because of rounding errors.

Table 2. Contribution to the improvement of the energy index 2017-2018 by changes of the stock.

Progress EI	Δ EI	%
New built dwellings	-0.007	-9,1 %
Purchase/administrative	-0.001	-0.7 %
Demolishing/sale/administrative	-0.004	-4.4 %
Total change	-0.012	-14.3 %

New built dwellings are responsible for -9,1% of the improvement of the energy index. The number of dwellings in the sample with a building year in 2017 (8,960) are however lower than expected. According to the supervision organisation of the social housing sector in the Netherlands on average about 15,200 dwellings are built each year [14]. The reason for the lower number in the sample could be that housing association rely on the EPC (obliged for a new built dwelling) rather than on calculating the EI (obliged for existing stock). It can be expected that if all new built dwellings would be present in the sample, a lower average EI would be measured.

When looking at the quality of the new built dwellings it can be seen that the average energy index of a new built dwelling in 2017 or 2018 is 0.70. The main characteristics of new built dwellings are:

- Insulation: floor Rc (m^2K/W) =4.1, envelope Rc=4.5, roof Rc=5.3, windows U (W/m^2K) =1.80
- Heating: HR107 boiler (59%), external heating (28%), heat pumps (11%), electrical heating (2%)
- Ventilation: Balance ventilation system (21%), mechanical exhaust ventilation system (72%)
- Solar: PV systems (58%, in case of single family dwellings 80%), solar heating (1%)

These characteristics can be seen as still quite traditional for Dutch standards. Only solar PV is becoming a standard solution on new built dwellings. To improve the impact of new built dwellings on the sectoral progress, the building quality of new built dwellings should improve further. This means more innovations should be applied in new dwellings. Also if the number of annually new built dwellings increases, the impact on the sectoral improvement of the energy performance could increase.

Besides new construction, dwellings can also be added to the stock by purchase or by administrative corrections. It is not possible to identify which dwellings are purchased or added by administrative corrections, but the impact is relative small. The reason for this is that the energy index of these dwellings is close to the sectoral average. The impact on the improvement of the EI is -0.001 or -0.7%.

Parallel to added dwellings, dwelling can be removed from the stock by demolishing, selling or for administrative reasons. It is not possible to identify how dwellings are removed from the stock, because this information is not given in the database. However, the impact of the dwellings not present in the database of 2018 compared to 2017 can be calculated by measuring the effect as if they would be present in both years. This effect is EI -0.004 and is therefore responsible for -4.4% of the improvement. The removal of dwellings has thus a significant impact on the improvement of the average energy index.

3.2.3. Changes within the stock. Besides changes of the sample and changes of the stock, the major part of the sectoral improvement of the energy index is due to improvements made within the existing stock. This share is calculated to be -87.7% or EI=-0.072. In this paragraph the relative importance of changes within the existing stock is explained, expressed in the contribution to the improvement of the energy index of applied energy saving measures. Table 3 shows the contribution of changes within the stock. Three groups of changes are responsible for a large part of the sectoral energy index improvement. These are improved insulation, heating systems and improved airtightness. The improved ventilation systems have a negative effect on the average sectoral improvement of the energy index, probably because these are mostly added mechanical systems which use electricity for operation. The effect of adding solar systems is responsible for -8.8%. Cooling and some minor other improvements which are not specified in the database are responsible for another -4.1% of the sectoral improvement.

Table 3. Contribution to the improvement of the energy index 2017-2018 by changes within the stock.

	Δ EI	%
Insulation	*-0.016*	*-19,1%*
Ventilation	*0.002*	*2.0%*
Airtightness	*-0.021*	*-26,5%*
Heating	*-0.021*	*-25,5%*
Hot water	*-0.004*	*-5,5%*
Solar systems	*-0.007*	*-8,8%*
Cooling	*0.000*	*-0.2%*
Minor other	*-0.003*	*-4.1%*
Total change within the stock	*-0.072*	*-87,7%*

Within the changes of the stock a group of measures can be described as more innovative, being innovative heating systems and solar systems. When looking at the impact of the innovative heating solutions (combined heat and power, biomass systems, external heating systems and heat pumps) the effect is respectively -0%, -3% -7% and -6% of the total effect of the changed heating installations. This adds up to -4% of the total sectoral change. Installed solar systems were found to be responsible for -8.8% of the energy index improvement. These innovative solutions used in changes of the stock together are responsible for -13% of the total sectoral energy index improvement.

3.3. *How do characteristics of housing associations explain the improvement of the energy index?*

3.3.1. Size of the housing association. When looking at the size of housing associations it shows the importance of the order of magnitude. Table 4 shows groups of housing associations with the same size, their average EI of 2017 and 2018, their average EI improvement between 2017 and 2018, the percentage of the non-profit housing stock they own and their absolute contribution to the improvement of the EI between 2017 and 2018. It shows that on average the large housing associations (XL>25000 dwelling) have an energy index of 1.70 in 2018, which is the only group higher than the average energy index of 1.65 of the total non-profit housing sector. However, because of their size they are also responsible for a large part (-0.036) of the total sector improvement (-0.082). To achieve sectoral improvement of the average energy index the large associations have the most potential and the most impact because of the size of the stock they own.

Table 4. Groups of housing associations with the same size, their average EI of 2017 and 2018, their average EI improvement between 2017 and 2018, the percentage of the non-profit housing stock they own and their absolute contribution to the improvement of the EI between 2017 and 2018.

	EI 2017	EI 2018	Av Δ EI	% stock	Abs Δ EI
XL (>25000 dwellings)	1.80	1.70	-0.099	36.12	-0.036
L (10000-25000 dwellings)	1.71	1.64	-0.073	28.69	-0.021
M (5000-10000 dwellings)	1.71	1.63	-0.081	21.77	-0.018
S (2500-5000 dwellings)	1.59	1.54	-0.051	9.25	-0.005
XS (1000-25000 dwellings)	1.59	1.53	-0.063	3.64	-0.002
XXS (<1000 dwellings)	1.62	1.58	-0.040	0.54	0.000
Total sector	1.73	1.65	-0.082	100	-0.082

3.3.2. Location of the housing association. When looking at groups of housing associations sorted by province, it also shows the importance of the order of magnitude. Table 5 shows that 44% of the total non-profit housing stock is located in two provinces (*Noord-Holland and Zuid-Holland*) and in both provinces dwellings have an average EI higher than average in the Netherlands. However, together the dwellings in these two provinces are responsible for 0.045 of the total progress between 2017 and 2018. In some other provinces, housing associations made good progress between 2017 and 2018 (*Zeeland and Limburg*) or good progress has been made in the past (in *Flevoland, Drenthe, Overijssel and Gelderland*), but because of the lower percentage of the total stock the impact of the housing associations in these provinces is overshadowed by the housing associations in *Noord-Holland* and *Zuid-Holland.*

Table 5. Groups of housing associations with the same location, their average EI of 2017 and 2018, their average EI improvement between 2017 and 2018, the percentage of the non-profit housing stock they own and their absolute contribution to the improvement of the EI between 2017 and 2018.

	EI 2017	EI 2018	Av Δ EI	% stock	Abs Δ EI
Flevoland	1.41	1.35	-0.063	1.20	-0.001
Drenthe	1.63	1.55	-0.074	2.33	-0.002
Friesland	1.80	1.72	-0.075	2.53	-0.002
Zeeland	1.68	1.56	-0.119	1.70	-0.002
Groningen	1.71	1.65	-0.052	4.23	-0.002
Overijssel	1.65	1.58	-0.073	5.89	-0.004
Utrecht	1.74	1.67	-0.070	6.85	-0.005
Gelderland	1.56	1.51	-0.050	10.38	-0.005
Limburg	1.84	1.73	-0.112	6.05	-0.007
Noord-Brabant	1.64	1.58	-0.065	14.38	-0.009

Zuid-Holland	1.81	**1.72**	-0.092	**22.83**	**-0.021**
Noord-Holland	1.79	**1.68**	-0.113	**21.63**	**-0.024**
Total	1.73	1.65	-0.082	100	-0.082

3.3.3. Degree of urbanity of the dwellings of the housing association. When looking at the degree of urbanity of the dwellings of housing associations in Table 6, the same pattern arises as with the size and location of housing associations. A large part (almost 70%) of the stock is located in dense cities which is also the part of the stock with the highest energy index. Housing associations operating in suburban areas have a better energy index, but still higher than the average goal of 1.40.

Table 6. Groups of housing associations located in areas with the same degree of urbanity, their average EI of 2017 and 2018, their average EI improvement between 2017 and 2018, the percentage of the non-profit housing stock they own and their absolute contribution to the improvement of the EI between 2017 and 2018.

	EI 2017	EI 2018	Av Δ EI	% stock	Abs Δ EI
Urban	1.83	**1.71**	-0.115	**35.47**	-0.041
Urban-suburban	1.72	**1.65**	-0.068	**33.23**	-0.023
Suburban	1.61	1.55	-0.058	12.36	-0.007
Suburban-rural	1.63	1.57	-0.058	15.83	-0.009
Rural	1.69	1.60	-0.092	3.12	-0.003
Total	1.73	1.65	-0.082	100	-0.082

3.3.4. Financial strength of housing associations. When looking at the financial strength of housing associations in Table 7, it is found that housing associations with a weak financial strength have a higher energy index, but they also make the largest average progress. However, the sector impact of this group of housing associations is modest, because they own a relative small percentage of the total housing stock. Housing associations with a higher financial strength are largely responsible for the progress of the sector.

Table 7. Groups of housing associations with the same categorization of the financial strength, their average EI of 2017 and 2018, their average EI improvement between 2017 and 2018, the percentage of the non-profit housing stock they own and their absolute contribution to the improvement of the EI between 2017 and 2018.

	EI 2017	EI 2018	Av Δ EI	% stock	Abs Δ EI
Above average	1.70	1.61	-0.088	38.78	-0.034
Average to weak	1.74	1.67	-0.067	48.67	-0.033
Weak	1.90	1.73	-0.173	5.21	-0.009
Total	1.73	1.65	-0.082	100	-0.082

3.3.5. State of the energy performance of the housing association. When looking at groups of housing associations with a low, average or high average energy index, it is found that housing associations with an high average energy index, make more progress than the housing associations which already have a low energy index. Table 8 shows that this is both relative and absolute. This can be understood because housing associations in the 1/3 low EI group are already closer to the sector goal of an average EI of 1.40, and secondly, for the 1/3 high EI group it is probably easier to improve the energy index because the dwellings with a lower energetic quality are easier to improve.

Table 8. Groups of housing associations with a low, average or high average EI, their average EI of 2017 and 2018, their average EI improvement between 2017 and 2018, the percentage of the non-profit housing stock they own and their absolute contribution to the improvement of the EI between 2017 and 2018.

Progress EI	EI 2017	EI 2018	Av Δ EI	% stock	Abs Δ EI
1/3 low EI	1.48	1.45	-0.037	20.08	-0.008
1/3 average EI	1.64	1.58	-0.054	35.02	-0.019
1/3 high EI	1.90	1.79	**-0.109**	44.90	**-0.050**
Total	1.73	1.65	-0.082	100	-0.082

3.4. How do the insights of the progress of the energy index between 2017 and 2018 relate to the goal of an energy index of 1.40 in 2020?

Figure 1 shows that the annual improvement of the energy index has followed a linear line for several years. With a linear extrapolation of the average energy index in 2017 and 2018, the goal of EI = 1.40 will probably be achieved in 2021. However, this study shows that several factors influence this rate of improvement. First the impact of new built dwellings could be higher in the future if more new built dwellings are reported or more are built in future years. Second the improvement of the average energy index is mostly done with traditional renovations. The effect of innovative solutions on the improvement of the average energy performance is still quite small. It is unclear whether this effect will increase in future years to come. Finally it can be deduced, that given the analysis of the characteristics of housing associations related to the sectoral improvement of the energy performance, the large urban housing associations play a big role in achieving sectoral progress and therefore the sectoral improvement to the goal of 1.40. If the large urban associations fail to improve the energy index as they did between 2017 and 2018, it cannot be expected that smaller housing association are able to compensate this.

4. Discussion

Several remarks can be made about the research method and results of this analysis.

First, this analysis looks at the progress of the energy index of the non-profit housing sector only between 2017 and 2018. This a short time period, but it is aimed to extend this in 2019 and 2020.

Second, this analysis only looks at the progress of dwellings of non-profit housing associations in The Netherlands. It cannot be expected that the described improvements are also applicable to the other two-third of the mostly privately owned dwellings in The Netherlands.

Third, when looking at the data of the dwellings in 2017 and 2018 it can be seen that not all improvements of the energy index are expected to be related to actual improvements of the dwellings. A share of the improvements shown in Table 3 are administrative corrections. This is the case with the improved energy index through improved airtightness. Is it not a measure which is actually measured in the dwelling and therefore could be, to a large extend, a change in the administration of this parameter. It could not be deduced what the actual effect of these administrative corrections are, but they could be considerable.

Fourth, this analysis gives a good insight in the measures taken in the Dutch non-profit housing sector and their effect on the average energy index. Other researchers present measures taken in other countries [15], [16], [17] and [18], but they do not add up to a comparable average energy index. However most of these researches use a theoretical indicator, just like the energy index. Several studies in the Netherlands and in Europe have shown that the results of the forecasting of the actual energy consumption with the energy index can deviate strongly from reality and lead to the systematic overestimation of potential energy savings [19], [20]. According to Filippidou et al. [21], a greater number of renovations at a single dwelling lowers the effectiveness of the individual measures and therefore the realized savings in actual energy consumption are lower than expected. More research is needed to close the knowledge gap on a sectoral level, between the measures taken and the actual energy savings.

5. Conclusion

In this paper, the progress of the improvement of the energy performance of the non-profit housing sector in the Netherlands between 2017 and 2018 is explained. This was done by looking at the relative contribution of sample changes, changes of the stock and changes in the stock to the improvement of the energy index combined with an analysis of characteristics of housing association themselves. Most important conclusions are that large urban housing associations with adequate financial positions drive the improvement of the average sectoral energy performance. The improvement of the average sectoral energy performance happens for a large part within the existing stock, mostly with traditional improvements like heating installations and insulation. Innovative solutions like PV, CHP, biomass, heat pumps and external heating in retrofits, are responsible for only a relative small part of the improvement. New built and demolishing are responsible for another relative small part of the annual improvement, but there is potential to improve this, if the quantity and energetic quality of new built dwelling improves in the future.

Acknowledgements and funding source
Aedes, the Dutch association of housing associations, is acknowledged for the data made available for the analysis. Aedes is a full sponsor of the research project.

6. References

[1] European Commission 2007 *Limiting Global Climate Change to 2 degrees Celsius The way ahead for 2020 and beyond* Brussels
[2] Ministerie van Volkshuisvesting Ruimtelijke Ordening en Milieubeheer 2008 *Convenant Energiebesparing corporatiesector* Den Haag
[3] Ministerie van Volkshuisvesting Ruimtelijke Ordening en Milieubeheer 2012 *Convenant energiebesparing huursector* Den Haag p 3
[4] Sociaal Economische Raad 2013 *Energieakkoord voor duurzame groei* Den Haag
[5] Nederlands Normalisatie Instituut 2011 *NEN 7120 - Energy performance of Buildings - Determination method* Delft
[6] European Commission 2003 *Directive 2002/91/EC of the European Parliament and of the Council of 16 December 2002 on the energy performance of buildings* in Official Journal
[7] European Commission 2010 *Directive 2010/31/EU of 19 May 2010 on the energy performance of buildings (recast)* Brussels
[8] Instituut voor Studie en Stimulering van Onderzoek 2015 *ISSO 82.1 opnameprotocol energie-index woningen op basis van het Nader Voorschrift* Rotterdam
[9] Blok S A 2016 *Kamerbrief energiebesparing gebouwde omgeving* Ministerie van Binnenlandse Zaken en Koninkrijksrelaties Den Haag
[10] Filippidou F, Nieboer N and Visscher H 2017 Are we moving fast enough? The energy renovation ate of the Dutch non-profit housing using the national energy labelling database *Energy policy 109* 488-498
[11] Filippidou F, Nieboer N and Visscher H 2016 Energy efficiency measures implemented in the Dutch non-profit housing sector *Energy and Buildings* **132** 107-116
[12] Thomsen A and Flier K v d 2009 Replacement or renovation of dwellings: The relevance of a more sustainable approach *Building research & information* **37** 649-659
[13] Nieboer N 2016 *Improving energy performance: many small interventions or selective deep renovations? in Conference Proceedings SBE16* Hamburg: ZEBAU p 690-695
[14] Autoriteit Woningcorporaties 2018 *Sectorbeeld 2017* Utrecht
[15] Loga T, Stein B and Diefenbach N 2016 TUBULA building typologies in 20 European countries-Making energy-related features of residential building stock comparable *Energy and Buildings* **132** 4-12
[16] Csoknyai T, et al. 2016 Building stock characteristics and energy performance of residential buildings in Eastern-European countries *Energy and Buildings* **132** 39-52
[17] Diefenbach N, Loga T and Stein B 2016 Reaching the climate protection targets for the heat supply of the German residential building stock: How and how fast? *Energy and Buildings* **132** 53-73
[18] Serghides D K, Dimitrou S and Katafygiotou M C 2016 Towards European targets by monitoring the energy profile of the Cyprus building stock *Energy and Buildings* **132** 130-140
[19] Itard L C M 2016 *Energie en COMfort MONitoring voor energiezuinige en comfortabele woningen* TU Delft Delft
[20] Itard L and Majcen D 2014 *Relatie tussen energielabel, werkelijk energiegebruik en CO2-uitstoot van Amsterdamse corporatiewoningen* TU Delft Delft
[21] Filippidou F, Visscher H and Nieboer N 2017 *Effectiveness of energy renovations: a reassesment based on actual consumption savings in ECEEE summer study proceedings* p 1737-1746

IOP Conf. Series: Earth and Environmental Science **329** (2019) 012036 doi:10.1088/1755-1315/329/1/012036

A Review of the Existing Residential Building Retrofit in Tianjin, China

Y S Liu[1], C H Liu[1,*] and W Li[2]

[1] School of Architecture, Tianjin University, Tianjin, China
[2] School of Architecture, Tianjin Chengjian University, Tianjin, China

[*] Corresponding author's e-mail: conghong_liu@163.com

Abstract. Tianjin, situates 120 km southeast of the capital city Beijing, is the largest open coastal city in northern China and has high economic status. From the beginning of the 20th century, the municipal government has done a lot on residential building retrofit. In China, the policies as well as standards of local governments play an important role in guiding and supporting the building industry. This paper clarify the characteristics of the existing residential buildings, and attempts to review the policies, standards and retrofit pilot projects implemented by governments since the beginning of 20th century in Tianjin. Four stages of retrofit process toward residential buildings in Tianjin are indicated: reconstruction of shanty housing, retrofit of basic function including building facade beautification, outdoor environment update and balcony reinforce, energy saving retrofit and comprehensive retrofit based on sustainable conception. In each stage, policies, retrofit principles and typical cases are analysed. We also shows that lots of retrofit technologies have been applied which can reduce energy consumption effective and make living environment more comfortable. But it also has many problems of residential retrofit, such as lack of local standards and financial strategy. Meanwhile, the trend of retrofit is predicted, which is based on sustainable concept, aimed at building performance, integrated with environmental protection, energy saving and high efficiency. At last, some suggestions are also presented for existing residential buildings retrofit in Tianjin.

1. Introduction

In China, the existing buildings that cannot meet the demand for use accounts for about 30-50% of the total building energy consumption [1]. In the National Science and Technology Support Program for the *11th Five-Year Plan*, the research and demonstration of key technologies for comprehensive building retrofit have been emphasized. Subsequently, the *Special Planning for Scientific and Technological Development of Green Building* in the *12th Five-Year Plan* period also specifies that green retrofit of existing buildings should be treated as a key task.

As the north economic center of China, Tianjin has its unique regional culture and architectural style. Since the impact of 1976 Tangshan (a city near Tianjin) earthquake, lots of buildings in Tianjin need reconstruction. But the buildings built in this period cannot meet the demand of today's people [2-3]. It is imperative to retrofit on account of the development for socio-cultural factors.

Tianjin has promulgated many policies and standards, and became one of the earliest cities to carry out large-scale existing residential buildings retrofit (ERBR) projects. There are many scholars focus on the ERBR in China, but most of them only pay attention to any one or two aspects of policies, financial model and pilot projects to get some conclusions. As we all know, it is essential to understand policies and standards clearly to complete the retrofit process. This paper summarized primary policies, standards and pilot projects of ERBR, pointed out achievements and shortcomings of different retrofit methods, in order to dissect policy trend and comprehensive the concept of retrofit. Finally, some comments for ERBR in future were given to provide reference.

2. Existing Residential Building Stock

According to the statistics, by 2016, urban residential buildings accounted for 43.89% of the total building area in China, which occupied 37.71% of the total building energy consumption [4]. Since 1986, Chinese government has attached great importance to building energy conservation, and promulgated relevant policies [5-6]. But most of the residential buildings built until the end of the last century cannot meet the standards because of unbalanced development of regional economy and society [7-8].

Tianjin is located in the North China Plain, 120 km away from Beijing. It is hot and rainy in summer, cold and dry in winter. On the basis of China's thermal design zoning, Tianjin belongs to cold climate (B) zone [9], which regulated that in the process of building design, it should meet the needs of thermal performance in winter and take into account the heat prevention in summer [10].

At the end of last century, the old residential buildings in Tianjin were demolished and reconstructed in a large scale due to earthquake disasters, also for meeting the requirements of population explosion. At that time, all reconstructed building were set up according to the design standards formulated by Tianjin Architectural Design Institute. These buildings are structurally stable and can continue to be used. [11-14]. Most envelop material of them are 370mm solid brick combined with the steel concrete as load-bearing components (Figure 1). These buildings have flat roofs, one or two rooms, and the typical floor plans are shown in Figure 2 to Figure 4 [15]. Because there is no insulation layer, caused lower indoor temperature in winter and higher in summer, resulting in cooling and heating loss.

(a) (b) (c)

Figure 1. Existing residential buildings in Tianjin

Figure 2. Floor plans of One-bedroom

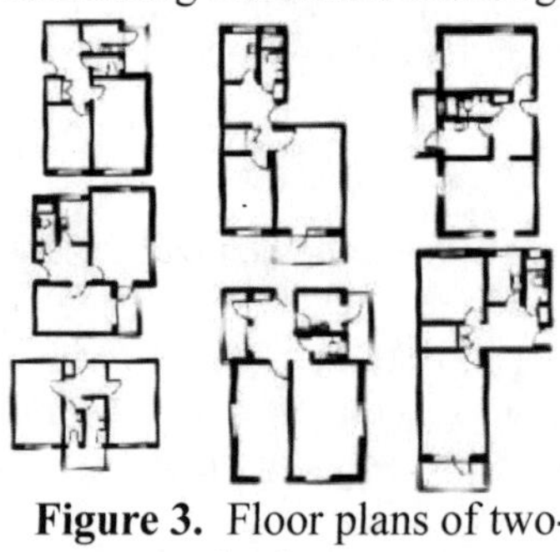

Figure 3. Floor plans of two-bedroom

Figure 4. Floor plans of three-bedroom

The existing residential buildings (ERBs) we mentioned above built in 1980 to 2000 are in large number and brings a series of problems, such as unreasonable division of indoor functions, high energy consumption and uncomfortable community environment. So corresponding measures should be taken since the development of society. Furthermore, there are many scholars indicate that retrofit is an environmentally better choice than the construction of a new building by use different research methods [16-18]. In this context, we have studied ERBR in detail from many aspects, summarize the experience both here and abroad, and point out the shortcomings in the following section.

IOP Conf. Series: Earth and Environmental Science **329** (2019) 012036 doi:10.1088/1755-1315/329/1/012036

3. The Experience of ERBR in Tianjin

3.1. Policies and standards
Initially, the boom in policies issued for residential buildings started after the 1976 Tangshan earthquake. From 2000, according to *Technical Specification for Energy Conservation Renovation of Existing Heating Residential Building (JGJ129)*, Tianjin municipal government began to retrofit the ERBs to do some exploration. At first, the documents called *Flat-to-sloping Roof Conversion Implementation Plans* promulgated by government provided retrofit content and process of the roofs. The projects alleviated the problem of heat loss through buildings' roof and beautifies the facades, as seen in Figure 1(c). This conservative retrofit is worth advocating rather than demolishing, raises by scholar Dongfan W [19].

From 2005, Tianjin has become the second city to implement *The Third Step Energy Saving* in China (i.e. the energy consumption level of newly designed heating residential buildings saved 65% on the basis of the energy consumption level of local general design in 1980-1981). Since 2008, Tianjin has gradually implemented some policies of heating system retrofit and planned to finish some comprehensive retrofit of typical ERBs. Relevant works during this period play an important guiding role in carrying out ERBR projects. In 2011, Tianjin took the lead in *The Fourth Step Energy Conservation Plan* that saved 75% energy consumption, expanding the area of ERBR gradually, supplementing and updating standards, pointing out the ERBR will become a continuous work. In 2018, the Tianjin government changed the heat source from coal to gas, and began to retrofit the community by the requirements of sponge city. Relevant data is shown in Table 1-3.

Table 1. Policies and works

Year	Name	Remark
2003	*Guiding Opinions on Pilot Work of Urban Heating System Reform* [20]	Initial exploration of heat metering.
2008	*Implementing Scheme of Heat Metering and Energy Saving Retrofit for 13 Million square meter ERBs in Tianjin*	Implemented heat metering retrofit, explored the organization pattern and financial problem.
2012	*Implementation Plan of Comprehensive Work on Energy Saving and Emission Reduction in Tianjin during the 12th Five-Year Plan*	Indicated that Tianjin completed 12 million square meters of energy saving retrofit of existing buildings.
2014	*Tianjin Green Building Action Plan*	Promoted energy saving retrofit and implemented heat metering retrofit.
2017	*Tianjin Construction Commission Implement the Concept of Green Building Design*	There are 43.74 million square meters of energy saving retrofit of ERBs have been completed in five years.
2018	*Notice on Strengthening the Management of Sponge City Construction* *Pilot Project of Sponge City Retrofit of ERBs in Tianjin has been Launched Extensive*	Including the retrofit of permeable pavement, parking lot, green spaces, rainwater collection system and pipe network.

Notes: Mentioned above come from Tianjin Municipal People's Government Network, http://www.tj.gov.cn/zw/

3.2. ERBR for Energy Saving
Tianjin has completed many projects of ERBR. The stage of demolition and reconstruction mentioned above targeted the shanty houses, has been completed basically. Projects of facade beautification mainly includes the flat-to-sloping roof conversion and some simple methods, such as repainting the building. There are a few projects about outdoor environment update including increasing green space, adding parking spot, repairing pavement and providing leisure space. ERBR for energy saving is the popular method that government promotes vigorously nowadays. Compared with the previous stages, energy saving retrofit has tended to diversify. This section mainly summarizes pilot projects of energy saving retrofit in Tianjin. [21-22]

Table 2. Guideline and atlas

Year	Name	Remark
2008	*Technical Guidelines for Heat Metering and Energy Saving Retrofit of Existing Residential Buildings in Tianjin* (MOHURD 2008 NO.442)	The main contents included heating system retrofit and roof retrofit.
	Guidelines for Heat Metering and Energy saving Retrofit of Existing Residential Buildings in Tianjin (MOHURD 2008 NO.939)	
2009	*Approaches for Acceptance of Heat Metering and Energy Saving Retrofit Projects of Existing Residential Buildings in Northern Heating Areas* (MOHURD 2008 NO.216)	
	Atlas of Flat Roof Retrofit Structure Diagram of Existing Buildings (03J203)	
2011	*Agreement on Heat Metering and Energy Saving Retrofit of Existing Residential Buildings in Northern Heating Areas*	

Table 3. Standards

Year	Name	Remark
2001	*Unified Standards for Constructional Quality Acceptance of Building Engineering* GB 50300-2001 (Old Version), GB 50300-2013	The formulation of standards provided comprehensive technical support, achieved technology integration, and formed a relatively complete system for the retrofit of ERBs in Tianjin.
2008	*Specification for Design of Heat Measurement-based Supply System in Residential Building with Central Heating of Tianjin* DB 29-26-2008 (Old Version), DBT 29-26-2017	
2010	*Tianjin Energy Efficiency Design Standard for Residential Buildings* DB 29-1-2010 (Old Version), DB 29-1-2013	
	Tianjin City Accepting Specification for the Quality of Energy-efficiency Engineering of Civil building DB 29-126-2010(Old Version), DB 29-126-2014	
2015	*Assessment Standard for Green Retrofitting of Existing Building* GB-T 51141-2015	
	Design Standard of Green Building for Tianjin DB29-205-2015	
2018	*Tianjin Assessment Standard for Green Retrofitting of Existing Building*	In editing

3.2.1. Envelope. Retrofit of envelope is the primary object because of the high heating and cooling energy consumption. Relevant research shows that the energy loss of external wall and window is the main factor of high building energy consumption, and the proportion of exterior wall is larger than window, so the effect is obvious after retrofit. [23]. In Tianjin, the principal method to retrofit the external wall is to increase the insulation layer. The thickness of insulation in the pilot projects are about 50-70 mm and the main material is EPS or XPS which are common used. Exterior insulation is used to reduce the impact of residents and condensation phenomenon of inside walls. The structure diagrams of exterior insulation used in Tianjin are shown in Figure 5 and Figure 6.

Most of the windows in ERBs are aluminum alloy with single glass. There are two difference methods of window retrofit. One is to replace the existing window with double glass (5mm+12mmAir+5mm) plastic steel window (Figure 7), the other is to retain the existing window and add a new layer of single glass (5mm) plastic steel window.

There are two methods for roof retrofit: flat-to-sloping roof conversion and insulation layer added. Flat-to-sloping roof conversion refers to adding a sloping roof on the basis of the existing flat roof, which can not only beautify the facade of the building, but also play an important role in heat preservation by the air interlayer (Figure 8). Steel structures can also be used for sloping roofs, but attention should be paid to their mechanical property and daylighting requirement [24]. In addition, upright roof is used generally for adding insulation layer, which structure is shown in Figure 9.

3.2.2. Staircase and balcony. As the component between indoor and outdoor space, staircase plays an important role in building thermal performance. Most of the staircases in ERBs were not designed the entrance doors at first, which directly led to the loss of heating and cooling. In most projects, insulation doors with self-closing function has been added to improve thermal performance effectively, and intercom system has been installed to ensure the safety of communities. Moreover, some pilot projects also using single glass (5mm) plastic steel window to replace the existing window of staircase.

For the open balcony of the ERBs, most of them are already sealed by residents using single glass windows. In the retrofit projects, all of the balcony were sealed by using single glass (5mm) plastic steel windows, reinforced for safety and added insulation for thermal performance.

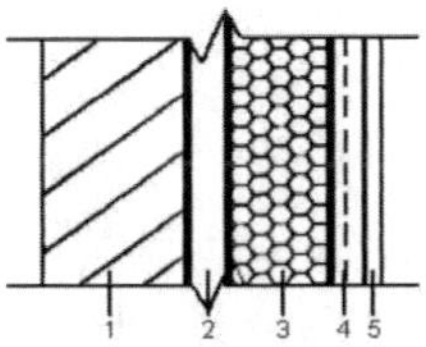

1. External Wall 2. Bonding Layer 3. Insulation Material 4. Protective Layer 5. External Plaster

Figure 5. Adhesive Structure

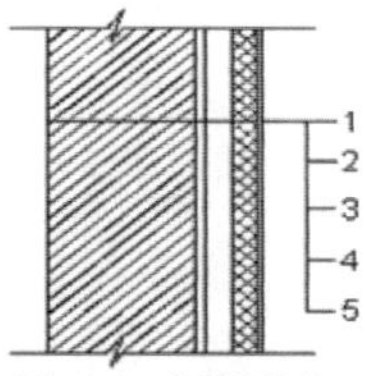

1. External Wall 2. Framework 3. Cavity 4. Insulation Material 5. External Plaster

Figure 6. Dry-hang Structure

Figure 7. Windows before and after replacement

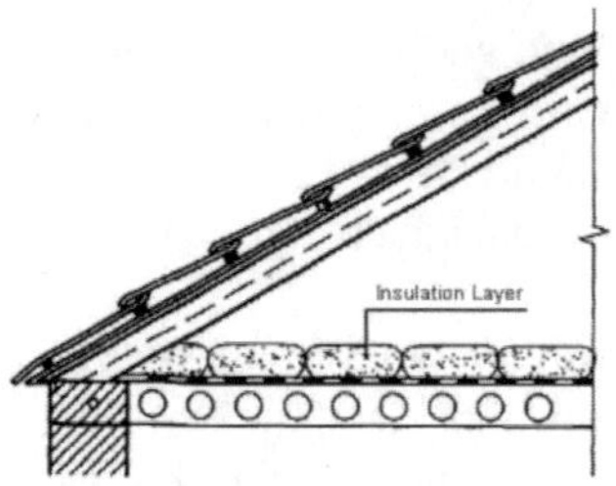

Figure 8. Structure of Flat-to-sloping Roof Conversion

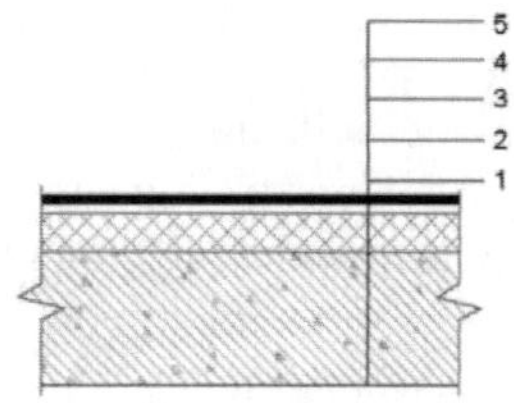

1. Original Roof 2. Insulation Material 3. Cement Mortar Screed 4. Waterproof Layer

Figure 9. Structure of Upright Roof

3.2.3. Heating system. The retrofit of heating system can divide into three parts: indoor heating system, heating pipe network and the heating source. The retrofit of indoor heating system included pipe network maintenance, temperature control system and heat metering system. The retrofit of heating source focused on improving boiler efficiency. In recent years, Tianjin has implemented the transformation of replacing coal with gas, which has greatly improved the efficiency of heating source. The retrofit of heating system can greatly reduce energy consumption in winter. Heat metering retrofit also enhanced the enthusiasm of households and popularized energy saving awareness.

3.2.4. Retrofit mode. In the process of ERBR, it has formed a variety of retrofit models in China, including government-led, heating enterprises-led, energy service companies-led, property rights units-led and other types. Most of the projects in Tianjin are dominated by municipal government who selected and determined the implementing company through bidding. After completion and acceptance, the projects shall be submitted to the superior management department for examination and evaluation. The cost of retrofit mostly funds by the government, and many pilot projects are welfare-oriented.

In summary, Tianjin has done many works of ERBR and formed a retrofit system. This section also summarizes some typical pilot projects, as shown in Table 4. According to the statistical data, the cost of ERBR in Hemu community is about 250 RMB/m^2. The retrofit can reduce about 40% of heating consumption and 43% of coal consumption. Flat-to-sloping roof conversion completed in Zijin community can also shorten the use time of air conditioning for top level residents by about 30% in summer, which is worthy although it is expensive. [25-28]

3.3. Other research of ERBR

Many scholars at home and abroad have studied about ERBR. These achievements can also provide reference for the future retrofit works in Tianjin.

Table 4. Pilot projects of ERBR

Community Name	Retrofit Item							
	External Wall	Windows & Doors	Roof	Building Structure	Indoor Heating System	Outdoor Heating Network	Heat Source	Outdoor Environment
Dongjiang Community	-	Δ	□	-	√	√	√	-
Taoyuan Community	√	Δ	□	√	√	√	√	-
Hemu Community	√	Δ	□	√	√	√	-	-
Zijin Community	-	-	■	√	-	-	-	√
Guanyun Community	√	Δ	□	-	√	√	-	-
Tanggu Street NO.1	√	▲	□	-	-	-	-	-

Notes : Δ. Window Replacement, ▲. Add Window, □. Add Insulation, ■. Flat-to-sloping

Numerous researches have shown that retrofit the windows and walls separately is uneconomical, it should combine with the heating system simultaneous. Furthermore, the scholar Shuqin C takes the ERBR projects in north China as an example, suggest that we should have specific target before retrofit. Aimed at reduction of space heating intensity, different retrofit packages can achieve 50% and 65% reduction by use simulation method. [23, 29-31]

On the other hand, many scholars have studied the retrofit of heating system, the results show that energy saving rate of 18.5% can be achieved by using of intermittent regulation of the heat exchange station [32]. Scholar Xin L researched the evaluation system for heat metering by using multi-level expert evaluation method, suggested rewards and punishment mechanism [33]. Another scholar figured out government financial subsidies should concentrate on supporting high-cost retrofits (building envelope measures) only. Heat metering, heating source and pipeline measures are more likely to attract funding in the retrofit market because of their lower cost and greater cost-effectiveness [34].

Due to various reasons, ERBR has not been widely popularized. Only pilot projects have been carried out in some cities. Actually, the effective retrofit methods proved by practice mainly include external wall and heating system retrofit.

4. Discussion and Conclusion

According to these works of ERBR, envelop and heating system retrofit are the most wildly used as well as effective methods in Tianjin. ERBR is a continuous project, from large-scale demolition to comprehensive retrofit, Tianjin has made many achievements. Social and cultural awareness has been improved than before, which indicate that it will turn to a sustainable retrofit aimed at building performance and integrated with environmental protection, energy saving and high efficiency. However, it also have some problems: a. There is no complete design standards for ERBR in Tianjin, results in most projects were limited to envelop and heating system and lack of retrofit for indoor functional space. b. Although some communities have finished environmental update project, there are some unsolved problems, such as the confined area of parking lot, lack of public space, etc. c. Because most of the cost funded by the Chinese and municipal government, the financing modes, incentive policies and reward mechanisms have not been generalized. For these reasons, ERBR cannot be popularized and still in the stage of pilot projects. d. Few residents participated in the ERBR caused the process slowed down, and the energy saving effect after retrofit was unsatisfactory, which may because the residents' motives, intentions and living habits were not taken into account during the retrofit projects [35].

Taking Tianjin as an example, this paper summarizes the development process of ERBR, points out the problems to provide reference of retrofit research. Indeed, implementation of the discussed methods is not always viable, ERBR is difficult in some respects. However, it's remarkable that a comprehensive review based on policies, projects and researches are essential obviously. Furthermore, suggestions are presented: a. In view of the development conception, ERBE should be a long-term works for our country devoted to sustainable development, to attain the goals of energy efficiency improvement and higher life quality for residents, to popularize awareness of energy conservation and environmental protection, to eliminate unnecessary waste of natural resources and to retain the historical value of these ERBs. b. Besides the retrofit funds, it is also an important issue that deficiency in energy saving awareness hindered development. Thus, we deem that highlighting the importance of public participation is contributes to the development of ERBR. c. Building reflects the development of society, economy and culture. It is not suitable to demolish all old buildings. When the structure of residential buildings is firm and safety, considering to the energy consumption, economy performance as well as the research results, it should be retrofit rather than reconstruction.

5. Acknowledgement

The work in this paper was supported by the National Key Research and Development Program of China (2016YFC0700200) and the Program of Introducing Talents of Discipline to Universities (B13011). We would like to thank the anonymous reviewers and the editors for their valuable suggestions and comments.

References

[1] Yutong L, Jun R, Zhou J, Jianping L and Qing Yea 2017 The Existing Building Sustainable Retrofit in China-A Review and Case Study *J. Procedia Engineering* **205** 3638–45.
[2] Tianjin Urban Housing Census Office 2001 *Data Collection of Residential Houses of Tianjin City* (Tianjin) Tianjin Urban Housing Census Office.
[3] Weixin L 1990 A bold exploration of urban renewal: objectives, stages, experiences and characteristics of old city renewal in Tianjin *J. Reform of Economic System* **05** 64-68.
[4] Special Committee on Energy Consumption Statistics of China Building Energy Conservation Association 2018 China Building Energy Research Report *J. Construction and Architecture* **02** 26-31.
[5] CABR 1986 *Energy Conservation Design Standard for New Heating Residential Buildings (JGJ26-86)* (Beijing) Ministry of Urban and Rural Construction and Environmental Protection.
[6] CABR 1995 *Energy Conservation Design Standard for New Heating Residential Buildings (JGJ26-95)* (Beijing) Ministry of Construction of the PRC.
[7] MOHURD 2012 *12th Five Year Special Plan of Building Energy Conservation.* Document of MOHURD http://www.gov.cn/zwgk/2012- 05/31/content_2149889.htm.
[8] Jinlong O, Chunyuan W, Haifeng L and Kazunori H 2011 A methodology for energy-efficient renovation of existing residential buildings in China and case study *J. Energy and Buildings* **43** 2203–10.
[9] MOHURD 2016 *Code for Thermal Design of Civil Building (GB50176-2016)* (Beijing) China Architecture & Building Press.
[10] Xiaotu L 2010 *Building Physics* (Beijing) China Architecture & Building Press.
[11] National Bureau of Statistics 1985 *China Building Almanac 1984-1985* (Beijing) China Architecture & Building Press.
[12] Zhonglin G 1990 *Modern Architecture in Tianjin* (Tianjin) Tianjin Science and Technology Press.
[13] Shuwei L 1993 *Modern Urban History of Tianjin* (Beijing) China Social Science Press.
[14] Department of Science and Technology for Ministry of Construction 1997 *A Collection of Demonstration Projects for Well-off Residential Buildings in China* (Beijing) China Architecture and Building Press.
[15] Xin L 2006 Research on Development and Renovation of the Old Multi-floor Residence in Tianjin, Tianjin University.
[16] Ireland D 2005 *How to Rescue a House: Turn an Unloved Property into Your Dream Home.*

(London) Penguin Books.
[17] Erlandsson M and Levin P 2005 Environmental Assessment of Rebuilding and Possible Performance Improvements Effect on A National Scale *J. Building and Environment* **40** 1459-71.
[18] Anne P 2008 Does Demolition or Refurbishment of Old and Inefficient Homes Help to Increase Our Environmental, Social and Economic Viability? *J. Energy Policy* **36** 4487–501.
[19] Dongfan W 2002 Historical and Cultural during the Reform of Old Areas: Reflections on Renovation of Old Residential Buildings in Tianjin (Tianjin) Urban Conditions 21, the 46th IFMP World Congress.
[20] Jiangde Y 2011 Policy and Models of Reconstruct ion of Energy Saving for the Existing Residential Building in Tianjin *J. Building Energy Efficiency* **39** 74-7.
[21] Yan W 2012 Study on Appropriate Strategies for Technologies Renovation of Existing Residential Building in North China: Cases in Tianjin, Hebei University of Technology.
[22] Hongwei Y, Li L, Xiaojun L, Conghong L and Phil J 2016 Tailored Domestic Retrofit Decision Making towards Integrated Performance Targets in Tianjin, China *J. Energy and Buildings* **140** 480-500.
[23] Shuqin C, Jun G, Mark D L, Linna X and Yowargana P 2013 Elaboration of Energy Saving Renovation Measures for Urban Existing Residential Buildings in North China Based on Simulation and Site Investigations *J. Building Simulation* **6** 113–25.
[24] Weiwei W 2008 Techno-economic Study on the Energy Efficiency Transformation of the Extant Residence Buildings, Changan University.
[25] Lifeng Y 2014 Study on Existing Residential Building Energy Saving Reconstruction in Tianjin, Tianjin University.
[26] Ling Z 2017 Study on the Transformation of Old Residential District, Tianjin University.
[27] Shengbo Z 2007 Analysis on Energy-saving Renovation of Existing Residential Buildings in North China: Taking Taoyuan Nanli Residential Renovation in Hongqiao District of Tianjin as an Example *J. science* **30** 595.
[28] Ying G, Yue F, Yiye B, Xiao W and Qiong Z 2016 Comparison of Existing Residential Buildings Retrofit toward Six Cities in North China *J. Housing Industry*, **05** 45-50.
[29] Chuanzhi L, Mengmeng L and Yong W 2012 Research on Indoor Thermal Environment in Winter and Retrofit Requirement in Existing Residential Buildings in China's Northern Heating Region *J. Energy Procedi* **16** 983-90.
[30] Yuming L and Weijia G 2013 Effects of Energy Conservation and Emission Reduction on Energy Efficiency Retrofit for Existing Residence: A case from China *J. Energy and Buildings* **61** 61–72.
[31] Shuqin C, Jun G, Mark D L, Haiying L and Yowargana P 2015 Evaluation on Retrofit of One Existing Residential Building in North China: Energy Saving, Environmental and Economic Benefits *J. Procedia Engineering* **121** 3-10.
[32] Xin L, Chenchen W, Chuanzhi L, Guohui F, Zekai Y and Zonghan L 2018 Effect of theEnergy-saving Retrofit on the Existing Residential Buildings in the Typical City in Northern China *J. Energy & Buildings* **177** 154-72.
[33] Xin L, Yan D, Yujia T, Neng Z and Zhe T 2014 Research on the evaluation system for heat metering and existing residential building retrofits in northern regions of China for the 12th five-year period *J. Energy* **77** 898-908.
[34] Shilei L, Wei F, Xiangfei K and Yong W 2014 Analysis and Case Studies of Residential Heat Metering and Energy-efficiency Retrofits in China's Northern Heating Region *J. Renewable and Sustainable Energy Reviews* **38** 765-74.
[35] Wenling L, Jinyun Z, Bettina B, Arthur P J M and Can W 2015 Public Participation in Energy Saving Retrofitting of Residential Buildings in China *J. Applied Energy* **147** 287-96.

IOP Conf. Series: Earth and Environmental Science **329** (2019) 012037 doi:10.1088/1755-1315/329/1/012037

Retrofitting near to zero energy homes in a Mediterranean climate

S De Marco[1] and **V M Buhagiar** [2]

[1] 47, Carmel Flats No 6, Psaila Street, Birkirkara, BKR 9072 Malta, Tel: (+356) 21 330804, Mob: (+356) 79056990, E-mail: ing.stefan@gmail.com
[2] Department of Environmental Design, Faculty for the Built Environment, University of Malta, Msida MSD 2080, Malta. Tel: (+356) 2340 2849, Mob: (+356) 99449930, E-mail: vincent.buhagiar@um.edu.mt

Abstract. The present local built environment has a common thermal comfort problem namely that most dwellings have a great reliance on electricity for environmental control indoors. The main objective of this research work was to offer a practical and cost-effective working solution to this problem. The feasible energy–saving measures that can be retrofitted to an existing dwelling were designed and applied to an existing building; a top third floor flat in Birkirkara, Malta, thus converting it into a thermally comfortable minimum energy home. The indoor climate of the subject flat, its mirror image apartment and the Birkirkara microclimate were monitored for one year and the necessary tools to analyse this data were utilised: a psychrometric chart analysis with Malta's defined thermal comfort zones. Compared to its microclimate and the mirror apartment, the results show that the subject flat managed to keep a constant and very comfortable indoor climate across both the hot and cold seasons. It is only for a small portion (a total of 9 out of 122 days – 7% in summer and 16 out of 121 days – 13% in winter) that the energy–saving retrofit measures did not fall within the thermal comfort zones limits. This case study also shows that the combined energy saving retrofit measures had a payback period of 15 years, which eventually pays off with a surplus of over €700.

1. Introduction

Vitruvius in his classic books, '*De Architectura*', cleverly coined Architecture as resting on three pillars, namely "*Commodity, Firmness and Delight*". This paper essentially deals with Commodity of dwellings in Architecture.

Today's motivation behind building dwellings is to provide a secure shelter, protect ourselves from adverse climatic conditions and to obtain a neutral thermal comfort level. The present local built environment has a common thermal comfort problem as most dwellings lack passive measures and thus have a great reliance on fossil fuels that come at a cost to both the individual and the government. In addition such poor thermal comfort conditions imply health problems, leading to another problem – an escalating national health bill (13.7% related deaths in 2012 up to 16.8% in 2017) [1]. What is certain is that comfort, up to now, has come at a price – high energy consumption because existing buildings are very inefficient energy wise and consume 39% of the national energy load.

All energy and environment stakeholders are very well aware of the 20–20–20 energy targets that all EU countries are bound to achieve by 2020 via the NZEB and relevant legislation [2]. Both EU and local policy follow such legislation via the relevant directives and legal notices – however the latter only focus on new buildings and renewable energy sources. Thus unless policy and legislation look into the possibility of energy retrofitting existing building stock, the EU targets will not be achieved to effectively reduce the evident problem of greenhouse gas emissions [3]. Such a situation, if unchanged, will continue to increase the fossil energy demand problem, increasing CO2 emissions. Thus considering this above scenario, the main objective of this work was to offer a practical working solution to this problem. The idea was to analyse and point out what are the feasible energy–saving measures that can be actually retrofitted to an existing dwelling without affecting the occupant's lifestyle and daily schedule. Such energy–saving retrofit measures must be based on our climatic conditions and existing building fabric to effectively reduce energy consumption.

Some valid information has been analysed both locally and overseas. However, most of these studies remain redundant or limited in dissemination or use. Thus rather than 're-inventing the wheel' and creating another bench study, the idea was intended to analyse these local and foreign studies in detail and utilise their results and suggestions via this hands-on project directly, for Malta. Thermal comfort can be achieved either by adapting to a building's climate or by changing the building's climate to one's comfort. The issue is that people can adapt or be comfortable to a wide range of climates [4]. As a matter of fact various studies quote different comfort temperature ranges and to date, even though an adaptive standard is being mostly considered, the 'ideal' standard comfortable temperature for all simply does not exist [5]. This is because thermal comfort is based on both the physiological aspects and psychological expectations, i.e. what may be ideal for a person might be uncomfortable to another – apart from social and economic constraints [6]. In addition to these factors, the utilisation of a particular building needs also to be taken into account in respect to the requirements of the specific group of people that will be occupying it – e.g. the requirements of a home are different than that of a work place.

In order to rectify this problem, it makes sense to investigate at a more practical level the energy–saving retrofitting solutions suitable for our climate and existing dwellings. The social and economic benefits of such an initiative could be quantified to encourage policy makers to look into them. The first step that needs to be taken before looking into how to design, build or alter a home in any country is to have a detailed look at its climate – in most cases the microclimate is even more important than the former [4], [6]. Once this data is collected and analysed, the relevant passive measures suitable for such a climate can be designed accordingly. If a building has a low thermal mass, adding external insulation, apart from the other benefits, is the key to increasing its thermal mass index. In fact placing thermal mass in between insulation is beneficial. This is indeed a possible solution for many local dwellings as they feature light thermal mass properties. Night time ventilation (when coupled with thermal mass) can be effectively utilised especially in the hot season. However we need to consider pollution and noise issues especially in Malta's urban areas. Dust is also another major issue (construction sites) and apertures must have insect screens. Some foreign case studies confirm that building near to zero energy dwellings that utilise all the prevailing climate conditions to our favour is indeed possible.

The reality is that electricity was considered a social commodity; hence its pricing was originally kept at bay by Government, running Enemalta, the only energy utility in Malta. Passive design solutions were therefore put aside for want for "modern" homes – albeit at a price. Today electricity tariffs were left to float as per international oil markets – often unpredictable. The repercussions we are facing due to this volatility in electricity tariffs are evident. Building *new* energy efficient homes is not going to solve the *old* problem of existing building stock – the solution is therefore to look into retrofitting, deploying energy–saving measures that are cost-effective and adaptive to a Mediterranean climate.

2. Project Methodology

The chosen methodology for this research work was to investigate various options for retrofitting to implement them onto an existing dwelling, thus potentially converting it into a near to zero energy home. An existing building (a top third floor flat in B'Kara), referred to as the subject flat, was used as a test bed for such conversions.

Before applying any energy–saving retrofit changes to an existing building, the said apartment needs to be thoroughly analysed to expose the main areas of heat losses and gains. This can be done by analysing the heat transfer process (HTP) of the building. Such an HTP must be carried out because the outside part of the building shell is strongly thermally influenced by outside air. The HTP can be a very complex analysis, as it involves the combined effect of all three heat transfer methods: convection, conduction and radiation [7]. However it can be safely assumed that buildings reach a steady state of heat transfer – such theory is the basis of all energy performance certification software across all European countries. Such a heat transfer model (HTM) yielded the following results – Figure 1.

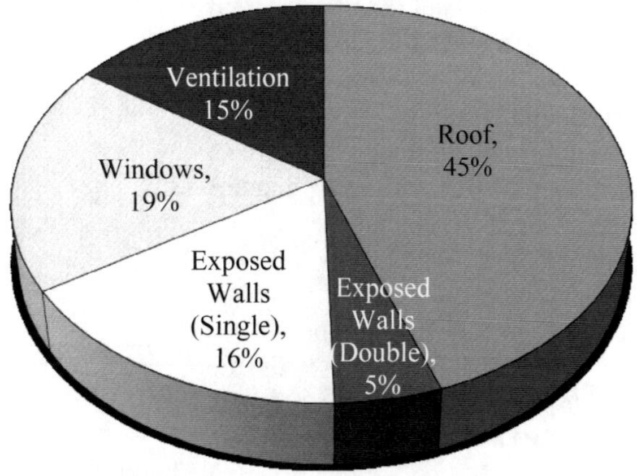

Figure 1: Steady State Heat Transfer Model of the Original Subject Flat

After a cost–effective analysis (based on the available project budget) and site considerations, the following design changes were applied to the HTM – these yielded the respective Heat Transfer Savings (HTS):

1. 75mm of expanded polystyrene (EPS) to the roof – 82% HTS
2. 50mm of rigid polyisocyanurate polyiso foam to the external walls – 74% HTS on double walls and 81% HTS on single walls.
3. Existing aluminium apertures replaced with PVC double-glazed and argon-filled windows – 81% HTS
4. All ventilators sealed – 90% HTS

Following the results obtained, Figure 2, the respective energy–saving retrofit measures mentioned above, including adjustable louvers on the south and west apertures were applied to the subject flat.

As previously described, thermal comfort (TC) is quite an extensive subject and thus requires a quantitative approach to verify if the energy–saving retrofit measures that were applied to the subject flat have managed to contain the indoor climatic conditions inside the standard thermal comfort zone (TCZ). The method of analysis adopted was to utilise the bioclimatic approach via a psychrometric representation [8], [9]. This was applied by first delineating the relevant local TCZ on the psychrometric chart and then superimposing the apartment's indoor climate parameters onto it to verify how many data points plotted were actually contained by such a defined TCZ.

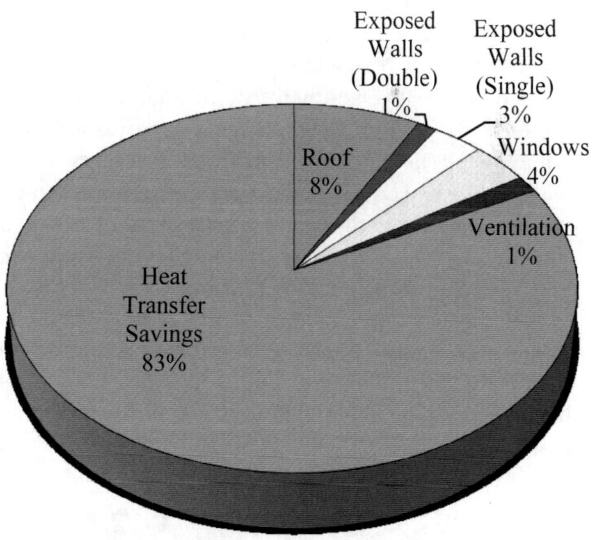

Figure 2: Steady State Heat Transfer Model of the Retrofitted Subject Flat.

A Microsoft Excel Tool (MET) was programmed with the necessary psychrometric chart parameters and a combination of Szokolay's and Givoni's algorithms for the TCZs were utilised based on the local climate. [10]. It was decided to use the 90% acceptability TCZs throughout the project as it reflects the best thermal comfort conditions needed for our local climate – Figure 3.

The pre-requisite to quantify if such energy–saving retrofit measures are effective from a thermal comfort point of view included a detailed analysis of the indoor climatic data, to check whether the temperature (T) and relative humidity (RH) readings fell within the defined TCZs. The hourly mean values of such T and RH readings need to be analysed to sum up the number of hours in the year when each specific value of T and RH occurs. Such data can then be plotted in a psychrometric chart with the number of hours (24 / day across a whole year) at each co-ordinate point [8].

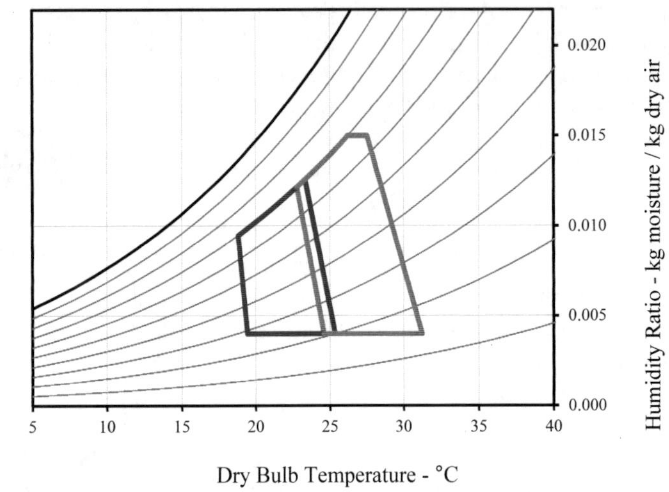

Figure 3: Local Thermal Comfort Zones with 90% Acceptability.

Unfortunately, before the retrofit changes were applied to the subject apartment, the indoor climatic data (for one year) was not recorded. However in order to have a good simultaneous comparison between the retrofitted subject flat and one that is standard, it was decided to also monitor the adjacent apartment (Flat 5) that happened to be a mirror image of the subject flat. As previously stated it was also important to monitor the B'Kara micro climate simultaneously with the subject's apartment readings. Thus T and RH hourly mean readings over a period of one year (June 2013 – May 2014) were recorded by using Lascar EL-USB-2 USB data loggers for both apartments and the B'Kara micro climate. The loggers (2 for each apartment) were placed at a 1.65m height in the living area sleeping area to record a home owner's thermal perspective.

In order to have another form of quantitative comparison and verification of such retrofit measures, the local EPC (energy performance certification) software for dwellings, EPRDM software was used, whereby the subject flat is considered as a single zone dwelling. In addition, the state-of-the-art software DesignBuilder (DB) was also applied since it gave the possibility of introducing adjacent dwellings, which may have some effect on the energy performance of the subject flat. Figure 4 shows the subject flat drawn in DesignBuilder, forming part of a whole block of 6 apartments, with two ground floor shops and adjacent blocks, as shown in Figure 5.

This scenario was created so as to reach as much as possible a close to reality simulation including shading effects and combined thermal masses from the adjacent apartment blocks. In both software, all the relevant building fabric parameters such as U-Values, wall thicknesses and heating / cooling schedules were carefully inputted to obtain a design model as close as possible to reality.

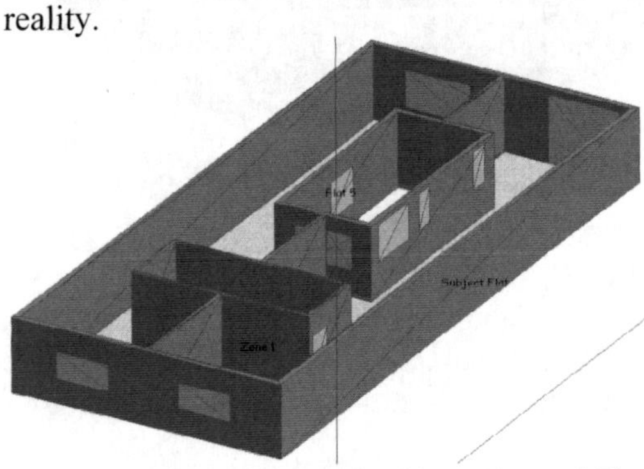

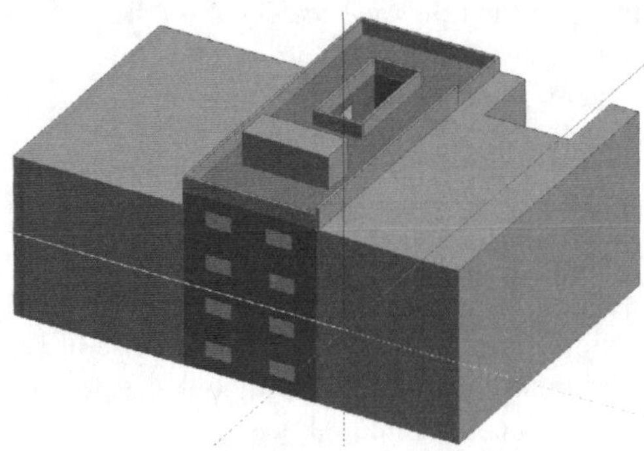

Figure 4: DB 3D view of Subject Flat and Flat 5. **Figure 5:** DB 3D view of Apartment Blocks

3. Results and discussion

IOP Conf. Series: Earth and Environmental Science **329** (2019) 012037 doi:10.1088/1755-1315/329/1/012037

3.1 Temperature Comparison

Figure 6 is a direct temperature comparison between the subject flat's (Blue) and flat 5's (Red) indoor temperatures, together with the B'Kara microclimate outdoor temperature (Green), following renovations to the subject flat.

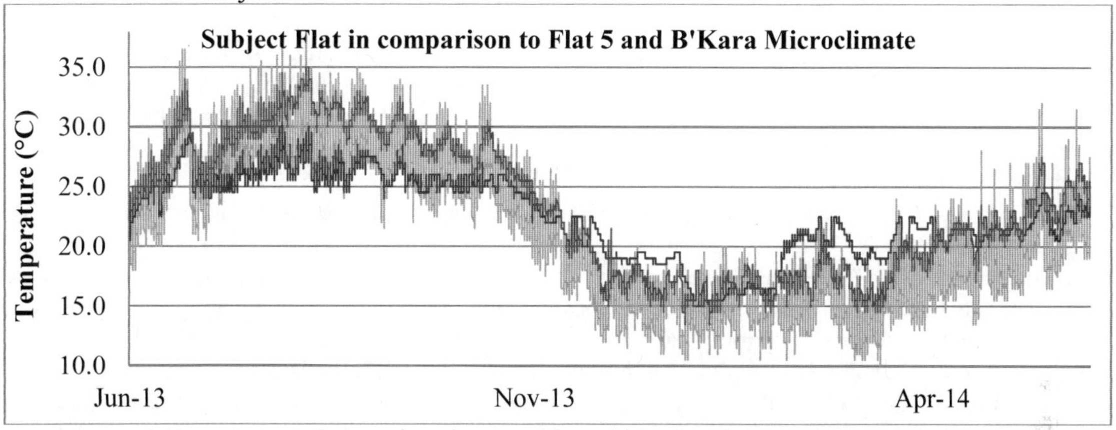

Figure 6: Temperature Comparison between the subject flat, flat 5 and B'Kara microclimate.

The subject flat doesn't make use of air conditioners (ACs), unlike flat 5 that makes extensive use of ACs. In flat 5 the Lascar temperature / humidity sensor was placed in an unoccupied room (spare bedroom) that does not make use of ACs.

Summer analysis: The subject flat was carefully controlled during most of the period July - August 2013 by:

1. Blocking off all sun rays via the adjustable louvers / shutters.
2. Windows were kept closed during most of the day (10:00hrs – 20:00hrs).
3. Night time ventilation was used accordingly to favour the prevailing climatic conditions offered during this period.

The result was that the apartment's inside temperature mirrored the lowest part of the microclimate during this period. On the other hand during the months of June, September and October 2013 the apartment was left unattended (closed up) and as predicated from various studies, it's inside temperature followed the microclimate mean temperature. In addition a fan had to be used for an evaporative cooling effect on the occupant (the author) during the heat wave periods.

Winter Analysis: Once the outside temperatures started falling (November) the subject flat indoor climate was controlled as follows:

1. The adjustable louvers were opened and retracted to allow the incident sun rays build up the internal solar gains.
2. All windows were kept closed at most times. They were only opened occasionally at noon to ventilate the apartment when the outside temperature was prevailing.
3. No form of artificial heating was used.
4. Internal humidity was kept to a minimum.

The results showed that the subject flat managed to keep quite a constant and very comfortable temperature of approximately 18°C throughout the whole cold season – as a matter of fact the occupant noted that unlike other dwellings, the clothing level was kept to a simple long sleeve top and trousers. On the other hand, during the month of January the apartment was left unattended (with closed shutters and louvers). A detailed look at the temperature hourly readings showed that the insulation helped to contain the internal solar gains within the subject flat for an 18-hour period. Once the solar gains were cut off (January), the apartment's temperature started falling towards the microclimate mean temperature.

Figure 6 also shows that flat 5 practically followed the highest temperature section of the B'Kara microclimate and when compared to the subject flat, the inside temperature swings are more frequent. This means that flat 5's thermal mass is very poor, as it did not offer sufficient dampening effect – unlike the subject flat (due to its insulated walls).

The occupants of flat 5 (a middle aged couple) stated that both summer and winter are unbearable without the continuous use of ACs for cooling and gas heating, respectively. Statistical analysis of flat 5 (room without any air-conditioning) showed that the internal temperatures reached up to 33 °C in summer and went down to 13.5 °C in winter. The apartment block featured the standard building practices of the 1950s that lead to a very poor thermal comfort. Apart from some plastering modifications, flat 5 is still in the original state as the subject flat was – both structurally and building fabric wise (230mm globigerina limestone).

The selected energy saving retrofit measures that were applied to the subject flat are very effective with an overall 3 °C (7 °C maximum) temperature difference in extreme hot and cold seasons. This difference comes at a cost; either via using ACs or by investing in such energy saving retrofit measures, thus one would need to analyse the cost effectiveness. However such a preliminary study already showed that if the selected energy saving retrofit measures are correctly installed and the dwelling is controlled well, then it is indeed possible to achieve minimum energy homes in Malta. However if such a retrofitted dwelling is left unoccupied it will simply follow the mean outdoor temperature swing and this will lead to the need of system heating and cooling.

3.2 Psychrometric Chart Analysis

The main scope of collecting the climatic data (T and RH) was to process it in the psychrometric charts to analyse if such energy–saving retrofit measures managed to contain the indoor climate within the TCZ limits. The measured climatic data was processed accordingly and inputted in the MET. The following plots are the results obtained – each black dot represents an average hourly reading of temperature and corresponding humidity ratio, the latter derived from the T, RH and atmospheric pressure (AP).

Figure 7 shows that at most times, the indoor climate is by far out of the TCZs and this means that flat 5 needs a considerable amount of heating and cooling – the latter being the greater load. In addition the humidity in winter is high – this might be due to the fact that the occupants use gas heating and they keep the apartment closed due to cold temperatures. The retrofitted subject flat model was simulated via the DB software with the ISE 2005 weather data. Ideally the DB simulations had to be carried out utilising the B'Kara microclimate weather. Unfortunately this was not possible since apart from the T, RH and AP, further detailed climatic data is necessary, such as: solar incidence (albedo, all direct and diffused components), wind (speed and direction), sky visibility parameters and precipitation. The indoor climate data obtained from the DB simulations was processed via the MET obtaining the respective psychrometric chart – Figure 8.

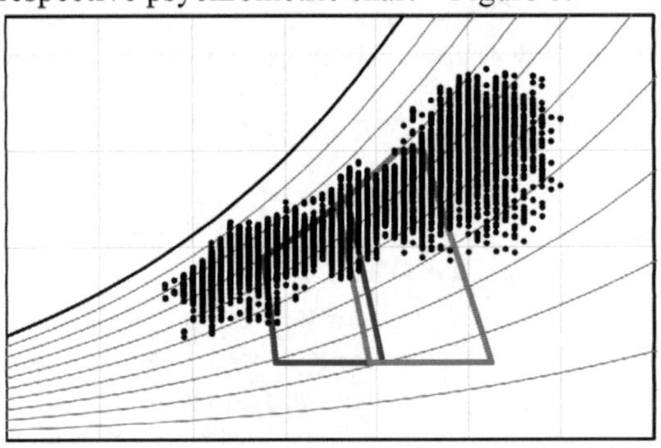

Figure 7: Flat 5 Indoor Climatic Data with 90% TCZs.

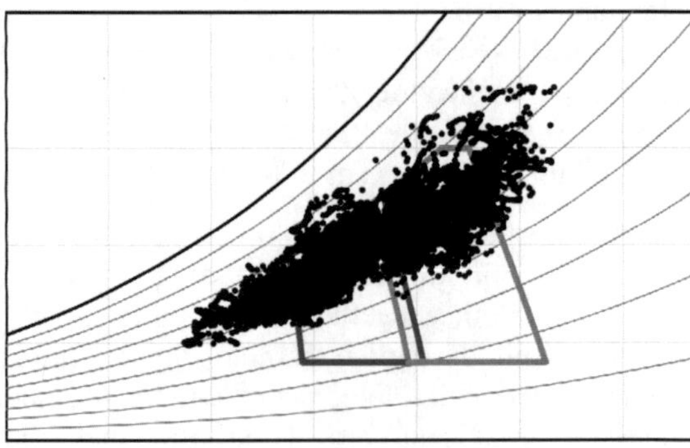

Figure 8: Retrofitted Subject Flat DesginBuilder Indoor Climatic Data with 90% TCZs.

Similarly the actual measured hourly indoor climatic data of the retrofitted subject flat was inputted in the MET and the respective psychrometric chart is shown in Figure 9. Comparing the two plots (Figures 8 and 9), the simulation and actual measured data showed a good correlation, even though they might look different. The differences in the extreme hot and cold periods are due to different humidity levels. This may be due to the fact that the DB software might not manage to accurately calculate the humidity levels – mainly in summer via night time ventilation. In fact the DB software only offers an OFF or ON option for natural ventilation and the air changes per hour (ach) - unlike the detailed heating and cooling schedules that the software can offer.

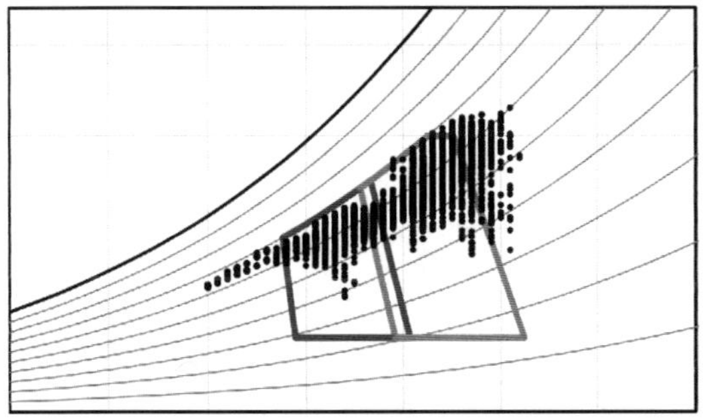

Figure 9: Retrofitted Subject Flat Indoor Climatic Data with 90% TCZs.

This means that it considers natural ventilation throughout the whole hot season. In reality a controlled schedule was used in summer as explained in Section 4.1, since the occupant of the subject flat realised that natural ventilation during the day increases the indoor temperature. In addition the occupant also used the site's prevailing climatic conditions to ventilate the apartment in winter (most often during midday), so as to reduce humidity levels too. All in all, even though the subject apartment actually performed better than the DB simulation, it shows that DB is a potential tool to carry out such climatic simulations as long as all parameters are correctly inputted.

With regards to Figure 9, apart from a few data points that fell out of the TCZs, most of the indoor climate is contained and this means that the energy–saving retrofit measures have successfully served their purpose. Most of the points that fall out of the TCZs are the ones when the subject apartment was intentionally left unattended in winter (no solar and internal gains during the period between December 2013 to January 2014) and in summer (when the apartment was closed up during June, all of September and October 2013). In fact it is only for this small portion (a total of 9 out of 122 days – 7% in summer and 16 out of 121 days – 13% in winter) that the energy–saving retrofit measures failed to satisfy the TCZ limits.

Such psychrometric chart analysis (Figure 9) and the temperature graph (Figure 6) show that energy efficient dwellings are indeed a possibility in our local climate if retrofitted with such measures. However the analysis also showed that if such a dwelling is left unattended or wrongly used, the tendency is that it will follow the mean outdoor temperature, thus the occupants would still have to rely on active measures such as the use of ACs to reach a thermal comfort level – the latter loads won't be as large yet still considerable.

3.3 Software Energy Analysis

Table 1 shows a comparison between the software results and the actual kWh readings over one year. Since no readings are available for the original state of the subject flat, flat No 5's results were used. It stands to reason that actual results for the modified state are not possible and were thus omitted.

Except for the subject flat cooling load, the EPRDM is nowhere close to the actual readings, however on the other hand the DB software and Flat No. 5 readings are close and this means that such a software, if carefully used (as there are many variables to consider), can be employed for relatively good simulations.

Table 1: Software and Actual Energy Analysis.

All Readings: kWh	EPRDM		All Readings: kWh	DesignBuilder (DB)		Actual	
	Original State	Retrofitted State		Original State	Retrofitted State	Flat No. 5	Subject Flat
Space Heating	734	626	Space Heating	2000	261	1671	0
Space Cooling	1266	161	Space Cooling	2307	1493	2433	164
Total	2000	787	Total	4306	1754	4104	164

Unfortunately neither of the two software managed to get close to the actual results obtained in the retrofitted subject flat. This may be due to the fact that even though the relevant natural ventilation parameters were carefully inputted, such software still relies a lot on system use (as per their defined system standards) rather than work around the adaptive comfort standards. This conclusion was reached since even though the DB temperature and RH readings were close to the actual ones, the DB software still recommended such cooling and heating loads.

3.4 Payback Periods of Energy–Saving Retrofit Measures

The most sought-after FAQ of such energy–saving retrofit 'investments' is "When will they eventually pay off?" The advantage of this project was that the subject flat and flat 5 were identical in size and layout (mirror image). Thus it was decided to utilise the subject flat and flat 5's electricity bills for such a payback calculation exercise. The subject flat electricity bills amounted to an average of 2,439 kWh (€266.41) while Flat 5's were 7,837 kWh (€1,023.88) per year. The cost breakdown was calculated utilising the Enemalta electricity residential tariffs as of April 2014.

The cost of each energy–saving retrofit measure was calculated in detail – Table 2. These costs reflected the actual installed cost as they included all purchased material, hiring of tools and heavy machinery, labour and the corresponding permits that were required. In order to carry out the right financial comparison, the cost of the installed AC units in flat 5 had to be calculated (3 AC Units at €1,012 each => € 3,036) and subtracted from the energy saving retrofit measure costs. This was done by dividing the AC cost in a ratio equivalent to the UA–value percentages (Figure 1) and then subtracting it from the corresponding retrofitted measure as shown in Table 2. Since the air tightness measure's cost is very low, it was decided to shift its ratio to the apertures cost as these are 100% draught-proof. In addition, since the aperture shades cannot be presented as a UA–value, no AC ratio cost was subtracted from the actual retrofit costs.

Table 2: Energy–Saving Retrofit Measures Costing, in Euros.

Retrofit Measure (UA-Value %)	Actual Installed Cost (€)	AC Cost Ratio (€)	Subtracted Cost (€)
Roof Insulation (45%)	2,033.26	1,366.20	667.06
External Insulation (21%)	1,170.00	637.56	532.44
PVC Double Glazing (34%)	3,450.69	1,032.24	2,418.45
Aperture shades	2,276.20	–	2,276.20
Air Tightness and Humidity Control	55.90	–	55.90

It stands to reason that only the cost of the heating and cooling section of flat 5 (4,104kWh) has to be used to calculate the energy–saving retrofit measures paybacks. This part amounts to a cost of €589.49 per year. However for a proper payback period calculation the subject apartment's heating and cooling part (164 kWh – €21.32) has to be subtracted from this amount. Thus this falls to €586.17.

Since the aperture shades cannot be represented in the respective UA-Value ratio, the DB simulation software was used to calculate the difference in the overall cooling load for the solar gains, with and without such shades across the hot period only (May – October). The difference (25%) was converted into the respective cost saving (€144.04) and thus subtracted from €586.17 to reflect the cost savings without the shades: €424.14. This amount was then divided according to the UA–value percentages as shown in Table 3. Rather than working out a simple payback period, a discounted payback period was utilised with a discount rate of 5%, as suggested by various financial institutions [11], [12], [13], [14]. The respective payback periods are shown in Table 3

Table 3: Energy–Saving Retrofit Measures Pay Back Period.

Retrofit Alteration	Subtracted Cost (€)	Yearly Cost Savings (€)	Discounted Payback Period (Years)
Roof Insulation	667.06	190.86 (45%)	3.9
External Insulation	532.44	89.07 (21%)	7.3
PVC Double Glazing	2,418.45	79.93 (19%)	35.9
Aperture shades	2,276.20	147.51 (Solar gains)	17.3
Air tightness and Humidity Control	55.90	63.10 (15%)	0.92

It is evident that the most effective energy saving retrofit measure is the air tightness and humidity control one, followed by the cost effective insulation (roof and external) measures. The last (yet most sought) is the double glazing one. Actually, this exercise shows that such a double glazing measure is not worth investing in. As a matter of fact, locally, there is a misconception that the best form of insulation measure is double glazing. In fact, such a measure comes at a high cost and with a very long payback period as opposed to the other beneficial measures. The shades, with a 17.3 year payback period (quite a long one) are still a more cost effective measure than the double glazing one. Thus it would make more economic sense to perhaps change single glazed windows to draught-proof ones as air tightness is more crucial than actual double glazing and install external shades.

Considering these payback periods, it would make more sense (from an economic point of view) for government to increase subsides on roof insulation and introduce a grant for external wall insulation – rather than the ongoing double glazing scheme. It is important to state that for the right economic analysis, only the discount rate was applied to this payback periods exercise. In reality; even though recently (March 2014) the electricity tariffs were revised downwards, the long-term tendency for energy prices is to rise, given our carbon tax disincentives. Such an outcome would decrease the payback periods and thus make such retrofits even more attractive.

Once the payback periods were calculated and eventually be reached by time, it would be interesting to use the same discount rate method to determine the additional cost savings (revenue) that one can get for the lifetime of the dwelling. Table 4 shows the obtained such cumulative results.

Table 4: Future Income of Energy–Saving Retrofit Measures, following the break-even point. Figures are in Euros.

Retrofit Alteration	5 Years	10 Years	15 Years	20 Years	25 Years
Roof Insulation	159.27	806.72	1,314.01	1,711.49	2,022.93
External Insulation	-146.82	155.83	392.06	577.55	722.89
PVC Double Glazing	-2,027.34	-1,655.20	-1,301.12	-964.23	-643.70
Aperture shades	-1,577.12	-911.98	-279.12	323.03	895.95
Air tightness and Humidity Control	219.54	435.36	604.46	736.95	840.76
Totals	-3,372.47	-1,169.77	730.29	2,384.79	3,838.84

This case study showed that after 15 years the combined energy saving retrofit measures pay off with a surplus of € 730.29. Considering the thermal comfort status achieved and the energy-cost analysis, stating that such energy saving retrofit measures aren't feasible, as most people think, is simply not correct. One has to appreciate that the study did not include any social benefits that may be enjoyed by the application of such retrofitting measures, such as better health and well-being. In fact this project has succeeded to achieve its objectives and its results can be used to aid policy direction and propose incentives regarding 20-20-20 targets for energy efficiency.

4. Conclusion

When considering the temperature distribution across a whole year (Figure 6), the psychrometric representation with the TCZ parameters (Figure 9) and the cost effective analysis carried out (Tables 3 and 4), this case study clearly showed that retrofitting our existing building stock via energy saving measures is indeed an achievable target and the outcome is a winning and positive situation from all aspects – such as:

1. A substantial reduction in energy use – both for the consumer and the national energy grid load.
2. A financial investment worth considering – especially if the payback period is surpassed thus making the investment render a profitable return for the remaining years.
3. A more thermally comfortable lifestyle in our existing dwellings and better well-being.

It is only for this small portion (a total of 9 out of 122 days – 7% in summer and 16 out of 121 days – 13% in winter) that the energy–saving retrofit measures fell outside the TCZ criteria. Such results, if utilised well, can open new business opportunities for an important sector of our economy – the construction industry – which has been on the decline due to the lack of demand and also due to the saturation of new buildings rising within the available land space.

Moreover the restraint on building permits outside development schemes – claimed as 'restricted' – has pushed developers to look inwards within building zones and possibly village cores to demolish and redevelop old houses to build new modern apartments – even if with a limited building height and floor area. These are unfortunately replacing the true houses of character, where most of the inherent physical features lie, including passive design unwittingly incorporated by our forefathers within the building fabric itself. Therefore retrofitting is surely one bold way forward. This not only eliminates the take up of new plots of green land and our finite resources (limestone), but moreover conserves embodied energy from construction as well as exploits the energy saving potential of such in-built features.

Hence retrofitting of existing dwellings into an NZEB home (near zero energy) home could be a potential for resuscitating the building sector. Apart from creating such an opportunity that will help increase our local economy due to new or modified skills and job take up, it will also help in reaching the EU energy efficiency targets.

However, it is of outmost importance that tradesmen need to be educated via adequate courses to improve their skills in retrofitting. In addition all relevant energy efficient products need to be certified and registered with the relevant authorities such as MRA and MCCAA. Quality assurance in the execution of such retrofits is key to it all.

5. References

[1] Ghirlando R., Fsadni M, "Traditional Building Features in a Passive House Concept for Malta.," *Energy Efficiency in the Mediterranean: 1st Passive House Conference, Malta,* 2011.

[2] European Commission, "The EU climate and energy package," [Online]. Available: http://ec.europa.eu/clima/policies/package/.

[3] V. G. V. B. Hens H., "Impact of energy efficiency measures on the CO2 emissions in the residential sector, a large scale analysis," *Energy and Buildings 33,* p. 275–281, 2001.

[4] Fergus N., "Adaptive Standards for Comfortable Buildings," *LEARN, London Metropolitan University – Commoncense project supported by the EU intelligent Energy Europe programme,* 2009.

[5] W. R. Darby S, "THERMAL COMFORT, Background document C for the 40% House report," *Environmental Change Institute, University of Oxford,* 2005.
[6] Buhagiar V., "Energy Efficiency in Building Design," *BICC report,* 2001.
[7] G. Z. Y. L. Y. L. Junli Zhou, "Coupling of thermal mass and natural ventilation in buildings," *Energy and Buildings 40* , 2008.
[8] Fsadni M, "Thermal Mass in a Bioclimatic Approach to Building Design in Malta," *PhD Thesis, Hertfordshire University, UK,* 2007.
[9] M. Tahbaz, "Psychrometric chart as a basis for outdoor thermal analysis," *International Journal of Architectural Engineering & Urban Planning,* pp. 95 - 109, 2011.
[10] A. A. Szokolay S. V, Thermal Comfort, Plea: Passive and Low Energy Architecture International, in association with Department of Architecture, The University of Queensland, Brisbane 4072, 2007.
[11] R. S. S. Chunekar A, "Discount Rates and Energy Efficiency," 2012.
[12] S. R. N. Jaffe A B, "The Energy-Efficiency Gap," *Energy Policy,* pp. 804-810, 1994.
[13] M. T. P. I. W. D. Isaacs N, "Benefit-cost Analysis of Thermal Insulation in Australian Dwellings," *ICEC Conference, Melbourne,* p. No. 103, 2002.
[14] D. B. L. G. K. B. M. Audenaerta A, "The inter-relation between heating systems, ventilation systems, insulation, energy price growth rates and discount rate for different dwelling types in Flanders (Belgium): a cost and E-level analysis," *HUB Research Papers – Economics & Management,* 2012.

IOP Conf. Series: Earth and Environmental Science **329** (2019) 012038 doi:10.1088/1755-1315/329/1/012038

Building monitoring protocol development for deep energy retrofit

E Perisoglou[1], M Ionas[1], J Patterson[1] and P Jones[1]

[1]Welsh School of Architecture, Cardiff University, Bute Building, King Edward VII Avenue, Cardiff CF10 3NB, Cardiff, UK

PerisoglouE@cardiff.ac.uk

Abstract. This paper presents the design of a monitoring protocol to support retrofit strategies and drive the evaluation of a low carbon house. Monitoring is used during pre-intervention stage to facilitate the decision-making process, as well as in the post-intervention to evaluate the performance of the building and systems. The first phase refers to monitoring as a diagnostic tool with a view to reduce the uncertainty of suggested low carbon solutions by providing real measurements and evidence on building conditions, comfort, systems and occupants behaviour and interactions. The second phase utilises monitoring as an evaluation tool to quantify benefits and challenges of the retrofitting by providing information and analysis on the supply and demand reduction, the difference in comfort and the effectiveness of each intervention. The study explains the protocol development and discuss different monitoring depths and requirements based on the performance indicators required. A typical pre-1919 Welsh end-terraced dwelling is used as a case study to demonstrate the two phases of the proposed protocol. The protocol was developed as part of the Low Carbon Built Environment (LCBE) project at Cardiff University and it was tested in various building types and research-led retrofit packages.

1. Introduction

It is well understood that buildings are a major contributor to greenhouse gas emissions [1, 2]. This is more significant in countries with old buildings such as the UK where emissions from buildings account for 37% of the total emissions and 2/3 of this coming from residential properties [1, 2] . The Welsh building stock is one the oldest in Europe with 75% of the houses built before 1980 and 36% before 1944 [3]. From 1976 building standards have been established to ensure high standards of energy efficiency in old and new-built properties. Improvements to the existing building stock is of a major importance in order to reduce emissions, energy bills and levels of fuel poverty and increase the standards in comfort and health [2, 4-6]. Governments push the market by providing incentives and schemes that offer a variety of mediums towards low carbon retrofitting. However, the design and implementation of technologies in existing occupied dwellings is a challenging process involving a variety of stakeholders and expertise with a high risk for performance gaps.

The SPECIFIC Low Carbon Built Environment (LCBE) project in Welsh School of Architecture (WSA), Cardiff University is a European funded project in collaboration with the industry, government, academia and the public on ten low carbon built environment demonstration projects. These demonstration projects combine appropriate renewable energy supply, energy storage, and energy demand reduction technologies to create a low carbon built environment that is both replicable and affordable. The project, amongst others, targeted 24 old residential buildings in west Wales and Valleys area and applied a deep retrofitting holistic process. This involves the identification and modelling of the most appropriate low carbon technologies, the implementation and demonstration of these technologies in practice and the monitoring, before and after, to provide evidence of the impact of the retrofitting solutions.

This study will describe how building monitoring fits in all the different stages of a deep retrofit process. From the pre-intervention to the post-intervention, a monitoring protocol will be described with the aim to propose a procedure replicable as a whole or partially depending on the depth of the retrofitting, the maturity of the implemented technology and the performance indicators set by the stakeholders. In this study the term building monitoring summarises all the measuring, monitoring and instrumentation processes and techniques to quantify parameters related to the building, its systems, occupants and environment as well as all the data processing procedures required to respond to the performance indicators established to quantitively describe the impact of the interventions. The protocol has been applied in full for research purposes, to a number of SPECIFIC LCBE case studies; one of the case studies, a semi-detached end-terrace house in the Wales valleys us used in this study to exemplify each monitoring step in a real-life setting.

2. Protocol development

The monitoring protocol was evolved to respond to the different steps of the retrofitting process and evaluation:

- To assist in decision making process of the interventions.
- To quantify the benefits of the interventions.

The first reason refers to the pre-intervention stage where monitoring is a design tool, whereas the second one is an evaluation tool and refers to both the pre and the post-interventions stage. In order to quantify the benefit of an intervention, the difference between the post and pre-intervention needs to be measured, meaning that monitoring actions must be taken before the intervention to make this comparison feasible. A brief overview on how monitoring processes scheduled before and after the retrofitting interventions are shown in figure 1.

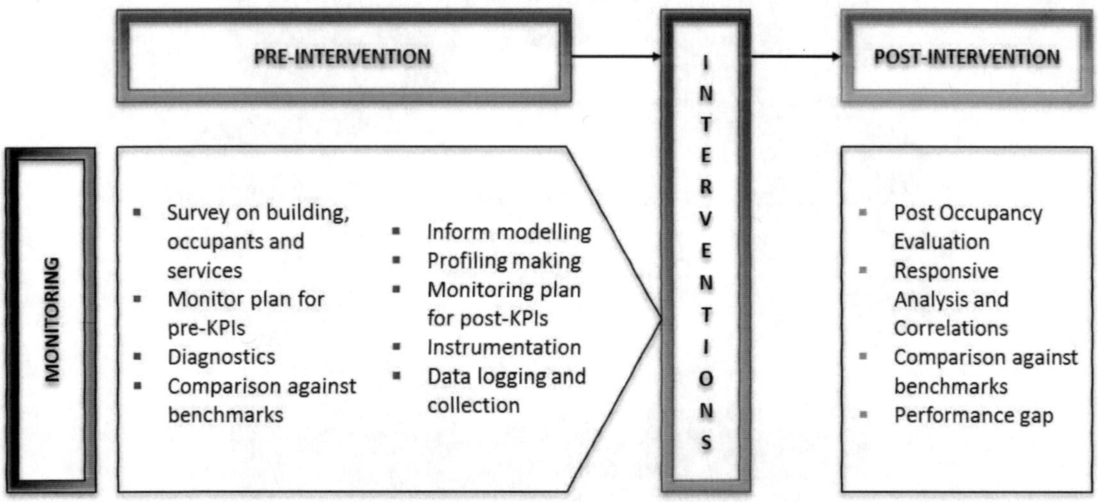

Figure 1. A summary of monitoring processes before and after the interventions

The very first step in a retrofitting process is the identification of the building related issues and concerns; this can be initiated by the occupants, housing associations or councils by gathering complains and by analysing more objective mediums such as Energy Performance Certificates (EPCs). These initiation routes include some risks as the occupants may not report issues or do not have the technical background to detect and recognise technical problems; in addition, EPCs (if exist) include some indicative information on potential retrofitting benefits but they lack information on comfort, dynamic modelling and accurate measurements.

Building monitoring as a powerful diagnostic tool is essential to recognise, quantify and characterise the building as a parametric framework where the environment, systems and occupants interact. Also, monitoring assists the design/sizing of the interventions responding to a specific building and occupants. However, one of the most crucial decisions to be made in the initial stage of the retrofit is if the interventions will respond to the existing occupants/tenants or to a typical occupancy profile for the house type and area. This will drastically affect the monitoring procedure as in order to create a typical demand or comfort profile, benchmark data may be enough; whereas in order to create real profiles, extensive monitoring processes are required. In this study, real life occupancy patterns were monitored; however, it is suggested that every profile should be compered against benchmarks to detect any extreme differences and understand if the interventions could be replicable.

Another vital decision that will drive the monitoring process as a performance evaluation tool is the identification of the key performance indicators (PIs). This is mainly depended on the stakeholders and the multiple perspectives and priorities on what the performance focus and depth is. An example on the different views on the performance indicators is illustrated in figure 2 where each stakeholder asks different possible questions on the implementation of an innovative low carbon technology. The building evaluation methodology used in this study was based in monitoring techniques and protocol developed in the Welsh School of Architecture as well as literature and guidance on building metrics and performance evaluation protocols [2, 6-13]. The focus and PIs are research driven and examine both the building as a whole and the impact of individual systems in response to real life demand profiles.

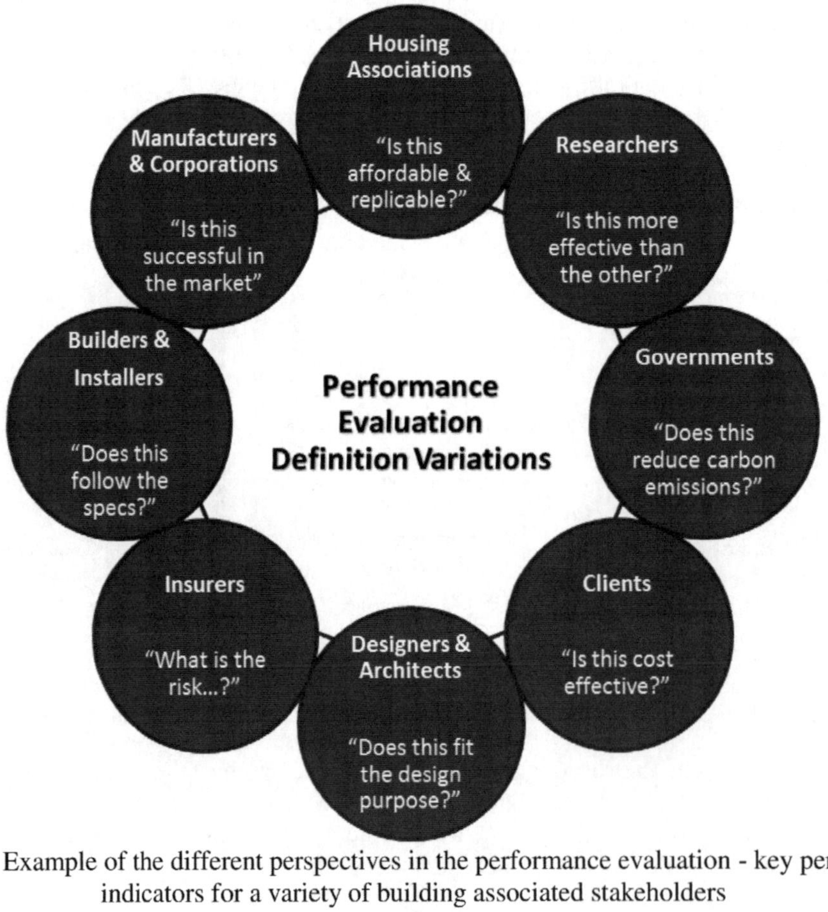

Figure 2. Example of the different perspectives in the performance evaluation - key performance indicators for a variety of building associated stakeholders

3. Dynamic and non-dynamic monitoring

By using the term non-dynamic monitoring, we refer to all the measuring processes and data collection that are not depended on the occupancy patterns or dynamic weather conditions. This type of data refers to the site, the building, the systems and the occupants' static (unchanging) information. There are a lot of tools to extract the information needed, described in table 1 below, such as a site survey, building survey, EPCs, occupants survey, historic weather data files, building envelope tests etc. The data analysis of this information collected is the output of the initial building's diagnosis which will inform the first stage of the decisions on passive and active retrofitting interventions.

Table 1. Non dynamic pre-intervention monitoring & data collection

Task Details	Action/Equipment/Technique
Site analysis and surroundings	Collect information of natural context such as nearby natural resources in abundance, topography, landscape, climate and weather trends, rivers, lakes, shading of surrounding and vegetation, mountains coordination etc. Tool: site survey, weather data files
	Collect information of artificial context such as surrounding buildings and neighbourhood context and patterns, artificial shading, noise, drainages, utility grids (gas, electricity, water etc) and future developments. Tool: site survey, DNOs, research
Building type, envelope, dimensions, construction, fabric details	Obtain information on building type, listed facades, age, orientation, access, views, number of storeys, roof inclinations, EPCs or other certificates Tools: building survey, research, occupants survey
	Record information about the building envelope and structure, openings and glazing ratio, type, frames, seals etc., fabric/roof materials and layering, insulation, loft boarding, U values Tools: Building survey, research, EPC U value tests (heat flux meters)
	Take dimensional measurements, internally & externally, fabric thickness, roof slopes, record flashing, ridge, eaves, soffits, downpipes, gutters, sills, floor joists, base course, air bricks, grills, trickle vents, chimneys etc. Tools: Building survey, occupants survey, EPC, dimensional instruments (tape, laser, ultrasonic meter)
	Inspect defects in fabric, air and water tightness, drainages (cracks, mould, dampness, degradation etc), air permeability. Tools: Building survey, thermography, Blower door test
Systems survey, existing capacity and nominal performance	Record energy mains, gas and water piping, consuming unit capacity, existing meters, suppliers and utility companies. Tools: Building survey, occupants survey,
	Collect nominal information on all the systems (electrical, mechanical), setting points, usage, number and capacity of appliances and lighting, specs and efficiency of boiler, electric shower, air conditioning units, electric heaters, mechanical ventilation units, renewables etc.

	Tools: Building survey, occupants survey, EPCs, research, wattmeter, air velocity meter
Occupancy	Collect subjective/objective information from the occupants on behaviour, set points, comfort, occupancy patterns, activities, bills, cost, ability to change and adjust. Tools: building survey, bills and specs data collection

The second stage of the decision-making process requires more information on the building performance as a response to weather/occupancy variation. This will then inform any modelling/projection in terms of systems feasibility, sizing and running costs. Also, long term performance outputs (for example monthly) can be produced in order to compare the effectiveness of the systems by comparing pre and post intervention indicators. The long-running monitoring usually refers to weather, comfort and energy data with a view to create profiles on different occupancies (weekdays / weekends) or heating patterns (heating / non-heating season) etc. Both, the monitoring indicators and tools and techniques are presented below in table 2.

Before applying a long-term monitoring plan, the duration and the time interval of the data collection needs to be identified. In the case study presented below, a 12-months pre-intervention and a 12-month post intervention monitoring period was decided to include a full heating and non-heating season. The interval was decided to be 5 minutes which provides a higher resolution than the suggested in the existing protocols [9, 11, 14]. This was decided for research purposes as the project investigated real time systems performance and efficiencies in response to transient and intermittent weather conditions. Extra attention was given to synchronise all sensors in order to represent simultaneous measurements despite any inertias. The data collection was grouped in one logger (CR1000 from Campbell scientific) and the information was sent back wirelessly to WSA servers every five minutes to enable live data visualisation and analysis.

Table 2. Dynamic pre-intervention and post intervention monitoring

Task Details	Action/Equipment/Technique
Weather data	Data logging of weather data on site: solar radiation, ambient temperature and relative humidity, barometric pressure, rainfall, wind direction and speed Tools: installation of weather station on site. Monthly average/total analysis and 24 hours profiles for heating and non-heating season
Comfort data	Data logging of temperature, relative humidity and CO^2 (in main or all occupied spaces). Tool: Installation of temperature, humidity & CO^2 sensors and loggers. Monthly average/total analysis and 24 hours profiles for heating and non-heating season
Energy data	Data logging of main electricity import (or export if exists), data logging of main generators (photovoltaics, air condition unit etc). Data logging of main gas meter and gas boiler. Tool: Installation of main meters and sub-meters (clamp-on or inline with pulse output) for electricity. Installation of a main or secondary meter for gas mains and gas boiler (induct with pulse output). Monthly average/total analysis and 24 hours profiles for heating and non-heating season
	Data logging of space heating and hot water

	Tools, heat meters if water based system, mass flow rates + temperature sensors if air heating system. Monthly average/total analysis and 24 hours profiles for heating and non-heating season
	Data logging of electricity sub-meters (most important or all circuits in the consuming unit). Tool: Installation sub-meters in the consuming unit (inline DIN-rail with pulse output) for electricity. Monthly average/total analysis and 24 hours profiles for heating and non-heating season
	Data logging of any other supplementary system or sub-system, For example DC circuits, solar thermal systems, heat exchanger units etc Tool bespoke approach depending on the system. Preferably meters with pulse output per energy unit, alternatively, installation of apparatus that monitors individual components of their energy delivery equation. Monthly average/total analysis and 24 hours profiles for heating and non-heating season

4. Case study: Protocol Application

A monitoring plan based on the pre and post intervention protocol as well as the non-dynamic and the dynamic monitoring suggested above was designed to fit the development priorities set in the LCBE project. The plan aimed to respond to two main principles: to assist the decision making process of the interventions and to facilitate the quantification of their impact. The case study building is a typical 100 years old, two storey 80m2, semi-detached end-terraced house in south Wales. During the monitoring period, the occupants were a family of two adults and two children, spending most of the time at home (social housing). The main heating system a gas fed combi boiler did not change, providing space heating and hot water. There were complaints on the building fabric in terms of air tightness and damp. The improvements on building envelope included external (EWI), internal (IWI), loft, and overlapping (anti-thermal bridging) insulation. Also, LED light bulbs were fitted and mechanical ventilation system with heat recovery (MVHR) was installed with a delivery of fresh air to all living spaces. On the supply side, photovoltaic (PV) panels were mounted on the double pitched roof (5.9kWp), electric batteries (13.5kWh) were fitted in the attic and transpired solar collectors (TSC) were installed on the south facade feeding solar heated warm air to the ventilation system during the heating period (figure 3). The system allows the export of excess renewable electricity generation to the grid.

The monitoring plan indicates the monitoring works in three categories: The first category includes non-dynamic studies on the building fabric such as, multiple U-values tests, air infiltration test and thermal imaging. In addition, it includes all the static data gathering on the building and site. Most of these measurements were taken before the interventions, however, some measurements were repeated after the interventions to identify changes of the building fabric and occupants' perception. The second and third box in the figure below contain long term monitoring processes refer to both pre and post intervention whereas some of them refer to the new systems installed. The plan also includes a brief indication of the position of the sensors in terms of location and in response to the systems of the building. The results of the monitoring informed the retrofitting strategy followed including the type of insulation and the type and size of the technologies. Also, the results included PIs such as total energy and operational cost reduction, changes in comfort, contribution of the individual system and payback times.

IOP Conf. Series: Earth and Environmental Science **329** (2019) 012038 doi:10.1088/1755-1315/329/1/012038

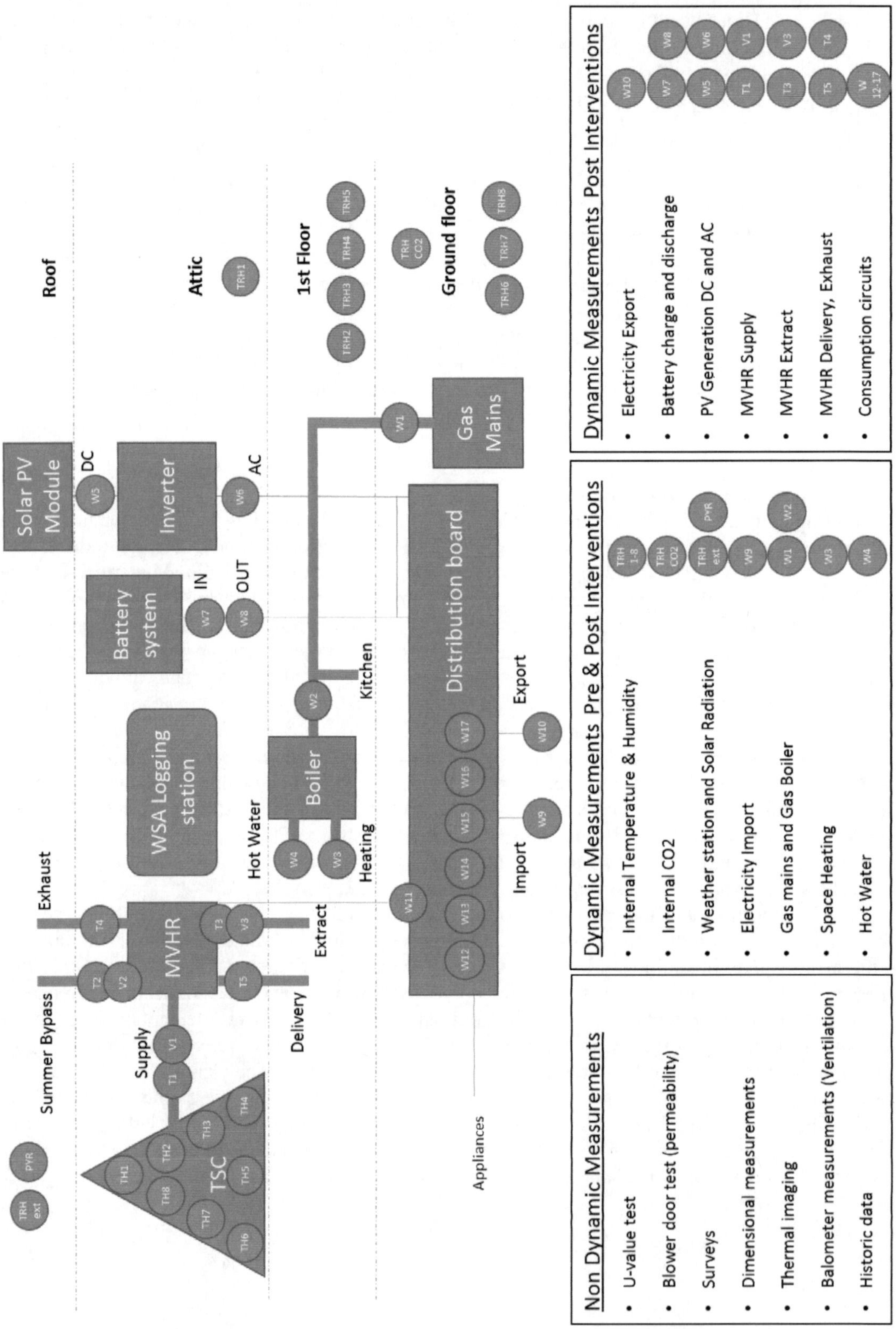

Figure 3. Monitoring Plan and Measuring indicators applied in a Welsh case study

5. Conclusion

This study introduces the development of a monitoring protocol to be used in a retrofitting of a dwelling. The process enhances the understanding of the building and enriches the interventions proposal with qualitative evidence. Also, the monitoring is used for building and system evaluation and for quantifying the benefit of the intervention in terms of energy, comfort and cost. This work splits the monitoring work into dynamic and non-dynamic monitoring. The non-dynamic is the pre-intervention initial monitoring that works as a diagnostic tool whereas the dynamic monitoring refers to long-term monitoring to explain and group variations in response to energy, comfort and weather. The dynamic monitoring assists the modelling process providing realistic profiles and allows for analytical building and systems performance evaluation. Tools and techniques are also analysed and used in a case study in a dwelling in south Wales.

Acknowledgments

Authors wish to acknowledge assistance and encouragement from colleagues from the LCBE SPECIFIC group especially Huw Jenkins for his technical guidance and Esther Tallent for her managerial support. The LCBE SPECIFIC project is delivered by the Welsh School of Architecture at Cardiff University. SPECIFIC is part-funded by the European Regional Development Fund (ERDF) through the Welsh Government, and also by Innovate UK and the Engineering and Physical Sciences Research Council (EPSRC). The case study retrofitting was partly funded by the West and Wales Housing Association.

[1] Committee on Climate Change. UK Emissions by Sector London: Committee on Climate Change; 2015 [cited 2015 December]. Available from: https://www.theccc.org.uk/charts-data/ukemissions-by-sector/.
[2] Jones P, Li X, Perisoglou E, Patterson J. Five energy retrofit houses in South Wales. Energy and Buildings. 2017 2017/11/01/;154:335-42.
[3] Welsh Goverment. Welsh Housing Conditions Survey 2017-18: Headline Report. Statistics for Wales, 2018.
[4] Jermyn D, Richman R. A process for developing deep energy retrofit strategies for single-family housing typologies: Three Toronto case studies. Energy and Buildings. 2016;116:522-34.
[5] Gupta R, Gregg M, Passmore S, Stevens G. Intent and outcomes from the Retrofit for the Future programme: key lessons. Building Research & Information. 2015 2015/07/04;43(4):435-51.
[6] Urge-Vorsatz D, Petrichenko K, Staniec M, Eom J. Energy use in buildings in a long-term perspective. Current Opinion in Environmental Sustainability. 2013;5(2):141-51.
[7] CIBSE. Building energy metering. London: The Chartered Institution of Building Services Engineers, 2009.
[8] SHC/IEA. Measurement and Verification protocol for Net Zero Energy Buildings. Bolzano: EURAC research, 2013.
[9] Guerra-Santin O and Tweed CA. In-use monitoring of buildings: An overview of data collection methods. Energy and Buildings. 2015;93:189-207.
[10] Ahmad MW, Mourshed M, Mundow D, Sisinni M, Rezgui Y. Building energy metering and environmental monitoring–A state-of-the-art review and directions for future research. Energy and Buildings. 2016;120:85-102.
[11] IEA/EBC. Total Energy Use in Buildings: Analysis and Evaluation Methods (Annex 53). Birmingham: 2016.
[12] Guerra-Santin O, Tweed C, Jenkins H, Jiang S. Monitoring the performance of low energy dwellings: Two UK case studies. Energy and Buildings. 2013;64:32-40.
[13] Sharmin T, Gül M, Li X, Ganev V, Nikolaidis I, Al-Hussein M. Monitoring building energy consumption, thermal performance, and indoor air quality in a cold climate region. Sustainable Cities and Society. 2014;13:57-68.
[14] IEA/IPEEC. Building Energy Performance Metrics. France: IEA, 2015.

IOP Conf. Series: Earth and Environmental Science **329** (2019) 012039 doi:10.1088/1755-1315/329/1/012039

Sensitivity energy analysis for the Saudi residential buildings envelope codes under future climate change scenarios: the case for the hot and humid region in Jeddah

M A Abuhussain[1,2], D H C Chow[1], S Sharples[1]
[1]School of Architecture, University of Liverpool, Liverpool, United Kingdom
[2]Corresponding Author: Mohammed.Abuhussain@Liverpool.ac.uk

Abstract. The residential building sector of the Kingdom of Saudi Arabia (KSA) is the primary consumer of energy consumption, using around 50% of the total amount of electricity KSA generates. Much of this is caused by the heavy demand for air conditioning due to the extremely high temperatures in the country. On the basis of KSA's government initiatives, it is compulsory that new residential buildings adhere to strict energy codes. This study places its focus on analysing the ability of KSA's new residential building envelope codes to successfully function according to the climate at present and in the future. In Jeddah, KSA, two single-family villas were examined with DesignBuilder simulation software and impacts of the new codes that were applied were assessed. Achieving the validation of the energy model requires the measurement of temperature indoors and outdoors. In addition, a long-term financial study was conducted to analyse the economic feasibility of the codes under the current and future climate change periods. Although the study shows the new codes produce a reduction in energy consumption by 14% to 40 % from the current cases, with the inclusion of wall and roof insulation being the most effective features of the new requirements of the codes. However, these codes cannot nullify the effect of future climate change.

1. Introduction

1.1. Background and context.

The Kingdom of Saudi Arabia (KSA) exclusively depends on natural gas and oil as primary sources of energy, as over half a million barrels of oil are used on a daily basis to provide electricity for the country [1]. In just the space of the year 2014, KSA saw an increase of 7.3% in oil consumption, and their rank changed to the eighth largest consumer of oil in the world because it domestically uses 25% of the crude oil and natural gas that it produces [2]. KSA has a population of 30.7 million people, 2 million km2 in area and 15.3 persons per square km density. It is also one of the fastest growing countries in the world and has 6.8% economic rate of growth per year [3]. There is therefore speedy expansion, urbanization and developmental growth within the construction industry allowing the standard of living to be fulfilled in a population that is growing alongside its demand.

Owing to its dependence on air conditioning, as well as the hot and dry climate of KSA, the building sector has been a significant contributor to the demand for energy. The building sector use up to 75% of the entire provision of electricity in KSA, and this grows every year at a rate of 7% [4]. Between the various type of buildings, 50% of the total use of electricity in the country is by the residential building sector, 70% of the entire national electrical demand is consumed by air conditioning [5].

In light of this issue, various researchers have shown that implementing an energy code and standard can perform a significant role in improving the energy efficiency in [1]. To respond to the issue of high consumption of energy caused by the building sector's expansion, the government in Saudi Arabia initiated Royal Decree No.6927 in 2012. This stated that thermal insulation is required throughout the building sector and was introduced to improve the energy efficiency of buildings. The analysis and description of the Saudi residential buildings envelope codes were described in an earlier paper [6].

1.2. Climate change in KSA.

In relation to KSA, one study suggested that owing to current trends, by 2041 Saudi's average warming will grow higher than the global average of [7]. Moreover, according to the study by Almazroui et al. (2012), Saudi Arabia's temperature will grow at a 0.72°C rate in mean temperature every decade. By 2050, a rise in temperature at a rate between 2.0–2.75°C was predicted [8]. The domestic building sector will therefore need to tackle substantial challenges regarding climate change in the future.

1.3. Research Objectives.

The main objectives of this study are following:

- To assess the capability of existing air-conditioned houses and newly built to the new residential building's envelope codes in the hot climate of KSA to perform under the current and the future climate change with the focus on energy usage.
- To examine how financially feasible the codes are for the long term with regards to the uncertain nature of climate change in the future.

2. Methodology

2.1. Case studies.

To assess the performance of the envelope codes for residential buildings, two detached single-family houses (villas) located in Jeddah city were chosen for this research, as shown in figure 1. These houses were chosen as a representation of Jeddah's mainstream domestic buildings. To ensure an acceptable level of representation of Saudi Arabia's typical, middle class residential buildings, both houses are in different locations and have varying building forms, architectural designs, building orientations, opening sizes, thermal insulation levels and number of inhabitants. This type of housing (villa) represents 20% of the number of total housing in KSA [9]. The house built first in 2008 before the code's announcement will be recognized as House A within this study. House B was built in 2015 after the code introduced. However, House B did not meet the requirements of the code because the code was not fully enforced. The full analysis of the study of House A can be seen in a previous paper [6] and an identical approach was conducted for House B. The general information and construction details of House B are shown in table 1.

Table 1. Characteristics of House B

	Construction
Total area	327.92 m^2
External walls	Mortar (20 mm) + Siporex block (200 mm) + mortar (20 mm)
Roofs	Terrazzo tiles (25 mm) + mortar (25 mm) + bitumen layer (4 mm) + expanded polystyrene (50 mm) + reinforced concrete (150 mm) + mortar (20 mm)
Window glazing	6mm/6mm Air double clear glazing, with 5.01 % window to wall ratio
Occupancy	0.0183 person/m2

2.2. Modelling.

The use of DesignBuilder simulation software as investigatory method is to principally assess the performance of the residential buildings` envelope code under future climate change. Three-dimensional models for the two buildings were constructed based on the architectural drawings using DesignBuilder simulation software. Based on physical site visits and survey, the buildings characteristics including construction materials, cooling system types, lights and installed appliances of the physical villas were applied and modelled in DesignBuilder with occupancy behaviour and activity profile.

2.3. Models validations.

The DesignBuilder models were validated with real measurements from the buildings using iButton dataloggers (DS1921H-F5 Thermochron) for interior rooms and ibutton dataloggers (DS1923-F5# Hygrochron) for exterior conditions, as described by Abuhussain et al. [6].The procedure of taking field measurements was initiated in June 2017 and concluded in August 2017. The research's intention behind using this period of time was due to the fact that these three months in summer are the hottest of the year, therefore air conditioner use is at its peak. In addition to the replicated buildings characteristics in the models for the empirical validation, the monthly average ground temperature was inserted in DesignBuilder and this was obtained from a previous study which conducted field measurements to record the actual ground temperatures at various depths in Jeddah city[10] The monitoring process also aimed to observe and extract the schedule and the occupancy of the house within this three-month period. The weather data files that were used for calibration derived from hourly observations recorded at each location for each building. Depicted by figure 2 is the hourly comparison of actual measured data and results calculated by the DesignBuilder of the living room in House B for five days whilst building was free running.

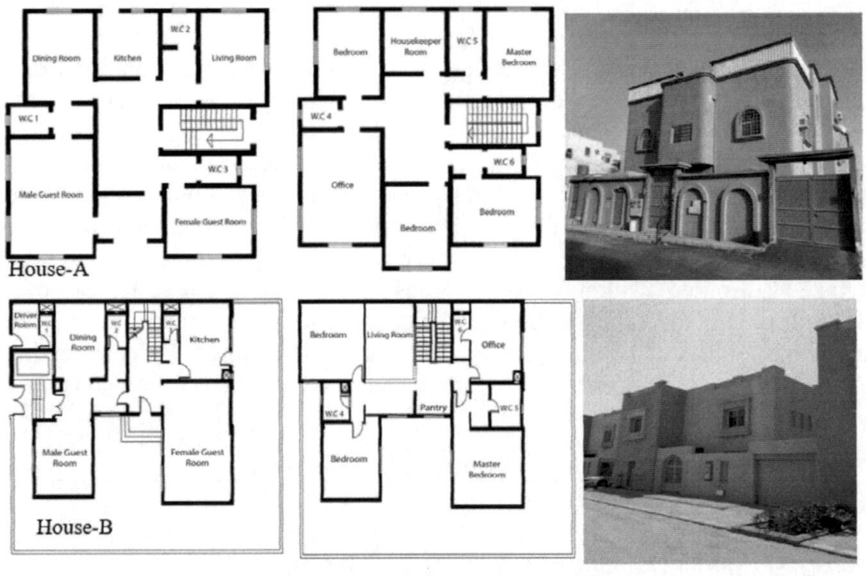

Figure 1. Buildings under study.

Additionally, the bill for electricity consumption per month and the villa's reports were acquired from the Saudi Electricity Company. The daily schedule of lighting, air conditioning, equipment, and occupant habits were observed and obtained from the monitoring process throughout the three-month period. A comparison of actual energy usage and that calculated by DesignBuilder are shown in figure 3 for House B. The highest energy consumption for the two houses is occurred due to requiring more energy to cool the interior through the summer. However, although House B is smaller and is better insulated than House A, House B uses as much energy as House A. The reason for this, observed through the monitoring process, was that the actual behaviour of the occupants of House B differs from the behaviour of the occupants of House A leading to more usage of energy. Moreover, House B was fully occupied but there are some rooms were not used in House A. Having validated the software made it possible to investigate how the Saudi codes might perform in different current and future climate scenarios.

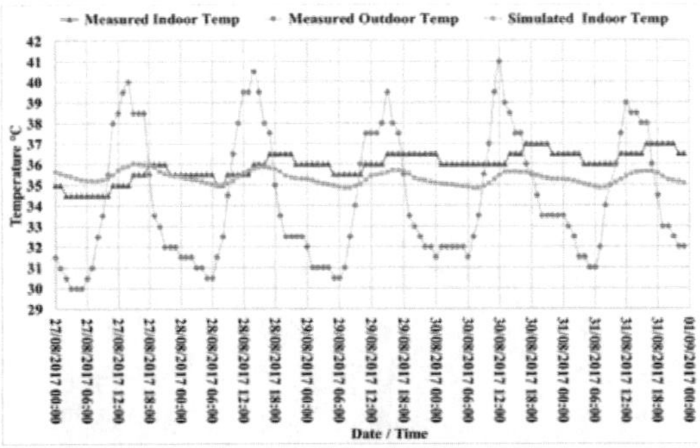

Figure 2. Hourly calibration between on-site measured temperature and calculated temperature for living room in house B.

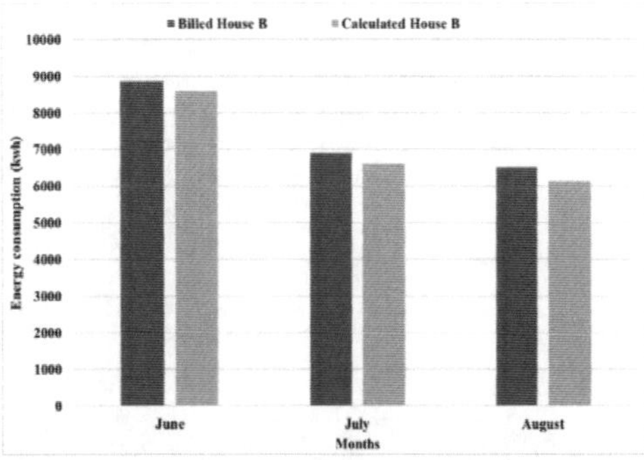

Figure 3. Comparison between utility bills consumption and DesignBuilder results for the three months for house B.

2.4. Assessments and simulations process.

To guarantee the constancy of the results of annual simulation analysis, typical daily schedules in KSA for lighting, equipment, cooling, and occupancy were specified from a prior study on energy conservation in residential buildings that exist in the country [11] . Weather data files for the current and future during the periods current, 2050 and 2080, as well as scenarios related to greenhouse gas emission, the A2, A1B and B1 of Jeddah were acquired from Meteonorm 7[12] .Table 2 draws a comparison between U-values of house A, B and the current codes which have been studied in this research.

Table 2. U-values (W/m2K) for existing houses and the envelope codes (code1 and 2).

U-Value (W /m2K)	House A	House B	Code1	Code2
Roofs	3.40	0.58	0.31	0.20
External walls	1.82	0.68	0.53	0.34
Window glazing	5.71 – SHGC 0.81	3.09 – SHGC 0.7	2.67 – SHGC 0.25	2.67 – SHGC 0.25
WWR %	14.5 %	5.01 %	maximum 25 % of wall for both codes	

2.5. Financial analysis.

Financial feasibility is among the most important concerns. The initial investment and running cost acted as the primary focus on completion of testing the residential building envelope codes in KSA under present and future climates, because they are of utmost significance when decision makers or people invest and inhabit the buildings. Thus, the emphasis has been placed on energy costs and economic feasibility of the codes for conserving energy in buildings as well as related costs varying from construction to running expenditures. The investigation goes beyond energy use and economic feasibility at present, however it reflects on scenarios in the future and consumption in the long term, due to economic assessment and long-term financial returns.

Table 3. Total initial costs for existing houses and the retrofitted houses.

Building	Base case initial cost	Code1 initial cost	Code 2 initial cost
House A	SR 81245	SR 124800	SR 136205
House B	SR 74590	SR 91186	SR 107942

2.5.1. Total initial costs. For the purpose of this study, the initial costs of the base cases (house A and B) were calculated by Saudi Riyal (SR). Also, the initial costs of improving walls insulation, roofs insulation and window glazing to achieve the codes requirements were calculated. Because of the fluctuation in the prices of thermal insulation and construction materials in KSA to predict the actual cost of constructing buildings, the research evaluated the cost of the construction labour and materials in the present market on the basis of the construction market survey. Table 3 presents the initial total cost of House A and House B, the base case buildings, and retrofitted villas to code 1 and code 2.

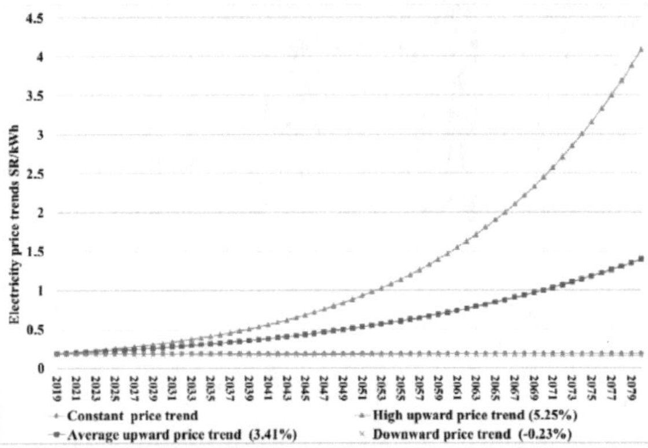

Figure 4. Predicted possible prices trends for electricity.

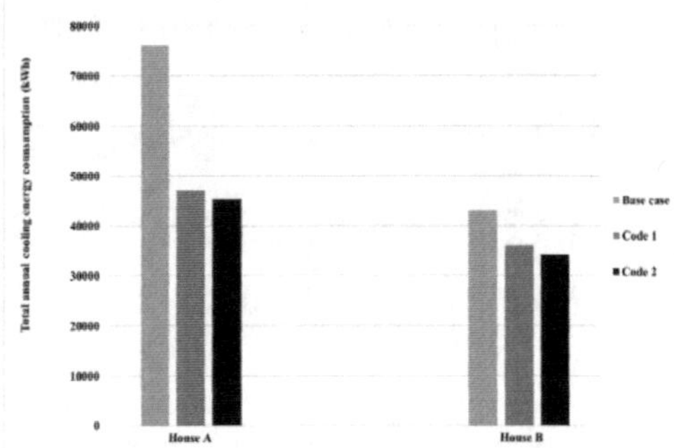

Figure 5. Total annual cooling energy consumption for the existing houses and retrofitted houses to code 1 and 2.

2.5.2. Energy running cost. An electricity consumption tariff was used for the calculation of actual running cost for buildings in the current year. The Saudi Electricity Company's average tariff used for residential buildings is 0.18 SR/kWh of electricity currently [13]. However, to study the financial feasibility of the Saudi codes in the long term over 61 years period from 2019 to 2080, the inflation rate of the energy price is considered in this study since electricity was the only running cost considered in this study, taking into consideration that, the increase in the electricity tariffs (the removal of subsidies by the government) will reflect increased financial benefits from applying code 1 and 2 in favour of this research. The escalation in energy prices is a significant factor to be considered when evaluating energy retrofit projects, as the energy price increase can be deemed an example of the rate of inflation.[14]. Four scenarios were chosen for future electricity prices to depict the uncertainty around the cost of electricity during the forecasting period on the basis of prior historical data provided by the Saudi Arabian Monetary Authority on the yearly percentage fluctuation of inflation rate over the previous 9 years between 2009-2018 in KSA [15]. In relation to prior data on inflation rates during these years,, this study considered high upward price scenario referred to the highest annual percentage variation of inflation with a rate of 5.25%, and the average upward price trend referred to annual percentage variation with a rate of 3.41%, and downward price trend referred to the lowest annual percentage variation of inflation with a rate of -0.23%. In addition, this study also considered a fourth scenario in which the price of electricity would be constant. Figure 4 illustrates the prediction of different electricity prices scenarios in each year up to 2080.To calculate the future yearly electricity price (FC), the present cost of electricity (PC) and the inflation rate (λ) over a life time (N) are required for the equation as follows [14]: $FC = PC\,(1 + \lambda)^N$

3. Results and discussion

3.1. Effect of applying the Saudi residential building envelope codes to the villas for current climate conditions.

Figure 5 shows the total annual cooling energy consumption for the existing Houses A and B and the retrofitted houses to code 1 and code 2 under the current climate of Jeddah. It has been demonstrated that applying code 1 and 2 to House A accomplished a significantly high reduction in the total annual cooling demands at respective rates of 38% and 40%. Meanwhile implementing code 1 and code 2 to House B as it stands meant that saving rates seen by reduction of total annual cooling requirements at just 16% and 20%. As has been shown, the difference of 2-4% in energy savings from code 1 and code 2 in both houses is negligible.

As mentioned by SEEC (Saudi Energy Efficiency Centre), this code has an expectation to deliver a reduction in energy of around 30-40% when implemented in the residential sector [4]. This estimation fulfils the results of this research which presents a reduction of 40% in the total annual cooling loads in base case A with the implementation of code 2. This finding is attributed to the fact that low insulation levels for walls and roofs and single glazed windows with high window to wall ratio of 14.5 %. However, the low rate of energy savings in the total cooling of 17% by applying code 2 to the base case B is due to relatively low window to wall ratio of 5%, double glazed windows and relatively low U-Values for external walls and roofs. As a result, implementing both codes to house B has a lower effect. based on the simulations performed, considerable reductions in energy seem to be accomplished already by retrofitting the villas to the standard of the new code. This shows how the standard for this type of housing in KSA can enhance the energy performance as well as minimize energy consumption in new builds of the residential building sector in the country.

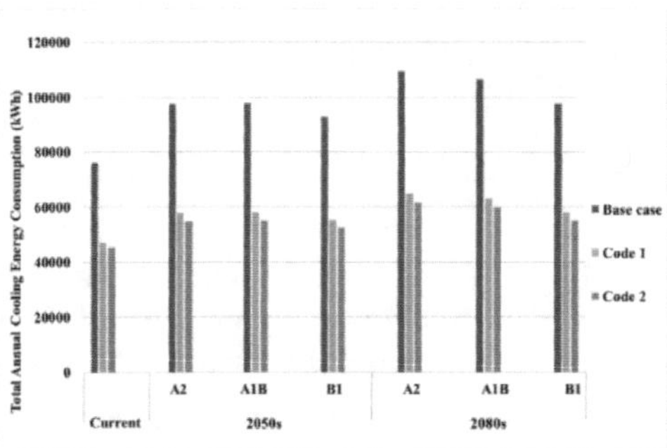

Figure 6. The total annual cooling energy consumption in the current and future climate periods under different climate change scenarios for the base case villa A, retrofitted villa A to code 1 and 2.

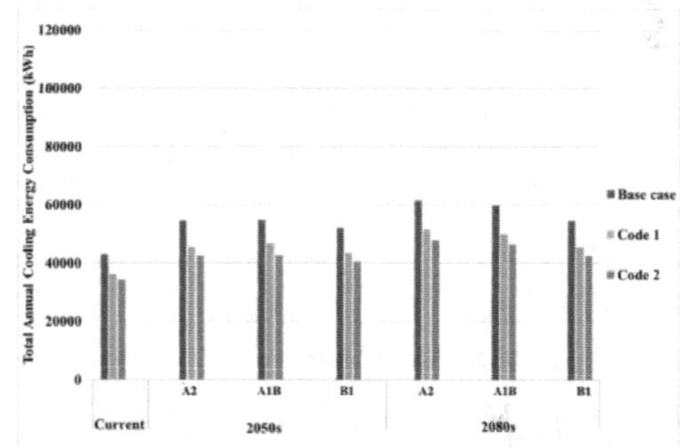

Figure 7. The total annual cooling energy consumption in the current and future climate periods under different climate change scenarios for the base case villa B, retrofitted villa B to code 1 and 2.

3.2. The potential impacts of the future climate change scenarios on the base case villas and retrofitted villas to the Saudi Residential codes.

Figures 6 and 7 show the total annual cooling energy consumption for house A and B in the current and future climate periods (2050s and 2080s) under different climate change scenarios (A1, A2 and B1) for the base case villas , retrofitted villas to code 1 and retrofitted villas to the standard of the code 2 .With the future climate data, the cooling demand for the base A rises sharply at a rate between 28 % to 44 % in 2080s depending on the future scenarios, whereas, for the retrofitted building A to code 2, this increase percentage falls in 2080s to 22-36 %. Similarly, for base B but less sever increase of cooling energy consumption could reach 27% to 43%, and for the retrofitted B to code 2 to 24 -40%. It can be seen that the total cooling energy consumption for the retrofitted houses to code1 and 2 in the 2080s will still be lower than the levels of cooling consumption with existing houses under the current climate.

3.3. Payback Analysis and Feasibility Study

3.3.1. House A. Figure 8 showing the cumulative cash flow that results from the electricity saving and the time required to recover initial costs of investment of applying code 2 to house A with constant, growing and decreasing electricity prices over 61 years for A2 climate scenario. The figure of (SR 54960) shows the initial investment cost of applying code 2 construction with corresponding accumulated saving in the electricity cost yearly. The cumulative saving flow of the retrofitted house to code 2 for various electricity prices trends shows that the payback time from electricity bill saving by improving wall and roof insulation and glazing to the standard of code 2 will be 7 to 8 years with all electricity future prices scenarios.

3.3.2. House B. The same assumptions for different electricity prices were made to observe the cost benefits of applying code 2 to the house B. By applying code 2 to the base case of house B, figure 9 shows that for this case the number of years needed to regain the investment on construction (SR 33351) will be 11 to 12 years alongside high and average upward prices and realistically may pay back after 16 years if the prices of electricity fall or remain constant. As a result, there will be a positive effect on the annual savings of electricity bill after 11 years, which will be even more beneficial due to the high upward price trend of electricity.

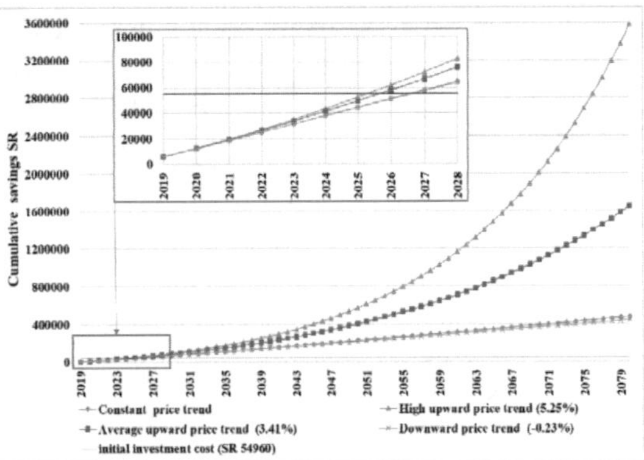

Figure 8. Cumulative yearly saving of applying code 2 to house A for current and future A2 climate scenario for different electricity prices.

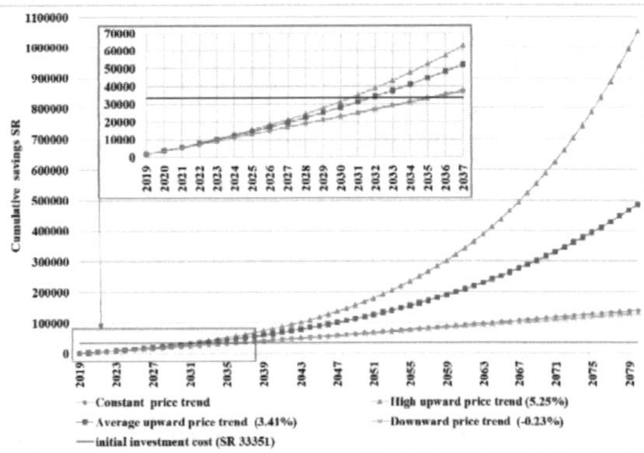

Figure 9. Cumulative yearly saving of applying code 2 to house B for current and future A2 climate scenario for different electricity prices.

4. Conclusions

This paper investigated the effectiveness of applying the Saudi residential buildings envelope codes for two detached single-family houses (villas) located in hot and humid climate Jeddah, KSA. Also, this paper has examined the potential impacts of future climate change scenarios on the existing houses and the retrofitted houses to the codes. DesignBuilder was the software used as the tool for investigating primarily the energy performance of the two buildings and that of the residential building envelope code in KSA under scenarios in the future climate change. Additionally, a financial assessment on the basis of results by the simulation was conducted in order to examine the financial feasibility of the codes in the long term while considering the uncertainty around climate change in the future. It becomes evident that the base case villas that were studied will not be able to tackle the impact of global warming in the future, with cooling being the main concern owing to the severe weather conditions in the country. Applying the standards by codes for residential buildings in existing villas caused a decrease in the total yearly consumption of cooling energy by a rate of 40 and 17% and can also reduce the impact of climate change in the future. However, there is still a rise in demand for cooling energy because of climate change. It is likely that global warming will cause an increase in the amount of electricity used in residential buildings for cooling in 2080s by 22-36%. Further research is necessary to enhance the current code by using codes 1 and 2 as a base case. A more thorough and passive strategy needs to be considered in future analysis, as they are not a part of the residential building envelope codes in KSA for various climate zones to neutralize the effects of climate change in the future.

Comparing the initial cost construction of applying code 1 and 2 to the base cases of house A and B was undertaken. Accordingly, based on the different future electricity prices, energy costs were calculated for the base case houses and for the retrofitted houses. It can be concluded that the payback periods to recover the initial investment and the financial benefits of applying the codes is very dependent on the growth of the energy prices and that energy prices are unpredictable. The most favourable electricity bill savings occurred by imposing the high upward rising prices which this gave a payback time of less than 13 years for all houses. There are low electricity bills, seeing as the current average price of electricity is low, thus there is no motivation towards consumption habits becoming more rationalize. Furthermore, the low cost of bills does not encourage higher investment into the efficiency of energy, and new houses are seen as more favourable to renovating old ones. As a result of this, this study explains that if the cost of electricity rapidly increases further, over the period of study, the payback time to regain earlier investment of implementing the code in residential building envelopes will be lower and the economic benefits reaped from savings on electricity cost will be deemed more favourable.

References

[1] M. Abdul and O. S. Alshamrani, "Prospects of energy conservation and management in buildings – The Saudi Arabian scenario versus global trends," vol. 58, pp. 1647–1663, 2016.

[2] B. Petroleum, "BP Statistical Review of World Energy June 2015''bp. com/statisticalreview," 2018.

[3] N. Al-Tamimi, "A state-of-the-art review of the sustainability and energy efficiency of buildings in Saudi Arabia," *Energy Effic.*, 2017.

[4] SEEC, "Saudi Energy Efficiency Center," Building energy sector, 2017. [Online]. Available: http://www.seec.gov.sa/en/buildings-en. [Accessed: 06-Jan-2019].

[5] Electricity and Cogeneration Regulatory Authority, "Annual statistical booklet for electricity & seawater desalination industries," KSA, 2016.

[6] M. Abuhussain, D. Chow, and S. Sharples, "Assessing the adaptability of the Saudi residential building`s energy code for future climate change scenarios," *in PLEA,Proceedings of the 34th International Conference on Passive and Low Energy Architecture,* 2018, pp. 73–78.

[7] A. Assiri and E. M. Darfaoui, "Response to Climate Change in the Kingdom of Saudi Arabia," *A Rep. Prep. FAO-RNE*, vol. 17, 2009.

[8] M. Almazroui, M. Nazrul Islam, H. Athar, P. D. Jones, and M. A. Rahman, "Recent climate change in the Arabian Peninsula: Annual rainfall and temperature analysis of Saudi Arabia for 1978-2009," *Int. J. Climatol.*, vol. 32, no. 6, pp. 953–966, 2012.

[9] M. Alhashmi *et al.*, "Energy Efficiency and Global Warming Potential in the Residential Sector : Comparative Evaluation of Canada and Saudi Arabia," vol. 23, no. 3, pp. 1–12, 2017.

[10] J. Hijazi and S. Howieson, "Displacing air conditioning in Kingdom of Saudi Arabia : An evaluation of ' fabric first ' design integrated with hybrid night radiant and ground pipe cooling systems," pp. 1–14, 2018.

[11] H. Monawar, "A Study of Energy Conservation in the Existing Apartment Buildings in Makkah Region , Saudi Arabia," *PhD Thesis, Sch. Archit. Plan. Landscape, Univ. Newcastle upon Tyne, United Kingdom.*, 2001.

[12] "Meteonorm," 2017. [Online]. Available: http://www.meteonorm.com/. [Accessed: 11-May-2017].

[13] Saudi Electricity Company, "Consumption tariffs," 2018. [Online]. Available: https://www.se.com.sa/en-us/customers/Pages/TariffRates.aspx. [Accessed: 04-Jan-2019].

[14] M. Krarti, *Energy Audit of Building Systems An Engineering Approach*, 2nd Editio. Boca Raton: CRC Press. PP.3-4, 2010.

[15] Saudi Arabian Monetary Authority, "Inflation Rate," 2019. [Online]. Available: http://www.sama.gov.sa/en-US/Indices/Pages/InflationRate.aspx. [Accessed: 18-Jan-2019].

Envi-MET validation and sensitivity analysis using field measurements in a hot arid climate

Y N Ayyad[1,2] and S Sharples[1]
[1] School of Architecture, University of Liverpool, Liverpool, L69 3BX, United Kingdom.
[2] Department of Architecture, Al-Ahliyya Amman University, Al-Salt, Jordan.

Abstract. Envi-MET is a computational fluid dynamics (CFD) modelling software that analyses interactions between the microclimate and the built environment through simulation, predicting parameters such as wind speed, air temperature and thermal comfort indices. The large number of variables and calculations involved makes validating Envi-MET models difficult. This study focuses on the validation of Envi-MET through sensitivity and calibration testing. A site was chosen in a hot arid climate location. The testing involved measuring air temperature, relative humidity and wind speed at the site. For the sensitivity testing a hypothetical site was modelled with three scenarios for four variables (wind speed, relative humidity, albedo and grid size) set as low, base and high values. The monitoring study concluded that the accuracy of Envi-MET varied with the variable in question, showing a high correlation factor with wind speed and air temperature but a poor correlation with relative humidity, especially between 04:00 and 10:00. Envi-MET displayed different sensitivity levels, with a good response to variables change in wind speed and relative humidity, but little response to albedo changes.

1. Introduction

Envi-MET has been described as "*a holistic three-dimensional non-hydrostatic model for the simulation of surface-plant-air interactions not only limited to, but very often used to simulate urban environments and to assess the effects of green architecture visions*" [1]. The software was developed by Michael Bruse and his team in 1993, and since then enhancements to the software have elevated it from a specialist research tool to mass market implementation [1]. Envi-MET as an urban simulation programme deals with large numbers of variables, and this makes the software potentially susceptible to calculation error. Considering that risk, this study investigated the accuracy of the outcomes of Envi-MET simulations by comparing them to field measurements made under hot arid conditions at Al Ahliyya Amman University in Jordan. The sensitivity of Envi-MET outputs to changes in the values of input variables was also analysed.

2. Previous validation studies of Envi-MET in hot climates

Early attempts to validate Envi-MET, for example [2][3], found adequate rather than strong levels of agreement between modelled and measured data. As Envi-MET developed and meteorological instrumentation improved, it became possible to undertake more detailed validation studies, Measurements of net-radiation, air temperature surface temperature and wind speed for an urban site in South Korea (hot humid climate) found significant differences between the logged and simulated results [4]. Conversely, monitored and modelled environmental conditions in the Al-Muizz street area of Cairo, Egypt (hot arid climate) found overall good agreement [5]. Wind speeds and other meteorological parameters were measured in Wuhan, China (hot summer climate) to assess the impact of tree type and distribution during a hot summer and cold winter. Envi-MET predictions showed good agreement for air temperatures whilst the findings for air velocities were less satisfactory – the measured wind speeds displayed more variability than the simulated values.

The results of this small literature review suggest that more validation studies of Envi-MET need to be performed under a range of climate types and urban/vegetative configurations. This need provided the motivation for the current work.

3. Methodology

3.1 Meteorological monitoring for Envi-MET comparison

The study site was Al Ahliyya Amman University in Jordan. This site offered a range of different urban elements, such as a wide variety of vegetation, tiled pathways, buildings and a car park (see Figure 1); the area surrounding the university is a hillside with empty plots and some scattered buildings. Trees at the edges acted as wind blockers. In addition, the University has 24 hour security that ensured the safety of the monitoring equipment. The Köppen climate classification system puts Jordan in two classifications - hot arid desert (BWh) and cold arid desert (BWk) [6], with a maximum temperature in August of 41.5°C and a minimum in February -4.5°C, Precipitation mostly occurs in winter, with an average of 55 rainy days. The cladding of most University buildings is white limestone with an average albedo of 60%, and the pathways that link the buildings have different material. However, the area where the loggers were placed was tiled with grey cement tiles with an average albedo of 30%. The locations of A and B allowed for different conditions, with B being near to sprinklers and A being under larger thicker trees. The vegetation at the site is mostly local and coniferous in nature - the only deciduous tree is Populus nigra. other coniferous are Pinus halepensis, Mediterranean cypress (Cupressaceae), Phoenix dactylifera and Cupressus macrocarpa 'Goldcrest' (Figure 2).

Figure 1. Rendering of the site showing the monitored locations.

Figure 2. Tree types found at the measurement site. Courtesy of Al Ahliyya Amman University.

For comparison with Envi-MET predictions, the meteorological parameters measured at the two sites identified in Figure 1 were wind speed, air temperature, relative humidity and mean radiant temperature. Two Kestrel 5400 Heat Stress Trackers were used to log these parameters. The Kestrel 5400 is a handheld, portable WBGT (Wet Bulb Globe Temperature) logger and weather station. The quoted accuracies of the Kestrel readings are ±3% for wind speed; ±0.5°C for air temperature; ±2% for relative humidity.

3.2 Envi-MET sensitivity analysis

In order to test the sensitivity of Envi-MET, a hypothetical site was constructed and modelled in three scenarios of low, base and high settings. The tested meteorological parameters were wind speed, relative humidity, albedo of the surrounding surfaces and grid size. The evaluation of the data was based on the change in air temperature for the different scenarios

3.3 Statistical analysis of the monitored and modelled data

Historically, one of the most widely used methods of model validation is Pearson's correlation coefficient r. It is defined as "*the measure of the strength of the linear relationship between two variables*" [7]. Pearson's correlation coefficient consists of the covariance of the data sets given and their standard deviations, with the resulting value of the correlation coefficient r ranging between +1 and -1, where 0 means no correlation within the data sets. Root Mean Squared Error (RMSE) is a dimensional metric used to identify how large the difference is between observed and predicted values of a parameter and to indicate how divergent the predicted data are from the observed values. RMSE consists of the square root of the sum of the squared variables difference divided by the number of the variables. Mean Absolute Error (MAE) is a dimensional metric used to measure the error between two sets of data. The formula consists of the sum of the absolute difference of the predicted and observed values over the number of variables. RMSE and MAE are widely used by climate researchers to test and validate predicted values. It has been argued that RMSE and other new measures, such as the index of agreement, were superior to Pearson's correlation coefficient for evaluating models [8]. The index of agreement ranges between 0 and 1, with 1 being perfect correlation and 0 no relationship between the predicted and observed values. Later research suggested that RMSE is not appropriate and replacing it with MAE should give a more accurate description of the model validation [9]. In recent years it has been argued that the replacement of RMSE with MAE is not an accurate representation of the model validation process, especially when there is a Gaussian error distribution. The use of a combination of model verification methods to reach accurate results was recommended [10].

4. Envi-MET modelling of the Al Ahliyya Amman University site

4.1 Spaces

A 3D model of the site was built using *spaces* —one of Envi-MET's components, where initial AutoCAD format drawings were used as a base for the model. Modelling in *spaces* is a grid system modelling with a default resolution of 2 metres. The 3D resolution was changed to dx=1 dy=1 dz=1 since the site in question is small enough to fit in the 60x60-pixel grid. *Equidistant* was chosen as the method of vertical grid generation as the highest point of the buildings only reaches 18 meters, which is enough to avoid any boundary issues that might occur due to the proximity to the upper limit of the model.

4.2 Vegetation

Envi-MET allows its users to create their own vegetation database through *Albero*. To be able to use *Albero* in Envi-MET, the leaf area density (LAD) must be determined. However, in this study there was no need to construct new vegetation since Envi-MET's library had similar trees to those on the site, and only some modifications on the existing Envi-MET's data base was made.

4.3 Wind

The wind speed values for the given simulated day at the monitoring site were extracted from the EPW file for the nearest weather station to the site. Usually, weather readings are taken in an open field 10 metres above ground level, which makes the readings potentially unreliable for an urbanized setting, It has been suggested that the wind speed values from a weather stations can be adjusted to suit an urban setting by using a ratio S - the ratio between the wind speed at a height H above the urban area (V_H) and the wind speed at 10m height in open flat country (V_{10}), where $10\ m \leq H \leq 150$ [11]. Table 1 shows the S factor values for urban and suburban areas.

Table 1. S factor S = VH /V10

Height	10	20	30	40	50	60	70	80	90	100	110	120	130	140	150
S (suburban)	0.6	0.73	0.82	0.89	0.94	0.99	1.04	1.08	1.11	1.14	1.18	1.21	1.24	1.27	1.29
S (urban)	0.36	0.47	0.55	0.62	0.68	0.73	0.77	0.82	0.86	0.89	0.93	0.96	0.99	1.02	1.05

Al Ahliyya Amman University is situated outside of the city Amman on a hillside; therefore, it does not fall under urban or suburban areas regarding to wind speed. The average wind speed was calculated as 4 m/s from a north westerly direction (291°).

4.4 Meteorological settings
The day chosen for the simulation was 1st October 2017- chosen for the mild conditions of the month, with average meteorological parameters. The values for air temperature and relative humidity were obtained from an EPW file, as shown in Table 2.

Table 2. Air temperature and relative humidity values for the 1st October

Hour	01:00	02:00	03:00	04:00	05:00	06:00	07:00	08:00	09:00	10:00	11:00	12:00
Air temp (°C)	17.4	16.2	15	14.3	13.8	13.8	15.3	17.4	19.5	21.4	22.9	24.1
Rel humidity (%)	47	49	50	52	50	50	50	41	34	28	24	22
Hour	13:00	14:00	15:00	16:00	17:00	18:00	19:00	20:00	21:00	22:00	23:00	24:00
Air temp (°C)	24.9	25.2	25.1	24.3	23	21.5	20.3	19.1	17.9	16.8	15.6	14.4
Rel humidity (%)	21	23	22	24	28	30	37	41	46	50	55	60

5. Results and discussion

5.1. Comparison of the measured and Envi-MET data
Figure 3 shows the predicted and measured values of air temperature at monitoring point A. Envi-MET values are generally lower than the logged ones by an average difference of 2.8 °C between the time intervals of 09:00-19:00 and 00:00-04:00. However, the predicted values show a slight increase for the time intervals 20:00-23:00 and 05:00– 08:00, with an average difference of 0.3°C. Overall, the two sets of data show a good correlation, with an index of agreement value of 0.886 and Pearson correlation coefficient of 0.933, as shown in Table 3.

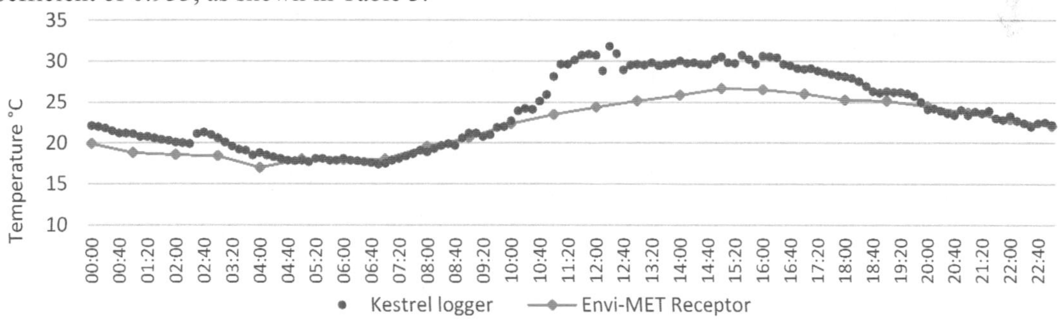

Figure 3. Data comparison between the observed and the predicted air temperature values, location A

Table 3. Model validation location A

Validation method	Temperature	Relative humidity	Wind Speed
Index of agreement	0.886	0.646	0.104
RMSE	2.602	18.797	0.424
MAE	1.974	14.4366	0.362
Pearson correlation coefficient	0.933	0.743	0.102

Figure 4 shows, for point B, the predicted temperature values for the time intervals of 09:00-19:00 and 00:00 – 04:00 are lower compared to the loggers' readings with an average change value of 2.9°C. For the rest of the day, the predicted values are similar to the observed ones, with a slight increase of an average change value of 0.2°C. Overall, the two sets of data show a good correlation, with an index of agreement value of 0.890 and Pearson correlation coefficient of 0.934 (see Table 4).

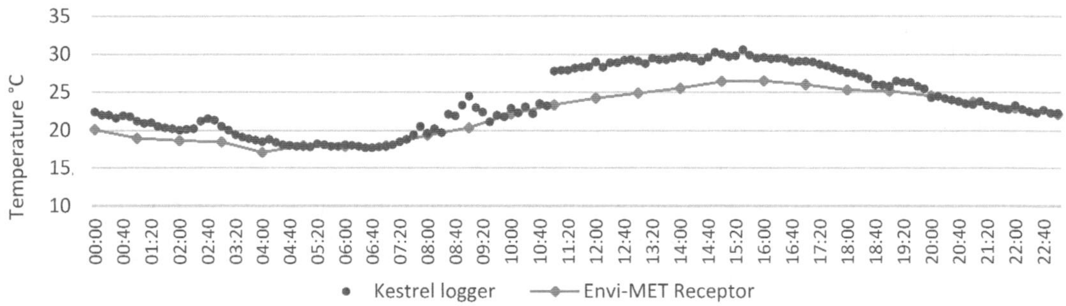

Figure 4. Data comparison between the observed and the predicted air temperature values, location B

Table 4. Model validation location B

Validation method	Temperature	Relative humidity	Wind Speed
Index of agreement	0.890	0.688	0.168
RMSE	2.455	16.624	0.819
MAE	1.889	12.814	0.593
Pearson correlation coefficient	0.934	0.768	-0.325

Figure 5 shows the predicted and measured values of relative humidity at monitoring point A. The divergence can be seen clearly for the time interval of 03:00-10:00, where the observed relative humidity value rises to its daily maximum due to the lower air temperature values at night. However, this impact is much smaller in the predicted values. For the rest of the day the predicted values seem to have a closer pattern behaviour to the observed ones with an average change value of 8.8%. Overall, the two sets of data present an adequate correlation, with an index of agreement value of 0.646 and Pearson correlation coefficient of 0.743 (Table 3).

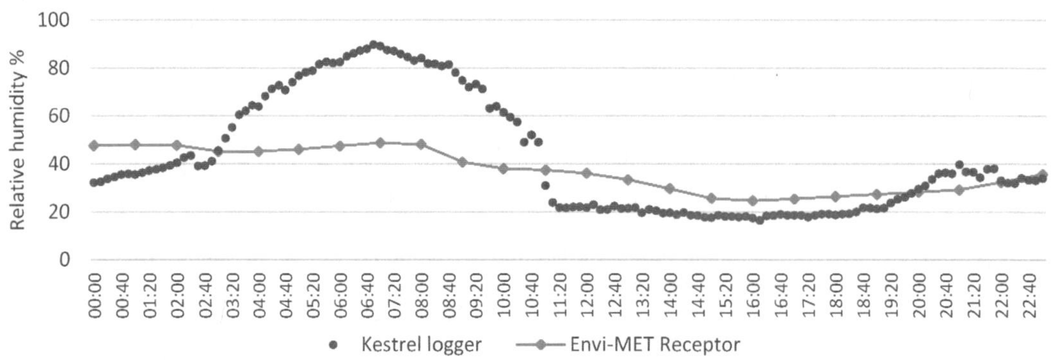

Figure 5. Data comparison between the observed and the predicted for relative humidity, location A.

As shown in Figure 6 for location B, RH values from Envi-MET at location B behave in a similar fashion to location A, with values having a large divergence between 03:00 and 10:00 with an average RH difference of 18.4%. compared to an average of 8.1% for the rest of the time. Overall, the data sets have an adequate correlation, with an index of agreement value of 0.688 and Pearson correlation coefficient of 0.768 (Table 4).

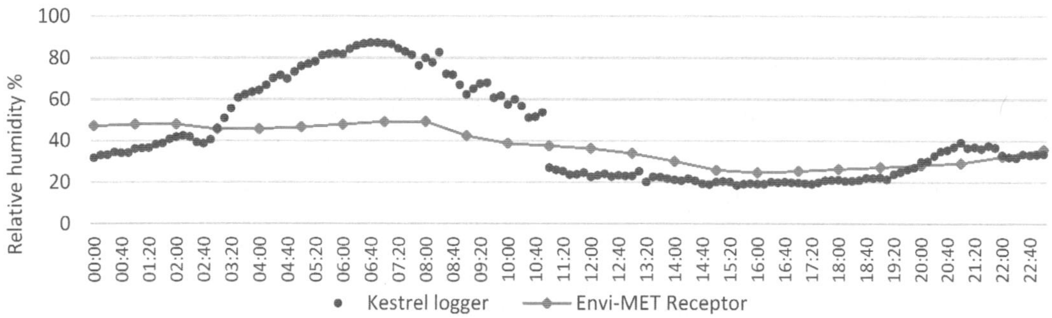

Figure 6. Data comparison between the observed and the predicted for relative humidity, location B.

Figures 7 and 8 indicate the predicted and observed values for wind speed; however, the predicted values for wind speed are averaged over an hourly basis, whereas the observed values were recorded every 10 minutes. The trend line shows great similarities with the predicted values, although there are large individual differences. In this case, even though the model validation methods in Tables 3 and 4 show a very low correlation, Envi-MET's results are still valid. This example shows the difficulty of trying to validate Envi-MET with a rapidly fluctuating parameter like the wind.

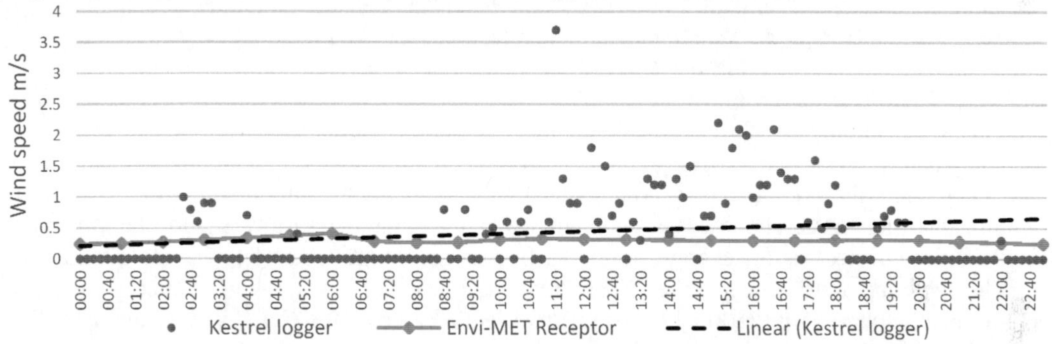

Figure 7. Data comparison between the observed and the predicted for wind speed, location A.

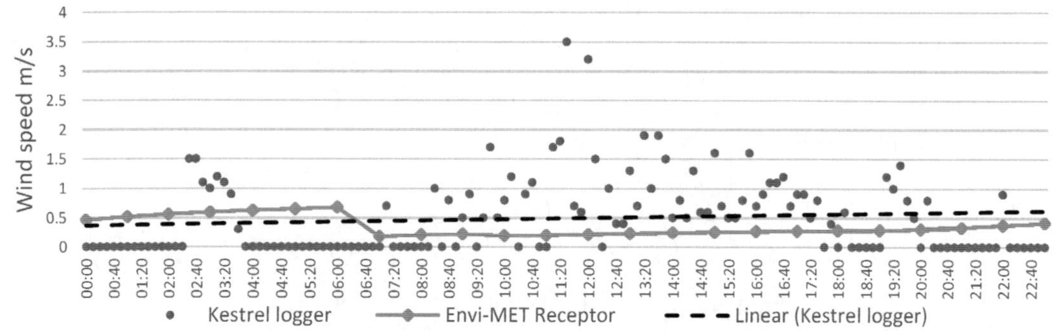

Figure 8. Data comparison between the observed and the predicted for wind speed, location B.

5.2. Envi-MET model sensitivity testing

For the purpose of testing Envi-MET's sensitivity, a series of model runs were made for different parameters. The tested area was comprised of 50 x 50 metre plots, with six 9-metre-high buildings surrounding the buildings. The model was run three times for every parameter, testing the base, the low and the high values of that said parameter. These parameters included relative humidity, wind speed, albedo and grid resolution. Air temperature was the derived parameter from running Envi-MET to test sensitivity to the range of inputs.

5.3 Wind Speed Sensitivity

Figure 9 shows the plotted data from the Envi-MET model testing the wind speed. A low wind of 1 m/s tended to lower the base model air temperature for 06:00-08:00 by an average value of 1.5°C, a maximum of 2.3°C and a minimum of 0.5°C. However, through the time interval of 09:00 to 17:00 the low wind speed showed a significant increase in air temperature, by an average of 1.8°C, a maximum value of 2.9°C and a minimum of 0.3°C. High wind speeds, on the other hand, showed a slight increase of air temperature for the same time intervals in the night time with an average of 0.4 °C, a maximum value of 0.8°C and a minimum of 0.1°C. Day time showed lower air temperature value, with an average of 1.1°C, a maximum value of 1.4°C and a minimum of 0.5°C.

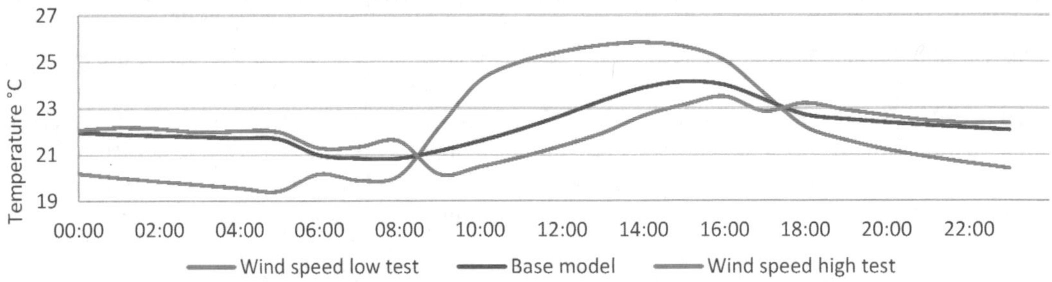

Figure 9. Wind speed sensitivity testing

5.4 Relative Humidity Sensitivity

Figure 10 shows the plotted data from the Envi-MET relative humidity testing. The low test value of 40% relative humidity showed a higher air temperature values compared to the base model for the time interval 11:0-17:00; the average change value was 0.3°C, while the maximum value was 1.0°C and the minimum value was 0.1°C. The low RH tended to lower the air temperature values for the time interval of 05:00-10:00, with an average change value of 1.0°C, a maximum value of 1.3°C and a minimum of 0.4°C. The air temperature values of the 90% relative humidity high test showed an average increase of 0.9°C for the time interval of 06:00 - 13:00 and a maximum value of 1.3°C and a minimum of 0°C. For the rest of the day, the high relative humidity tended to lower the base model's air temperature by an average value of 0.4°C, a maximum value of 0.8°C and a minimum of 0.3°C.

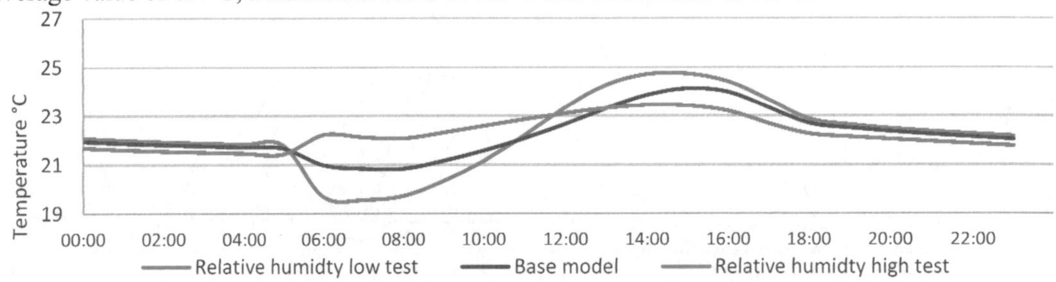

Figure 10. Relative humidity sensitivity testing

5.5 Albedo Sensitivity

Figure 11 shows the plotted data from the Envi-MET model testing the albedo. The albedo of the buildings' cladding and the pavement area were modified as low albedo of 10%, base model of 50% and high albedo of 90%. The high albedo test show virtually no change in the air temperature compared to the base model. However, the albedo test showed a slightly higher change in air temperature values compared to the high albedo test, with an average change of 0.2°C and a maximum value of 0.3°C.

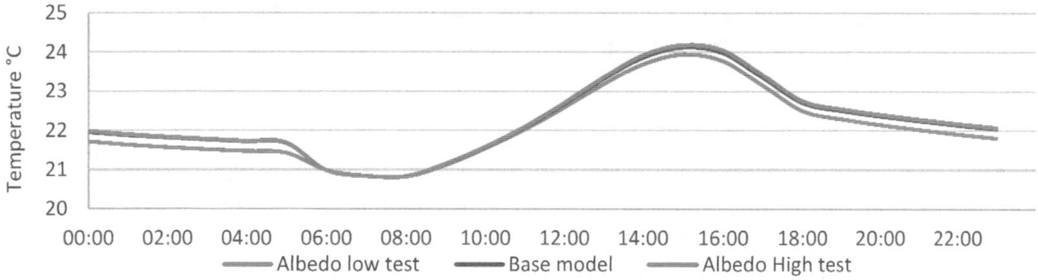

Figure 11. Albedo sensitivity testing.

5.6 Grid size

Simulating in Envi-MET relies heavily on the size of the actual project and in some cases the studied parameter. A higher resolution grid or a thicker mesh would give more accurate results that is if the site is relatively small. However, increasing the resolution of the grid will raise the hardware requirement and the simulation time. The base model that was used in the testing was a 50 x 50m plot with a resolution of 1x1m. In order to test Envi-MET's sensitivity two other mesh counts were introduced: 2x2 and a 3x3. As Figure 12 shows, the 3x3 grid's air temperature values had a large increase compared to the base model for the time interval of 13:00-03:00, with an average change value of 4.8°C, a maximum change value of 12.2°C and a minimum change value of 0.4°C. For the rest of the day the variations showed a decrease in air temperature values of an average change value of 2.8°C, a maximum change value of 4.2°C and a minimum change value of 0.1°C. The 2x2 grid had better accuracy than the other grids. However, air temperature values showed a decrease from the base model for the time interval of 14:00-09:00, with an average change value of 1.9°C, a maximum change value of 2.9°C and a minimum change value of 0.1°C. For the time interval 10:00-13:00 Figure 13 shows a slight increase in air temperature values of an average change value of 0.5°C, a maximum change value of 0.7°C and a minimum change value of 0.5°C.

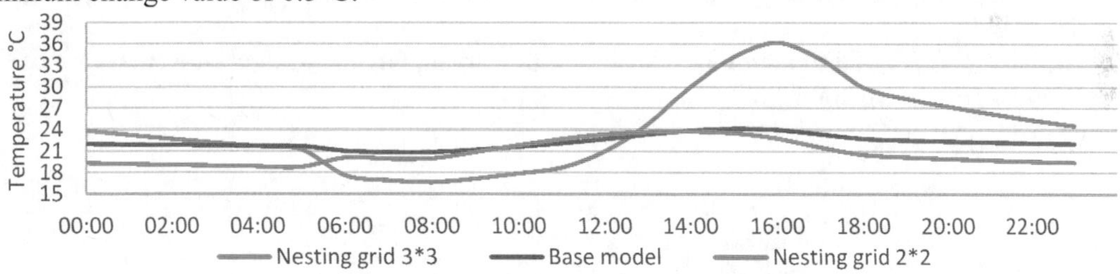

Figure 12. Grid size sensitivity testing.

6. Conclusion

Envi-MET showed different sensitivity levels across the tested parameters. The relative humidity change showed a good response to air temperature, especially at the highest value of the day, where the rise in relative humidity produced lower air temperature values and vice versa. For the wind speed Envi-MET also showed an effect in moving the air temperature compared to the base model. Air temperature had a significant increase when reducing the wind speed to a minimum but showed a smaller value change when increasing the wind speed to double the base model. Lowering the grid count was shown to produce inaccurate results compared to the base model, particularly when simulating small urban plots in Envi-MET using a small resolution, as seen for the 3x3 grid. However, the 2x2 grid showed less error percentage compared to the base model. The albedo test had the least impact on air temperature compared to the other parameters where it only showed a slight decrease in values in the low test.

Though Envi-MET has some unsatisfactory results, as with the relative humidity testing, it showed that the data reacted to air temperature change for night time to some extent, and it had a good correlation value for the rest of the day. The air temperature testing showed the most accurate results with the highest index of agreement values. As for the wind speed, Envi-MET showed a very good correlation with the trend line of the observed values since it averaged the hourly data. Overall, the study showed that Envi-MET as a simulation software is better suited to analyse the relative change in parameters rather than the change in absolute values, as it shows a great response to the modification of microclimatic parameters.

References

[1] M. Bruse, “Envi-MET official website,” 2019. [Online]. [Accessed 17 February 2019].
[2] E. Lahme and M. Bruse, “Microclimatic effects of a small urban park in densely built- up areas: measurements and model simulations,” ICUC5, Lodz, 2003.
[3] R. Emmanuel and H. J. S. Fernando, “Urban heat islands in humid and arid climates: role of urban form and thermal properties in Colombo, Sri Lanka and Phoenix, USA,” climate research, vol. 34, 241-251, 2007.
[4] B.-G. Song, K.-H. Park and S.-G. Jung, “Validation of ENVI-met Model within Situ Measurements Considering Spatial Characteristics of Land Use Types,” Journal of the Korean Association of Geographic Information Studies, vol. 17, no. 2, pp. 156-172, 2014.
[5] M. Elnabawi, N. Hamza and S. Dudek, “Numerical Modelling evaluation for the microclimate of an outdoor urban form in Cairo, Egypt,” HBRC Journal, vol. 11, pp. 246-251, 2015.
[6] M. Kottek, J. Grieser, C. Beck, B. Rudolf and F. Rubel, “World Map of the Köppen-Geiger climate classification update,” Meteorologische Zeitschrift, vol. 15, 259-263, 2006.
[7] D. M. Lane, M. Hebl and R. Guerra, “Introduction to Statistics,” University of Houston, Houston, 2013.
[8] C. J. Willmott, “on the validation of models,” Physical Geography, pp. 184-194, 1981.
[9] C. J. Willmott and K. Matsuura, “Advantages of the Mean Absolute Error (MAE) over the Root Mean Square Error (RMSE) in assessing average model performance,” climate research, vol. 30, pp. 79-82, 2005.
[10] T. Chai and R. Draxler, “Root mean square error (RMSE) or mean absolute error (MAE)? – Arguments against avoiding RMSE in the literature,” Geoscientific Model Development, pp. 1247-1250, 2014.
[11] M. Nikolopoulou, Designing Open Spaces in the Urban, Greece: Center of Renewable Energy Sources (C.R.E.S), 2004.

IOP Conf. Series: Earth and Environmental Science **329** (2019) 012041 doi:10.1088/1755-1315/329/1/012041

Scoping Low-Cost Measures to Nowcast Sub-Hourly Solar Radiations for Buildings

L Chen[1], **H Du**[1] **and Y Li**[1]
[1] Welsh School of Architecture, Cardiff University, Bute Building, King Edward VII Avenue, Cardiff, CF10 3NB, Wales, UK

Abstract. Solar energy, as a clear, renewable and promising resource, has brought tremendous value for carbon emission reduction in the last three decades. However, the uncertainty of solar radiation brings significant challenges to the renewable heat and electrical system for buildings. Thus, many different solar radiation forecasting methods, such as Numerical Weather Prediction Methods (NWP), Statistical Methods, Top-Down Methods, Bottom-Up Methods and Hybrid Methods have been developed to make the best guess on future solar radiations. Based on the meteorology definition, nowcasting refers to the forecast of the temporal horizon from the next few seconds up to six hours. Predicting solar radiation within this range are extremely challenging, but crucial for solar energy system operation. This paper firstly reviewed the state of art of solar radiation forecasting methods and compared the key features of each method. According to the advantage and costs of each method, this paper then proposed a methodology to generate high accuracy low-cost solar radiation nowcast data for optimising building solar thermal system performance and related control strategies. Note that this paper focused on the scoping study of methodology rather than field experiments, and further research with experiments data will be reported in a journal publication.

1. Introduction

Solar energy, as a clear, renewable and promising resource, has attracted a lot of researchers' interest in the last three decades. A large number of researches have explored the utilisation of solar energy in the built environment. However, the uncertainty of solar energy brings significant challenges to its utilisations. As a result, solar radiation forecasting methods have been developed to achieve a better understanding of solar radiation. Nonetheless, it's worth noting that there are very few pieces of research focusing on near-future forecasting of solar radiation and its significant potentials for building performance optimisation. In the meanwhile, there are increasing opportunities, such as the development of the Internet of Things and 5G internet infrastructure, to enable the development of building optimisation through using real-time solar radiation forecasting.

Therefore, this paper pays attention to study a new nowcasting method of solar radiation, which combines the advantages of various solar radiation forecasting methods and discuss its application on building performance optimisation. According to the definition of nowcasting in meteorology term, this research aims to explore a solar radiation nowcasting method with the temporal horizon from the next few minutes up to six hours.

IOP Conf. Series: Earth and Environmental Science **329** (2019) 012041 doi:10.1088/1755-1315/329/1/012041

2. The state of art of solar radiation forecasting methods

2.1 Diversified types of solar radiation forecasting methods

The first step of studying solar radiation forecasting methods is to explore their classifications. Previous reviews [1], [2], [3] have discussed the classification of solar forecasting methods. Based on previous studies, solar forecasting methods can be classified into five clusters: Numerical Weather Prediction Methods (NWP), Statistical Methods, Top-Down Methods, Bottom-Up Methods and Hybrid Methods.

2.2 Numerical Weather Prediction Methods (NWP)

Numerical weather predictions (NWPs) use mathematic models of the atmosphere and oceans to predict the weather in the future. In general, NWPs methods are suitable for the macroscopical and long-term forecast. Representative NWP models involve into two types: global models, such as Global Forecast System (GFS), or regional models including European Centre for Medium-Range Weather Forecasts (ECMWF), Weather Research and Forecasting (WRF) for North America. In recent years, a lot of researchers have expressed their interests in the WRF model because it is open-source and can be configured by the user to a high resolution over a specific region. The representative studies of NWP methods are listed in table 1.

Table 1. NWP Methods for Solar Radiation Forecasting

Article	Parameter	Frequency	Data Source	Method
Mathiesen and Kleissl. [4]	GHI	Intra-day	Public Meteorological Observations	NAM/GFS /ECMWF
Yang and Kleissl. [5]	GHI	Intra-day	Public Meteorological Observations	WRF
Mathiesen et al. [6]	GHI	Day Ahead/Intra-day/Intra-hour	Public Meteorological Observations Meteorological Satellites	WRF-CLDDA
Lara-Fanego et al. [7]	GHI/DNI	Hourly	Public Meteorological Observations	WRF

GHI: Global Horizontal Irradiance, DNI: Direct Normal Irradiance, DHI: Diffuse Horizontal Irradiance

2.3 Statistical methods

Statistical methods for solar radiation forecasting have been very popular in the past double decades because of the rapid development of computer techniques. Statistical methods can be divided into two categories: model-driven methods and data-driven methods. Model-driven methods utilise statistical model to determine the interdependent quantitative relationship among variables. Typical model-driven methods include an autoregressive integrated moving average (ARIMA), exponential smoothing (ETS), etc. Unlike model-driven methods which tend to require researchers' mastery of empirical models, data-driven methods are based on machine learning (ML) and thus emphasise self-learning of models based on ample data sample, where learning implies classification, regression, and prediction. The most typical data-driven methods include artificial neural network (ANN), support vector machine (SVM). In general, statistical methods have good elasticity in spatial and temporal resolution. However, the main challenges involve the selection of input variables, models, algorithms, and validation methods. Table 2 lists remarkable researches of statistical methods for solar radiation forecasting.

Table 2. Statistical methods for solar radiation forecasting

Article	Parameter	Frequency	Data Source	Model-Driven Method
Al-Alawi and Al-Hinai. [8]	GHI	Monthly	Public Met Observations	ARMA
Yang et al. [9]	GHI/DHI /DNI	Hourly	Public Met Observations	ARIMA
Dorvlo et al. [10]	GHI	Monthly	Public Met Observations	RBF/ ANN (MLP)
Renno et al. [11]	DNI	Daily /Hourly	Privately-owned Measurement	ANN (MLP/BP/LM)
Jiang and Dong. [12]	GHI	Hourly	Public Met Observations	SVM
Chu and Coimbra. [13]	DNI	5/10/15/20 Mins	Privately-owned Measurement	kNN

2.4 Top-Down methods
The principle of top-down methods is to analyse satellite cloud images obtained from the atmosphere above. Through examining two consecutive images captured by the meteorological satellite, cloud motion can be tracked based on statistical algorithms. Cloud motion accompanied by cloud cover information is translated into the solar radiation forecast based on the specific mathematic models. The working features of satellite determine its good performance on regional and middle-term solar radiation forecasting. Typical researches of top-down methods are listed in table 3.

Table 3. Top-Down Methods for Solar Radiation Forecasting

Article	Parameter	Frequency	Data Source	Method
Perez et al. [14]	GHI	6 Hours	Met Satellites	Satellite Image Analysis
Nonnenmacher et al. [15]	GHI	1/2/3 Hours	Met Satellites	Satellite Image Analysis
Lorenz et al. [16]	GHI	30 Mins-6 Hours	Met Satellites	Satellite Image Analysis
Hammer et al. [17]	GHI	30 Mins-2 Hours	Met Satellites	Satellite Image Analysis

2.5 Bottom-Up methods
In recent years, researches of bottom-up forecast methods have presented promising potentials to overcome the limitations induced by the low spatial and temporal resolution of NWP and top-down methods. As similar as top-down methods, the key of the bottom-up method is to analyse advection of clouds through consecutive cloud images observed with ground-based all-sky cameras called "Total Sky Imager" (TSI) for achieving solar radiation forecast. Based on the ability of the ground-based camera, these methods can provide advance cloud information at a lead time of several minutes to hours for forecast solar radiation in the very short-term temporal horizon. Table 4 presents the typical researches of bottom-up methods.

Table 4. Bottom-Up Methods for Solar Radiation Forecasting

Article	Parameter	Frequency	Data Source	Method
Richardson et al. [18]	GHI	Intra-hour	Public Met Observations + Privately-owned Measurement	Ground Cloud Image Analysis
Yang et al. [19]	GHI	30 s-15 Mins	Public Met Observations + Privately-owned Measurement	Ground Cloud Image Analysis
Marquez et al. [20]	DNI	3-15 Mins	Public Met Observations + Privately-owned Measurement	Ground Cloud Image Analysis
Cervantes et al. [21]	DNI	5 Mins	Public Met Observations + Privately-owned Measurement	Ground Cloud Image Analysis
Peng et al. [22]	GHI	1/5/10/15 Mins	Public Met Observations + Privately-owned Measurement	Ground Cloud Image Analysis

2.6 Hybrid methods
Hybrid Methods usually consist of any two or more of the methods described previously. In recent years, several Hybrid Methods have presented their superior performance for forecasting high-quality solar radiation. Through taking advantage of the strengths of each methodology, hybrid methods can increase the forecasting accuracy efficiently. Another reason to develop hybrid methods is to overcome the limitations of each forecasting methods in their specific spatial and temporal horizon. Remarkable researches of hybrid methods are presented in table 5.

Table 5. Hybrid Methods for Solar Radiation Forecasting

Article	Parameter	Frequency	Data Source	Method
Lima et al. [23]	GHI	24 hours	Public Met Observations	ANN + WRF
Marquez et al. [24]	GHI	30/60/90/120 Mins	Public Met Observations + Met Satellite	ANN + Satellite Image
Voyant et al. [25]	GHI	Hourly	Public Met Observations	ANN + ARMA + Time Series
Wu and Chen. [26]	GHI	Hourly	Public Met Observations	ARMA + TDNN
Chu et al. [27]	DNI	5/10 Mins	Public Met Observations + Privately-owned Measurement	ANN + TSI

2.7 Comparative analysis
Table 6 lists various spatial-temporal horizons, where different methods can achieve good performance. All of them have respective advantages and limitations according to their spatial-temporal horizon. In this case, hybrid methods can integrate the strengths of different methods that thus can perform well in widest spatial and temporal horizon once the suitable combination of different methods is utilised. As a result, the spatial and temporal horizon of hybrid methods is contributed to achieving good performance of solar forecasting according to flexible application purposes, such as building simulation.

Table 6. Performance of Different Methods in Various Spatial-Temporal Horizon

Methods	NWP	Statistical	Top-Down	Bottom-Up	Hybrid
Spatial Horizon	5-20 km	1 m-2 km	1-10 km	1 m-2 km	1 m-20 km
Temporal Horizon	4-36 hours	1 second-1 month	30 mins-6 hours	5-30 mins	1 second-1 month

2.8 Data acquisition for solar radiation forecasting

2.8.1 Public Meteorological Observations. The database of public weather observations, such as National Oceanic and Atmospheric Administration (NOAA) and UK Met Office, which can provide the most comprehensive and long-term meteorological data involving temperature, relative humidity, etc. In general, all solar radiation forecast methods are more or less dependent on this data source.

2.8.2 Privately-owned Measurement. This way of data acquisition usually is conducted by individuals or teams comes from the laboratory of academic and research institution, such as the Welsh School of Architecture. Through utilising specific equipment for different measuring purposes, such as BF5 Pyrheliometer for measuring solar radiation, this way can obtain high accurate historical and current data.

2.8.3 Meteorological Satellites. This way can provide the uppermost data – satellite cloud images – for top-down methods. Different meteorological satellites, such as GOES-West Satellite, are operated by various meteorological organisations, such as National Oceanic and Atmospheric Administration (NOAA). The quality of satellites cloud images is dependent on the characters of a specific satellite, such as coverage area, movement orbit, image resolution, and photograph intervals.

2.8.4 Ground-based Total Sky Imagers. The most important data source of bottom-up methods – ground-based cloud image – can be obtained from total sky imager. The performance of different "total sky imagers" are relevant to various characters including equipment composition, resolution, environmental adaption, size, cost and associated software application.

2.8.5 Weather Application Programming Interface (API). Recently, an ascendant approach, which utilises Application Programming Interface (API) to provide meteorological data, has expressed its potentials as an accurate and accessible data resource. Table 2 in the authors' previous study [28] showed the available data types from various APIs.

3. Scoping study into the methodology

According to the comprehensive and comparative analysis of existing solar radiation forecasting methods, this study proposes a mixed nowcasting method based on the representative data-driven model - artificial neural network (ANN) and several significant ways of data acquisition and processing including TSI methods, Privately-owned Measurement and publicly available Weather API. Unlike conventional studies using direct meteorological data from public weather observation or privately-owned measurement as inputs for ANN model without consideration of actual physical conditions, this study introduces several data resources to acquire both meteorological data and physical cloud information as the model inputs which have potential to improve the accuracy of ANN forecasting model in very short temporal horizon. The proposed working flow of this mixed solar radiation nowcasting method is illustrated in figure 1.

IOP Conf. Series: Earth and Environmental Science **329** (2019) 012041 doi:10.1088/1755-1315/329/1/012041

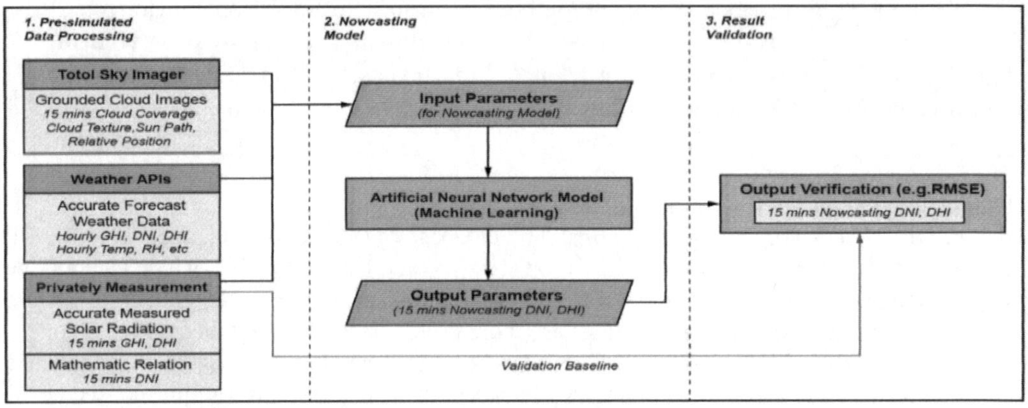

Figure 1. Proposed Nowcasting Method of Solar Radiation

3.1 Data acquisition and processing for nowcasting model

For this scoping study, it is necessary to demonstrate a good understanding of available and reliable data sources for nowcasting model development. This section presents three ways of data acquisition and processing, which generate crucial input parameters for the ANN nowcasting model.

TSI (Total Sky Imager) Method. Through analysing cloud cover information based on consecutive grounded cloud images, TSI methods can generate the forecast of physical sky conditions. The major steps of image processing in TSI method include Exposure Compensation, Undistorted and Cropping, Masking (Removing useless information), Segmentation (Generating grayscale image based on "Red to Blue Ratio") – Filtering (Transforming the grayscale image into a binary image), Optical Flow & Cloud Motion Vectors (Generating associated vectors based on feature points in consecutive images for forecasting cloud motion). The specific content of these steps can refer to the paper by Richardson et al. [18]. Following above steps, this study operates TSI to take cloud image in the very short time interval for cloud image analysis to acquire the nowcasting of cloud physical information involving cloud coverage, cloud texture, sun path, the relative position between cloud and sun, etc.

Privately-owned Measurement. In this study, accurate historical and current data, which are measured by specific equipment, are used as important inputs for nowcasting model and baseline. The typical equipment for solar radiation, such as BF5 Pyrheliometer, are used to measure the precise value of GHI and DHI in the very short time interval and calculate the simultaneous value of DNI.

Weather Application Programming Interface (API). With the development of weather APIs, some APIs, such as Weatherbit and Solcast, have started to provide not only general meteorological data, such as humidity, temperature etc., but also hourly historical, current and forecast solar radiation data. Although the data accuracy of weather APIs still needs to be improved and verified, these data undoubtedly can be used as an essential reference that has significant application potentials. In this study, DNI and DHI obtained from Weatherbit API are used as inputs for nowcasting model.

In summary, TSI methods could provide second/minute-ahead cloud physical information; Privately-owned Equipment provides minutes-ahead accurate value of DNI and DHI; Weather APIs provides the hourly average value of DNI, DHI and other meteorological parameters. All these parameters as inputs will be introduced to train the ANN nowcasting model.

3.2 ANN nowcasting model

The principle of ANN model is to operate classification, regression, and prediction automatically through machine learning. Once the training and testing stages of machine learning are completed, ANN, as a kind of "black box" model, can apply historical data to forecast future data. The basic architecture of ANN model mainly consists of input layers, hidden layer and output layer. In this research, minute-ahead cloud physical information from TSI methods, hourly average value of DNI, DHI and other meteorological parameters from Weather APIs are introduced into input layer; Minute-ahead accurate

IOP Conf. Series: Earth and Environmental Science **329** (2019) 012041 doi:10.1088/1755-1315/329/1/012041

value of DNI and DHI from Privately-owned Equipment is introduced into output layer. As a result, the training of ANN nowcasting model can establish a connection among physical cloud information, general meteorological data and accurate solar radiation data. Representative researches of ANN model are presented by Renno et al. [11].

3.3 Result Validation
A range of metrics can be used for validating time series data prediction. The commonly used metrics include the Root Mean Square Error (RMSE), the Mean Absolute Error (MAE) and the Mean Absolute Deviation Percent error (MADP). The study plan to test all metrics and make a sensible judgement on how good they are to suit the validation process for solar radiation forecasting. Crucial comparative data are forecast solar data from nowcasting method and actual solar data from measuring equipment.

3.4 Software choices
The study has considered a range of software environment to implement data capturing, data and image processing, algorithm development. Python 3.4 and OpenCV 4.0 are cost-free open-sources software that could act as crucial roles for data classification, function calculation, and image processing because its flexibility to link hardware equipment with software programs. In further, Python can also be integrated with building simulation engines, such as Modelica, to implement optimisation and control.

3.5 Hardware requirement
Research equipment of Conventional TSI methods tends to be very professional and expensive, such as the European Union's Seventh Programme DNICAST project (€3m, 2013-2017). The costs of such equipment are in a range of £25k-30k per set which is not affordable for general application at a domestic scale. Therefore, this study proposed alternative equipment considering middle/low-cost option. Table 7 listed the cost, major equipment and key features of recommended hardware option.

Table 7. Options of hardware for nowcasting

Cost Options	Combination Types	Parameters	Camera Resolution	Camera View	Environmental Adaption	Built-in Software
High (~£25k)	ASI-16 All Sky Imager SPN1 Sunshine Sensor	Cloud Image GHI/DHI	5 Mega Pixels	180°	Strong (Case + V&H System)	√
Middle (~£3.5k)	ASI183MC Pro Camera BF5 Sunshine Sensor	Cloud Image GHI/DHI	20 Mega Pixels	180°	Middle (Case + V&H System)	√
Low (~£50)	Raspberry Pi Camera Weather API	Cloud Image GHI/DNI/DHI	5 Mega Pixels	175°	Weak (Case Only)	X

3.6 Future work - nowcasting application
This research will focus on building simulation & control once the solar radiation nowcasting is completed. A simulation engine, such as Modelica, can be used to construct a Model Predictive Control (MPC) model for building system optimisation. By introducing DNI and DHI generated from solar radiation nowcasting methods and other weather data obtained from measurement as input parameters, the MPC model can conduct self-adjustment of indoor environmental parameters including temperature, relative humidity, pressure, etc. to optimise the solar thermal system. Potential control units of the MPC model includes thermal storage, valves, switches, etc. The specific content of these steps will be based on the work conducted by Oldewurtel et al. [29]. The proposed nowcasting application method is illustrated in figure 2.

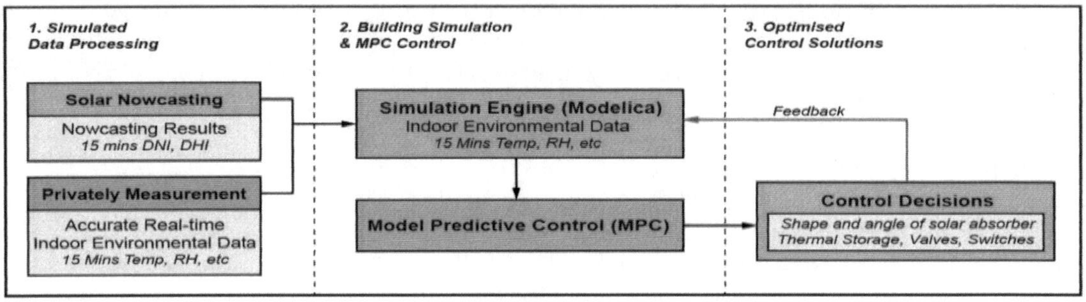

Figure 2. Proposed Nowcasting Application Method

4. Conclusion and discussion

Through reviewing previous solar forecasting methods, this study proposed a new nowcasting method of solar radiation with consideration of its practice and future work. Significant contributions of this research include four parts. Firstly, this study presents a comprehensive and comparative analysis of various solar forecasting methods. Secondly, a mixed nowcasting method is constructed by combining the advantages of various forecasting methods and various way of data acquisition that thus can provide the nowcasting of both DNI and DHI. Thirdly, lower-cost hardware and software are discussed which will be welcomed by further researchers. Finally, this study presents the plan to use nowcasting solar data to operate building simulation and control for optimising building performance.

Acknowledgement

This research has received funding from the Welsh Government's Sêr Cymru (Stars Wales) fellowship programme and the European Union´s Horizon 2020 research and innovation programme under grant agreement No 768735.

References

[1] Inman, Rich H., Hugo TC Pedro, and Carlos FM Coimbra. "Solar forecasting methods for renewable energy integration." *Progress in energy and combustion science* 39.6 (2013): 535-576.
[2] Antonanzas, J, et al. "Review of photovoltaic power forecasting. " *Solar Energy* 136 (2016): 78-111.
[3] Yang, Dazhi, et al. "History and trends in solar irradiance and PV power forecasting: A preliminary assessment and review using text mining." *Solar Energy* 168 (2018): 60-101.
[4] Mathiesen, Patrick, and Jan Kleissl. "Evaluation of numerical weather prediction for intra-day solar forecasting in the continental United States." *Solar Energy* 85.5 (2011): 967-977.
[5] Yang, Handa, and Jan Kleissl. "Preprocessing WRF initial conditions for coastal stratocumulus forecasting." *Solar Energy* 133 (2016): 180-193.
[6] Mathiesen, Patrick, Craig Collier, and Jan Kleissl. "A high-resolution, cloud-assimilating numerical weather prediction model for solar irradiance forecasting." *Solar Energy* 92 (2013): 47-61.
[7] Lara-Fanego, V., et al. "Evaluation of the WRF model solar irradiance forecasts in Andalusia (southern Spain)." *Solar Energy* 86.8 (2012): 2200-2217.
[8] Al-Alawi, S. M., and H. A. Al-Hinai. "An ANN-based approach for predicting global radiation in locations with no direct measurement instrumentation." *Renewable Energy* 14.1-4 (1998): 199-204.
[9] Yang, Dazhi, Panida Jirutitijaroen, and Wilfred M. Walsh. "Hourly solar irradiance time series forecasting using cloud cover index." *Solar Energy* 86.12 (2012): 3531-3543.
[10] Dorvlo, Atsu SS, Joseph A. Jervase, and Ali Al-Lawati. "Solar radiation estimation using artificial neural networks." *Applied Energy* 71.4 (2002): 307-319.
[11] Renno, C, et al. "ANN model for predicting the direct normal irradiance and the global radiation

for a solar application to a residential building." *Journal of Cleaner Production* 135 (2016): 1298-1316.

[12] Jiang, He, and Yao Dong. "A nonlinear support vector machine model with hard penalty function based on glowworm swarm optimization for forecasting daily global solar radiation." *Energy conversion and management* 126 (2016): 991-1002.

[13] Chu, Yinghao, and Carlos FM Coimbra. "Short-term probabilistic forecasts for direct normal irradiance." *Renewable Energy* 101 (2017): 526-536.

[14] Perez, Richard, et al. "Validation of short and medium term operational solar radiation forecasts in the US." *Solar Energy* 84.12 (2010): 2161-2172.

[15] Nonnenmacher, Lukas, and Carlos FM Coimbra. "Streamline-based method for intra-day solar forecasting through remote sensing." *Solar Energy* 108 (2014): 447-459.

[16] Lorenz, Elke, Annette Hammer, and Detlev Heinemann. "Short term forecasting of solar radiation based on satellite data." *EUROSUN2004 (ISES Europe Solar Congress)*. 2004.

[17] Hammer, Annette, et al. "Short-term forecasting of solar radiation: a statistical approach using satellite data." *Solar Energy* 67.1-3 (1999): 139-150.

[18] Richardson, Walter, et al. "A low cost, edge computing, all-sky imager for cloud tracking and intra-hour irradiance forecasting." *Sustainability* 9.4 (2017): 482.

[19] Yang, Handa, et al. "Solar irradiance forecasting using a ground-based sky imager developed at UC San Diego." *Solar Energy* 103 (2014): 502-524.

[20] Marquez, Ricardo, and Carlos FM Coimbra. "Intra-hour DNI forecasting based on cloud tracking image analysis." *Solar Energy* 91 (2013): 327-336.

[21] Cervantes, et al. "Utilization of Low Cost, Sky-Imaging Technology for Irradiance Forecasting of Distributed Solar Generation." *2016 IEEE Green Technologies Conference*. IEEE, 2016.

[22] Peng, Zhenzhou, et al. "3D cloud detection and tracking system for solar forecast using multiple sky imagers." *Solar Energy* 118 (2015): 496-519.

[23] Lima, Francisco JL, et al. "Forecast for surface solar irradiance at the Brazilian Northeastern region using NWP model and artificial neural networks." *Renewable Energy* 87 (2016): 807-818.

[24] Marquez, Ricardo, Hugo TC Pedro, and Carlos FM Coimbra. "Hybrid solar forecasting method uses satellite imaging and ground telemetry as inputs to ANNs." *Solar Energy* 92 (2013): 176-188.

[25] Voyant, Cyril, et al. "Numerical weather prediction (NWP) and hybrid ARMA/ANN model to predict global radiation." *Energy* 39.1 (2012): 341-355.

[26] Ji, Wu, and Keong Chan Chee. "Prediction of hourly solar radiation using a novel hybrid model of ARMA and TDNN." *Solar Energy* 85.5 (2011): 808-817.

[27] Chu, Yinghao, Hugo TC Pedro, and Carlos FM Coimbra. "Hybrid intra-hour DNI forecasts with sky image processing enhanced by stochastic learning." *Solar Energy* 98 (2013): 592-603.

[28] Du, H., et al. "Nowcasting Methods for Optimising Building Performance." *Proceedings of Building Simulation 2019: 16th Conference of IBPSA*, Rome, Italy, 2-4 September 2019

[29] Oldewurtel, Frauke, et al. "Use of model predictive control and weather forecasts for energy efficient building climate control." *Energy and Buildings* 45 (2012): 15-27.

IOP Conf. Series: Earth and Environmental Science **329** (2019) 012042 doi:10.1088/1755-1315/329/1/012042

A parametric analysis of the influence of wind speed and direction on the thermal comfort performance of a Passive Downdraught Evaporative Cooling (PDEC) system – field measurements from a Saudi Arabian library

M A Alshenaifi[1,2] and S Sharples[1]

[1]School of Architecture, University of Liverpool, Liverpool, UK
[2]Corresponding Author: m.alshenaifi@liverpool.ac.uk

Abstract: Building energy consumption in the desert climate of Saudi Arabia is dominated by cooling demand. Electricity for this cooling is generated predominantly from finite fossil fuel reserves. To improve resource efficiency and decrease carbon emissions, reducing this demand by using more passive cooling approaches is desirable. One system is the passive downdraught evaporative cooling (PDEC) tower. PDEC captures hot, dry winds at the top of a tower and then cools the air by passing it through or over water. This cooler air then flows out from the base of the tower into the building. In this study, a PDEC system in a small Saudi public library was monitored for two summer months. A key aim of this study was to investigate the relationship between local wind speed and direction and the performance of the PDEC towers. A thermal comfort analysis investigated the acceptability limits of indoor temperature using the adaptive thermal comfort model. A parametric analysis of the wind effects was conducted by grouping wind data in to ranges of wind speed and direction and then correlating them against environmental conditions in the library. The results indicated that the PDEC towers could deliver significant cooling for library users. However, the towers' effectiveness was influenced by changes in wind speed, and in a counter intuitive way – stronger wind speeds tended to reduce the tower cooling efficiency.

1. Introduction
In Saudi Arabia, around 75% of the country's total electricity generation is used in buildings, with air conditioning being accountable for most of that consumption [1]. Most electricity generation is from the burning of crude oil - around 900,000 barrels/day in the summer months [2]. Reducing or replacing air conditioning use in buildings with passive cooling systems could have a major influence on Saudi Arabia's energy consumption and greenhouse gas emissions. This study investigated one such cooling system – the Passive Downdraught Evaporative Cooling (PDEC) tower. The performance of an actual PDEC system was monitored in a real building. The analysis of collected data revealed that this passive system did provide cooling, although the system's performance was negatively affected by the wind.

2. Literature and Background
Passive Downdraught Evaporative Cooling (PDEC) towers is a direct evaporative cooling technique. When hot, dry air passes through a water medium, sensible heat is converted into latent heat, by the evaporation of the water, and the air temperature decreases as the relative humidity increases. A PDEC tower contains a wind catcher located at the top of a tower, an evaporative/water medium, and a shaft to deliver the caught, cooled air to an occupied space via openings at the bottom of the tower. Hot and

arid climatic regions provide an ideal environment for PDEC systems, which could create a significant decrease in cooling energy consumption [4]. Contemporary applications of PDEC towers may use shower towers (large droplets of spray); wetted porous ceramic; wetted pads or water misting nozzles [5].

2.1. PDEC Case Studies
The Torrent Research Centre (TRC) in Ahmadabad, India was the first large scale application of a misting PDEC system, which was positioned above a central atrium that separated the offices from the laboratories. PDEC reduced the interior temperatures compared to outside by between 10 and 15°C, with a 64% savings in cooling demand, compared to full air conditioning, was achieved [5]. The Zion National Park's Visitors' Centre is in Utah, where the outdoor summer daytime temperatures range between 35°C and 37°C. The Centre incorporates two cooling towers and clerestories to circulate the cooled air within the spaces. The evaporation method used in this case study is wetted pads. The building was monitored over two years and its energy consumption was found to be approximately 70% less compared with a similar building built to the applicable Federal codes[6][7].

3. Thermal comfort models
Two dominant thermal comfort models have been developed. The Fanger heat balance model considers occupants as a passive recipient of thermal stimuli, while the adaptive model recognizes occupants as active users interacting with their environment. In the Fanger model heat balance principles are defined by several factors including metabolic rate, clothing insulation and environmental conditions. Controlled climate chamber studies led to the formulation of the average comfort score (on a seven-point scale from hot to cold) that a group of people would choose – the Predicted Mean Vote (PMV) [8], [9]. The adaptive model was developed by conducting field-studies in 160 buildings from 9 countries [8]. To apply the adaptive model, the investigated building (occupied space) must be exposed to the outdoors, conditioned naturally, and mechanical cooling/heating should be avoided. However, if the mean monthly outdoor temperature is less than 10°C or greater than 33.5°C, this model may not be used and the only model available is the PMV [8]. A thermal comfort model for mixed-mode buildings, with both passive and active cooling systems (as was the case in this study) is difficult to choose. Researchers have suggested using the adaptive model in mixed-mode, saying that the occupants have some control over some local thermal conditions, which is an adaptive feature [10], [11]. People with some control over their building conditions have been found to tolerate a wider comfort temperature range, which is similar to the adaptive model range in naturally ventilated buildings [10].

4. Case study of Dar Al-Rahmaniah library

4.1. The Studies Building and its PDEC towers Characteristics
The PDEC building assessed in this study was the Dar Al-Rahmaniah library, which has two PDEC towers that use wetted pads. The library is in Alghat city, central Saudi Arabia. Its climate is hot and arid, with external dry bulb temperatures (DBT) in summer reaching 45°C. The annual average DBT and wet bulb temperatures (WBT) are 36.5°C and 18.8°C respectively. The library was monitored for over 70 days during the summer of 2018. The daytime relative humidity was typically below 20% during this period, and the prevailing wind directions during the summer season were north and north-west. Figure 1 shows the library and its two PDEC towers. The design of the Library respects the traditional architecture of the surrounding environment. The main entrance is located on the north-west side of the building between two PDEC towers, with the left-hand tower designated as Tower A in this study and Tower B on the right side of the entrance. The two towers are approximately 10m high with four openings around the top. The bottom of each Tower has a large opening to provide the cool air to the occupied space. Clerestories are placed in the centre of the roof facing north-eastern and south-western side. The leeward clerestory openings in the roof were designed to assure the circulation of the air inside the building. Data loggers were installed in the library from 21st June to 30th August 2018, recording for

IOP Conf. Series: Earth and Environmental Science **329** (2019) 012042 doi:10.1088/1755-1315/329/1/012042

24 hours a day with a logging interval of 10 minutes, giving a total of 1688 recorded hours. The library working time is divided into two shifts from Sunday to Thursday. The first shift starts from 09:00 to 12:00 while the second shift is from 16:00 to 20:00. The PDEC Towers worked 24 hours each day.

Figure 1. The main entrance of Dar Al-Rahmaniah library and the two PDEC towers.

4.2. On-site monitoring

Four different data logging equipment were used for the monitoring of the building. The recorded parameters included external and internal dry-bulb and wet-bulb temperatures, external and interior relative humidities, external wind speed and wind direction, and internal air velocity. The data logger used were a Kestrel 5500, Kestrel 5200, EXTECH SDL350 thermo-anemometer, and Rotronic HL-1D. The Kestrel 5500 mini weather station was installed on the roof of the Library to record the outdoor weather conditions. Two Kestrel 5200 data loggers were installed at the towers supply air openings to measure the conditions of the delivered air. The thermo-anemometer was used to measure temperature and air velocity within the tower. Seven compact Rotronic HL-1D were installed at different locations inside the building and the tower to record temperature and relative humidity (RH). All the loggers were new and unused. Factory calibrations were checked by running the loggers in a controlled space for 12 hours to ensure consistent readings. Note that the minimum starting speed for the Kestrel loggers is 0.6m/s, which meant that external wind speeds below 0.6m/s will be recorded as zero. The data loggers were installed in the library and the PDEC towers as shown in Figure 2.

5. Results and discussion

5.1. Cooling impact

The temperature difference between the external DBT and that delivered at the bottom of the PDEC towers ranged from 6°C in the early morning to 22.5°C during the hottest parts of the days (~3.00pm). The hottest recorded period, from 5th to 15th July, is shown in Figure 3, and describes the measured hourly external dry-bulb temperature (DBT), external wet-bulb temperature (WBT), and the indoor temperature (Tin) observed by a Rotronic HL-1D data logger placed in the middle of the library. The PDEC towers provided a significant amount of cooling to the space, considering that the mechanical air conditioning was on for only 4 hours from 16:00 to 20:00 on the 11th July. On July 14th, the maximum external DBT peaked around 46°C mid-afternoon while the WBT was around 20°C. The supply air temperatures for Tower A (Ta) and B (Tb) were recorded at 24.2°C and 23.5°C respectively at the same time while the internal temperature Tin was around 25.8°C. At the same peak hour, the indoor relative humidity (RH) increased rapidly when compared to the external RH due to the evaporation process occurring within the PDEC tower. The recorded external RH was around 8% while the internal RH was approximately 65% at tower A, 70% at tower B, and 61% in the middle of the library. Despite the PDEC towers providing cooling for most of the time, there was still a need for mechanical cooling as the PDEC towers could not provide enough cooling all the time. For instance, the maximum DBT reached around 43°C on July 11th. The WBT was around 18°C while the external RH was 7% during the same time. Given these suitable conditions for such a passive cooling system, the result, however, shows a higher indoor temperature although with the relatively lower DBT compared to the previous case. The Tin was observed around 28°C at 15:00 while the indoor RH was about 43%. The total reduction from the DBT to delivered temperatures was around 16°C, while it was around 22.5°C in the first case, leading to the mechanical cooling being used after 16:00 when the building was occupied in line with the PDEC towers. It was noted that under certain weather conditions, the performance was less effective. The wind speed played significant roles in the overall performance of the PDEC. As a result, analyses of wind direction and wind speed effects were undertaken.

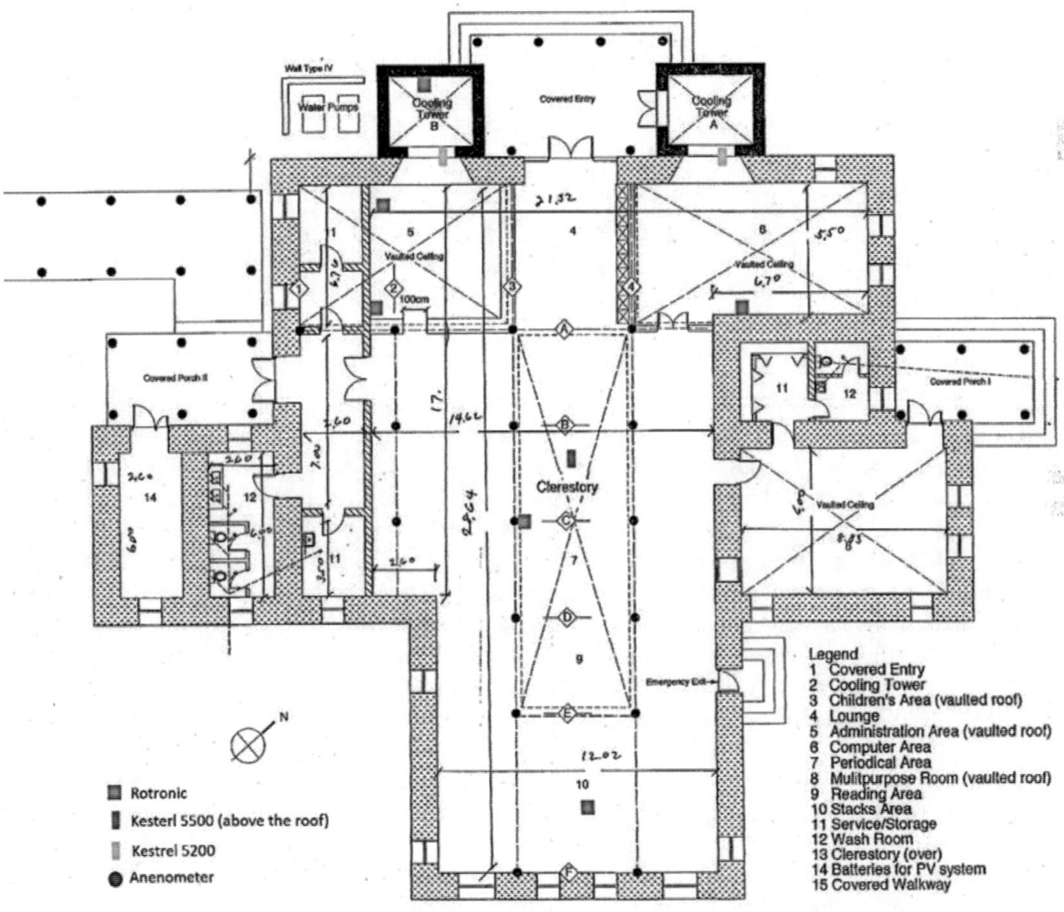

Figure 2. Library floor plan showing the location of the data loggers.

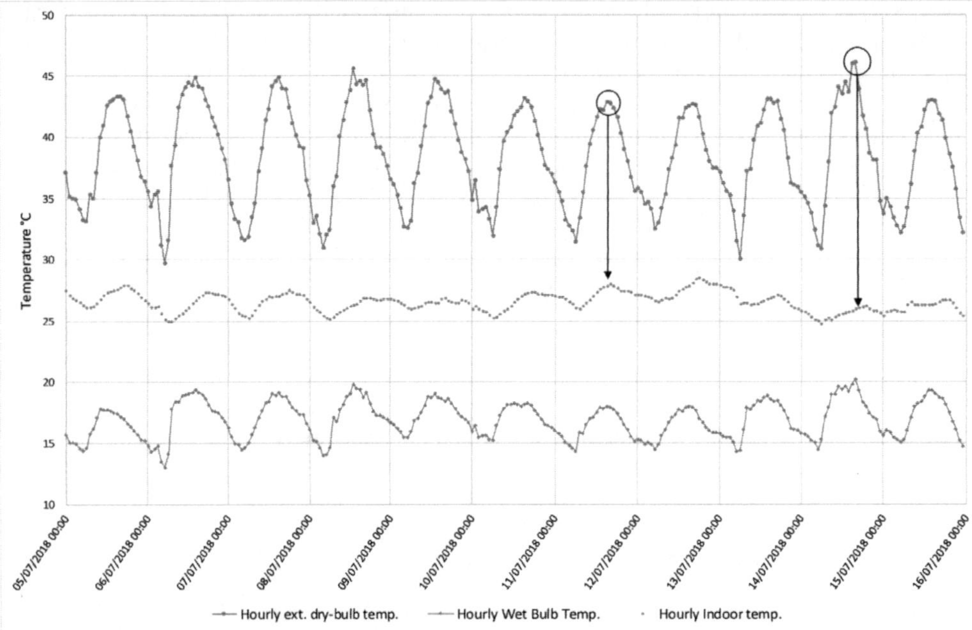

Figure 3. 10 days from 5th to 15th July, representing the hottest days of the recorded period.

5.2. Wind direction effect

In this case study, it was observed that the wind direction has a minimal impact on PDEC performance. This is attributed to four main reasons. First, the building was oriented in a way that placed the two cooling towers towards the prevailing wind direction, which was north-west. This minimised the effect of wind direction for most of the time. Secondly, it is strongly expected that the design and placement of the leeward clerestory openings within the roof significantly minimised the effect of other wind directions, which agrees with a previous computational study conducted by the authors [12]. Third, the wind catcher, at the top of each tower, was designed with four openings facing each direction. Then, an X-shaped wind barrier was placed inside the wind catcher directing the coming winds form any direction towards the airstream within the tower. Last, most of the collected data were at low wind speed while the data that was during higher wind speed was mostly coming from the prevailing wind direction. As a result, the impact of wind direction was neglected when investigating the wind speed effect.

5.3. Wind speed effect

It was apparent from the collected data that the wind speed had a direct influence on the performance of the PDEC towers. Hence, a parametric analysis of the wind speed was conducted by grouping wind data into ranges of wind speed and then correlating them against environmental conditions in the library. The investigation was performed for hours from 9:00-18:00 daily. This was the time of the day representing the higher DBT when there was a big difference between the DBT and WBT, known as wet-bulb depression (WBD). Another justification was that the higher wind speeds were recorded during daytime while nighttime was mostly calm. The results indicated that the PDEC towers' effectiveness was influenced by changes in wind speed, but in a counter-intuitive way as stronger wind speeds tended to reduce the efficiency of the towers. These findings from measured data support previous PDEC simulation analyses by the authors [12]. Figure 4 shows the observed average temperature reduction (external-internal air temperature difference ΔT) plotted against wind speed for the PDEC towers (A and B). ΔT ranged from approximately 18°C during near calm conditions to 13.5°C at the highest wind speed recorded. A simple linear correlation analysis between the ΔT and wind speed showed a strong negative relationship. A possible explanation for this is that turbulence increased around the tower inlet opening at the top due to the higher wind speeds, as discussed by [13].

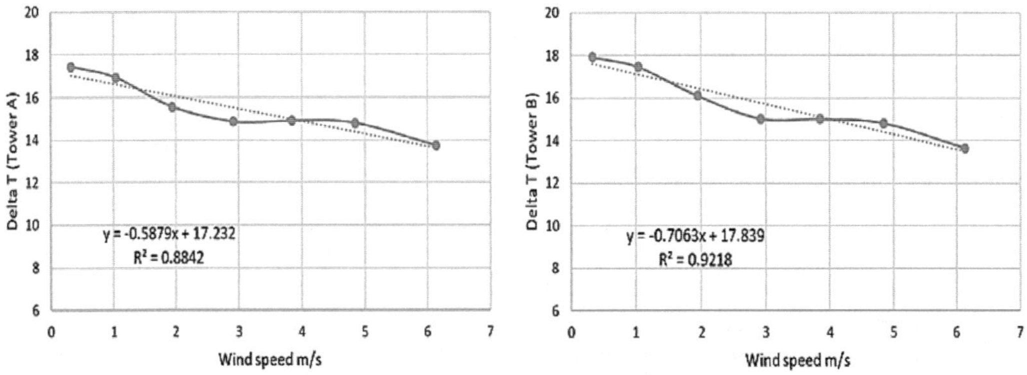

Figure 4. Influence of wind speed on the temperature reduction in Tower A (left) and B (right).

5.4. Overall Performance

The measured data from the PDEC towers demonstrated their ability to provide useful levels of cooling. Both towers performed well, but it was clear that Tower B was generally more effective than Tower A. This could be because the W and SW wind directions played a role, as discussed previously, or the layout of the Library could be a factor. It can be seen in the floor plan (Figure 2) that the supply opening of Tower A is facing a partition wall that forms a computer zone within the main area of the Library. This obstruction could hamper the airflow from Tower A, which would lead to reduced performance under certain circumstances. On the other side, Tower B has a direct unobstructed connection with the Library's open space.

6. Thermal comfort analysis

6.1. Passive cooling mode - adaptive comfort model

Of the 1688 monitored hours in the Library, the building was occupied for 410 hours. The mechanical cooling was working in conjunction with the PDEC towers (mixed-mode) for about 19% of the total occupied time, while the remaining 81% of the occupied time was just in PDEC passive cooling mode (i.e. a mixed-mode arrangment). Following the recommendation of [10] for mixed-mode buildings, the thermal comfort performance of the PDEC towers was analysed using the adaptive approach during the 410 occupied hours and the entire 1688 monitored hours (~70 days). The adaptive comfort model, as developed by de Dear and Brager [8] uses the following equation to calculate the comfort temperature:

$$Tcomf = 0.31Tm + 17.8 \qquad (1)$$

Where Tcomf is the comfort temperature, and Tm is the monthly mean outdoor temperature. The model can represent two comfort zones – an 80% acceptability limit is used for typical applications, whilst a 90% acceptability limit is used when a higher standard of thermal comfort is desired. These two ranges can be defined by adding ±3.5°C to the comfort temperature to determine the 80% acceptability or ±2.5°C for the 90% acceptability [6][8]. Following the adaptive comfort model limits stated in the ASHRAE Standard [8], the higher end point of the mean monthly outdoor temperature (33.5°C) was considered in Equation (1), giving a Tcomf of 28.2°C. Consequently, the 90% acceptability limits were 25.7°C and 30.7°C while the 80% acceptability limits were between 24.7°C and 31.7°C. Using these comfort levels, the number of comfort hours experienced in the Library can be derived from the measured data.

6.2. Comfort hours using the adaptive comfort model

The adaptive model was used to predict thermal comfort hours during both the occupied working hours (410 hours) and the total recorded hours (1688 hours) - see Table 1 and Figure 5. Most recorded indoor temperatures fall within the comfort zone, ranging from 71% of acceptable temperatures within the narrower *90%* range of the total monitored period to 98% for the wider *80%* range during occupied hours only. 1606 hours where recorded during the passive cooling mode only, including occupied and non-occupied hours. Approximately 92% of the recorded indoor temperatures fell within the 80% thermal comfort range. Within the 90% comfort range, it was found that 70% of the monitored period was considered thermally comfortable. The rest of the measurements were recorded below this limit.

Table 1. Thermal comfort analysis using the adaptive comfort model for both occupied hours only and the total monitoring period for 80% and 90% acceptability limits.

	PDEC + mixed-mode (hours)	Comfort range 80% (°C)	No. of Comfort hours 80%	Comfort hours 80% (%)	Comfort range 90% (°C)	No. of Comfort hours 90%	Comfort hours 90% (%)
Occupied hours	410	24.7-31.7	401	97.8%	25.7-30.7	357	87.1%
Total monitored hours	1688	24.7- 31.7	1565	92.7%	25.7-30.7	1203	71.3%

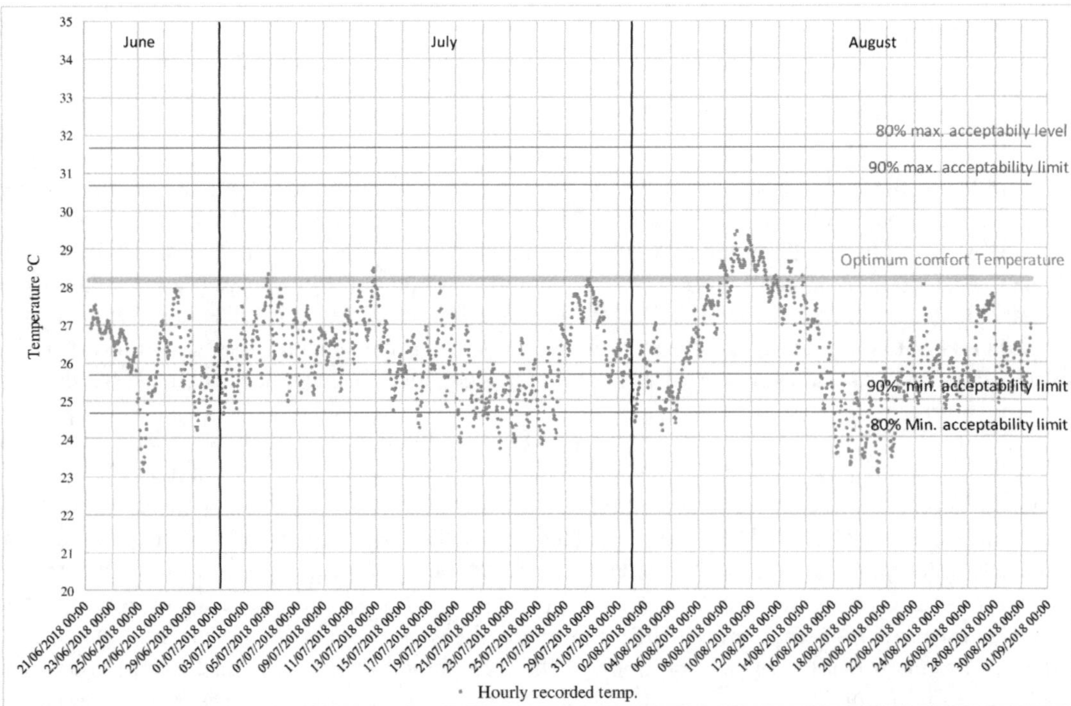

Figure 5. Measured summer internal air temperatures and 80% and 90% adaptive comfort ranges.

7. Conclusion

This paper has analysed the performance of an existing PDEC building – the Dar Al-Rahmaniah Library in Saudi Arabia. The case study provided detailed information about the ability of the PDEC towers to provide effective passive cooling, although the degree of cooling was affected by prevailing wind speeds. It was apparent from the finding that higher wind speeds had a negative impact on the performance of the towers, leading to higher supply air temperatures. The effect of wind direction was found to be minimised by the overall design and form of the library. Limitations of the current study include the fact that very low wind speeds could not be measured as the weather logger only recorded speeds above 0.6 m/s. The second part of the analysis used the adaptive comfort approach to determine levels of thermal comfort in the library. Results indicated high levels of comfort could be delivered by the PDEC towers for most of the occupied time. Further work could include detailed further parametric analysis of wind speed effect by linking the external wind speed to the tower and supply air velocities.

Acknowledgments

The authors acknowledge Abdulrahman Al Sudairy Cultural Centre for providing full access to Dar Al-Rahmaniah library and sharing all the information and documents needed to conduct this research.

References

[1] Abuhussain M, Chow D and Sharples S Assessing the adaptability of the Saudi residential building`s energy code for future climate change scenarios *Proc.PLEA 2018*, p 73
[2] U.S. Energy Information Adminstration (EIA), Country Analysis Brief: Saudi Arabia 2017.
[3] Lechner N *Heating, Cooling, Lighting Sustainable Design Methods for Architects* 4th ed. 2009.
[4] Ford B, Schiano-Phan R and Francis E *The Architecture and Engineering of Downdraught Cooling*. Bologna, Italy: PHDC Press, 2010
[5] Ford B, Patel N, Zaveri P and Hewitt M Cooling without air conditioning: the Torrent Research Centre, Ahmedabad, India *Renewable Energy* **15**, pp. 177–182, 1998
[6] Torcellini P, Judkoff R, and Hayter S Zion National Park visitor center: significant energy savings achieved through a whole-building design process *ACEEE Summer Study on Energy Efficiency in Buildings* July, 2002
[7] Torcellini P, Deru M, Griffith B, Long N, Pless S, and Judkoff R, Lessons learned from field evaluation of six high-performance buildings *ACEEE Summer Study on Energy Efficiency in Buildings* July, 2004
[8] ASHRAE Thermal environmental conditions for human occupancy, ASHRAE Standard 55-2010. American Society of Heating, Refrigerating and Air-Conditioning Engineers, Atlanta, Georgia, 2010
[9] CIBSE The limits of thermal comfort : avoiding overheating in European buildings, CIBSE TM52: 2013
[10] de Dear, R. and G. S. Brager G S Thermal comfort in naturally ventilated buildings : revisions to ASHRAE Standard 55, 2002
[11] Paul M, John R, and de Dear R Mixed-mode buildings : a double standard in occupants ' comfort expectations *Build. Environ.* **54**, pp. 53–60, 2012.
[12] Alshenaifi M and Sharples S Investigating the impact of architectural form and wind direction on the performance of a passive downdraught evaporative cooling tower in Saudi Arabia *Proc. PLEA, 34th International Conference on Passive and Low Energy Architecture*, 2018, p 294
[13] Kang D and R. K. Strand R K Significance of parameters affecting the performance of a passive down-draft evaporative cooling (PDEC) tower with a spray system *Appl. Energy*, **178**, pp 269-280, 2016.

IOP Conf. Series: Earth and Environmental Science **329** (2019) 012043 doi:10.1088/1755-1315/329/1/012043

The Cooling Efficiency of Urban Greenery Coverage in a High-density City

Wanlu Ouyang[1*], Tobi Eniolu Morakinyo[2] and Edward Ng[3]
[1] Ph.D. student, the Chinese University of Hong Kong, wanlu.oy@link.cuhk.edu.hk
[2] Postdoctoral Fellow, the Chinese University of Hong Kong, tobimorak@cuhk.edu.hk
[3] Professor, the Chinese University of Hong Kong, edwardng@cuhk.edu.hk
*Corresponding author

Abstract. Urban greenery especially trees has been identified as one of the most effective measures for urban climate moderation through the mechanism of evaporative cooling and shading. For subtropical high-density cities like Hong Kong, intense heat stress arises from the compact urban form, poor ventilation, and the high percentage of impervious urban cover. Thus, concerted actions are being taken to harness the cooling potential of urban greenery for improved microclimate. However, the cooling effects of urban greenery vary based on multiple factors, including background climates, urban morphology, anthropogenic heat emission, vegetation types and quantity, etc. This study aims at quantifying and comparing the cooling efficiency of different urban greenery coverage for a typical residential neighbourhood in Hong Kong. The questions guiding our research are: what is the relationship between tree coverage and ambient environment, including air temperature and thermal comfort during a typical summer day in the selected neighbourhood? To achieve this objective, we employed a previously validated ENVI-met model setting of a typical summer day to conducted parametric studies of cooling benefit of variable tree coverage ratio from 0 to 30% at the interval of 2% on a typical summer day. The results verified the greenery had a higher cooling potential during period of maximum temperature than the whole diurnal time, 1 °C and almost 0.8 °C respectively for air temperature. This study also showed that increasing greenery coverage indeed lowers diurnal temperatures, but this relationship is logarithmic instead of linear. The ratio of change in cooling effects to greenery increase and then decreased gradually. Between the coverage ratio of 8%-35%, the cooling potential would be underestimated in a linear relationship. Greenery measures may lead to ineffective results if greening principles have not been comprehensive understood, such as greenery abundance, composition and types. The present study allows urban planners and managers to understand more in greenery amount for urban micro-climate in high-density cities.
Keywords: Urban greenery; ENVI-met; High-density cities

1. Introduction

Urban heat island (UHI) describes a phenomenon of higher temperature within urban areas is apparently higher than surrounding rural areas [1], which is attributed as the outcome of urbanization, abundant impervious surfaces and increasing anthropogenic heat emission [2]. Extensive and rapid urbanization also increase both the city size and urban density. With compact urban form and concentrated population, multiple benefits have been brought up, such as maximizing the potential of land resources, reducing transportation costs, etc. However, compact urban development without proper guidance in planning and design can lead to serious thermal stress and damage residents' health and wellbeing [3].

Hence, there is a pressing need to evaluate mitigation and adaptation strategies for UHI in the context of high-density.

As one of the main strategies to mitigate UHI effects, there is a growing interest in applying urban greenery into urban planning and design for climatic regulation [4]. Previous studies have proved urban greenery can attenuate urban overheat by reducing daytime heat storage and facilitate night-time heat cooling [5]. Through the mechanisms in surface shading, evapotranspiration, and wind pattern alteration [6], temperature regulation effect of urban greenery have been reported to be 1-7 degrees [7, 8]. Nevertheless, our knowledge in the most optimum types, abundance and distribution of urban greenery that is necessary to alleviate heat at neighbourhood and street canyon levels is still insufficient [9]. In addition, the effectiveness of urban greenery to certain urban and neighbourhood contexts has not been fully assessed [10]. Specifically, in high-density cities like Hong Kong, most of the land is covered by buildings and roads, where there is limited land resources for greenery. In hot-humid climate, the main contribution of urban greenery is shading, and increasing humidity and reducing speed at pedestrian level are undesirable from the thermal comfort perspective [11].

Therefore, the main research question guiding this study is: what is the relationship between tree coverage and thermal environment during daytime in summer, for a typical residential neighbourhood in Hong Kong? To answer this question, parametric studies based on a validated ENVI-met model setting were conducted. The contribution of this study lies in extending our understanding in urban greenery quantity and its benefits in climatic regulation. Although this study was taken in Hong Kong, the result of it can be a reference to other cities in high-density and subtropical regions.

2. Methods

2.1. Model's description

ENVI-met is a holistic, three-dimensional, non-hydrostatic, and grid based microclimate model, which is designed for simulating the complex interactions among surface, vegetation and air in urban environment. Been constantly evaluated as acceptable accuracy for simulating bio-meteorological conditions, it has been widely using in micro-climate and greenery studies [12]. ENVI-met V4 was firstly released in 2010 [13], and was applied for numerical simulations in this study. The main inputs for ENVI-Met are buildings and vegetation elements, ground and wall surface properties, initial soil temperature and wetness profiles, as well as meteorological data. The outputs of this model include variety of climatic parameters, such as air temperature, relative humidity, mean radiant temperature, and thermal comfort index like PET, PMV, and UTCI, etc.

2.2. Model validation

Numerical models need observation to initiate and validate the simulations for synthetic weather conditions [14]. In this study, we have applied the validated settings of a typical extreme summer day in Hong Kong in which the simulated and measurement air temperature (AT), and mean radiant temperature (MRT) comparison where reported [15]. Their results reveals a good correlation of R^2 = 0.79 - 0.81 and 0.70 – 0.74 for AT and MRT respectively between the measured and simulation results which corresponds to a low RMSE and MAPE[1] of 3.7% (for tree-shaded Ta), 5.1% (for unshaded Ta), 7.7% (for tree-shaded MRT) and 13.2% (for unshaded MRT).

[1] RMSE: Root Mean Square Error; MAPE: Mean Absolute Percentage Error

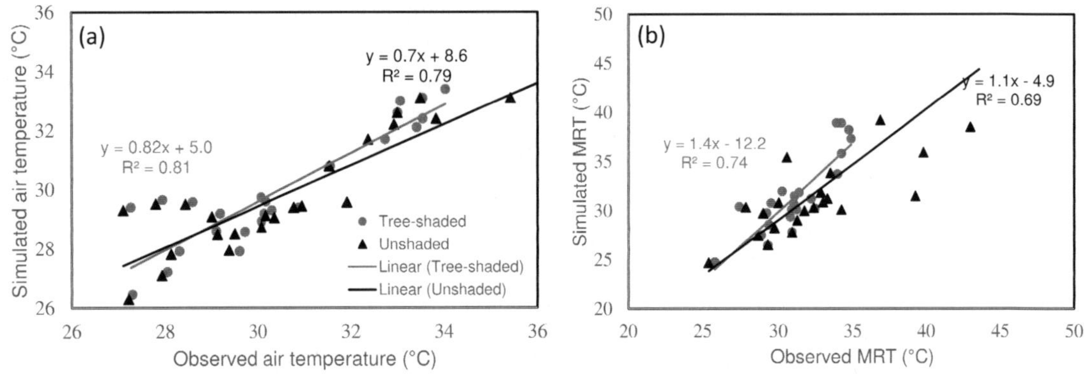

Figure 1.Relationship between ENVI-met simulated and observed (a) air temperature and (b) mean radiant temperature, MRT at tree-shaded and unshaded location on a typical summer day (Source: [15])

2.3. Experimental design

2.3.1. Model initialization. Mong Kok is a typical densely built district in Hong Kong with a high population density of 130,000 /km^2 [16]. This area is featured with high commercial towers with residential buildings and is characterized as LCZ-1 [17]. This study was based on a simplified neighbourhood where the building height is 30 meters on average, street width is 20 and 32 meters, and building coverage is 44%. The simulation domain contains 170 × 200 horizontal grids at 4-meter resolution, which equal a 680 × 800 m^2 domain. And there were 27 vertical grids were 1-m resolution with 15% telescoping after 2 meters building height. In addition, 10 nesting grids were added around the simulation area to ensure enough space between buildings and model border. A schematic layout is shown in Figure 2.a.

2.3.2. Experiment scenarios. The parametric studies include 16 scenarios -one reference consisting of only street and building elements without no greenery, while the remaining contains greenery coverage increasing from 2% to 30% at 2% interval. The building coverage ratio was set as 44% and constant in all scenarios. The initial meteorological conditions came from Hong Kong Observatory[2], which is a government department for weather forecast and meteorological information provision. In this study, a typical extreme hot day was chosen as the simulation day for the meteorological conditions because 2016 has the 38 very hot days (daily maximum temperature >= 33.0 degrees), which has been the highest records since 1884. Therefore, a typical extreme hot day with meteorological conditions in initial air temperature as 29.6 °C, initial relative humidity as 87%, wind in east direction and speed of 3.75 m/s, which was similar to that of the validation study, was adopted and used as the initial condition for parametric simulations.

In this study, *Aleurites moluccana* was selected for the main tree species for the following reasons: 1) it has a high air cooling capacity in the previous studies [15]; 2) It is a common landscape tree in the residential sectors in Hong Kong [18]. The tree height, trunk height, crown height, crown width and leaf area index (LAI) were obtained from a study [15], which was listed for details in Table 1. The tree model was built in the 3D vegetation module in ENVI-met at the basis of pixel. Besides, in order to exclude the effects of the spatial composition of urban greenery, the trees were located evenly throughout the whole domain (Figure 2.b). To show the experimental setting as well as the scenarios, Figure 2 shows the configuration setting as well as the analysis domain.

Table 1. Tree parameters in the model

Tree height	Trunk height	Crown Height	Crown Width	LAI
9 meter	3 meter	6 meter	7 meter	2.77

[2] HKO website: https://www.hko.gov.hk/

2.3.3. Statistical analysis. The analysis domain cut the surrounding building pixels and minimize the edge-effects furtherly (Figure 2.b). The simulation results were extract from LEONARDO and the analysis parameters include air temperature, relative humidity, mean radiant temperature, and thermal comfort index PET. The pre-processing was implemented with a VBA code and visualized in R software. The statistics showed distribution pattern of the climatic parameters within the neighbourhood and across both diurnal period (09:00-17:00 local time), and period of maximum temperature (14:00-15:00 local time).

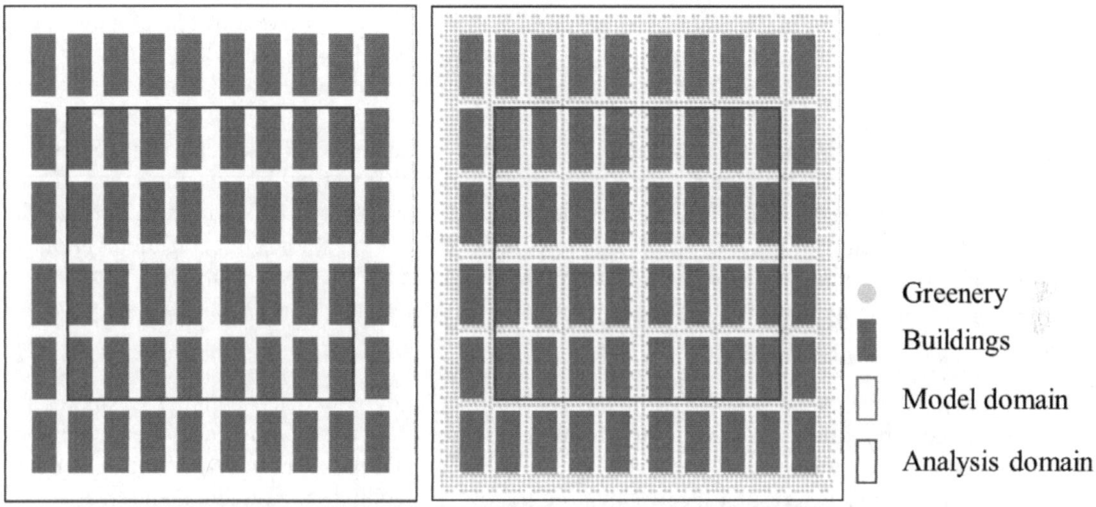

(a) Reference scenario (b) 20% Greenery coverage scenario

Figure 2. Model schematic layout and scenarios examples (Building coverage ratio = 44%)

3. Results and Discussion

3.1. Cooling Effects of urban greenery

Extracting the data within the analysis domain (Figure 2), the average values for air temperature (AT), Universal Thermal Climate Index (UTCI), physiological equivalent temperature (PET) were calculated for different greenery coverage ratios. The cooling magnitude during the period of maximum temperature (14:00 – 15:00) was larger than that in the whole diurnal time (9:00 – 17:00). The maximum AT reduction could be 1 °C for the whole area compared with the base case without greenery, which fit the range indicated in a previous study in Hong Kong [3]. As for outdoor thermal comfort, the thermal benefits reached over 2 °C and 8 °C for UTCI and PET, respectively. All these highest cooling potentials were achieved by the 56% coverage, which means that the resting areas besides buildings were covered as greening.

3.2. The relationship between greenery coverage and cooling effects

Based on the trend pattern of R^2, the relationship between greenery coverage and its cooling benefits was logarithmic than linear as shown in Figure 3. The pattern increased sharply as the greenery coverage ratio increased within the lower end of the coverage range, and the rate slowed down gradually with the coverage increasing. When greenery coverage reached between 20-30%, the cooling benefits seemed to level off. The results in this study kept corresponding to Grimmond's findings [19]. They simulated and compared surface energy balance among a dozen cities in North America, noting that proportional latent heat flux came to be stable when the vegetated fraction reached 0.2-0.3. If there were less coverage classes, the relationship would be recognized as linear and the cooling effects would be overestimated between the coverage of 8-35%. Therefore, our study illustrated the necessary to conduct parametric study at fine interval of greenery coverage.

However, there are also other studies indicating that there was linear relationship between greenery coverage and air temperature. In a study for hot and arid climate, a parametric study was conducted for air temperature at 3:00 pm local time and tree coverage ratio [10]. This study found there was a strongly linear relationship between these two factors. The reason for the contradiction between findings of our study and reference study is possibly because the difference background climate. In the arid climate, greenery provides cooling effects through both shading, evapotranspiration, while for subtropical climate in our study, higher humidity decreases the cooling rate and effects.

In addition, the trend of PET was not identical as AT and UTCI, it showed generally no pattern among different coverage ratio. One possible reason is that the PET is more sensitive with wind speed than UTCI. Another explanation for this different is that the standard clothing index (clo = 0.9) for PET, and UTCI adjusts clothing based on meteorological conditions accordingly. These should be tested in the future studies.

Besides, only one species of trees was used for simulation and analysis. The selection of species was referring to a previous study in Hong Kong, which compared different cooling performance of tree species with different height and LAI [15]. Considering the objective of this study is to explore the cooling efficiency for street trees, it is acceptable to conduct parametric studies in controlled variables. Further studies could explore the sensitivity in tree species with different features, as well as boundary conditions, grid resolution and time-interval for simulations for uncertainty analysis and testing.

4. Conclusion

Based on previously validated ENVI-met model setting, this study conducted parametric studies for exploring the cooling effects regarding to different greenery coverage ratios. In the context of subtropical climate background and high-density urban form, a neighbourhood in Hong Kong was selected as a case study. The results showed that the maximized cooling magnitude of the trees was 1 °C and 2 °C and 8 °C reduction in air temperature, UTCI and PET, respectively.

With coverage ratio increasing, the cooling effects of urban greenery presented a logarithmic trend instead of linear as some previous studies indicated. The results in this study emphasizes the necessary of investigating the cooling potential of urban greenery within the lower end of the vegetated coverage ratio range. This is especially significant for a high-density context that is an irreversible trend, and for some places where land resources are limited.

In the planning standards and guidelines published by the Planning Department of Hong Kong, the green coverage is suggested to be 20-30% in the residential development. This study corresponded to this regulation and illustrated the scientific consideration for environmental design, as this range utilizes and maximizes the cooling efficiency of greenery.

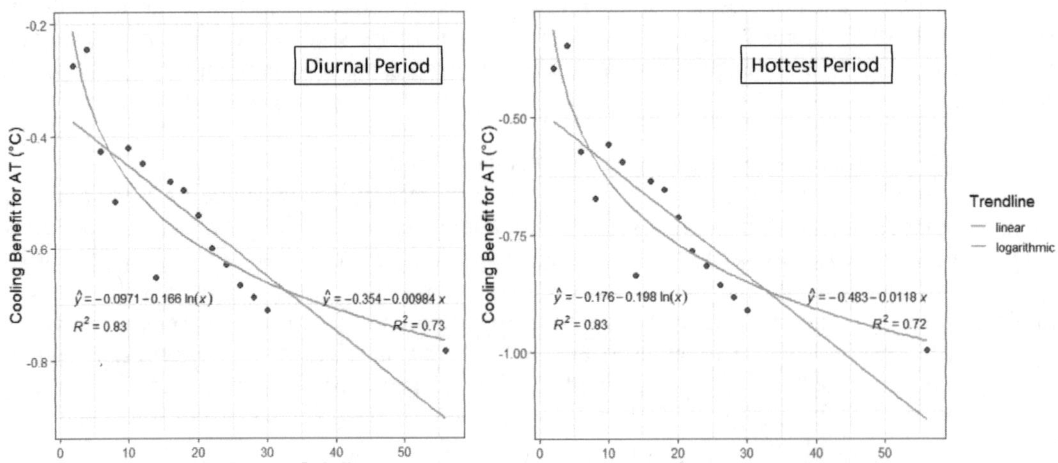

(a) AT during the day (b) AT during the solar noon time

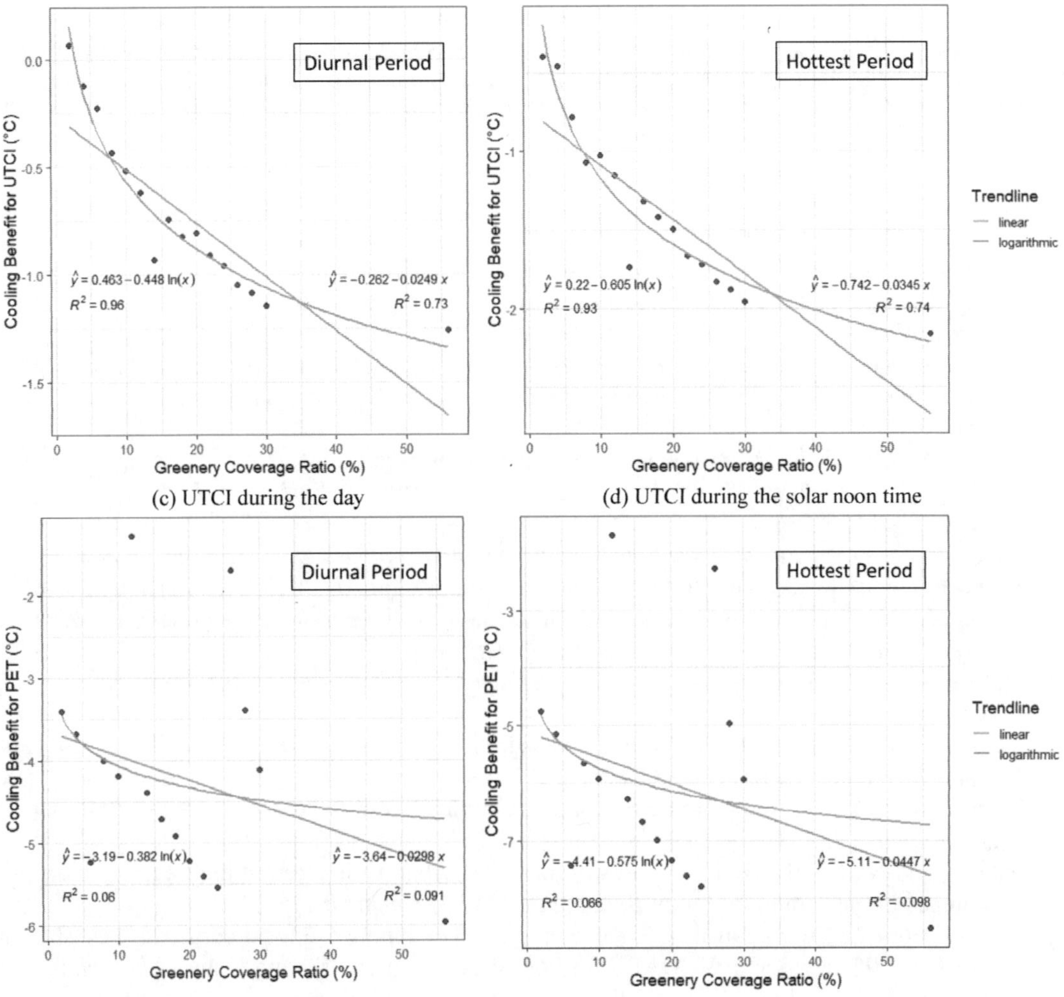

(c) UTCI during the day (d) UTCI during the solar noon time

(a) PET during the day (b) PET during the solar noon time

Figure 3. Logarithm patterns between climatic parameters and greenery coverage ratio (during the day refers 09:00-17:00 local time, solar noon time refers 13:00-14:00 local time)

Acknowledgments
This study is supported by the Post Graduate Scholarship (PGS) grant from The Chinese University of Hong Kong. Any opinions, findings, and conclusions or recommendations expressed in this article are those of the authors and do not necessarily reflect the views of the sponsoring agency.

References

[1] Oke T R 1982 THE ENERGETIC BASIS OF THE URBAN HEAT-ISLAND *Quarterly Journal of the Royal Meteorological Society* **108** 1-24

[2] Monteiro M V, Doick K J, Handley P and Peace A 2016 The impact of greenspace size on the extent of local nocturnal air temperature cooling in London *Urban Forestry & Urban Greening* **16** 160-9

[3] Ng E, Chen L, Wang Y N and Yuan C 2012 A study on the cooling effects of greening in a high-density city: An experience from Hong Kong *Building and Environment* **47** 256-71

[4] Jeong S-Y and Yoon S-H 2012 Method to quantify the effect of apartment housing design parameters on outdoor thermal comfort in summer *Building and Environment* **53** 150-8

[5] Gober P, Brazel A, Quay R, Myint S, Grossman-Clarke S, Miller A and Rossi S 2009 Using Watered Landscapes to Manipulate Urban Heat Island Effects: How Much Water Will It Take to Cool Phoenix? *Journal of the American Planning Association* **76** 109-21
[6] Oke T R, Crowther J M, McNaughton K G, Monteith J L and Gardiner B 1989 The Micrometeorology of the Urban Forest [and Discussion] *Philosophical Transactions of the Royal Society of London. Series B, Biological Sciences* **324** 335-49
[7] Shashua-Bar L, Potchter O, Bitan A, Boltansky D and Yaakov Y 2009 Microclimate modelling of street tree species effects within the varied urban morphology in the Mediterranean city of Tel Aviv, Israel *International Journal of Climatology* **30** 44-57
[8] Chang C-R and Li M-H 2014 Effects of urban parks on the local urban thermal environment *Urban Forestry & Urban Greening* **13** 672-81
[9] Bartesaghi Koc C, Osmond P and Peters A 2018 Evaluating the cooling effects of green infrastructure: A systematic review of methods, indicators and data sources *Solar Energy* **166** 486-508
[10] Middel A, Chhetri N and Quay R 2015 Urban forestry and cool roofs: Assessment of heat mitigation strategies in Phoenix residential neighborhoods *Urban Forestry & Urban Greening* **14** 178-86
[11] Givoni B 1998 *Climate considerations in building and urban design*: John Wiley & Sons)
[12] Salata F, Golasi I, de Lieto Vollaro R and de Lieto Vollaro A 2016 Urban microclimate and outdoor thermal comfort. A proper procedure to fit ENVI-met simulation outputs to experimental data *Sustainable Cities and Society* **26** 318-43
[13] Huttner S and Bruse M 2009 *Numerical modeling of the urban climate - a preview on ENVI-met 4.0*
[14] Velasco E 2018 Go to field, look around, measure and then run models *Urban Climate* **24** 231-6
[15] Morakinyo T E, Lau K K-L, Ren C and Ng E 2018 Performance of Hong Kong's common trees species for outdoor temperature regulation, thermal comfort and energy saving *Building and Environment* **137** 157-70
[16] Yuan C and Ng E 2012 Building porosity for better urban ventilation in high-density cities – A computational parametric study *Building and Environment* **50** 176-89
[17] Wang R, Ren C, Xu Y, Lau K K-L and Shi Y 2018 Mapping the local climate zones of urban areas by GIS-based and WUDAPT methods: A case study of Hong Kong *Urban climate* **24** 567-76
[18] Zhang H and Jim C Y 2014 Contributions of landscape trees in public housing estates to urban biodiversity in Hong Kong *Urban Forestry & Urban Greening* **13** 272-84
[19] Grimmond C S B and Oke T R 2002 Turbulent heat fluxes in urban areas: Observations and a local-scale urban meteorological parameterization scheme (LUMPS) *Journal of Applied Meteorology* **41** 792-810

Awareness assessment of biophilic design principles application

Zeinab Nasr Aly Tahoun
Architecture Department, MTI (Modern University for Technology and Information), Cairo, Egypt. Email: zeinab_tahoon@yahoo.com

Abstract. The green architecture has developed in the last few years, since the sustainable system have evolved specially in buildings construction. The aim is to keep sources of non-renewable energy. With this evolution, a new concept related to the green architecture has been emerged called biophilia, or the living nature which is related to the human inheritance of loving nature. Many studies have proved that the existence of humans in the external environment reduces physiological and neurological distress and makes them feel better. This research deal with definition of architectural biophilic design with relation to the physical and physiological human health; as well as the merging of building interior with nature and their effects on the employee in terms of health and productivity. Moreover, it investigated the applications of the biophilic design inside work place to make it more comfortable and effective for the employee to enhance their productivity. The aim of this paper is to assess awareness about the biophilic design and its impact on employees and their wellbeing in Egypt because it's too important to apply biophilic design especially at workspaces here in Egypt which had big percentage of unsuitable places to work full of stress and anxiety.

Keywords: Awareness, human health, Sustainability, Biophilia, Biophilic design, Workplace.

1. Introduction

Human health and psychology are one of the most important factors that must be considered in creating an efficient working environment. This building impact on human shall be taken ecosystem in sustainable systems. Even though, biophilia is in consideration as much as we consider about building impact on physical and physiological human health. Although the biophilia has been considered as a new lifestyle and an upcoming trend in the architecture, the foundation of it is not recent at all; since, the nature of the human has always been with preference of being surrounded by and part of nature. However, the awareness of the application of the biophilic design and the way it can influence the human's health is considered to be limited is a specific part. This paper discusses the warness of enhancing human health and recovery using the elements of biophilic design that should be at workplace among professionals in the Egyptian building industry environment. The aim of this paper is to present the application of biophilic design at work spaces; this has been traced through a survey that measured architects impact when applying biophilic design inside work places.

2. Research methodology

The research investigates the awareness about the biophilic design, this can be reached by examining the architects' knowledge of it as a modern trend and whether they apply it in their work or not. This investigation has been performed through a qualitative applied approach, depending on questionnaire to collect the needed information to be analysed and measured and the literature review of biophilic design and its benefits. Online survey has been publicized over a population of comprising of student of architecture department, developer, government officer, project executives and also academicians within the construction discipline. The total number of respondents participated in this survey is 70.The research is designed as follow; sections discuss the Biophilia and the Biophilic Design academically, the third section discusses the direct experience. Whereas the forth section represents the benefits of implementing biophilic design in workspace, and the fifth section. The form of the survey is discussed in the sixth section whereas the last two sections show the conclusion and recommendation.

3. Biophilia

A concept first popularized by Edward O. Wilson in 1984 [1], describes the innate relationship between humans and nature, and concerns the need we have to be continually connected to nature. Plenty of research confirms this human preference for the natural, rather than built, environment [2]. For example, in a 2004 study, when asked to describe their ideal city, people more often chose non-urban characteristics, greenery in particular [3], and in other studies it has been shown that a pleasant and natural view can raise the price of a house considerably.

4. Biophilic design

Biophilic design is a method of designing the places in which we live and work in such a way that satisfies our deep and fundamental need to be connected with nature. Biophilic design encourages the use of natural elements and processes as design inspiration in the built environment [2]. The idea behind this is that exposure to natural environments and features have positive effects on human health and wellbeing, which has been supported in a wealth of research [4]. According to the Biophilia hypothesis, these positive effects of exposure to nature originate in a biological bond between humans and the natural world [2]. These ideas have been taken forward in two theories developed in the Environmental Psychology literature: Attention Restoration Theory [5] and Stress Recovery Theory [6]. Both theories suggest that some environments are stressful, others are not and yet others can actively help people recover from stress and mental fatigue. Environments that evoke positive moods, have properties that draw people's attention without being stressful or demanding, can help people recover more quickly and fully from mental fatigue and stress are known are restorative environments. According to Kaplan and Kaplan [5] and Ulrich and colleagues [8] natural environments in particular contain elements that promote renewed attention by providing a sense of being away, fascination, extent and compatibility [9]; and by containing elements that promote survival and therefore positive appraisal. Urban environments, on the other hand, tend to be full of demanding, stressful, under stimulating or boring features. Biophilic design then suggests that built environments could be made more restorative by incorporating natural elements in their design. Gifford and McGunn [7] suggest that Biophilic design can be viewed as belonging under a larger restorative design umbrella. Much of the small but growing peer-reviewed literature on Biophilic design often cites research on restorative environments to support the health and wellbeing benefits of Biophilic design [10–11]. Although the concept of Biophilic design is relatively new, the plethora of research on nature and restorative environments makes a strong case for the health and wellbeing potential of incorporating.

5. Experiences and attributes of biophilic design

The results of the psychological literature search on the three Biophilic design experiences; some overlap exists between the independent variables (the 24 attributes) explored in research and this was mentioned in the overlapping attributes. For example, water will be seen as being highly restorative in the built environment from both direct and indirect nature experience.

Moreover, different natural elements often feature together in environmental design, making it difficult to distinguish clearly between different aspects of Biophilic design.

Table 1: Experiences and attributes of Biophilic design By Kellert and Calabrese [18].

Direct Experience of Nature	Indirect Experience of Nature	Experience of Space And Place
Light Air Water Plants Animals Weather Natural landscapes and ecosystems Fire	Images of nature Natural materials Natural colours Simulating natural light and air Naturalistic shapes and forms Evoking nature Information richness Age, change, and the patina of time Natural geometries Biomimicry	Prospect and refuge Organized complexity Integration of parts to wholes Transitional spaces Mobility and wayfinding Cultural and ecological attachment to place

5.1. Direct Experience of Nature
The direct experience of nature means having a direct contact with nature and natural processes. The eight Biophilic design attributes that are used in this experience are listed in Table.1. The most researched attributes in this experience category in psychological literature are the use of plants in the built environment and natural landscapes and ecosystems.
5.1.1. Natural Light has been promoted by various disciplines as being beneficial for wellness of building occupants, including psychology. From a Biophilic view, humans evolved under natural, diurnal light conditions and therefore natural light and natural light processes should be preferred and most beneficial.
5.1.2. Water has also been found to be restorative, both through views of water [8] and sounds of water. Kaplan and Kaplan [5] note how preferred environments often have a view of water and this was also found in various other studies. In fact, images of the urban environment that contained water were found to be even more preferred than nature images containing no water.
5.1.3. Natural Landscapes and Ecosystems. The view of greenery from a window, which is listed under the natural landscapes and ecosystems attribute, has been shown in several studies as having beneficial effects on the wellbeing of building occupants.

5.2. Indirect Experience of Nature
As direct contact with nature may not be possible in every design situation, such as in certain medical environments, and looks at representation of nature in the built environment. This experience has ten attributes, which can be found in Table 1.
5.2.1. Images of Nature can even be more restorative than the view of real nature, depending on the content of the image and the view of nature. Therefore, images of nature may have a large role in environments where exterior views with lush nature may not be possible, due to seasons, adjacent buildings, or external environment.
5.2.2. Natural Materials The amount and type of material is important for perceived restorative quality and preference. In a study that was made by Nyrud and colleagues [17] the amount of wood in a patient room that was most preferred was an intermediate amount, with the floor, one wall and furniture being made of wood. An entire wood surfaced room and a room with no wood were the least preferred amongst respondents.
5.2.3. Natural Geometries refer to mathematical properties commonly encountered in nature. These include hierarchically organized scales, sinuous rather than rigid artificial geometries, self-repeating but varying patterns, and more. For example, fractals are a geometric form often encountered in the natural world, where a basic shape occurs in repeated but varied and predictable ways that contribute both variety and similarity to a setting. Other prominent natural geometries include hierarchically ordered scales such as the "Golden Ratio" and "Fibonacci Sequence."

6. Benefits of implementing biophilic design in the workspace

While employers hope their employees are productive every hour, functioning at 100% efficiency is unlikely. Understanding the drivers of absenteeism and the resulting direct and indirect costs of deficient employee productivity however, can positively influence strategies for workplace design and strengthen an organization's ability to attract and retain workers. The incorporation of biophilic design elements into the built environment has been demonstrated to produce an abundance of physical and physiological benefits, which, in turn, powers a more productive and innovative workplace.

6.1. Economic benefit of biophilic design in the work space.

With biophilia originating in the human-nature connection, this is justifiable, and Browning et al. supported this further by pointing out that "today productivity costs are 112 times greater than energy costs in the workplace," and that by daylighting schemes in offices can "save over $2000 per employee per year in office costs" [18]. The economic gains to be made from environmental benefits such as reduced energy costs, extended building life, and decreased water management costs are apparent.

6.1.1. Increased worker productivity. The biophilic design further seeks to sustain the productivity, functioning, and resilience of natural systems over time. Alteration of natural systems inevitably occur as a result of major building construction and development. Moreover, all biological organisms transform the natural environment in the process of inhabiting it. The question is not whether ecological change occurs, but rather will the net result over time be a more productive and resilient natural environment as measured by such indicators as levels of biological diversity, biomass, nutrient cycling, hydrologic regulation, decomposition, pollination, and other essential ecosystem services. The application of biophilic design can alter the environmental conditions of a building or landscape in the short term, but over the long run, it should support an ecologically robust and sustainable natural community. The successful application of biophilic design should also result in a wide spectrum of physical, mental and behavioural benefits. Physical outcomes include:

- Enhanced physical fitness
- Lower blood pressure
- Increased comfort
- Satisfaction
- Fewer illness symptoms
- Improved health

Mental benefits range from increased satisfaction and motivation, less stress and anxiety, to improved problem solving and creativity. The positive behavioural change includes better coping and mastery skills enhanced attention and concentration, improved social interaction, and less hostility and aggression.

6.1.2 Health promotion programs. The willingness of employers to invest in programs that are oriented to prevention, to enhancing wellness, and to educating employees with respect to health issues is hardly surprising. The assumption is that such health promotion efforts will generate cost savings, and that the savings will be greater than the expenses incurred in offering the programs [19-20-21].

6.2 The psychological benefit of biophilic design in the work space.

First, the mental benefits to those who work in nature-focused offices are worth noting. When a person's surroundings are congruent with their biological desires, the following results occur:

- Stress levels decline
- Motivation increases
- Creativity flows
- Ability to concentrate improves

Furthermore, physical improvements also take place, including speedier recovery times after sickness and fewer stress related illnesses. If all that isn't enough to convince people that biophilic design is worth the investment, think about running an organization where employee's quality of life improves – as a direct result of the environment they work in.

6.2.1 Wellbeing. It seems clear that urban life, with its disconnection from the natural world, stimulates a desire for contact with nature that needs to be satisfied. A key factor in maintaining positive well-being is reducing levels of stress. Research has identified that visible connections to nature can have a positive effect on an individual's reported stress levels. In a review of numerous studies looking at the effects of different landscapes on health, it was found that natural landscapes had a more positive effect compared to urban landscapes [26]. In fact, in some cases, urban landscapes had a negative effect.

6.2.2 Productivity. Feeling good often equates to being able to do more. In addition to the abundance of research that confirms the relationship between well-being and productivity [27] there is also clear evidence directly linking biophilia with an organization's output. One of the more recent, and most relevant, research studies looking at these effects was carried out in the UK (a study called 'The relative benefits of green versus lean office space: Three field experiments) [28], where university researchers in Cardiff compared the levels of productivity of two groups of office workers who were exposed to different levels of nature contact. They found that those who worked in offices with natural greenery saw a 15% rise in productivity over a three month period, in comparison to those working with no greenery or natural elements within their immediate environment.

6.2.3 Creativity. The impact of biophilic design on the individual's ability to act, behave and perform creatively within the context of their role is something that has received less focus within the realm of research into biophilic design.

7. Questionnaire

This section discusses the results of the online survey that aims to understand how the professionals in construction industry react toward four of the most important biophilic design attributes that can help to achieve a successful biophilic designed work space (natural light, quite workspace, indoor planting and colours indoor).

Online survey is selected as it is a fast and easily reachable method to get feedback from the professionals. The survey was posted online for 44 days; the survey is divided into 4 sections. The first the demographic section which collects data related to the gender of the respondents, age, educational level, occupation, and the years of experience, as well as the geographic location of the respondents. The three other sections examine the independent variables through different examples about the biophilic design. Each of them divided into 6 subsections each covers an attribute of the biophilic design dependent variables. The professionals in built environment industry were chosen as they are the major parties involved in applying biophilic design philosophy in workspaces. Therefore, their opinions may affect the direction of biophilic design development in future. The professionals were categorized based on their field to understand the different views among them. The targeted population of this research comprise of student of architecture department, developer, government officer, project executives and also academicians within the construction discipline. Since it's really hard to reach every single one of the population; the research is based on a sample of the 70 respondent, and are categorized into groups shown in Table 2.

Table 2: Classification of respondent of the sample

Occupation	Sample size	Percentage of Sample (%)
Interior Designer	14	20
Architect	10	14.3
Academic	29	41.4
Developer	12	17.1
Student	5	7.2

The number those who answered survey questions in accordance to their respective fields is unequal. Based on the findings, the highest respondents group is the academician 29 respondents in the Academician discipline forming 41.4% of the responses. Then, come the developer and the architect groups in the second and third ranks, with number of respondents 12 and 10 respectively; forming 17.1% for the developer and 14.3% for the architects. Whereas the lowest number off the respondents group is student with only 5 forming 7.2% of the responses.

7.1. Form

The survey form has a total of 56 questions; 8 are open-ended questions while the rest are close-ended. The questionnaire is divided into four sections. The first section of the survey questionnaire is the demographic section designed to identify and categorized respondents. This section aims to determine whether the respondents are familiar with Biophilic design and whether they have already implemented the design elements in their previous project. The rest three sections of the survey questionnaire aim at identifying whether our local professionals are aware of the benefits of applying Biophilic design elements into work-space, by comparing three different case study qualities. Each section consists of four sets of questions that will deal with the presence of the following Biophilic design elements (daylight, quiet workspace, planting indoor and colours indoor). Respondents were also asked to share their opinion about how to implement these elements more efficiently within the case studies if possible.

7.2. Result and discussion

Table 3 indicates the response percentage rate for each biophilic design element either if it is desirable in space or not and the factors that may affect them.

Table 3: The response rate for each biophilic design element

	Yes	No
Natural illumination	96.6%	3.4%
Effect of space plan design in light efficiency	83.5%	16.5%
Building orientation	73.1%	26.9%
Façade treatment	83.5%	16.5%
Workstation	34.5%	65.5%
Cubicle Office	55.2%	44.8%
Plants existence	100%	0%
Colours appropriate	100%	0%
Effect of colours in work and productivity	84.4%	15.6%

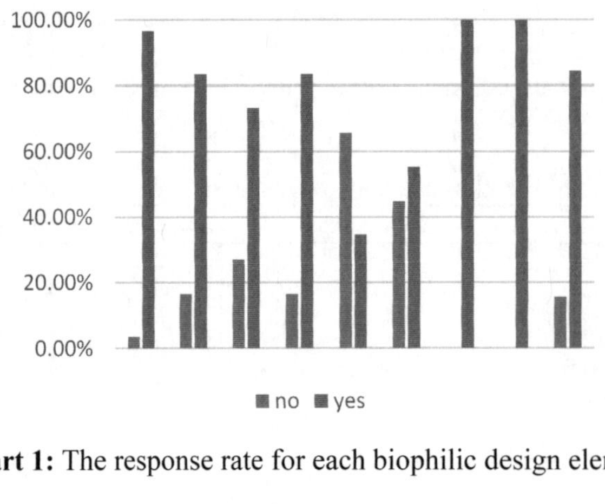

Chart 1: The response rate for each biophilic design element.

- 96.6% of the respondents preferred natural illumination in workspace, while only 3.4% preferred artificial illumination. This could be due to the fact that professionals' desire to keep the illumination level standard.
- As for factors that affect natural illumination quality in space, 83.5% of the respondents agreed that space plan design and façade treatment can improve natural light in space while 16.5% disagreed.
- 73.1% of the respondents agreed about building orientation role in controlling light quality versus 26.9% who thought not.
- 55.2% of the respondents preferred cubicle office on the contrary 34.5% who didn't mind workstation as a quiet workspace.
- All respondents (100%) have preferred indoor planting in work-space, as well as the existence of appropriate light colours in workspace. However, only 84.4% of the respondents thought it can affect positively on employees productivity in compared with 15.6% who found it unnecessary.

8. Conclusion

The compilation of 30 green roof projects from the survey shows that intensive green roof is the most common green roof being constructed in the country. While the common type of building where most intensive green roof are used is residential. The survey findings suggest that most of Malaysian professionals in built environment industry are aware of green roof technology. However, due to certain barriers including lack of knowledge and technical expert, expensive cost of green roof and green roof application techniques are not widely disseminated or spread among professionals led to less interest to use the system. Malaysian professionals involved in the construction of green roof need to import construction materials from abroad, in particular, the western countries. However, the cost of green roof construction could be decrease if we have our own green roof supplier and manufacturer in the country.

Research must be done to find potential local resources to be tested and further utilize usage for green roof construction in Malaysia. The product may be more durable to the harsh tropical climate condition and also be more compatible with the nature of our native plants. Therefore, all parties in the industry and government body have the responsibility to promote green roof system as one of the approaches for green technology application in the country. Thus in return, it could help to accelerate the marketability and progress of green roof in the country.

Although the biophilic design is considered as modern trend in architect in specific and a modern life style in general, the awareness about it is extremely wide and not only among those on the architect discipline but among the public as well. However, the knowledge about the biophilic design and how it can be applied effectively and how it can be applied expensively or in an efficient cost reduction way is only available to a specific party within the architect discipline (as mentioned earlier the academician). The research has focused on the application of the biophilic design within the workspace and its impact on the employees' health and productivity. The main measure used in this research is a qualitative measure through an online survey. The results show that the appearance of the Biophilic design elements in the completed projects can be applied to other workspaces. The survey findings reflect that academicians are the highest group of the respondents that has awareness about the Biophilic design movement. Professionals involved in the construction of workspace need to be aware more about importing Biophilic design in workspaces and how it can impact employees' health. Government body has the responsibility to promote Biophilic design movement as one of the approaches for sustainability in the country. Thus in return, it could help to accelerate the marketability and progress of Biophilic design in the country.

1. Biophilic design is the missing link between a healthy built up environment and human need for nature.
2. A review of Biophilic design from a restorative environment perspective yielded many results demonstrating the benefits of Biophilic design for human wellbeing.
3. There are three applications to apply biophilic design:
    - Direct experience of nature
    - Indirect experience of nature

- Experience of space and place

4. The majority of work on Biophilic design refers to environmental restoration theory as an underlying basis.
5. This paper focused on psychological literature examining the evidence for the restorative qualities of being Biophilic attributes, or the extent to which the presence of such elements in a work-space can help foster recovery of stress and mental fatigue, and how it can impact economically.
6. The survey shows that most of respondents found the Biophilic design elements were useful to make a workspace more efficient and liveable.
7. Like many design philosophies, Biophilic design strategies need to be employed in consideration with the building occupants, location, and function.

References

[1] Wilson, E.O. Biophilia; Harvard University Press: Cambridge, MS, USA, 1984.
[2] Kellert, S.R. Dimensions, Elements, and Attributes of Biophilic Design. In Biophilic Design: The Theory, Science, and Practice of Bringing Buildings to Life; Heerwagen, J., Mador, M., Eds.; Wiley: Hoboken, NJ, USA, 2008
[3] Félonneau, M. L. (2004). Love and loathing of the city: Urbanophilia and urbanophobia, topological identity and perceived incivilities. Journal of Environmental Psychology, 24, 43–52.
[4] Kellert, S.R. Building for Life: Designing and Understanding the Human-Nature Connection; Island Press: Washington, DC, USA, 2005.
[5] Bowler, D.; Buying-Ali, L.; Knight, T.; Pullin, A. The Importance of Nature for Health: Is There a Specific Benefit of Contact with Green Space?
[6] Kaplan, R.; Kaplan, S. The Experience of Nature: A Psychological Perspective; Cambridge University Press: Cambridge, UK, 1989.
[7] Joye, Y.; van den Berg, A.E. Restorative environments. In Environmental Psychology: An
[8] Introduction; Steg, L., van den Berg, A.E., de Groot, J.I.M., Eds.; Wiley: Hoboken, NJ, USA, 2012.
[9] Ulrich, R.S.; Simons, R.F.; Losito, B.D.; Fiorito, E.; Miles, M.A.; Zelson, M. Stress recovery during exposure to natural and urban environments. J. Environ. Psychol. 1991, 11, 201–230.
[10] Kaplan, S. The restorative benefits of nature: Toward an integrative framework. J. Environ. Psychol.1995, 15, 169–182.
[11] Ryan, C.O.; Browning, W.D.; Clancy, J.O.; Andrews, S.L.; Kallianpurkar, N.B. Emerging nature-based parameters for health and well-being in the built environment. Int. J. Archit. Res. 2014, 8, 62–75.
[12] Joye, Y. Architectural lessons from environmental psychology: The case of biophilic architecture Rev. Gen. Psychol. 2007, 11, 305–328.
[13] Kellert, S.R.; Nature by Design: The Practice of Biophilic Design.
[14] Bringslimark, T.; Hartig, T.; Patil, G.G. The psychological benefits of indoor plants: A critical review of the experimental literature. J. Environ. Psychol. 2009, 29, 422–433.
[15] Larsen, L.; Adams, J.; Deal, B.; Kweon, B.S.; Tyler, E. Plants in the workplace: The effects of plant density on productivity, attitudes, and perceptions. Environ. Behav. 1998, 30, 261–281.
[16] Ulrich, R. View through a window may influence recovery. Science 1984, 224, 420–421.
[17] Nyrud, A.Q.; Bringslimark, T.; Bysheim, K. Benefits from wood interior in a hospital room: A preference study. Archit. Sci. Rev. 2014, 57, 125–131.
[18] Browning B, Garvin C, Ryan C, et al., The Economics of Biophilia-Why Designing with Nature in Mind Makes Financial Sense. 2012.
[19] Aberth, J., 1986. Worksite wellness programs: An evaluation. Manage. Rev., 75: 51 -53.
[20] Adams, J.D., 1988. A healthy cut in costs. Personnel Administrator, 33: 42-47.

IOP Conf. Series: Earth and Environmental Science **329** (2019) 012045 doi:10.1088/1755-1315/329/1/012045

Analysing the relationship between urban form and urban heat island effect at a city-block scale

Xi Deng[1, *], Shunqian Zhang[2], Phillip Jones[1]

[1] Welsh School of Architecture, Cardiff University, Cardiff, CF10 3NB, UK
[2] Institute of Plateau Meteorology of China Meteorological Administration, Chengdu, 610072, China
* dengx4@cardiff.ac.uk

Abstract. Over the last twenty years China has witnessed rapid urbanisation process. The modification of urban land surface due to urbanisation changes not only affects the geometry of an urban settlement but also its thermal properties, which may give rise to the urban heat island (UHI) affect. Although there has been a large amount of research on mitigation strategies of UHI at the urban scale, there is still a lack of research that concerns the influence of urban geometrical settlement on UHI at a city block scale. Chengdu, a megacity in south-west China, has been chosen as the research subject in this study, to identify the existence of the UHI phenomenon, to measure its magnitude, and to investigate the relationship between UHI and urban design variables, including Floor Area Ratio, Building Coverage Ratio, Green Coverage Ratio, Building Height, Plot-layout Form, and Project Type (function). In this study, urban maps of UHI had been derived from MODIS satellite data for 103 selected projects and 28 urban zones. Correlations were carried out between UHI intensity and urban design variables. The study has developed design guidelines for improving outdoor thermal performance at an city block level. Moreover, a new approach to observe UHI at multi-scales in a long-time period had been conducted, providing a new method to analysis UHI at a city-block scale.

1. Introduction

It has been widely recognized that urban form is closely related to urban sustainability and local climate policy [1][2][3], which plays a significant role in influencing solar access on building façade [4] and altering the thermal properties of urban surface [5], thereby determining the variation of temperature in microclimate and creating the urban climate [6][7], and in specific, causing the urban heat island (UHI) effect. On the other hand, urban form alters the urban wind conditions and determines the pattern of population distribution, and both of the features are critical factors of UHI intensity [8][9]. However, existing literature focuses on the features of UHI at macro scales; research at the city-block level is relatively low. In this study, the high-definition of UHI images obtained from satellite remote sensing was processed and overlaid on the selected 103 sites and 28 urban zones within the city of Chengdu in ArcGIS- a tool of geographic information system (GIS), it investigates the relationship between each design variables and UHI value for each site through long-term observation and data processing.

1.1. Remote sensing satellite imagery

The derived MODIS data of land surface temperature has been widely implemented investigating the UHI effect [9][10][11]. In this paper, the remote sensing data was obtained and processed through the receiving and processing system of DVB-S system broadcasts of China National Satellite Meteorological Center and Huayun MODIS satellite [12]. Ninety-nine days with the clear sky from 2005 to 2010 were selected for observation and data collection.

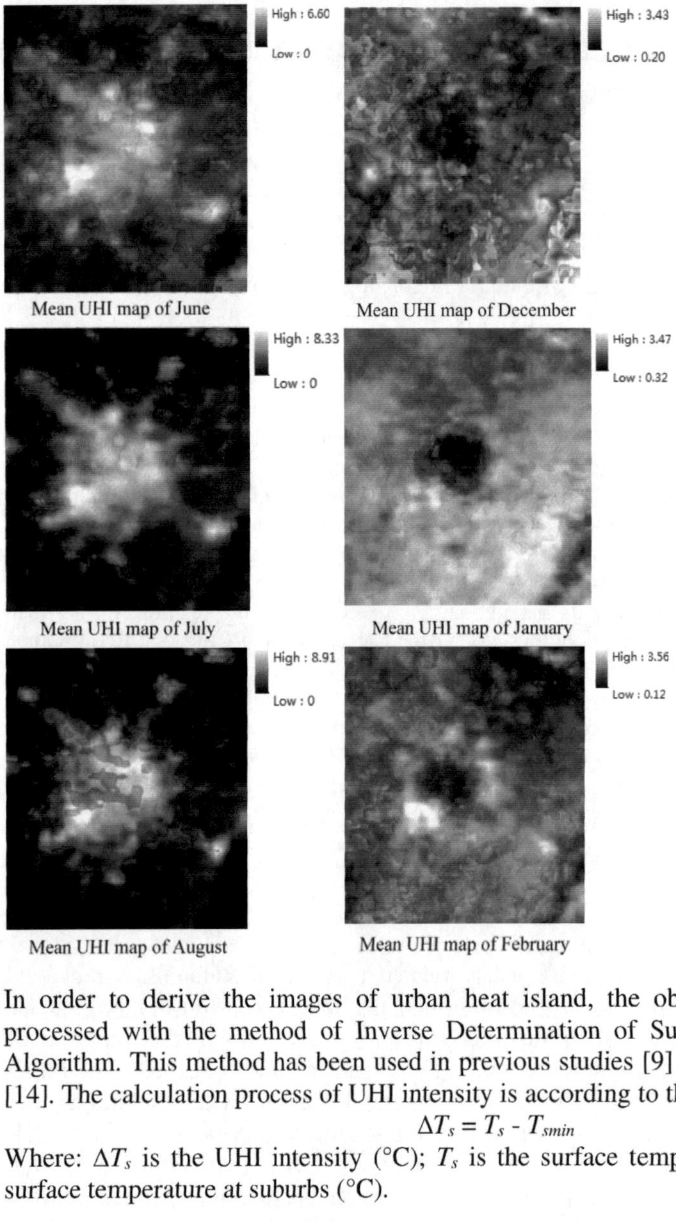

Figure 1. UHI in summer and winter months (°C)

In order to derive the images of urban heat island, the obtained raw data of remote sensing is processed with the method of Inverse Determination of Surface Temperature- the Split Window Algorithm. This method has been used in previous studies [9] [13], being verified with high accuracy [14]. The calculation process of UHI intensity is according to the equation:

$$\Delta T_s = T_s - T_{smin} \quad (1)$$

Where: ΔT_s is the UHI intensity (°C); T_s is the surface temperature at downtown (°C); T_{smin} is the surface temperature at suburbs (°C).

Due to the characteristics of UHI, there is a temperature interval with an abrupt change at the urban boundary. Moreover, this pattern of temperature distribution can be applied to locate the 'real' temperature boundary between urban and suburban areas, thereby identifying rural temperature. Accordingly, the calculation of surface temperature in suburban areas (T_{smin}) is conducted by using this novel automatic extraction method of suburb temperature [12]:

$$\Delta T_s \leq \Delta \bar{T}_s + m\delta_{\Delta T_s} \quad (2)$$

Where: ΔT_s is the difference of temperature between two belts next to each other (The city core is the first belt); $\Delta \overline{T}_s$ is the mean value of the difference of temperature between belts, $\delta_{\Delta T_s}$ is the standard deviation of the difference of temperature between belt; m is a constant that is usually assigned 3 [12].

Therefore, the raw images in typical summer months (June, July, and August) and winter months (December, January, and February) are grouped and recalculated to produce monthly mean images for each of six months, accordingly (Figure 1). All images are processed with 376 × 368 pixel working display (139.25 metres × 139.25 metres for each pixel), corresponding to the rectangular area of 52,358 metres × 51,244 metres urban region of Chengdu.

1.2. The city and site selection

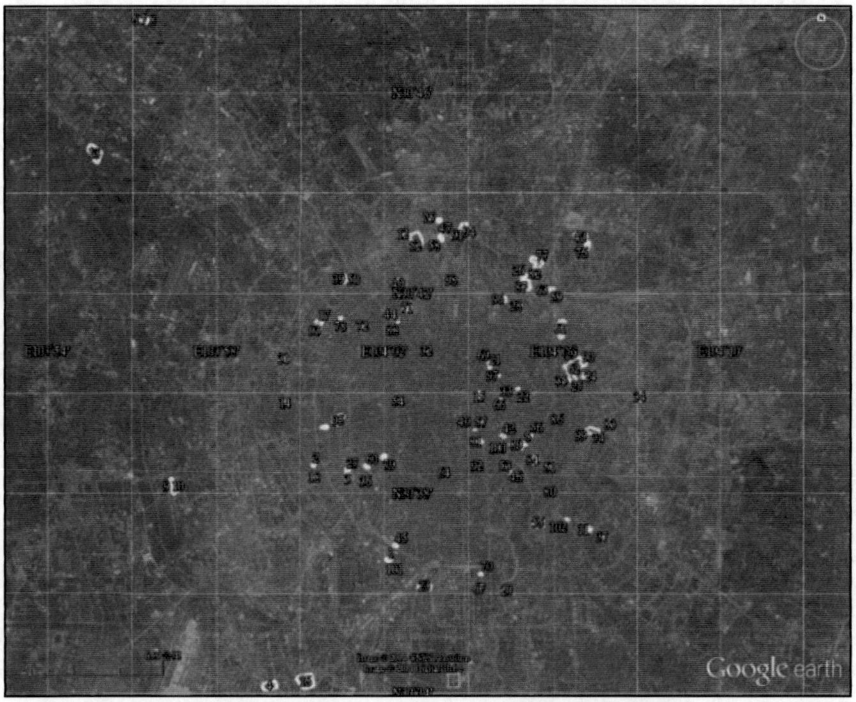

Figure 2. 103 sites in Chengdu City Map

Chengdu is located at Chengdu Pain and Sichuan Basin within the region of 102°54′E to 104°53′E and 30°5′N to 33°26′N. The city has a climate of hot summer and cold winter, and it is humid (above 69% in humidity) in all seasons. The annual rainfall is around 1000 mm (with a monthly value of 110 mm to 230 mm in summer), while the yearly average evaporation is 840 mm to 1100 mm (with a monthly value of 115 mm to 153mm in summer). The average wind speed is low (around 1.2m/s) with many cloudy and foggy days in winter. According to Deng's [15] study, till 2010, the urban core of Chengdu had been shaped with tense urban constructions in the forms of modern concrete high-rise buildings and skyscrapers. In the meanwhile, above 6 °C UHI intensity (UHII) has been continuously observed since 1988. However, local research intention has been more focused on the urban macro-scale, which leads to the missing of fundamental and theoretical data about the relationship between planning variables and UHI at a micro-scale. Therefore, studies of quantitative research on correlating multiple parameters of urban form with on-site UHI values are needed, in order to provide theoretical support and database for the practice of planning and designing a city block, whereas the large-scale building group has become the mainstream of the urban development [15].

In this paper, 103 the new-constructing city-blocks were selected as target projects; all projects were completed before 2010. The 103 projects are finely distributed within the area of 3 rings development of Chengdu city (Figure 2). Among the 103 projects, 62 are commercial-residential

IOP Conf. Series: Earth and Environmental Science **329** (2019) 012045 doi:10.1088/1755-1315/329/1/012045

mixed, one is office-residential mixed, 30 are residential, and 10 are single commercial in use. Due to the restriction of construction height in an urban area, more than 85% of the target projects are limited below 100 metres height. Only one skyscraper, in the central business district, is constructed greater than 150 metres. Moreover, the site area for the 103 projects varies from 3,000 m² to 356,000 m² with an average area of 30,900m². More than 50% are built with 30% building coverage ratio. The floor area ratio in three-quarters of all projects is restricted no more than 5. The green coverage ratio reaches 25% for most of the projects. Nearly half of the 103 projects are designed with the enclosed plot, while there are 24 sites are semi-opened.

1.3. Variable selection

There are several control parameters for planning any construction project in China, among which Floor Area Ratio (FAR), Building Coverage Ratio (BCR), Building Height (Height), and Green Coverage Ratio (GCR) are mostly concerned in codes and regulations. Moreover, variables of Project Type and Plot Layout (PL) are restricted accordingly due to the urban planning policies and local codes. Table 1 summarises the six key variables that have been selected in this study. In this study, seven types of functionality are summarized, which are the different combinations of residential, commercial, office, and HOPSCA. Moreover, as shown in Figure 3, the target projects are categorized into five groups: tower, linear, semi-closed, interspersed and court, according to the geometric form.

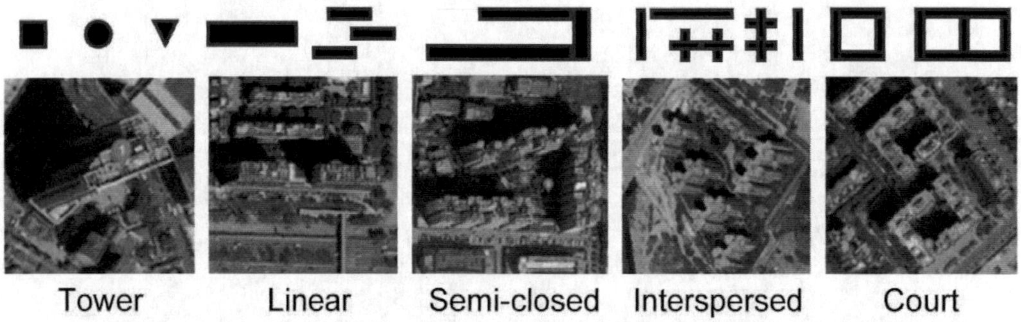

Figure 3. Examples of five types of plot layout with plan (top) and aerial view (down)

Table 1. Variable selected

Variables	Abbreviations	Properties
Project Type	TYPE	Residential (R), Commercial (C), Office (O), HOPSCA
Plot Layout	PL	Tower, Linear, Semi-closed, Interspersed, and Court
Green Coverage Ratio	GCR	The ratio of green area to the total site area
Floor Area Ratio	FAR	The ratio of total building floor area to the total site
Building Coverage Ratio	BCR	The ratio of building ground floor area to the total
Building Height	HEIGHT	Building height

1.4. Mapping UHI onto 103 projects and GIS-based (ArcGIS) analysis

Through the Spatial Analysis Tools of the Esri's ArcMap, the 103 selected sites are symbolised as a shapefile according to their geo-coordinates, and the UHI intensity value within satellite-derived images are input into this tool and are assigned into the different layers, thereby obtaining the mean value of monthly UHI intensity over the target projects.

2. Data analysis

2.1. Mapping Plot Layout vs UHI

According to the plot plan, the projects are categorised into five groups (Table 1): Tower, Linear, Semi-closed, Interspersed, and Court. The result of the mean value of UHI intensity in each type of projects is shown in Figure 4. In general, sites with less openings in terms of plot layout (Interspersed and Court) are recorded lower summer-UHII and higher winter-UHII, while the 'fully open' projects (tower) suffer the UHI nearly 0.9°C higher in summer and 0.7°C lower in winter than the projects in more closed plot layout (Semi-closed and Courts). When comparing to the other four types of plot layout, the Tower layout suffers highest summer-UHII (4.45°C) and lowest winter-UHII (1.01°C).
The Interspersed Group, which has multi openings on the plan, inner courtyards, and Interspersed buildings, records the lowest summer-UHII and the highest winter-UHII. It reveals the most appropriate form of plot layout for city blocks in Chengdu in terms of plot layout form of building group with minimum the thermal impacts of building group on outdoor spaces. Moreover, the majority of the building groups with the three types of plot layout (Tower, Linear, and Semi-closed) are located within areas of 5km away from the city centre, and the Interspersed Group is more frequently planned in outer urban areas. Therefore, the location of the buildings has a significant impact on plot layout, which indicates the possible influence of background effect on on-site UHI, which is provided by the surrounding constructions.

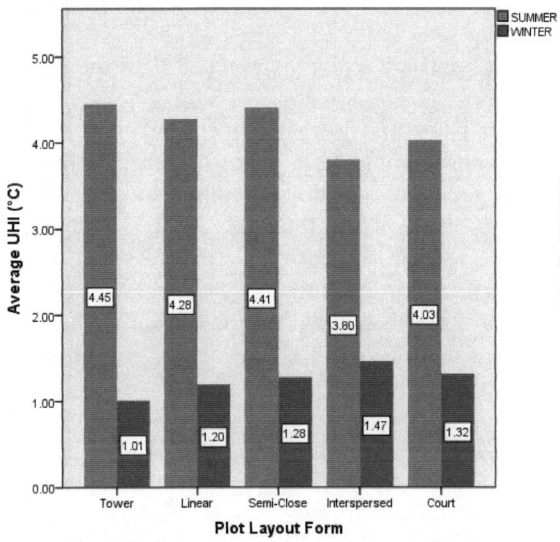

Figure 4. 103 Plot layout (Table1) vs UHI

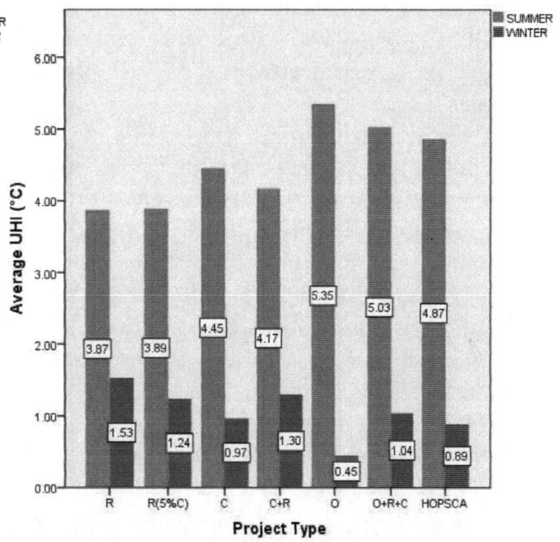

Figure 5. Project Type (Table 1) vs UHI

The interspersed form is often more homogeneously designed in terms of buildings height, buildings within the group provide over-shading to each other, while this form allows as much openings as possible on the layout of the site to let the breeze take away a considerable amount of heat stored within site generated by high-angle sun in the summertime. The courtyards within the interspersed sites are designed with more significant green area, sometimes even with water bodies, which potentially provide efficient UHI mitigation measures. In the wintertime, the interspersed group benefits from its wide-spread and highly free building settlement to receive and store solar radiation, while wind speed is lower than its annual average (below 1.2 m/s), thereby leading to the highest UHII among all layout forms. However, the towers receive relatively high amounts of solar radiation due to little over-shadowing in summer, resulting in the highest UHII. Moreover, their relatively more compact layout avoids the towers receiving high levels of solar radiation compared to other layout forms with much larger site area. In general, the analysis above explains the different potential causes of UHI in summer and winter across different city blocks.

2.2. Function vs UHI

IOP Conf. Series: Earth and Environmental Science **329** (2019) 012045 doi:10.1088/1755-1315/329/1/012045

As stated in section 1.3, four basic uses- Residential (R), Commercial (C), Office (O), HOPSCA (complex with hotel, office, park, shopping mall, convention, and apartment)- consists of different types of combination of project functions. Accordingly, the 103 the projects are assorted into seven groups based on their design functions: Residential (R), Residential with 5% Commercial (R(5%)), Commercial (C), Commercial + Residential (C+R), Office (O), Office + Residential + Commercial (O+R+C), and HOPSCA. The most popular forms of dwelling in Chengdu are single residential, and residential with limited commercial (small retails with up to 5% of total building area), while the other five types of developments are more public-oriented, including shopping malls, office, hotels, and the mix of them.

As shown in Figure 5, the projects of single residential are observed to have the lowest summer-UHII (3.9°C) while their winter-UHII(1.5°C) is the highest among all project types. On the other hand, compared to the residential city blocks, the average UHII of the public building types is higher in summer and lower in winter. The project type- Office- records the highest value (5.4°C) in summer and lowest (0.5°C) in winter. Introducing other types of use will alter the on-site UHII pattern. The integration of a 5% portion of commercial buildings raises the UHI-summer and lowers the UHI-winter in the residential building projects. Compared to the single commercial projects, the mix of commercial and residential projects tends to have lower values in summer and higher values in winter. The HOPSCA and Office+ Residence Hotel+ Commercial sites have an average proportion of each building type, which decreases the UHII in both summer and winter when compared to the single office sites.

In general, office project sites suffer highest summer-UHII and lowest winter-UHII among the three types of single functions (other two are residential and commercial). The location of a project determines building type. HOPSCA and Office + Residence Hotel + Commercial are located within city core area of less than 3km away from the city centre, whereas residential types are frequently built in locations more than 6km away from urban core. Lastly, the majority of commercial and office projects are located within the middle areas, between 3km and 6km from the city centre. Projects with office functions are mainly high-rise towers. Their "disadvantageous" layout forms enhance the UHI effect on office sites in winter, while they reduce the solar heat stored in winter, thereby resulting in low winter-UHII.

2.3. *Performances of design variables on site and point basis*

2.3.1 *Floor Area Ratio vs UHI*

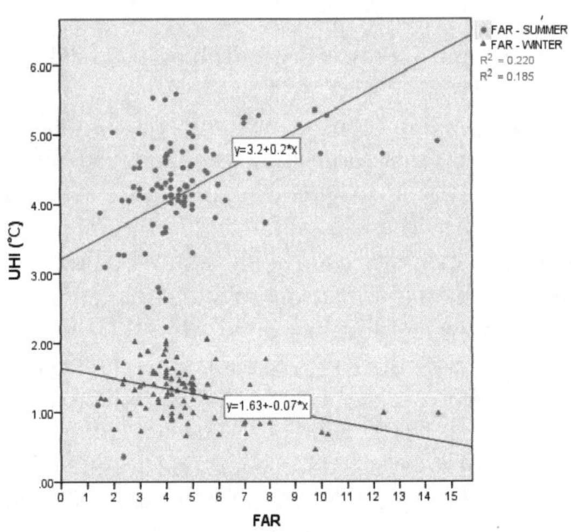

Figure 6. Floor Area Ratio vs UHI

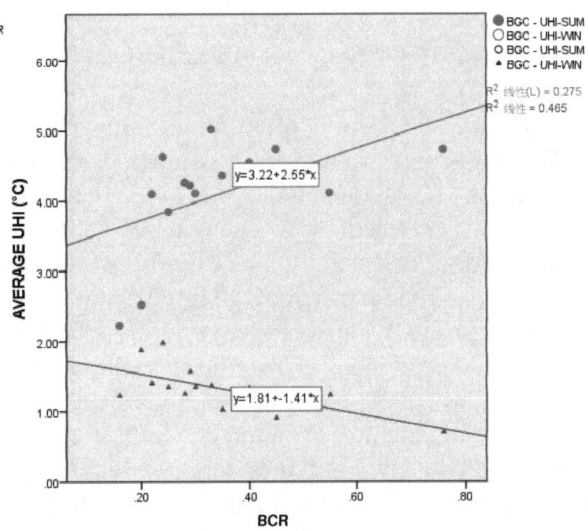

Figure 7. Building Coverage ratio vs UHI

As shown in scatter plots and correlation figure (Figure 6), FAR is positively related to summer-UHII, while negatively related to winter-UHII. FAR has a significant linear correlation with both summer-

UHII and winter-UHII (significance being regarded as an R^2 greater than 5% level [16]). A project with larger FAR tends to suffer from higher summer-UHII and lower winter-UHII in winter. A decrease in a value of 1 FAR will reduce summer-UHII by 0.20°C and will benefit from 0.07°C higher winter-UHII. In general, the distribution of FAR values is relatively uneven: more than two-thirds of the points fall within the range of 1 to 8, which is restricted by the local statues for urban planning in the urban area of Chengdu.

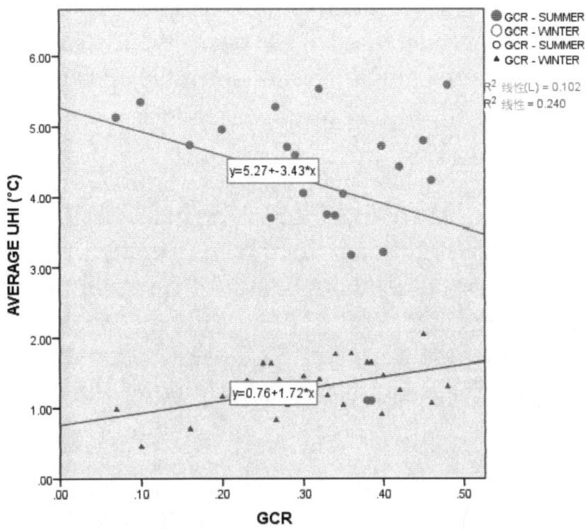

Figure 8. Green Coverage Ratio vs UHI

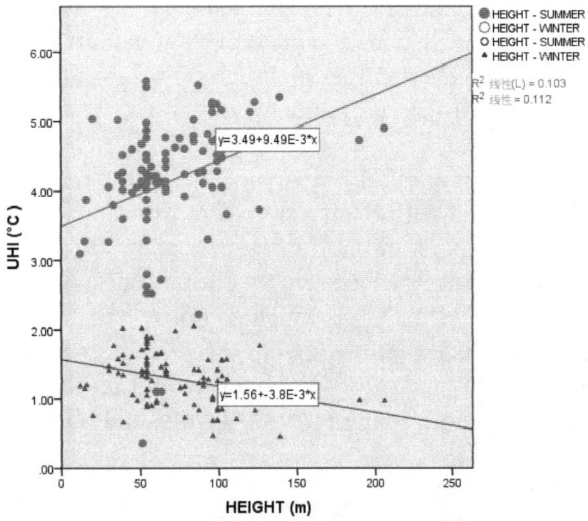

Figure 9. Building Height vs UHI

2.3.2 Building Coverage Ratio vs UHI. Figure 7 shows reasonable linear correlations of BCR with average summer-UHII and average winter-UHII. In general, the denser plot in plan raises summer-UHII and lowers winter-UHII. It is estimated that every 0.1 (10%) increase in the value of BCR will increase summer-UHII by 0.26°C. While in contrast, a decrease of 0.1 (10%) in BCR will increase winter-UHII by 0.14°C. The distribution of BCR values is also restricted by local planning policies with certain levels; most projects were designed with the BCR 30%, 35% or 40%.

2.3.3 Greening Coverage Ratio vs UHI. Figure 8 indicates that increasing GCR is an effective way to mitigate UHI in summer months and to rise outdoor surface temperature in winter: an increase of 0.1 in BCR will reduce summer-UHII by 0.34°C, and will raise winter-UHII by 0.17°C.

2.3.4 Building Height vs UHI. Figure 9 indicates that building height has a positive relationship to summer-UHII and a negative relationship to winter-UHII. An increase of 50 meters in average building height of a project will result in raising summer-UHII by 0.48°C and decreasing winter-UHII by 0.19°C.

3. Discussion and conclusions

3.1. Selection of Design specific variables: FAR, BCR, GCR, HEIGHT, PL, and TYPE

In this study, all the correlations of parameters with both summer-UHII and winter-UHII are higher than 5% of the significant level. However, no single variable can independently determine the on-site UHII, which is due to the multi-factor effect to which all variables contribute, as reported by other research [16]. From the results on Floor Area Ratio, the denser plots tend to have a higher summer-UHII and lower winter-UHII. Increasing Green Coverage Ratio and openness in plot layout (less enclosure in plot layout on site) are effective to reduce summer-UHII and to benefit from higher winter-UHII. Both increasing Building Coverage Ratio and Building Height raises summer-UHII and reduces the winter-UHII. Residential projects tend to have lower summer-UHII, and public projects, especially in office projects, tend to have lower winter-UHII. For a given density, a careful mix of buildings of various heights has been previously reported to help improve local ventilation as compared to a homogeneous height setting [17].

3.2. Recommended design strategies of city-blocks in Chengdu

The choice of building density should be based on the overall impact on urban heat island effect. The study shows that the site with higher building density tends to have higher outdoor surface temperature, especially for public (non-residential) buildings. It was found that the taller buildings could provide over-shadowing within a limited range, however, the average outdoor surface temperature of the entire site still increases as the buildings get taller. The choice of plot layout should be considered in relation to the microclimate. The study shows that the building groups with interspersed and court plot layout were achieved with a reduced UHI intensity in summer and increased UHI intensity in winter. The study suggests that the modification of ground surface with greening has a considerable impact on the outdoor thermal environment.

3.3. Summary of the Top-Down process

This Top-Down process carried out in this study is a practical approach to investigate the long-term urban temperature at different scales. In comparison to traditional methods (weather stations), processing the surface temperature through MODIS is a more practical method for research on urban heat island effect [18], which can provide substantial images in a relatively short time. Moreover, due to the high resolutions of the images obtained in this study, the derived LST images not only reflect the general pattern of UHI over the entire city at a macro scale, but also precisely describes the UHI values at micro scales (blocks, streets, and buildings).

Acknowledgement

The procedures of collecting and processing of remote sensing data were operated by Chengdu plateau Meteorological Institute of China Meteorological Bureau under China Meteorological Bureau's Integration & Application Project (national level) of Key Meteorological Technologies-CMAGJ2014M47. We would also acknowledge that Dr. Xinshuang Zhang made a significant contribution to this study.

References

[1] Korthals Altes W K and Tambach M 2008 Municipal strategies for introducing housing on industrial estates as part of compact-city policies in the Netherlands *Cities* 25(4) pp 218–229
[2] Tian Y and Jim C Y 2012 Development potential of sky gardens in the compact city of Hong Kong *Urban Forestry and Urban Greening* 11(3) pp 223–233
[3] Wolsink M 2016 "Sustainable City" requires 'recognition'- The example of environmental education under pressure from the compact city *Land Use Policy* 52 pp 174–180
[4] Ratti C, Baker N, and Steemers K 2005. Energy consumption and urban texture *Energy and Buildings* 37(7) pp 762–776
[5] Lee S and Lee B 2014 The influence of urban form on GHG emissions in the U.S. household sector *Energy Policy* 68 pp 534–549
[6] Futcher J A and Mills G 2013 The role of urban form as an energy management parameter *Energy Policy* 53 pp 218–228
[7] Wong N H et al 2011 Evaluation of the impact of the surrounding urban morphology on building energy consumption *Solar Energy* 85(1) pp 57–71
[8] Oke T R 2002 Boundary layer climates 2nd Edition (London: Routledge)
[9] Tran H et al 2006 Assessment with satellite data of the urban heat island effects in Asian mega cities *International Journal of Applied Earth Observation and Geoinformation* 8(1) pp 34-48
[10] Pongracz R, Bartholy J, and Dezso Z 2006 Remotely sensed thermal information applied to urban climate analysis *Advances in Space Research* 37(12) pp 2191–2196
[11] Yao R et al 2017 Temporal trends of surface urban heat islands and associated determinants in major Chinese cities *Science of the Total Environment* 609 pp 742–754
[12] Zhang S and Zhou C 2013 Spatiotemporal distribution characteristics and causes of sunny days'

heat island effect in Chengdu City of Southwest China *Chinese Journal of Applied Ecology* 24(7) pp 1962–1968

[13] Wan Z and Dozier J 1996 A Generalized Split- Window Algorithm for Retrieving Land-Surface Temperature from Space *IEEE Transactions on geoscience and remote sensing* 34(4) pp892–905

[14] Wan Z, Zhang Y, Zhang Q and Li Z 2004 Quality assessment and validation of the MODIS global land surface temperature *International journal of remote sensing* 25(1) pp 261–274

[15] Deng X 2018 Microclimate and building energy in the built environment: A study of planning high-rise building groups at a city-block scale in China (Cardiff, PhD: Architecture)

[16] Yang F, Lau S S Y and Qian F 2011 Urban design to lower summertime outdoor temperatures: An empirical study on high-rise housing in Shanghai *Building and Environment* 46(3) pp769–785

[17] Ng E, Wong N H and Han M 2005 Parametric studies of urban design morphologies and their implied environmental performance *Tropical Sustainable Architecture: Social and Environmental Dimensions* (London: Architectural Press)

[18] Tomlinson C J et al 2012 Derivation of Birmingham's summer surface urban heat island from MODIS satellite images *International Journal of Climatology* 32(2) pp 214–224

The Influence of Urban Green Systems on the Urban Heat Island Effect in London

Hashem Taher[*], **Heba Elsharkawy and Darryl Newport**

School of Architecture, Computing and Engineering (ACE),AVA Building, Room 1.08, University of East London (UEL), Docklands Campus ,London, UK, E16 2RD

*u1630377@uel.ac.uk

Abstract. Urban areas are typically warmer than rural ones. This is mainly due to denser configuration dominated by impermeable surfaces such as buildings and roads, compared to rural areas which are less densely built and mainly dominated by open spaces. Rapid urban expansion in dense cities bares direct impact on surface and air temperature patterns within street canyons; a phenomena which is known as the Urban Heat Island (UHI) effect. Thus, several UK city councils such as Birmingham, Manchester, and London have started to develop strategies aiming at enhancing urban green systems (UGS) through trees, green walls and green roofs. Some of those strategies include considering the green space factor, and increasing green areas within the cities to improve street canyon microclimate and reduce UHI. The Mayor of London has adopted a strategy for London 2050 aspiring to transform it to be the greenest city in the world by increasing the green areas up to 50%. This paper investigates the influence of increasing the UGS percentage which is considered as a key solution to mitigate UHI effect which will, in turn, provide thermally comfortable outdoor environments for pedestrians. The investigation is undertaken by comparing the morphology of precincts and streets in relation to air temperature, mean radiant temperature and surface temperature within Oxford Street canyons in London city centre; being one of the world's busiest streets. The results from this research demonstrate that different UGS interventions with varying percentage are required depending on particular canyon orientations and geometries. The study found that, in general, more trees would have significant thermal comfort effect followed by living façade, while high albedo pavement (HAP) came last. However, HAP had high influence on improving thermal comfort in North-South orientated streets with minor variance to trees and living facades which, changing their percentage levels was insignificant.

1. Introduction

Global cities are currently inhabited by almost half of the world's population, which is expected to increase by 70% by 2050 [1]. The UK has more than 60% of its population living in cities, while most European countries have almost 50% living in cities [2]. Thus, the higher population densities, the higher the impact of climate change [3]. Climate change and associated heat waves may increase the number of heat-related deaths in Europe from 152,000 to 239,758 a year by 2080, leading to 50 times death rise, while in the UK nearly 11,000 persons may die every year as a result [4].

Depending on relative vulnerabilities of the population, infrastructure, ecosystem, etc.; the impact on cities may be diverse [5]. Since London's streets and pathways contribute by up to 80% of London's public spaces [6]. Thus, policy makers are establishing climate change programs that require architects, planners and urban designers to integrate climate resilience in refurbishment, retrofitting and newly built developments [7]. There have been various calls for interventions initiated in the UK, in response to climate change adaptation and mitigation on the urban scale level as Healthy Streets for London by Transport for London (TfL) [8], London Environment Strategy, London Urban Heat Island, London Infrastructure Plan 2050, and Urban Green Factor by Greater London Authority (GLA) [9–14]. On the other hand, there have also been some initial design proposals to address this issue; such as Walkable London by Zaha Hadid Architects [15]. Notably, Adaptation and Resilience in a Changing Climate (ARCC) research network for different scales and level projects, has evolved to develop adaption analysis tools for urban areas, focusing on the built environment, infrastructure, transportation system as the main objective for effective climate change mitigation initiatives [16, 17].

The current study addresses an important topic of assessing the influence of UGS on mitigating UHI since UGSs may help minimise cooling and heating loads and overall operational energy demand within buildings. The research investigates how green roofs, trees and vertical green systems (green walls and living façades) may potentially mitigate the adverse effects of climate change, by lowering the UHI effect and improving pedestrians' thermal comfort.

2. Research context

2.1 Urban Heat Island (UHI) in London

UHI is the phenomenon at which air temperature in cities would tend to be higher than the rural areas of close proximity during warmer seasons particularly during night time [18]. The UHI occurs due to solar energy storage within the urban fabric accumulating during the daytime, which is then released back to the local environment during night time. When green spaces are being replaced by buildings and roads, the thermal, radiative, moisture and aerodynamic properties of the surface and the atmosphere change (GLA, 2006). UHI bears a direct effect on London's climate leading to warmer winters, earlier spring season and less snow. The UHI effect leads to approximately 2°C warmer temperature during the night and -0.2°C cooler temperature during the day leading to UHI intensity reaching up to 9°C in 2003 compared to only 4-6°C by 1960 in London compared to surrounding rural areas [10]. It is expected that summer air temperatures in the South East of the UK may rise by 3.5°C, and 5°C by 2050 and 2080, respectively [19]. London City centre may also face up to 9°C in air temperature compared to the surrounding greenbelt with expectations to the frequent increase of heat waves [10]. London UHI mitigating strategies must be developed coherently between key stakeholders; urban planners, climate change scientists, and policy makers. Thus, in developing mitigation strategies for London's UHI, it must be considered that the UHI is a city scale phenomenon and the outcome of the combination of the vast range of microclimates that occur across London.

2.2 Urban Greenery Systems for UHI mitigation

A classification system of UHI mitigation strategies has been recently established [20] grounded on grouping the measures according to selection criteria of action-oriented adaptation which aligns with the shared and mutual view of policy makers. This system led to categorising UGS interventions into 4 main scales: urban landscape, street geometry, building envelope and pavements. Consequently, 11 mitigation strategies have been identified to mitigate UHI; namely: cool building envelope, green roofs, green facades, ground vegetation, shading trees, cool pavements, water bodies, water-retentive pavements, built environment orientation, built environment prevailing winds, built environment orientation built environment typical section and built environment orientation to the sun [20].

Urban greening is much more than a luxury or an aesthetical intervention for cities. It needs to be considered as a vital part of urban planning [21]. Thus, the GLA recommends a few measures to manage UHI in London [10]. Firstly, urban greening as trees, parks is an effective way of enhancing and improving harsh urban climates throughout individual buildings within the neighbourhood scale, due to evapotranspiration process and reducing surface temperatures between 5-20°C due to shading. It is also very important to consider the type of tree canopy in order to avoid trapping air pollutants from the street [18]. Secondly, living façades, green walls and green roofs (GR), which include a growing medium

IOP Conf. Series: Earth and Environmental Science **329** (2019) 012046 doi:10.1088/1755-1315/329/1/012046

planted above a waterproof membrane, might potentially have an influence on the atmosphere around it and on the building fabric (walls) insulation, while GR has an influence on the upper floors of buildings in the green roof case [22–24]. Lastly, fixing cool pavements with high reflective solar material and water preamble will help in high urban temperature mitigation through rain storage and reflecting high solar energy rates [25]. The current study will be mainly focusing on trees, living walls and high albedo pavements (HAP).

3. Research Methodology

The main aim of this study is to investigate the extent to which UGS may play an effective role in the mitigation of the UHI, in the current and future climate scenarios using Oxford Street in London as the case study. The research methodology is primarily quantitative; applying ENVI-met software for modelling and on-site measurements. The paper focuses on ENVI-met modelling and data analysis. Envi-met was developed at Ruhr-University Bochum in Germany [26]. ENVI-met software package facilitates modelling the microclimate effect of vegetation and its influence on UHI, Physiological Equivalent Temperature (PET), Air temperature (Ta), Radiant Temperature (Tmrt) and Predicted Mean Vote (PMV). Thus, the tool helps undertake the investigation of the impact of UGS on the microclimate.

3.1 Model Geometrical Specifications/inputs

The main criteria for deciding the case study is to choose and model some of the most polluted and highly populated streets in the UK [27] as a worst case scenario. Hence, central London's Oxford Street was chosen as the research case study where many of the aforementioned challenges exist. The canyon between Orchard Street and Park Street, characterised by Height: Width of 1:1 while the length of the building block is 120m and the street canyon is 140m as illustrated in Figure 1 with a total neighbourhood area of (520mX520m).

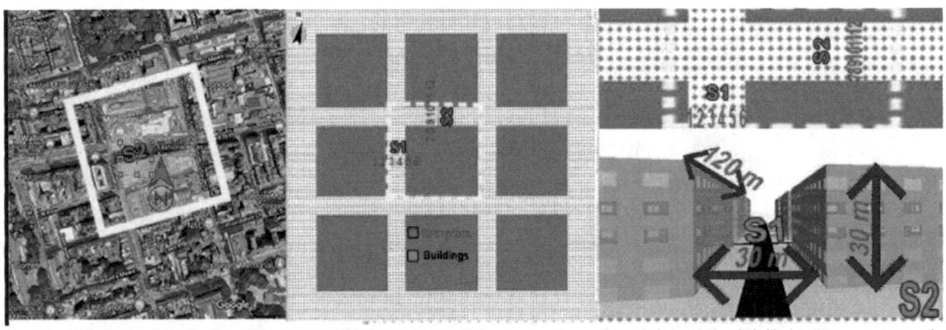

Figure 1. Oxford street (left), Street 1&2 layout with receptors locations (middle), Receptors locations (up) & Model dimensions bottom (right) (Source: Google map, ENVI-met)

Several studies [28, 29] have asserted that similar canyons dimensions need to be investigated for the purpose of improving pedestrians' thermal comfort (PTC), especially during the summer season[28, 29]. Thus, measurements are taken within the North-South (NS) street (S1) and East-West (EW) street (S2) as shown in Figure 1 (middle picture) which represents the Oxford street case. This would also be for the middle canyon within our middle neighbourhood. Since it represents the real canyon within central city urban fabric figure. Measurements are taken at the location of 6 Receptors (R) located in the middle of each canyon (total of 12 Rs in both canyons as in figure 1 middle). The physiological equivalent temperature (PET) is used to determine PTC as it uses the skin of the human body to calculate all energy exchanges, core temperature and sweat rate while considering clothing as a secondary variable [30].

3.2. Modelling scenarios
In order to evaluate the effectiveness of the proposed UHI mitigation scenarios in relation to PET, the real empirical model will be testing the following scenarios as illustrated in Fig 2. Two different green system types; trees, living façade (LF) are then incorporated at two 25 and 50 percent. High pavement albedo (HPA), such as white concrete (Albedo=0.8), is also tested with 66.6% of canyon area, which represents all pavement area. It must be noted that the vegetation in central London is almost negligible in main streets, whilst in the case study canyons is negligible, hence the base case setting was fixed at 0% vegetation.

The simulations are then undertaken for the four different scenarios; base case, trees, living facade (LF) and pavement albedo using 2018 Met Office climatic data in order to evaluate their influence on UHI mitigation by using the 12 Receptors. The two different ratios for each vegetation type is then applied, by increasing the current case by (25% and 50%). The Greater London Authority (GLA) [9, 12] plans to upsurge the green areas in London from 38.4% to 50%, hence applying 50% as the maximum level of vegetation.

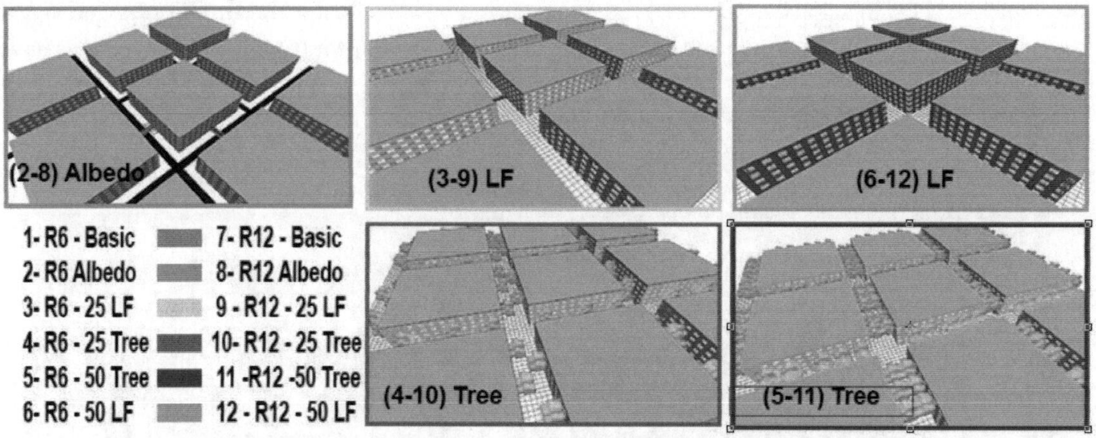

Figure 2. UGS alternatives (legend colours represents different UGS alternatives')

4. Results and Discussion
After running simulations for Oxford Street canyons previously identified, results were extracted from ENVI-met in the form of graphical illustrations and numerical excel files from the Receptors within both canyons. Street 2 (S2) represents Oxford Street while Street 1 (S1) represents Regent's Street which intersects S1. The different vegetation and albedo strategies were applied to determine which intervention(s) are more appropriate to be used within the case study canyons to improve pedestrians' thermal comfort. Thus, PET, PMV, Predicted Percentage of Dissatisfied (PPD) have been the focus of the study. Further analysis is undertaken to explore how Ta, Tmrt, Surface Temperature (Ts) Wind Speed (Ws) and Relative Humidity (RH) influence pedestrians' thermal comfort (PTC).

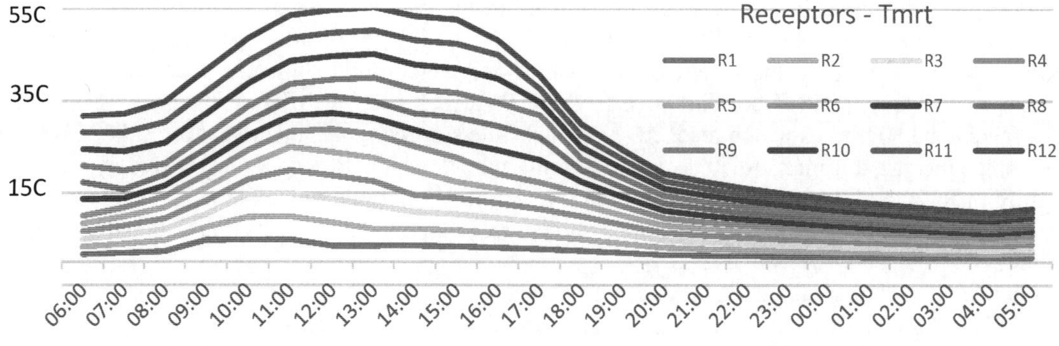

Figure 3 Mean Radiant Temperature (Tmrt) Measurements of the 12 receptors (Source: ENVI-met)

As shown in Fig. 3, the Mean Radiant Temperature (Tmrt) at all 12 Receptors have been extracted for the average of the summer season in 2018. However, as there have not been significant differences between Tmrt and Tair at each receptor, the highest value of each street receptors based on PET and PMV has been chosen to represent the canyon (S1, and S2), those being R6 and R12. Consequently, PMV and PET were measured to determine which time of the day has the maximum heat stress, which was found to occur at 16:00. Thus, further heat stress indicators (PET, PMV and PPD) and calculations regarding UGSs and albedo effects are then undertaken.

4.1 Thermal Comfort Values (PET, PMV and PPD) at Street 1 and Street 2

Figure 4 demonstrates the calculations at R6 and R12 concerning the PET, Ta, Tmrt and Ts. It was clear that the Ta had no notable variance with the changing strategies. However, UGSs had a significant influence on both Ts and Tmrt which have reflected on PET. Thus, PET was used as a clear indicator of pedestrians' thermal comfort within street canyons based on other outputs as Ts, Tmrt, RH and Ws. Overall, in S1 (north-south) UGS was not as effective as in S2 (east-west) due to higher solar radiation and higher sun hours associated with S1 orientation and canyon geometry. Hence, altering pavement albedo would be a feasible and cost-effective solution, particularly for the Ts. Shading pavements have been the main reason for lowering temperature due to blocking access of solar radiation to the pavement surface and hence improving pedestrians' thermal comfort level.

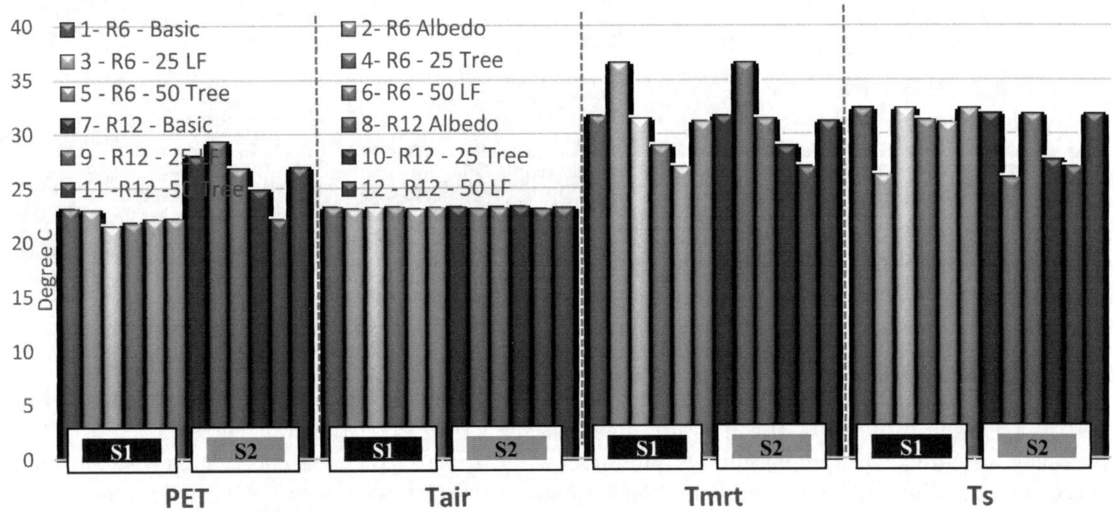

Figure 4. PET, Air Temperature, Mean Radiant Temperature and Surface temperature comparison (Source: ENVI-met)

It was not as effective to change the UGSs percentage for either trees or Living Facades (LF) where the maximum reduction in PET was -0.3 °C which means, increasing trees up to 50% may not be an effective (nor feasible) solution. While it was -0.67 °C for LF where walls have been receiving higher solar radiation and hence more temperature variations were found. For S2, trees have been more practical in improving PET comfort levels where there has been a difference which was observed between 25 and 50% trees which reached -2.76 °C. While surprisingly PET increased +0.14 °C when LF percentage increased from 25 to 50%, where RH might be the main driver for that, however, Ta lowered slightly in R12 and Ts and Tmrt had a similar pattern to R9. High surface albedo was increased heat stress by +1.36 °C for PET than the basic case, however, it has decreased Ta by -0.20 °C and by – 4.91 °C for Ts and increased Tmrt by +5.42 °C.

Interestingly, the percentage of reduced PET due to UGSs in S1 street had a similar pattern were using 50% of either trees or LF had same -1.04% while similar reductions found 1.06 and 1.07 from trees and HAP. While for S2 Street, both LF of 25 and 50% had almost the same PET reduction of -1.04% and -1.03% respectively. The higher saving percentage was found for trees with 25% and 50%

reached -1.12% and -1.26% correspondingly. These percentages could be used to give us indications about how UGSs act to heat stress in the cloudy would or a cooler day or in winter for instance based on S1 Street.

4.2 Relative Humidity (RH) and Wind Speed (Ws)

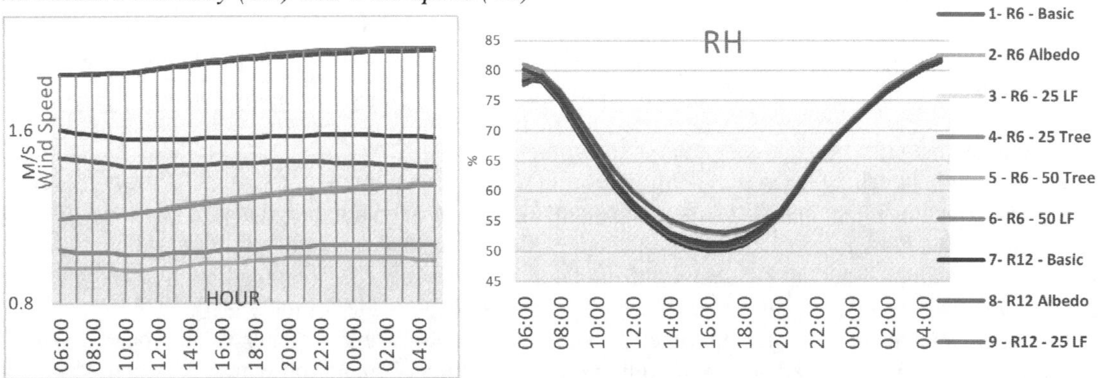

Figure 5. Wind Speed (left) and Relative Humidity (right) comparison (Source: ENVI-met)

In Figure 5 for S1, wind speed ranged from 1.1-1.3 m/s across the day while with negligible changes for both LF percentages', which was 1.12- 1.28 m/s. For the two tree percentages, the wind speed reduced to 0.99-1.01 m/s for R4 and 0.92-0.95 m/s for R5 which clearly shows that trees have blocked even the low wind flow.

For S2, wind speed ranged from 2-2.2 m/s for all cases except the R4,5 with trees at which trees have significantly reduced the Ws in R5 from 1.6-1.55 m/s and 1.43-1.38 m/s in R5. This could show us great evidence that in-case a higher wind speed, a higher thermal comfort would have been expected during the hot season. Although trees have lowered Ws, the highest thermal comfort was met in PET. For RH in S1 and S2, it had the same pattern across different UGSs however Rs near trees or vegetated walls might have felt a slight difference. Higher RH was noticed in S2 than S1 especially for R12 followed by R11, R10, R9 and with the same vegetation sequence at S1.

4.3 Predicted Mean Vote (PMV) and Predicted Percentage of Dissatisfied (PPD)

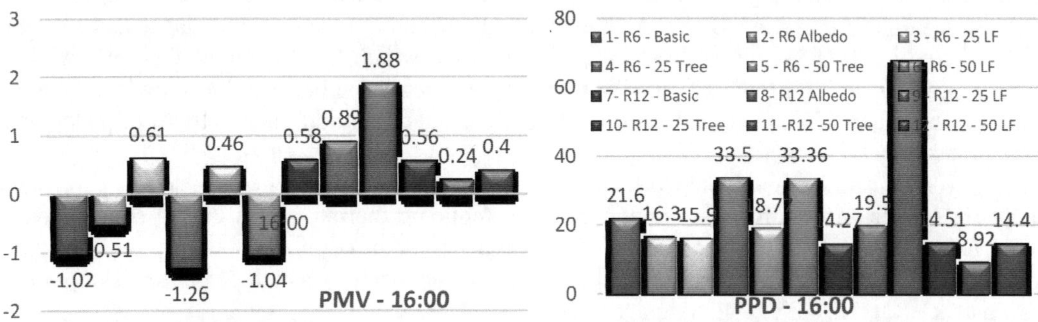

Figure 6. PMV (left) and PPD (right)comparison (Source: ENVI-met)

Figure 6, PMV and PPD have a linear relationship where PMV calculations are based on (Ta, Tmrt, Ws and vapour pressure) where its assumptions were mainly for steady-state indoor situations with extending the energy fluxes with long and short wave radiations in order to use it outdoors [30]. For S1, using higher albedo material, 25 LF and 50% tress have to lead to change PMV positively from cold to slightly cool and slightly warm for R4 and R5. While it was cooler when added 50% LF and cooler for 25% trees. For S2, UGS has improved PMV from warm to slightly warm, except at R2 and R3 where R3 has increased massively to be hot.

PMV seems not the following PET, Ta and Tmrt follow a similar trend, which might be due to PMV considers the effect of clothes and human activity into consideration in a scale, not in a numerical number, which might have several interpretations. For instance, adding lighter clothes within the warmer street canyon would lead to higher discomfort levels, that's due to increasing the skin surface facing the sun, however, it is more logical that it should be more thermally comfortable. As for PPD; in S1 around 33.5% of people felt discomfort at R10 and R12 however it is still acceptable. While in S2, 67.6% of people have been dissatisfied which also demonstrates that there is a linear relationship between PMV and PPD [30].

5. Conclusion

A relationship between UGS and UHI mitigation in the current climate scenarios of summer 2018 is being established using numerical methods through addressing this knowledge gap in this field in order to determine the most influential factors affecting pedestrian thermal comfort. Being able to decide whether increasing UGS percentage would likely to increase the PTC or not. On another hand, the location and the way of UGS placement are more vital and responsible for increasing PTC.

UGSs implementation within London city centre would be different in terms of the type of UGS, its percentage, where it is applied and that would be based on street orientation where S1 streets does not need much vegetation as S2 streets that is mainly due to solar radiation and the buildings height within the canyon limits sunlight rays to reach street level due to building shading. For S1 street, applying high albedo surfaces for pavements would increase comfort through reducing PET by 0.21°C. The latter was due to its high effect on lowering surface temperature while its reflectivity had increased Tmrt. while other vegetation percentages, PET improved by -0.35 to -1.05°C. With regards to S2 canyon, a very high thermal stress was found compared to S1, HPA increased thermal stress by 1.3°C due to its high Tmrt, while trees lowered PET by 3.08 and 5.84°C for 25% and 50%, espectively. While for LF 25% and 50%, -1.17°C and -0.09°C was found as a Tr which is almost negligible for LF 50%.

Concerning Tr percentages in PET, it could be concluded that with cooler temperatures across the year seasons and days, similar savings in Tr percentages were similarly found based on S1 and S2 comparison where S1 has a well-oriented canyon and has low direct sun hours. PET was a more precise way to indicate thermal stress and thermal comfort since it considers the effect of several variables on human skin: Ta, Tmrt, Ts, Ws, RH, and solar radiance. Trees reduce Ws slightly but do not influence thermal comfort as Ws is already too low. However, they work better in higher solar radiance and challenging orientations of streets. Increasing LF and trees percentages raise RH levels, especially LF.

Notably, HAP type has a huge influence on surface temperature while it has a higher Tmrt as well, thus it is advised to be applied in NS streets which lacks direct sun light and high radiance as well. PPD has been lowered in all UGSs alternatives even in HPA case which had higher PET before, while it increased in 25% LF case up to 67.6% in EW Canyon. That might be because PMV and PPD do not take solar radiation and Ws into consideration, while it considers humidity and activity of pedestrians and their clothes.

Overall, HPA and low vegetation work better for NS canyons and higher vegetation percentage especially trees which are more pedestrian friendly in reducing thermal stress during summer days. While for EW canyons, trees are more advised for higher thermal comfort especially with a higher percentage. Thus, it might be advised to use HPA with another strategy which offers to shade such as trees for instance.

Trees have been more effective during extreme heat condition, due to their shading effect in the first place as well as their direct influence on pedestrians rather than LF. This has more influence on building fabric and it only makes a difference when solar radiation falls over it which is not occurring in our case in central London with narrow canyon. The results of this study provides an in-depth understanding of the potential of UGSs to mitigate the UHI effect in a sustainable, energy-efficient and cost-effective way. However, it is recommended to carry out further numerical simulations to estimate the extent to which UGSs may help mitigate climate change in future scenarios of 2050s and 2080s.

Acknowledgements
The authors would like to acknowledge Mr Peter Massini from the (GLA), Dr Katherine Drayson and Mr Charles Snead from (TfL) for their guidance and time to help facilitate this research.

References
[1] UN. World's population increasingly urban with more than half living in urban areas | UN DESA | United Nations Department of Economic and Social Affairs, http://www.un.org/en/development/desa/news/population/world-urbanization-prospects-2014.html (2014, accessed 10 June 2018).
[2] UN. The World's Cities in 2016, http://www.un.org/en/development/desa/population/publications/pdf/urbanization/the_worlds_cities_in_2016_data_booklet.pdf (2016, accessed 10 June 2018).
[3] Ibarrarán ME. Climate's Long-term Impacts on Mexico's City Urban Infrastructure Case study prepared for Cities and Climate Change: Global Report on Human Settlements 2011 Climate's Long-term Impacts on Mexico's City Urban Infrastructure, http://www.unhabitat.org/grhs/2011 (2011, accessed 10 June 2018).
[4] Hajat S, Vardoulakis S, Heaviside C, et al. Climate change effects on human health: Projections of temperature-related mortality for the UK during the 2020s, 2050s and 2080s. *J Epidemiol Community Health* 2014; 68: 641–648.
[5] HuntP A, WatkissP P, U ecsasph HT, et al. Climate Change Impacts and Adaptation in Cities: A Review of the Literature, http://opus.bath.ac.uk/22301/1/Hunt_ClimateChange_2011_104_1_13.pdf (2011, accessed 10 June 2018).
[6] New London Architecture NLA. Public Realm in London - ideas for the future, https://www.newlondonarchitecture.org/whats-on/events/2017/december-2017/public-realm-in-london--ideas-for-the-future (2017, accessed 22 June 2019).
[7] Shaw R, Colley M, Connell R. Climate change adaptation by design: a guide for sustainable communities. TCPA, London, www.acclimatise.uk.com (2007, accessed 10 June 2018).
[8] TfL. Healthy Streets for London: Prioritising walking, cycling and public transport to create a healthy city, http://content.tfl.gov.uk/healthy-streets-for-london.pdf (2017).
[9] GLA. LONDON 2050 BIGGER AND BETTER.
[10] GLA. London's Urban Heat Island: A Summary for Decision Makers, https://www.puc.state.pa.us/electric/pdf/dsr/dsrwg_sub_ECA-London.pdf (2006, accessed 10 April 2018).
[11] Greater London Authority. London Infrastructure Plan 2050: Improving Quality of Life as London Grows. 2015; 3: 1–9.
[12] Mayor of London. *London Environment Strategy*, https://www.london.gov.uk/sites/default/files/london_environment_strategy.pdf (2018).
[13] Gary Grant. *Urban Greening Factor for London*, https://www.london.gov.uk/sites/default/files/urban_greening_factor_for_london_final_report.pdf (2017, accessed 25 June 2019).
[14] GLA. Policy G5 Urban greening | Draft New London Plan, https://www.london.gov.uk/what-we-do/planning/london-plan/new-london-plan/draft-new-london-plan/chapter-8-green-infrastructure-and-natural-environment/policy-g5 (2019, accessed 25 June 2019).
[15] Zaha Hadid. PRESENTATIONS — Walkable London, https://www.walkablelondon.co.uk/presentations (2018, accessed 22 June 2019).
[16] Scott S. Adaptation and Resilience to a Changing Climate Collaborative research to enable the design of resilient and sustainable systems in the UK built environment and infrastructure sectors (ARCC), www.ukcip-arcc.org.uk (2011, accessed 10 June 2018).
[17] ARCC. Adaptation and Resilience to a Changing Climate (ARCC) - Coordination Network (ACN), http://gtr.ukri.org/projects?ref=EP%2FG036586%2F1 (2018, accessed 10 June 2018).
[18] Zupancic T, Westmacott C, Bulthuis M. The impact of green space on heat and air pollution in urban communities : A meta-narrative systematic review. 2015; 1–68.

IOP Conf. Series: Earth and Environmental Science **329** (2019) 012046 doi:10.1088/1755-1315/329/1/012046

[19] Hulme M, Lu X, Turnpenny Tim Mitchell J, et al. Climate Change Scenarios for the United Kingdom Climate Change Scenarios for the United Kingdom: Climate Change Scenarios for the United Kingdom, http://artefacts.ceda.ac.uk/badc_datadocs/link/UKCIP02_tech.pdf (2002, accessed 10 April 2018).

[20] Aleksandrowicz O, Vuckovic M, Kiesel K, et al. Current trends in urban heat island mitigation research: Observations based on a comprehensive research repository. *Urban Clim* 2017; 21: 1–26.

[21] James P, Tzoulas K, Adams MD, et al. Towards an integrated understanding of green space in the European built environment. *Urban For Urban Green* 2009; 8: 65–75.

[22] Bianco L, Serra V, Larcher F, et al. Thermal behaviour assessment of a novel vertical greenery module system: first results of a long-term monitoring campaign in an outdoor test cell. *Energy Effic* 2017; 10: 625–638.

[23] Afshari A. A new model of urban cooling demand and heat island—application to vertical greenery systems (VGS). *Energy Build* 2017; 157: 204–217.

[24] Bass B, Krayenhoff S, Martilli A, et al. Mitigating the Urban Heat Island with Green Roof Infrastructure.

[25] Leal Filho W, Echevarria Icaza L, Neht A, et al. Coping with the impacts of urban heat islands. A literature based study on understanding urban heat vulnerability and the need for resilience in cities in a global climate change context. *J Clean Prod* 2018; 171: 1140–1149.

[26] Envimet. ENVI_MET – Decoding Urban Nature, https://www.envi-met.com/ (2018, accessed 16 September 2018).

[27] ISOBEL HAMILTON. One of the world's most polluted streets could eliminate vehicles in 2018. *Mashable*, https://mashable.com/2017/11/06/oxford-street-pedestrianisation-2018-pollution/?europe=true (2017, accessed 27 June 2019).

[28] Shafey AM. *Urban form to achieve the thermal comfort application on New Cairo*. Cairo University Faculty, 2016.

[29] Shalaby AS, Shafey AM, SaadShalaby A, et al. Optimizing the Thermal Performance of Street Canyons in New Cairo , Egypt , Using. *Int J Adv Res Sci Eng Technol* 2018; 5: 35–43.

[30] Envimet. apps:biomet_pmv [A holistic microclimate model], http://www.envi-met.info/doku.php?id=apps:biomet_pmv (2019, accessed 25 June 2019).

IOP Conf. Series: Earth and Environmental Science **329** (2019) 012047 doi:10.1088/1755-1315/329/1/012047

Simulation-Informed Urban Design: Improving Urban Microclimate in Real-World Practice in a High Density City.

Jianxiang Huang[1], Tongping Hao[1], Shan Shan Hou[2], Phil Jones[2]

[1] Department of Urban Planning and Design, Faculty of Architecture, Hong Kong University

[2] Welsh School of Architecture, Cardiff University, Cardiff, Wales, CF10 3NB, UK

Corresponding author email: JonesP@cardiff.ac.uk

Abstract. In many dense cities, urban heat and the interaction of buildings with their immediate urban environment emerges as a pressing issue due to growing urban heat island effect and climate change. Informed evidence based design decisions to mitigate heat stress becomes a priority for urban planning and design practitioners. The aim of the study is to develop informed design and development decisions using computer simulation tools concerning urban microclimate performance. In this study, academic researchers have worked with industrial partners in an urban renewal project in Hong Kong's high density urban area. In-house developed simulation software such as CityComfort+ and HTB2-Virvil were applied to assess urban microclimate conditions and risks of pedestrian thermal stress throughout key seasons. Simulation results were provided as feedback to project designers and managers at early stage, allowing timely design modification to improve performance while maintaining code compliance and design and fiscal priorities. The procedure is iterative until performance attributes converge. Preliminary results show that the informed design can deliver significant microclimate benefits compared with "business-as-usual scenarios". By shaping building mass, orientation, and strategic placement of shading and vegetation, the improved design is expected to reduce summer-time outdoor heat stress by 1°C measured in UTCI equivalent temperature, thus bringing the average conditions for the hot season into the "comfort zone" for the local community. Energy simulation can predict overall energy demand and the potential for renewable energy supply at an urban scale. The simulation-designer workflow shows promising potentials to improve urban microclimate performance of design outcomes and the potential for zero carbon urban blocks. The early-stage action, forward-looking partnership, and computing efficiency of the simulation tools are the keys.

1. Introduction

The study consists of computer simulation of environmental qualities of the development of urban form for the KC-AA1 site in To Kwa Wan, optimising for external environmental comfort conditions. Simulated data were provided as feedback to project designers and managers at early stage, allowing timely design modification to improve performance while maintaining code compliance and the Authority's development priorities.

The Sustainable High Density Cities Lab (SHDC) of the HKUrbanLab, Faculty of Architecture has developed a framework software and hardware to conduct computer-based urban environment assessment. The team has a number of in-house developed software tools. CityComfort+, a computer simulation tool that is uniquely equipped to assess urban climate and human responses at micro-scale. It has collaborated with the Cardiff University in the further development of HTB2 and VIRVIL, to model urban scale thermal and energy performance.

The modelling framework has been engaged by the URA (Urban Renewal Authority) Hong Kong to provide quick computer simulation tools that can be more readily applied for environmental analysis to enhance planning and design works at early design stage, particularly for the district-based redevelopment project in To Kwa Wan, namely KC-AA1. The overall aim is to apply a holistic planning approach. The objectives include creating a smart, sustainable and walkable neighbourhood and to enhance amenity value of public open space within the high density urban context of Hong Kong. The existing site to be redeveloped has an area of 57,800 ㎡ is shown in Figure 1. It mainly consists of residential dwellings of 10 floors or below, mostly built at the middle of 20th century. The initial scheme as proposed by the design team features a high-rise (100m), high-density (FAR=9.0), mixed use development as illustrated in Figure 2.

Figure 1. Existing condition of the To Kwa Wan site in Hong Kong

Figure 2. Initial scheme proposed by the design team.

The research aimed to investigate:

(i) How simulation tools can inform measurable urban environmental performance instead of merely adding a "green label?

(ii) How much can a passive design strategy improve key environmental indicators for an urban redevelopment project in Hong Kong's high-density neighbourhoods?

(iii) To what extent can a high density urban block generate energy using building integrated renewables?

This paper presents some preliminary findings.

2. Method

The primary method employed is computer simulation using software tools such as CityComfort+, FlowDesigner and DIVA-for-Rhino, together with HTB2-VIRVIL for energy simulation. Performance criteria for pedestrian thermal comfort, wind speed, daylighting, and visual comfort (glare) were established for key open spaces. Simulation runs were conducted for the existing condition and a range of design options. The software used is summarised below.

CityComfort+ Rhino Software Plugin is a design and planning tool jointly developed at the University of Hong Kong and Harvard University [3], [2]. To model micro-scale environmental parameters, including human perception of thermal comfort, day-lighting, wind speed, solar radiation, air temperature, air quality, and noise, etc. [1].

FlowDesigner is a 3D simulation software for air flow and thermal/contamination distribution developed by Advanced Knowledge Laboratory (AKL). The software uses Computational Fluid Dynamic (CFD) models developed to assess airflow in urban areas. FlowDesigner is user-friendly, adaptable, and practical CFD software with a high-precision and high-performance algorithm based on SIMPLE-C with automatic mesh generation, which allows designers to rapidly simulate, achieve high-accuracy result and visualize wind flow / temperature distribution in urban areas.

HTB2 – VIRVIL is a building energy model, HTB2 [5], which is able to simulate the annual hourly thermal performance of the building, using local weather data, building construction details and occupancy profiles. HTB2 has been developed at Cardiff University over a period of nearly 40 years, and has undergone extensive testing, validation, including the IEA Annex 1 [8], IEA task 12 [6] and the IEA BESTEST [7]. HTB2 – VIRVIL is an extension of single building energy modelling to apply to multi-building scale developments, typically up to a few hundred buildings, which could be a new or existing development, or a mix [4].

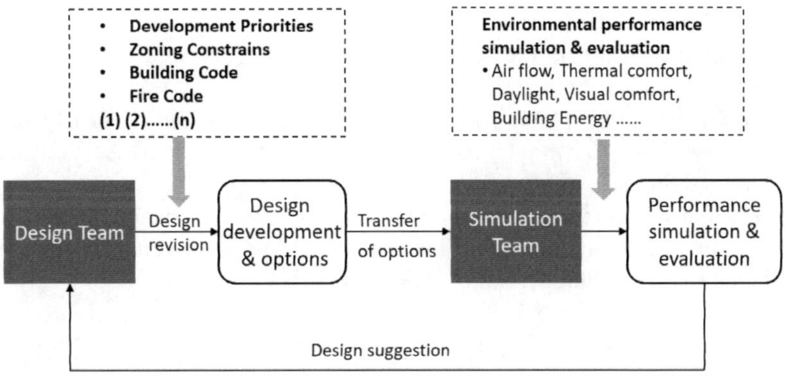

Figure 3. Simulation interaction between the design and simulation team

3. Results

Pedestrian Simulation software were applied to predict performance indicators, i.e. urban microclimate, pedestrian thermal comfort, and building energy use, for various design options. Results were provided as feedback to project designers, allowing timely modification. The procedure repeats itself until performance gains converge. A summary of the iterative design revision informed by simulation results is provided in Figure 4 below.

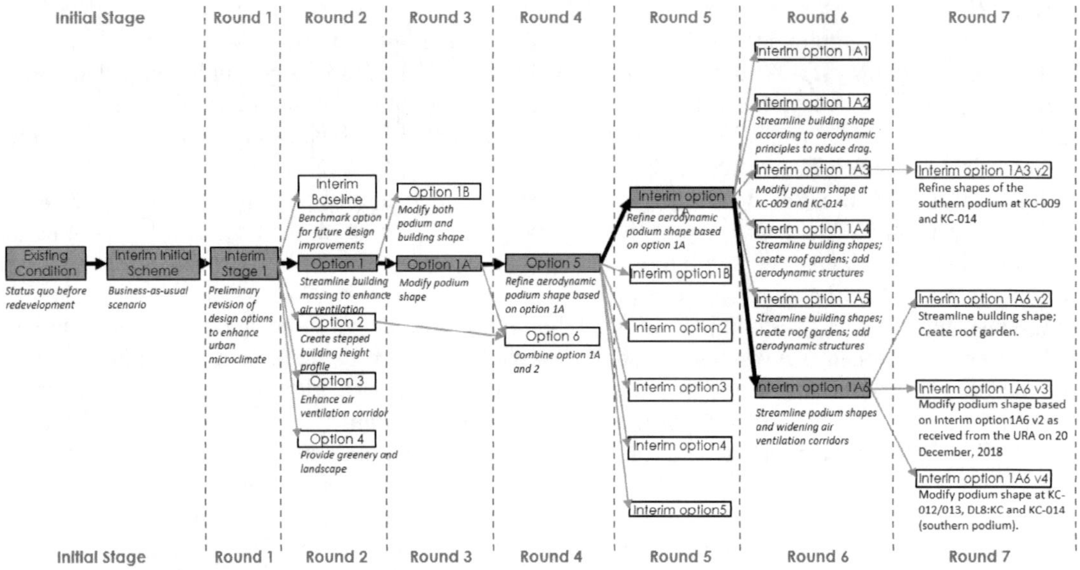

Figure 4. Roadmap for applying the simulation-design workflow to evaluate and revise intermediate options

3.1. Pedestrian-level wind assessment

Figure 5 shows the pedestrian-level (1.5m) wind speed by open space and streets are for the Existing Condition, Initial Scheme and Interim Baseline Scheme.

Table 1 shows significant improvements of pedestrian-level wind speed are expected for the Interim Baseline Scheme in comparison with the Initial Scheme and the Existing Condition.

- On-site mean wind speed is 1.07 m/s for the Interim Baseline Scheme, a nearly 70% improvement compared with those of the Existing Condition (0.63 m/s) and 40% increase compared with those of the Initial Scheme (0.89 m/s)
- The percentage of site area with wind speed above 1.1 m/s, the acceptable threshold for hot season, is 41% for the Interim Baseline Scheme. This is a significant improvement compared with those of the Initial Scheme (35%) and the Existing Condition (19%)

Table 1. Statistics of simulated wind speed results for existing, initial and interim baseline scenario

Scenarios	Mean wind speed on-site (m/s)	% of area of wind speed above 1.1 m/s	Mean wind speed at Urban Square (m/s)*	% of area of wind speed above 1.1m/s at urban square *	Size of Urban Square (m^2)	Building Footprint Area(m^2)
Interim baseline	1.07	41%	1.19	65%	2,092	31,656
Initial scheme	0.89	35%	1.45	98%	388	35,192
Existing condition	0.63	19%	N/A	N/A	0	34,380

* The size of Urban Square varies by schemes, therefore the mean wind speed and area percentage of wind speed >1.1 m/s cannot be simply compared across the board.

Table 2. Statistics of simulated outdoor thermal comfort results for existing, initial and interim baseline scenario

Scenarios	Yearly	Spring	Summer	Autumn	Winter	On-Site Mean UTCI during hot season (May-Oct.)
Interim Baseline Scheme	50.96%	59.46%	6.46%	39.00%	98.90%	28.8°C
Initial scheme	51.83%	61.04%	6.13%	40.63%	99.50%	29.3°C
Existing Condition	50.56%	55.22%	5.05%	42.14%	99.83%	29.7°C

A main contributor to pedestrian-level wind improvement is the reduced building ground coverage in the Initial Scheme and the Interim Baseline Scheme compared with the Existing Conditions, despite the fact that the former two features more building floor area than the later. The dense, medium-height building fabrics of the Existing Condition blocked wind from penetrating the neighbourhood. Figure 6a to b shows the percentage of comfortable hours that falls into acceptable thermal conditions (UTCI 15.2-28.8°C) for spring, summer, autumn and winter, for the Existing Condition, the Initial Scheme and the Interim Baseline Scheme. Table 2 shows that the Interim Baseline Scheme is expected to reduce summer time heat stress compared with the other two On-site Mean UTCI equivalent temperature during the hot season (May.-Oct.) is expected to be 28.8 °C, which is 0.9 °C cooler from those of the existing condition.

Figure 5. Simulated wind speed contours for Existing Condition, Initial Scheme and Interim Baseline Scheme

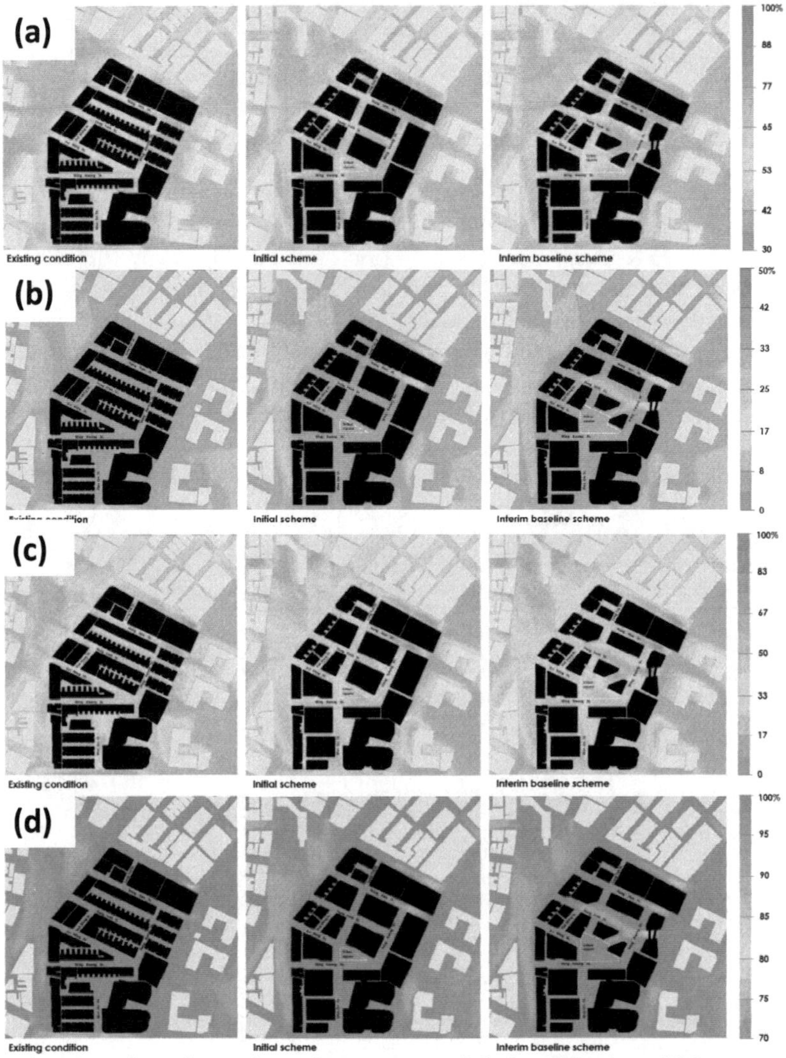

Figure 6. Simulated outdoor thermal comfort contours (%) for (a)Spring, (b)Summer, (c)Autumn, (d)Winter for Existing Condition, Initial Scheme and Interim Baseline Scheme

3.2. Passive design principles

A comparison of the mean thermal comfort conditions aggregated across the core KC-AA1 site shows a varied performance level across a range of building form and layout options (Figure 7). The options are: existing condition, initial scheme, interim baseline, interim option 2, interim option 3, and interim option 1A6.

While the existing condition features a challenging mean UTCI equivalent temperature of 29.7 °C, the Interim Option 1A6 is expected to be 28.7 °C, a one-degree reduction from the status quo (in line with the benchmark set previously). The improvement is largely due to the acceleration of on- site wind, with Option 1A6 being the only option with a satisfactory on-site mean wind speed above 1.1 m/s.

3.3.

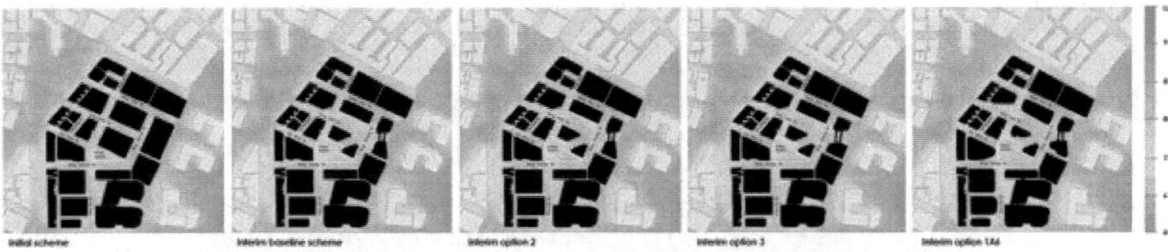

Figure 7. Simulated thermal comfort conditions across design options

Thermal comfort performances by option (Figure 8) show improvements from the interim baseline scheme in hot season. This suggested that streamlined podium shapes and reduced building footprints have marked effects on moderate thermal heat stress. Compared with other options, interim option 1A6 achieved the best performances in relation to thermal comfort. The second best is option 2. The increment, although small in absolute percentages, is statistically significant and is expected to deliver sizable impact at block-scale.

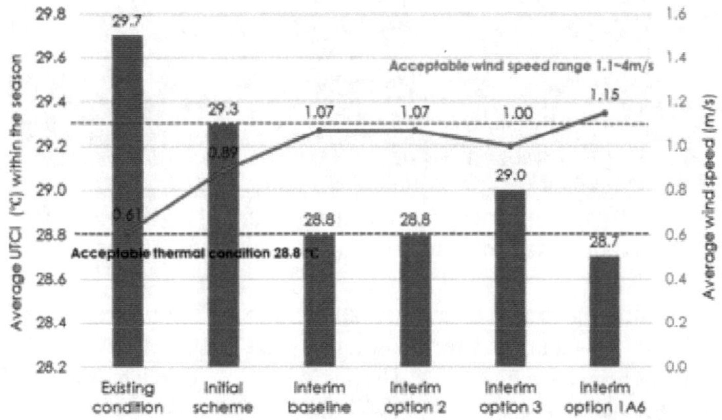

Figure 8. Summary of on-site average wind speed and equivalent temperature during hot season across design options.

Wind ventilation corridor is an effective approach to introduce cooler wind into built areas. Suitable disposition of building blocks could help effective air flows around buildings in desirable directions. To minimize obstruction of airflow, the axis of the building blocks should be parallel to the prevailing wind.

1. Wind speed at 50% of the site area falls within 1.1~ 4.0 m/s under site prevailing wind conditions.
2. 80% of the areas within key open spaces, i.e. the Urban Square, has wind speed within 1.1~ 4.0 m/s under annual prevailing wind conditions.
3. Reduction of the site-wise mean UTCI equivalent temperature in hot season by 1.0 °C, from 29.8 °C of the existing condition to 28.8 °C, within the range acceptable by Hong Kong's local communities.

4. Energy use and renewable energy generation

Future urban blocks may be required to generate renewable energy, which for solar energy, will be affected by overshadowing. Computer simulations of building energy use for the KC-AA1 development was conducted using Virvil-HTB2 software (Figure 9). The hourly values of solar energy generated from the roof top and wall surfaces of buildings on-site were predicted on an annual basis. The results indicate that the renewable energy generation is sufficient to cover the buildings cooling demand, but not small power appliances (lighting, computers, refrigerators, etc.) which dominates building energy use (Table 3). Despite relative small roof-to-volume ratio in tall buildings which tend to limit the application of on-site renewal energy in high-rise high-density cities, the findings indicate the potential for low-carbon high-rise communities.

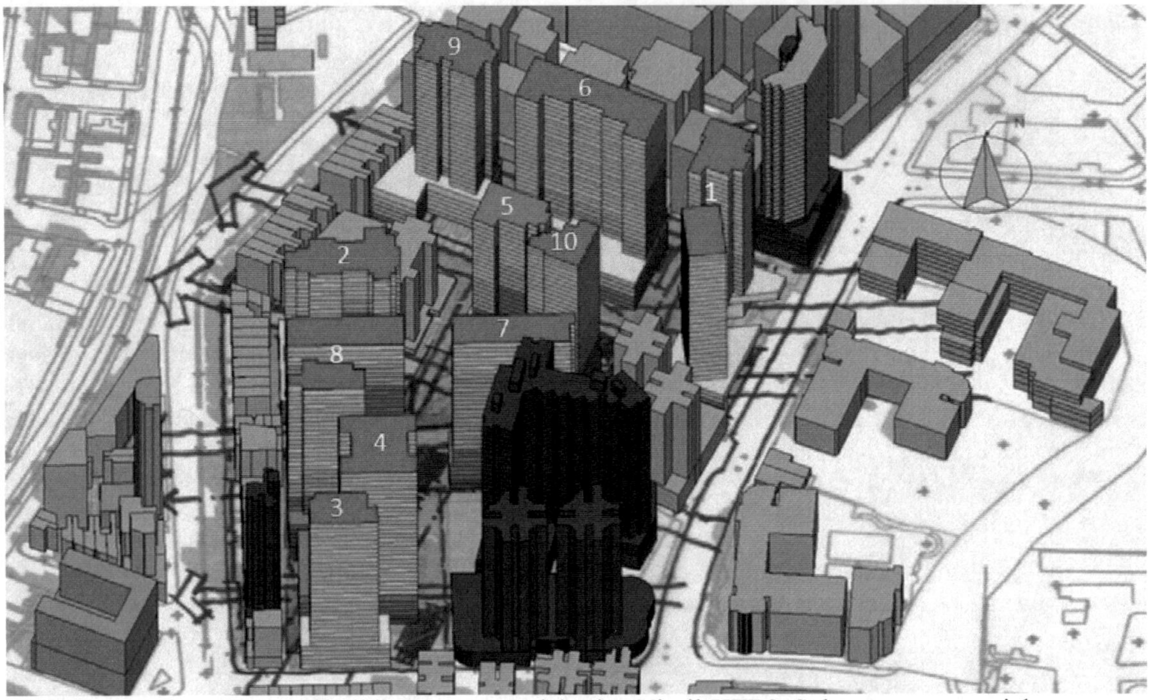

Figure 9. Visualization of building energy model using Virvil-HTB2. Solar energy potentials are color-coded for building façade and roof surfaces.

Table 3. Simulated building annual energy demand and building integrated renewable energy generation for the interim stage 1 option

Buildings	Floor area (m2)	PV generation on all the walls and roof (kWh/floor area/year)	Heating load (kWh/floor area/year)	Cooling load (kWh/floor area/year)	Small power load (kWh/floor area/year)
1	161,000	18.7	0.6	-23.7	43.7
2	31,100	26.3	1.0	-25.0	43.7
3	13,600	37.4	1.1	-27.7	43.7
4	33,500	22.0	0.8	-24.2	43.7
5	13,000	32.1	1.4	-25.0	43.7
6	36,400	25.0	0.8	-25.5	43.7
7	25,000	22.5	1.1	-24.4	43.7
8	39,400	17.2	1.1	-24.0	43.7
9	21,700	32.2	1.0	-26.6	43.7
10	12,700	27.4	1.7	-27.0	43.7

5. Conclusion

Several key findings can be drawn from the verification of design refinements:

- Simulation results show that the informed design can deliver significant microclimate benefits compared with "business-as-usual" scenarios. The improved pedestrian-level wind speed due to streamlined podium shapes and reduced building footprint has marked effects on UTCI reduction and moderate thermal heat stress in hot season;
- Greenery benefits the outdoor thermal comfort but limits in the magnitude of improvements in the thermal condition in hottest summer season;
- Artificial shading canopies are effective in in reducing UTCI and mitigating thermal heat stress in hottest summer season;

- Passive measures and devices cannot guarantee the thermal conditions stay 100% within the acceptable range during extreme hot days. Therefore, there is a need for active device such as evaporative cooling or mechanical ventilation to mitigate thermal heat stress.
- Computer simulation, if integrated in early stage planning and design, shows potentials to enhance urban microclimate performance of design schemes. It can be integrated with energy simulation and the potential for generating renewable energy on the building facades. The early-stage action, forward-looking partnership, and computing efficiency of the simulation tools are the keys.

References

[1] Huang, J. *et al.* (2012) 'a Gis-Based Assessment Method for Mean Radiant Temperature in Dense Urban Areas', in *SimBuild 2012 Conference*. IBPSA-USA, Madison, Wisconsin,USA, pp. 246–254.

[2] Huang, J. *et al.* (2016) 'Outdoor Thermal Environments and Activities in Open Space: An Experiment Study in Humid Subtropical Climates', *Building and Environment*. doi: 10.1016/j.buildenv.2016.03.029.

[3] Huang, J., Cedeño-Laurent, J. G. and Spengler, J. D. (2014) 'CityComfort+: A simulation-based method for predicting mean radiant temperature in dense urban areas', *Building and Environment*. Elsevier Ltd, 80, pp. 84–95.

[4] Jones, P. et al. (2013) 'Intensive Building Energy Simulation At Early Design Stage', in Proceedings of BS 2013: 13th Conference of the International Building Performance Simulation Association.

[5] Lewis, P. T. and Alexander, D. K. (1990) 'HTB2: A flexible model for dynamic building simulation', *Building and Environment*, 25(1), pp. 7–16. doi: 10.1016/0360-1323(90)90035-P.

[6] Lomas, K. J. et al. (1994) Empirical validation of detailed thermal programs using test room data, Volume 1: Final Report. Available at: http://www.ecbcs.org/annexes/annex21.htm.

[7] Neymark, J. *et al.* (2011) 'Multi-zone non-airflow in-depth diagnostic cases', in *Building Simulation 2011 Conference*. Sydney, Australia: IBPSA. Available at: http://www.nrel.gov/docs/fy12osti/51589.pdf.

[8] Oscar Faber and Partners (1980) IEA Annex 1 Computer Modelling of Building Performance: Results and Analyses of Avonbank Simulation. St. Albans, UK.

Climate change and emerging risks. Innovative urban climate services as a strategy to improve resilience of human systems.

Chiara Bernardini[1,3], Giancarlo Paganin[2] and Cinzia Maria Luisa Talamo[1]

[1] Department ABC, Politecnico di Milano, Via G. Ponzio 31, 20133 Milano, Italy
[2] Department DASTU, Politecnico di Milano, Politecnico di Milano Via Edoardo Bonardi 3, 20133 Milano, Italy
[3] Corresponding Author: chiara.bernardini@polimi.it

Abstract. Climate change and extreme natural events are increasingly threatening human systems. Even if greenhouse gases emission were reduced to zero today, global warming and the consequent climate change would continue to affect future generations: as climate is changing, human systems must learn to adapt. This study focuses on the role, the use and the management of climate information to overcome the barriers to climate change adaptation action. The article first addresses the relationship between sustainable development and climate change, regarded as "two mutually reinforcing sides of the same coin". After a short section on the complementarity of mitigation and adaptation strategies, it analyses the barriers to investments in climate adaptation action by public authorities: the uncertainty about the magnitude and the frequency of climate change-connected events, the common wrong perception of unlikelihood of global warming consequences and the intrinsic complexity characterizing human systems are identified as major constraints. These considerations lead to introduce the concepts of soft-resilience and low-regret adaptation strategies. The discipline of climate change adaptation based on soft-approach climate resilience strategies is closely linked to information management, information technologies, communications, logistics, geographic information services and good governance. Indeed, information management is here seen as the pillar of soft-approach strategies. Low-regret strategies are defined as measures that would generate net social and/or economic benefits both under current climate and a range of future climate change scenarios. Section 5 explores potential synergies between Climate Services (understood as any service, equipment, techniques, practical knowledge and skills that can be used to increase climate resilience of human system) and Urban Facility Management services. This study identifies Innovative Urban Climate Services, an integrated system of services addressing both climate and urban issues simultaneously, as an effective, soft-approach and low-regret strategy to overcome the barriers to adaptation action. These solutions provide a general framework allowing a flexible and customized support for human systems in adapting to unavoidable impacts of climate change.

1. Introduction

Earth's climate has been relatively stable for the past 12000 years, and modern life is tailored to that climate [1]. In the last century, human systems face emerging physical and social risks arising from climate change, one of the biggest global challenges of modern world. Climate change has become a central theme in many political and public debates: in the last decade international and intergovernmental institutions and scientific community attention was increasingly drawn to this issue.

Climate change impacts on human systems are beginning to change how urban infrastructure is designed and located, how populations are settled, and their organization [2].

This paper summarizes the initial phase of a three-years research work on climate adaptation strategies for built environment funded by the Italian Ministry of Education (Ministero dell'Istruzione dell'Università e della Ricerca). After a rationale on the link between climate change and sustainable development, the issue of complementarity between mitigation and adaptation measures is approached, followed by an investigation on the barriers to climate adaptation action. Section 3 introduces the concepts of low-regret adaptation strategies and soft-resilience. The aim of the research is then stated: identifying effective strategies centred on the implementation of services to support human systems in coping with climate change. The next sections outline the potential synergies between Climate Services and Urban Facility Management. Lastly, the concept and implementation of Innovative Urban Climate Services, intended as an effective, soft-approach and low-regret strategy to improve resilience of human systems, is discussed.

The interest of this study, in line with the multidisciplinary nature of the issue itself, includes all the aspects of built environment management ranging from tangible to intangible infrastructures management, risk management, digitization and innovation in products and systems. It offers an original standpoint on the issue of sustainable built environment and climate change, exploring possible soft-adaptation solutions to improve resilience of human systems centred on the development of a set of urban services based on typical Urban Facility Management approaches.

2. Complementarity between climate change, sustainable development and built environment

The definition of sustainable development provided by the Brundtland Report in 1987 reads: "Sustainable development is development that meets the needs of the present without compromising the ability of future generations to meet their own needs." [3]. Even if greenhouse gases emission would be reduced to zero today, global warming will continue to affect future generations in the next centuries [1][4]. Climate change is the result of the unsustainable industrial production, unsustainable land use, unsustainable energy consumption, unsustainable resource use globally fostered and accepted by various stakeholders in the last two centuries.

In the UN agreement addressing climate change and promoting sustainable development are regarded as "two mutually reinforcing sides of the same coin" [5]: sustainable development cannot be accomplished without climate action and, vice versa, the Sustainable Development Agenda Goals tackle the main drivers of climate change by reducing greenhouse gases emissions. Climate change is the ultimate sustainable development issue [6].

Which role could built environment play in mitigation and adaption to climate change? The building sector is the main energy-consuming sector worldwide: almost 41% of final energy consumption and comparable contributions to greenhouse gases emissions are attributable to buildings life cycle [7]. Yet, this sector has great potential to significantly reduce GHG emissions [8], enable climate adaptation and enhance resilience of human systems.

3. Emissions mitigation and climate adaptation: adoption of national and local climate plans among European countries

As reported below, NASA stated that even if greenhouse gases emission would be reduced to zero today, global warming will continue to affect future generations in the next centuries [1]. Global climate models and IPCC climate projections support this statement [4]. Even assuming the efforts to reduce emissions demonstrate their effectiveness, climate change effects are inevitable, and complementary actions to adapt to its impacts are needed.

Climate adaptation, understood as "the process of adjustment to actual or expected climate and its effects" [4], seeks to mitigate or avoid climate-related impacts and, in the best-case scenario, to exploit their beneficial opportunities. The Paris agreement proposes a similar and inspiring definition of climate adaptation, identifying it as "changes to the socio-technical systems that support human living in light of inevitable climate change" [9].

So, ultimately, climate changes human systems must learn to adapt. Despite this evidence, data show that this commitment continues to be poorly implemented in practice: a voluntary survey launched by Eurostat and up-to-date as of January 2018 about adoption of national adaptation strategies and plans across 33 European countries reveals that climate change adaptation efforts have increased, particularly over the past decade, but only the 50% of the respondent countries dispose of a national adaptation plan at present (Eurostat, 2018).

What then are the barriers to adaptation action for public administrations and, more in general, for stakeholders? The next section seeks to examine the factors that interacts to impede adaptation planning and implementation.

4. Barriers to climate adaptation action

The IPCC identifies the uncertainties about projected climate change impacts as a major constraint, among others (e.g. limited financial and human resources; limited integration or coordination of governance; different perceptions of risks; competing values; absence of key adaptation leaders and advocates): they interact to impede adaptation planning and implementation [4].

Climate change is an extremely uncertain source of hazard and, thus, difficult to describe and model quantitatively. We need only think of the fact that an integral feature of IPCC reports is the communication of the strength of uncertainties in scientific understanding underlying assessment findings: the "IPCC AR5 guidance note on consistent treatment of uncertainties" was published, where a qualitative level of confidence is expressed for assessment findings (from very low to very high) and, when possible, they are probabilistically characterized with a quantified likelihood (from exceptionally unlikely to virtually certain).

Limitations of available measurements, especially for rare events, make the probability, the frequency of occurrence and the magnitude of such climate change-connected events still partially unknown. There is evidence from the literature of the lack of reliable statistics based on the past, as we are currently experiencing a transition phase of ecosystems. Limited knowledge leads to the common wrong perception of unlikelihood of global warming consequences.

Human systems are complex networks with very specific framework conditions, so the implementation of adaptation strategies is embedded in a complex frame that is influenced by internal and external factors. The issue of complexity should be therefore introduced: multiple communities of people, tangible and intangible infrastructures (built environment) and services characterized by high interdependency make up what we call a "human system". Human systems are characterized by different geographical location, city structure, size and density, urban-rural relationships, socio-economic aspects, cultural habits and backgrounds, human resources, interests of the involved stakeholders, different exposure and vulnerabilities to climate change impacts, level of awareness of climate change, operational capability.

In conclusion, the uncertainty surrounding climate change impacts scenarios, the consequent common wrong perception of unlikelihood of global warming consequences and the intrinsic complexity of human systems are major obstacles to long-term, significant investments in adaptation actions by stakeholders.

In addition to the above constraints, information management-related barriers should be mentioned. Despite the growing availability of climate data, lack of actionable information can be observed: it may require tailoring, interpretation, or integration with other information before it can be useful [10]. In this paper, we seek to identify effective information management-based strategies that facilitate the implementation of adaption measures.

5. Soft-resilience and low-regret strategies to overcome barriers to adaptation action

Facilitating the investments in climate adaptation action is a major challenge for the international scientific community. According to the outcomes of many studies carried out by the World Energy Council, addressing the issue of the reticence of policy makers to invest in climate adaptation measures would require a shift from the concept of hard-resilience (defined as the direct strengthening of structures or institutions in relation to specific stresses, obtained through specific measures to reduce their probability of collapse) to an innovative soft-resilience approach [11]. Soft-resilience is here referred to as the ability of systems to absorb and recover from the impact of disruptive events without fundamental changes in function or structural-measures, relying on preparedness and systemic interventions [12].

The discipline of climate change adaptation based on soft-approach climate resilience strategies is closely linked to information management, information technologies, communications, logistics, geographic information services and good governance. Indeed, information management is here seen as the pillar of soft-approach strategies: it is the basis for the operation of Climate Services, as further explained in the next section.

Hard adaptation strategies rely predominantly on tangible built infrastructures, are resource-consuming and lack flexibility and adaptability to sudden changes in projections of climate change [13]. Given the high grade of uncertainty characterizing climate models, soft adaptation strategies present a great advantage considering their flexibility, reversibility and cost-effectiveness.

Table 1. Proprieties of soft-approach adaptation strategies and hard-approach adaptation strategies.

Propriety	Soft – approach adaptation strategies	Hard – approach adaptation strategies
Flexibility	Flexible, adaptable to sudden changes in climate conditions or projections	Inflexible in the short-term, non-adaptable to sudden changes in climate conditions or projections
Reversibility	Reversible in the short term without adverse impacts; do not narrow future choices	Reversible in the long time with impacts; narrow future choices
Technologies	Rely on simple, low-impact technologies and intangible infrastructures	Rely on complex technologies and human-built infrastructures
Time horizon	Short-medium term investments	Long-term investments
Scale	Small-scale measures	Large-scale measures; large-scale disturbances
Compatibility	Allow coexistence and synergies with other measures	Trade-offs with other measures; narrow future choices
Resources	Resource-saving; cost-effective	Resources-consuming; capital intensive
Type of interventions	Deal mainly with organizational, operational and institutional capacity	Mainly provide for physical interventions
Adjustability	Adjustable to stakeholders needs	Difficult to proportion on the basis of uncertain data
Environmental impact	No environmental drawbacks	Medium to high environmental drawbacks

Soft adaptation strategies may be regarded as low-regret solutions. It is worth introducing here the concept of low-regret adaptation: the IPCC defines it as measures that would generate net social and/or economic benefits both under current climate and a range of future climate change scenarios. Summarizing the UK Adaptation Sub-Committee definition this type of solutions can be described as adaptive measures for which the associated costs are relatively low and for which the benefits, although primarily realized under projected future climate changes, may be relatively large. The key is maximizing the positive impacts of climate adaptation measures.

Table 2. Proprieties of low-regret and high-regret adaptation strategies

Propriety	Low-regret adaptation strategies	High-regret adaptation strategies
Resources	Relatively low; resource-saving; cost-effective	High investments; resource-consuming; capital-intensive
Purpose	Often multi-purpose; address more than one societal objective	Specific purpose in specific conditions
Benefits	Benefits under current climate and a range of future scenarios	Benefits under current specific climate

Adjustability	Incrementable; often modular solutions	Rigid solutions

Typical examples of low-regret strategies are restrictive land use planning; insurance, early warning and evacuation scheme; green infrastructures; improved building standards. As can be seen from the tables, soft solutions possess many proprieties that make them potential low-regret strategies.

The aim of this study is here declared: providing a methodology, a conceptual framework for the development of a set of strategies to enhance human systems resilience, taking into consideration the principles of soft-approach and low-regret solutions. The proposed strategies are centred on the enhancement of urban services. The next section first defines what Climate Services and Urban Facility Management are to then explore potential synergies arising from their integration.

6. Climate Services and Urban Facility Management

6.1 Climate Services

Interest in Climate Services has grown over the past decade, particularly with the launch of the Global Framework for Climate Services, a global partnership of governments and organizations that produce climate information and services, in occasion of the 3rd World Climate Conference. The aim was to turn scientific information into operationally available information and services that would help society to better cope with climate variability and change [11]. Later on the EU declared in the European Roadmap for Climate Services, published in 2015, its will to support the development of a market for Climate Services with and for diverse categories of users. In this direction numerous knowledge platforms and initiatives were launched, to foster the application of climate information into effective adaptation and mitigation actions: Copernicus Climate Change Services (C3S), that aims at providing open source climate information; Climate-ADAPT, presented as "a climate knowledge intelligent portal"; climate-knowledge and innovation communities like Climate-KIC, supported by the European Institute of Innovation and Technology (EIT); and many others.

As already mentioned above, Climate Services show limitations in their effectiveness when it comes to deliver usable information for adaptation measures implementation; making climate information applicable may require tailoring, interpretation, or integration with other information [10].

The definition of Climate Services provided by the European Commission includes a broad number of applications obtained from "the transformation of climate-related data - together with other relevant information - into customized products such as projections, forecasts, trends, economic analysis, assessments, counselling on best practices development and evaluation of solutions and any other [...] that may be useful for the society at large". It introduces the possibility to integrate climate-related information with other relevant information from different sources. Kathleen, Swartling et al. broadly conceptualize the developments of Climate Services as 'climate knowledge systems" [10], stressing the importance of the three-phases transformation from data to usable information to knowledge.

For the purpose of this study it resulted extremely useful interpreting a definition provided by the CTCN (Climate Technology Centre and Network) that identifies climate technologies, in which the emerging idea of a new, innovative set of Climate Services generated by the synergy with Urban Facility Management services is well represented by the following adapted definition: any service, equipment, techniques, practical knowledge and skills that can be used to increase climate resilience of human system. This concept will be further explained in section 6 of this work, after having delineated what we intend with Urban Facility Management.

6.2 Urban Facility Management

UNI 11447:2012 defines Urban Facility Management as an integrated management service for the operation, the serviceability and the valorisation of urban facilities. [14] Urban facilities include: i) Services for the Territory: monitoring and maintenance of urban green, plants, road network; ii) Services for buildings, plants, infrastructures: monitoring and maintenance of water systems, sewers, infrastructure for public transport, security technology installations, telephone and data transmission networks, availability and emergency intervention; iii) Services for the environment and the community:

cleaning and environmental hygiene, waste collection and disposal, space management; iv) Government services: technical and inventory registers, implementation and management of information systems, operational centres.

Urban Facility Management is increasingly required not only to satisfy the primary requirement – guaranteeing an efficient and effective management of urban facilities through conditions monitoring and maintenance operations – but also to ensure a quick reaction and practical response in case of extreme natural events. In the light of raising hazard magnitude and frequency due to climate change, this aspect becomes of fundamental importance.

6.3 Potential synergies between Climate Services and Urban Facility Management

Optimizing Urban Facility Management by integrating it with climatic information through the use of current ICTs complies with the definitions of soft-resilience and low-regret strategy: this strategy relies on information management practices and gives back benefits (a better level of service and users comfort) whatever the climate conditions.

The table below provides some examples of adaptation strategies classified by the definitions introduced in section 4 of this paper, showing the overlaps between soft/hard regret strategies and low/high regret strategies.

Table 3. Matrix of adaptation strategies: examples and classification by definitions.

Strategies	Soft-approach adaptation strategies	Hard-approach adaptation strategies
Low-regret adaptation strategies	UFM integrated with CS: monitoring and maintenance practices for urban facilities	Green infrastructures; Improved building standards; Improved drainage systems
High-regret adaptation strategies	Soft-approach adaptation strategies are not high-regret strategies by definition	Flood protection walls; Improved irrigation systems; Improved urban infrastructures

The heterogeneity of services covering the urban scale requires the formulation of an overall framework in which each individual service is correctly identified through an accurate mapping of the package of goods and services and the relationships among them. This allows to identify and analyse services among which synergies are necessary or convenient.

This is the case of Urban Facility Management and Climate Services. The serviceability of existing urban facilities can be enhanced by linking the information provided in the form of climate models by scientific bodies, and elaborated by services providers to deliver Climate Services, with systems of Urban Facility Management. This allows to anticipate harmful events, raise the grade of preparedness, enable a quick response in case of emergency and, consequently, improve the resilience degree of human systems. Section 6 presents the outcome of this integration.

7. Innovative Urban Climate Services as an effective soft-approach and low-regret strategy to improve resilience of human systems

Despite Climate Services have been broadly investigated in the scientific literature, their grade of integration in human systems management is still insufficient. The lack of operative tools for their application is a main limiting factor of such operationalization [15].

The previous section explores potential synergies between Climate Services and Urban Facility Management systems: integrating the information provided in the form of Climate Services with operations and procedures typical of Urban Facility Management approaches gives rise to the implementation of a new model of urban services based on the effective information sharing to support systematic monitoring and interventions. Human systems needs and requirements for improving their resilience to the impacts of climate change are highly context specific: much meta-information is needed to transfer adaptation strategies [16]. Meteorological information as well as maps, risk and vulnerability analyses, impacts assessments (Climate Impacts Models), and long-term projections and scenarios (Climate Models) crossed with non-meteorological data (such as settlement in high-risk areas, road and infrastructure maps, urban facilities conditions), may be combined depending on specific user needs [17].

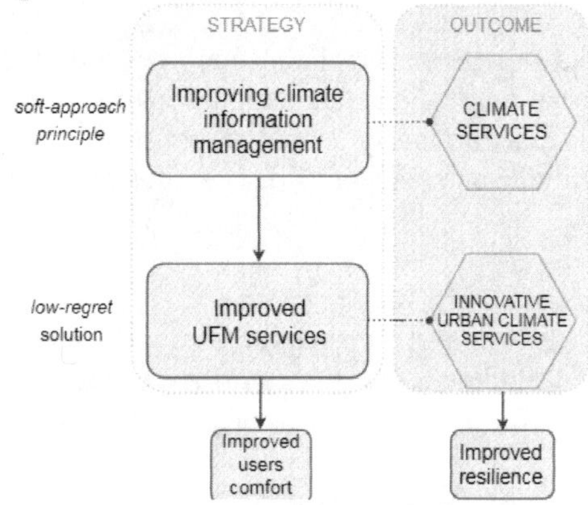

Figure 1. Synergies between Climate Services and Urban Facility Management.

The incorporation of a new operational entity responsible for the processing of multiple information should be envisaged: it will elaborate information making it usable and applicable for various purposes, tailoring it to specific users' categories needs.

The outcomes of this work will include improved conventional urban services that enhance the utility of existing urban facilities and a set of new urban services to cope with climate change trends and extreme events; these services will allow a constant monitoring of climate conditions and urban facilities conditions. The set of integrated Urban Facility Management systems may then be "fed" with geo-spatial and real-time status information extracted from the numerous ICTs systems incorporated in urban space: this would allow to take a step forward from the conventional management method that usually relies on manual maintenance following periodical inspection reports [15].

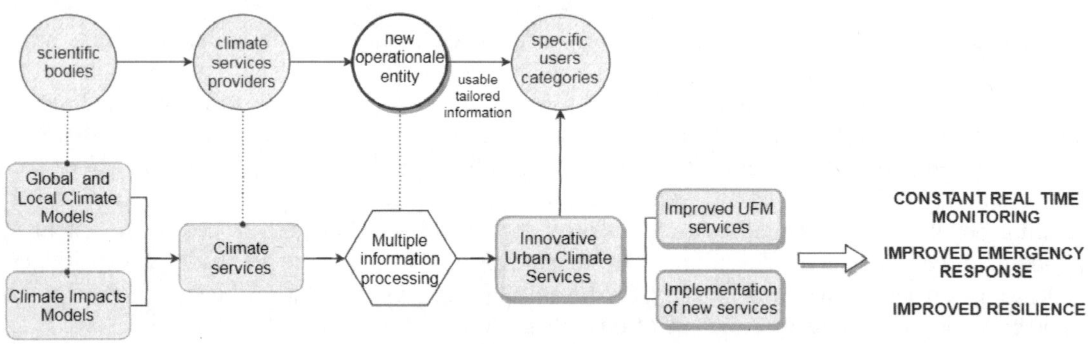

Figure 2. Innovative Urban Climate Services generated from the synergies between UFM and Climate Services.

This paper does not provide by choice a list of Innovative Urban Climate Services due to the fact that one-fit-all solutions would be in contrast with the logic of this research itself. The intrinsic complexity of human systems makes it necessary to consider the specificities of each context and tailor services on user's needs. We leave the presentation of a repertoire of case studies and applications for next publications.

8. Conclusions

This paper seeks to provide an original standpoint to address the issue of how to facilitate the investments in climate adaptation by stakeholders. It introduces the concepts of soft-approach adaptation measures and low-regret solutions and it applies them in the identification of effective short-medium term strategies to be included in the mix of strategies to improve resilience of built environment and human systems. It identifies Innovative Urban Climate Services, an integrated system of services

IOP Conf. Series: Earth and Environmental Science **329** (2019) 012048 doi:10.1088/1755-1315/329/1/012048

addressing both climate and urban issues simultaneously, as a strategy to overcome the barriers to adaptation action (uncertainty, wrong perception of unlikelihood of global warming consequences and intrinsic complexity of human systems) thanks to their flexibility, reversibility and cost-effectiveness. Maximizing the positive impacts of climate adaptation measures means promoting sustainable development. The forecasted side-benefits arising from the implementation of the solution presented are improved efficiency and effectiveness of urban facilities management, which may entail an improved urban facilities users' comfort level and a constant monitoring of climate conditions and urban facilities conditions; this may ensure an improved emergency response that will lead to improved resilience of human systems to climate change hazards. This study paves the way to the development of a set of Innovative Urban Climate Services allowing a flexible and customized support for human systems in adapting to climate change.

References

[1] National Aeronautic and Space Administration – NASA (no-date) *Global Climate Change: Vital Signs of the Planet*. www.nasa.gov
[2] Intergovernmental Panel on Climate Change – IPCC (2012) *Managing the Risks of Extreme Events and Disasters to Advance Climate Change Adaptation. A Special Report of Working Groups I and II of the Intergovernmental Panel on Climate Change.*
[3] United Nations – UN (1987) *Report of the World Commission on Environment and Development: Our Common Future*. Brundtland Report.
[4] Intergovernmental Panel on Climate Change - IPCC (2014) *Summary for policymakers*. In: Climate Change 2014: Impacts, Adaptation, and Vulnerability.
[5] United Nations – UN (2016*) Transforming our world: the 2030 Agenda for Sustainable Development.*
[6] United Nations - United Nations Framework Convention on Climate Change – UNFCCC (2006) *Climate Change and Sustainable Development: an international workshop to strengthen research and understanding*. Keynote speech by Richard Kinley, Officer-in-Charge, UNFCCC, New Dehli, 7 April 2006.
[7] European Commission – EC (2015) *A European Roadmap for Climate Services Roadmap.*
[8] United Nations Environment Programme – UNEP (2013) *Sustainable Buildings and Climate Initiative.*
[9] United Nations - United Nations Framework Convention on Climate Change - UNFCCC (2015) Adoption of the Paris agreement. In: United Nations Framework Convention on Climate Change (UNFCCC). United Nations, Paris, France.
[10] Kathleen, E., Swartling, M. et al. (2019) *Identifying climate service production constraints to adaptation decision-making in Sweden.* Environmental Science and Policy.
[11] World Meteorological Organization – WMO (2016) *Climate Services for Supporting Climate Change Adaptation. Supplement to the Technical Guidelines for The National Adaptation Plan Process.*
[12] Proag, V. (2014) *Assessing and Measuring Resilience*. Procedia Economics and Finance.
[13] Sovacool, B. (2011) *Hard and soft paths for climate change adaptation.* Climate policy.
[14] UNI11447:2012 (2012) *Urban Facility Management Services. Guidelines to set and program contracts.*
[15] Lee, J., Jeong, Y. et al. (2013) *An integrated approach to intelligent urban facilities management for real-time emergency response.* Automation in Construction.
[16] Cortekar, J., Bender, S. et al. (2016) *Why climate change adaptation in cities needs customised and flexible climate services.* Climate Services.
[17] Global Framework for Climate Services – GFCS (no-date) www.wmo.int

IOP Conf. Series: Earth and Environmental Science **329** (2019) 012049 doi:10.1088/1755-1315/329/1/012049

A feasibility study to evaluate the potential replication of an energy positive house in the UK

X Li, J Patterson, E Coma Bassas and P Jones

Welsh School of Architecture, Cardiff University, Bute Building, King Edward VII Avenue, Cardiff CF10 3NB, United Kingdom

Junglix@cardiff.ac.uk; patterson@cardiff.ac.uk

Abstract. This paper presents an evaluation of the economic and technical feasibility of a renewable-led low carbon house in the UK. A holistic systems-based approach to achieve energy positive house has been taken. Long-term economic and technical feasibility analysis have been carried out based on a validated thermal and energy model of the house. The economic analysis employs the Return on Investment (ROI) method and considers changes to government financial support and technology progress over time. Results show that the extra investment on the house, compared with that for building a standard social house of similar size, can be paid back within the system lifespan under both the old Feed-in Tariff and its proposed replacement with reduced financial support. Variants examined in the technical feasibility analysis include housing type, orientation and location. Results show that the house can be replicated to achieve an energy positive performance for all variant combinations. Among the variants, location has the highest impact on building performance including annual electricity import, CO_2 emission and electricity self-sufficient ratio, with South UK performing better than North UK. This implies that more efficient or advanced technologies would be needed to achieve similar building or system performance in areas of unfavourable climate. The outcome of the research has demonstrated the affordability of the energy positive house, and the technical feasibility of its replication with different housing types, orientations and locations in the UK. This study supports the wide scale replication of this affordable systems-based approach in domestic building design and construction when incorporating appropriate technologies.

1. Introduction

There have been worldwide efforts to employ renewable energy systems for tackling the fuel poverty issue, either at a community scale [1] or for single domestic building [2]. Around 291,000 households, equivalent to 23% of all households in Wales, still live in fuel poverty [3], and for England the estimated proportion is 11%, approximately 2.5 million [4]. In the UK, renewable technologies have been employed since the 1980s at community scale, for example in the Linford Solar Court (1979), the Rainbow Housing project (1980), the Giffard Park Housing (1984), the BedZed Development (2001), the Hockerton Housing (2002) or the Lammas Community (2009). Also at the single building scale, with innovative house projects such as the Bradville Solar House (1972), the projects presented in the Homeworld 81 exhibition, the House for the future (2000) or the Ty Unnos (2011). There have been UK government supported schemes to provide financial support for small scale renewable energy generation, such as the Feed-in-tariff (FIT) and its potential replacement, the new supply-led 'Smart Export Guarantee' (SEG) scheme [5], and the Renewable Heat Incentive (RHI). Driven by government incentive mechanisms and the demand on minimizing fuel poverty, low carbon domestic retrofit in the UK has been moving towards a systems-based approach with regards to renewable energy integration. There have also been attempts to integrate a systems-based approach in new houses. The SOLCER house, built in Bridgend (South Wales) in 2015, was designed and constructed by the Welsh School of Architecture (WSA) at Cardiff University to be an energy positive house [6]. The SOLCER house is a new build house that employs a whole house systems approach, integrating high performance building fabric, an efficient service system, renewable energy generation and energy storage [7]. Unlike most renewable energy applications in domestic buildings, where renewable energy generation only serves as an auxiliary energy source, the 'all electric' SOLCER house can meet around 70% of its total energy needs and produce more power export than import over an annual period [8]. The total floor area of the building is 100m^2 and is designed to meet social Welsh Housing Quality Standard (WHQS).

Figure 1. The SOLCER house by WSA.

The design of SOLCER house incorporates existing and emerging technologies and design approaches developed through the LCRI Low Carbon Buildings Programme [6]. The technologies have been integrated and optimized through a systems-based approach to meet the house's reduced energy demand using renewable energy supply and energy storage. The house is built with high performance Structural Insulated Panels (SIPs) for walls and roof (U-values = 0.12 W/K/m^2), a highly insulated floor slab (U-values = 0.10 W/K/m^2) and composite windows (U-values = 1.12 to 1.51 W/K/m^2). The measured air leakage after completion was 2.91 m^3/h/m^2 at 50 Pa. Space heating and domestic hot water (DHW) are provided through Integrating Transpired Solar Collector (TSC) and a GENVEX Combi-unit which combines Mechanical Ventilation Heat Recovery (MVHR), an Exhaust Air Heat Pump (EAHP) and thermal water storage. The dark coloured TSC is located on the upper floor of the south-facing facade (Figure 1), which preheats the outdoor air supply prior to the MVHR, but also reduces heat losses through the wall by providing an additional external layer on the south facing external wall. A 4.3kWp integrated solar photovoltaic roof and a 6.9kWh lithium battery storage are assisted by grid import to meet the electricity demands. The system is designed to import electricity from grid only when there is no PV generation and the battery is fully discharged.

The performance of the SOLCER House has been fully monitored since its construction and has proved to perform as expected, however, evidence is required to prove its cost-effectiveness compared with new build social houses, and whether the systems-based approach works for different housing types, orientations and locations in the UK. This research studies the economic and technical feasibility for the SOLCER House's replication across the UK. Work is carried out from the following aspects, 1) a long-term economic analysis to produce a Return on Invest (ROI) graph of the system compared to a best practice new build social house; 2) a technical feasibility analysis to apply the system to different housing types, orientations and locations in the UK.

The UK government has reduced its financial support for renewable energy generation. FIT closed on 31st March 2019 and the alternative SEG scheme proposes to pay only for power export. RHI is not applicable for the SOLCER house as it uses an exhaust air heat pump. The economic analysis in this paper will discuss impacts resulting from changing financial support over time. By examining a worst-case economic scenario with limited income from power export and no RHI gain, it will provide an unbiased and referable case for international researchers living in countries where there is limited or no financial support for domestic renewable system application.

2. Methodology

In preparation for the economic and technical feasibility analysis, a thermal and energy model was developed in the dynamic tool HTB2 [9], to predict energy performance of the as-built SOLCER house. The model has undergone several calibration processes based on survey and monitoring data collected from the SOLCER House over a two years period, proving to be reliable for scenario modelling [10]. HTB2 is typical of the more advanced numerical models, using input data including hourly climate for the location, detailed building materials and construction, spatial attributes, system and occupancy profiles, to calculate the energy required to maintain specified internal thermal conditions. It can also deal with most thermal related systems such as Mechanical Ventilation Heat Recovery (MVHR) and Transpired Solar Collectors (TSC) when relevant technical parameters are available. As a result, HTB2 provides more complete and dynamic prediction of the performance of individual system components. Its results can be further processed to calculate the resulting energy supply, CO2 emission, operating cost and electricity self-sufficiency.

2.1. A long-term economic analysis

The long-term economic analysis is carried out using the Return on Investment (ROI) method. ROI measures the amount of return on a particular investment, relative to the investment's cost. A simple formula is presented in Equation (1). The ROI varies over time, along with the increasing of gain or loss from investment and the adding of extra investments if applicable.

ROI = (gain from investment –cost of investment) /cost of investment (1)

For this research, the investment's cost will be the extra money needed to build the SOLCER house compared with building a best practice social house of the same size, which complies with Welsh Building Regulations [3]. Differences between the two houses in terms of fabric, system element and total cost are presented in table 1. MVHR is employed in the social house to achieve similar indoor air quality as that of the SOLCER house. The total cost of the social house refers to the published cost statistics for Welsh new build dwellings, mainly social houses [11].

Table 1. A summary of the differences between the SOLCER house and a best practice new build social house of the same size in Wales, UK.

	SOLCER house	A best practice new build social house
Building fabric element	Wall U-value: 0.12 W/K/m^2	Wall U-value: 0.18 W/K/m^2
	Roof U-value: 0.12 W/K/m^2	Roof U-value: 0.13 W/K/m^2
	Floor U-value: 0.10 W/K/m^2	Floor U-value: 0.13 W/K/m^2
	Window U-value: 1.12-1.51 W/K/m^2	Window U-value: 1.40 W/K/m^2
	Air permeability: 2.91 m^3/h/m^2	Air permeability: 5.0 m^3/h/m^2
Service and renewable system	TSC area: 14.0m^2;	No TSC;
	GENVEX Combi-unit, including MVHR, ASHP (COP 3.21) and hot water tank;	Gas fired system boiler with hot water tank, radiators;
		With MVHR;
	4.3kW$_p$ building integrated solar PV;	No solar PV;
	6.9kWh lithium battery.	No battery storage.
Total cost	£1380 /m^2	£ 1240 /m^2 [11]

Financial benefits from the extra investment on the SOLCER house, compared with the best practice social house, are mainly associated with the operating energy cost savings from the systems based approach of integrating reduced energy demand, renewable energy supply and energy storage. Income from government schemes such as the old Feed-in Tariff or the potential new SEG scheme are considered, together with costs for maintaining the two properties. The total gain from investment can be obtained using equation (2). Expenses for maintaining the components in SOLCER house are added, and taken as a negative gain. While any maintenance cost of the social house is taken as a return of the investment. Replacement expenses are included in the maintenance costs for components with lifespans shorter than the designed system lifespan, which is assumed to be 25 years for this case study. A summary of the lifespans, maintenance and replacement costs of individual components is presented in Table 2. The costs for maintenance and replacement were obtained from market searches and suppliers.

Table 2. A summary of the low carbon technologies.

Components	Specifications	Maintenance (£/yr)	Replacement (£)	Lifespan
External wall insulation	SIPs Panels	0	0	>= 50
TSC	An area of 14.0 m^2	0	0	>= 30 [12]
Combi-unit	HP: COP 3.21; MVHR; DHW tank: 185L.	40 (filters)	0	20-25 [13,14]
PV	4.3kWp BIPV, module efficiency 15.0%.	20 (cleaning)	Inverter replacement every 15 years for £300 per kWp PV.	25-30 [13,15,16]
Battery	Lithium, 6.9kWh	0	6,500	10-20 [17,18, 19]
MVHR	SPF 0.56W/l/s, 91% efficiency	40 (filters)	2,500	15 [20]
Gas boiler	System boiler with tank (90% efficiency)	100 (inspection)	4,000	15-20 [20, 21]

The research also considers the degradation of component performance over time, such as annual degradation rates of 0.7% for PV capacity [22], 2-3% for lithium battery capacity if well managed [23]and for every 10 year, 6-10% for boilers efficiency and 10-24% for air cooled chillers COP [24]. A standard UK discount rate of 3.5% [25] is used to calculate the present value of future cash flows, therefore enable a comparison of investments and benefits in the long term. This discount rate indicates how rapidly the value today of a future real pound decreases by time, and it applies to real values with the anticipated effects of inflation already removed. In this research, it is assumed that inflation will affect all prices equally, in which case the real values of future cash flows will equal to the current prices, so the inflation rate is not needed for the calculation.

$$C_{gain} = \Sigma_1^t (C_{si}/(1+r)^{i-1}) + \Sigma_1^t (C_{pvi}/(1+r)^{i-1}) - \Sigma_1^t (C_{mi}/(1+r)^{i-1}) \quad (2)$$

$$C_{si} = \Sigma_1^h (E_{zi}P_z) - \Sigma_1^m (E_{yi}P_y) \quad (3)$$

$$C_{pvi} = C_{pv} (1-\alpha)^{i-1} \quad or \quad C_{pvi} = C_{pvEi} \quad (4)$$

$$C_{mi} = \Sigma_1^n M_x - \Sigma_1^a M_b \quad (5)$$

C_{gain} – total gain from the extra investment on SOLCER house, compared with that for building the social house;

r –discount rate;

C_{si} – operating energy cost savings on year i;

C_{pvi} –income from Solar PV of year i when i is within the tariff timescale, 20 years;

C_{mi} – net expense from maintenance on year i;

t – the last year of the selected time period, for example the designed system lifespan;

h – total number of fuels used in the social house;

E_{zi} – use of fuel z in the social house on year i, obtained from the calibrated model with consideration to system performance degradation over time;

P_z – price of fuel z;

z – fuel No., ranging from 1 to h;

m – the total number of fuels used in the SOLCER house;

E_{yi} –use of fuel y in the SOLCER house on year i, obtained from the calibrated model with consideration to system performance degradation by time;

P_y – price of fuel y;

y – fuel No., ranging from 1 to m;

C_{pv}– 1st year generation and export incomes from PV according to the old Feed-in Tariff;

α – performance degradation rates for PV;

C_{pvEi}– income from power export based on the potential SEG scheme (only metered power export would be paid, so would vary with PV and battery degradation over time, and can be obtained from the calibrated energy model);

M_x, M_b –maintenance fees of SOLCER house component x and the social house component b, respectively (components here and after include only those presented in Table 1);

n, a – the total number of components in the SOLCER house and the social house, respectively.

With consideration to the discount rate which account for the devaluation of pound over time, a more reliable discounted ROI is developed and presented in Equation (6). The year when the total gain from investment is equal to the total cost of investment is taken as the payback time. The payback time illustrates the amount of time needed to recover the investment. A positive value of ROI indicates a good investment, while a negative value indicates a bad one. And the steeper the ROI line the faster the return.

$$ROI_i = [\Sigma_1^t (C_{pvi}/(1+r)^{i-1})+\Sigma_1^t (C_{si}/(1+r)^{i-1})- C_{in}-\Sigma_1^t (C_{mi}/(1+r)^{i-1}]/[C_{in}+\Sigma_1^t (C_{mi}/(1+r)^{i-1})] \quad (6)$$

ROI_i – the ROI of the investment on year i; C_{in} –the extra investment on SOLCER house, compared with cost for building the social house.

2.2. *A technical feasibility analysis*

To have a complete examination of the adaptability and replicability of the SOLCER house, the following variants have been selected and studied in the technical feasibility analysis:

- housing type (detached, semi-detached and terrace),
- orientation of the main roof (west, southwest, south, southeast and east) assuming that PV is installed on the main roof only,
- location (Cardiff, Glasgow, Manchester and Southampton).

Four typical cities are selected to include the majority of climate conditions in the UK, with one in the north, one in the south and two in the middle. CIBSE test reference year (TRY) weather data of individual cities were employed for the simulation.

The technologies installed at the SOLCER house were reviewed to accommodate any performance improvement since the SOLCER house was built. The major potential improvements include the increased module efficiency of the Building Integrated PV (BIPV) according to the latest market data, and the employment of a big TESLA battery with a capacity of 13.5kWh and a round trip efficiency of 90%. The adapted PV module efficiency is assumed to be 19.75% for this case study, much higher than that of the SOLCER house, which is 15%.

A single-variant analysis and a multi-variant analysis were used to understand how the SOLCER house performs in different scenarios, including examining house performances after adaptation to the efficiency of PV module, and comparing sensitivities of different variants in relation to energy consumption, CO_2 emission, operating cost, electricity self-sufficiency. The sensitivity of each variant in relation to a specific performance is determined by the performance range achieved to include the average performances of individual case groups employing the same variant value.

3. Results and discussion

3.1. The long-term economic analysis

Three scenarios are examined in the economic analysis, including:

- Scenario 1: As-built SOLCER house with the Feed-in Tariff from 2015;
- Scenario 2: As-built SOLCER house with only power export income under the SEG scheme;
- Scenario 3: Adapted SOLCER house with only power export income.

The first two scenarios are to enable the consideration of changing of external financial support over time, while the last one is to account for technology performance improvements over time, such as an increased PV efficiency.

Results from the ROI analysis are summarized and compared in Figure 2. It indicates, compared with a best practice new built social house,

1) the extra investment on the SOLCER house can be paid back in 12 years under the 2015 Feed-in Tariff when it was built;

2) the extra investment on the SOLCER house cannot be paid back within the designed system lifespan of 25 years under the new SEG Scheme with income only from power export;

3) with consideration to the relevant technology improvements over time, extra investment on the adapted SOLCER house can be paid back in 17 years at the first time, thereafter 25 years with replacement costs over time taken into account;

4) in all the scenarios, the steeper increase at year 16 is because of a higher replacement cost of the social house compared with that of the SOLCER house, while the drop at year 21 is due to the big investment on replacing the battery of the SOLCER house;

5) for each scenario, the payback speeds can be ranked as, year 1-15 >=year16-20>year 21-25, and the big decrease from year 21 results from the finalisation of the Feed-in Tariff which lasts only for 20 years for each application.

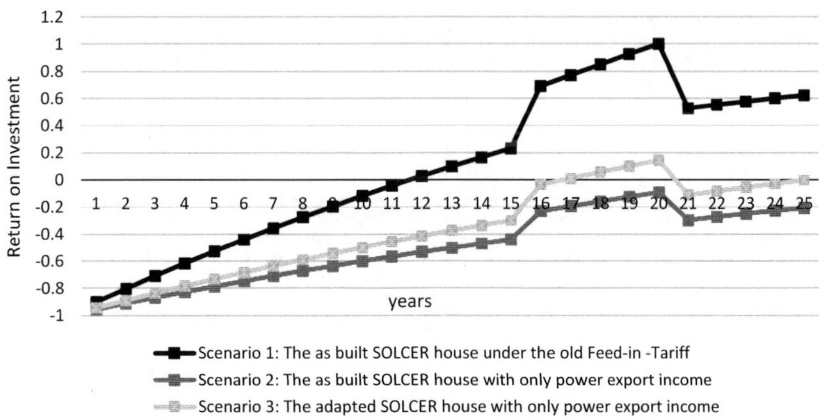

Figure 2. A comparison of ROIs for different scenarios.

3.2. The technical feasibility analysis

Simulation results from HTB2 for the technical feasibility analysis are summarized below in Table 3 and Figures 3 and 4. Table 3 presents the electricity export to import rates, with values above 1 being energy positive, and values close or below 1 being near-zero energy performance. Figure 3 compares the saving ratios of electricity import from single-variant changes on the basis of SOLCER house in Cardiff, and Figure 4 illustrates the multi-variant's sensitivities in relation to electricity import, CO_2 emission and electricity sufficient ratio.

Table 3 shows the majority of the replication cases are energy positive except the east and the west facing semi-detached houses and the east-facing mid-terrace house in Glasgow, which only achieve near-zero or zero energy performance. However, these can be improved by installing PV on the east and west facing roof, as the annual solar potentials and electricity yields on both sides of the roof will be similar. This would not increase the total peak PV power as the peak power outputs from both sides of the roof would not occur at the same time, and therefore would not require additional capacity accommodation from the grid, which would be an additional cost. Therefore, an energy positive target can still be met by these houses.

Table 3. A summary of the predicted electricity export to import rates for replication for different housing types, orientations and UK locations.

	Export to import rates											
	Cardiff			Glasgow			Manchester			Southampton		
	Detach	Semi	Mid	Detach	Semi	Mid	Detach	Semi	Mid	Detach	Semi	Mid
W	1.8	2.0	2.2	0.9	1.0	1.1	1.2	1.3	1.4	1.6	1.7	1.9
SW	2.5	2.8	3.1	1.2	1.4	1.5	1.6	1.7	1.9	2.1	2.4	2.6
S	2.9	3.2	3.7	1.4	1.6	1.8	1.8	2.0	2.2	2.5	2.8	3.1
SE	2.4	2.6	3.0	1.2	1.3	1.5	1.5	1.7	1.9	2.1	2.3	2.6
E	1.7	1.8	2.1	0.9	1.0	1.0	1.1	1.2	1.4	1.5	1.6	1.8

Detach--Detached house, Semi--Semi-detached house, Mid--Mid-terrace.

According to the single variant analysis in Figure 3, alternative housing types result in a decrease of electricity import by up to 22%, while the changing of orientation or location increases annual electricity import, with the case in Glasgow being the worst, requiring a 54% increase in electricity import from the grid.

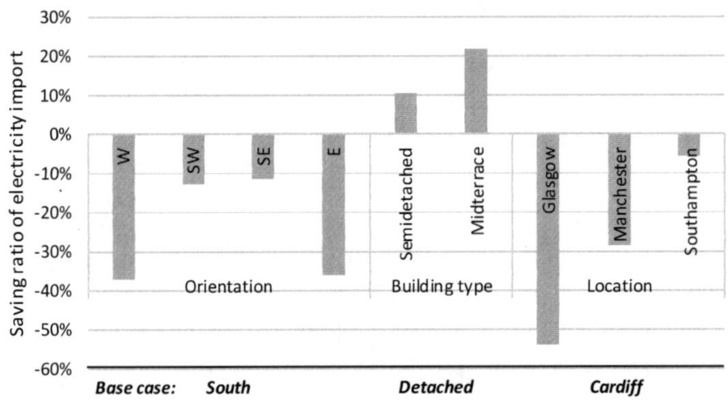

Figure 3. Saving ratios of annual electricity import from single variant changes on the basis of the SOLCER House.

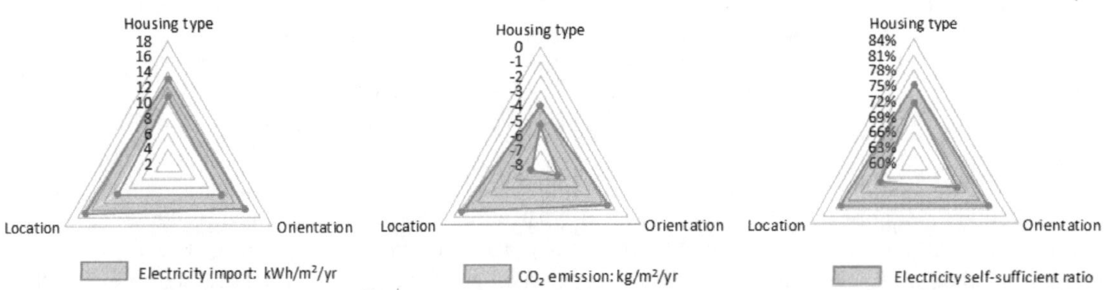

Figure 4. A sensitivity analysis on housing type, orientation and location in relation to electricity import, CO_2 emission and electricity sufficient ratio.

The results of a sensitivity analysis in Figure 4 indicate that the location has the highest impact on annual electricity import, CO_2 emission and electricity self-sufficient ratio, while housing type has the lowest. The results imply that there would be big performance differences between geographical areas if the SOLCER house was replicated across the country, and the trade-offs between housing type and location, orientation and location will not minimise the differences. Therefore, besides the favourable housing type and orientation, more efficient or advanced technologies would be needed in the less favourable locations, such as cities in the north, to achieve similar performance target.

4. Conclusion

This paper has presented a long-term economic analysis and a technical feasibility analysis of the SOLCER house, a renewable-led energy-positive house in the UK. Analyses have been carried out based on a validated thermal and energy model of the SOLCER house, also taking into account the changing of financial support and technology improvements over time. Results from the economic analysis show that the extra spending on building the SOLCER house can be paid back in 12 years under the old Feed-in Tariff, when compared with a best practice social house of similar size. Under the new SEG scheme, where only power export would be paid, pay back of the extra investment on the adapted SOLCER house can still be achieved during the system's lifespan. Results from the technical feasibility analysis show the replication of the SOLCER house with different housing types and orientations and in different locations across the UK can be energy positive, although east-west facing houses in northern areas might require additional PV installation to boost generation due to low solar potentials. The technical feasibility analysis also indicates location has the highest impact on building performance, such as annual electricity import, CO_2 emission and electricity self-sufficient ratio; implying that more efficient or advanced technologies would be needed to achieve similar building or system performance in the less favourable locations. The outcome of the research has demonstrated that the affordability of the SOLCER house, and the technical feasibility of its replication of different housing types and orientations in different areas in the UK. The payback within lifetime of the system implies that occupants can benefit from improved internal conditions, namely healthier life quality with the same cost they can afford in social houses, therefore will reduce fuel poverty.

The research also highlights the importance of government schemes for promoting renewable energy application in domestic buildings, although at some point, the reducing or withdrawal of this support can be offset by foreseeable technology improvements. As the highest impact on building performance is a result of location, financial support for renewable energy application should be tailored according to location, such as providing more support to less favourable areas to justify expense and increase affordability. Future work could investigate innovative or more advanced technologies for further improving building performance of the SOLCER house, in particular its replication in the less favourable areas in the UK.

Acknowledgments

The research presented in this paper was funded through the Cardiff University EPSRC IAA project 'SOLCER house - from demonstration to real world'. It is also part of the Wales Low Carbon Research Institute (LCRI) research programme, funded through the European Regional Development Fund (ERDF).

References

[1] Sreeraj E S, Chatterjee K and Bandyopadhyay S 2010 Design of isolated renewable hybrid power systems *Solar Energy* **84** 1124-36
O'Flaherty F and Pinder J 2011 *The Role of Micro-generation Technologies in Alleviating Fuel Poverty* (Sheffield: Sheffield Hallam University)
Martiskainen M, Heiskanen E and Speciale G B 2017 Community energy initiatives to alleviate fuel poverty: the material politics of Energy Cafes *Local Environment* **23** 20-35
[2] Zubi G, Fracastoro G V, Lujano-Rojas J M, Bakari K E and Andrews D 2019 The unlocked potential of solar home systems; an effective way to overcome domestic energy poverty in developing regions *Renewable Energy* **132** 1425-35
[3] Welsh Government 2016 *The Building Regulations 2010, Conservation of Fuel and Power:*

Approved Document Part L -2014 L1A-New Dwelling (Cardiff: Welsh Government)
[4] BEIS 2017 *Annual Fuel Poverty Statistics Report 2017 (2015 data)* (London: BEIS)
[5] BEIS 2019 *The Future for Small-scale Low-carbon Generation: A Consultation on a Smart Export Guarantee 2019* (London: BEIS)
[6] Jones P, Pearson P, Irvine S, Guwy A, Masters I and Bowen P 2015 *LCRI 2015: Overview of LCRI Research 2008 to 2015* (Cardiff: LCRI Cardiff University)
[7] Jones P, Li X, Patterson J L, Coma Bassas E and Lannon S C 2016 *Proc. 32nd Int. Conf. on Passive and Low Energy Architecture*, ed P La Roche and M Schiller (Los Angeles: PLEA) pp 1095-1101
[8] Jones P, Li X, Coma Bassas E and Patterson J L 2017 *Proc. 15th Conf. of Int. Building Performance Simulation Association* (San Francisco: IBPSA) pp 1334-8
[9] Lewis P T and Alexander D K 1990 HTB2: a flexible model for dynamic building simulation *Build Environ* **25** 7-16
[10] Perisoglou E, Coma Bassas E, Lannon S C, Li X, Jenkins H, Patterson J L, Jones P and Hou S 2018 *Proc. 4th IBPSA-England Conf. on Building Simulation and Optimization* (Cambridge: IBPSA) pp 223-9
[11] BCIS 2015 *Housing development: the economics of small sites – the effect of project size on the cost of housing construction* (London: BCIS)
[12] Perisoglou E, Patterson J, Stevenson E V and Jenkins H 2017 *Proc. Int. Conf. on Sustainability in Energy and Buildings (*Chania: SEB)
[13] McCanne and Parteners 2017 *Low and Zeron Carbon Technology Feasibility Study* (Cardiff: McCann and Partners)
[14] EHPA 2013 *European Heat Pump Market and Statistics Report 2013* (Brussels: EHPA)
[15] Salasovich J, Geiger J, Mosey G and Healey V 2013 *Feasibility Study of Economics and Performance of Solar Photovoltaics at the Standard Chlorine of Delaware Superfund Site in Delaware City, Delaware* (Golden: NREL)
[16] Hammond G P, Harajli H A, Jones C I and Winnett A B 2012 Whole systems appraisal of a UK building integrated photovoltaic system: energy, environmental, and economic evaluations *Energy Policy* **40** 219-230
[17] Tesla 2017 *Tesla powerwall limited warranty (European Warranty Region)* (Amsterdam: Tesla)
[18] May G J, Davidson A and Monahov B 2018 Lead batteries for utility energy storage: A review *Journal of Energy Storage* **15** 145-157
[19] Pena-Bello A, Burer M, Patel M K and Parra D 2017 Optimizing PV and grid charging in combined applications to improve the profitability of residential batteries *Journal of Energy Storage* **13** 58-72
[20] BSI 2017 *BS EN 15459- 1:2017 Energy performance of buildings – Economic evaluation procedure for energy systems in buildings - Part 1: Calculation procedures, Module M1-14* (London: BSI)
[21] Vignali G 2017 Environmental assessment of domestic boilers: A comparison of condensing and traditional technology using life cycle assessment methodology *Journal of Cleaner Production* **142** 2493-2508
[22] Jordan D C, Smith R M, Osterwald C R, Gelak E and Kurtz S R 2010 *Proc. the 35th IEEE Photovoltaic Specialists Conf.* (Hawaii: IEEE)
[23] Smith K, Saxon A, Keyser M, Lundstrom B, Cao Z and Roc A 2017 *Proc. 2017 American Control Conf.* (Seattle: IEEE)
[24] Waddicor D A, Fuentes E, Sisó L, Salom J, Favre B, Jiménez C and Azar M 2016 Climate change and building ageing impact on building energy performance and mitigation measures application: a case study in Turin, northern Italy *Building and Environment* **102** 13-25
[25] HM Treasury 2018 *The Green Book: Central government guidance on appraisal and evaluation* (London: HM Treasury)

IOP Conf. Series: Earth and Environmental Science **329** (2019) 012050 doi:10.1088/1755-1315/329/1/012050

Design of a Self-Shading Mass as a Function of the Latitude for Automatic Seasonal Adjustment

Shady Shawky Saifelnasr

Assistant Professor, October University for Modern Sciences and Arts "MSA", Cairo, Egypt

ssaifelnasr@msa.eun.eg

Abstract. The main objective of this paper is to propose a simple tool for architects to design a self-shading mass, resulting in an automatic seasonal adjustment, i.e., full shading in summer but allowing solar heat gain in winter, within the low latitudes. Commonly, self-shading masses might have inverted pyramidal forms or inward terracing. But in the proposed design, crystal-like forms are generated on the eastern and western façades and an inward sloping form is generated on the equator-facing façade. Those generated forms are not only used from an aesthetical point of view, as might be done in some contemporary buildings, but also ensuring external shading when needed. The different dimensions of the proposed self-shading mass could be obtained by using a design chart that was previously designed and developed by the author. The obtained dimensions are function of the latitude, indicating the building location, the shading height, indicating the building height and the selected cut-off times, indicating the selected hours of shading. A number of smaller crystalline forms, stacked above each other, could have the same effect in blocking the sun rays as a single large one, which might be essential to overcome the elongated depth or protrusion of a self-shading mass.

1. Introduction

Shading design could be considered as one of the most important passive design strategies, needed within the overheated zone, i.e., warm or hot climates. It involves the exclusion of a large amount of the direct solar radiation, emitted from the sun.

Self-shading masses signify those masses that shade themselves and they have been used in both traditional and contemporary buildings. Commonly, self-shading masses might have inverted pyramidal forms or inward terracing, e.g., ST Diamond Building, in Malaysia, whose self-shading inverted pyramid protects itself during the overheated period (Figure 1) [1]. But in the proposed design, crystal-like forms are generated on the eastern and western façades and an inward sloping form is generated on the equator-facing façade.

Those generated forms or elements might have been included in some contemporary buildings for conceptual or aesthetical reasons. Conceptual reasons could be exemplified in the extension to the Royal Ontario Museum (ROM), in Canada, whose intersecting crystal-like forms were inspired by the crystalline forms in the ROM's mineralogy galleries (Figure 2) [2]. Whereas aesthetical or formalistic reasons could be seen by the growing number of buildings with crystal-like forms within different contexts (Figure 3).

But in the proposed design, the generated forms are not only used from an aesthetical or conceptual point of view, but also ensuring external shading when needed.

Figure 1. Self-shading inverted pyramidal form of the ST Diamond Building, Putrajaya, Malaysia (Source: M. Boubekri, 2014).

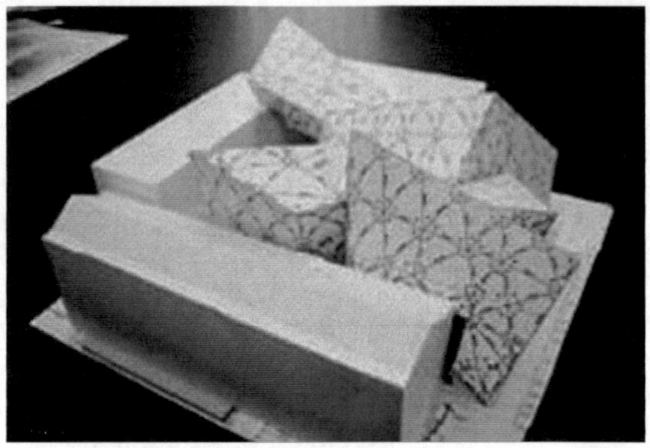

Figure 2. Crystalline forms of the extension to the Royal Ontario Museum, Toronto, Canada (Source: Boake, Terri Meyer, 2012).

Figure 3. Growing number of buildings with crystal-like forms, Cairo, Egypt (Source: the author).

Nomenclature

HSA	horizontal shadow angle [°]
VSA	vertical shadow angle [°]
LAT	latitude [°]
h	shading height [m]
l	self-shading mass length [m]
d	depth of self-shading mass exceeding the original depth [m]
p	protrusion of self-shading mass exceeding the original length [m]
n	number of crystalline forms

2. Main Objectives and Hypothesis

The main objective of this paper is to propose a simple tool for architects to design a self-shading mass, resulting in an automatic seasonal adjustment, i.e., full shading in summer but allowing solar heat gain in winter, within the low latitudes.

The main hypothesis for this paper is the possibility of determining the different dimensions of a proposed self-shading mass having one or more crystalline forms, as a function of the latitude (LAT) and shading height (h) at the selected cut-off times.

3. Methodology

In this paper, the different dimensions of the proposed self-shading mass could be obtained by using a design chart that was previously designed and developed by the author [3].

IOP Conf. Series: Earth and Environmental Science **329** (2019) 012050 doi:10.1088/1755-1315/329/1/012050

To use this measuring technique, the scope and limitations of the paper, associated with some simplifications and assumptions are shown as follows.

3.1. Scope and Limitations

The paper mainly focuses on shading design within the low latitudes, where shading is needed as an essential passive design strategy. It provides a simple method for architects to generate a proposed self-shading mass having one or more crystalline forms. This method is simple and generic, as it involves the use of a design chart to determine the different dimensions of the proposed self-shading mass as a function of the latitude (LAT) and shading height (h) at the selected cut-off times, resulting in an automatic seasonal adjustment.

The paper does not concern about a number of aspects or parameters, which in turn could be addressed in future researches, such as the cost and the stability of such unique structures or their effect on the quantity or quality of daylighting within the different spaces at the different façades.

3.2. Simplifications and Assumptions

Some simplifications and assumptions were considered while designing the proposed self-shading mass concerning its applicability, orientation of inward sloping form, effect of the shadow angles and specifications of the used base case.

3.2.1. Applicability of the Proposed Self-Shading Mass. The proposed self-shading mass is applicable for forms with rectilinear plans, whose edges are oriented towards the main orientations, located at lower latitudes, with temperate, hot-dry or warm-humid climates where shading is a need [4].

3.2.2. Orientation of Inward Sloping Form. The inward sloping form of the proposed self-shading mass is facing the equator (due south in the northern hemisphere and due north in the southern hemisphere). The used design chart is based on the fact that an equator-facing horizontal shading device, can give an automatic seasonal adjustment: full shading in summer but allowing solar heat gain in winter if the (VSA) = 90° - (LAT) [5], e.g., an equinox cut-off for the city of Cairo, Egypt ((LAT) = 30° N), it will be (VSA) = 90° - 30° = 60°.

3.2.3. Effect of the Shadow Angles. The proposed self-shading mass is handled as a horizontal shading device that is not only characterized by a vertical shadow angle (VSA) that specifies its depth (d) in elevation, but also is affected by the horizontal shadow angle (HSA) that specifies its protrusion (p) resulting in its trapezoidal shape in plan.

3.2.4. Specifications of the Used Base Case. The form, dimensions, selected latitudes, cut-off times and number of crystalline forms are shown as follows.

- Form of the original mass: cube.
- Dimensions of the original mass: external length = 9.00 m, external depth = 9.00 m and external height = 9.00 m.
- Selected Latitudes: LAT 10°, LAT 20°, LAT 30° and LAT 40°.
- Selected cut-off times: 0900-1500 hr., 1000-1400 hr. and 1100-1300 hr.
- Used number of crystalline forms: 1, 2, 3, 4, 5 and 6.

3.3. The Design Chart to Determine the Different Dimensions of the Proposed Self-Shading Mass

The design chart shown in Figure 4 is a combination of two graphs; the first graph on the right and the second graph on the left.

The first graph (on the right) determines the ratio (d/h) between the required depth (d) and the shading height (h), as a function of the latitude (LAT), and it is drawn using the formula:

$$\frac{d}{h} = \tan(LAT)$$

It consists of two axes and a curve; the horizontal axis represents the latitude (LAT), ranging from 0° (Equator) to 40° N and S, while the vertical axis represents the ratio (d/h) between the required depth (d) and the shading height (h). That ratio (d/h) is obtained on the vertical axis after intersecting the curve from the marked latitude (LAT).

The second graph (on the right) determines the ratio (p/h) between the required protrusion (p) and the shading height (h), as a function of the latitude (LAT) and cut-off times, and it is drawn using the formula:

$$\frac{p}{h} = \tan(LAT) \times \tan(HSA)$$

It consists of two axes and three curves; the vertical axis represents ratio (d/h) between the required depth (d) and the shading height (h), while the horizontal axis represents the ratio (p/h) between the required protrusion (p) and the shading height (h). That ratio (p/h) is obtained on the horizontal axis after intersecting one of the three curves according to the selected cut-off times from the marked ratio (d/h), obtained from the first graph.

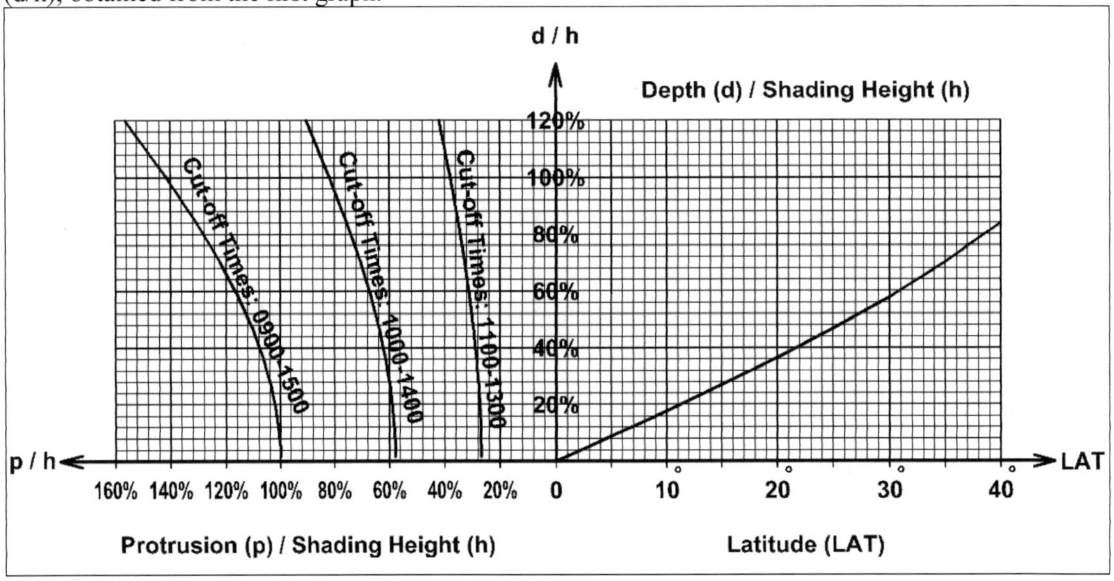

Figure 4. The design chart to determine (d/h) and (p/h) of the proposed self-shading mass (source: Saifelnasr, 2015).

4. The Proposed Self-Shading Mass

4.1. Concept of the Proposed Self-Shading Mass

Shading could be achieved either through shading devices or through building form, i.e., self-shading methods [6]. Accordingly, in this research, the proposed self-shading mass is based on the idea of designing a fixed horizontal shading device, that functions best for equator-facing façades [7], and then integrating it with the original building mass within a single form. That combination would result in the generation of crystal-like forms on the eastern and western façades with dividing ridge lines, and an inward sloping form on the equator-facing façade.

The different dimensions of the proposed self-shading mass could be obtained by using a design chart that was previously designed and developed by the author [3]. The obtained dimensions are function of the latitude (LAT), indicating the building location, the shading height (h), indicating the building height and the selected cut-off times, indicating the selected hours of shading. Thus, they could be determined as ratios with (h); (d/h) and (p/h) for a given latitude at the selected cut-off times (Figure 5).

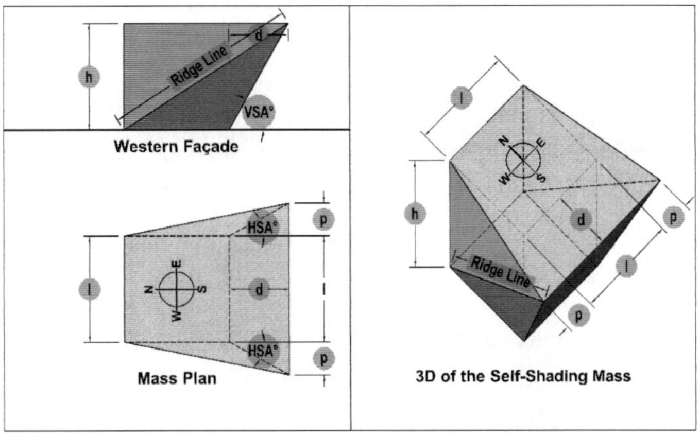

Figure 5. The different dimensions needed to construct the proposed self-shading mass could be obtained from the design chart as ratios with (h) (Source: the author).

4.2. Formation of the Proposed Self-Shading Mass

To construct the proposed self-shading mass, follow the shown steps in Figure 6 as follows. First start with the original rectilinear mass, oriented towards the main orientations; north, south, east and west. Determine the depth (d) from the first graph (on the right) of the design chart and apply it towards the equator-facing orientation. Determine the protrusion (p) from the second graph (on the left) of the design chart and apply it towards the eastern and western orientations. Join the upper equator-facing corners to the upper non-equator-facing corners that would lead to the formation of the trapezoidal shape of the roof. Join the upper equator-facing corners to the lower equator-facing corners that would lead to the formation of the trapezoidal shape of the inward sloping equator-facing façade. Construct the dividing ridge lines by joining the lower non-equator-facing corners with the upper equator-facing corners that would lead to the formation of crystal-like forms on the eastern and western façades. The constructed self-shading mass would result in an automatic seasonal adjustment, i.e., full shading in summer but allowing solar heat gain in winter.

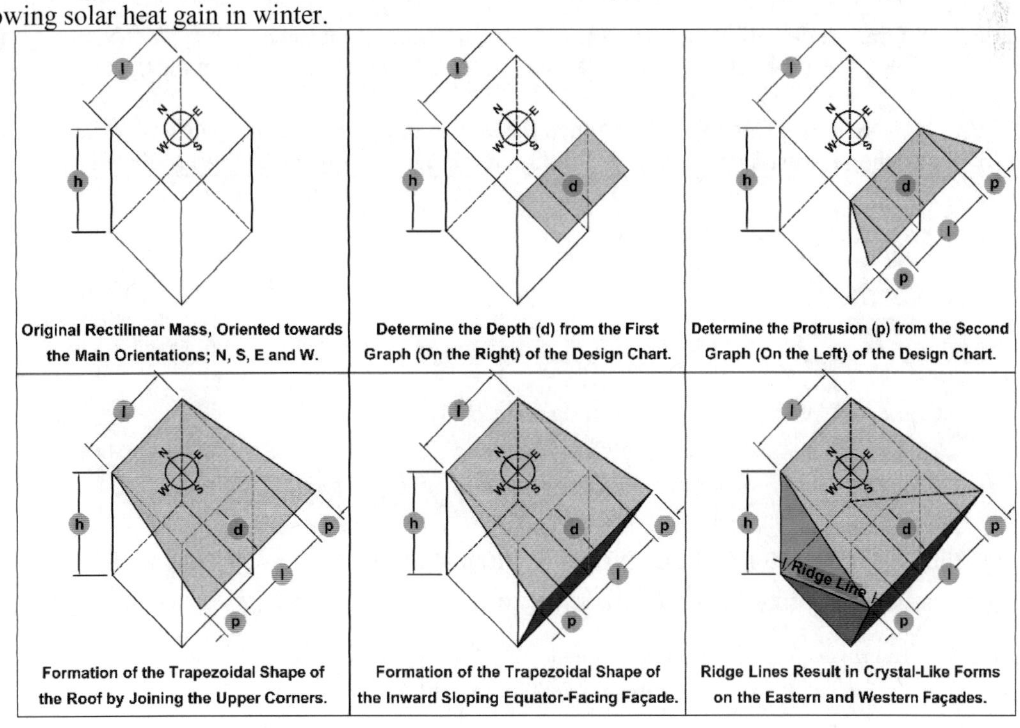

Figure 6. The different steps to construct the proposed self-shading mass (Source: the author).

IOP Conf. Series: Earth and Environmental Science **329** (2019) 012050 doi:10.1088/1755-1315/329/1/012050

4.3. Notes on the Proposed Self-Shading Mass and Its Variables

4.3.1. Effect of the Latitude (LAT). The change in the latitude is correlated with change in both shadow angles; vertical (VSA) and horizontal (HSA). Consequently, when the latitude (LAT) increases, both of the depth (d) and the protrusion (p) increase. The increase in the depth (d) leads to increasing the slope of the inward sloping form on the equator-facing façade, while the increase in the protrusion (p) leads to elongating the equator facing façade (Figure 7).

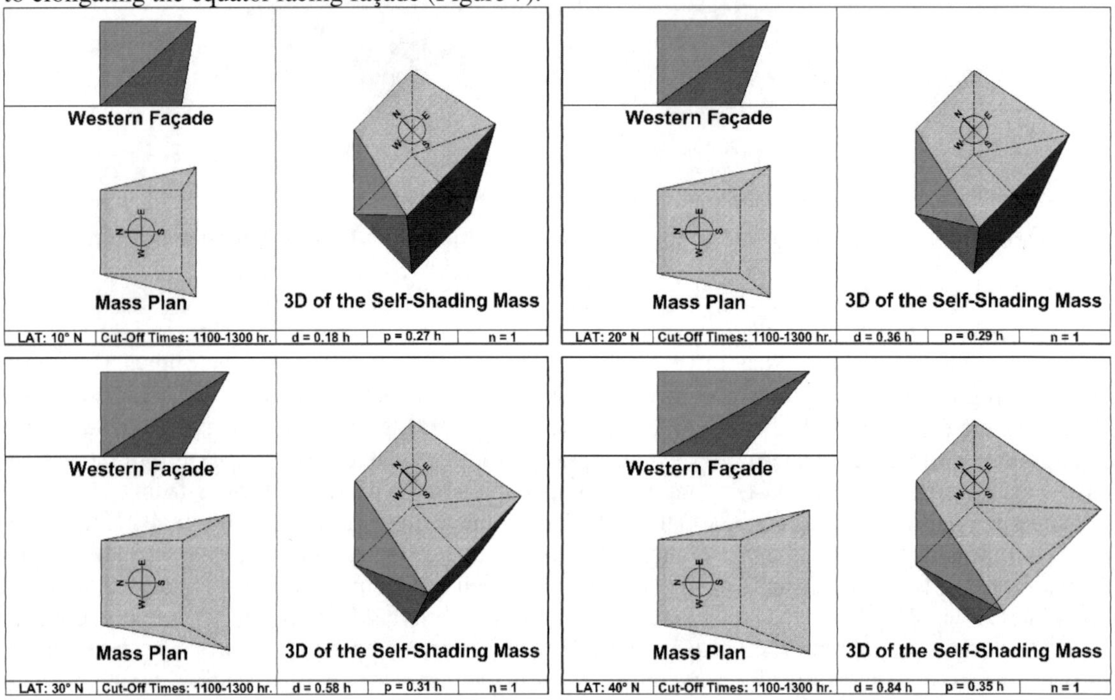

Figure 7. Different self-shading masses constructed for different latitudes; 10° N, 20° N, 30° N and 40° N, at the selected cut-off times: 1100-1300 hr. (Source: the author).

4.3.2. Effect of the Selected Cut-Off Times. When the self-shading mass is designed over a larger period of cut-off times, the protrusion (p) increases, leading to elongated equator facing façade (Figure 8).

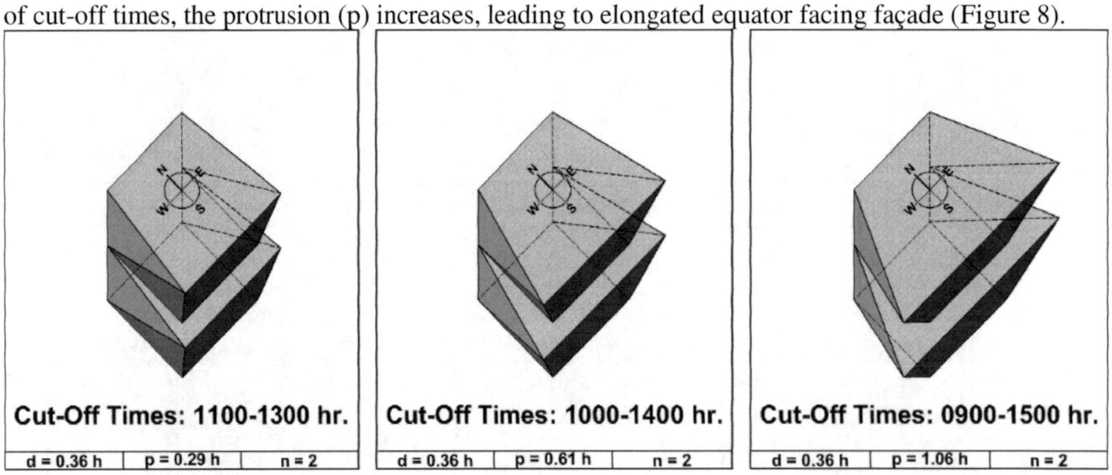

Figure 8. Different self-shading masses constructed for latitude 20° N, at different selected cut-off times (Source: the author).

4.3.3. Number of Crystalline Forms. A number of smaller crystalline forms (n), stacked above each other, could have the same effect in blocking the sun rays as a single large one, which might be essential to overcome the elongated depth or protrusion of a self-shading mass. In that case, the shading height for each crystalline form is equal to the original mass height divided by the number of crystalline forms (n) (Figure 9).

$$shading\ height\ for\ each\ crystalline\ form = (original\ mass\ height)/n$$

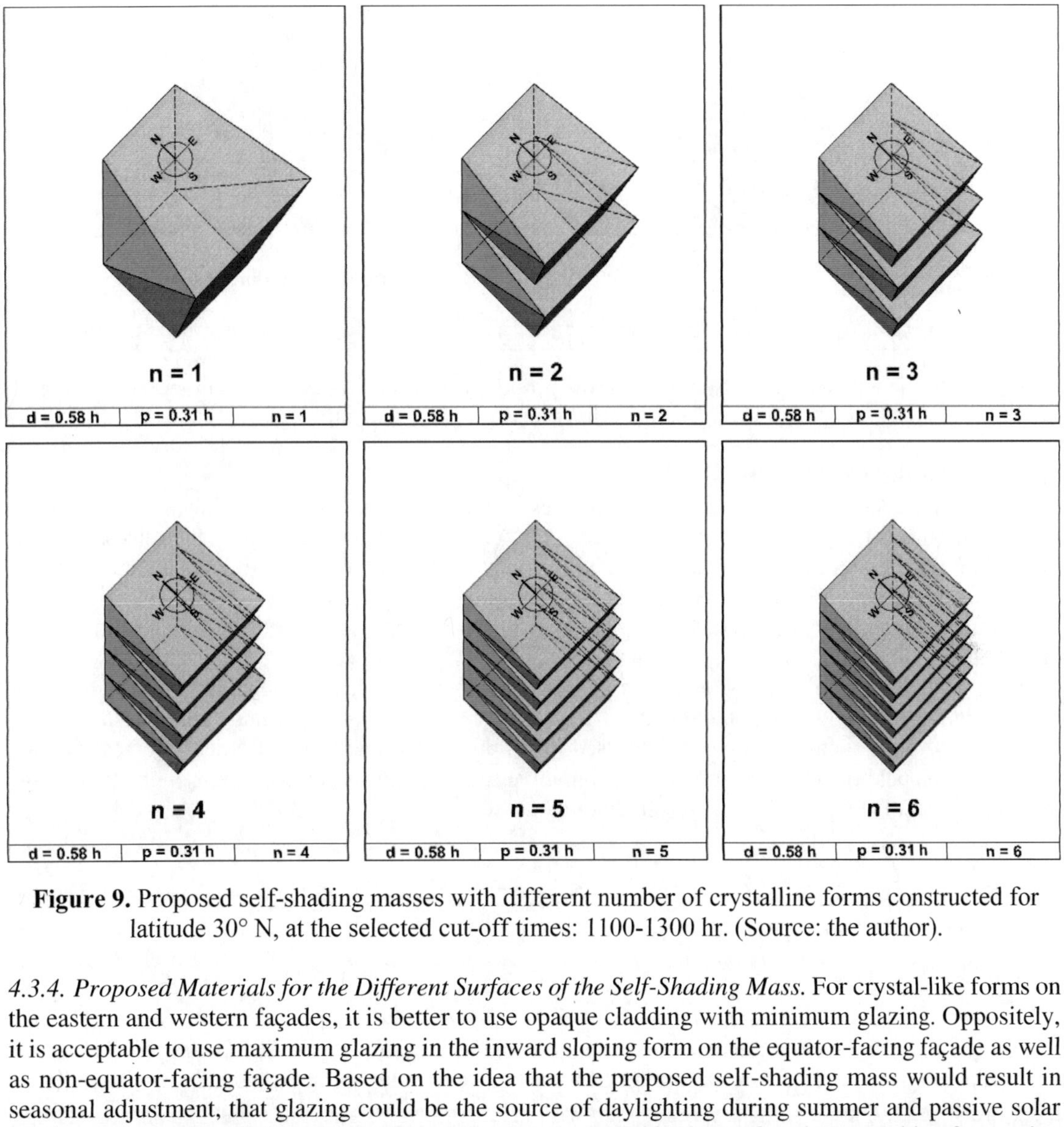

Figure 9. Proposed self-shading masses with different number of crystalline forms constructed for latitude 30° N, at the selected cut-off times: 1100-1300 hr. (Source: the author).

4.3.4. Proposed Materials for the Different Surfaces of the Self-Shading Mass. For crystal-like forms on the eastern and western façades, it is better to use opaque cladding with minimum glazing. Oppositely, it is acceptable to use maximum glazing in the inward sloping form on the equator-facing façade as well as non-equator-facing façade. Based on the idea that the proposed self-shading mass would result in seasonal adjustment, that glazing could be the source of daylighting during summer and passive solar heating during winter [8]. As for the flat surfaces, represented in the roof or those resulting from using more than one crystalline form, they acquire maximum solar radiation. For that reason, they could be either being applied as green roofs for insulation or applying photovoltaics on them with suitable slopes for generating energy (Figure 10).

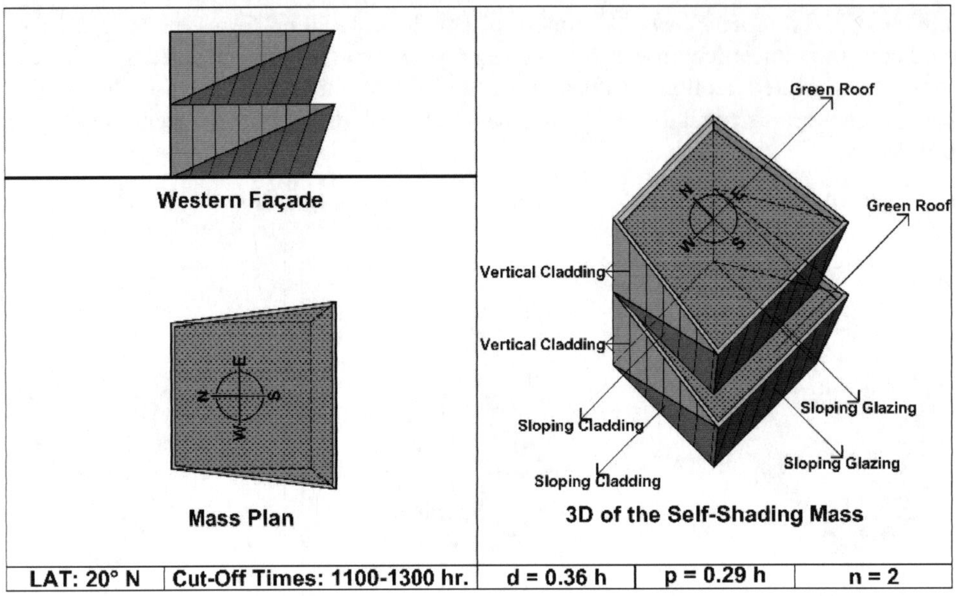

Figure 10. Schematic drawing showing some proposed materials for the different surfaces of the proposed self-shading mass (Source: the author).

5. Discussion and Conclusions

It is possible to determine the different dimensions of the proposed self-shading mass having one or more crystalline forms, as a function of the latitude (LAT) and shading height (h) at the selected cut-off times. Those generated forms are not only used from an aesthetical point of view, but also resulting in an automatic seasonal adjustment, within the low latitudes.

In future researches, the proposed self-shading mass could be upgraded to be suitable for masses other than rectilinear ones, e.g., masses with curves, or masses whose edges are not oriented towards the main orientations. It could be upgraded to find specific dimensions or details that might be suitable for a particular task, e.g., when applying photovoltaics to the surfaces with maximum solar radiation, i.e., the roof. The same methodology could be used for finding the dimensions of other volumes such as, climatic envelopes, etc. Other researches might involve studying the cost effectiveness or energy savings that might result from the use of such self-shading masses.

References

[1] M. Boubekri, *Daylighting Design: Planning Strategies and Best Practice Solutions*, Basel: Birkhäuser, 2014.

[2] T. M. Boake, *Understanding Steel Design: An Architectural Design Manual*, Basel: Birkhäuser, 2012.

[3] S. S. Saifelnasr, "*A Design Chart to Determine the Dimensions of a Horizontal Shading Device over an Equator-Facing Window as a Function of the Latitude and the Shading Height*," in Urban Planning and Architecture Design for Sustainable Development (UPADSD), Lecce, Italy, 2015.

[4] V. Olgyay, *Design with Climate: Bioclimatic Approach to Architectural Regionalism*, Princeton, N.J.: Princeton University Press, 2015.

[5] S. Szokolay, *Introduction to Architectural Science: The Basis of Sustainable Design*, Oxford: Architectural Press, 2008, pp. 60-62.

[6] A. Aksamija, *Sustainable Façades: Design Methods for High-Performance Building Envelopes*, Hoboken, N.J.: John Wiley & Sons, 2013.

[7] A. Athienitis and W. O'Brien, Eds., *Modeling, Design, and Optimization of Net-Zero Energy Buildings*, Berlin: Ernst & Sohn, 2015.

[8] N. Lechner, *Heating, Cooling, Lighting: Sustainable Design Methods for Architects*, Hoboken, N.J.: Wiley, 2008.

IOP Conf. Series: Earth and Environmental Science **329** (2019) 012051 doi:10.1088/1755-1315/329/1/012051

Regional Space Strategy and Zero-energy Engineering Strategy: An Analysis of "In-Between Nature" for Solar Decathlon China 2018

WANG Shaosen[1], SHI Feng[1], YAN Shurui[1], and CHEN Hong[1]
[1] School of Architecture and Civil Engineering, Xiamen University, 182 Daxue Road, Siming District, Xiamen, Fujian, China 361005.

ymcai@xmu.edu.cn; 347605132@qq.com

Abstract. With an analysis of the project "In-Between Nature" for SDC2018 that got 3rd place in the competition, this paper explored a combined approach to achieve zero-energy consumption building: space strategy learning from local architecture prototypes which is mainly passive, and engineering strategy which covers life-cycle of the building. During the competition, statistics such as temperature, humidity, energy produced and consumed are measured, proving that this combined strategy is feasible and the house has achieved the goal of zero-energy in Dezhou during early August, when the temperature and humidity is highest of a year.

1. Project Background

During both urban construction and rural renovation, we must respond to the problem of reserving traditional culture while fulfilling the demand of current living method. Traditional and regional architecture often means old in public opinion, but old doesn't necessarily mean outdated. On the contrary, it contains experience accumulated by multi-generations that finally combines architectural form, building function, climate characteristics and geographical environment. Nowadays, hi-tech and modernization rather than traditional architecture are many people's first impression of green building. However, simple and efficient wisdom are embedded in traditional constructions in wide-spreading regions. That's the reason we use tradition prototypes, local materials and combine traditional wisdom to modern technology to build a zero-energy residence.

Team JIA+, which means Home Plus, consists of faculty and students from Xiamen University, Shandong University and Team Solar Bretagne in France. Taking the opportunity of 2018 Solar Decathlon China, we placed our site in a village inside the city of Xiamen to reflect the building environment and climate features of South Fujian Province. Each team is required to build a 100-200 ㎡ zero-energy consumption residence within 20 days.[1] In order to meet the demands of the competition, we built a 1:1 frame model of surrounding houses in the competition site in Dezhou, illustrating the spatial relationship between old and new. The whole process from design, management to construction was carried out by students under the instruction of the faculty, and was completed in 2.5 years. In the final competition, we achieved the third place with a total score of 915.82.

IOP Conf. Series: Earth and Environmental Science **329** (2019) 012051 doi:10.1088/1755-1315/329/1/012051

Figure 1. Current situation of site

Figure 2. Bird's-eye view

Figure 3. Southeast view

2. Space Strategy: Prototype from Regional Architecture

2.1. Two Spatial Prototypes: Dacuo and Shoujinliao

South Fujian DaCuo[2] , a kind of big shady house widely distributed in this region, has been proved as an effective way of adaptation to the local climate. Overhanging roof shades the porch in front of the house, providing a heat-buffering zone between natural and artificial environment. Rather than stone which is also commonly used in the roof construction in Fujian, the roof of Dacuo is built with wood structure and covered with tiles, both of which are low thermal conductive materials. Sky-windows and building-gaps on the slope roof enhance the chimney effect of the interior space, making full use of natural ventilation.

Shoujinliao, which means Handkerchief House because of its narrow shape, consists of a series of small-scale yards connecting different building blocks. Houses and yards work together as a self-shading system, sheltering each other from the sun. Space is linear distributed to minimize the heat gained by radiation, and a typical 4-meter wide Shoujinliao is usually 20-30 meters in length. It can contain as many as 9 yards and 31 rooms in a single group that are diverse in size and shape. Variation of space dimension also leads to imbalanced air density, which accelerates the interior wind speed to keep the air fresh and cool down the temperature.

IOP Conf. Series: Earth and Environmental Science **329** (2019) 012051 doi:10.1088/1755-1315/329/1/012051

Figure 4. Bird's-eye view of Dacuo

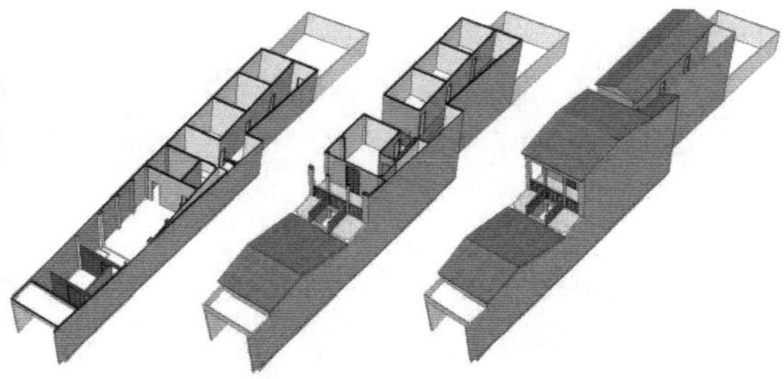

Figure 5. Layout of Shoujinliao

2.2. Embracing Traditional Wisdom: Passive Space Strategies

Learning from these two types of traditional architecture, we researched into the relationship between interior space and exterior environment, especially the design of in-between space, and try to adopt climate adaptive strategies of traditional architecture in modern technical design, making full use of passive method to reduce cost of energy.

In the plan, we introduced 3 space to promote the thermal performance. Firstly, a 3-meter-wide porch space to the south of the building is covered with louvers, insulating the rooms from the heat and meanwhile guarantee the natural lighting. Secondly, inside of the new building, a heat-buffering inner-yard, which is narrow and high enough to block its floor-ground from the radiation, is placed along with the porch and adjusts the micro-climate and further advances the ventilation. Nearby rooms open to the small yard, which uses its skylight window to work as a heat buffering space. In the winter, the skylight windows are closed, converting the yard into a greenhouse which warms up surrounding spaces by heat convection. Filler materials in the yard floor helps to accumulate the heat and keep a stable temperature. In the summer, the skylight window is open with louvers blocking the sunlight, assisting the ventilation of neighbouring space with chimney effect. According to previous research on Shoujinliao, in such yards located in South Fujian, the average air temperature near the floor-ground can be 2.1°C lower than the outdoor space, and temperature range per-day is also reduced from 10.7°C to 6.2°C. Compared to its modern opponents, average temperature of the bedroom in a typical Shoujinliao is also 1.1°C lower than that of a concrete-structure house in the same district. [3]

The overhanging attic increases the aspect ratio of the yard, and further shelters the living-room underneath. Finally, the new building places itself just 6 meters from the old one, leaving a gap space consist of two 1.5-meter corridors adjacent to each house and a 3-meter-wide yard between them, so that the old house and the new can work together as a yard-block self-sheltering system, just like the case in a traditional dwelling. These three spaces all derive from the study of traditional prototypes and form a series of hear-buffering space between indoor and outdoor.

In terms of section, the south and north roofs are separated, leaving space for motor-driven skylight windows facing the north which enhances the natural ventilation by heat and wind pressure and fills the space with gentle lighting. We choose a slope roof based on two factors. Firstly, the site is surrounded by slope-roof South Fujian DaCuo. By creating formal similarity between new and old, the visual consistency was acquired. [4] Secondly, for a higher generation rate of PV panels, slope roof takes advantage of the angle between solar panels and incident sunlight, achieving high solar energy gathering efficiency. Considering annual incidence angle and height limit, we adopted a 20° slope to balance among various aspects. In addition, the 15cm cavity between PV panels and roof structure allows the wind flow to reduce solar heating of the roof and cools down the panels, improving their electricity generation efficiency.

IOP Conf. Series: Earth and Environmental Science **329** (2019) 012051 doi:10.1088/1755-1315/329/1/012051

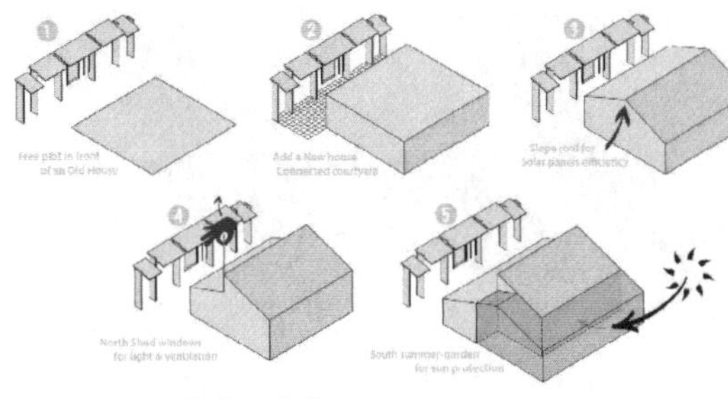

Figure 6. Spatial strategy

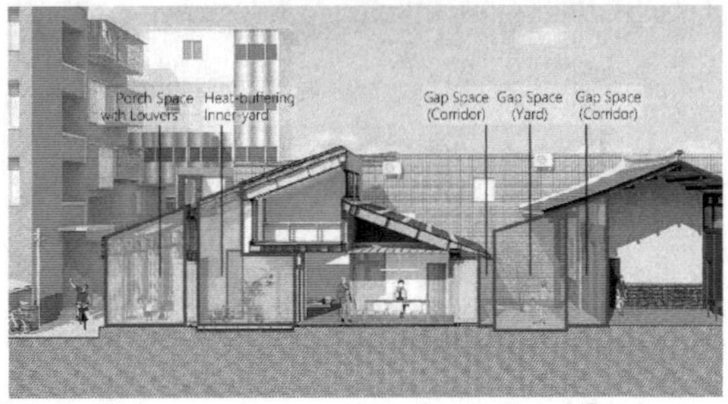

Figure 7. Spaces promoting thermal performance

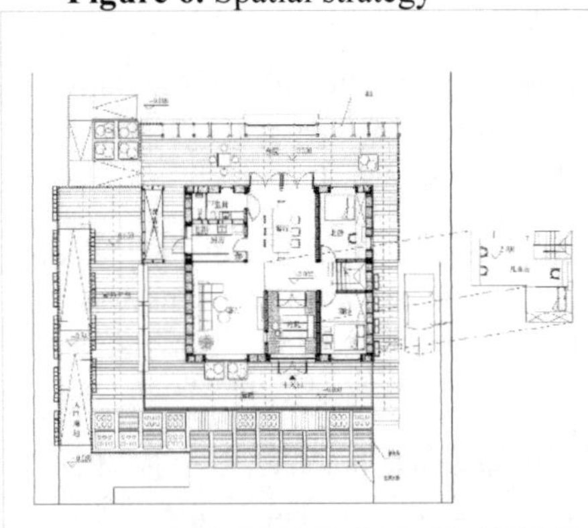

Figure 8. Ground floor plan

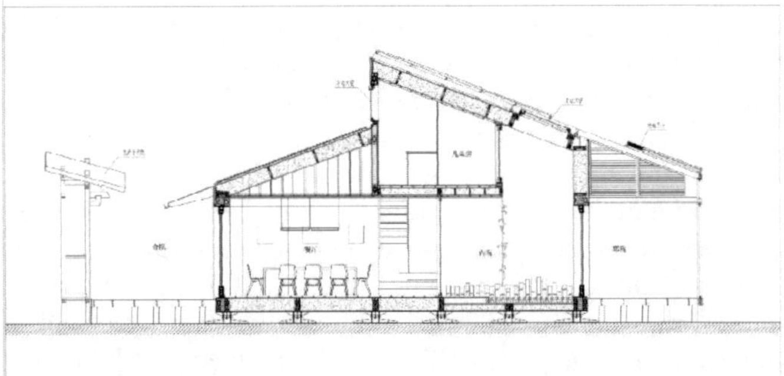

Figure 9. Section

Figure 10. Space between neighbourhood

Figure 11. Front porch space with louver

3. Engineering Strategy: Zero-energy Techniques

The goal of zero-energy consumption is a full-cycle process, from material selection, construction management, daily operation to maintenance. There are 3 technical strategies that we proposed for this project, each focusing on different aspects of the life-cycle of building.

3.1. Prefabrication and Fabrication

An eco-responsible house has to be built quickly with a dry and healthy process. The team works on a modular solution to optimize the assembly time through simple and accessible construction modes. The wet blocks are bundled with integrated networks to prevent plumbing on site. The bearing structure, that is to say, the perimeter walls, the ground floor slab and the roof, have been designed with the same prefabricated method: 2D box including wood frame, insulation and also interior facing. Structural elements are prefabricated, so that only interior decorations need to be made on site. Thus, even untrained students or novices can take over the system and participate in the construction.

IOP Conf. Series: Earth and Environmental Science **329** (2019) 012051 doi:10.1088/1755-1315/329/1/012051

Building equipment and pipelines are also highly compacted for the ease of modular construction. The cavity under the north roof works as equipment space for air conditional fans and fresh air ventilator. Air pipes are distributed between decorative boards and kneels. Partition walls of both toilet and kitchen contain water supply and drainage system. Electric wires are also hidden inside wall units. We took full advantage of prefabrication system and integrate structure with equipment.

Figure 12. Prefabrication and assembling

3.2. 90% Eco-materials

In a logic of sobriety, we choose bio-sourced materials, resulting from the biomass, for an architecture respectful of the environment. Four local biomass materials including straw, wood, bamboo and sawdust are the major materials of the building.

We choose straw board insulation because this material is available in large quantities in Shandong, humidity regulator and inexpensive, which is usually burned as agricultural by-product and causes serious air pollution. Wood is mainly used as the structural units, and bamboo as decorations, both natural, recyclable, non-toxic and carbon sink materials. With no chemical treatment in the whole manufacture process, it brings no risk for the health to the builders or the inhabitants.

Figure 13. Straw-wall Configuration

MATERIAL	DENSITY	THERMAL CONDUCTIVITY	HEAT CAPACITY	GREY ENERGY	PHASE SHIFT
	kg/m3	W/mK	J/kgK	kWh/m3	H
Wood Fiber	50	0,039	2100	60	13
Glass Wool	35	0,039	1030	470	3
Cork Panel	125	0,049	1560	450	10
STRAW	230	0,080	1330	0	16

Figure 14. Physical Qualities of Straw

IOP Conf. Series: Earth and Environmental Science **329** (2019) 012051 doi:10.1088/1755-1315/329/1/012051

3.3. Combination of passive and active equipment system

The team's driving line is "the best energy is the one you do not consume". We prefer the use of the "free" that are the sun, wind and water in a passive design. However, the climate in Dezhou is not always temperate, and when the passive system is not enough to meet the comfort requirements in harsh climatic conditions, the active system can take over the house regulation for some time.

The panels installation on the south roof has been designed to be as efficient as possible. The roof slope (20°) has been set to have the best orientation during summer and produce energy to balance house energy needs. The installation consists of 44 PV panels and one water heating solar panel.

A flexible skin system is used as the envelope of the building.[5] Outer louvers and skylight windows are all electric controlled and adjusted according to the weather for a better micro-climate. Taking the south façade as an example, it is equipped with motor-control aluminium lover and manual-control bamboo louver. The aluminium louver adjust itself automatically to create sunshade, while the bamboo louver is controlled by human to conduct the airflow. Building intelligent control system is introduced to link the building equipment of different brands and different control methods. By integrating and optimizing active and passive methods, environment control measures become more intelligent and energy-saving.[6][7] The whole design process is optimized by BIM techniques.

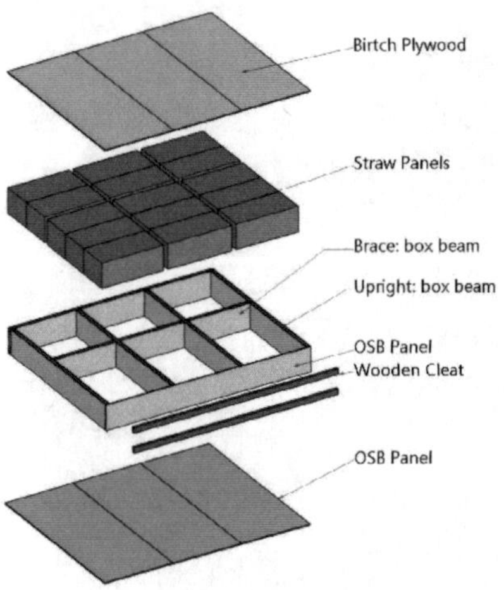

Figure 15. Wall-block configuration

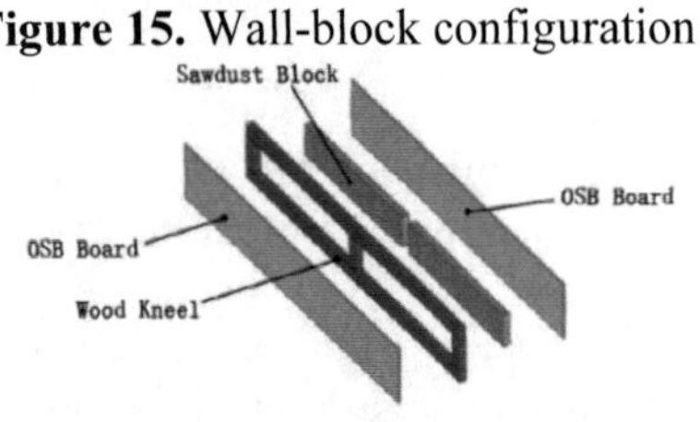

Figure 16. Louver configuration

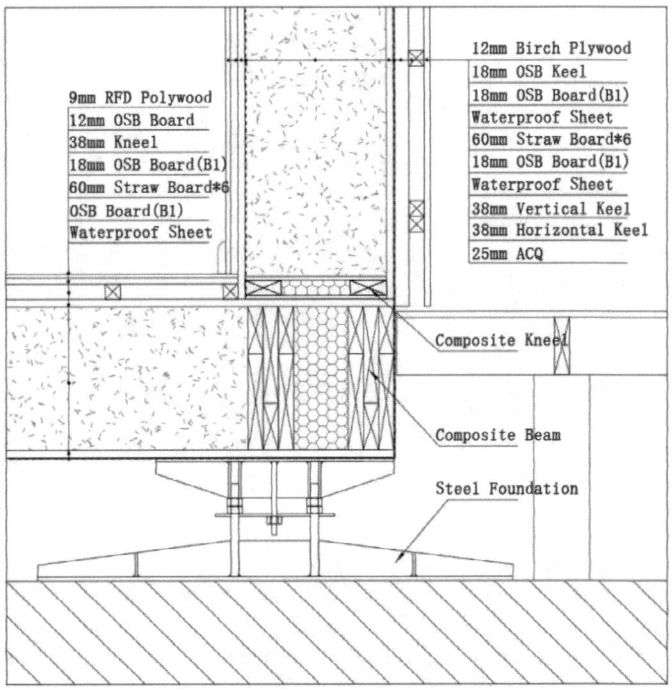

Figure 17. Detail drawing

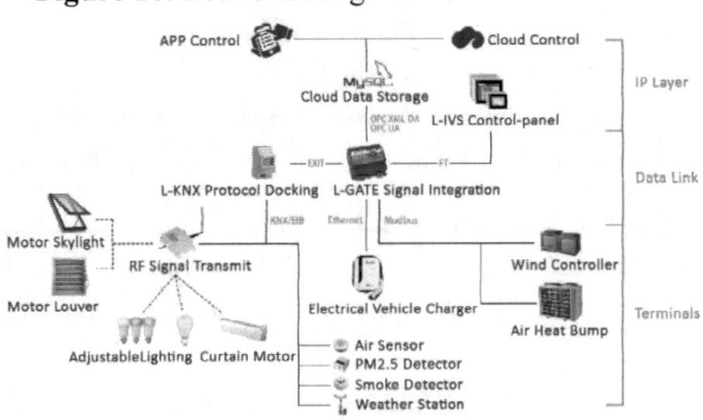

Figure 18. Intelligent control system

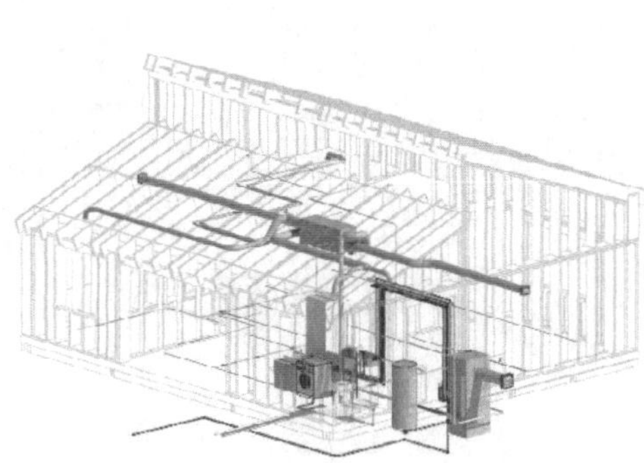

Figure 19. BIM model of building equipment

4. Testing Data and Conclusion

Climate data during the competition were measured using a meteorological station. Air temperature, humidity and solar radiation during the competition were recorded.

Fig. 20 shows the air temperature and Fig. 21 shows the humidity during the competition indoors and outdoors. The outdoor air temperature often exceeded 40°C at noon and 30°C at night, and the air humidity was also very high. Thus, heat gain had to be reduced and a great deal of heat had to be removed to reach the 22-25°C indoor temperature standard of the competition. The HVAC system was used together with passive strategies, such as creating heat-buffering spaces, using dynamic shading to reduce heat gain from solar radiation and opening the skylight of the atrium for ventilation at night. Meanwhile, the temperature at inner-yard was kept around 30°C without the help from the HVAC system during the test, significantly lower and much more stable that the outdoor temperature, which proves the passive strategies effective.

Solar radiation intensity on the competition site was measured from August 2nd to August 16th. It was sunny and extremely hot, except for a heavy rain on August 14th.

Fig. 22 shows the solar radiation intensity and generation power of PV panels in early August. The peak intensity of solar radiation during those days was approximately 800 W, and the peak power production was approximately 11 kW. Fig. 23 compares the electric power generation and consumption during the period. By the end of the test, there was a total energy surplus of 37.00 kWh, which means that the house can meet the zero-energy building standard, and its energy efficiency strategies are feasible.

For a dwelling building, it's of vital importance to achieve balance between living comfort and energy efficiency. Passive shading measures shouldn't obstruct the natural lighting. The simulation of day lightning is made with Ecotect, and we revised the distribution of louvers several times so that the house is correctly lighted except the machine room (housing requires an average of 200 to 400 lux). Thus, during the construction, we guaranteed living quality for resident along with energy performance.

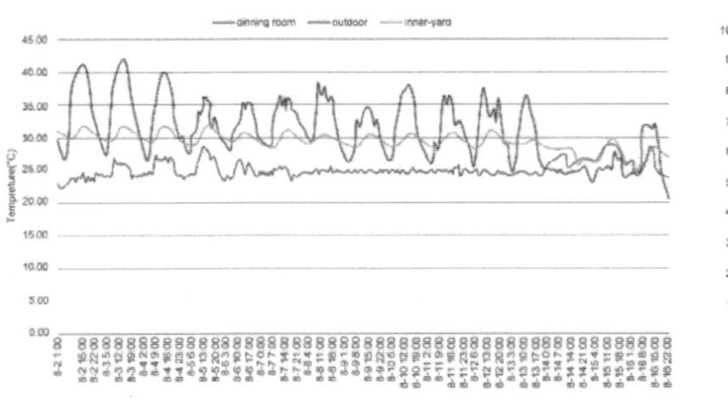

Figure 20. Air temperature during the contest

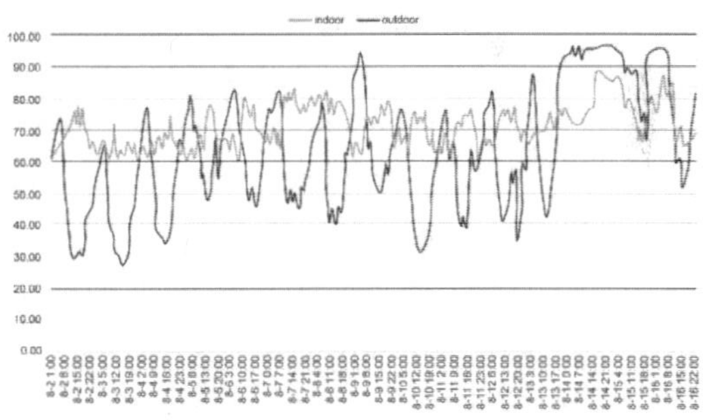

Figure 21. Humidity during the contest

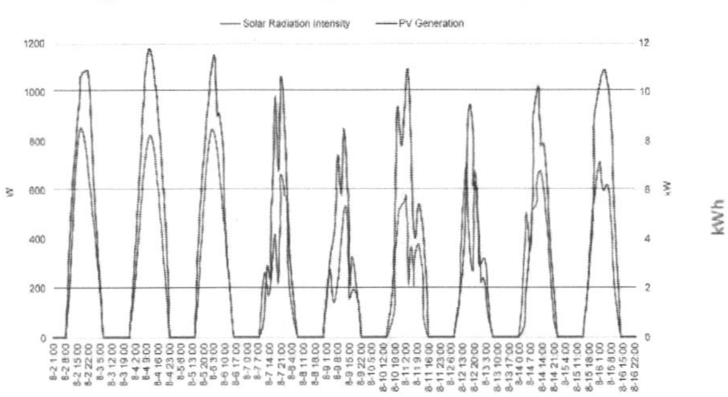

Figure 22. Solar radiation intensity and PV generation

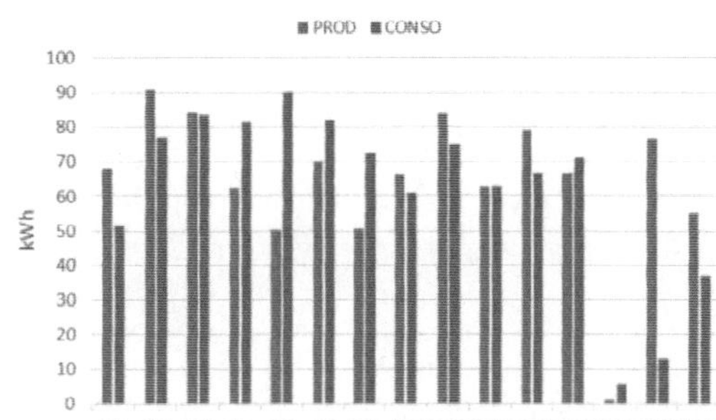

Figure 23. Power production and consumption

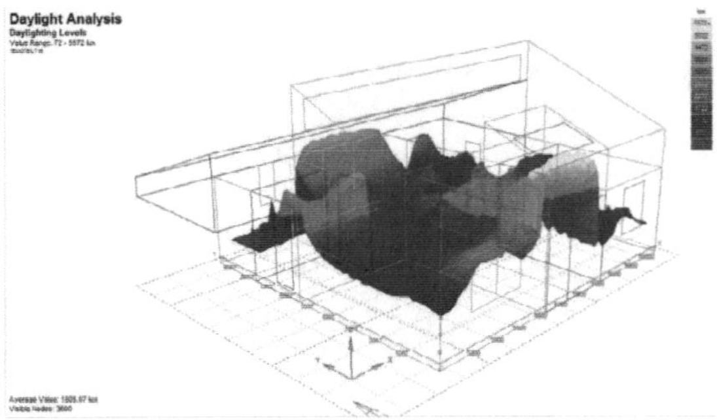

Figure 24. 3D Daylight Analysis

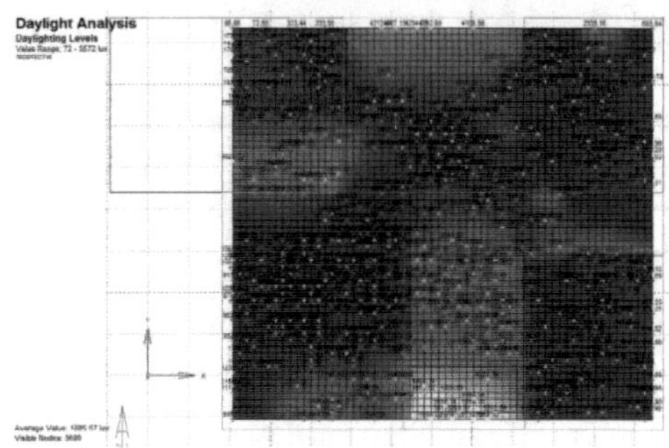

Figure 25. Daylight Analysis Plan

References

[1] Solar Decathlon China Committee, Solar Decathlon China [EB/OL]. [2018-07-01]. http://www.sdchina.org.cn.

[2] Wang Shaosen. The Perspective of Architecture: An Introduction for the Art of Architecture [M].Beijing:Science Press，2000：88-89.

[3] Chen Xiaoyang, Xuejiawei, Zhen Bin. Field Study on Thermal Environment of Quanzhou Traditional Shoujinliao Residence in Summer*[J]. Architectural Journal,* 2010(s1).

[4] Wang Shaosen. Study on the Regional Expression of Contemporary Architecture in South Fujian [D].South China University of Technology,2010.

[5] Shi Feng, Hu Chi, Zheng Weiwei. An Analysis of Design Strategies of Operable Building Skins with Dynamic Adjustment on Environment Parameters: The Example of the Solar Decathlon*[J]. New Architecture,* 2017(2).

[6] Biloria N, Sumini V. Performative Building Skin Systems: A Morphogenomic Approach Towards Developing Real-Time Adaptive Building Skin Systems*[J]. International Journal of Architectural Computing,* 2009, 7(4): 643-675.

[7] Loonen R C G M, Trcka M, Cóstola D, et al. Climate Adaptive Building Shells: State-of-the-art and Future Challenges*[J]. Renewable & Sustainable Energy Reviews,* 2013, 25(5): 483-493.

IOP Conf. Series: Earth and Environmental Science **329** (2019) 012052 doi:10.1088/1755-1315/329/1/012052

Design of Earthship for Climate Conditions in Macedonia

Todorka Samardzioska[1,3], **Nena Salih**[2], **Vesna Grujoska**[1] **and Milica Jovanoska**[1]

[1] Ss. Cyril and Methodius University in Skopje, Faculty of Civil Engineering, 24, Partizanski odredi blvd, 1000 Skopje, Macedonia
[2] New Vision, Skopje, Macedonia
[3] corresponding author: samardzioska@gf.ukim.edu.mk

Abstract. The Earthship, as a sustainable nearly zero house, built of natural and recycled materials, represents one of the solutions for the numerous environmental issues. A specific location is chosen by the Lake Ohrid in Macedonia. The house is designed for a single family, with dimensions and materials selected according to recommendations in the literature. It is not connected to the public infrastructural network, reducing its dependence on public services and the use of fossil fuels. It is built of natural and recycled materials, having thermal solar heating and cooling, collection and storage of rain water, controlled wastewater treatment, power from sun or wind and food self-production. Static and dynamic analysis has been performed and stability in this highly seismic region is confirmed, followed by the analysis of the energy efficiency and bill of quantities. This design meets the passive house criteria and the planned investment is considerably lower in comparison to a passive house built of "standard" building materials. The potential benefits of this designed Earthship might have an impact in the future building construction and architecture in Macedonia, especially for suburbs. They might help raise people's awareness of the benefits of sustainable architecture: reduction of the greenhouse gas emission, saving fresh water reserves by harvesting rainwater and reusing it as grey water, controlling the micro climate of the suburban areas by using energy efficient materials and choosing the right location for building.

1. Introduction

1.1. Earthship building

Responsible behaviour towards nature will be one of the most important challenges in the future. Construction of energy efficient buildings incorporates harmonious design of the building, solar orientation, precise positioning of glass surfaces in the building envelope, adequate volume of the building according to the planned built area, durability of materials, use of natural and local materials, reuse of already used materials and economic viability.

Republic of Macedonia has great potential for using renewable energy sources, according to the Strategy for energy development, [7]. Namely, the unused potential of renewables, which can be utilized by 2030, is 1565 GWh annually for heat generation and 3795 GWh annually for electricity production. In the past, Macedonian traditional houses were built with local materials, characteristic for the surroundings where the houses were to be built.

Earthship is a passive solar house built of natural and recycled materials. The specific location for this building is chosen in the settlement of Lagadin, along Lake Ohrid. Buildings built on a location along water surface are exposed to abrupt climate change, wind, sunshine, extreme heat, water damage, etc. In addition, these site types are sometimes not connected to an electricity grid, waste, pipeline or plumbing. The construction of passive houses at such locations is a combination of the necessary solutions of the natural conditions of the site, with technologies that maximize energy efficiency and minimize the impact of the building on the environment. This house is designed to create conditions for forming a sustainable community, which will offer opportunities for further development. As Freney et al [3] states, each of the houses of that community creates its own electricity, supplies it with its own water, has its own sewerage and treatment of wastewater, produces bio fuel, and raises a good portion of the food needed on a daily basis. Furthermore, the buildings are heated and cooled naturally and independently. Since Earthship houses are designed as structures that exist in harmony with their surroundings, it is important to have their own independent systems and to use materials that are sustainable and ready to use. These houses, in order to be completely self-sustaining, need to manage the three main systems: Water, Power and Climate.

The first Earthship house was built over forty years ago in Taos, New Mexico by Michael Reynolds, [9]. Design is constantly evolving and improving, opposing the conventional architectural concept. The first Earthship in Europe was built in 2004 at Kinghorn Loch in Fife, Scotland, while the second one was built in 2005 in Stanmer Park, Brighton, England, both in UK, [6].

Following the recommendations in the literature [1], [4], [6], [9], a design for an Earthship at a specific location near the Ohrid Lake is proposed in this paper. The advantages and potential benefits of the proposed house are summed and pointed out.

1.2. State of the art in energy in Macedonia

Total electricity production of 5646 GWh in Macedonia in 2015, divided regarding to the source, is presented in figure 1, according the official statistics of the Macedonian Energy Agency, [2]. Additional 2656 GWh were imported, and negligible 143 GWh were exported in the same period. The domestic energy production in Macedonia is insufficient to satisfy the increasing energy consumption.

Regarding the renewables, due to the hydrological conditions in the region, hydroelectricity is the most harvested among the renewable sources, following by biomass, whilst solar power offers a minimal contribution. Hydropower can cover 10% to 20% of the total demand for electricity. There are 9 big hydropower plants with total production of 1392 GWh per year, [2]. According to the energy balance, the biomass participates with 6% of the total primary energy in the Republic of Macedonia, i.e. 9.5% of the total energy consumption. The wood is the most used resource from biomass. Biomass, of wood or coal, is almost exclusively used in the domestic sector. Local geothermal waters are of insufficient temperature, and they are not used for electricity generation. Currently, there are 18 known geothermal fields, with more than 50 geothermal sources with total outflow of 1000 l/s with temperature of 20°C - 78°C. Macedonia was the first country in the Western Balkan region to put into operation a sizeable wind facility, wind farm of 36.8 MW at Bogdanci. Solar PV capacity is also the highest compared to neighbouring countries, but the potential is hardly used, [2]. The energy consumption in the Macedonian residential sector, according to the official statistical data for the period of 1996 to 2006, grew by 0.47% annually. However, the growth of the household energy consumption is assessed at nearly 2% annually, taking into account the unrecorded consumption of firewood and electricity, [7]. Figure 2 represents the final consumption of electricity in the residential sector as 47.4% of the total final consumption of electricity in 2015 in Macedonia, according to IEA. On the other hand, the annual growth of construction of new homes in the last six years is 5.25%, [7]. The average consumption by types of energy commodities per household and per capita at the national level, as well as average energy consumption in households that consumed the particular energy commodity in 2014 are: fuel, wood and other plants with 61.59%, electricity 28.5%, LPG 0.44%, derived heat 8.33%, coal 0.07%, heating oil 0.1%. Regarding the CO_2 emissions, 3.48 tons per capita were calculated in 2015.

Given the current environmental problems and the increase of energy consumption in Macedonia, it is of great importance that adequate actions are taken. They include preservation of the vegetation, implementation of renewable sources of energy, construction of energy efficient buildings, building with local and sustainable building materials etc.

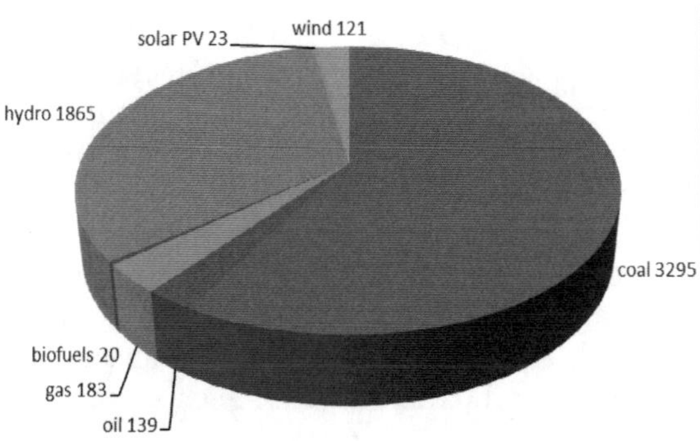

Figure 1. Total electricity production in MK 2015.

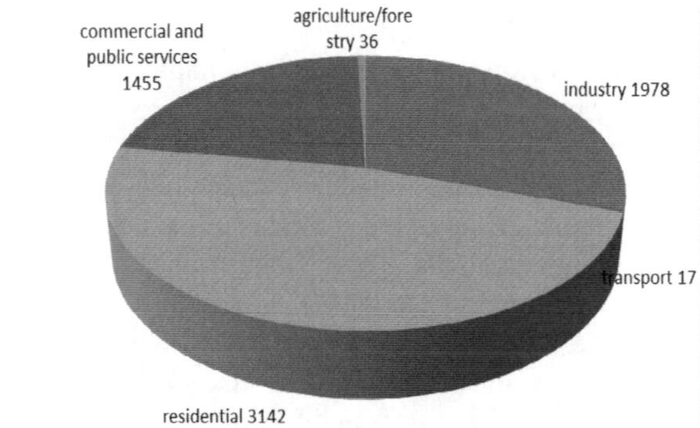

Figure 2. Final electricity consumption in MK 2015.

2. Design of Earthship house

2.1. Location and climate conditions

The location chosen for this Earthship house is in the settlement of Lagadin, with geographical coordinates of 41.1°N 20.8°E. The house is located by the Ohrid Lake and is 9 km away from the Ohrid city centre. The settlement of Lagadin has been developed in the second half of the 20th century and its construction is still on-going. The building plot has an area of 1200 m^2, while the maximum building area is 216 m^2, with maximum permitted height of 9 m. All the houses are designed to have south-west orientation, receiving maximum solar gains and visibility to the lake. Figure 3 represents the solar radiation in Ohrid during the year; south, west and horizontal directions only, as the most representative ones. The climate in Macedonia is mixed, continental with Mediterranean, cold in winter and hot in summer. The country has no outlet to the sea and it is separated from the Adriatic and Aegean Seas by mountain ranges. The monthly average temperatures in °C for Ohrid are given in figure 4. Location of the building and its orientation in the neighbourhood is presented in figure 5, while its designed 3D view fitted in the environment is shown in figure 6.

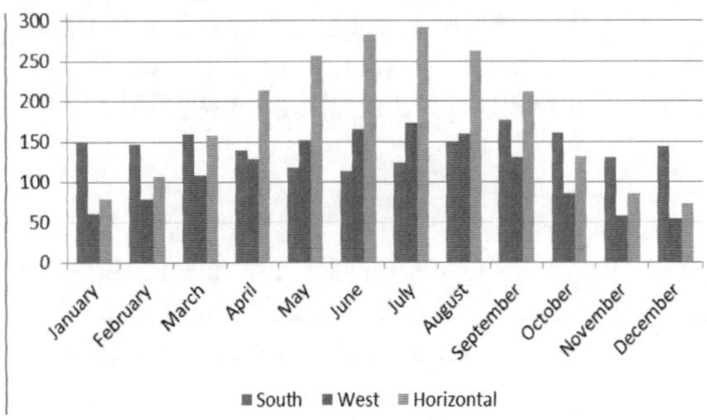

Figure 3. Solar radiation in Ohrid, for south, west and horizontal direction, in [W/m^2].

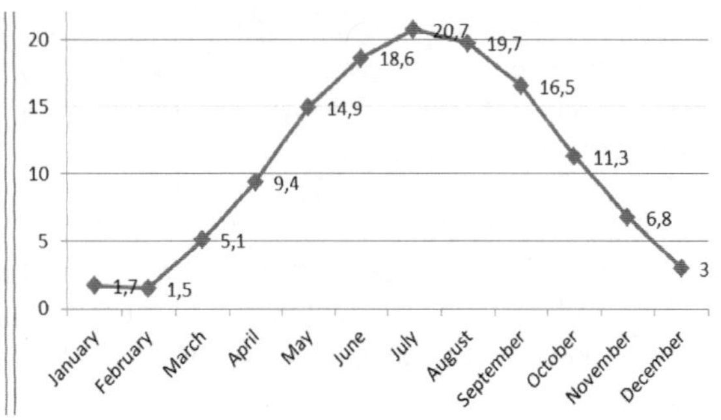

Figure 4. Average monthly temperature in Ohrid.

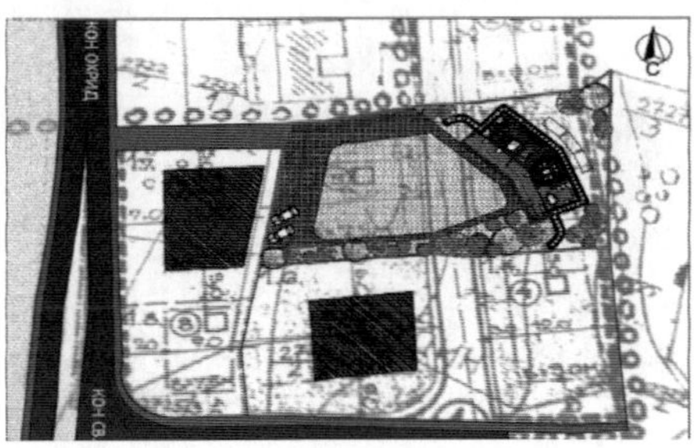

Figure 5. Location of Earthship.

Figure 6. 3D view of the building and its environment.

2.2. Materials for the Earthship

One of the ways to integrate sustainable architecture in buildings is through the use of recycled and natural building materials. Given the already mentioned energy problems, as well as the energy used in the manufacturing process of conventional building materials, air and water pollution, the use of these materials is an imperative in the house like Earthship.

The most impressive material used, which represents the main bearing part of the Earthship, are actually used car tires filled with rammed earth, cans and tins as partition walls, glass bottles as fillers and decoration of exterior walls. In addition to these materials, due to the stability and durability of the building, natural materials are used in a certain parts, such as wood, glass, and also conventional materials such as concrete for buttresses on the supporting wall of tires and other segments, materials for thermal and hydro insulation etc.

2.3. Geometry of the building

The gross dimensions of the base are 26.2 m x 11 m, including the thermal layer of 10 cm expanded polystyrene with air barrier and 100 cm thick fill of earth, figures 7 and 8.

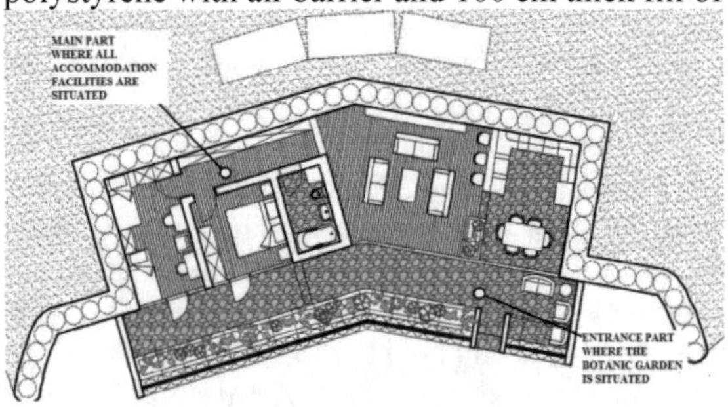

Figure 7. Layout of the building.

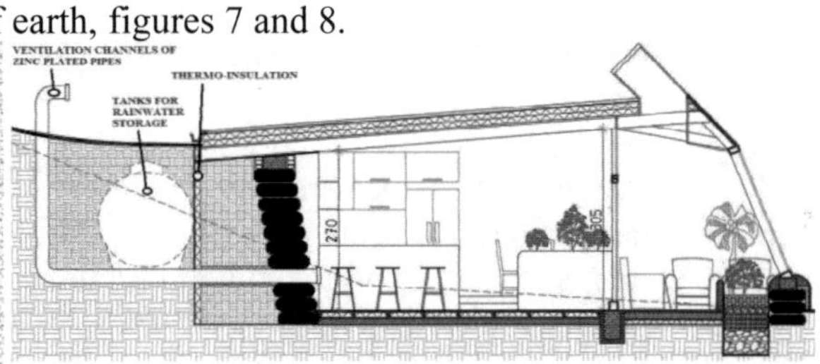

Figure 8. Cross section of the building.

One row of the wall of tires, filled with a compacted soil, consists of 46 tires R15 in total. Tires' dimensions are taken as d = 76 cm and height h = 24 cm for the analysis. It has 11 rows, with a total height of 264 cm, which does not exceed the recommended Earthship Biotecture of 300 cm, [9]. The recommendations of Reynolds [9] and Griepentrog [4] are followed regarding the adopted dimensions of the other elements, as well.

2.4. Solar and wind energy

The sunbeams fall at a 22.8° angle on the shortest day of the year - December 21th, dictating the slope of the wooden structure oriented to the southwest side to be at a 67.2° angle, see figure 9. In this way, the total solar insulation is gained during the winter, while the hot summer sun is prevented to enter the house.

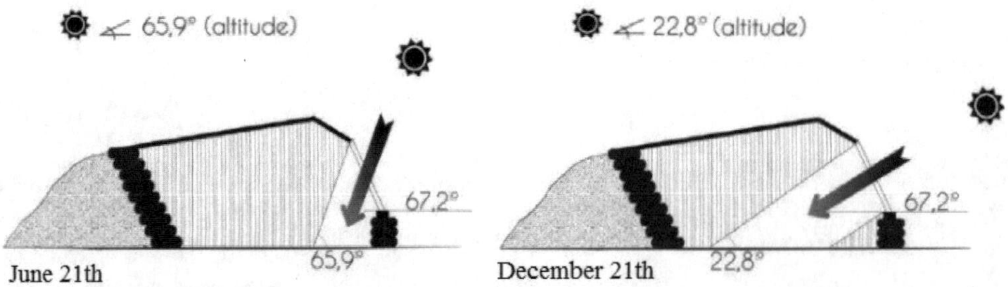

Figure 9. Position of the sun to the southwest facade in June 21th and in December 21th.

The particular building has unfavourable position in terms of generating wind energy; the mean wind speed in this region is relatively low, 3.4 m/s, while the minimum required for the turbine to start generating energy is 3.5 m/s. Furthermore, the wind has no constant continuity, and Ohrid is located in the geographical zone I regarding the intensity of wind (area of moderate winds).

Therefore, in terms of energy supply, the total energy is designed to be provided with solar photovoltaic systems. The biggest solar radiation in the Republic of Macedonia is determined in the south-west part of the state as 1530 KWh/m^2. Produced electricity can be used directly or stored in batteries and used according to the needs. Six modules of photovoltaic of 240 W, with dimensions 980 x 1660 x 32 mm are provided for this Earthship house. Their produced energy is stored in 6V (325 Ah) batteries, grouped into two groups of 5 batteries mutually connected, which manage to meet the minimum needs of 5 kWh per day designed by the architect. Similar recommendations are given by the group of architects which originally created Earthship house, [4].

2.5. Collection and storage of water

Earthship houses are designed to be autonomous, both in terms of water supply and communal networks. Therefore, water is provided by taking, directing and storing rainwater through the roof. Table 1 represents the average annual and monthly precipitation (in mm) in the Ohrid valley (National plan for management with National park Galichica 2011).

Table 1. Average monthly and annual precipitation in mm.

Months												Per
I	II	III	IV	V	VI	VII	VIII	IX	X	XI	XII	year
76.4	72.6	59.8	50.3	63.7	36.4	23.2	21.1	47.4	71.9	98.4	78.9	708.3

In the vicinity of Lake Ohrid, where the building is estimated to be, the annual amount of precipitation is slightly larger than the Ohrid valley and it is 759 l / m^2. On the other hand, estimated water consumption was 270 l per day per resident for 2015. Given that the Earthship house uses the water four times, it is calculated that 70 l per day per inhabitant have to be provided (270 l : 4 ≈ 70 l).

The roof area is 150 m^2, and then the total amount of water is about 114 000 litters per year, which meets 70 l of the total quantity of water needed per day per inhabitant, for a family of four members.

Regarding the installation of the entire system, 3 tanks of 6000 l are provided. The water is gravitationally directed from the tanks in a water organization module, whereby filters are used to purify the rainwater from bacteria and other pollutants. Then, the water is brought into a standard pressure vessel in order to achieve water pressure as in a conventional housing building. This water is further used for any activity in the household, except for the rinsing of the toilet, where grey water is used, which is actually waste water from the bathtub or sinks in the house that also passes through a particle filter and continues into the botanical garden in the front of the house. In some cases, at the request of the investor, the water from the sink in the kitchen is directed directly to the black water in order not to affect the grease of the filter from the botanical garden. When passing through the botanical garden, the water is filtered through the gravel and through the roots of the plants, whereby oxygen is added and the nitrogen is taken off the water. In addition, the water flowing through the plants and evaporating from them, maintains the humidity of the air in the room.

2.6. Controlled wastewater treatment

The waste water problem is solved independently of the city sewerage system, like other systems within the Earthship house. The grey water, after being used to flush the toilet, becomes black water that is directed to a septic tank for anaerobic bacteria decomposition. The septic tank is dug into the ground, it is coated with additional insulation and has a south-oriented window through which the pit is additionally heated by the sun and thus accelerates the anaerobic process. When the black water comes to the septic tank, the liquid part is separated from the solid, and then the treated water is directed through an external filtration and trenching field to the outer cell with decorative plants, similar to the inner botanical cell.

3. Analysis of the building

3.1. Static analysis

The analysis of the supporting wall is made in correlation with the paper of Zimmerman [12], which offers explanation of the problem through a more comprehensive analysis.

A realistic model has been used for static analysis, with 3 tires in length and 8 in height, filled with earth and connected to each other. The modification regarding the standard method of stacking and compacting the tires with earth, consists of cutting off the side of the tire from the open sides facing to the upper part, thus facilitating the fulfilment and compacting of the ground.

The coefficient of friction between two tires filled with soil and between tires and the ground, has been determined. For the R15 type of tires, the coefficient of friction "tire – tire" is 0.588, and "tire – ground" is 0.697. Its effect has been simulated by application of a force at the centre of the earth pressure diagram. The results of the analytical analysis include analysis of the stability of the structure, in terms of overturning and slipping, ground - tire and tire-tire, as well as in relation to the height of the wall, and expressed through the allowed active earth pressure.

The analysis has been performed for two types of structure:

- without inclusion of the bearing simulated by the reinforced concrete connecting beam, on which the roof beams are mounted, which additionally increases the stability, and
- with its inclusion (laterally restrained tire walls).

The incorporation of the roof into overall stability substantially better reflects the real behaviour of the wall when used specifically in Earthship, although it is noted that the results obtained by the analysis are quite conservative and on the side of reliability, since they do not include the buttresses and the placement of the structure in Π - form, which also additionally increases stability.

This wall is the most magnificent element of the Earthship, regarding the insulation characteristics. About 800 car tires R15 are needed to build the wall of this house. Tires are organized and placed in the same way as bricks, with each row of tires being set 3.5-4 cm behind the row below it. In that way, the wall is tilted to the outside and supported by the compacted ground. Galvanized tubes are placed in two places above the third row of tires, presenting ventilation ducts, coated with 5 cm of EPS, through which the natural ventilation of the interior is carried out. A small mesh is placed at each end of the tube for protection against insects and small animals. Expanded polystyrene thick 10 cm for thermal insulation on a surface coated with a vapour membrane is placed behind the back side along the entire length of the tire wall at a distance of 100 cm, figure 10. The rainwater collection tanks are placed behind the thermal insulation and are completely insulated with 5 cm EPS.

The roof is designed as a timber structure, with external roofing of metal panels that enable maximum collection of the rain water. Its cross section is given in figure 11. On the top of the house, there are ventilation openings that improve the natural movement of the air together with the ventilation pipes. The heat transfer coefficient for the roof structure is calculated as 0.13 W/m^2K.

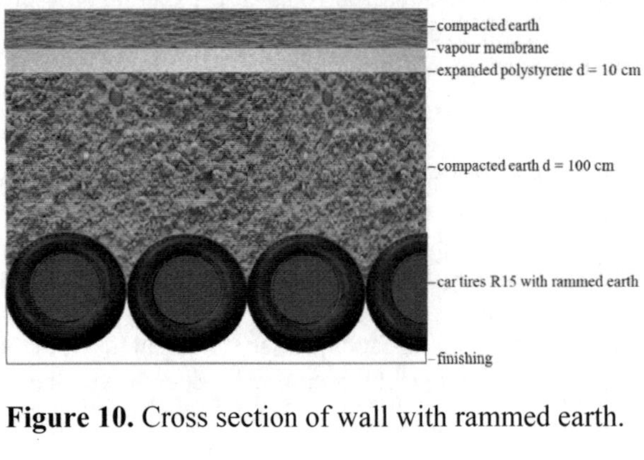

Figure 10. Cross section of wall with rammed earth.

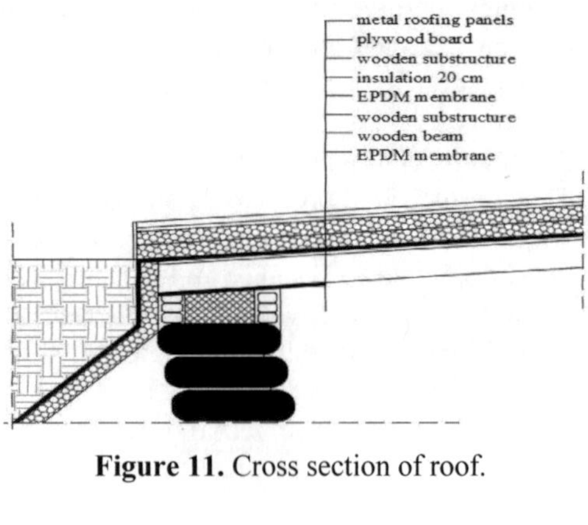

Figure 11. Cross section of roof.

Thermal barrier, made up of the wall of tires, compacted earth and expanded polystyrene foam have the greatest role in maintaining the thermal comfort and maintaining the desired indoor temperature. It has the best thermal insulation characteristics compared to other structural elements. Calculated U-value for this wall is 0.12 W/m^2K, which is a value typical for passive houses.

The partition walls between the bedrooms and between the bedroom and the bathroom are built of two layers of cans (0.33 l) with cement and 5cm thermal insulation between the layers. The use of cans helps to form a wall mass and reduces the necessary amount of cement. The wall of cans is easy to build, and on the other hand recycling of the cans helps in the waste reduction. The partition wall between the bathroom and the living room, and the walls in the corridor of the house are built in the previously described manner with cans in the lower part, and in the upper part with "glass bricks" connected with cement. One glass brick is made of two glass bottles of the same kind, cut into the throat, washed and glued with adhesive tape. All the walls are plastered and painted. The U-value of the wall made of cans is calculated as 0.09 W/m^2K.

Only the floors in the house are designed of reinforced concrete slab, placed on a levelled ground soil and the vapour barrier on it. Given the climatic conditions at the location where this house is planned, a 10 cm thermal insulation should be installed above the vapour barrier. Above it there is another layer of vapour barrier, on which the mesh reinforcement is arranged and concrete is placed. The floor in the rooms is with finishing of brown wooden flooring, with U-coefficient of 0.21 W/m^2K; while the flooring in the toilet, kitchen and at the entrance hall is made of stone tiles and its U-coefficient is U=0.23 W/m^2K.

3.2. Energy consumption analysis

After determining the coefficients of thermal conductivity of all elements of the building envelope, the ENSI Key Number software is used to calculate the energy consumption of the building. The software includes the location of the building and the type of building, [11].

The schedule of using the premises in the house is entered as 16 h/day per working day, according to the standard working hours of 8 hours, and 20 h per weekend days, because an absence of a part or all members of the family of 4 hours a day is assumed. The Earthship house does not use electricity for further heating, but a value greater than zero must be entered in the heating schedule fields, even though it is not necessary. Furthermore, regarding the systems for heating, cooling, ventilation and sanitary hot water, most of the input values are zero, because the Earthship house uses natural ventilation, generates its own electricity through solar energy, uses solar energy to heat water and uses the thermal properties of the earth. Finally, the calculated energy consumption for heating and cooling of the house is 12.5 kWh/m^2 annually. The results obtained with the ENSI Key Number Software for the total annual energy consumption of the house, including the lighting, electrical equipment and pumps, is about 45 kWh/m^2. The electricity used in the house is the energy obtained through the photovoltaic panels, and without the use of electricity from the city's electricity system. The additional heating of the water from the solar collectors, if required, is carried out with the aid of a gas, and the additional heating of the premises is done with a fireplace or a wood / pellet stove.

Unlike the ENSI Key Number software, a simple 3D model of the building is drawn in Design Builder software. Materials of each constructive element are input data, as well as the calculated U values and structural characteristics of each of the elements. The model is divided into zones according to the activities taking place in each room and the respective parameters are entered for each zone. Results from the conducted analysis show the thermal balance during the year and adequate comfort in the house.

4. Conclusions

The Earthship designed for a specifically chosen location in the settlement of Lagadin, along Lake Ohrid, is a sustainable passive solar house. Built of natural and recycled materials, it represents one of the solutions for the numerous environmental issues. The house is designed for a single family, with dimensions and materials selected according to the recommendations in the literature. The house is an off grid house, i.e. it is not connected to the public infrastructural networks, reducing its dependence on public services and the use of fossil fuels. It is built of natural and recycled materials, having thermal solar heating and cooling, collection and storage of rain water, controlled wastewater treatment, power from sun or wind and food production. Static and dynamic analysis of the building determined the dimensions of the structural bearing members. All the necessary checks were performed and controlled.

The results of the conducted analysis for energy consumption of the building show that the Earthship house meets the passive house criteria, [8]. Heating energy demand is 12.5 kWh/m^2 and it does not exceed 15 kWh per square meter of net living space. Its total energy consumption to be used for all domestic applications (heating, cooling, domestic hot water, lighting and electricity for household appliances) is 43.5 kWh/m^2, i.e. less than 120 kWh per square meter of treated floor area per year. Furthermore, according to the results from the Design Builder Software, the thermal comfort in the house is met for all living areas during the all seasons, with not more than 10% of the hours in a given year over 25 °C. The financial estimations show that the construction of the Earthship house requires 20% less investment funds for construction compared to a passive house built of reinforced concrete in an urban area, with the same living space, with classical HVAC systems incorporated, [10]. The main advantage of the Earthship house is that it is an off-grid house.

According to the conducted research, the appropriate location for the construction of the Earthship house is in the suburbs or periphery of the larger cities, due to its construction and because of the space it occupies. There is an opportunity to build buildings for collective living of this type, but they will still have the same limit in terms of space and number of floors. Finally, these types of houses are environmentally friendly, affordable and energy efficient alternative to conventional houses, without reduction of the living comfort.

References

[1] Bauer M 2010 *Green Building – Guidebook for Sustainable Architecture* Springer-Verlag Berlin Heidelberg New York
[2] Energy Agency of Republic of Macedonia at http://www.ea.gov.mk/index.php?lang=en
[3] Freney M, Soebarto V and Williamson T 2013 Thermal comfort of global model Earthship in various European climates *Proc. of BS2013: 13th Conf. of Int. Building Performance Simulation Association* Chambéry France August 26-28
[4] Griepentrog T E 1990 *Engineering Evaluation of Earth-filled Tire Construction* Consulting Engineers & Geologists, Colorado
[5] Hoyle B 2011 Low *Energy Building Engineering* The English Press Delhi
[6] Miller A and Ip K 2005 *The Brighton Earthship: Evaluating the Thermal Performance* University of Brighton Centre for Sustainability of the Built Environment
[7] Ministry of Economy of Republic of Macedonia 2010 Strategy for energy development in the Republic of Macedonia until 2030 http://www.ea.gov.mk/projects/unece/docs/legislation/Macedonian_Energy_Strategy_until_2030_adopted.pdf
[8] Passive House Institute 2014 *Active for more comfort: Passive House, International Passive House Association* available at http://www.passivehouse-international.org
[9] Reynolds M 1990-1993 *Earthship* vol 1 2 and 3 Earthship Biotecture
[10] Salih N 2015 MSc thesis: Comparative analysis between modern passive house made of new energy efficient materials and passive house made of natural and recycled materials. Ss. Cyril and Methodius University Faculty of Civil Engineering in Skopje Macedonia
[11] Software and User Guide 2012 *ENSI Tools on Energy Auditing in Buildings* Energy Saving International AS
[12] Zimmerman A 2011 *Testing and Analysis of Modified Rammed Earth Tire Walls* E90 Project Swarthmore College Philadelphia

IOP Conf. Series: Earth and Environmental Science **329** (2019) 012053 doi:10.1088/1755-1315/329/1/012053

Assessment of carbon foot print of Jaffna Town, Sri Lanka

S Raveendran[1,2], and V Srikaran[1]

[1] University of Jaffna, Sri Lanka,
[2] Corresponding Author

Abstract. Jaffna town is one of the emerging urban centers in Northern Sri Lanka which attract more people, resources and activities in the post war scenario. This study assesses the carbon foot print of the town. Cities play a key role to top up greenhouse gases into the atmosphere worldwide which lead to global warming and consequent climate change and related impacts. The carbon footprint of Jaffna town has not been assessed yet and the awareness of low carbon development is not common in the development domain which are essential to make decisions to carry out development consistent with Sustainable Development Goals. The purpose of the research to assess the gross carbon emission of Jaffna town and sectoral composition of the emission. The study shows that the city emits 312,096,821 Kg Co2e per day. More emission comes from domestic activities which account for 60% and second largest emission comes from transport sector. Nearly 60 % of the commuters are using private vehicles that account for nearly 20% of the total emission. Among the domestic emission, light and ventilation consume more energy than other uses. Solid waste management also contributes to emit nearly 6 – 10 % of the emission of the city. More houses consume relatively higher energy for light and ventilation because of the poor design of house that do not utilize the natural light and ventilation. The study suggest that the building design must adhere to the low carbon building designs which is not present in Jaffna. Appropriate building design that minimize the energy consumption need to formulated and implemented. Use of centralized air conditioning system in public building will minimize the energy consumption compared to distributed system. Integrated waste management system and sanitary land fill facilities will minimize the emission from solid waste disposal. Suitable urban design will encourage walking and cycling of residents and commuters. Deliberate planning of routes and facilities of public transport will promote the use of public transport between Jaffna and other suburban towns in the Northern Region.

1. Introduction

This research has assessed the level of carbon emission into the atmosphere from various activities in Jaffna town, Sri Lanka. Jaffna town is a first order city in the Northern Province of Sri Lanka and growing rapidly after the end of civil war. Though, Jaffna town is a smaller town center compare to the cities around the world, all urban areas depending on the size of the city and functions, they emit greenhouse gasses into the atmosphere. Based on the general principles of sources of green house gas emission, possible sources of emission were studied in this research. Data on transportation activities of residential population, retail and wholesale business activities, government and private institutions, energy use in residence, business places and institutions, goods and services consumed by residential people and institutions were collected using questionnaires and waste generated and disposed from the city were collected from statistical records of Jaffna Municipal Council. In each sector, per unit emission was calculated based on the survey data using carbon calculators and total emission of the city was calculated considering the total number of entities.

Jaffna town is located at the Northern most part of Sri Lanka, nearly 400 kms away from the capital city of Colombo. The administrative areas of the town cover nearly 20 square kilometers and actual urban areas have spread beyond the administrative boundary. Jaffna town is mostly serving the Jaffna Peninsula and the Northern Province.

2. Research Problem

Cities around the world account for 80% of the green house gas emission and each city in the world has stake in the total emission[1]. More than 70% of the energy consumed by cities in the world which are burned and emitted as greenhouse gases into the atmosphere[2]. Cities proportionately emit greenhouse gases, depending on the total population, urban form and functions. Number of people living in Jaffna town has been increasing rapidly during the last ten years. There are more than 100,000 people live in the city and nearly 65,000 people are commuting to the city daily. Number of automobiles used by the residents and commuters are increasing day by day. There are larger number of construction projects undertaken in various parts of the town. Land use and land cover changes are taking places at different scales in different walk of life of the city. More than 63% of the land use is allocated for residential activities. Residential areas consist of home garden with greeneries. There are 25238 housing units in the town area. Green land cover and natural environments are converted into unnatural and constructed surfaces. Since there are activities related to energy consumption and methane productions, Jaffna town also contributes to add carbon into the atmosphere which has been unaccounted in the past[3]. Every aspect of Jaffna Town add carbon into the atmosphere. They are land use conversion, waste disposal, increasing dependency on automobiles, waste water disposal, energy consumption at households, institutions and business places. Though the global Warming and climate change are global issues, created by local activities, mitigation requires local response to them by minimizing carbon emission from various activities locally. What is measured get managed is true in carbon emission as well. Therefore, the current condition of the emission from various activities of the town needs to be assessed in order to take decisions with respect to low emission urban design.

3. Objectives

This research has four objectives to;

- Assess the total carbon emission of Jaffna Town
- Assess sectoral contribution of carbon emission
- Assess per capita emission
- Formulate guidelines to control low carbon development in Jaffna Town

4. Research Design

This research has been conducted based on survey data and quantitative analysis to assess the total and sectoral carbon footprint of the town. The aim of the research is to assess the level of carbon emission and devise methods to create low carbon environment. It required to assess per activity emission and sectoral emission which are the fundamental level functional understanding necessary for planning a low carbon environment. The activities that emit carbon were mainly considered transport, energy, goods and services and solid waste. Questionnaire survey was conducted to collect data on activities that have the nature of carbon emission. The questionnaire consisted of open ended and closed questions under four sections. The transport sector consisted questions of what type of transport the people live in the city use, type of transport used by the government and private institutions, used by retail and wholesale business people, distanced travelled, type of vehicles used, capacity of motorbikes, what type of vehicles used for goods transport etc.

Questions in energy sector consisted the following questions electricity use of households, private and government institutions, retail and whole sale business places, type of energy used, industries in the town, utility equipment used such as refrigerators, washing machines, cookers, bulbs, fans, iron boxes, air conditioners, electric water pumps. Primary data was collected from questionnaire survey and summarized and analyzed using SPSS v25. Total of 120 questionnaires were filled and stratified random sampling method was adopted to select the samples. The number of questionnaires surveyed in each sector are as follows; 40 questionnaires for householders, 20 for retail business establishments. 20 for wholesale establishments, 20 for government institutions and 20 for private institutions. The carbon emission from each activity was calculated based on the data collected from questionnaire survey and using carbon calculator software[4] available online and standalone. Per head emission in each sector was calculated and was multiplied by number of people or activity or institutions in the town to get total emission.

5. Results and Discussion

The performance of each sector that contribute to carbon emission has been taken into account and assessment were carried out. Main functions and sectors that are liable for carbon emission have been studied and data were collected. In this section, sectoral contribution such as transport, energy use, goods and services consumption, waste generation and disposal have been assessed and emission from those activities were calculated. Further, functionalities also have been assessed and their carbon emission were worked out.

5.1 Activities based Carbon Emission

5.1.1 Transport in Jaffna Town Assessing the carbon foot print of a town, assessment of transport sector emission is crucial. Transport sector contributes to the emission substantially. In this research, transport uses in different sectors were considered and data on the distance travelled collected. Each respondent was asked the distance travelled per day and the mode of transport they used. The data were summarized and analyzed. The table 1 shows the distance travelled and the mode of transport used in different sectors.

Table 1: Distanced Travelled Per Day (Kms) By Various entities in the Town

Transport Mode	Total Travelled Distance (Km)
Bicycle	315
Motor Bike	10,175
Lorry	3,417
Private Bus	3,296
Public Bus	1,716
Car/ Auto	3,799
Van	2,769
Land Master	27
Total	**22,697**

Source: Questionnaire Survey, 2014

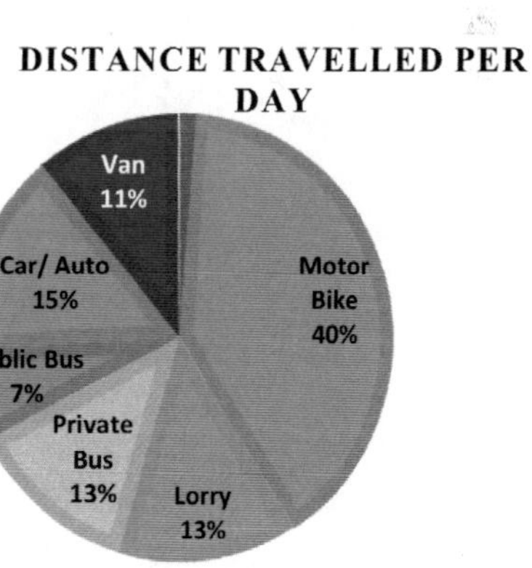

5.1.2 Motorbikes Use and Capacities In Jaffna, most people use motorbikes, which have different capacities, for transport needs. The capacity of those motorbikes used by different users in various places was also considered to assess the carbon emission. More journeys made using motorbikes by people in government and private institutions. Most motorbikes are in the 100 cc category.

Table 2: Distance Travelled Using Motorbike Types in Business Places, Institutions and Settlements (Km)

MB Capacity	Business Places
90 cc	30
100 cc	7099
125 cc	2361
135 cc	267
150 cc	195
180 cc	315
200 cc	06
Total	**10273**

Source: Questionnaire Survey, 2014

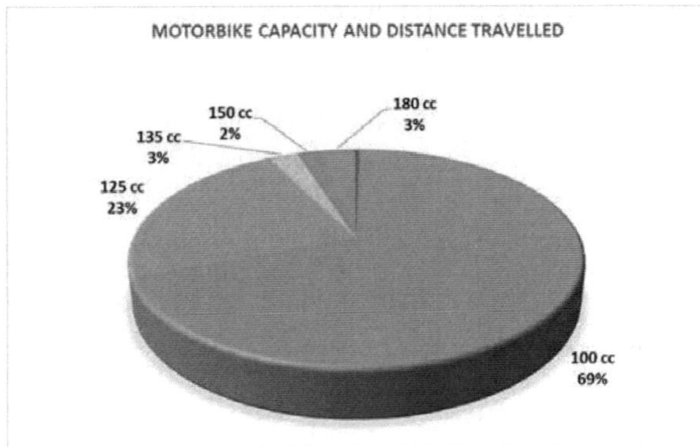

5.1.3 Usage of Electrical Equipment Emission from energy use also has been studied by assessing the number of electrical equipment used and electricity units consumed by different sectors. The total energy consumed from different functions and sectors were assessed from secondary data on electricity unit consumption. The following table shows the electrical equipment used in different sectors.

Table 3: Electrical Equipment Used in Business Places, Government and Private Institutions, and Settlements

Electrical Equipment	Business Places	Govt. Institutions	Private Institutions	Dwelling Units	Total
Bulb	451	26,260	119	608	27,506
Fan	105	13,760	1,246	164	13,879
Computer	14	340	241	51	581
Air Conditioner	08	312	183	-	495
Fridge	04	-	15	25	19
Water Pump	04	15	06	43	68
Printers	04	156	14	-	174
Cookers	-	-	-	24	24
Washing Machines	-	-	-	12	12
Iron Boxes	-	-	-	40	40
Industrial Equipment	-	-	-	04	04

Source: Questionnaire Survey, 2014

5.1.4 Electricity Consumption Electricity use contributes to carbon emission when they are generated from fossil fuel-based sources. In Sri Lanka, more than 60% of the electricity is generated from thermal power. Data on energy usage, which was represented by electricity units utilized by every separate entity, were collected from each sector. Almost all people use electricity for light, ventilation and operating household equipment in the town area. The following table 4 shows the amount of electricity consumed per day in sectors.

Table 4: Electricity Consumption in Sectors

Sector	Electricity Units Used
Retail and Wholesale Business Places	3,461
Government and Private Institutions	14,309,904
Settlements	5,046
Total	14,318,466

Source: Questionnaire Survey, 2014

5.1.5 Consumption of Goods and Services Consumption of goods and services also contribute to emit carbon since they are manufactured and processed using fossil fuel energy. Food and stationery were accounted for the estimates of carbon emission. Data on goods and Services consumed by different sectors in Jaffna Town were collected. The amount of papers consumed by users in different sectors were considered in calculating the carbon emission. The following table 5 shows the amount of papers consumed by sectors.

Table 5: Stationery Consumption by Sectors

Sector	Paper Consumed in Kg
Retail and Wholesale Business Places	65
Government and Private Institutions	388
Settlements	13.5
Total	**466.5**

Source: Questionnaire Survey, 2014

5.1.6 Disposal of Solid Waste in Jaffna Town Jaffna urban area disposes 30.2 metric tons of wastes from various activities, settlements, markets, hospital, commercial and business activities. The entire wastes are disposed in open dump site in Kallundai which lies 2 kilometers away from the town at the North Western direction. The dumping area inundates during rainy season from October to February. Open dumping of solid waste liable for emission of methane by anaerobic process and carbon dioxide by burning activities.

5.2 Carbon Emission from Activities and Sectors

The carbon emission for each activity in the town was calculated on the basis of unit emission using carbon foot print calculators. Carbon emission from transport, energy use, stationery consumption and waste disposal were calculated for the data collected using questionnaire survey and secondary sources.

5.2.1 Emission from Transports Activities If looked at the carbon emission due to the transport activities of retail and wholesale business in the town, one business place emits 31.425 Kg of carbon per day and there are 8326 business entities in the town. All business entities emit total of 261, 644 Kg of carbon per day.

The analysis shows that the transport activities of government and private institutions as follows; carbon emission of one institution is 98.34 Kg and there are 783 institutions in the town which all together emit 77,003 Kg of carbon per day.

Settlements due to transport activities emit carbon in the following manner. One dwelling unit emits 10.81 Kg of carbon per day. There are total of 25,238 dwelling units in the town area which emit 273,077 Kg of carbon per day. The table 6 below shows the distance travelled and total carbon emitted by transport activities in all three sectors in the town.

Table 6: Carbon Foot Print of Transport Activities of Business Places, Institutions, Settlements

Transport Mode	Distanced Travelled (Kms)	Carbon Foot Print (Kg)
Bicycle	315	00
Motorbike	10,175	1,240
Car / Auto	3,799	680
Lorry	3,417	1,168.86
Van/ Ambulance	2,769	703.82
Land Master	27	2
Private Bus	3,296	1,252
Public Bus	1,716	587
Total	**22,697**	**5,633.68**

Source: Questionnaire Survey, 2014

Considering the sectoral assessment of transport activities, carbon emission has been calculated. This has been done by summing up the sectoral transport performance. The following table 7 shows the total carbon emission from transport in three sectors

Table 7: Carbon Emission from Transport Activities of Business Places, Institutions and Settlements

Sectors	Carbon Emission (KgCo2e)
Retail and Wholesale Business Places	262,644.55
Government and Private Institutions	77,003.15
Settlements	273,077
Total	**611,724.7**

Source: Questionnaire Survey, 2014

5.2.2 Emission from Energy Usage. When people consume energy, it contributes to add carbon into the atmosphere. In this research, Carbon Foot Print from energy sources used in various sectors in the town was calculated. The following table shows the sectoral emission from energy use.

Table 8: Sectoral Carbon Emission from Energy Use (Samples)

Sector	Monthly Energy Use (Units)	Daily Energy Use (Units)	Carbon Emission per day (KgCo2e)
Retail and Wholesale Business	3,461	115.36	120.44
Government and Private Institutions	14,309,909	476,996.96	497,984.88
Settlements	5,096	169.86	177.34
Total			**498,282.66**

Source: Questionnaire Survey, 2014

Total carbon emitted from energy per month is 14,948,480 KgCo2e. The emission per day was calculated dividing the monthly emission by 30 which is 498,282.99 KgCo2e.

Table 9: Total Sectoral Carbon Emission from Energy Use Per Day

Sector	Emission per Entity (KgCo2e)	No of Entities	Total Emission (KgCo2e)
Retail and Wholesale Business Places	3.011	8326	25,069.58
Government and Private Institutions	12,449.62	783	9,748,052.46
Settlements	4.43	25,238	111,804.34
Total			**9,884,926.38**

Source: Questionnaire Survey, 2014

5.2.3 Emission from Consumption of Goods and Services. The following table shows the carbon emission from Consumption of Stationery particularly papers. Manufacturing of papers consume energy that leads to carbon emission. Based on the total paper consumed by various entities, emission per kg of papers and total emission were calculated. Table 9 shows the emission from consumption of papers.

Table 10: Carbon Emission from Stationery Use in Sectors

Sector	Emission
Retail and Wholesale Business Places	18,759,310.6
Government and Private Institutions	10,530,880
Settlements	11,808,860
Total	**41,099,050**

Source: Questionnaire Survey, 2014

Production, processing, transporting of foot stuff also involve energy consumption and carbon emission. Based on the amount of food consumed by an individual, Carbon Foot Print one kg of Food has been calculated. Table 10 shows the amount of emission from consumption of food.

Table 11: Carbon Foot Print of Consumption of Food Per Day

Carbon Foot Print Per Head / Day (KgCo2e)	Total Population in the Town	Total Carbon Emission Per Day (KgCo2e)
2740	98,647	247,364,460

Source: Questionnaire Survey, 2014

5.2.4 Emission from Solid Waste Disposal. Emission per metric ton of waste has been calculated using carbon calculator. Table 11 shows the amount of emission from the waste produced in the town area.

Table 12: Carbon Foot Print of Solid Waste Per Day

Carbon Foot Print Per Kg of Solid Waste (KgCo2e)	Total Solid Waste Per Day (Metric Ton)	Total Carbon Emission Per Day (KgCo2e)
441.2	30	13,236,669

Source: Questionnaire Survey, 2014

5.2.5 Sectoral Carbon Emission. Total emission from various sectors have been calculated summing up the sectoral emissions. The following table 12 shows the sectoral emission. In considering the sectoral emission, goods and services contributes more.

Table 13: Sectors Carbon Emission Per Day

Sector	**Emission**	**%**
Transport	611,724	0.2
Energy/ Power	9,884,926.38	3.59
Goods and Services	251,474,411	91
Solid Waste	13,236,660	4.8
Total	**275,207,722.01**	**100**

Source: Questionnaire Survey, 2014

5.2.6 Carbon Emission from Functions

Carbon Foot Print from Different Functions have been calculated for the Town considering the data collected and analyzed for different functions in the town. Table 13 shows the functional type emission. Higher amount of carbon emitted from goods and services which is 91%.

Table 14: Carbon Emission Per Functions Per Day in the Town

Sector	**Emission (KgCo2e)**	**%**
Business	19,046,024.73	6.5
Government and Private Institutions	20,355,935.66	6.5
Settlements	259,458,201	83.1
Solid Waste	13,236,660	4.2
Total	**312,096,821**	**100**

Source: Questionnaire Survey, 2014

In considering the functionality-based emission, larger amount of carbon is emitted from settlements which is 83.1 percent. Per capita carbon emission has been calculated based on the data and analysis carried out for the sectoral and functional activities of the town. The per capital emission is 718.66 KgCo2e per day.

6. Mitigation Measures for Low Carbon Environment

Since most people use private vehicles, promoting non motorable and public transport required to be promoted in order to minimize the emission from transport sector. Currently the public transport is not efficient and attractive therefore; people prefer to use private transport. Making public transport accessible and more comfortable is important to reduce private vehicles. Walking and cycling could be promoted among residents and commuters, if facilities and conditions for cycling and walking are provided. Arterial roads, A9 road, Point Pedro road, Palaly road, KKS road, Manipay road and Karainagar roads are mostly used by residents and commuters. Improving internal roads also will reduce the strain on the major roads. Most roads in Jaffna town are narrow and do not have right of way. Land acquisition depending on the grade of the road need to be undertaken in order to provide space for the provision of cycling and walking[3].

The designs of residential and public buildings are not energy efficient and consume substantial amount of energy and emitting carbon into the atmosphere[5]. Green Building Standards similar to LEEDs are adopted in Sri Lanka but those standards are not mandatory now. Making the Standards of Green Building Council are mandatory for all buildings including residential buildings, building designs would be more energy efficient[6]. Energy efficient design must be incorporated into the built environment that will utilize natural light and ventilation. Most of the residential buildings are owner designed and did not consider the elements of energy efficiency. Awareness about the importance of building design and energy efficiency will encourage people to seek the assistance of architects in designing the buildings. Production and use of renewable energy are not adopted in Jaffna town now. Less awareness and poor economic conditions prevent people from installing and using non renewable sources such as solar photovoltaic and wind power generation.

There are many ways to reduce the emission from solid waste disposal in Jaffna Town. At the moment, recycling, reuse, composting of wastes are not effectively adopted[7]. Most wastes end up into dumping site unsegregated. People also do not have the habit of any of these good practices. Awareness among community members on reduction, reuse, recycle and composting will minimize the waste coming into the waste stream from the city. Sanitary landfill site for the final nondegradable waste will minimize the emission of methane and other toxic and green house gases. Promotion of urban agriculture also will absorb substantial amount of organic waste and contribute to minimize the waste volume.

In conventional methods of business, activities consume more papers. Jaffna town has not adopted the digital technologies for main activities. Use of digital technologies and electronic communication will minimize the usage of papers in various places.

Water purification, water pumping and distribution consume energy. Rain water harvesting, waste water reuse and recycle are not practiced in Jaffna. Rainwater harvesting and waste water reuse will reduce water demand. Planning and building regulations to promote rain water harvesting and waste water recycling and reuse need to be formulated and adopted for Jaffna town. Adopting water efficient fixtures will minimize the water usage and reduce demand for supplied water.

References

[1] G. Chen, T. Wiedmann, M. Hadjikakou, and H. Rowley, "City carbon footprint networks," *Energies*, 2016.
[2] Z. Lin, "Ecological urbanism in East Asia: A comparative assessment of two eco-cities in Japan and China," *Landsc. Urban Plan.*, vol. 179, pp. 90–102, Nov. 2018.
[3] Urban Development Authority, "Jaffna Development Plan (Draft)," Jaffna, 2015.
[4] T. Letete, N. Mungwe, M. Guma, and A. Marquard, "Carbon footprint of the University of Cape Town," *J. Energy South. Africa*, 2011.
[5] C. S. Ho, Y. Matsuoka, J. Simson, and K. Gomi, "Low carbon urban development strategy in Malaysia - The case of Iskandar Malaysia development corridor," *Habitat Int.*, 2013.
[6] GBCSL, *Green Building Council - Sri Lanka*. Green Building Council of Sri Lanka, 2017.
[7] J. M. Council, "Statistical Hand Book," Jaffna, 2014.

A review on coal-to-electricity policies into practice in northern China

Zhaoyang Li[1], Zhaowei Xu[2], Yuying Sun[1], Zhichao Wang[2], Shiming Deng[3,1], Yiming Cui[1], Wei Xu[2,]*, Wei Wang[1,]*

[1]Beijing Key Laboratory of Green Built Environment and Energy Efficient Technology, Beijing University of Technology, Beijing, 100124, China
[2]Institute of Building Environment and Energy Efficiency, China Academy of Building Research, Beijing 100013, China
[3]Department of Building Services Engineering, The Hong Kong Polytechnic University, Kowloon, Hong Kong SAR, China

*Corresponding author's e-mail: xuwei19@126.com
*Corresponding author's e-mail: mrwangwei@bjut.edu.cn

Abstract. China has been promoting clean energy heating in an all-around way since recent year to reduce air pollution and energy consumption as a result of using coal for winter heating. Supported by a series of Chinese government policies, a coal-to-electricity project become a significant measure for clean energy heating in Beijing-Tianjin-Hebei region and the surrounding areas, which has obviously improved air quality in northern China. The coal-to-electricity project can involve engagement from multiple levels of government and social companies, wherein government is uniquely positioned to provide leadership. However, factors contributing or hindering policies into practice are largely unexplored. This paper presents a review on coal-to-electricity policies into practice through summarizing government policies and practical investigation, including its planning, scope, funding, process, and technical types analysis. In addition, a prospect of coal-to-electricity project in northern China is also given. It is found that the coal-to-electricity project holds a meaningful contribution and promising potential of development given the urgent need for clean energy heating in large areas of northern China. Government financial subsidy is driving force and efficient technical types are carrier. It was further demonstrated that a large amount of government financial subsidy and the application of efficient technology will contribute to the implementation of coal-to-electricity project, which may provide reference for transition of other energy and low carbon policies into practice.

1. Introduction

As the world's second-largest economy, rapid industrialization and urbanization in China has caused significant air pollution [1-2]. The mean annual $PM_{2.5}$ concentration in China's major cities reported to be 43 μg/m^3, had exceeded 4 folds based on the threshold of 10 μg/m^3 annual mean concentration set by the World Health Organization in 2017, and the Air Quality Index (AQI) has been in "red alerts" for multiple consecutive days [3-5]. The sources of $PM_{2.5}$ mainly include town coal heating boilers, rural coal heating, coal-fired power plants and automobile exhaust [6], many studies indicated that the large amount of coal-fired heating in winter were the main cause of the poor air quality. The annual air quality report released by the Ministry of Environ-mental Protection of China in 2014 showed that cities requiring heating had an average of 168.3 days during which the air quality standard was not reached, which was 59.8 % higher than that of cities which did not require heating [7]. ALMOND et al. evaluated TSP concentration in 76 cities in China and observed that the average value of TSP concentration in heating cities was 300 mg/m^3 higher than that in non-heating cities [8]. Zhang et al. reported that the contribution of coal-fired district heating to $PM_{2.5}$ concentration in winter was up to 53 % [9]. As seen, although the high $PM_{2.5}$ concentration in China is synthetically caused by industrial, domestic, traffic, and heating emissions, during the heating season, coal-fired heating emissions are the main cause of the poor environmental quality. Therefore, in northern China, where heating is commonly needed in winter, air pollution caused by coal-fired emissions would be more severe as a result of tremendous rising in building space and residential heating demand.

Air quality of northern China during heating season in 2013 has been monitored and shown in Figure 1. As seen, large-scale air pollution in northern China, especially in Beijing-Tianjin-Hebei region and its surrounding areas, has become an urgent issue that jeopardized the health of residents. Many studies have proven that $PM_{2.5}$ has serious adverse effects on the human health. Because of its small particle size, $PM_{2.5}$ can reach the lungs and alveolar regions [10], and there are positive and statistically significant relationships between the short-term exposure to $PM_{2.5}$ and cardiovascular and respiratory diseases [11]. To this end, the Chinese government has begun to call for clean energy heating and established a series of policies to alleviate air pollution. Various provinces and municipalities in northern China have adopted different alternative energy sources according to local conditions. At present, a coal-to-electricity project has become a main measure of clean heating conversion in Beijing-Tianjin-Hebei region, which has played a positive role in improving air quality. The coal-to-electricity is a project of transforming the users who use coal-fired heating into those who use electric drive equipment such as air source heat pump (ASHP), ground source heat pump (GSHP), solar energy, electric heater and so on. It is a complex engineering that requires a lot of manpower, financial resources and technical support, involving engagement from multiple levels of government and social companies. However, due to differences in policy guidance among regional governments, the implementation effect of coal-to-electricity project was different in northern China. Therefore, to ensure the better promotion of the coal-to-electricity project, it is necessary to explore the factors that contribute or hinder the policies of the coal-to-electricity project into practice. As a result, a review on coal-to-electricity policies into practice about its planning, scope, funding, process, and technical types were detailed in this paper. Followed by a prospect of coal-to-electricity project in northern China. Finally, a conclusion is given.

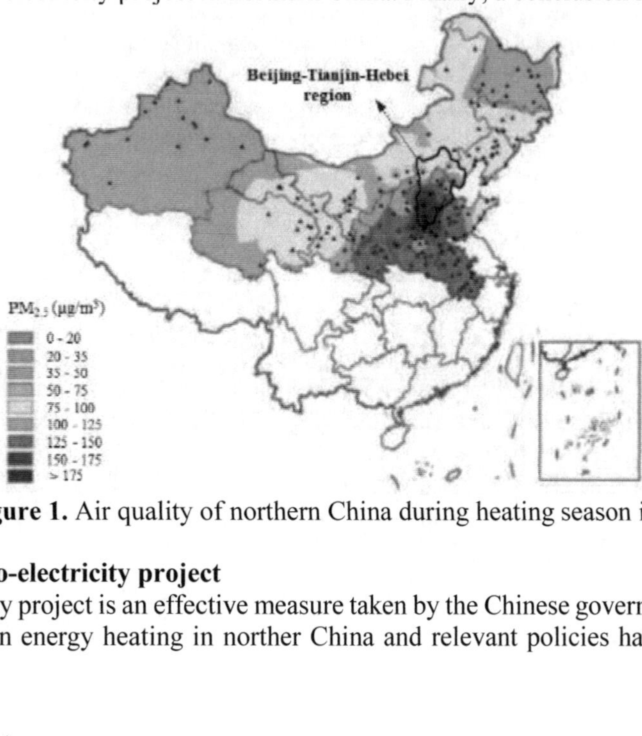

Figure 1. Air quality of northern China during heating season in 2013

2. Policies of coal-to-electricity project

The coal-to-electricity project is an effective measure taken by the Chinese government based on macro-requirements of clean energy heating in norther China and relevant policies have been introduced in succession. The contents of coal-to-electricity policies in various regions will be sorted out based on the background of clean energy heating, and specific objectives are also listed in this section.

2.1. Background

2.1.1. Heating Status in northern China. By the end of 2016, the heating area of buildings in northern China was about 20.6 billion square meters. Among them, the heating area of urban buildings was 14.1 billion square meters, while that of rural buildings was 6.5 billion square meters. Coal was the main energy for heating and coal-fired heating area accounted for 83 % of the total heating area, resulting in annual consumption reached about 400 million tons of standard coal. The proportion of clean energy heating in northern China was lower, as shown in Figure 2.

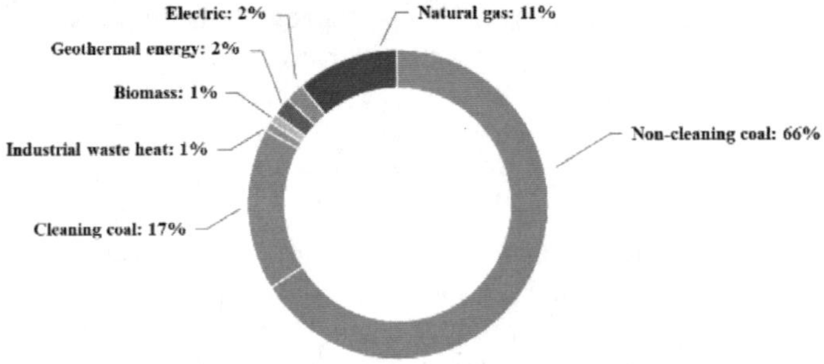

Figure 2. Distribution of heating sources in northern China in 2016

2.1.2. Development process of clean energy heating. Before 2016, the development of clean energy heating was sluggish in northern China, and there was a lack of overall planning and management. Except for the support measures introduced in the Beijing-Tianjin-Hebei region, most of the regional support policies were few.

On December 21, 2016, President Xi Jinping proposed to speed up the increasing of the proportion of clean energy heating, so that the work of heating conversion has received extensive attention. On March 5, 2017, Premier Li Keqiang reported that all levels of government should devote themselves to the planning route of clean energy heating, emphasizing the importance of "Defending the blue sky". At this point, cleaning energy heating have become China's key livelihood project, which have been highly valued by the government. Subsequently, the government promulgated <the Clean Energy Heating Plan for Winter in North China (2017-2021)>. This plan required that the goal of clean energy heating rate in northern China will reach 70% by 2021, and heating sources should be selected according to local conditions, including clean energy such as electric, solar and geothermal energy. The conversion project will start from the Beijing-Tianjin-Hebei region and expands to its surrounding areas. Meanwhile, the government has put forward 36 key cities and invested a lot of money to ensure the implementation of clean energy heating conversion, as shown in Table 1.

2.2. Subsidies of coal-to-electricity policies

The coal-to-electricity project has become a main measure of clean energy heating conversion in Beijing-Tianjin-Hebei region due to the policies guidance of the government, and will reach the goal of 500 million square meters of electric heating in northern China in 2021. Coal-fired heating was mainly used in rural areas, in thus **Beijing** has issued the <Plan for Clean Energy Heating in Winter in Rural Areas of Beijing in 2018>, which proposes to realize "coal-free" in plain areas of the whole city by 2019. **Tianjin** has issued the <Tianjin Residents' Winter Clean Energy Heating Program> and planned to implement clean energy heating conversion for 1.23 million residents, including a total of 543,000 coal-to-electricity. **Hebei Province** promulgated <Hebei Province Winter Clean Energy Heating Program in 2018> that 400,000 households were converted from coal to electricity. Shanxi, Shandong and Henan provinces have also introduced relevant policies and Table 2 shows the subsidies for coal-to-electricity project in key cities of northern China.

IOP Conf. Series: Earth and Environmental Science **329** (2019) 012054 doi:10.1088/1755-1315/329/1/012054

Table 1. Key Cities and Funding of clean energy heating in norther China

Beijing-Tianjin-Hebei region atmospheric transmission channel	Key cities	Beijing; Tianjin; Shijiazhuang; Tangshan; Langfang; Baoding; Cangzhou; Hengshui; Xingtai; Handan; Taiyuan; Yangquan; Changzhi; Jincheng; Jinan; Zibo; Jining; Dezhou; Liaocheng; Binzhou; Heze; Zhenzhou; Kaifeng; Anyang; Hebi; Xinxiang; Jiaozuo; Puyang
	Funding	Municipality directly under the Central Government: **1 billion RMB per year**; Provincial capital city: **700 million RMB per year**; Prefecture-level city: **500 million RMB per year**.
	Promulgation date	17/02/2017
First batch of central financial support cities	Key cities	Tianjin; Shijiazhuang; Tangshan; Langfang; Baoding; Hengshui; Taiyuan; Jinan; Zhenzhou; Kaifeng; Hebi; Xinxiang
	Funding	Central Finance: **20.1 billion RMB**; Local Finance: **69.5 billion RMB**; Social Capital: **211.6 billion RMB**.
	Promulgation date	05/06/2017
Second batch of central financial support cities	Key cities	Handan; Xingtai; Zhangjiakou; Cangzhou; Yangquan; Changzhi; Jincheng; Jinzhong; Yuncheng; Linfen; Lvliang; Zibo; Jining; Binzhou; Dezhou; Liaocheng; Heze; Luoyang; Anyang; Jiaozuo; Puyang; Xi'an; Xianyang
	Funding	Central Finance: **30.4 billion RMB**; Local Finance: **65.3 billion RMB**; Social Capital: **221.4 billion RMB**.
	Promulgation date	22/08/2018

Table 2. Subsidies for coal-to-electricity project in key cities of northern China

City	Technology	Equipment subsidy (RMB/household)	Tariff subsidy
Beijing	ASHP; GSHP	100 yuan/m^2	20:00 – 8:00 0.1 RMB/kWh
	Others	12,000	
Tianjin	Electric radiator	1200 – 4400	21:00 – 6:00 0.2 RMB/kWh
	ASHP	25000	
Shijiazhuang	——	7400	0.2 RMB/kWh
Baoding	——	7400	0.2 RMB/kWh
Langfang	——	7400	0.2 RMB/kWh
Chengde	——	25000	2000 RMB/household
Hengshui	——	7400	0.2 RMB/kWh
Taiyuan	ASHP	27400	0.2 RMB/kWh
	Electric boiler; Electric radiator	14400	
Datong	——	8000	0.2 RMB/kWh
Lvliang	——	24000	2000 RMB/household
Yuncheng	——	2000	0.21 RMB/kWh

2.3. Analysis of coal-to-electricity policies

The coal-to-electricity policies introduced in the Beijing-Tianjin-Hebei region was more clear than other cities, in terms of conversion target. These policies have formulated the area of coal-to-electricity project and the number of converted households. Under the mandatory goal, the progress and efficiency of coal-to-electricity was higher. On the other hand, subsidies for Beijing, Tianjin, and Hebei were higher so that residents can basically buy clean energy heating equipment for free. Meanwhile, the subsidies of local governments on electricity prices have reduced the operating costs of heating equipment, making it easier for residents to accept conversion.

3. Practice of coal-to-electricity project

The coal-to-electricity project has reached the final stage of the Beijing-Tianjin-Hebei region, where the air quality has improved significantly. The application of various technologies makes electric heating more efficient and energy-saving however the practical effect was different. In this section, the practice of various technologies in coal-to-electricity project was detailed though field measurement and data investigation.

3.1. Application Technology of Coal-to-Electricity project

In <the Clean Energy Heating Plan for Winter in North China (2017-2021)>, the types of electric heating technology were also defined, including electric radiator, ASHP, GSHP and so on. In order to investigate the practical application of coal-to-electricity project, a total of nearly 20,000 coal-to-electricity households have been monitored, in terms of their equipment operation cost and performances.

3.1.1. Economy of technology. Each technology can satisfy the requirements of residents' comfort in winter and the main difference was reflected in the cost. Figure 3 shows original investment and operating cost for different technologies. As seen, the original investment of combination of ASHP and solar energy system was the largest due to its complex forms while the simplest electric heating technology has the smallest investment. The original investment of ASHP was in the middle. On the other hand, the cost trend has changed in practical operation, the combination of ASHP and solar energy system was the lowest and the electric heating technology was the largest. Taking 120 square meters of households as an example, the electricity cost of ASHP (Air/Air) was 13.2 RMB per square meter, and the total cost of heating season was 1584 RMB. The operating cost of electric heating, electric radiator, electric boiler, solar energy, GSHP and ASHP-Solar system were 3468, 2220, 3360, 1812, 1656 and 972 RMB in heating season, respectively. Taking into account the original investment, the economic advantages of the ASHP were obvious. Meanwhile, the ASHP was recognized by 97% of users in the survey of user satisfaction and thus it became the main utilization technology of coal-to-electricity project in northern China.

3.1.2. performances of technology. ASHP has the advantages of environmental protection, energy-saving and flexibility, but there were still inevitable problems in its application, which influenced the operation efficiency of the ASHP. The practical operation performances of the ASHP in coal-to-electricity project was monitored and the average *COP* of heating season in Beijing from 2016 to 2018 was 2.19, 2.24 and 2.2, respectively. The *COP* distribution of the ASHP in 2017 is shown in Figure 4, during which the ambient air temperature was averaged at -1.1 °C and the relative humidity was averaged at 36.7%. It can be seen that only 27% of the ASHP was in a good performances range during the whole heating season. Wang et al. reported that frosting will occur on an outdoor coil of a space heating ASHP when its outdoor coil metal surface temperature is below both air dew-point temperature and freezing temperature of water, which lead to the output heating capacity of the ASHP unit decreased by 30 % to 57 %, the *COP* decreased by 35 % to 60 %. Furthermore, mal-defrosting will also lead to a significant drop in the output heating capacity and operating efficiency of ASHPs [12-13]. Therefore,

IOP Conf. Series: Earth and Environmental Science **329** (2019) 012054 doi:10.1088/1755-1315/329/1/012054

developing good technologies to improve the operation efficiency of the ASHP is an effective way to promote the practice of coal-to-electricity policies.

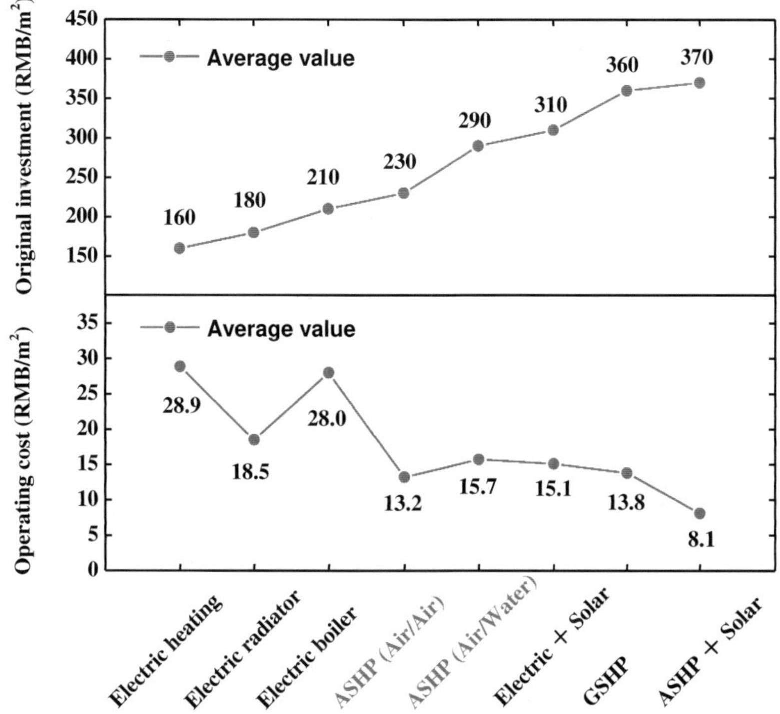

Figure 3. Original investment and operating cost for different technologies

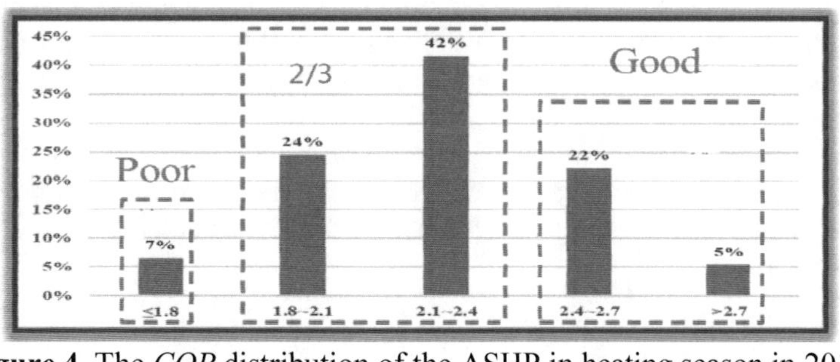

Figure 4. The *COP* distribution of the ASHP in heating season in 2017

3.2. Effect of coal-to-electricity project

From 2016 to 2018, more than 3 million households were converted from coal to electricity in northern China. The proportion of coal consumption decreased by 8.1%, and that of clean energy consumption increased by 6.3%. The conversion of 700 villages in Beijing has been completed, reducing coal-fired emissions by 900,000 tons and $PM_{2.5}$ by about 1,260 tons per year. Air quality of northern China during heating season in 2017 is shown in Figure 5. As seen, at the beginning of the implementation of the coal-to-electricity project, the environmental pollution during heating season in northern China has been significantly improved. In addition, according to the data of Ministry of Environmental Protection, the average concentration of $PM_{2.5}$ in 12 cities supported by first batch of central financial such as Tianjin and Baoding et al. decreased by 33% in the heating season of 2017, and the number of heavy pollution days decreased by 57% compared with last year. A number of data showed that coal-to-electricity project is an effective way to alleviate air pollution in northern China.

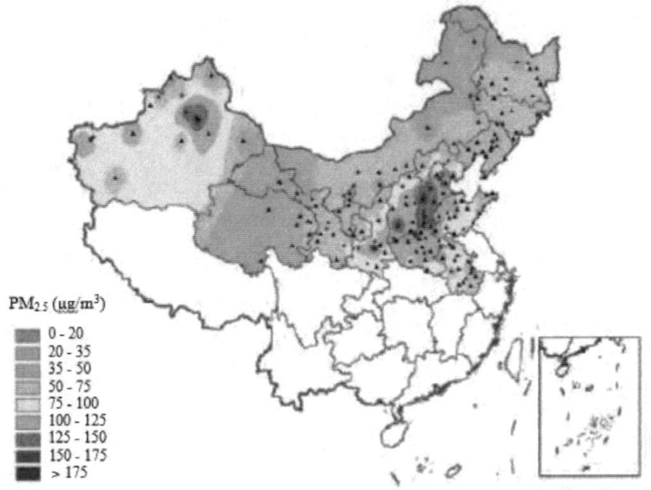

Figure 5. Air quality of northern China during heating season in 2017

4. Prospect of coal-to-electricity project in northern China

Clean energy heating was still the key implementation project of the Chinese government. In 2019, it will increase 15 billion RMB for clean energy heating. Coal-to-electricity project was rapidly being implemented in the surrounding areas of Beijing-Tianjin-Hebei region, and the Fenwei Plain will be the focus of conversion in the government's planning. It can be inferred that the coal-to-electricity project holds promising potential of development given the urgent need for clean energy heating in large areas of northern China.

The development of commercial ASHP was also predicted and shown in Figure 6. As the continues to increase in demand for urban building conversion, the ASHP were in thus vigorously promoted. The concomitant problem was how to improve the operating performances of equipment so that it can be used efficiently in coal-to-electricity project. To solve these problems, it is necessary to give play to the joint role of universities and enterprises to carry out technical research and the technology must be adapted to local conditions.

Finally, the experience of coal-to-electricity project in Beijing-Tianjin- Hebei region should be fully used for reference, and local governments should speed up the introduction of policies. Defining the target, scope and financial subsidy of coal-to-electricity policies can actively promote conversion.

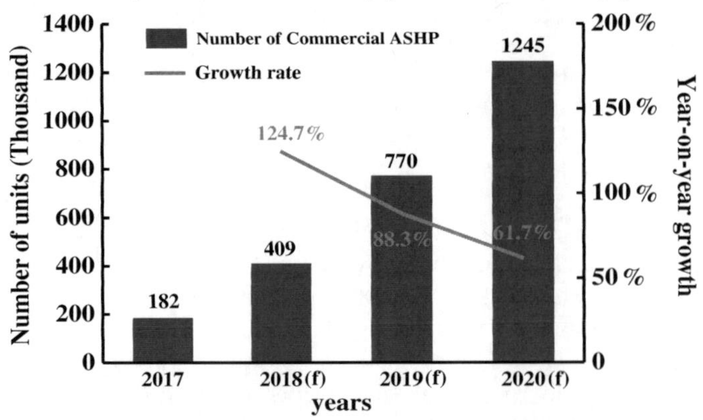

Figure 6. Development forecast of commercial ASHP in China

5. Conclusions

In this paper, a review on coal-to-electricity policies into practice is reported. Its planning, scope, funding, process, and technical types are analyzed in detail and a prospect of coal-to-electricity project in northern China is also given. It is found that the coal-to-electricity project has become a main measure of clean energy heating conversion in Beijing-Tianjin-Hebei region and holds a meaningful contribution and promising potential of development. Meanwhile, due to the advantages of energy-saving and environmental protection, ASHPs has become the main utilization technology of coal-to-electricity project, which could be recommended for large-scale application in northern China. Lacking of policy subsidies and technical deficiencies are the factors that hinder the coal-to-electricity policies into practice and should be given attention in coal-to-electricity project in different regions. With clear funding policies by government and further research on suitable technology by social companies, it is contributing to the coal-to-electricity project and providing reference for transition of energy and low carbon policies into practice.

Acknowledgements
This work was supported partially by the China National Key R&D Program (Grant No. 2016YFC0700100), the Young Top-Notch Talents Team Program of Beijing Excellent Talents Funding (Grant No. 2017000026833TD02) and the International Research Cooperation Seed Fund of Beijing University of Technology (Grant No. 004000514118557).

References
[1] Lim L, 2007. Air pollution grows in tandem with China's economy. NPR. https://www. npr.org/templates/story/story.php?storyId=10221268.
[2] Zeng YY, Cao YF, Qiao X, Seyler B.C, Tang Y, 2019. Air pollution reduction in China: Recent success but great challenge for the future. Science of The Total Environment. 663, 329-337.
[3] Reuters, 2014. China to 'Declare War' on Pollution, Premier Says. 4 March. http://www. reuters.com/article/us-china-parliament-pollution-idUSBREA2405W20140305.
[4] Bai RQ, Lam Jacqueline CK, Li Victor OK, 2018. A review on health cost accounting of air pollution in China. Environment International. 120, 279-294.
[5] Tilt B, 2019. China's air pollution crisis: Science and policy perspectives. Environmental Science & Policy. 92, 275-280.
[6] Beijing Municipal Bureau of Environmental Protection (BMBEP), 2014. The Beijing $PM_{2.5}$ source resolution was officially released. http://www.bjepb.gov.cn/bjhrb/xxgk/jgzn/jgsz/jjgjgs zjzz/xcjyc/xwfb/607219/index.html.
[7] MEP, 2015. Air Quality Report 2014 of Major Areas and Major Cities in China. Beijing, China (in Chinese).
[8] Almond D, Chen Y, Greenstone M, Li H, 2009. Winter heating or clean air? unintended impacts of china's huai river policy. American Economic Review, 99(2), 184-190.
[9] Zhang RJ, Jing JS, Tao J, et al., 2013. Chemical characterization and source apportionment of $PM_{2.5}$ in Beijing: seasonal perspective. Atmos. Chem. Phys. 13 (14), 7053e7074.
[10] Kampa, M., Castanas, E., 2008. Human health effects of air pollution. Environ. Pollut. 151 (2), 362e367.
[11] Pascal, M., Falq, G.,Wagner, V., et al., 2014. Short-term impacts of particulate matter (PM 10, PM 10e2.5, PM 2.5) on mortality in nine French cities. Atmos. Environ. 95, 175e184.
[12] Wang W, Xiao J, Guo QC, Lu WP, Feng YC, 2011. Field test investigation of the characteristics for the air source heat pump under two typical mal-defrost phenomena. Appl. Energ. 88 (12), 4470-4480.
[13] Wang W, Feng YC, Zhu JH, Li LT, Guo QC, Lu WP, 2013. Performances of air source heat pump system for a kind of mal-defrost phenomenon appearing in moderate climate conditions. Applied Energy, 112, 1138-1145.

IOP Conf. Series: Earth and Environmental Science **329** (2019) 012055 doi:10.1088/1755-1315/329/1/012055

Impact of low-carbon technologies on the power distribution network

A Rafi [1], T Lee [1], and W Wu [1]

[1]School of Engineering and the Built Environment, Birmingham City University, City Centre Campus, Millennium Point, Birmingham B4 7XG, United Kingdom

Abstract. Over recent decades, there has been a growing concern about how modern technologies would impact on an electrical grid. Whilst the UK is set to meet the target of 15% energy demand from renewable sources by 2020 to ensure energy security and decarbonisation objectives, there has been a greater concern over how the increasing deployment of low carbon technologies (LCTs) would affect an electrical grid, in particular the residential low voltage distribution network. While LCTs can provide clean energy and decreases the dependence on oil and natural gas stocks, their impact on distribution networks is unknown and the operators are visibly blind as they use a 'fit and forget' approach. Consumers' connections are uncertain and stochastic and the LCT uptake poses a potential threat for distribution network operators. A review is presented in this paper, illustrating the potential impacts the LCTs have on a local electrical grid such as voltage regulation, thermal limits, power quality and harmonics. Potential solutions, such as Deregulation, Soft Open Points, On Load Tap Changer, and Active Power Filter currently available for Distribution Network Operators and power system planners before costly network reinforcement work is undertaken.

1. Introduction

The dynamic role of energy in day-to-day life is irrefutable, with fossil fuel being the primary resource [1]. However, the significant negative impact from this form of energy on the environment is the emission of greenhouse gases like carbon dioxide (CO_2) which lead to the question for policymakers, as to how to ensure low carbon economy. Being eco-friendly and non-depletable, renewable energy has surpassed as the means of energy of choice and various strategies have been implicated for its generation [2]. With an ambitious target for CO_2 reductions, these are leading to more low carbon technology adoption, which threatens the electricity network stability.

2. Literature review

According to the UK Renewable Energy Roadmap [3], renewable sources in the UK needs to contribute 15% to the energy demand by 2020. Significant increment has been seen in energy demand from renewable sources to contribute to energy security and decarbonization objectives. To meet these targets, low carbon technologies (LCTs) development and deployment gain a considerable amount of importance over the past decade. UK has aimed to provide clean energy, decrease the dependence on oil and natural gas stocks and meet the greenhouse gas emission targets of 80 % reduction on 1990 figures by 2050 [4]. To accommodate LCT more efficiently on the electrical grid, DECC and Ofgem are joined together to establish a smart electrical grid which facilitates LCTs in the generation, supply, and consumption, while ensuring the security of supply [5]. This will give the chance for the domestic consumers to get a return on what they generate from their distributed generators.

Recent research studies [6] suggest that LCT growth at the domestic level, may have a deep threat to the UK distribution network, though no certain reports have been recorded due to the consumer's connections are uncertain and stochastic. It is even more challenging to assess the impact of forthcoming scenarios when there is a lack of understanding and monitoring on the LV. Despite many researches [6] emphasized and sited potential solution to such problems, still, uncertainty exists as to what likely to happen in the future to the distribution network notably with respect to the growing electric vehicles (EVs), photovoltaics (PVs) and heat pumps (HPs) at the domestic level [7].

2.1. Electric vehicles

Electric Vehicles (EVs) have been around in the market for over a century since 1990s after the vast development in power electronics, new battery technologies, and advancement in its design and testing [8]. Very soon the importance of EV has been noticed on its low carbon emission making a priority compared to normal Internal Combustion Engines [3]. It is reported in [3] that the government has committed £400 million up to 2015 to support ultra-low emissions vehicles.

Most of the current EV owners find the home to be the most convenient location to recharge their EVs, but their load cumulates in a local domestic area which may cause an increase in the number of voltage violations [8]. According to Dubey [9] the impact is particularly on three areas which are, increases the substation load demand, transformer overheating, and the violation of under-voltage limits, and three-phase power supply unbalance.

2.2. Photovoltaics

Compared to EV, photovoltaics (PVs) system became more popular due to Feed-in Tariff (FiT) scheme which was launched in April 2010 and very soon closed in April 2019 due to the numbers of installations made have exceeded the requirement. This shows the UK is on a fast track to reach the target to reduce carbon emission [10].

Network issues due to the increasing number of PVs was addressed in [11] report as well. It was mentioned that distribution network operators will have to face severe consequence to manage their operation if this issue is not resolved earlier. The impact may occur during the off-peak time as the large number of PVs cumulatively exporting electricity back to the grid would result in voltage violation issues such as the voltage will rise about its threshold which typically regulates within +10% and -6% of 230V phase-neutral. Other issues related are harmonics and power quality [12].

2.3. Heat pumps

The use of heat pumps (HPs) by a Domestic Renewable Heat Incentive (Domestic RHI) scheme would help to reduce its carbon emissions and meet its renewable energy targets stated in [10]. Moving from fossil fuel heating to heat pumps will make a significant amount of reduction in greenhouse gas emission, although the high level of HPs penetration poses a new challenge for the low-voltage networks. Since HPs typically use induction motors the effect could be on power quality which includes voltage variation (dips and surges), harmonics and frequency variation [13].

3. Future prognosis

According to National Grid plc which is multinational company for transmission and distribution of gas and electricity, summaries in their Future Energy Scenarios (FES) [6] about how the energy consumption will take place on their transmission system. These scenarios are credible pathways for the future of energy, from 2018 to 2050 but not considered as a forecast. Four potentials scenarios mentioned in the FES 2018 reports are, Community Renewables (CR), Two Degrees (TD), Steady Progression (SP), and Consumer Evolution (CE). Instead of going into each scenario it is worth looking at analysis made by LCTs on them.

3.1. Expected domestic electricity demand with heat pump

Appliances alone makes up to more than 61% of domestic electricity demand in which Heating accounts for almost a quarter. The significant amounts of electricity for heating is used by traditional heating elements mainly resistive for storage heaters and showers, heat pumps are also used to consume less energy per unit. Domestic electricity demand scenario profile for annual energy demand is being illustrated in figure 1 excluding EVs.

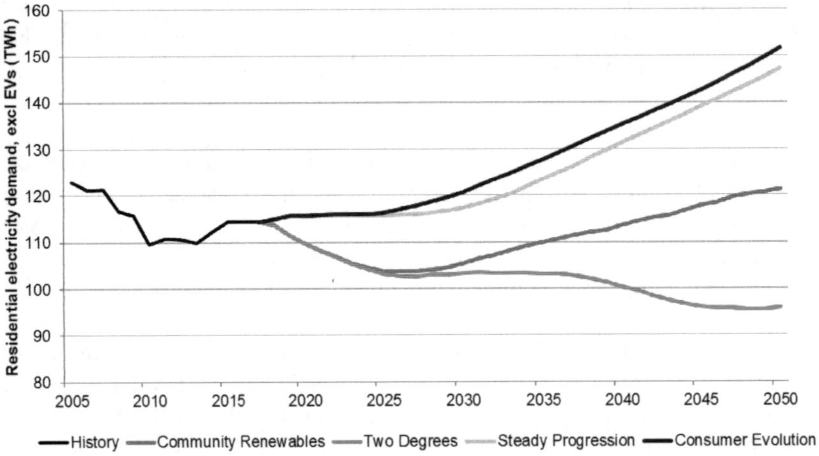

Figure 1. Domestic electricity demand with HP [6]

A high demand rise can be seen in consumer evolution and the steady progression scenarios. In consumer evolution based scenario, the rise is due to the increasing number of heat pumps used at home, whereas, in steady progression based scenario the growth of population exceeds slow efficiency savings. In two degrees scenario, the drop is due to the alternative heating method used such as the hydrogen networks will start and also used improving insulated housing stock whereas in community renewables scenario, the rise after 2030 is due to growth in Air-Source Heat Pump which also mainly depends on at certain times and seasons.

3.2. Expected domestic electricity demand with electric vehicles

Similarly, the electric demand in the transport matters the most when the EVs being charged at a peak time as shown in figure 2. Those four scenarios are made in a way that some have this factor that the avoidance of peak charging such as in consumer evolution and steady progression scenarios the net electric demand for charging EVs went above 13GW and 10GW respectively.

The other two scenarios that is two degrees and community renewables the net demands to charge the EVs rises until it reaches the point when vehicle to Grid (V2G) connections are widely available shows a study line until it drops as the growth in V2G continues and pulls down net peak demand. The net demand is not as steep in community renewables scenario as compare to two degrees because it's considered that in community renewables people are avoiding making charging during the peak time still it is 10 times more than the current consumption.

IOP Conf. Series: Earth and Environmental Science **329** (2019) 012055 doi:10.1088/1755-1315/329/1/012055

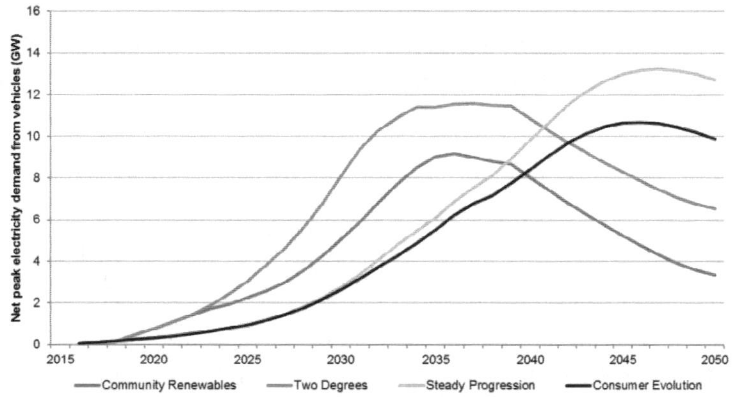

Figure 2. Net peak electricity demand from EV [6]

3.3. Expected domestic electricity demand with photovoltaic

The supply of electricity using micro solar like PVs growth found from FES report is at least doubled from the date till 2050 as confirmed in figure 3. All four scenarios show an increase in the supply due to the reduction in the cost of PVs technology, promotion of these by government policies and the fascination consumers sees from them as they can import less during daytime and export what they save. In community renewables scenarios the solar capacity goes more than 66GW by 2050 from which 33GW is due to micro-solar used at the residential level. The necessity of storage capacity will be significantly observed when PV generates during the day whereas EV charges mostly during the night at home.

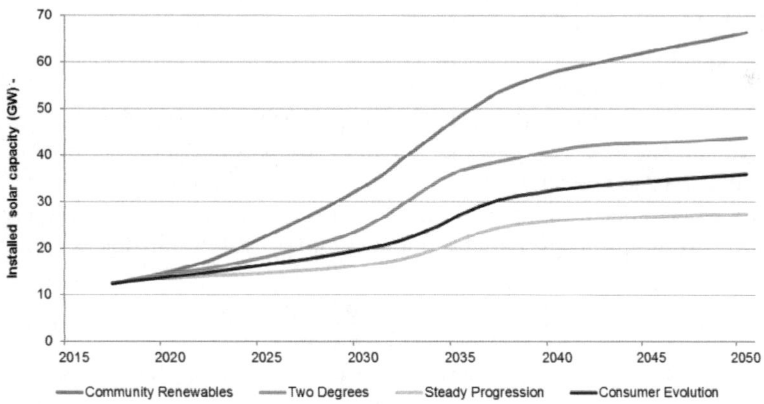

Figure 3. Solar capacity scenarios PV [6]

4. Impact of increased demand & supply

The effect on the low voltage networks are associated to the increasing number of residential demands in electric vehicles, heat pumps and photovoltaics, hence they are outlined. The type of network constraints is illustrated by Fishbone Diagram in figure 4.

IOP Conf. Series: Earth and Environmental Science **329** (2019) 012055 doi:10.1088/1755-1315/329/1/012055

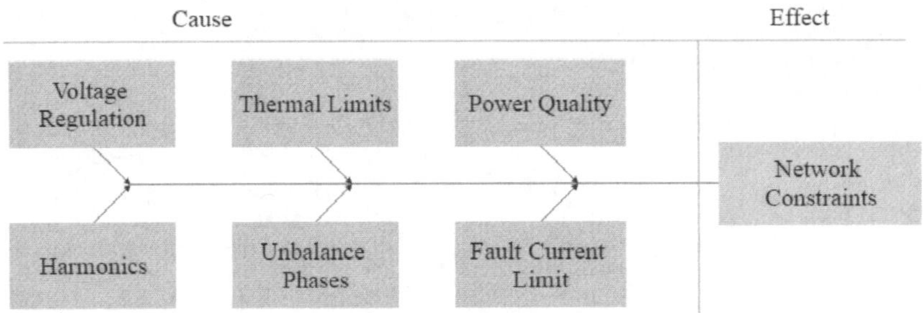

Figure 4. Network Constraints using Ishikawa diagram

The normal voltage needs to be regulated in its range that is +10% and -6% of 230V phase-neutral as illustrated in [14]. When the voltage is not regulated within the limit, a potential threat is generated to the whole low voltage network, since the power could flow in either direction [15]. Due to high PV penetration, the power could flow in the reverse direction causing back feeding, it is a situation where personnel working on equipment may not be aware and this could cause hazardous situation. However, the thermal limit constraint is far greater than the other two because electric vehicle would require high amount of power (typically 3kW or 7kW) which could not be delivered from the existing cables [16]. This also creates the fourth constraint that is imbalance in the phases. Most of the residential units are on single-phase system in the UK and an unbalance condition would occur especially when one of the three loads is either consuming more or in a worst case generating high [15].

Harmonic current or voltage on an electrical system is considered as multiple of the fundamental frequency caused by non-linear electric loads. [16]. Lastly a fault may occur on the load side as a short-circuit or in between two phases which would results in a high amount of current flowing from the power source, hence an uptake of solar panels on the system would require another protection method to disconnect the downstream as well as the fault site [17].

5. Solution to network constraints

Reinforcement of the distribution network is the conventional way to address the issues alleviated by changes in the loads. Some of the options that are classified as a network reinforcement are, increasing the thermal capacity on the network, selecting transformer to accommodate an increase in demand, considering other ways to deliver power instead of using longer routes, building advance substations, and using reactive power compensation techniques. Unfortunately, all these options are costly and disruptive [18].

Table 1. Solutions to list of problems

Solutions	Network Constraints: Voltage Regulation	Thermal limits	Power quality	Harmonics	Unbalance phases	Fault current limit
Reinforcement	✓	✓	✓	✓	✓	✓
De-regulation	✓		✓		✓	
SOP	✓	✓	✓		✓	✓
OLTC	✓				✓	
Energy Storage	✓	✓				
APF		✓	✓	✓		

The table 1 shows list of solutions along with their respective problems. Firstly, de-regulation is one of the approach addressing the effect of voltage regulation, power quality and eventually unbalance phases. This concept provides solution by storing energy (which is costly similar to energy storage) and dispatching it when the demand is high [8]. Secondly, soft open points (SOPs) covers all the constraints apart from harmonics. Advantages of SOP is that it can be used for any network system such as radial or loop [19]. Thirdly on load tap changer (OLTC) method proposed by [10] can accommodate high PV penetration. This method is effective but costly and mainly applicable to be proposed at the planning phase of the network. Lastly, active power filters (APF) which are widely used to provide a solution to the harmonics issues in the system. Active filters long exist in the literature since the 1970s [20] [21], whereas APF resolves harmonics issues caused by loads of the future.

6. Conclusion

In this paper, a broad review has been presented illustrating various types of impact existing due to the enormous implementation of low carbon technologies, particularly from electric vehicles, photovoltaics and heat pumps. A brief summary is also presented of different solutions available to reduce them including reinforcement, de-regulation, soft open points, on load tap changer, and active power filters. Although the impact is typical for electrical distribution system, their combined effect would require initial assessment of the present system and provide an economical, ecological and sustainable solution.

Analysing the active occupancy, appliance consumption, weather forecast, irradiance, and power consumption would help to predict these types of impact. In future work, we are evaluating the type of domestic loads, creating a technique to implement different types of low carbon technologies into those loads, and developing a forecasting system to analyse such impact effectively.

References

[1] BEIS, "Digest of UK Energy Statistics (DUKES) 2018 Chapter 5: Electricity," 2018. [Online].

[2] European Commission. (2010b). Europe 2020: A strategy for smart, sustainable and inclusive growth, COM (2010) 2020.

[3] DECC, "UK Renewable Energy Roadmap," *Carbon N. Y.*, vol. 5, July, pp. 293–298, 2011.

[4] Godlee, "Climate change: permission to act," BMJ, vol. 336, no. 7637, Jan. 2008.

[5] Ofgem, DECC, and Smart Grid Forum, "Smart Grid Vision and Routemap Smart Grid Forum," *Rep. Number URN 14D / 056*, February, 2014.

[6] National Grid, "Future Energy Scenarios," July, pp. 1–133, 2018.

[7] T. Lee, R. Yao, and P. Coker, "An analysis of UK policies for domestic energy reduction using an agent based tool," *Energy Policy*, vol. 66, pp. 267–279, 2014.

[8] F. Marra *et al.*, "EV charging facilities and their application in LV feeders with photovoltaics," *IEEE Trans. Smart Grid*, vol. 4, no. 3, pp. 1533–1540, 2013.

[9] A. Dubey, S. Member, and S. Santoso, "Electric Vehicle Charging on Residential Distribution Systems : Impacts and Mitigations," pp. 1871–1893, 2015.

[10] T. Aziz and N. Ketjoy, "Enhancing PV Penetration in LV Networks Using Reactive Power Control and On Load Tap Changer with Existing Transformers," *IEEE Access*, vol. 6, pp. 2683–2691, 2017.

[11] DECC, "Electricity System: Assessment of Future Challenges," August 2012, 2012.

[12] D. Fischer, J. Scherer, A. Flunk, N. Kreifels, K. Byskov-Lindberg, and B. Wille-Haussmann, "Impact of HP, CHP, PV and EVs on households' electric load profiles," *2015 IEEE Eindhoven PowerTech, PowerTech 2015*, pp. 2–7, 2015.

[13] M. Akmal, B. Fox, D. J. Morrow, and T. Littler, "Impact of high penetration of heat pumps on low voltage distribution networks," *PowerTech, 2011 IEEE Trondheim*, pp. 1–7, 2011.

[14] Y. Li and P. A. Crossley, "Monte Carlo study on impact of electric vehicles and heat pumps on LV feeder voltages," *12th IET Int. Conf. Dev. Power Syst. Prot. (DPSP 2014)*, no. 1, p. 12.13-12.13, 2014.

[15] T. Schlösser, S. Stinner, A. Monti, and D. Müller, "Analyzing the impact of home energy systems on the electrical grid," *2014 Power Syst. Comput. Conf.*, pp. 1–7, 2014.

[16] M. R. Hans, "Identification of Underground Cable," pp. 5–8, 2017.

[17] J. Keller, B. Kroposki, R. Bravo, and S. Robles, "Fault current contribution from single-phase PV inverters," *Conf. Rec. IEEE Photovolt. Spec. Conf.*, June, pp. 001822–001826, 2011.

[18] Z. Hu and F. Li, "Distribution network reinforcement utilizing active management means,"

IEEE PES Gen. Meet. PES 2010, pp. 1–3, 2010.
[19] W. Cao, J. Wu, N. Jenkins, C. Wang, and T. Green, "Operating principle of Soft Open Points for electrical distribution network operation," *Appl. Energy*, vol. 164, pp. 245–257, 2016.
[20] H. Sasaki and T. Machida, "A new method to eliminate AC harmonic currents by magnetic flux compensation — Considerations on basic design," *IEEE Trans. Power Appar. Syst.*, vol. PAS-90, no. 5, pp. 2009–2019, 1971.
[21] A. Ametani, "Harmonic reduction in thyristor converters by harmonic current injection," *IEEE Trans. Power Appar. Syst.*, vol. 95, no. 2, pp. 441–449, 1976.

IOP Conf. Series: Earth and Environmental Science **329** (2019) 012056 doi:10.1088/1755-1315/329/1/012056

Growing Greener: Creating a New Values-based Environmental Engagement Toolkit for SME Intermediaries

Sam Hampton[1,4], Richard Blundel[2], Tina Fawcett[1], Chris Shaw[3]

[1] Environmental Change Institute, University of Oxford, South Parks Road, Oxford, OX1 3QY, UK.
[2] The Open University Business School, Milton Keynes, MK7 6AA, UK
[3] Climate Outreach, The Old Music Hall, 106-108 Cowley Rd, Oxford OX4 1JE, UK.
[4] Corresponding Author: samuel.hampton@ouce.ox.ac.uk

Abstract. This paper explores a radically different way of facilitating energy and environmental initiatives in small and medium-sized enterprises (SMEs). In terms of energy policy, smaller firms in Europe are exempted from most of the major fiscal and regulatory mechanisms that are applied to larger organisations. Policies to reduce energy demand and associated carbon emissions in SMEs are largely based on providing incentives, such as face-to-face support and grants for energy efficiency in buildings. Energy advisors are therefore key intermediaries, providing advice and encouraging the uptake of low carbon technologies and practices by SMEs.

Previous studies have found that advisors often find it difficult to engage effectively with SME owners and managers, and that traditional 'win-win' messaging can have limited impact, resulting in implementation problems such as under-investment in energy-saving technologies, reluctance to adopt new environmental practices, and a tendency to revert to previous ways of operating once the incentive is removed. Recent research also suggests that SME owners' and managers' personal values play an important mediating role in their response to environmental issues, acting in combination with more established factors such as educational background, access to resources and the views of customers and suppliers. The implication is that policy interventions in this area could be delivered in more cost-effective ways if accompanied by a more nuanced, values-based approach to engagement.

This paper reports findings from 'Growing Greener' a UK multi-disciplinary project that aims to equip advisors and other types of intermediary with the skills, knowledge and understanding they need in order to incorporate a values-based approach into their existing interactions with SMEs. It opens with an overview of the policy context and a brief overview of the relevant research literature.

The main section explains how the research team co-produced a values-based engagement toolkit in conjunction with a group of energy advisors and external specialists. The design process included a series of facilitated 'narrative workshops', where advisors shared their experiences and experimented with early versions of the engagement tools. This is followed by an outline of the completed engagement 'toolkit', which includes free-to-access online course, a communication guide and an interactive engagement tool. These three inter-related components are designed to help advisors to engage SMEs beyond a narrow, cost-benefit framework, and in turn help SME owners and managers to connect low carbon choices with the personal and business values that are important to them as individuals.

Our findings indicate the potential for more effective, longer-lasting interventions beyond the low hanging fruit of building efficiency measures.

IOP Conf. Series: Earth and Environmental Science **329** (2019) 012056 doi:10.1088/1755-1315/329/1/012056

1. Introduction

There are over 5.7 million small and medium-sized enterprises (SMEs) in the UK, including a wide variety of organisations, ranging from multi-sited manufacturing companies to small firms, social enterprises and micro-businesses. In the UK, they collectively use over 50% of energy in the commercial and industrial sectors [1] .

A government report found evidence from local energy efficiency support programmes that suggested the average SME could save up to 25 per cent of its energy use through relatively simple, low cost measures, like upgrading lighting or installing occupancy sensors. However, take up of these opportunities is systematically low. In the UK, the 'missed opportunity' of energy savings for SMEs is estimated to equate to between £1.26 bn and £2.63 bn per year [2].

The need for government intervention in this market failure is widely recognised [3]. In the UK, this has conventionally included financial incentives such as grants and loans [4], alongside the provision of expert advice for SMEs, to help them select the most appropriate energy efficiency measures for their buildings. Given the scale of SME energy use, and the potential for savings, the energy advisors that deliver such advice are critical actors in the transition to a lower carbon economy. However, they have been the subject of little research, and while they are funded, sometimes generously, by the taxpayer, they are given little other support as a community of practitioners [5].

This paper reports on a 12-month project aiming to address this deficiency. Growing Greener was a project funded by the Economic and Social Research Council's Impact Acceleration Account. A collaboration between Oxford University, Open University Business School and Climate Outreach, the project held a series of 5 workshops around England (Birmingham, Manchester, Oxford and two in London) with energy and sustainability advisors. The workshops gathered views from these practitioners on the topic of 'greening' the SME population, identifying insights into the practice of engaging with SMEs on the topic of the environment. Drawing on evidence collected in these workshops, and from interviews with selected advisors, Growing Greener went on to produce a 'toolkit' for supporting these important intermediaries to increase their potential for influence amongst SMEs. This paper is the first academic report from the project. The rest of this article is structured into four sections. The next section presents a brief review of literature on SMEs, energy and the environment, and the role of intermediaries in promoting energy efficiency and pro-environmental action. Section three then describes the design and content of the five workshops put on as part of the project. Section four summarises the content of the Growing Greener 'toolkit'; and the final section discusses the broader implications of this work.

2. Literature review

Academic literature on SMEs, energy and environmental impact is large, but somewhat fragmented. For example, there is a significant literature from the business studies tradition which has sought to develop and implement environmental management tools [6]; whilst contributions from outside of Business Schools have been made from geographers [7,8] and public policy scholars [9,10]. Perhaps the most influential body of research has focused on identifying and evaluating the barriers to the uptake of energy efficiency measures by SMEs [11–13]. As well as being popular amongst academic researchers, this approach has dominated the approach of policy-focused 'grey' literature, which includes research sponsored by the UK and Australian Governments [14,15], and the European Commission [16]. Many contributions from this tradition, having summarised the key barriers to action, highlight the importance of intermediaries in helping SMEs to navigate a complex array of energy-saving options.

However, there has been significantly less research conducted on the practices and experiences of these business advisors themselves. Katy Janda and colleagues have sought to place intermediaries in the centre of their research focus and highlight their role as more than simply go-betweens, operating between policy makers and SMEs [17–19]. In doing so, they use the term 'middle-actors' to differentiate their approach from literature which insufficiently addresses the agency and capacity of intermediaries such as builders [20,21], landlords [18] and data providers [19].

Contributing to this small and emerging 'middle actor' literature within energy research, Hampton's [5] auto-ethnographic study of low carbon SME advisors shed light on their everyday practice. It highlighted the dominance of the economic growth narrative within business support networks, and the associated prevalence of financial arguments used by advisors when seeking to engage with and 'sell' energy efficiency technologies and other pro-environmental measures to SMEs. Hampton argues that such approaches are insufficient to address the multiple market failures, which already lead SMEs to miss opportunities for compelling cost savings. Interviews with UK based advisors reveal that this professional community typically hold strong environmental values, but tend to suppress these when talking to SMEs; making a set of assumptions about the values and motivations of business owner-managers, including that they have little consideration for environmental impact.

Hampton points to research on SME owner-managers values including work from researchers at the Open University Business School [22–25], who have worked extensively to apply Schwartz' Theory of Basic Human Values [26]. Schwartz's framework, also known as the Schwartz Values System (SVS), categorises basic values into ten different types, and four broad groupings: Self-Transcendence; Conservation; Self-Enhancement; and Openness to Change (see Figure 1). In a recent contribution, Schaefer et al. [23] develop a typography of SME managers based on their personal and professional values. Challenging the assumption that business leaders are motivated primarily by profit, they identify a range of values based on in-depth interviews. These include power values, where motivations include wealth and influence, to achievement values, where managers strive for positive, visible markers such as accreditation and peer-recognition, and universalism, where concern for social equality or climate change are primary motivators. The Growing Greener project leads directly from this work on SMEs, middle-actors and values.

3. Workshops with SME advisors

The Growing Greener project (2017-2018) follows on from an ESRC funded 'kick-starter' impact project called Growing Green in 2016. With similar aims of exploring how basic human values come into play for SME owner managers with respect to their environmental impact, this project convened two workshops in Oxford. Co-hosted with Oxfordshire Business Support, the workshops were heavily promoted through existing SME networks, but nonetheless struggled to attract significant attendances. Participating SMEs were largely made up of those already working in the environmental sector, including charities and non-profits. It was felt by the project team that the values-based discussion was skewed towards universalism and benevolence as a result [27].

Learning from the limitations of the kick-start project, Growing Greener instead turned its attention to energy and sustainability advisors. A series of five workshops were conducted in Autumn 2018, in Birmingham, Manchester, Oxford and two in London. Recruitment of participants was aided by the involvement of Sam Hampton, who holds a part-time role with Oxfordshire Local Enterprise Partnership, and had established contacts through his previous research [4,5].

The workshops were led by project partner Climate Outreach, who are a charity specialising in climate change communications. They have unique experience of engaging with conventionally 'hard to reach' audiences on climate change, such as centre-right voters [28], and faith groups [29], and have developed a methodology of 'narrative workshops' [30]. Following this methodology, the workshops were broadly structured into two parts. They began with an introduction to the Theory of Basic Human Values, and a discussion about how values come into play in our personal and professional lives. Participants were asked to consider which of the values they most strongly identified with, and to plot well-known brands such as Lloyds Bank, Facebook and Ben & Jerrys against the diagram (Figure 1). Advisors were invited to share stories of how they had encountered values at play in their work with SMEs.

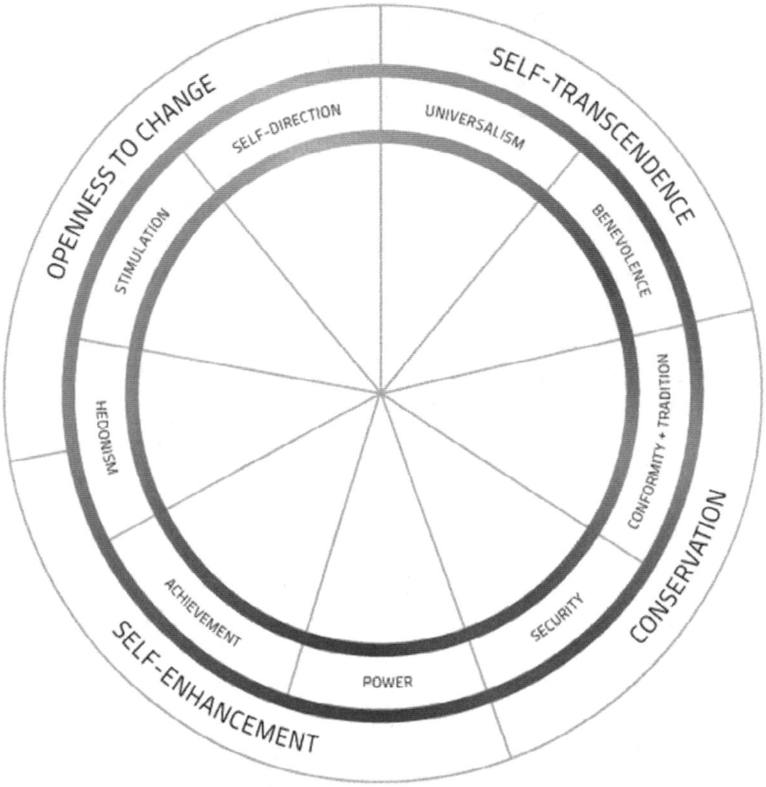

Figure 1: Diagram of the Theory of Basic Human Values [26]

Having explored the theory behind values-based engagements, in the second part of the workshop, participants applied these ideas in a series of role-playing exercises. In pairs, one participant would play the role of an SME owner-manager with a strong set of guiding values, and the other, an advisor, who would attempt to discover the key values at play, and to select the most salient language and arguments for their client. The workshops concluded with a plenary discussion of the challenges and opportunities associated with values-based engagements with SMEs.

4. A toolkit for energy and sustainability advisors

The Growing Greener toolkit consists of three resources. Firstly, Climate Outreach have produced a 'communications guide' for a broad audience of organisations and individuals wishing to communicate environmental messages to SMEs [31]. This 10-page publication includes six principles for effective communication; summarised in Table 1.

Table 1: Six principles for effective communication with SMEs

People do not just define themselves by who they are –they also define themselves by who they are not.	Show your clients that taking effective action on climate change is something 'we' do, and isn't just something 'those' environmentalists worry about. Sharing case studies that can tell a story about why and how people like themselves have cut their carbon footprint will help your audience identify with your message.
Communicate from the 'values up', not the numbers down'.	People make sense of the world on the basis of their values; when it comes to issues such as climate change, people's world view is more important than facts and targets. Use stories that connect the business case with the values of

	the person you are speaking to, and their concerns as a human being. The vision of clean and affordable energy, for example, is an idea with wider appeal than the need to meet targets set for the year 2050.
Connect climate change with other environmental concerns, and the desire for a less wasteful society.	When talking about energy efficiency, use the idea of our common wish to avoid being wasteful. Research has demonstrated that the desire to use resources carefully is widely held, and provides a strong rationale for reducing our energy use.
Highlight how action on climate change is an issue of security.	Security matters. Businesses want a sense of security, just as we want to feel secure on a personal level. Climate change is itself a driver of a more uncertain future for us and our families. Using home grown renewable energy sources offers the potential for greater energy security.
Highlight continuity with what is familiar and trusted.	A sense of tradition and continuity with the past are values that sit side-by-side with business leaders' desire to have a certain and knowable future. Talk about renewables in terms of the nation making best use of its natural resources. Show how new technologies speak to the best of British ingenuity and resourcefulness.
Stress the inevitability of a low carbon future.	Momentum for the switch to a low carbon energy system is growing amongst the public and within the business sector. Ambitious targets are locked into energy policy. When businesses feel this is the normal thing to do they will have less reason for postponing action.

The second resource is an 'Open Educational Resource' (OER), 'Promoting sustainability in business: a values-based toolkit', which is hosted on the Open University's 'OpenLearn Create platform' (www.open.edu/openlearncreate). This online course is designed for SME advisors, intermediaries and people working within organisations to effect pro-environmental change. Taking approximately five hours to complete, it first introduces the key theoretical ideas behind values-based engagement with business, including the Schwartz Theory of Basic Human Values, and some of the literature cited above. We have incorporated six videos into the course, which provide insights from business advisors who attended the consultative workshops and from members of the Growing Greener team. The video content provides the basis for several learning activities. There are also interactive exercises based on key concepts, and a series of 'thinking points' and short end-of-session quizzes that encourage learners to reflect on their own practice, and how they have encountered personal, professional and organisation values in their careers to date.

For example, in one of the interactive exercises, learners are asked to 'drag and drop' a set of key words on to a diagram that summarises the Schwartz Theory of Basic Human Values, the aim being to identify which of the values they consider them to be most salient. Having completed the task, the authoring team's version is displayed (Figure 2), along with a short commentary.

IOP Conf. Series: Earth and Environmental Science **329** (2019) 012056 doi:10.1088/1755-1315/329/1/012056

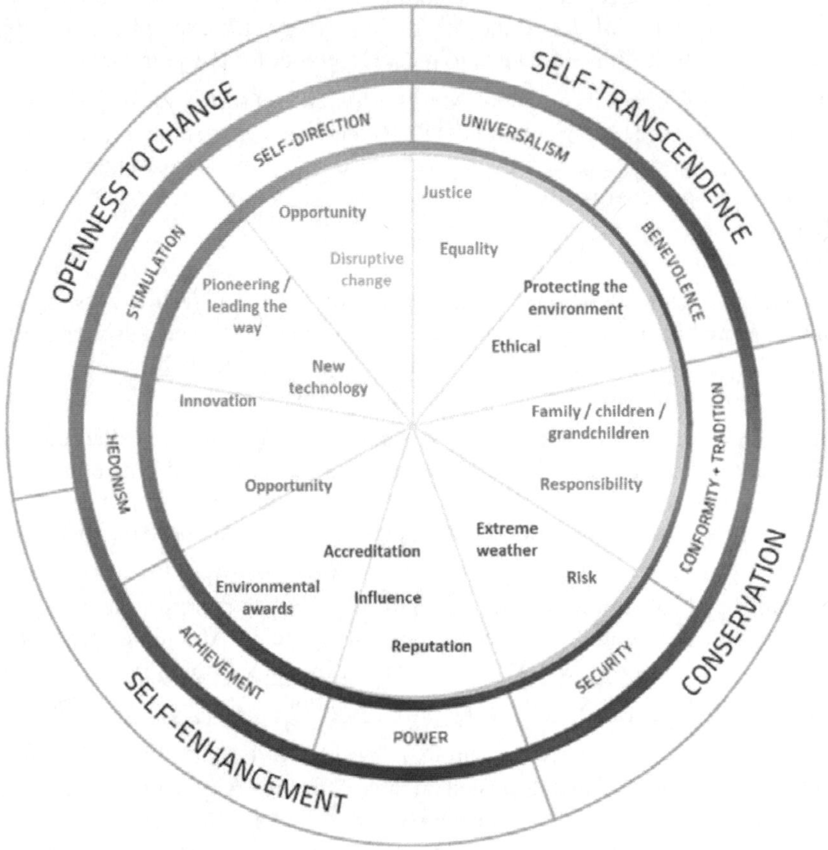

Figure 2: Key words relating to environmental responsibility which are likely to resonate with SMEs holding different values-sets

In another exercise learners are asked to consider how sets of values can be derived from different sources, such as those embedded in organisational cultures and norms; those demanded by particular job roles; and those brought into the SME by individual staff members (Figure 3). The resulting complexity was one of the issues raised at the consultative workshops, and we have designed the learning activity to prompt similar reflections amongst those completing the online course.

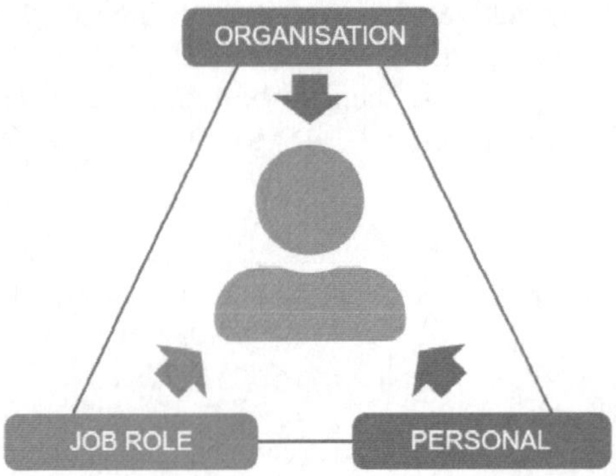

Figure 3: Diagram illustrating the different sources of values which may influence an individual working within an SME.

The final part of the Growing Greener toolkit is a short 'step-by-step engagement guide'. This 3-page document is designed as a series of prompts for advisors when preparing to meet an SME client. Structured into *before, during,* and *after* sections, it suggests ways to anticipate the kinds of values that are likely to be at play when meeting an SME owner-manager; offers tips for asking questions and steering conversations; and gives ideas for how to maximise impact by following up after an initial meeting. This resource is both included as part of the online course and available as a printable document.

5. Discussion

The toolkit brings together insights from academic literature; professional experience as advisors; climate outreach's previous work on climate narratives; and the series of workshops conducted as part of the Growing Greener project. Whilst the workshops provided an opportunity for the academic team to test out a series of ideas and approaches, these were refined and adapted as we received feedback and input from low carbon advisors who meet with SMEs on a daily basis to talk about energy and environmental impact.

One key contribution from workshop attendees was to emphasise that *all* segments of Schwartz's Value System are represented in SME owner-managers. By instinct, the advisers are already highly adaptive. Meeting a wide variety of businesses in their day-to-day roles, they tailor the language they use and the advice they provide. They might find themselves talking about the efficiency of an industrial kiln with one SME in the morning, and giving advice about crowdfunding to an environmental charity in the afternoon. They valued the framework as a way to articulate and visualise what was already familiar to them. However workshops also revealed how advisors might commonly avoid talking about questions of environmental ethics and responsibility if they gleaned that their client had self-enhancing values. The Growing Greener toolkit therefore aims to celebrate and enhance good practice, but has also been developed to support advisors to develop the language, concepts and approaches which might help them avoid steering clear of these crucial topics.

Our primary research focuses on environmental messaging and has found that understanding and appealing to the full range of Basic Human Values can help to engage a wider set of people. Some businesses, for example long established family firms, may be mainly concerned with maintaining tradition and continuity and minimising risks (conservation values). For these businesses, we have found that common approaches used to discuss climate change fail to motivate them. They are likely to respond better to messages on climate change which make links to risk, responsibility and family [28].

Conversely, some SMEs consider themselves to be drivers of innovation and disruption. Leaders of these businesses are more likely to identify with sets of values in the upper-left quadrant of Schwartz's framework. They are motivated by newness and change, and respond positively to environmental messaging that focuses on new business opportunities, innovative technologies and the potential to disrupt incumbent corporations.

Schwartz tells us that it is impossible to change anybody's basic values. But, rather than focus on certain types of SMEs, or individual business leaders with particular values, the Theory of Basic Human Values allows the advisers to adapt their language, messages and approaches so that they can work most effectively with all types of organisation. All value-sets have merit and a successful transition into a low carbon economy will not only require engaging businesses with all types of values, but will also benefit from the range of different skillsets associated with each.

6. Conclusion

The Growing Greener project is part of broader movement seeking to reduce the energy consumption and environmental impact from the SME population. Too long a neglected sector of the economy, the urgency of the need to cut carbon emissions across the world means that they can no longer be neglected by policy makers.

Energy and sustainability advisors are already playing a crucial role in supporting SMEs to mitigate their impacts, but they are limited by a set of assumptions about SME motivations which is underpinned by an impoverished appreciation of the multiple values at play in smaller organisations. Largely assuming that SME owner-managers hold power and achievement values-sets, advisors' practice is dominated by economic narratives. This leads them to pursue engagements with SMEs with a narrow focus on investment-payback periods and the promise of increased brand reputation.

In recent years, there has been a significant increase in the number of social enterprises and community interest companies being established in the UK [32], as well as ethical accreditation schemes such as B-Corps [33]. The idea of 'start-up culture' has entered public discourse, conjuring images of modern workplaces with indoor putting-greens and swings [34], and associated with explicitly stated, extrinsically focused values [35]. These trends indicate that values are coming to play a more significant role in how business is conducted, and that the assumptions made by advisors and policy makers about what motivates and drives SMEs must be updated accordingly. Correcting assumptions alone is not enough however. Learning to communicate effectively using values-based techniques is not easy, as the participants in our workshop role-plays would attest.

The Growing Greener toolkit provides a set of resources for helping energy and sustainability advisors become equipped with the skills and knowledge to be able to use values in their working practice, however this is just a start. It is too early to evaluate the take-up of the toolkit by advisors, although several indicated that the workshops alone had inspired them to adapt their practice.

Policy makers must begin to recognise the strategic importance of this community of practitioners with more comprehensive forms of professional support and development. There are some signs of promise in this regard. The Growing Greener team has been invited to present the toolkit to civil servants in both the UK and Ireland, and to brief senior figures in a leading political party. Given the universal nature of the SVS, and the shared challenges faced by policy-makers in including SMEs in sustainability transitions, the toolkit has potential for global reach.

References

[1] Sykes R 2009 *Engaging SMEs to improve their Energy Efficiency: A Market Appraisal* (DECC)
[2] DECC 2016 *Potential of Smart Technologies in SMEs* (London: Department of Energy & Climate Change)
[3] DCLG 2015 *European Regional Development Fund Operational Programme 2014 to 2020* (London: Department for Communities and Local Government)
[4] Hampton S 2018 Policy implementation as practice? Using social practice theory to examine multi-level governance efforts to decarbonise transport in the United Kingdom *Energy Research & Social Science* **38** 41–52
[5] Hampton S 2018 'It's the soft stuff that's hard': Investigating the role played by low carbon small- and medium-sized enterprise advisors in sustainability transitions *Local Economy* **33** 384–404
[6] Johnson M P and Schaltegger S 2016 Two Decades of Sustainability Management Tools for SMEs: How Far Have We Come? *Journal of Small Business Management* **54** 481–505
[7] North P 2016 The business of the Anthropocene? Substantivist and diverse economies perspectives on SME engagement in local low carbon transitions *Progress in Human Geography* **40** 437–54
[8] Eadson W 2014 Towards a spatially and socially embedded approach to SME support for carbon reduction *People, Place and Policy Online* **8** 129–38
[9] Andrews R N L and Johnson E 2016 Energy use, behavioral change, and business organizations: Reviewing recent findings and proposing a future research agenda *Energy Research & Social Science* **11** 195–208
[10] Mallaburn P 2016 A new approach to non-domestic energy efficiency policy
[11] Sorrell S, Mallett A and Nye S 2011 *Barriers to industrial energy efficiency: A literature review* (United Nations Industrial Development Organisation)
[12] DeCanio S J 1993 Barriers within firms to energy-efficient investments *Energy Policy* **21** 906–14

[13] Crocker R 2012 SME Perceptions of Low Carbon Options: Drivers, Barriers and Business Sector Differences *Earth and the Environment* **7** 232–259
[14] DECC 2014 *Research to Assess the Barriers and Drivers to Energy Efficiency in Small and Medium Sized Enterprises* (London: Department of Energy & Climate Change)
[15] EEC 2017 *SMEs and community organisations: enabling best practice energy efficiency* (Energy Efficiency Council)
[16] European Commission 2016 *Improving Resource Efficiency in SMEs*
[17] Janda K B and Parag Y 2013 A middle-out approach for improving energy performance in buildings *Building Research & Information* **41** 39–50
[18] Janda K B 2014 Building communities and social potential: between and beyond organizations and individuals in commercial properties *Energy Policy* **67** 48–55
[19] Janda K B, Bottrill C and Layberry R 2014 Learning from the "data poor": Energy management in understudied organizations *Journal of Property Investment and Finance* **32** 424–42
[20] Killip G 2013 Products, practices and processes: exploring the innovation potential for low-carbon housing refurbishment among small and medium-sized enterprises (SMEs) in the UK construction industry *Energy Policy* **62** 522–30
[21] Owen A, Mitchell G and Gouldson A 2014 Unseen influence—The role of low carbon retrofit advisers and installers in the adoption and use of domestic energy technology *Energy Policy* **73** 169–179
[22] Blundel R, Monaghan A and Thomas C 2013 SMEs and environmental responsibility: a policy perspective *Business Ethics: A European Review* **22** 246–62
[23] Schaefer A, Williams S and Blundel R 2018 Individual Values and SME Environmental Engagement *Business & Society* 0007650317750134
[24] Williams S, Schaefer, Anja and Blundel R Williams, S., Schaefer, A. and Blundel, R. (2017). Understanding value conflict to engage SME managers with business greening *Perspectives on Philosophy of Management and Business Ethics, Ethical Economy. Studies in Economic Ethics and Philosophy* ed J D Rendtorff pp 73–92
[25] Williams S and Schaefer A 2013 Small and Medium-Sized Enterprises and Sustainability: Managers' Values and Engagement with Environmental and Climate Change Issues *Business Strategy and the Environment* **22** 173–86
[26] Schwartz S 2012 An Overview of the Schwartz Theory of Basic Values *Online Readings in Psychology and Culture* **2**
[27] Blundel R, Fawcett T, Shaw C, Hampton S and Westall A 2017 Growing green?: co-creating an evidence-based model of SME engagement ISBE Annual Conference (Belfast)
[28] Climate Outreach 2016 *Communicating effectively with the centre-right about household energy-efficiency and renewable energy technologies*
[29] Climate Outreach 2016 *Faith & Climate Change: A guide to talking with the five major faiths*
[30] Shaw C and Corner A 2017 Using Narrative Workshops to socialise the climate debate: Lessons from two case studies – centre-right audiences and the Scottish public *Energy Research & Social Science* **31** 273–83
[31] Climate Outreach 2019 *Effective communication guide: using values to promote sustainable ways of doing business*
[32] DCMS and BEIS 2017 *Social Enterprise: Market Trends 2017*
[33] B-Corps Certified B Corporation
[34] Workspace Design & Build 2016 What can we learn from Google's offices about workplace design? *Workspace Design & Build*
[35] Kulturenvy C M 2013 How Do You Define Startup Culture? *Wired*

AUTHOR INDEX

AUTHOR INDEX